# A DARK and DIS (A Bright and Shining Life)

A STORY
about
COAL,
COALMINERS
and
LOCHHEAD  COLLIERY

by

# DAVID MORRIS MOODIE

*Cover Photograph:*
A rather weary looking Jimmy Hutton is seen at the controls
of a 15-inch 'Bluebird' coal-cutting machine.
Note the tension on the steel driving cable and the rubberised power cable.

# A Dark and Dismal Strife
## (A Bright and Shining Life)

by

David Morris Moodie

Published by
Dauv.Miner Publications
Grantully, Perthshire, PH15 2QY

ISBN 0 9541120 0 8

Printed by Danscot Print Limited, Perth.

LOCHHEAD COLLIERY

J. DRYBURGH.

*This line drawing of Lochhead Colliery is reproduced here by courtesy of Mr. J. Dryburgh, Buckhaven, Fife.*

# CONTENTS

|  |  | *Page* |
|---|---|---|
| **ACKNOWLEDGEMENTS** | . . . . . . . . . | I |
| **DEDICATION** | . . . . . . . . . | V |
| **BELATED RECOGNITION** | . . . . . . . . | VI |
| **HISTORICAL FOREWARD** | . . . . . . . . | VII |
| **INTRODUCTION** | . . . . . . . . . | LVIII |
| **CHAPTER I** | Insight (Part I) . . . . . . | 1 |
| **CHAPTER II** | Black Bloody Bings . . . . | 25 |
| **CHAPTER III** | Lochhead Pit Bottom . . . . | 37 |
| **CHAPTER IV** | A Not So Wicked Youth! . . . . | 45 |
| **CROSS SECTION** | of the Dysart Main Seam . . . . | 54 |
| **CHAPTER V** | The Taking of the Dysart Main . . . | 55 |
| **CHAPTER VI** | "Tally Carbide!!". . . . . . | 82 |
| **CHAPTER VII** | Dauvit's Progress I . . . . . | 86 |
| **CHAPTER VIII** | The Denbeath Rows . . . . | 91 |
| **CHAPTER IX** | The Victoria Pit . . . . . | 111 |
| **CHAPTER X** | The Getting of the Coal . . . . | 131 |
| **CHAPTER XI** | Band Aid . . . . . . | 159 |
| **CHAPTER XII** | Old Dook Extractions . . . . | 162 |
| **CHAPTER XIII** | Dauvit's Progress II . . . . | 186 |
| **CHAPTER XIV** | Banana Boy! . . . . . | 190 |
| **CHAPTER XV** | The Hugo Mine . . . . . | 196 |
| **CHAPTER XVI** | A Day in the Life of the Dipping 'Tailsman!' . | 199 |
| **CHAPTER XVII** | Pit-Head Baths – A Clean Environment . . | 209 |
| **CHAPTER XVIII** | Dauvit's Progress III . . . . | 216 |
| **CHAPTER XIX** | The New Dook . . . . . | 222 |
| **CHAPTER XX** | Diesel Davie . . . . . | 244 |
| **CROSS SECTION** | of Lower Dysart Seam . . . . | 247 |
| **CHAPTER XXI** | The Lower Dysart Coal Taking! . . . | 248 |
| **CHAPTER XXII** | The Road to Dusty Death! . . . | 272 |
| **CHAPTER XXIII** | Dauvit's Progress IV . . . . | 278 |
| **CHAPTER XXIV** | The Coaltown of Wemyss – The Middle Village | 289 |
| **CHAPTER XXV** | It's An Absolute Gas! . . . . | 307 |
| **PHOTOGRAPHS** | . . . . . . . | 325 |
| **CHAPTER XXVI** | The Surface Dipping . . . . | 331 |
| **CHAPTER XXVII** | The Gentle Giant! – or Local Bogie Man? . | 343 |

| CHAPTER XXVIII | Dauvit's Progress V | . | . | . | . | . | 348 |
| CHAPTER XXIX | Harry's Stutter! | . | . | . | . | . | 353 |
| CHAPTER XXX | Out of the East! | . | . | . | . | . | 356 |
| CHAPTER XXXI | Pitch and Toss! | . | . | . | . | . | 366 |
| CHAPTER XXXII | The Black Art! | . | . | . | . | . | 370 |
| CHAPTER XXXIII | Dauvit's Progress VI | . | . | . | . | . | 375 |
| CHAPTER XXXIV | Mary's Transgressions!. | . | . | . | . | . | 379 |
| CHAPTER XXXV | A Miner's Piece! | . | . | . | . | . | 382 |
| CHAPTER XXXVI | The Lochhead Syphon | . | . | . | . | . | 388 |
| CHAPTER XXXVII | The 'Baund' Section! | . | . | . | . | . | 393 |
| CHAPTER XXXVIII | Dauvit's Progress VII | . | . | . | . | . | 406 |
| CHAPTER XXXIX | Happy Elder | . | . | . | . | . | 411 |
| CHAPTER XL | Pithead Tables! | . | . | . | . | . | 416 |
| CHAPTER XLI | Old Harry – an Institution! | . | . | . | . | . | 422 |
| CHAPTER XLII | Escape Routes! | . | . | . | . | . | 429 |
| CHAPTER XLIII | The Iron Men | . | . | . | . | . | 434 |
| CHAPTER XLIV | Why Goest Thou! | . | . | . | . | . | 439 |
| CHAPTER XLV | Dauvit's Progress VIII | . | . | . | . | . | 443 |
| CHAPTER XLVI | Tally Ho! | . | . | . | . | . | 451 |
| CHAPTER XLVII | Insight!  (Part II) | . | . | . | . | . | 456 |
| CHAPTER XLVIII | Miners at Play! | . | . | . | . | . | 481 |
| CHAPTER IL | The Outcrop | . | . | . | . | . | 485 |
| CHAPTER L | Zeitgeist! | . | . | . | . | . | 494 |
| CHAPTER LI | Army Rations | . | . | . | . | . | 511 |
| CHAPTER LII | Willie's Pirouette! | . | . | . | . | . | 514 |
| CHAPTER LIII | Panzer Power! | . | . | . | . | . | 518 |
| CHAPTER LIV | Hoch Dutch! | . | . | . | . | . | 546 |
| CHAPTER LV | Dauvit's Progress IX | . | . | . | . | . | 550 |
| CHAPTER LVI | The Old Dook Re-visited | . | . | . | . | . | 560 |
| CHAPTER LVII | Death of a Coalmine! | . | . | . | . | . | 577 |
| CHAPTER LVIII | The Passing! | . | . | . | . | . | 579 |
| THE GÖTHENBURG EXPERIMENT! | | . | . | . | . | . | 585 |
| ADDENDUM | I Tiddley-Winks Clubs. | . | . | . | . | . | 590 |
| | II  Miner's Wedding Cake | . | . | . | . | . | 592 |
| | III  Recipe for 'Po Toast' | . | . | . | . | . | 593 |
| | IV  Goodness Gracious Me! | . | . | . | . | . | 594 |
| IN MEMORANDUM | . | . | . | . | . | . | 595 |
| BIBLIOGRAPHY | . | . | . | . | . | . | 599 |

# Acknowledgements

During the long process of searching out old Lochhead miners and gathering information for this book in 1999 and 2000 - my wife and I spent most of our weekends travelling to various destinations in the Kingdom of Fife to interview many old ex-miners who had 'wrocht' in Lochhead pit or the Surface Dipping.  Some of these ex-coalminers were actually interviewed more than once and without exception we were made very welcome with the only time limit being imposed being the lateness of the day and the need to return home.  The ex-coalminers ranged from colliery managers' to oversmen to deputies to shot-firers, through every class of face-worker, brusher/developer and oncost miner, all the way to the Pithead workers.  These people gave unstintingly of their time and memories - and I can but only apologise to those who spent many sleepless nights re-living old memories in anticipation of my impending visits.  One such ex-miner's wife confided to me a few weeks after my visit that it had been several years since she had seen her husband so alive, animated and unable to sleep.

I am deeply indebted to the following ex-Lochhead miners who suffered my persistent questioning - and to their wives who sometimes added more than just a touch of reality to the proceedings.

Bob Rodger, Methilhill ................................................ Collier/Miner for 49 years.
*(Late of the Coaltown of Wemyss.)*

Tom Mathers, Coaltown of Wemyss ...................................... Developer, Collier.
*(Late of West Wemyss)*

Jackie Dryburgh, Buckhaven ..................................................... Mine Electrician.
*(Late of Coaltown of Wemyss)*

Laurie Gibb, Kirkcaldy .............. Roadsman, Rope-splicer, Face-worker, Deputy.

Smith Anderson, Kennoway ...... Versatile Coalminer,  First Aid Gold Medallist.

Jim Allen, Glenrothes, (Young 'Lots o' Coal') ....... Coalminer/brusher/developer.

Bob Ross, West Wemyss ...................................... Stripper/developer, Coalminer.

David Davidson, Coaltown of Balgonie ............................ Engineer at Lochhead.
*(Lastly supervised the final filling-in of the Lochhead Pit Shaft.)*

George Halley, Leslie, Fife ... Shotfirer, Deputy, Oversman, Undermanager/Manager.

David Black, Coaltown of Wemyss ...................................................... Coalminer.
*(Late of West Wemyss)*

David Rodger, Coaltown of Wemyss .............. Collier & Deputy at Lochhead Pit.

Jackie Black, Aberhill/Methil ........................ Stripper/developer in Lochhead Pit.

Ted Paslowsky, Glenrothes .............. Stripper/brusher/developer in Lochhead Pit.

Bob Kinnear, Glenrothes .................... Life-long coalminer in Lochhead Colliery.

Jackie Dryburgh, Glenrothes ................................................. Versatile Coalminer.
*(Formerly of Coaltown of Wemyss.)*
*(Mentor to this author in 'Stoop and room' coal-mining.)*

Jackie Penman, Coaltown of Wemyss ...... Long term Coalminer in Lochhead Pit.

John Abbot, Coalminer .... Deputy and then Oversman in Lochhead's New Dook.

Bill Shields, West Wemyss/Coaltown of Wemyss ............ Miner in Lochhead Pit.

Bobby Grubb, Buckhaven ........... Long-term versatile Coalminer in Lochhead Pit.


The following people did not work in Lochhead pit, but were never-the-less generous
with their particular knowledge of related mining subjects.

Mr W. Kerr, M.I.Min.E., Pitteuchar, Glenrothes . Coalminer to Colliery Manager.

Mr G. Fry, C.Eng.,M.I.Min.E., Leslie, Fife ..... Scientist with N.C.B. In Fife Area.

Bobby Allan, Glenrothes .. Coalminer & Power-loading team leader. Francis Pit.

Belle Smart, Coaltown/Dysart ........... Mother and Former employee at Lochhead.

Brian Smart, Coaltown of Wemyss ........................................... Gen. Sec. & Treas.
Coaltown Soc.& U.S.Club *(Now defunct.)*

I am also indebted to:- David Clarke, Mining Records Office, Burton on Trent, Staffs., who put up with my several requests for specific items, to which he freely and generously responded.

George Archibald, Volunteer at the Scottish Mining Museum, Newtongrange.

Kevin Brown and volunteer staff at Methil Heritage Centre, Methil, Fife.

Frank Rankin, also at Methil Heritage Centre, Methil, Fife.

Members of Staff at the Reference section of Kirkcaldy Central Library, Fife.

Bill and Betty Hay, Proprietors, Amateur Radio Emporium, Woodside, Glenrothes. *(Who provided the welcome cups of coffee en route!)*

The very largest 'thank you' is reserved for Mr. Bob Stevens and his lady wife Jessie of Woodside, Glenrothes. He, a versatile coalminer and raconteur 'extra ordinaire' who encouraged me greatly throughout the whole of my research in 1999, who fired my imagination with hair-raising stories of 'unmentionable things' and 'unbelievable doings' amongst the miners of Lochhead Colliery, both above and below ground - and who did so much voluntary 'leg-work' in the successful searching out of old ex-miners in the Kingdom of Fife! All keenly done on my behalf ! Thank You Bob!

During my sustained travels in and around the East coast of Fife, one of my keenest pleasures was to meet up with and converse with the present Master of Wemyss, Mr. Michael Wemyss of that Ilk. I made the initial approach with the help of David Black and his lady wife who were kind (and bold) enough to point me in the right direction. My approach to Mr. Wemyss at the Coaltown of Wemyss annual Gala day was rather tentative, in that I had already written most the opening Historical Foreword to this book, but felt that it was not altogether quite correct. This then this was the person to whom I should address any request. I described that which I had written and made suitable apologetic noises as to its probable content in the hope that I could solicit some further gems of information. I was completely outdone! Not only did the Master of Wemyss offer a few hitherto unknown facts, he actually invited me down to the Castle library where were deposited the one set of quarto volumes* that would enable me to produce a limited, but factual record of some of the coal dealings of nearly all of the previous Laird's since 1428. The Master of Wemyss very generously allowed me the long loan of these three volumes of his families' long history.
Thank you sir!

* see Bibliography.

I also wish to acknowledge the help and guidance freely given by Mrs Jean Black, the manageress at Danscot Print in Perth - also Trish and Valerie whose painstaking job it was, to convert the whole written project into readable book format. My grateful thanks to you, and all concerned at Danscot!

David Morris Moodie.

In such parts of this narrative where I have reverted to the use of the local vernacular, mining slang or the Auld Scots tongue, I have followed this expression almost immediately with an explanation of the word or phrase. On those few occasions where I have not done so, it is because I feel that the word or phrase is in the context of the narrative - and therefore should be easily understood. The use of such words, phrases and the Scots tongue has been kept to an absolute minimum, except in the case of original materiel where I feel that originality enhances authenticity. In this pass (instance) I have done my best to translate the original auld 'Scots Tongue' into readable text. I hope that I have succeeded!

D.M.M

# DEDICATION

This book is dedicated to the memory of Harry Moodie,
coal miner, who was my father. He was an extraordinary quiet man who,
because of a speech impediment caused men to listen patiently to him in
the event they missed something. He very rarely raised his voice.

He knew only the coal-mines and his back-garden joinery and his
knowledge was profound. His free time was spent exclusively on his
secondary hobby to which he successfully initiated me. He worked
extremely hard for all of his life and was often the only miner to go
to the pit on a Saturday night.

As an individual, he cherished nothing but a few ancient carpentry tools.
He had no materiel wealth and wished for none.

His capacity for patience was unlimited, although his frustrations were often
suppressed. He was seriously minded, would not suffer fools and was
disinclined to frivolity.

Once, when I was thirteen years old, he took me to a Miner's Gala in
a large park in Edinburgh, where I was given a brown paper bag containing
two sandwiches, a sticky bun and a sponge cake - along with
a small bottle of tepid milk. All for free!

My first journey in a train - and he took me.

I have never forgotten that!

D.M.M.

# In Belated Recognition!

The very first ex-coalminer whom I interviewed in the course of my research for this book, was Bob Rodger, formerly of the Coaltown of Wemyss, but now living in a newish semi-detached bungalow in Methilhill. It also may be of interest to note that these dwellings were built on the site of the original Methilhill pit and bing of previous notoriety!

As this interview progressed, I became aware that even though Bob had spent all of forty-seven continuous years in one coalmine, i.e. Lochhead pit, there did not seem to be very much in the way of happenings, events or anecdotes that could be attributed or linked to Bob. I soon discovered that he was very knowledgeable in the working of coal - had experienced the many facets of primitive and exploratory mining techniques, and had also witnessed and been part of the mechanical revolution that slowly pervaded the Fife coalmines in the Forties, Fifties and Sixties. But, he modestly disparaged my futile attempts to acknowledge his ever having been involved in any event, or anything of particular note!

It was not until I had interviewed a full dozen more ex-coalminer's - and listened to their parts in the development of Lochhead colliery, that the simple truth concerning and describing the great majority of Lochhead miner's began to dawn on me! For the most part, there was simply no exciting stories to tell. At least, not any that would take readers to the heights of suspense or enthralment!

Bob Rodger, along with many other unnamed collier's who worked in Lochhead colliery, most assuredly, epitomised the doughtiness of the great unwashed majority of coalminer's, who were the unspoken mainstay of what was described even then, as a 'family' pit. One has but to peruse the lists of common surnames that proliferated the employment roll at Lochhead, to realise that whole families were employed at any one given time! Most of whom, came from either West Wemyss or the Coaltown of Wemyss. These coalminer's went to the pit every working day of their lives, completed their daily tasks and returned home, having given a fair day's work for sometimes, an altogether inadequate wage.

The majority of Lochhead's miner's were of this ilk. Steady, dependable, punctilious, taciturn and unassuming! They were the unacknowledged backbone of the Lochhead family colliery!

This is a story about some of those miners!

. . . And how they took the coal!

D.M.M.

# An
# Historical Foreword

## A Gey Auld Occupation? *or, How it All Began!*
## *(A binding of Miners, Wives & Bairns!)*

Beneath the surface of the Kingdom of Fife, there are two completely separate, but viable coal measures.  They are contained in what was sometimes referred to as the Wemyss field and the Dunfermline field.  The first here mentioned, known as the Wemyssfield, *(East Fife coalfields)* is contained in the **Productive or Upper coal measures**, while the second, the Dunfermline field, *(West Fife coalfields)* takes completely different and much deeper coals.

As described in greater detail elsewhere in this narrative, but may require introduction at this point, there are fully twenty different seams *(inc. the heavily faulted Lethamwell seam)* of coal in the upper coal measures *(sometimes also called the* **'True Coal Measures'***)* of the Wemyss field area, as proved and worked at various times during the last millennium. This is the coalfield that I wish to concern myself within this part of the narrative, in order to describe several 'workings' that took place from the Fifteenth century onwards, that discusses many of the 'ordinary' trials and tribulations faced by the coal-owners and some of those early coalminers in east Fife, from that time to the mid-twentieth century.

These twenty separate coal seams are contained within strata, *(in mining parlance - metals!)* that rise on a broad front in an irregular ascent, of between one in four and one in three, in a northwesterly direction from under the wide waters of the Firth of Forth.

This multiple-layered 'tongue' of sandwiched coals, intrudes into the Kingdom of Fife from under the Firth of Forth, between the coastal towns of **Kirkcaldy** *(Seafield colliery)* and the old seaside burgh of **Leven.** *(Two miles north-east of the Wellesley colliery.)*

The depths of these seams do vary considerably at different geographic locations along the coast, as shown by individual borings and test-drillings carried out at markedly different dates.  For example: - At Denbeath, near the site of the Wellesley pit, the Dysart main seam *(No. 18 in depth)* was to be found at a vertical depth of 280 fathoms, *(approximately 512 metres)* while at the site of the Victoria pit four miles along the coast, the Dysart* Main seam was to be found at a vertical depth of 130 fathoms *(approximately 239 metres)*.

*(The average thickness of this seam at both sites, was found to be very much
the same at approximately 32 feet/10 metres).*
*See chapter on Victoria pit.*

If all of the twenty coal seams to be found in the upper coal measures that
constitute the Wemyss coalfield, were to rise fairly consistently at the above-
mentioned slope *(angle)* and continue to do so without interruption, then it is fair
assumption *(barring any great geological upheaval)* that they would all eventually
surface somewhere inland. That is seemingly what transpired and to a greater and
lesser degree. The bottommost workable seam **the Lower Dysart** and the second
bottom seam **the Dysart Main,** actually surfaced several miles inland to the
Northwest of the A915, *(Standing Stone road)* between **Wellsgreen, Thornton** and
the **Boreland**. This meant that all of the intervening seams from No. 1 to No. 17
actually extruded to the surface and were probably 'wrocht' as coal-heughs at one
time or another, *(that is, if they had not petered out!)* somewhere between the coast
and beyond the line of the Standing Stone road, *(A955)* and along a broad semi-
circular front.

Before the industrial use of the steam-engine, electricity and the Davy safety
lamp, there were several factors that severely inhibited coal mines from being taken
to any great underground lengths or depths. Lack of adequate iron or steel tools, lack
of ventilation, ignorance of mine gases and inadequate roof supports, to name but a
few. One of the inhibiting factors *(amongst others)* that decided the fate of some of
the developing coal-heugh's* and most of the early deeper sinks, *(vertical shafts)*
was the appearance of running water from different parts of the above ground
workings. Water could be got rid of providing that there was somewhere lower that
it could gravitate to, initially by the driving of slightly dipping levels to lower
ground, or even by the use of a primitive 'syphon construction', if it could be made
to work! The build-up of 'standing water' in early sinks with subsequent flooding,
therefore severely curtailed many of the earlier attempts to go deeper underground
in search of greater volume of coals in potentially rich coal-fields or known coal
reserves.

*

The first mention of coal having been being worked on the Wemyss lands *(that
I can discover)* was during the time of the recognised laird, **David of Methil and
Wemyss.** (1428-1430) David the laird, entered into a series of 'bands' *(agreements, -
sometimes secret)* with **Robert Livingstone,** the laird of Drumry. *(Husband to one of
the co-parceners (coheirs) following the break up of the Wemyss estates at the death
of Sir Michael Wemyss in 1342, there being no surviving male issue.)* Both agreed to
separate the lands of *Wemyss at a place called the Dean Burn. The agreement was
that Robert, laird of Drumry, was to have all of the Wemyss lands including Macduff's

Castle to the east of Dean Burn, while David Wemyss of that Ilk was to have all of the lands including Wemyss Castle, to the west of the Dean Burn.

As part of the band, it was also agreed that either of them would have the freedom to work 'coal and stone' out-with each other's respective Manor lands, but within each other's respective lordships and without limit, the subsequent profit *(if there was any)* to be divided, with two parts retained by the originator *(to cover initiating and production costs)* and one part to the other non-participating partner. All other 'working' or 'small industry', *(coal-taking, salt-panning and harbouring of boats)* to be let or rented out at full value, with the profits being equally shared.

**This means that coal was being worked and sea salt was being manu-factured in the parish of Wemyss as early as 1428!** *(These two industries were synonymous in that the foreshore 'coal-heughs' supplied the immediately continuous and copious amounts of coal needed to maintain the necessary heat at the boiling of the salt pans!)*

\* This could have been the emergence and separation of **Easter Wemyss** and **Wester Wemyss**.

Author's Note: - Coal-heughs are not coal-mines *(per se)*. A description of a coal-heugh appears later on in the narrative of this chapter.

*

In 1448, **John Wemyss of Wemyss**, (1430-1502) son and heir to David Wemyss of that Ilk, married Margaret Livingstone, daughter to Robert Livingstone of East Wemyss. This marriage seemingly served to strengthen the terms of original 'band' concerning the lands of Wemyss-shire entered into by their respective fathers, thereby helping to ensure it's permanency. *(Or as much of a permanency as could be realised in those troubled times!)* At the coronation of **James III, King of Scots** in 1460, John, Laird of Wemyss was raised to Knighthood taking the title, **Sir John Wemyss of Wemyss of that Ilk.**

In the year of 1475, a dispute arose between **Sir John of Wemyss** and **Sir Michael Livingstone, Vicar of Wemyss,** to the effect that **Sir Michael** as Vicar of Wemyss was not receiving his 'due monies' in **teinds** from the coal and salt industries at **Wemyss**.

*(I would, with good reason, suppose that Sir Michael was closely related to Sir Robert of that Ilk, family nepotism in lucrative appointments being a way of life in those days!)* Sir Michael took his grievance and claim to the Ecclesiastical Court at St. Andrews, where, after 'evidence' was heard from both parties, a decree was pronounced setting forth the following deliberation: -

First, that the teind coals of the coal-heughs of the Laird of Wemyss should be levied on the multure of the coal-heughs in the place from which they were led to the sea; (second) that as the teind of salt, the Vicar should have the true tenth of each pan in the tank paid wholly to him at the (salt) pan.

*For teinds - read 'Church tithes.'*
*For 'teind coal'- read 'levied tithes paid to church!'*
*For 'multure of coal heughs' - read 'tithes taken at the point of production.'*
*For the 'salt tithe', - the Vicar was to recover a full tenth of the sea-salt recovered from each individual salt pan.*

By virtue of the fact that the Vicar of Wemyss deemed it necessary to approach the Ecclesiastical court in St. Andrews to gain his due teinds, showed that **Sir John of Wemyss** had developed the coal-heughs and the salt pans to such an extent, that the Church *(in other words, the vicar)* felt obligated to press for its due teinds at St. Andrews.

The Vicar of Wemyss was not the only Livingstone to harbour a difference of opinion with Sir John of Wemyss and, who also took complaint to the Lords of Council for considered arbitration. In or around 1501, **Sir Robert Livingstone** *(of Easter Wemyss)* complained to the Lords of Council, that he had not received his due 'one-third' of the profits *(as per the conditions of the 'band' agreed in 1428)* accrued from the produce and profits of the Coal-heughs and saltpans at Wester Wemyss. After hearing partial evidence from both sides, the noble **Lords** directed that Sir Robert should receive his one-third of the 'profits' accrued from the working of the coal and the salt at Wester Wemyss.

*(This judgement was seemingly arrived at, irrespective of the expenses incurred by Sir John of Wemyss in the development and running of both coal-heughs and the saltpans!)*

*

Sir John Wemyss of that Ilk died in 1502. He was succeeded by his fifty-four years old son also named **Sir John of Wemyss,** but also of **Strathardale.** (1502-1508) *(The late Sir John's eldest son who had worked with him at their coal-heughs and saltpans at Wester Wemyss.)* On his succession to the estates, Sir John re-involved himself in this dispute with Sir Robert Livingstone, *res* the claim Sir Robert had made against the Wemyss estates for his one-third shares in the profits accrued from the coal-heughs and saltpans at Wester Wemyss. On this occasion, the arbitration was heard in St. Mary's aisle of St. Giles church in Edinburgh, before the **Lords of Council** as arbiters. The Lords of Council called for the coal-grieve *(superintendent of the Coal-Heughs)* and other skilled witnesses. After further due consideration, the Lords decreed that the one -third of the 'profits' from both the coal and the salt supposedly due to Sir Robert, were to be retained by Sir John, to compensate him for the overall expenses incurred by Sir John in the further development of new coal-heughs.  In other words, the so-called 'profits' from both the coal and the salt were being swallowed up by development costs and running expenses, such as wages in the form of coals and monies in the form of silver merks to the coalliers and salt-panners, along with due teinds to the Vicar of Wemyss.  The Lords of Council further decreed, that Sir Robert Livingstone should in future, order that one third of the workers needed to work the coal-heughs and the saltpans at Wester Wemyss be directed from his estates and, that he pay one-third of the five shillings weekly wage of the coal-grieve!

*(As far as Sir John of Wemyss was concerned, justice was seen and heard to have been finally done!)*

*

In the spring of 1508, **David Wemyss of Wemyss** (1508-1513) succeeded to the estate.  He was knighted in 1510 by his grace, **King James IV.**  In the year of 1511, **James, King of Scots,** caused the estates of Wemyss to be erected into a Barony and named the **Barony of Wemyss**.  The great castle of Wemyss was also assigned as the principal *****messuage** of the new Barony and additional powers was given *(to Sir David Wemyss)* to erect the new **Haven town of Wemyss**.

Sir David Wemyss of that Ilk (1508-1513) was killed at the Battle of Flodden Field on September 9th, 1513.

* A principal dwelling house or 'seat' of Barony.

*

In the year of 1513, **David Wemyss of Wemyss** (1513-1544) succeeded to the Barony of Wemyss-shire.  In or around 1515, **Robert Livingstone**, the last male heir of the Livingstone's of Drumry, died without male issue.  He predeceased his widow Janet Betoun and their daughter Margaret Livingstone, his heiress.  In 1516, Janet Betoun married **James Hamilton**, the First Earl of Arran, while immediately prior to that, **Margaret Livingstone** *(heiress to the estates of Livingstone)* married **Sir James Hamilton** of Finnart, a natural son of **James,** the first **Earl of Arran**. *(Here, we have the not so strange circumstance, of the mother marrying the father of the son, who had married her natural daughter!)*  This meant that the whole of the lands of **Livingstone of (East) Wemyss** passed through the heiress Margaret Livingstone and were now bespoken for by her spouse, Sir James Hamilton, son of the Earl of Arran.  In that same year, Sir James *(on behalf of his lady wife)* leased to **David of Wemyss** the one-third *(ex-Livingstone)* part of the previous 'band' for the coal-

XI

heughs and saltpans at Wester Wemyss. The agreed yearly rental to be paid by David of Wemyss to Sir James Hamilton was to be Fifty pounds *(Scots)*. *(It would appear that Sir James Hamilton did not think that very much of a profit could be accrued from the coal production and the salt-panning at Wester Wemyss, hence the gentlemanly 'peppercorn' rental agreement!)*

In 1530, **Margaret Hamilton** *(nee Livingstone)* having first obtained the consent of her husband Sir James, exchanged all of her lands of Easter Wemyss for the lands of **James Ochiltree of that Ilk** in Ayrshire. The family of **James Colville** *(of Ochiltree)* therefore took the place of the Hamilton's/Livingstone's at Easter Wemyss and of course, acquired the rights of the one-third share of the coal-heughs and the saltpans at Wester Wemyss.

In 1534, David of Wemyss again gained the lease of **Colville's** one-third share for the same yearly rental of fifty pounds, *(Scots)* with the additional amount of ten pounds to be added on an annual basis, if all twelve saltpans were to be kept in production. David of Wemyss did eventually within his lifetime, acquire all of the rights to the coal-heughs and saltpans at Wester Wemyss from the **Colville** family. *(Thus, the rights to all of the coal-heughs and all of the saltpans at Wemyss, lay with the Lairds of Wemyss.)*

*

**Sir John Wemyss** of that **Ilk,** (1544-1572) succeeded to the Wemyss estates in 1544, before **Queen Mary** had come of age. I can find no record of his ever being involved with anything to do with coal-getting or salt-panning during his stewardship of the estates. It would seem however, that in the mid 1560's, the castle at Wemyss was where **Mary, Queen of Scots** met **Henry, Lord Darnley** for the first time. The rest of course, is bloody Scottish history!

*

**David Wemyss** of that **Ilk** (1572-1597) succeeded his father in 1572. He, like his father became heavily involved as a member of the judiciary in that they both presided over trials for criminals held in the **Barony of Wemyss**. This had come about because of a joint commission thrust upon them by the then Justice-general for Scotland, the Earl of Argyll.

One of the 'state' duties encumbrant upon the Laird as Baron of the Manor, was to 'entertain' certain uninvited guests thrust into his care by the King and Lords of Council. These house-guests were there by order of the King, to stand surety for the continued good behaviour of their unruly border kinsmen, who seemingly found it extremely difficult to refrain from pillage and cattle reiving on both sides of the Scottish/English border. That sort of murderous behaviour being a way of life with

XII

border peoples. This forced, but altogether civilised incarceration of rebellious lords and nobles and their sons and heirs with peaceful nobles and barons throughout the Country, was in keeping with the status of such lords and nobles and of course, their upkeep did not fall on the (King's) 'Privy Purse!'

In addition to this civilised incarceration and the so-called 'gentlemen's word' of the mischievous lords, the onus was on the encumbered lord or baron as 'keeper' to ensure that the wayward 'guests' did not attempt to depart or escape custody. The penalty for permitting any 'captive guest' to escape was a forfeiture of two thousand pound's Scots. *(This practice of course, all helped to fill the King o' Scots empty coffers!)*

In the year of 1573, David of Wemyss as owner of the saltpans at Wemyss, was summoned to appear before the Privy Council along with other Salt producers from around the Country, and called upon to restrict the export of local salt to other countries, until such times as the whole of this land had been supplied with a sufficiency of salt!

David Wemyss of Wemyss of that Ilk departed his life in February of 1597.

\*

**Sir John Wemyss of Wemyss** succeeded to the Lairdship of Wemyss in 1597, on the death of his father, David Wemyss of Wemyss. Sir John had previously married Margaret, daughter to William Douglas of Lochleven in 1574. Margaret died shortly afterwards having produced no heirs. Sir John remarried in 1581, to Mary, daughter of **Sir James Stewart of Doune**. Sir John seemingly immersed himself in politics during his time as Laird of Wemyss and I can find little mention of his ever getting too much involved with the coal-taking and salt-panning at Wester Wemyss. Obviously, these two trades were being well managed, probably by other members of the Wemyss family with the help of their factors - along with the local coal and salt grieves.

In the year of 1610, **Ludvick, Duke of Lennox,** granted to Sir John and his heirs, a commission as **Admiral of the Forth** with a fiefdom stretching from **Dysart** town to **Leven** burgh. *(Both on the same north side of the Forth)* In addition to this commission, by reason of heredity, Sir John was also 'Bailie' of the River Leven. In this capacity, Sir John was 'requested' by King James (VI of Scotland, I of England) to ensure the prevention of a ⋆lax-net at the mouth of the River Leven. At this time, that part of the river was in the hands of one Andrew Wood, resident in Largo. Mr Woods complaints were, that certain persons, the Laird's son included, had thrown a wide-meshed net across the mouth of the river to take only the large and heavy salmon heading upstream. The miscreants were controlling the net from both sides of the River mouth and were slaying the salmon as they were being hauled ashore on either bank. The upshot of this 'request', was that the miscreants had to find the sum of 400 silver merks each, to stand surety for their future good behaviour.

Sir John of Wemyss died in 1622, saddened by the fact that his oldest son David by his second marriage had died in 1608.

*

**Sir John Wemyss** of that **Ilk,** (1622-1649) knighted by King James VI around 1608, succeeded his father, also named Sir John Wemyss of Wemyss in 1622. In 1628, King Charles I. in recognition of services to the crown displayed by Sir John of Wemyss, conferred on Sir John, the dignity and rank of a Lord of Parliament with the title Lord Wemyss of Elcho. In 1633, in the palace of Dunfermline where **Charles I.** was born, a 'patent' which had been signed at Holyrood in June of that year, was presented to Sir John of Wemyss by the King, creating him **John, Lord Earl of Wemyss** - and **Lord Elcho of Methil.**

In spite of his elevation to the nobility, the Earl of Wemyss still found enough time to devote himself to the affairs of his estate, which by all accounts, he immersed himself to a great degree. He was fully aware that his estate was but one of three separately owned estates on the wide lands of Wemyss-shire - and he seemingly determined to somehow unite these three estates into one greater estate.

In 1630, the Earl made representation to Lord Colville of Culross, *(the then owner of the Easter Wemyss lands)* with a view to an immediate purchase of those lands. The offer was seemingly accepted and the Earl gained immediate access to the Easter Wemyss estate. The Earl was so pleased with his acquisition that also included the 'Macduff's' castle, that he took up residence in the historic castle that had for so long been associated with his families' name. *(Parts of the red stone remains of this once proud edifice stands to this day at the edge of the cliffs immediately above the old glass cave - and behind the Eastern extension to East Wemyss cemetery.)*

In addition to his gaining the lands of Easter Wemyss, the Earl now managed to gain control of the lands of Lochhead, Little Raith and other adjoining lands. The Earl knew, or strongly surmised that his own lands were rich in minerals and things - and now that he had acquired these additional lands he now sought to prove at least part of their estimated mineral potentiality. He did this by bringing from England a reputable mining engineer, to carry out 'drilling' tests and probable limited excavations over his extensive new lands, with a view to estimating the potentiality of the underground minerals over the whole of his properties in Wemyss-shire. The Earl estimated that he had upwards of twelve different seams of coal under his lands and hoped that this imported mining engineer would uphold, verify and expand on his own conclusions. The Earl had actually attempted to break into the 'minerals' at a point between the 'Glesse

Cove' *(glass cave)* and the Blair burn, *(stream)* and was convinced that he had uncovered 12 seams of coal. *(It may have been that he actually 'discovered' less than that, his workers probably uncovering the same seam more than once, at different locations.)* The Earl himself was credited with having actually discovered the existence of coal at a place called Lochhead near Lochgelly, which probably gave rise to his subsequent actions in 'importing' a mining engineer from England.

*

# An Act of Council

After the Winter of 1620-21, the Lords of Council passed a Statute that prohibited the sale of coal to foreign vessels as a matter of preference, and stated that the owners of the coals and coal-heughs upon the river and waters of the Forth, *(Firth of Forth)* should 'prefer' the sale of coals to Home vessels at the first instance - and furthermore, the coals should be sold at a price concurrent with that of the past three months. If any coal-owner or his agent were to be found guilty of contravening this stature, a penalty of One Hundred Pounds (Scots) would be imposed.

*This state of affairs had come about because of the very unseasonable weather that had been experienced in the summer of 1620. The weather had seemingly been so bad that the usual and very essential stocks of winter forest wood had not been cut in most parts of the land. This had happened before, but previously, the coal-stocks and the coal-heughs at the Lothians and at Wemyss had compensated for this deficiency. However, at the winter of 1620-21, the coal-stocks at the Lothians and at Wemyss had been severely depleted due to the 'preferred' exports of coal via foreign vessels. Things were so bad in the country at that time, that it was decreed to be* 'ane extreme skarsetie of fewell and fire' *in town and country, with the trades of baking and brewing having been brought to a standstill! What had brought about this calamity the people were informed by this statute, was,* 'the uncharitable doing of the owners of the said coale-heughs vpon the River and water of the Firth of Forth'. *This charge by the statute, was that foreign vessels were being ranked as 'preference' vessels by the coal-owners, to the great hurt of his majesty's subjects and it was to the* 'scandale and reproache of the kingdome that it should be tolerate and overseene!' *The Lords of Council had made their decision and promulgated their decree by statute. Now, it was up to the coal-owners to conform to the new statue and act accordingly!*

*

# An Act of Combination

The coal-owners did act, but not in the way that was envisaged and hoped for by the Lords of Council. There was obviously much discontent and mutterings of mild rebellion amongst the ranks of the coal-owners, but as yet, they were a disorganised voice with little individual 'clout' against the will of the Lords of Council. However, one coal owner decided that something could be done, if the owners of the coal-mines and the coal-heughs presented a united front to the demands of Council. **Lady Faced** *(Janet Lawson)* of Faced castle at Inveresk, issued an invitation to the coal-owners on the south side of the Firth of Forth to a dinner at the castle, ostensibly to entertain them, but in fact, to discuss their possible reaction to the Council's decree. The result of this gathering of coal-owners, was to immediately raise the price of a load *(possibly, one imperial ton)* of coal from the current price of three shillings, to four shillings per load. This was heartily approved amidst great applause by most of the coal-owners present.

*This then, must rank as one of the earliest Act of Combination on record at that time. Not only did this 'consortium' agree wholeheartedly to raise the price of every load of coal, they also agreed that no individual coal-owner would attempt to lower the price of their coals without the specific permission of the 'combination' as a whole!*

*The coal-owners then 'sat back' to await the reaction of the Lords of Council, if indeed, there was to be any response from their Lordships at all!*

\*

# An Act of Retribution

The Privy Council *(Lords of Council)* reacted fairly swiftly. They quickly condemned this 'act of combination' and indignantly charged the coal-owners with **'Resolving on matters injurious to the commonwealth,'** - declared the 'band' of agreement unlawful and null – and **'their persons and goods liable to condign (deserved, suitable) punishment,'** ordained them to return the coals to the former price, and **'prohibit the sale of coals to strangers!'** The coal-owners were further warned that any further acts of obstinacy or further retaliation, would place them at great risk or even peril, and that, directly from the Lords of the Privy Council.

Author's Note :- *I can find no mention anywhere of the co-operation or collusion of the Earl of Wemyss in this Act of Combination, that originated on the south side of the Forth. (I do understand however, that the Wemyss family did have a permanent residence in Edinburgh town at that period.) However, once this decree was issued by the Lords of Council and confirmed onto the statute books, its implications were to affect every coal-owner in the land, including the coals taken from the mines and coal-heughs on the Earl's lands of Wemyss-shire.*

The first Earl of Wemyss died on 22<sup>nd</sup> November, 1649. He outlived his lady wife by approx ten years. He departed this life with the knowledge that he had succeeded in re-uniting the lands of Wemyss-shire and with the profound knowledge that he would be succeeded on the Wemyss estates by his firstborn son **David, the Lord Elcho.**

\*

At the beginning of 1660, **David, the second Lord Earl of Wemyss,** determined on a great scheme for the continued and expansive development of the increasing mineral production in Wemyss-shire. He applied to King Charles II. for permission to build an all-weather harbour at Methil on the coast of Fife. Again, the Earl David seemingly committed his thoughts to his ongoing diary in that he recorded:-

The King, God bless him, did giue me a new gift to bould a herbure at Methil, 1660, and the Bishope of St Androis did erect itt in a free Brught of Barronrie, 1662, called MethiL, wt. a wickly markitt one the Weadnesdays ; and tu publick feaires in the year --- to witte, ane one the 22 of June, St Jo. Day, and the 27<sup>th</sup> December, also St Jos. Day in winter, yt yeir 1662, and so for euir, houldine of him and his suckssr. Bishops of St Andrews, peing him yeirly 20 shillings Scotts as a feu deutie for euir.

The above paragraph basically translates as follows: -
*In the year of 1660 AD. I did receive from King Charles, permission to build a harbour at a place called Methil. At nearly the same time, the Bishop of St Andrews (and of Fife) did raise the village of Methil to the status of a free burgh of the Barony, (of Wemyss) with permission to hold a weekly market every Wednesday throughout the year. Permission was also given to hold two public fairs during the course of every year, with one fair being held on the 22<sup>nd</sup> of June, St. Joseph's Summer's day - and the second fair being held on the 27<sup>th</sup> December, being St. Joseph's winter day. Permission to hold these market days and twice yearly fairs, being given of the Bishop and his successors in perpetuity and paying him the yearly feu duty of 20 shillings Scots (also in perpetuity) for the privilege.*

The Earl, having got royal assent for the building of his 'mineral' harbour at Methil, set about its construction with great will and energy. He was however, thwarted on three separate occasions by the forces of nature, in the form of great and sustained storms, that threatened to tear up the very foundations of the expensive project. It was reckoned that upwards of ten thousand Merks *(Scots silver pounds)* was wasted through rebuilding, caused by the sustained severe weather at that time.

Of the costs to himself incurred by the frequent storm damage during the building thereof, the Lord Earl records: -

> I must show you that this work has been being to me since 2 May 1662
> that I badged the harbor or Peine of Methil to this 2 February 1677,
> being many years. The stone herbure uas thrisse ouir throwin or I gott
> itt to any perfectione, and it hes beine to me 40,000 lib. Scotts to this day,
> 2 Feby. 1677.

*("I make it known just how much trouble I have encountered in the building off this pier or harbour, since its commencement on 2<sup>nd</sup> February 1662. The foundation stonework has thrice been overthrown (uprooted) by extremely bad weather during the building process, before the building thereof was successfully completed. This project to date (2 February, 1677.) has cost me the total sum of 40,000 lib. Scots.")*

The Earl however, was not completely thwarted in his great efforts and the harbour was finally opened for commercial traffic on Sept. 6th 1664. The Earl David recorded the opening ceremony for posterity as thus: -

> I was one 6 September 1654, 54 yeirs of Eaydge. One 15 September 1664,
> Andrew Thomsone in Leiuen did leade his Botte in the new Herbure of Methil,
> wt colles from the colle of Methil, being 60 leades of colles, and he did tak them
> to Leith one 17 of September 1664, which was the first Botte that did leade
> wt colles att that Herbure. The colles uas well loued att Leith, and since
> thorrow all sea ports in Scotland. I sould them att 5 lb. the 12 lodes and
> 2 sh to the griue. I give them 22d for uining them to the coller and 1 sh 2d
> to the caller (driver) of them from the colle pitte to the Herbure.

"On September 6, 1654 I celebrated my fifty-fourth birthday! *On the 15<sup>th</sup> of September of the same year, Andrew Thomson of Leven (ship-owner/Captain) moored his vessel in the new Methil harbour and loaded his boat with sixty tons of coal from the stocks on the dockside. On the 17<sup>th</sup> of the month, he set sail for the port of Leith on the other side of the Firth of Forth with a full cargo of coals.*

*The coal was successfully landed at Leith amidst great cheering and pleasure. Since then, coals have been landed at different ports all around Scotland.*

*I pay the coal-miners 22d per ton to howk (extract) the coal, I pay the carters (with horses and carts) one shilling and two pennies to transport the coal from the pits to the harbour - and I sell the coals at the dockside at five pounds for 12 tons (imp.)  The dockside loader earns two shillings per 12 tons load."*

In the year 1662, the Lord David, second Earl of Wemyss, also committed the following entry to his diary: -

It is my intention to bring a mind from the Sea Bank, on the full sea mark, to drye all the colles, as i have begunne itt already at the Deane Burne in the banks, and so to work just a croping north-west to yt sink (at Methillhill) which this watter work is sett on, and by mynding in stone and setting dounne caire holles one the mynd ; as ye work N.W. ye will drye all thir colles by working the mynd still forward till ye ditte the undermist Colle.

This then was the intention of the Lord David.  There was an existing pit or 'sink' at Methilhill that had been 'sunk' by the first Earl of Wemyss, but that had been abandoned in 1612 because the 'Egyptian wheel or chain and bucket method' could not cope with the rising/increasing levels of standing water.  The first Lord Earl could find no close level outlet, or deep glen into which the standing water could gravitate to drain his pit! *(A favoured method at this time was effectively to drain the mine into a lower lying level or into a burn (stream) within a deep glen. (Valley or wide, deep cutting.)*

The Lord David described to his miners the exact named coals in sequence, that they would intersect as they initiated and commenced this level mine, and also detailed the final seam that they would strike into to complete their immediate task. The target seam was the 'Chemiss' coal that he determined as being 120 feet *(36.5 metres)* below the ground surface at that point. *(Near Methilhill)*  He accordingly gave his contracted miners the following instructions: -

Work the mind first by ane opine cast till ye cum to the Brae yt tak one stone or rock one yr head. Then goe in att ye mind. Work itt five quarters brode, and highe. Itt most have carie holls all along as ye find itt neids ; if ye be oppressed wt water in the sinks or carie holles, work the mind under them, and bore doune to itt, which done will lett the water goe away.

Accordingly, and to specific instructions, his contracted miners began a small opencast working on the lower ground above the high water mark at a place named Deane burn.

*(This was a narrow stream that ran onto the foreshore near the bottom of what*

*is now called the Swan brae and adjacent to the site of what was to become the Denbeath pit, later the Wellesley colliery.)* The miners commenced this dead level working, initially through soft aggregates on a given accurate bearing *(direction)* which was set to follow a true straight line. *(Except for slight deviations to initiate the near vertical air-holes.)* They were to continue this short level 'cutting' inland, until such times as they had a least one metre of solid rock above their heads, where this opencast cutting would change and develop into a straight and level enclosed mine, aimed in the same specific direction and cut through the intervening metals. *(Not only coal, but also different sloping rock formations.)* At the point where the opencast working developed into a mine proper, they were instructed to drive the mine to a width of five *quarters and at a height of five quarters.

*\*At this time I have been unable to recall the modern equivalent of a 'quarter'? Common sense however, dictates that the Lord Earl would not pay out good money on a 'multi-lane highway' to reach these coal seams, even though the 'mine' had to be wide enough to take, or lay, two sets of parallel rails, (either of iron or wood) and be high enough to accommodate the heads of working horses. My estimation would therefore be a 'square tunnel' measuring approx seven feet wide by seven feet high. (approximately 2.15 metres square) Miners can work to this height without benefit of portable platforms.*

{Late entry: - I have returned often to the question of what the Earl meant, when he instructed the miners to drive this tunnel 'five quarters wide and five quarters high'. I have now 'fathomed-out' what he meant!  The system of measurement used then when dealing with depth *(and length)* was the fathom!, a measurement of six imperial feet. *(1.92 metres)*  A 'quarter' therefore, is a fourth part of six feet, i.e. 18 inches *(1.5 feet, or 0.457 metres)*. Five quarters is therefore five times 18 inches, which equals' seven and a half feet! (90 inches, or 2.29 metres).}

As the level mine continued, the miners were instructed to drive near vertical connecting air holes between the lengthening mine and the surface above.  The need for fresher air would determine the frequency and spacing of these air holes at the mine-face, which in turn, would be dictated by the increasing difficulty in breathing experienced by the miners.  Therefore, the frequency and spacing of subsequent air holes to the ground surface would be at the sole discretion of the miner's themselves! *(I would also hazard a guess that the stone miners were not paid for the driving of these air holes!)*

Author's Comment: - *My interpretation of the Earl's instructions with regard to the air holes is that the air holes were not to be placed directly above the line of advance. (This would make sense in that water or debris falling directly into the air hole might seriously compromise the mine.) I have therefore formed the opinion that they may have driven the air holes to a steep angle from next to, or at right-angles to the straight line of the mine. Regarding the miners being able to drive this mine on an absolutely level plane, this should not have presented any problem! The miners did have the help of the most primitive, yet most accurate means of level planing. 'Free water shall always find its own level!'*

In the year 1670, this level mine had been *wrocht* to a linear distance of approx. 1000 fathoms *(6,000 feet/1827.5 metres)* underground. I do not know just how many air holes were actually *wrought* to the surface, but the 'stone' mine was lengthened until it had reached a distance of approximately two underground miles, where it supposedly struck into the targeted 'Chemiss' coal seam. The driving of this level mine had deliberately intersected and 'opened-up' dry access to seven different, rising seams of coal. As the mine had progressed through the first intersecting seam, this gave the Earl the option of working this coal to the rise, from levels cut to the left and right at the point of intersection. These levels could be as long as was economically or practically viable, and could be aspirated through the same air holes already cut to the surface. These workings would therefore all be 'dry' extractions, but only if they were worked to the rise! At this time, the Earl also writes of employing 18 colliers within the mine, excluding the 'stone-miners!' He was therefore, having colliers taking the coals from the intersected seams as the mine progressed inwards!

Regarding the collier's remuneration agreed with the Earl, the following bears reading: -

At Martinmes 1665 A.D. i have agried wt my collers at Methill colle to get 22 pennies Scotts for every leade of colles they shal heu of gritte colle to the Herbure, and 20d for land leade, and 12d for lime colles.

*As far as I have gathered from broad interpretation of the above: - On St. Martin's day, the 11th of November 1665, (the November term-day) the Earl and his colliers agreed on the following rates to be earned for hewn coal. The sum of 1/10d per load for large sized clean coal delivered to Methil docks, 1/8d for every load of fist-sized nuggets of 'land-sale' coal and 1/- for every load of small coals (small chirles) or 'dross!' The latter having only limited commercial or domestic use.*

*The 'stone miners' who were contracted to drive the level mine, were paid in wages with small perks! They were paid ten shillings Scots per day, (or night) with their drawers (putters) being paid six shillings (Scots) per day or night. (Underground 'shifts' were fully 12 hours long at this time). The Earl provided them with candles (tallow) and allowed them free sharpening of their work tools, (mainly hand picks) for as long as they were on fixed wages. There was, only one proviso in the contract that the Earl recorded - and I quote: -* '; and when one faldoms they work men pies all.' *This translates:- 'where the colliers are working the sloping restricted galleries and working at hewing coal at piece-rates, they shall pay for all of their work tools and sharpening thereof - and candles used!'*

This 'Happy Mine' as it was described by the second Lord Earl David was by any measure a very successful mine in it's day. To have driven and gained access to seven different seams of coal and taken coal from all of them, speaks volumes for the far sightedness and daring attributed to the Earl, considering the great cost of the single venture.

Of the recorded expenditure, the Earl noted:-

The Happie Mynd for to drye the 7 colles uas 30,000 lib; then the boulding of 7 pans and ther patts 20,000 ; then the gritte doubell housse, and the horsse work that uas 5 yeirs one colle att the Hill of Methille or the mynd was wrought, cost 10,000 lid, so do trewly declare all was 100,000 lib.

This translates as follows:- *To drive the level stone mine to it's full distance in order to drain all seven seams of coal to the sea, the sum of 30,000 lib. The cost of the materials used in the construction of seven different levels, (one for each struck seam of coal) including the transport system used to get the extracted coal out, - 20,000 lib. The construction and maintenance of the simple coal separation buildings and the purchase and upkeep of enough horses to work the mine/coal for all of five years, - 50,000 lib. The total cost amounting to 100,000 lib. (I cannot say for certain the approximate value of a Lib, but would hazard a guess that it's value was that of a Scots pound, which was worth approx one-twelfth of an English pound. (sterling) That would fix the Earl's cost for the 'Happy Mine' at 8,330 Pounds sterling. The value of a Scots 'Merk', a silver coin peculiar to Scotland, was valued at thirteen and one-third pennies (13.3d imp.) against the English pound sterling).*

N.B. This was designed the ' Happie Mynd' after it's striking of the Chemiss coal at Methilhill! This epithet, no doubt signified the Earl's pleasure at the undoubted success of what was in those days, a very risky venture!

*

At a sometime in the 1620's, the first Lord Earl of Wemyss had initiated a level mine, that commenced in the solid strata at the Barncraig *(seam)* just above the sea high water mark in West Wemyss, to cut into and take *(extract)* as much coals as was possible from the rising strata, with the mine *(level)* having natural land-water drainage and without any part of the level mine being flooded by sea water.   This basically meant that all of the taken coals had to be mined on a plane, that was level with, or above the level of the highest sea tide.  This also meant, that all of the coals in the mine that lay below the level of the high tide mark, had to be left insitu on pain of being constantly flooded!  This was the limiting factor that bedevilled the first Earl so much, that it forced him to literally abandon many such incursions into the cliff-faced, sea-front, coal seams.

Fully fifty years later, **David, the second Earl** returned to the site of his father's Barncraig mine with an altogether new revolutionary *(risky?)* approach!  He knew of his late father's problems of probable flooding with both land water and salt water, but he had what he thought was a possible solution, which might partially overcome this potentially dangerous hazard!

The following passage is again an extract from Andrew Cunningham's 1909 volume on the House of Wemyss:-

> I am still working the 4 fitte colle N.N. East drift geate, allways caring the leuill of the lowest sea tyde wt me, and ther at the motte neir the barn Craige ye will find a new deuisse of a mynd and clousse (not in timber as is the ordinar custome is in the world) but a leanchard or mynd cutt in stone from the loue watter all in rock till ye cum to the motte yt is bilt wt stone and lime.  In yt motte ye will find a littill sink, in uhich sink I caused bore 4 holles with my Eiron rods in the natural stone wall yt is betwix the leuille yt cumes from the sea and yt sink touards the land, in uhich bores I have 4 timber doucks of 3 inches diamiter yt houlds out the watter of the sea when it cumes in and hould in the watter of the colle till the sea be out againe, and a loue watter every tyde.  And the lick is not in Scotland or evir sein befor in the world (yt I heir of) and it is the sourest fastione of any clousse to kipe in watter either of colle or othir occasiones.

As far as I have gathered, or indeed am able to translate from the above paragraph, it seems that David, the second Earl came up with a possible solution to the problem faced by his father, in having to leave untouched, all of the coals that lay below the high tide mark, but above the low tide level. *(This applied to all of the seams encountered and wrought in the original level mine at the high water mark, bearing in mind that this level crosscut mine cut its way through seven seams of coal to reach the Four-foot seam. i.e. starting with the Barncraig, the Coxtool upper, the Coxtool lower, the Den coal, the Chemiss coal, the Bush coal, the Wemyss parrot and then the Four-foot seam).* David, the second Earl, described *(above)* in his own words, how in the 1670's he tackled this watery problem.  Below, is my own belaboured interpretation of the above paragraph: -

"*I have restarted working (extracting) coals from the four-foot seam at the north, northeast mine gate. I have commenced a level gully, cut in the pavement stone from the low tide mark at the water's edge, to a point on the cliff-face directly underneath the entrance to the level mine at the high water mark, worked by the first Earl in the 1620's. I have caused the sidewalls of this cutting to be constructed with stone and lime and not of timber (as is the custom in the world?).*

*At the inside end of this cutting near to the cliff face, I have had built a moat cum sluice-gate. Between the moat and the rock face, I have excavated a short intrusion into the face of solid stone, in which I have had drilled with iron rods, four separate, three-inch diameter holes, to link up with a 'sink' that I had commenced at the other side of the natural rock (barrier). In the three-inch diameter holes, I have placed wooden plugs to prevent the ingress of the sea water when the tide is high, which they remove when the tide is low, to allow the drainage of land water (and saltwater seepage) to the sea at low tides. I believe that the likes of this has never been seen before in Scotland, or indeed, ever before anywhere in the world! (That I have heard of!) It is the surest fashion of a moat /sluice-gate to hold out the water of the sea, or of the land, whether it be from the wet coal (wet coal seams) or other means.*

David, the Lord Earl, was now able to drive this lower mine into the strata at the level of the low tide mark, knowing full well that as long as the solid face of rock held firm and the plugs were not removed prematurely, the standing water in the new deeper mine would be 'allowed' to drain away at Low water! He did not take too many chances with this innovative arrangement, in that he caused a man to be paid *(or fee'd)* to 'attend' the plugs on a daily basis.

\*

In the year of 1654, during the month of May, the Earl caused a 'sink' *(vertical shaft)* to be commenced near to the Blair Burn at a point above the high-water mark. He contracted with two miners to sink this shaft to sixty feet deep *(approximately 18 metres)* and at a width and breadth of 16 feet and 10 feet respectively. *(4.9 metres by 3 metres).*

The terms of the contract as recorded by the Lord Earl were as follows: -

I to give them Twenty pounds for every fathom of the said sink from the grass or strike-board till they set me down 10 fathoms, also 4 stones of iron, a loan of my quarry mell, and also 4 bolls of oat meal (2 at the first and the other two at the 10 fathoms end) and after that I am to agree anew with them or others ; But till it be 10 fathoms down, although they should meet with never so much water or hard stone, this is all they will get from me ; I cradling the sink and furnishing all windlass works.

It would appear that the Lord Earl struck a hard, uncompromising bargain with his two contracted miners. The Earl would initiate the contract by supplying the two men with approximately 12 bushels of oatmeal, (one 'boll' equals approximately six bushels in dry measure) and a half hundredweight (approximately 25 kilo's) of iron, (from which to have metal tools fashioned by a blacksmith). The miners were to be paid twenty pounds (Scots) for every 1.83 metres depth achieved. The Earl would lend the miners a heavy quarry mallet for the duration of the sink - and the Earl would install a pulley cradle and a windlass, with which to remove the mined waste from the sink bottom. The Earl further stipulated that he would not change the terms of the contract even if the miners struck water, or were further delayed by striking through hard stone. The only 'get-out' clause seemed to be, that when and if the miners did strike down to ten fathoms, then he could negotiate any other working with them, or any other miners. The carrot at the end of the stick also seemed to be, that when the miners struck the sink bottom at 60 feet, they would receive another 12 bushels of oatmeal!

Strike-board :- In this application, the strike-board would comprised of a flat rectangular wooden platform, built around the periphery of the hole at ground level, to act as a stable working platform and to stop the edges from crumbling in!

\*

**David, the third Earl of Wemyss,** (1705-1720) succeeded to the Lordship of Wemyss in the year 1705. He succeeded to the Earldom on the death of his mother, Margaret, the Countess of Wemyss, who, in the absence of a living male heir had succeeded to the 'Earldom' as **Countess of Wemyss** on the death of her father David, the second Lord Earl. (*Thus we have a first-hand example of a title being passed down through the distaff side of an ennobled family in Scotland!*)

Around the early 1700's, David, the Lord Elcho, before he had succeeded to the Earldom, did engage himself in an industrial enterprise on the foreshore of the Wemyss estates. He used the great cave between Easter and Wester Wemyss and had it fitted-out for his proposed venture. He had already approached the Lords of Parliament, to have himself and his business partners protected (*an early form of patent rights perhaps?*) from any possible competition in the making of certain types of glass. He needn't have bothered! After considerable expenditure by the partners in the installation of furnaces and interior building work, the project was abandoned.

(*This was not the first attempt at glass-making in this cave, but it seemingly was the last. These operations gave future description to this large cave, which from then on was described as the 'Glesse Cove'. In later years, the structural integrity of this cave was severely compromised by the underground workings between the Victoria pit, Lochhead pit and the Michael Colliery, in that part of their interconnecting levels lay directly underneath the foreshore in this area. The underground 'Michael Level' in the Dysart main seam of coal, ran from the Victoria pits' Sea Dook, through the bottom of Lochhead's Old Dook and on to one of the Michaels' pit shafts.)*

It was then that the young Lord Elcho turned his attentions to the mineral workings on the Wemyss estates. He seemingly employed a certain Dr Melville to advise upon and overseer the new Wemyss-shire mining development at Kirkland near Methil. Dr. Melville had seemingly hit upon the idea of using horses and oxen to pull the loaded hutches, thus improving and speeding-up the underground transport system. This introduction of working horses and oxen into the 'level parts' of the early coalmines, must have been rated as a great success, in that it was a practice that was to endure and be improved upon, for at least another two and a half centuries. In, or around 1705, the young **Lord Elcho,** who now succeeded to the Earldom on the death of his mother the Countess of Wemyss, must have felt sufficiently confident of his abilities as a coal-owner/overseer in that he took over the management and running of the 'coal-pit' at Kirkland. He was probably twenty-seven years of age on his succession to the Earldom, after the death of his mother. He was now **David,** the third **Lord Earl of Wemyss**. (1705-1720)

*(In the year of 1704, the then Lord Elcho entered into an arrangement where he 'lent' to Mr Christopher Seton, (brother to the Earl of Winton) the use of six colliers and eleven 'bearers' (or drawers) for labour at a coalmine in Tranent. The arrangement was that Seton would deliver up the said colliers and bearers 'on demand' if David, the Lord Earl required their immediate return.)*

Author's Note: - This arrangement shows that miners/colliers *(and their immediate families?)* were still 'owned or bound' by the coal-masters, who obviously felt free to do with them as occasion or opportunity demanded!

*

**James, the fourth Earl of Wemyss,** assumed the Earldom of Wemyss on the death of his father David, the third Earl of that name in 1720. He held to the Earldom for a total of thirty-six years until 1756, where the Earlship was infefted on his second son Francis, now resident on the lands of his 'good' father on the south side of the River Forth. He Francis, had now assumed/adopted the name of Charteris, his wife's family name and styled himself Francis Wemyss Charteris. Thus, on the death of Earl James, the Earlship left the original branch and lands of the Wemyss family seat in the Kingdom of Fife, never to return.

To the best of my knowledge, James, the fourth Earl of Wemyss, did not leave any published diaries, but he seemingly did leave of wealth of categorised correspondence dealing with both the labour troubles he experienced with his workers at the salt pans, and with the coalminers at his Wemyss pits. In one letter he wrote to his factor of the time, *and underneath, I have quoted from the original:* -

> Since these tenants are so stubborn that they won t coall the pans without their own price, I know no other way than first to protest against them for damnadges done me by their not working, and then to cause Baily Malcolm hold a Court on Munday, and any who stand indebted to me by the lists of rests to throw him in prison untill he pay d, and to break one of their tacks to deterr them from doing so in future. I think that the salt that s lost by the pans not going should be stated to their accounts,

The above communication is most illuminating in several aspects, not the least of which, is the very clear anglicisation of parts of the earlier Scots tongue within two generations, both in changed spelling and in syntax! The meaning thereof should be abundantly clear, except the following: - For 'coall the pans' *read 'fuel the underneath of the salt water boiling pans with coal'*, for 'Baily' *read 'Magistrate'*, for 'lists of rests' *read 'roll of company debtors'*, for 'stated to accounts' *read 'deducted from their credited wages'.* To 'break a tack' *means to 'dispossess them of their company house'.*

*Also, note the intent to extract from his workers, the probable profit from the lost production of sea salt!*

The Lord Earl also had problems with some of his '*bounden'* salt-workers, his miners and their families, in that a few of them were forever trying to 'break' the bonds, that still existed between the coal-owners and their bound workers *(and their children).* If a salt-panner or a miner decided to make the 'break', there were very few options open to him, as readers will conclude from the text of the two undermentioned *letters: -

> Don t forgett to write the name of the coallier and his wife which run away from Methill a few days before I left home, and desire William Forbes to search for them at Pinkie, and for Lindsay, and gett them over.

*It would seem that William Forbes was the name of the 'Bounty-hunter' employed to apprehend the 'runaway' miner and his wife! (and their child Lindsay?) It also seems from original correspondence, that when the arrest of one of the salt-panners was imminent, the arresting officer was 'deforced' by several of his co-workers, which was why the bounty-hunter was called-in! Local history does not seemingly record as to whether they actually apprehended the collier and his wife!*

A full year later, another letter in the same vein was also sent to his factor/manager. This communication is also quoted in full: -

I do not see that you had any occasion to delay requireing back stragled coalliers till you advis d with the commisioners, for that was a strict charge given you to look after them, and in consequence of the coall proprietors meeting at Edinburgh ; therefore the moment a coallier leaves his work he ought to be sent after immediately, otherwise it gives him time to get into England, where he can never be recover d. And when the grieves don t represent this to you in time, they ought to suffer for it. Besides the coalliers, thier children should be all look d after and sett to work below ground when capable, and not be allow d to hirr d cattle, or go to service, as many of them have done, and I wish may not be the case as yett. And if you see it for my benefitt, and that there s work and room for more people below ground, why don t you gett some of Balbirny s coalliers, who are now in different parts of the country, and nobody s property. Pray are Alexander Leslie s and Thomas Lumsden s children now working at the coal work ?

*As can be observed from the above original text, the tone and meaning of the correspondence is clear, although some of the sentences are very long with a syntax that could be termed quaint. This however, does detract from the implications that in the Earl's view, (and those of other coal owners) these men and their wives and families were wholly tied to him, the estate, his salt pans and his coalmines. It is apparent that the 'binding' conditions included the 'tack' of the tied house, the commitment to toil in the Earl's employment (whether it be on the saltpans or underground) and the promise of the bounden miner to take his offspring underground when they had 'come of age', or when deemed appropriate by the Earl. It is also obvious, that the Earl and his factor/manager were fully aware of the ages and 'maturity' of all the youngsters within the mining families!*

*It would also seem that the Earl was fully aware, that if miner's runaway (spirited-away?) adolescent children were not apprehended before they could gain employment on a farm, (herding cattle) or by going into service with the landed gentry, they would then be lost to him into the foreseeable future. The Earl also intimates that when his under- managers (grieves) are guilty of not informing on a runaway miner 'tut suit', then the recalcitrant grieves should be punished somehow. The Earl was also always on the lookout for other 'untied' workers, in that he extols and behoves his factor, to search out those ex-Balbirny coalminers who are seemingly roaming the countryside searching for food and shelter (and 'untied' work?)*

*The letter also shows that the Earl had an awareness and knowledge of his 'resident' worker's families, in mentioning the names of two of his workers by name, along with the pertinent knowledge of the children's 'work potentiality!' The Earl looked to his manager and through him to his grieves, to keep him well informed as to the 'well-being' and the potentiality of all of his bound families.*

*

The **Hon. James Wemyss of that Ilk,** *(1756-1786)* succeeded to the Wemyss estates in 1756, after the death of his father, the fourth Lord Earl. The Hon. James was not the natural inheritor of the Wemyss estates, bearing in mind that he was the third son of the late Earl. The right of succession should have gone to the Earl's first born son, the Lord Elcho, but fate *(and Parliament)* decreed that he spend the rest of his natural life in exile on the continent, due to his active support for the 'young pretender', Bonnie Prince Charlie at Culloden Moor. The Earl's second son Francis, eventually succeeded to the Earlship on the death of the exiled Lord Elcho and also to his estates in the Lothians, but did not succeed to the lands of Wemyss-shire due to the Earl having settled the Wemyss estates on his third son, the Hon. James Wemyss of Wemyss of that Ilk.

The Hon. James, Laird of the Manor, did have an interesting career before he involved himself in the affairs of the estate. He joined the Royal navy as a mid-shipman in 1741 and in September 1745, attainted the rank of lieutenant R.N. In 1757, he married Lady Elisabeth Sutherland his cousin and in doing so, decided to give up his navy career in order to promote himself as a politician and potential parliamentarian. In 1762, they returned him to parliament for one term as the member for Fife until 1768, when he was ousted by Colonel Scott of Balcombe. He did not however, stay out of Parliament for very long because he was adopted by the good folk of the constituency of Sutherland as their M.P. Where he remained as their representative until his death in May 1786!

James the Laird, did involve himself a great deal in the running and developments of both coal-pits and salt-works at Wester Wemyss. He was responsible for the sinking of what became known as the 'Engine-pit' at West Wemyss. This shaft was sunk to a depth of 40 fathoms, *(240 feet, or 73 metres)* to strike into the Chemiss seam of coal that was approximately 8 to 9 feet *(2.5 to 2.75 metres)* thick at that depth. In striking to this depth, the sinkers had bypassed six shallower seams of coal, the next thickest being the Barncraig seam at approximately six feet thick, but which had probably been 'taken' at an earlier time.

Having struck into the Chemiss seam, the immediate task was of course, to drive longish side levels to the left and right of the sink bottom, *(following the level of the coal)* in order to commence a series of separate 'headings' up into the rising seam, to simultaneous take coals at different locations. This breaking into the Chemiss seam was so much of a success, that the Laird decided to commence the taking of the seam to the dip! This involved the colliers having to do an about-turn at the sink-bottom levels and commence to drive several dippings or 'dooks' into the dipping seam, out and under the sea-waters of the foreshore. This then meant, that incoming water, whether it be land water or sea water, would eventually, if not sooner, gather at the bottommost part of the dippings or dooks and render them unfit to be worked. This water would quickly rise, fill the dooks and then the 'sink/shaft' up to mean sea level. The pit would then become flooded and unworkable!

The laird overcame this potential water problem, by installing a great steam pump on the 'pithead', that was powerful enough to control the level of the ingressing water and raise it the to the pithead, where it would be allowed to run down into the sea. This powerful steam engine cum water pump was quite a wonder for its time - and was probably the first of it is kind to be employed for such a purpose! This pit therefore became quite famous for this unique and inspired innovation and thus of course, became known as the 'Engine Pit!'

The Hon., James Wemyss of Wemyss died in Edinburgh in May 1786. He was sixty years of age.

*

**William Wemyss,** (1786-1822) the second oldest surviving son, succeeded to the Lairdship on the death of his father, the Hon. James. He was 26 years of age. William Wemyss seemingly had a glittering Army career in which he rose to the rank of 'General' officer. He also had an early secondary career, where, like his father before him, he entered Parliament first as the Member for Sutherland, then secondly as the member for Fife, on the death of the sitting M.P., Lt. Gen. Robert Skene.

I can find little record of General Wemyss association with any of the estates coal-mines, but obviously, the coal-heughs, pits and mines as they were now being named were being run and managed just as before. It is also a matter of record that General Wemyss was a great lover of woods and forests, and was responsible for the planting of the great wood that lay between the West side of Easter Wemyss and the Wemyss castle.

In 1815, the General made an approach to the Government, to request the loan or the grant of a goodly sum of money to carry out major reconstructions at the docks of Methil. Seemingly, such was the volume of coal and salt that was now flowing through the harbour /docks that the existing facilities were being hard pushed to cope. The request to the Government was turned down at that time, but the need for further increased handling capacity at the docks was pressing. The seeds however, had been sown and it was left to another future member of the Wemyss family dynasty to bring the vision of an extended Methil dock to fruition.

General William Wemyss died in February 1822, at the family home in Wemyss.

*

By the time **James Erskine Wemyss** succeeded to the Lairdship of Wemyss in 1822, he already held the rank of acting Captain in the Royal Navy.  He had retired from the Navy in 1814, with the understandable resolution to embrace a political career.  In 1820, he was chosen and returned to parliament as the Liberal representative for Fife.  His political career lasted from 1820 until 1847, where he did not seek re-election.

In 1836, the Baronies of Torrie and Lundin were added to the holdings of the Laird of Wemyss, being inherited through the laird's mother, wife to General William Wemyss.  The barony of Lundin was later disposed of by the then laird.

In 1837, long before he was promoted to Rear-Admiral in 1850, *(obviously, he was still on the Navy list!)*  Captain R.N., the Laird of Wemyss, was appointed Lord-Lieutenant of the County of Fife.  He died in April 1854 at Wemyss Castle.  He was sixty-five years of age!

*

The following is an edited extract from the 'New (second) Statistical Account of Scotland' MDCCCXLV, (1845) with regard to the coalmines in the Wemyss area:-

There are four coal pits in this parish of Wemyss.  The Wemyss pit (WestWemyss) employs 140 men, 24 boys and 42 girls.  (Could this have been the original Victoria pit shaft sunk in 1824?) The annual output from this pit is approx 40,000 tons at a selling price of 8/6d per ton. (42.5p) The coal taken is the main coal that is 9 feet thick. (This could only have been the Chemiss coal at that thickness and, was the deepest seam to be tackled in those days at 100 yards below sea level and worked in a widening area under the whole of West Wemyss.) The other pit/working? - in West Wemyss, took coal that was two seams below that of the Chemiss - namely the Wemyss parrot coal. But, this coal is described as being taken level-free , which may have meant that it was wrocht as an outcrop considering that it was an inherently gassy coal. Working this seam as a coal-heugh would have obviated the need to burn-of the exuded gasses! There were 20 men employed at this mine where the extracted coal sold for Ten shillings per ton. (10/- or 50p.)

XXXI

The other two coal pits within the parish that are mentioned in this extract are described as being wrocht exclusively for land-sale coal. The two pits employs About 50 men, 20 boys and 7 girls. (As to the exact location of these other two pits I would hazard a guess (considering the date) that they might just have been the original Lochhead pit taking shallower seams - and, the Duncan pit, not as yet sunk down to the 39 fathoms required to work the Dysart Main seam!) At this time, the Account mentions that there were some very powerful engines employed in this very extensive coal establishment. And also, that all the recent improvements in mining machinery here have been very successfully introduced and applied and, are all under the very active and efficient management and direction of Mr David Landale - mining engineer.

It may also be deduced that since no mention has been made of any other mining engineer or, colliery official, that Mr Landale s responsibilities actually extended to the other three pits in the Wemyss area! Mining engineers not being too much of a commonality at this time!

The original author of this part of the 'Account' was the 'Rev. John McLachlan, Minister, Presbytery of Kirkcaldy, Synod of Fife.

\*

**James Hay Erskine Wemyss,** (1854-1864) succeeded to the Lairdship of Wemyss and the baronial estates of Wemyss and Torrie in April 1854, whilst he was still in his twenty-fifth year. He had seen service in the Royal Navy, first as a twelve years old midshipman and then as a young lieutenant. In 1848, he left the Royal Navy on grounds of medical advice, after having contracted a tropical fever while on service around West Africa.

In 1859, the laird, Mr James Wemyss M.P., was returned to Parliament as the member for Fife, having beaten his opponent Lord Loughborough *(son of Lord Rosslyn of Dysart)* by a near thirty percent margin. This period also saw the rise and early development of the county volunteer force in Fife, in which Mr Wemyss became heavily involved, so much so in fact, that Mr Wemyss became the first Captain of No. 8 Company, that included the Leven and Wemyss contingent of the volunteer force. A few years later when the volunteer Company had expanded, the Wemyss contingent was separated from the larger body, taking the new name, the No.10 Company. The Laird, Captain J.H.E. Wemyss, took formal command.

Capt. James Wemyss' ability as a competent, capable and efficient commander was further recognised with his promotion and appointment, as Major-Commandant of the County of Fife Artillery Volunteers.

Captain Wemyss' interests did not only include the local militia, he also took a very keen interest in the agricultural and mining aspects of the Wemyss estates. It was said that cattle-breeders, farmers and agricultural minded gentry rode/coached into Wemyss from all over the country to see, inspect and examine the Laird's polled *(lopped or cropped horns)* herd of Aberdeen Angus cattle beasts. There was also and probably, three separate coal-mines being operated on Wemyss land at this time. The Victoria pit, *(long before it's deepening to take the Dysart main seam)*, the Lady Emma pit *(now long abandoned)* and the Barncraig. *(This latter could have been a level mine. The Barncraig coal was at a relatively shallow depth!).*

Author's Note: - *At this time, I can find no mention being made of either the Lady Lilian or Lochhead colliery.*

In January 1864, the Laird of Wemyss, James Hay Erskine Wemyss, was appointed Lord Lieutenant of the county of Fife. However, he was never destined to appreciate the appointment. Previously, during the past month of November, the laird had somehow contracted an ailment that quickly manifested itself into something more deep-seated, so that his health and strength deteriorated rapidly. Plans were made to take the ailing laird to Nice in the South of France in order to better his health - and towards recuperation, but he did not even manage to cross the channel. The laird succumbed to his illness in March 1864, at Buckingham Gate in London.

\*

On the death of his father, the young master, **Randolph Gordon Erskine Wemyss,** was but in the sixth year of his childhood. He could not therefore truly aspire to the Lairdship of the Wemyss estates until he had reached his majority in July 1879. The estates meanwhile were managed under the trusteeship of his mother Mrs Wemyss and several other titled notables. The late laird, James Hay Erskine Wemyss, had shown implicit faith in the managerial abilities of his lady wife, in that in 1860, he had executed a trust-settlement in favour of Mrs. Wemyss and a few other notables, probably knowing full well that if anything should happen to him, the management of the estate and the upbringing of their children would indeed be in very capable hands.

During the pupilage of the young laird and after the death of her husband, Mrs. Wemyss did not remain idle for very long. Being the woman of vision that she was, somewhere around 1871, she seemingly came to the conclusion, that if the mineral *(coal)* wealth of the parish *(West Wemyss?)* were to develop and expand, a positive attitude must needs be taken regarding improvements to the harbour at West Wemyss. Mrs. Wemyss then initiated a plan of action and commissioned Messrs.

J & A Leslie of Edinburgh, to produce practical plans for a wet dock at West Wemyss. Planning and preparation soon got underway and work was commenced around August/September in the following year. (1872) Within one year, the wet dock was opened at a terminal cost of around £10,000 sterling. At the opening ceremony, the first boat that cut the blue ribbon that stretched across the dock gate, was the vessel owned by the young laird, the Millicent!

In the mid part of the 1870's, the parish of Wemyss was struck by the scourge of a cholera epidemic, that many good folks including Mrs. Wemyss, put blame to the totally inadequate fresh water supply serving the parish. Before this, Mrs. Wemyss had given much thought to improving the supply of fresh water to the parish of Wemyss, knowing quite well that one of the keys to future expansion and subsequent development of the parish, lay in the supply of a consistent wholesome supply of clean fresh water. Up to that time, the supply of common water to the village was hopelessly inadequate, non too clean and commonly flowed from partially contaminated sources. In the aftermath of this dreadful catastrophe, the persistent agitation and promise of additional help from Mrs. Wemyss, were finally sufficient to encourage the Wemyss Parochial Board *(the local authority)* to introduce a constant supply of fresh clean water, to every miner's dwelling place in the parish. This was achieved by the creation of a gravity-fed supply from a controlled reservoir on the higher ground, well above and away from the villages and hamlets of the parish.

Because of this constant agitation from Mrs. Wemyss and the young laird - and the promise of every possible facility at their disposal, the Wemyss Parochial Board finally adopted a comprehensive scheme to build a large waterworks, to take and deliver clean fresh water from Carriston Reservoir, near Star, to the villages and hamlets in most parts of the parish of Wemyss.

The water scheme was adopted at the beginning of 1877, and the resulting work lasted for approximately twenty months. During October 1878, the fresh water supply was finally turned on in the parish, much to the great delight of its inhabitants. The young Randolph Wemyss though still technically in his pupilage, played his full part in these proceedings.

Mrs. Wemyss for her part, probably felt that this was her final gesture as head of the board of trustees, and something that she felt was bound to encourage that old feeling of sympathy, understanding and attachment, that had always lain between the Wemyss family and the people of the parish of Wemyss.

\*

In July 1879, the young laird **Randolph Gordon Erskine Wemyss,** attained his majority and almost immediately enthused his parishioners with the news, that within a few years, there would be a railway connection between the villages of

Thornton and Buckhaven. Parliamentary permission had already been granted to the laird to instigate the construction of this much needed extension, so tirelessly promoted by the laird's mother, Mrs. Wemyss. This railway scheme, promulgated and enthusiastically undertaken by the young laird, *(at a proposed cost of £25,000 sterling)* continued with alacrity and was completed and opened in August 1881. This short two-year construction time for the building of this railway line, surely exhibited the early zealous and fervid, industrial trait that was to underpin nearly all of the laird's future undertakings during his lifetime. *(One of the ironies of this railway connection between Thornton and Buckhaven, was that the railway line did not touch the village of West Wemyss at all! West Wemyss did have a 'named railway station', but that was situated within the 'Firs' to the north side of the Standing-stone road (A915). The distance from West Wemyss to its named railway station, was fully one and a half miles as the crow flies!)*

As mentioned earlier in this narrative, the underground coal reserves in the Kingdom of Fife, fall into two distinctly separate fields. Not only that, but at this time, the quality *(burning or steam navigation quality)* of the coals from the two coalfields fell into different categories, that partially served to decide their eventual destination. The 'steam' coals from the West Fife areas including the 'Dunfermline splint', were classified as grade I, while the coals from the Wemyss-field, *(upper coal measures)* were classified as grade III. This anomaly resulted in not only a difference of approximately one to two shillings per ton in the selling price of the different coals, but also as a bedevilling classification that unfortunately helped determine the disposition of the coals.

*{The Dunfermline splint coal was rated much 'superior' as a steaming coal, than that of any of the mixed coals from the Dysart Main seam. The problem (if indeed, a 'problem' did exist?) with the Dysart main \*named coals, was that they were very rarely worked in isolation (i.e. - Grounds & Nethers with Myslen, - Toughs & Clears with Sparcoal - etc.) with the marked exception however, of the thicker Coronation coal.}*

One obvious result of the difference in the steam grading of the coals, was that more of the West Fife coals was retained for 'home' market consumption, *(around 50% of total production)* while approximately 75% to 80% of the supposedly lower graded, Wemyss coalfield output was shipped for export. At this time, most of the output from the Wemyss coal-fields was being handled through the docks at Burntisland, where the handling/conveyance charges amounted to 1/- *(one shilling)* per ton in Wemyss Coal Company wagons - and 1/6d *(one and sixpence)* per ton in Railway Company wagons. This expenditure obviously gave the young laird much food for thought, especially because of the sobering knowledge that the mineral wealth that lay under the Wemyss estates was much, much more than he had ever realised. He also realised that the output from his

Wemyss coalmines, along with that of his two coal tenants, Messrs. Bowman & Coy, and the Fife Coal Coy, could only but increase in the long term - and that another practical and ultimately cheaper means must be found, to handle the steadily increasing outputs from all sources in the East Fife coalfields.

* See chapter on 'The Taking of the Dysart Main'.

To this end, the young Laird made overtures to both of his tenant coal companies, to solicit their views on the possible construction of a dock at Methil, to handle the combined outputs of their three enterprises. Because of these overtures, both of the tenants' companies entered into an agreement with the laird, in that they would guarantee to pay dues on a fixed amount of yearly tonnage, if the laird could provide them with shipping and docking facilities at Methil - and furthermore, they also agreed that if their yearly tonnage did fall short of the agreed amount, they would reimburse the laird in cash monies to make up the shortfall! At this time, the Bowman partners were shipping a yearly average of approx. 35,000 tons from the harbour at Methil, - while the Fife Coal Coy. were shipping approximately 40,000 tons from the loading dock at Leven.

The only possible obstruction that could thwart the young Laird's proposals, was the fact that the **Leven Harbour Board** had began negotiations with some local traders, with a view to improving the facilities at Leven, in the form of an extension to the existing basin. The one damning disadvantage to L.H.B's proposal, *(that seemed to work in the young laird's favour)* was that the mouth of the River Leven was forever silting-up with the abundant shifting sands, that stretched for miles around this part of the Fife coast.

*(Not for nothing was this part of the Fife coastline described as the Golden Fringe!).*

This was perhaps, one of the main reasons that many local traders fought shy of such a potentially risky investment.

Even though the Leven harbour board's plans for an extension to the basin did not come to fruition, the fact that they were still in business could prove to be a stumbling-block to the young laird's plans. He then did the one thing that would remove this competition, or possible opposition to his proposed venture - and that was to initiate and enter into negotiations for the probable buy-out of the Leven Harbour Company. Negotiations were commenced and proved to be very successful, with the Leven Company parting with their holdings for approximately £12,000 sterling. This, for a business that had cost around £40,000 sterling to establish!

This obstacle removed, the Laird in 1883, then applied for parliamentary powers to construct a 'wet' dock to the north side of the existing harbour at Methil. This proposed wet dock would cover a 'ground' area of approx. 22,000 square yards, or approximately four and a half acres. *(In today's terms that would amount to approximately 18,400 square metres or 1.84 Hectares.)* Parliamentary permission was soon granted to the Laird, who lost no time in appointing Messrs. Cunningham, Blyth & Westland as contractual engineers for the project.

On the 28th of July 1884, the Laird of Wemyss, Randolph Gordon Erskine Wemyss, married the Lady Lilian Mary Paulet in St. Paul's Church, Knutsbridge. *(Here, we now have the origin of the previous name given to the second Lochhead Colliery, - the Lady Lilian!)*

The Laird of Wemyss at this time, had no empirical knowledge regarding the building of a harbour, or an all-weather sea dock - and could only but guess at the problems that were likely to beset the planners and builders of this large undertaking. He therefore, for possible guidance or even inspiration, turned to the diaries of one of his illustrious ancestors, David, the second Lord Earl. He then read the original passage that gave instructions to the then builders of that harbour, in the year 1664.

I ordine you to bould a harbour in that place called in ould the Quarrille of Methil. Ye shall bould itt as follows :-- Let it be all of stone, for ye have abundance ther. It must be very brode (yt ye may cause it slope or botter the more). Bould it like to a half moune. When ye bould this herbure, boulde it so brode yt horses wt ther lodes may cum wt the colls (to the ship s seids, Bark s or Botts), and this will scaue the bearing and mak gritter despatche then they were led dounne at the full sea one land. Be anything bould itt high that the seas ouerflowit not in great stormes . . . Bould the peart yt is fardest in the sea first, for the entire of the herbure must be first diggit and made cline. . . . Above all, remember yt what ye found in March yt it be fully ended in October or winter stormes cum one, otherwise all your sumber s work is lost, and euery yeir ade to it so much as will be ended or winter cum one. I find by the aduice of the most juditius of maroners there most be a little kie.

*To the uninitiated, or to non-Scots readers, I shall again venture a close translation of the original script, as printed above: -*

"I ordain (commit) you to build a harbour in that place known of old as the Quarry haven (heaven?) of Methil. You are instructed to build it as follows:- Build in completely in stone of which there is a great abundance to hand. Build it high enough so that the winter storms with their high seas shall not pervade its walls. Make the walls broad enough, so that horse-pulled coal-carts may be driven directly to the side of any ship moored alongside, thereby saving time and labour, than when the horses and coal-carts were driven overland to a ship's side at high tide. Shape the base and the foundations of the harbour walls in a curving outwards direction, so that they may add strength to the building thereof and break-up the mounting seas.

Build the outer walls and that part of the quay that is farthest out into the water first, in order to protect those and the work that needs must be undertaken to dig out the deep basin of the dock.  Above all, do remember that what is commenced in the month of March, must be completed by the month of October - lest the rough winter storms undo all that has been done in the Summer months.  Work only between the months of March and October in each year, with each phase wrought to completion before the onset of Winter.  I am also advised by the most experienced of Mariner's, that there must be a seaward quay! "

As to how much the young laird followed the advice gleaned from the writings of his famous fore-bearer?  I have no way of knowing!  However, it is a matter of record that the wet dock at Methil *(later to be identified as No. 1 dock.)* was officially opened amid great rejoicing in good weather on 5th May 1887, less than four years on, from when it was first mooted. *(The young Laird was still but 29 years of age.)*

This opening ceremony was reportedly attended by up to 15,000 persons from all over the Kingdom of Fife, who wished to be part of the monumental proceedings, many of whom were coalminers who had been given a single day's holiday, in order to partake of the celebrations.  The first coal-boat that entered the new dock was the S.S. Newhaven.  It was soon moored to the dockside next to a fully working wagon hoist, that almost immediately began pouring coal down the great chutes and into the boat's coal-hold.  The opening of this dock at Methil, actually produced a domino effect that presaged the opening-up and development of many shallow coalmines in the Wemyss-field areas, and to an extent that would be almost unbelievable.

In looking at the date of the opening of this dock, and at the start date of the sinking of the Lady Lilian shaft down to take the Dysart main seam of coal in the 1890's, it would seem that one event was closely followed by another, especially considering the additional fact that Randolph Wemyss then came to the conclusion, *(regarding coal and coalmines)* that since there was no gathering stock of any type of coal at any of the Wemyss Pits-head, the coal markets both at home and abroad probably could and eventually would, consume much greater quantities, of yet to be extracted Fife coals.

*(The mined coal was being immediately transported for export and local consumption, as soon as it was extracted and surfaced.  Coal does not need to 'mature' before being consumed!)*

In the early 1890's, the Laird seemingly took stock of his position, with his thoughts turning to the opportunities to be gained from owning a great coalfield that had the potential to produce much more coal, and a docking facility from which to export nearly all that could be extracted from his working pits and mines.   This seemed a most propitious time to expand and develop his underground resources - and to the extent where he visualised many more coalmines with a greater number of colliers being employed, and with his new dock working at full capacity. What was the point of having this great expensive dock facility running at reduced capacity? Besides, what had been accomplished once, could be added to and expanded!

To this end, the Laird gave his full attention about how best to continue with coal developments in other areas of Wemyss-field. In the past, the coal development had been restricted to West Wemyss and the Coaltown of Wemyss, with the extracted coals mostly taken from seams that were in reality, quite shallow. The Laird knew that the thicker and probably more productive seams lay at a greater depth than had previously been worked, and if they were to be tackled, it would take time, planning and inevitably, much capital investment.

Randolph Wemyss then made three interlinking decisions, that in their implementation and execution, they were to have great bearing on the future prosperity and social development of the mining communities of the three villages' Wemyss, the embryonic Wellesley colliery, the model village of Denbeath and the future of Methil Docks.

The first decision was probably made out of necessity for the future of the second Lochhead pit, *(and Surface Dipping?)* in that, either most of the shallower coals had been taken with the original Lochhead pit with the seams worked to exhaustion, or the Laird having decided that he wanted to tap into the as yet unknown, underground bounty that awaited the miners in the breaking onto, and the subsequent development of Fife's greatest seam of coal, the Dysart Main! The Laird's next two decisions *(and that of his board of management)* were inextricably interlinked, and formed the basis for an under-taking that was to result in the initiation and development of the deepest, and probably the most prolific coalmine in the Scottish coalfields, the Michael colliery!

The Laird, had for some time, pondered the idea of sinking a new pit somewhere on the Wemyss lands, preferably in the local vicinity and at least within striking distance of his yet to materialise mineral railway. The site the board settled on was near the foreshore at East Wemyss, where there was ample space on near level ground, that would accommodate the necessary buildings and machinery, to manage and develop such a gigantic operation.

While Randolph Wemyss realised that the development costs for such an enterprise would overwhelm his financial means at this time, he felt that this could be a project, where the future gains in triple terms of job prospects, future employment and financial reward would far outweigh the original initial investment. To this end, he was instrumental in the flotation of a new company to be named the **'Wemyss Coal Coy. Ltd.\*'**, (1894) with a capital of approximately £225,000 sterling. The shares available for purchase, would consist of fifteen hundred (1,500) ordinary shares at one hundred pounds (£100 sterling) each, and seven thousand, five hundred (7,500) preference shares *(cumulative at 6%)* at ten pounds (£10 sterling) each.

*Directors of the New Wemyss Coal Company Ltd in 1894 were: -
Mr. R.G.E. Wemyss.  Chairman of the Board.
Mr. Joseph Budge.   Mr. John Gemmel.   Mr. John Oswald.   Mr. W. Nocton
Mr. A. Bowman.*  Mr. V.L.Gordon, General Manager. Mr. G.F. Underwood:
Secretary.    Mr. R. Anderson: Cashier.    Mr. J. Davis: Principal book-keeper.

*Mr A. Bowman was the same Archibald Bowman who, in 1864, and in partnership
with two brothers James and David Cairns, obtained from the then Laird, James Hay
Erskine Wemyss, a long lease on the Muiredge pit. This pit was at a 'temporary'
standstill, due to the presence of much land water in the shaft and the fact that the
exposed seam of Chemiss splint at 80 fathoms showed so much evidence of
calcification, that it was rendered almost unworkable. {The damage having been
caused many, many centuries before, by the much earlier upheaval of the Buckhaven
Fault (Dyke)}.  This pit was ultimately successful in that it produced coal at a profit,
and was operated and worked under the forceful management of Bowman &
Company, but that unfortunately, is not part of this narrative!*

The chosen site for the new Michael pit lay approximately one mile along the
shore to the north-east of Wemyss castle, immediately adjacent and to the south of
the small village of East Wemyss.   The decision was then made to sink two vertical
shafts within two hundred yards of each other, each shaft to be sunk to strike the nine
to ten feet thick Leven Main Coal seam, *(the Chemiss splint)* at approximately 140
fathoms deep. *(840 feet, or 256 metres.)*  In the year of 1898, both shafts successfully
struck into the thickish Chemiss splint seam at the Michael 'sinks', while down
Lochhead pit shaft, the much thicker and shallower Dysart Main seam *(struck at
sixty-three fathoms deep)* had been successfully broached.   The thick seam was
now slowly being opened-up, through mostly taking only that part of the thick coals
sloping to the rise, in a broad, ever-widening north-westerly direction.   The Dysart
Main seam could now be said to have been opened-up in the Wemyss coalfield, and
would now continue to produce increasing quantities of quality coal from Lochhead
Pit, *(as well as several others!)* far into the foreseeable future.

By this time, Randolph Wemyss was now employing reputable mining engineers,
whose sole task it was to produce informed and comprehensive estimations, based
on drillings and bore holes, as to just how many workable seams there were and how
great were the reserves of coal that lay under the Wemyss-field estates.   The
engineers in turn, came up with a figure that seemingly defied all previous
expectations. A conservative estimate put the figure at the staggering approximation
of two hundred million tons of coal! *(200,000,000 tons)*  From this, it was calculated
that future probable coal production in terms of one million *(1,000,000)* tons per
year, for the next one hundred years, was entirely feasible.  Randolph Wemyss had

been correct in his belief that up until now, no one had really been aware of just how much coal lay underground at Wemyss-field, and now, it was there for the taking.

To this end, the Laird of Wemyss now promulgated a deliberate policy to open-up and exploit the barely touched mineral wealth of the region. In conjunction with his board of management, the decision was made to initiate a second flotation with a projected overall capital, much more than that of the first. The second flotation, named the **'Wemyss Collieries Trust Ltd.'**, (1899-1900?) proposed to raise the capital sum of Five hundred thousand pounds sterling, (£500,000) with the issue of 25,000 ordinary shares, each at Ten pounds (£10) sterling, and 25,000 preference *(cumulative)* shares, also at Ten pounds (£10) each. This done, the Laird *(and board of management)* now had the capital with which to proceed with their plans for the opening up of the Wemyss coalfield. The Laird also recognised that there was no point in producing vast quantities of coal a different pit-heads, unless there was a viable means of transporting it. Transport by road was out of the question and was probably not even considered. The answer lay in the construction of a full-gauge mineral railway linking Thornton and the W.C.C. coalmines directly to the new docks at Methil! *(Not to be confused with the Thornton to Buckhaven thence Methil passenger railway connection!)*

*

In the year of 1898, during the month of October, a storm, the likes of which had never been experienced within living memory, struck the wide estuary of the Firth of Forth and in particular the south coast of the Kingdom of Fife. The storm raged for approximately four days, with storm force winds and tides that threatened to engulf the haven town of West Wemyss. The violence of the storm wrecked the harbour erected by Mrs. Wemyss, completely wrecked four of the seven ships that lay within the harbour and the entry roads, and did very considerable damage to most of the miners' dwelling houses in the village. The storm also severely damaged the offices of the Wemyss Coal Company, but more devastating damage had been caused to the pithead winding houses and mechanical gearings of both the Victoria and Lady Emma collieries.

In the aftermath of the storm, the scenes of carnage and destruction at West Wemyss defied belief. Nothing had been spared, wreckage of every description lay everywhere with everyone hard put to identify anything that once belonged to them. The harbour was in ruins, the dock was unusable, the coal gangway had disappeared and the two coalmines were brought to a halt! The wreckage and flotsam from four sailing ships lay scattered over the devastated foreshore! The haven town of West Wemyss was all but a total ruin!

The Laird then very quickly made a bold decision, that could only but endear him to the fishing folk and coalmining families of West Wemyss. After consultation with his resident architect Alexander Tod, he came to the conclusion, that he should put his mineral developments plans on hold for the time being, in order to concentrate on rebuilding the village, the harbour and most importantly, the pit-head buildings and winding gear for both the Victoria and the Lady Emma pits! In a short space of time, *(days perhaps)* plans were produced for the reconstruction of the dwelling houses, the harbour walls and gangway and the re-building of the two colliery pit-heads and winding gears.

*(To date, I have not been able to discover just how long all of this re-construction work took, but considering the determination of the Laird and the reputed phlegm of the Fife miners and fisher-folk, I can imagine that the time taken may have been counted as a limited number of months, rather than years!)*

*

Back in 1888, just after the grand opening of the dock at Methil, an approach was made to the Laird *(and by implication, to his board of management)* by the North British Railway Company, who had, up the opening of the Wemyss/ Buckhaven railway and Methil docks, a monopoly of all the railway traffic in the Kingdom of Fife. The North British Railway Company, wishing to retain that monopoly, proposed, that not only the railway, but the whole of the Methil docks be acquired by the Company to maintain it's overall monopoly. Negotiations were entered into between the two parties, which dragged on for several months, the outcome of which was that ownership of Methil docks and it's rail network was acquired by the N.B.R.C. for the total sum of £225,000 sterling. One of the conditions attached to the sale, was that Randolph Wemyss should join the board of the Railway Company as a full director. This actually happened, with the deal being finalised in January 1889!

Previously, at the opening of the dock at Methil, the Laird had openly promised, that if the coal trade in Fife showed signs of the expansion that he had previously predicted, he would then take immediately steps to increase the handling capacity at Methil by building an extension to the existing *(No.1)* dock. The Laird, mindful of his earlier promise, also 'committed' the N.R.B.C. to this undertaking, claiming that it was also a condition of the original purchase. Within one year of joining the board, the Laird found himself pressing, even agitating with his fellow board members for an expansion of the railway and loading facilities at Methil. He argued that the board members were lacking in imagination and forward thinking, in not anticipating the need for increased handling and loading capacity at Methil docks.

In 1891, the North British Railway Company did obtain parliamentary powers to proceed with the construction of what was later to become the No.2 dock. However, developments did not proceed at Methil according to plan, or indeed to the benefit of the increasing coal production of the Wemyss pits. *(In the opinion of the Laird of Wemyss!)* Also, in or around 1892, there occurred the death of Mr John Walker, the general manager of the N.B.R.C. At almost the same time, a strong proposal for an extension to the dock at Burntisland, was also on the business agenda of the N.B.R.C. The result of this one occurrence and the board's consideration of the other, meant that the proposed building of the second dock at Methil was put on indefinite hold, much to the chagrin of the impatient Laird of Wemyss.

In 1894, *(three years after they obtained parliamentary permission)* the N.B.R.C. finally went ahead with the construction of the second dock at Methil. The Laird, confident that this dock extension work would now proceed unreservedly, got on with the delayed commencement of the sinking the two new pit shafts *(the Michael Pit)* at East Wemyss. After work had started on the Methil dock extension, word reached the ears of the Laird that the N.B.R.C. had also decided to press ahead with the enlargement of Burntisland dock. The Laird of Wemyss was very strongly opposed to the extension at Burntisland for several reasons, not the least of which was that the facilities at Burntisland, were quite useless as a means to handle the output of the Wemyss coalfield, and that better and cheaper facilities could be had at Methil. The Laird also argued that repeating the same, but inferior facilities at Burntisland was an utter waste of shareholders money. *(However, it seemed that the members of the N.B.R.C. board were not to be moved!)* The upshot of the whole sorry affair, was that the Laird of Wemyss resigned his seat on the board of the N.B.R.C. in 1899.

Sometime during the summer of 1900, the new No.2 dock at Methil opened for traffic. This No. 2 dock was slightly larger than the No.1 dock at fully six and a quarter acres in capacity. *(30,625 square yards, 2.56 hectares, or 25,600 square metres.)* This No. 2 basin may have been a separate entity with it's own dock-gates, but it was actually an elongated extension to the No.1 dock and having the same approximate width of basin. That is to say, that any ship or boat that wished to berth in No.1 dock had now to pass through the sea-gates of No. 2 dock, past the berthed ships in that basin and then through the dock-gates of No.1 dock. *(It was often the case that the sea-gates of No. 1 dock were left open, which had the effect of equalising the level of the water between the two docks, whilst the sea-gates of No. 2 dock were closed to the sea. This may have had something to do with water seepage at lowered tides through the No. 2 dock-gates to the sea!)* This opening of the No. 2 basin certainly doubled the handling and loading capacity of Methil docks, but would it prove to be enough?

\*

Sometime during the year of 1893, the Laird of Wemyss became embroiled in a litigation against the Crown, that involved the mining and extraction of coals from under the sea in the Firth of Forth, in those areas that stood adjacent to the Baronies of West Wemyss, East Wemyss and Methil. For many years in the past, from as early as the 1810 to 1820, coal had been taken from under the sea by the previous Lairds, in those sea areas that lay adjacent to their Baronies. The respective Lairds had always assumed that this coal was theirs by right of charter, supported by Act of Parliament. However, previously to this action in the Court of Session, in or around 1874, agents for the Crown had maintained, that all minerals under the sea belonged to her Majesty Queen Victoria and that royalties would have to be paid on minerals *(in this case - coal!)* taken *ex adverso* the Baronies of Wemyss and Methil. At the time of the claim made by the Crown agents, the Laird was but sixteen years of age and therefore in no position to act for himself. In light of this, the decision was made by the then board of trustees, *(headed by Mrs. Wemyss)* to take a lease on the 'submarine' minerals *(coal reserves) ex adverso* of the Baronies of West Wemyss, East Wemyss and Methil - and paid a fixed 'royalty' on each ton of coal extracted. This course of action was taken by the trustees at that time, to avoid possible costly litigation and also, to avoid the possibility of a 'hold' being put on the undersea extraction of coal at Wemyss and Methil.

In 1887, thirteen years after the 'royalties agreement' had been reached, the Laird applied for and was granted a 50% reduction in the royalties to be paid for undersea coal, that brought the 'tax' down to 2d *(imp.)* per ton extracted. However, this was not enough to satisfy the Laird o' Wemyss. He also in 1890, applied for and successfully obtained further amendments to the terms and conditions of the 1874 lease.

Thus, in 1893, the Laird found himself in litigation in the Court of Session against the Crown. In having the whole contentious question of the coals *ex adverso* the three Baronies of Wemyss meticulously researched and examined, he had discovered that the three Baronies had been united into one Barony, *i.e. the 'Barony of Wemyss'* during the time of David, the second Lord Earl. This had been mooted by Crown charter in July 1651, in favour of the Second Earl and then later confirmed by an Act of the Scottish Parliament in 1661. The Laird maintained that the coals under the sea, adjacent to the three united Baronies were pertinent too and part of the combined Baronies, subject only to the rights of the general public having use of the foreshore down to the low tide mark. The Laird further maintained in light of this discovery, that he could not be bound by an agreement entered into with the board of trustees during his minority, even though he himself had obtained a reduction in the royalties and further amendments to the agreement. He did not see or feel, that these past happenings should prejudice his appeal with this present litigation.

In court, the Lord Advocate acting for the Crown, did not dispute the Laird's rights to the foreshore of the combined Baronies - but argued that the Laird's right to the undersea coals that lay opposite to the Baronies, had been negated by the agreement entered into with the Crown by the board of trustees in 1874, -

and further, by the Laird's own dealings with the Crown's representatives in 1890. In other words, the Laird had unwittingly prejudiced his own case by those actions.

The Lord Ordinary, with the consent of the Lord Advocate, also in the Court of Session, affirmed that the foreshore of the combined Baronies did belong to the Laird of Wemyss, but also confirmed that the agreements of 1874 and 1890 'set aside' any claim to the undersea minerals *ex adverso* the Barony of Wemyss.

For a while, the Laird of Wemyss accepted this judgement, but within three years, *(in 1896)* had requested a formal notice of appeal against the earlier judgment. This time, the Judges of the First Division recalled the 'interlocutor'*(a judgment exhausting the points immediately under discussion, in a cause that becomes final if not appealed against in due time!)* of the Lord Ordinary - and then made pronouncements of a declaratory decree in favour of the Laird. The judges declared that with respect to the foreshore and the adjacent undersea coals, that the Laird of Wemyss had full rights to the coals *'ex adverso'* the Barony of Wemyss. This judgment lasted all of three years!

Previous to 1899, the Lord Advocate had appealed to the House of Lords, who then, after due consideration, set aside the decision of the Judges of the First Division of the Court of Sessions - and practically reversed their appealed decision. Their Lordships decision could be summarised as follows: - 'After having concluded that the Laird, on coming of age in 1879 could have challenged the lease agreed to by the board of trustees on his behalf, but did not! And further compounded the agreement by obtaining amendments to the lease in 1890, by that, further reducing his right of challenge.'

Their Lordships opinions can be summarised as follows: -

1. *The barony title to East Wemyss contained the mineral rights down only to the low tide mark.*

2. *The barony title to Methil, did not refer to any undersea mineral rights at all and had insufficient description.*

3. *The undersea minerals adjacent to East Wemyss and Methil belonged to the Crown - and with regard to West Wemyss, the trustees had been entitled to enter into agreement with the Crown agents for the lease of the undersea minerals opposite that barony.*

The Laird of Wemyss fully accepted their Lordships final decision and almost immediately entered into negotiations with the Woods & Forests Commission, to gain a lease on the coals *ex adverso* the united Barony of Wemyss. Obviously, he was quite successful, as local history proves!

\*

On 28th November 1899, the Laird of Wemyss married Lady Eva Cecilia Wellesley, only daughter to the second Earl of Cowley. *(He had previously obtained a divorce from his first wife, the Lady Lilian Paulet.)* As from January 1900, the Laird of Wemyss and his new wife spent nearly all of the next one and a half years out of the Country, where he became involved in the Boer War in South Africa. That episode however, is not part of this narrative.

Authors' Notes: -

1. *To date, I can find information regarding the death or otherwise, of the Laird's first wife, the Lady Lilian.*

2. *Did the Laird change the name of the coalmine at the Coaltown of Wemyss to 'Lochhead' after being divorced from his first wife, the Lady Lilian?*

3. *Did the Laird name the old Denbeath Pit for his second wife?*

4. *Did the Laird re-name the present Cowley street in Denbeath after his second marriage? (Re his new father-in-law, the Earl of Cowley?)*

At the end of 1901, after the Laird's remarriage and his sojourn in South Africa and after their subsequent return to the Kingdom of Fife, the Laird again re-involved himself in the business of coal-getting. He soon realised, that while the Wemyss collieries were still well-manned and producing coal, they were not at the state of progress that he had expected!

He therefore set-to with renewed vigour and immersed himself in plans for further developments at the Michael pits and Lochhead colliery. He also went much further than that, and initiated the opening-up *(literally)* of the north section of the Wemyss coalfield at Thornton, where it was apparent that there was a confluence of several coal seams including the Dysart Main seam, that seemed to form a heightened apex at this location. These workings became well known as the Earlseat day mines!

Author's Note: - *These 'day mines' at Earlseat were absolutely unique, in that there was never such a confluence of coal seams anywhere in the whole of the Country, that surfaced in such a relatively small given area. There were fully five different seams of coal.*

*The one seam that was of the most interest to the W.C.C. at Earlseat, was the outcropping Dysart Main seam, that seemed to stretch out and dip into the surrounding fields in nearly all directions, and for considerable distances. The seam soon proved to be between 12 to 15 feet thick depending on which direction was looked at, with the known prospect of its thickening out as it dipped to the south, east and west. These day mines, or 'In-going eyes' as they were once named, were destined to be driven (or sunk) up to distances approaching 1,800 to 3,000 yards.*

*The overall beauty of the scheme, was that in deciding the exact location of the pithead construction, meant, that one set of buildings would serve all five mines as the 'pithead!' At the area centre of the horseshoe-shaped confluence, a huge platform was erected, onto which all of the coals drawn from the five mines was delivered, cleaned, screened and separated, and where the graded and different sized clean coals was poured into the waiting wagons below.*

*The method used in the initial taking of the coal was to be the tried and tested 'Stoop & Room' approach, where the need for general roof supports were negated and all of the mines could be worked with only the most primitive of equipment. i.e. Poker drills, low explosives, and colliers supplied with fillers and drawers! The general incline of the strata varied between 1 in 6 and 1 in 3, which was fairly normal for the area. This underlying area of the Dysart Main seam was also typical of most other worked areas, in that the multi-layered seam formed a large unbroken blanket of coal that varied from 12 feet in thickness at Earlseat, to almost 20 feet thick in nearly all directions to the Earlseat Boundaries. When all of the mines had been fully developed and worked to near maximum, the estimated output was expected to be in the region of 2,000 to 3,000 tons daily. The estimated overall yield from all of the day mines at Earlseat working the Dysart Main seam, was reckoned to be in the area of fifteen to twenty million tons of good coal, over their expected lifetime.*

*These mines were deemed to be dry mines, with any small amounts of standing water being shovelled up by the colliers in the normal course of filling. To the best of my knowledge, there were no water-pumps in use at the Earlseat day-mines! In the No. 3 mine however, there is record of quantities of water being brought to the surface in what could be described as a sealed 'wooden box', this being a lined and undamaged wooden hutch with a near water-tight lid! This 'water-hutch' would have been a coupled part of a normal race of up-coming coal. The extracted coal from these mines was transported direct to Methil docks via the Laird's mineral railway, down past Lochhead pit, past the north of East Wemyss and the Rosie, looped around behind the Muiredge pit, down through Stark's wood, past Cowley street back gardens - and under the Denbeath bridge into the extensive sidings at Methil Docks.*

Shortly after the opening-up and development of the Earlseat area around 1904, the Laird in 1905, completed the purchase of the extensive pithead works of Messes, Bowman & Co., *(Archibald Bowman, James & David Cairns)* whose leases at Muiredge, Cameron and the Denbeath pit had expired at the death in 1905, of the last surviving partner/founder, David Cairns. With these newly purchased established workings, Randolph Wemyss had now nearly doubled the number of working coalmines operated by the still newish Wemyss Coal Company. For the next three years, the Laird threw himself into developing all of the collieries now owned by the W.C.C. If increased coal output was a measure of how well the Laird threw himself into the colliery business, it is surely reflected in the coal production figures for the W.C.C. over the years between 1894 and 1909. The coal output from the W.C.C. for the year ending 1894, *(the year of the founding of the Wemyss Coal Co.)* was approximately 140,000 tons from all sources. The coal output at the turn of the century for the year ending 1900, amounted to approximately 278,500 tons, again from all sources, but with further coal developments. An overall output that all but doubled over a six-year period.

For the year ending 1909, *(a full year after the death of Randolph Wemyss)* the approximate total output from all Wemyss sources stood at 1,490,000 tons of coal. This was a tenfold increase over a fifteen-year period claimed by the Wemyss Coal Company. *(It must also be remembered, that output from the newly named Wellesley colliery was severely curtailed during the years 1906 to 1908, due to the deepening of the shaft to strike into the Dysart Main seam at 260 fathoms!)*

In 1894, the approximate number of miners of all types employed within the W.C.C. was approximately 500. The weekly wage bill to the W.C.C. then was nearly £750 sterling. Therefore, the average weekly wage paid to each miner was around thirty shillings (30/-, or £1-50p) for a six-day week. *{This makes sound (historical) sense at around five shillings (5/-d, or 25p) per diem!}*

In 1908, the number of coalminers of all types *(including pithead workers)* employed by the W.C.C. in all of its coalmines, amounted to approximately 4,700 in total. The weekly wages bill at this time amounted to approximately £9,000 sterling, which meant that over a 14-year period, the average wage for a coalminer had risen to around £1-18s-6p (£1.92) for a six-day week. *(This therefore amounts to approximately 6/5d, or 32p per diem.)* If the combined wages of all of the workers employed on the Wemyss-field estates were added to the weekly wage bill, this would amount to another £500 sterling per week. *{Assuming the weekly wage of an estate general worker to be in the region of 20/-d to 22/6d, (100p to 110p) this would approximate to between 450 and 500 employees?}*

Authors' Note: -

1. *This daily increase of around one and fivepence (1/5d) per diem for coalminers over a fourteen-year period, again makes sound sense in that the coalminers average wages over the next thirteen years, up to the time of the 1921 National Strike, would see a slow increase to around 6/6d to 6/9d per diem! (32.5p to 33.75p).*

2. *It has often been argued that the measure of a coalmine's efficiency and work-related output can be measured in the ratio of \*'coal produced per number of men employed'. In 1894, the total output from all sources within the Wemyss fields was approximately 140,000 tons of coal, from a total of 500 employed miners of all types.* **This would relate to approximately 0.94 tons (19 cwts.) of coal, per man, per shift.**

*In 1909, the total output from the W.C.C mines in the East Fife coalfield amounted to approximately 1,490,000 tons of coal, from a total of 4,700 miners of all types employed within the company.* **This therefore, equates to approx. 1.05 tons (21 cwts.) of coal, per man, per shift.** *This slightly increased, but seasonally fluctuating\* ratio seems to have been steadily maintained over this fifteen-year period!*

*\*This ratio could not be liberally applied to every working coalmine! Each and ever mine was a separate entity in its own right, with much depending on the depth of the shaft, the age of the coalmine, how deep and how thick were the coal seams and how far underground did the extracted coal have to travel? The ratio however, could have been used as a yardstick to determine or compare, one year's output to the last, or next!*

\*

With the deepening of the Wellesley shaft down to strike the Dysart Main seam, the intention of Randolph Wemyss, *(ergo the W.C.C.)* was for the miners to break into this thick rich seam and work it level and sideways in both directions, before mining to the dip, out and under the sea and on an ever-widening front. This would involve the W.C.C. in a capital investment of around £150,000 sterling in development costs, with the additional obligation to pay the Woods & Forests Commission the royalties as they become due, with the extraction the undersea coals *ex adverso* the Barony of Wemyss. *(I am under the distinct impression, that at that time, the Laird was actually given the choice of paying either a fixed Royalty per ton, or a fixed yearly rental of £1,000, while the undersea coals were being taken. I have no information what so ever on which option the W.C.C. settled for!)*

As soon as the initial developments at the Wellesley colliery were under way, the Laird then turned his attentions back to the Victoria and Lady Emma pits at West Wemyss. These two pits had been in operation for approximately seventy-five years, but as yet, neither of them had been extended downwards to take the potentially richer and deeper seams. It was at this juncture that the Laird/W.C.C. decided to sink
the Victoria shaft deeper to take the rich bounty of the Dysart Main seam, known to have a thickness of approximately 32 feet (9.75 metres) deep under the foreshore at West Wemyss.

*{The deepening of the Victoria shaft did not actually strike into the Dysart Main seam (per se) as had been done in the Wellesley pit. The W.C.C. had obviously given this particular operation much consideration as to longevity (and safety) and forwarded a diabolical plan for its development! This unusual, but successful development is fully described in the chapter: - The Victoria Pit!}*

The Laird also realised, *(and fervently hoped)* that as soon as the coalminers had broached the Dysart main in the Victoria pit - and then opened it out with widening developments, that the subsequent extracted coals coming up the Victoria pit could only but increase in volume. *(The Victoria shaft (not the pit!) was only 72.5 fathoms or 132.5 metres deep, with a twin cage, single hutch winding system.)* The Laird then estimated that with the potential output of the Victoria pit set to double, or even triple, he reasoned that the small harbour at West Wemyss would not be able to cope with the increased coal traffic. Therefore, in order to have the local shipping facilities match the proposed coal output from the Victoria pit, the Laird promulgated an order for an extension scheme at West Wemyss harbour to double the shipping/loading capacity, at a probable cost of £20,000 sterling. *(Randolph, the Laird o' Wemyss, did not live long enough to see this extension finished!)*

Authors' Note: - *With the striking of the Dysart Main seam at Lochhead's Nicholson's Dook, No. 1 level (west), but within the precincts of the Victoria pit, this intrusion had also opened up a possible secondary exit means of delivering the expected coal from the Victoria's pit embryo sea dock, to the ground surface via the Lochhead pit shaft. (It also could have been that the Lochhead shaft was winding coal to maximum capacity over one daily working shift!) There is also, worth of note to readers, to ponder the reason behind the decision to drive the Lochhead (Hugo) tunnel from the beach at West Wemyss, up to the Coaltown of Wemyss? (See the chapter on 'The Victoria Pit!')*

The Laird of Wemyss was also aware that in attempting to man his pits, that it was pointless to import coalminers and their families into the coalmining villages, if there was no worthwhile accommodation available to house them. He was also very aware of the fact that coalminers wives were especially minded of the need for extensive washing and bathing facilities - and the means to provide hot water. To this end, the Laird charged his resident architect Alexander Tod, to design and have built, houses and terraced cottages with separate bed-rooms and adequate facilities, to include running water and flushing toilets. These houses were not to be situated in any single village, but built to accommodate coal-miners families in the areas where the miners were needed. As a result, many new '*up-to-date' dwelling houses were built in Methil, Denbeath, Buckhaven, the Rosie, East Wemyss, Coaltown of Wemyss and a few in West Wemyss.

In the one area where there was a concentration of miners *(and their families)* that were not all employed at the local colliery, the Laird organised a special passenger train to run on the W.C.C. private railway, to transport and return miners from Denbeath to the mines at Earlseat and the Lochhead pit at Coaltown of Wemyss.
* *(Description of the building of these village extensions is contained elsewhere within this narrative.)* See Chapter on Denbeath Rows!

\*

In the closing months of 1903, it became hugely apparent to the Laird that the continued future prosperity of the parish of Wemyss lay mostly in the working and mining of coal. He also came to believe that if this were the case, then it should be apparent *(to those in the business)* that the existing docking and loading facilities at Methil docks would soon become inadequate. (Again!) The Laird reasoned that since the coal output from the W.C.C. coalmines had regularly increased on a yearly basis since the inception of the Company in 1894, then, there was every reason to surmise that it would continue to do so in the foreseeable future! The Laird therefore approached the N.B.R.C. to explain his point of view and asked them to seriously consider another possible extension to the existing two docks at Methil. Unfortunately, the board of management of the N.B.R.C. did not adhere to the Laird's point of view. Thus, he was met with a very negative response!

Undeterred, the Laird, in the spring of 1903, wrote personally to the manager of the N.B.R.C., hoping that a direct appeal might prevail if he described his arguments, detailed his concerns and produced some firm proposals for the project. His letter covered the following points: -

'*Mr Wemyss argued that there was insufficient width of gate and depth of water at the present double dock, that limited the size and capacity of the coaling ships currently loading. He envisaged the need for wider and larger capacity coaling ships, with additional berthing space to include greater and more efficient handling capacity for coal loading.*'

*He also reminded the board, that when he sold the original dock to the N.B.R.C., he undertook not to construct another docking facility on the assurance of the then manager Mr Walker, that the N.B.R.C. would extend the docking facilities at Methil when the increase in coal traffic warranted it! Mr Wemyss also said, that when the No. 2 dock extensions were completed in 1900, it was apparent at the time, that the dock would soon prove to be inadequate to cope with the potentially increasing coal traffic in the near future.*

*Mr. Wemyss also made clear, that if the N.B.R.C. were unwilling to provide additional docking facilities at Methil, then he himself, was prepared to finance an additional dock at Methil exclusively for W.C.C. traffic. Mr Wemyss was prepared to go to these lengths (and expense) to gain the extra shipping facilities, but hoped that favourable agreement could be reached between the N.B.R.C. and himself.*

*Mr. Wemyss also expressed a desire to meet with the manager of the Board of the N.B.R.C., with a view to having a frank and open discussion, along with an earnest request to lay this letter before the full Board of directors, so that they could have the opportunity to make comment on his three proposals, which were as follows: -*

1. That the N.B.R.C. builds the necessary additional facilities.

2. That he, the Lairds build the necessary facilities.

3. That both the N.B.R.C. and the Laird combine for the specific purpose of creating the necessary extra docking facilities required by the Laird for his Wemyss coals.

On being informed that his letter had indeed been conveyed to the board, Mr. Wemyss quickly followed this with a second letter containing statistical detail of past coal shipments at Methil since the inception of the first dock. He then went on to describe his actions since 1896.

*(The following is a summary of the Laird's second letter to the Board!)*

1. *In 1896, and in view of the congestion at Methil docks, I informed the board that unless something was done soon to ease this congestion, I would apply for an Act of Parliament to construct a dock at Buckhaven, to handle my Wemyss coal. I did not move quickly with this application, in order to give Mr Conacher a chance to settle into his new appointment. (With the board.)*

2. *I fully intend to sink the shaft of the Denbeath (Wellesley) pit deeper, to 'strike and take' the Dysart Main seam of coal, for which I intend to expend up to £200,000 in developing these coals 'ex adverso' the Barony of Methil.*

3. *I have discussed with the Woods & Forests commission, the necessity for further dock accommodation at Methil, to cope with a projected annual output of 250,000 tons of coal from the Wellesley colliery, along with an estimated annual extraction of approximately 500,000 tons from the Earlseat mines. In addition to the present 800,000 tons being produced annually from the Wemyss collieries, I also predict a further annual increase of 100,000 tons from the same source. I therefore estimate with sound reasoning, that both dock handling facilities and shipping capacity at Methil, shall need to be able to cope with Two million (2,000,000) tons of coal annually!*

4. *I feel at this time, that the answer to my dilemma, lies in the construction of an entirely new dock, having wider gates and greater depth of channel, to allow the passage of larger capacity, deeper draught coaling ships. This needs to be given great consideration, in that with the building of the No. 2 dock, no increase in depth of channel or width of gate, was embodied in the construction.*

5. *I trust that you will lay this letter containing my specific views before your board. If the N.B.R.C. are unwilling to consider my earnest requests, I stand ready to initiate that which I have already described! I firmly believe*

*that the hard-won trade of this part of the county must not be allowed to suffer - and that the N.B.R.C has a moral duty in ensuring that the facilities at Methil are extended to cope with positive increases in output from the W.C.C. coalmines!*

6. *In my opinion, the estimated cost to either of us working independently for the construction of an entirely new dock could be in the region of approximately £350,000 to £500,000. If however, we together, can come to an accommodation, it may be possible through the deepening of the existing channel and using the present entrance (sea roads), to modify this cost to around anything from £200,000 to £350,000.*

The answer to Mr. Wemyss's letters was not long in coming. In June, barely a month after the board's discussion and dissection of both letters, Mr Jackson communicated to Mr. Wemyss, the board's considered decision. Mr. Jackson had been instructed by his board to convey and report to Mr. Wemyss, that after a full investigation of the known relevant facts by the board, that in their opinion, they were already providing the necessary accommodation at Methil docks, with more than ample margin for any possible future increases in traffic. Mr. Jackson's report went on to state, that in their view *(the board)*, Mr. Wemyss had in 1889, expressly agreed not to construct any other dock or harbour in the county of Fife. Since then, the N.B.R.C. had in fact, more than doubled the dock accommodation (at their own expense) at Methil with the building of the No. 2 dock. Furthermore, regarding any possible future expansion at Methil, it would obviously be to the good and benefit of this company to prepare for further enlargement of the dock facilities, if and when, future traffic should warrant same! It must also not be overlooked, that a large percentage of the coal coming from West Fife could just as easily be shipped from Burntisland as from Methil!

Mr.Wemyss pondered this letter from Mr Jackson giving what amounted to the railway board's final decision, and decided that this indeed, was their last word on the matter. Randolph Wemyss also decided not to let the matter rest with this seemingly final refusal from the board of the N.B.R.C. - and began to reiterate and document his official complaints against this tardy illiberality.

He decided that his position was as follows: -

1. That the N.B.R.C. was putting a restraint on his trade.

2. That the agreements the N.B.R.C. were basing their object-ions upon, were applicable only to the protection of the N.B.R.C. from possible aggressiveness on the part of the Caledonian Railway Company.

3. That an agreement had been entered into between himself and Mr Walker (the then General Manager of the N.B.R.C.) that stipulated, that future Dock extensions should be 'kept abreast of the times!' Mr. Wemyss's contention was that the N.B.R.C. in refusing his plea for further dock extension with greater handling capacity, put them in breach of this original agreement.

It was now Mr. Wemyss's contention, that the refusal on the part of the N.B.R.C was in breach of his original agreement with the Railway management, and it was his intention to promote a parliamentary bill asking for powers to construct a W.C.C. dock at Buckhaven for the shipping of Wemyss coals. The proposal was for a dock with a quay length of 1800 feet (*approximately 550 metres*), capable of the simultaneous handling of eight coaling ships of between 6,000 and 8,000 tons each, at a proposed cost to himself of £260,000 sterling.

On 25th July 1905, the House of Lords committee handed down the following decision: -

"The Committee find the preamble not proved, but they have been much impressed by the evidence regarding the congestion of the district. They do not consider that it has been proved, that no other party can relieve the congestions except the promoter - and therefore, do not at present see sufficient reason for relieving him from the agreements he has entered into with the N.B.R.C., while there are hopes of the necessary accommodation being provided from another source."

Mr. Wemyss again had pause to ponder, this time as to the underlying meaning of their Lordships deliberations. He interpreted their findings as meaning: -

*'If the N.B.R.C. did not immediately promote a bill for a new dock at Methil, he himself would be granted powers to proceed with his Buckhaven proposal!*
*He immediately busied himself with modified plans, into which would shortly develop the proposed 'Buckhaven Dock Bill Scheme'.*

Inevitably and again, the board of the N.B.R.C. soon realised that Randolph Wemyss was a man of his word and retaliated with a new scheme of their own. The management of the N.B.R.C. had seemingly also read into their Lordships findings and come up with the same conclusions as had Mr. Wemyss. They then launched their own scheme for the construction of an entirely new additional dock at Methil, that surfaced in November 1905. Their plans were deposited under the Scottish Private Bill Procedure Act during March 1906, but when the Provisional Order was converted into a Bill, it was found to be too late to be carried in the 1906 session of the parliament. This Bill therefore, was not called forward until the month of April 1907.

(This Bill, promoted by the N.B.R.C., provided for the construction of a new dock at Methil between the high and low water marks, to cover an area of approximately 15 acres (73,500 square yards, 61,365 square metres, or 6.137 hectares), at a proposed cost of £700,000 sterling.  This Bill also provided for the twin-tracking of the Leven to Thornton railway, at the probable cost of a further £50,000 sterling.)

The Laird of Wemyss soon became aware of this rival Bill promoted by the N.B.R.C., and looked hard at both the financial implications and the proposed facilities that were to be offered him for the shipment of Wemyss coals.  Mr. Wemyss did not like the provisions afforded, nor did he like the financial impositions that could be hoisted on the W.C.C.  He therefore decided to press on with the promotion of the W.C.C's Buckhaven Dock scheme.

It was therefore inevitable, that when the schemes of both protagonists came before the appropriate committee in the House of Lords, that they would be considered '*inter alia*', since each in their own way wanted the same thing.  It being due, to where, when, at what cost and to whom, and would it satisfy Randolph Wemyss?  There were apparently deep divisions, if not an outright animosity between the two companies, with each side vying to outdo the other about their own stated priorities.  Mr. Wemyss, when called to give evidence before their Lordships, admitted that he had no real desire to continue with his dock scheme at Buckhaven, if the N.B.R.C and the W.C.C. could agree on favourable financial terms - and also recognise Mr. Wemyss's concerns regarding the future potential coal tonnage, soon to flow from both the new and upgraded coalmines belonging to the W.C.C.  This admission by Mr.Wemyss was accepted by their Lordships, that then formed the basis for further negotiation between the two protagonists.  Their respective agents soon got together and hammered out a form of agreement that was acceptable to both sides.  Thus, terms were agreed upon and Mr. Wemyss then quietly dropped the proposed Buckhaven Dock scheme.

On 27th August 1907, an Act authorising the construction of a new dock at Methil was given Royal Assent, much to the pleasure of the Laird o' Wemyss and the Wemyss Coal Company.

*

*In retrospect, it seems the Mr. Wemyss was absolutely right to have called for the building of No. 2 dock in 1893, and even more so during the spring of 1904, when he called for a new wider, deeper, larger dock to be constructed at Methil for the handling of W.C.C. coal. With hindsight, it could be said that he forced the hand of the N.B.R.C. with his threat to construct a private dock at Buckhaven. It can also be said that if the N.B.R.C. had not 'come to its senses' regarding the large extension at Methil, there might just have been a rather large dock and coal-handling facility at Buckhaven!*

*

The following extract from **Fife's County coal statistics** would seem fully to reflect the prophetic words of the late Mr. Randolph Wemyss: -

| Year | No. of Persons Employed | Coal produced | Coal Shipped at Methil. |
|---|---|---|---|
| 1887 | 7,558 | 2,585,410 tons | 219,880 tons |
| 1892 | 11,308 | 3,573,818 tons | 810,545 tons |
| 1897 | 11,945 | 4,077,820 tons | 1,090,320 tons |
| 1902 | 16,933 | 6,134,171 tons | 1,759,041 tons |
| 1907 | 23,100 | 8,530,040 tons | 2,823,720 tons |

From the above table the following statistics can be extracted: -

In 1887, the percentage of W.C.C. coal exported from Methil was approximately 8.5% of the total produced.

In 1897, the percentage of W.C.C. coal exported from Methil was approximately 26.8% of the total produced.

In 1907, the percentage of W.C.C. coal exported from Methil was approximately 33.1% of the total produced.

N.B.  Construction of the third and largest dock at Methil commenced around 1907-1908.  The construction of the basin, the quay, the pier *(and a small lighthouse)* and the fully protected fairway, took all of five years.  The No. 3 dock was situated immediately to the northeast of No.1 dock and tight to the shoreline.  The basin was approximately 1300 feet long, *(approximately 400 metres)* 825 feet wide, *(approximately 250 metres)* with two 150 metres long, 25 metres wide quay projections into the 16.25 acres sized basin, *(approximately 6.65 hectares).*  This brought the total quays length at Methil docks to something more than 10,000 feet. *(More than 3,000 metres in length.)*

The approach to No. 3 Basin was by means of a 1,640 feet *(approximately 500 metres)* long fairway *(protected channel)* built immediately next to - and along the outside length of both No.1 and No. 2 docks.  The new channel to dock No. 3 was deeper and wider than that to Docks No. 2 and No.1 - and the wider dock gates could pass certain ships up to 8,000 tons. *(The S. S. Woodfield at 7,907 tons for one.)*  The sea-roads into the entrance of the new dock remained as that for Nos. 1

and 2 docks, but were widened and deepened to take ships and boats of greater draft. The new outer sea wall, armoured and graded, stretched all of 4,125 feet *(1,250 metres)* in length, from the new small lighthouse at the outermost projection of the new pier, to the far end of No. 3 Basin. The new and much larger No. 3 dock was completed around 1913 and opened up for shipping almost immediately. Unfortunately, the start of the Great War in 1914, severely curtailed what might have been an even greater exporting enterprise!

Author's Note:- *During the construction of this chapter I have drawn heavily upon the work contained in: - 1. Andrew S. Cunningham's book - R. G. E. Wemyss - An Appreciation., published in 1909. (Although I am 'at odds' with some of the statistics contained therein!) 2. Memorials of the Family Wemyss of Wemyss by Sir William Fraser. K.C.B. LL.D. (Published in 1888). Most of the originals writings quoted in this chapter are taken and quoted verbatim from these publications. I have however, only quoted some of those parts that are relevant to the Wemyss coalfields, the development thereof, the working and taking of the coals, the subsequent Laird's dealings with colliers, the panning of salt upon the lands of Wemyss-shire and in particular, any information dealing with the developments of the three village's Wemyss!*

\*

# A
# DARK
## and
# DISMAL STRIFE
## *(A Bright and Shining Life!)*

## An INTRODUCTION

This book is essentially about a small coalmine in the Kingdom of Fife and some of the coalminers who 'wrocht' therein. This coalmine originally belonged to the Wemyss Family, ergo the Wemyss Coal Company before Nationalisation on January 1st 1947 - and is situated approximately half a mile to the North of the Coaltown of Wemyss in the middle of open country. It was singularly accessed along an open track locally known as the 'pit-road'. *(The short roadway between the main street and the start of this single track to the old colliery site is now officially designated 'The Lochhead pit Road'.)* This road was wide enough only to accommodate one single vehicle with no passing places *(miners with motor-cars were few and far between).* The coal mine was originally named the Lady Lilian, which was later changed to Lochhead Colliery. *(There is an open area of farmland between the Pit-road and the near parallel Check-bar road, that is named Lochhead on British Ordnance Survey maps.)*

The Colliery was one of several mines owned at that time and operated by the Wemyss Coal Company, that included the Michael Pit at East Wemyss and the Victoria pit at West Wemyss. *(The W.C.C. did at one time own many more Pits and Mines in the East Fife area, most of which have been named, but they play little part in this narrative. The Wellesley Colliery is the exception, for the reason that it was for a time owned by the W.C.C. - and the on-going aftermath of it's subsequent deeper extension coupled with the waste output from the Baum washer, was the cause of great social and economic upheaval and forced change to the local area!)*

Each of these three pits had their own vertical shafts and winding houses, with the Michael pit having its own separate management. The manager of Lochhead Colliery also had responsibility for the smaller Victoria pit, which had a resident under-manager who reported to Lochhead's manager. This make good economic sense in that the Victoria pit was quite a small pit, with all of the major decisions with regard to its on-going development being taken by the manager, under-manager and planners at Lochhead.  Before the driving of the sea dook in the Dysart Main coal in the Victoria pit, the mined coals were drawn up the shaft, wheeled around the small dock and loaded directly on coal ships.  When this trade fell away, the company constructed Hugo tunnel linking West Wemyss to the Coaltown of Wemyss, was used to transport the Victoria pit's output up and onto the small screening plant at the Coaltown ,before being transported onto the mineral railway line through the Coaltown of Wemyss, then past Lochhead's Surface Dipping, taking the Company's coal to the Methil docks via the Wellesley washer after its installation.

  With the driving of the level cross-cut mines from the deepened Victoria shaft, *(pit bottom)* and the subsequent linking-up with the Old Dook's long No. 1 level to the South-west of Nicholson's Dook. *(This underground place where the Victoria's cross-cut mines struck into the Dysart Main seam at approx 625 metres from the Victoria pit bottom, was the commencement point for the long, dipping, Victoria sea dooks.  See chapter on Victoria Pit!)*  The underground level connection was then established through to Lochhead pit's Nicholson's Dook, whence the output from the Victoria pit's workings was diverted to be integrated with its parent coals *(all Dysart main)* and wound up the Lochhead shaft.  The two pits were now truly economically connected and administratively combined, even though the Victoria pit's miners living in West Wemyss, still descended the Victoria pit shaft to go to work. *I have no evidence to suggest that any Lochhead miner from outside West Wemyss (except officials) descended the Victoria pit!*
  In the first four decades of the twentieth century and in a colliery like Lochhead, the methods of coal-getting especially in the Dysart Main seam progressed very little over that time, as did the simple machinery used to help take the coal.  Holing picks, shovels, poker drills, wooden hutches and powdered low grade explosives, were the mainstay of the miners tools.  Wooden roof supports in the form of 'trees and straps' were 'de rigueur' as was pillar-wood for packs, for the reason that 'steels' had not as yet made any inroads, were still expensive and not really viable within the methods used to take the Dysart main coal.  Larch bars *(railway type sleepers)*, heavy-weight, wooden supports were also expensive and not too readily available, even though they could be had.  Underground illumination for the miners was by cap worn 'tallow-lamps' which, even if they were of good quality would only produce a meagre* light source.   Store bought carbide lamps which produced better illumination, were a luxury that few miners could afford. *(An American made 'Autolite' with a large reflector plate cost six and sixpence (L.s.d.) in the 1940's.)*

*\*The one or two candle-power illumination given of by a tallow lamp was barely*

*sufficient for the underground miner to work by - bearing in mind that light reflection was nil, because of the enclosed coal black surroundings. The miner's eyes however, very quickly became accustomed to the near darkness surrounding them, after all, illumination was only needed at the point of labour and when a miner turned, so did his cap lamp! (Some readers may find it difficult to appreciate that ninety-nine percent of the underground area of a working coalmine was enveloped in and operated in total darkness!)*

This book also attempts to describe in some detail, the taking of the Dysart Main seam of coal and later, the Lower Dysart seam of coal. The Dysart main seam in Lochhead pit was intruded into in the latter part of the 1890's, with the deepening of the shaft to approximately sixty-three fathoms - and was worked continuously* until it's abandonment in 1960.

* Except during the weeks of the 1912 Minimum Wage Strike, and the months of both the 1921 National Strike and the 1926 General Strike.

The seam was however, by no means exhausted, with approximately 85 percent of it's coal still standing. That in itself, is still a contentious issue and continuing, but lost debate. The commencement of the taking of the coal from the deeper Lower Dysart seam in the late Forties/early Fifties was better planned, much faster and thoroughly mechanised. In fact, some of the coal was quite savagely taken! The output from the thickish, Lower Dysart, split seam, sustained Lochhead pit for approximately another twenty years.

Lochhead Colliery also had a Surface Dipping, which was sunk at a pre-determined descending angle to intrude into and take coal from three different seams *(neither of which was the Dysart main nor the Lower Dysart),* at the 240 level, the 550 level and finally the 350 level. *(Measured in imperial feet from ground level.)* The 350 level was the last to be worked was abandoned in 1954, with the remaining miners being transferred to the main pit, where they were successfully integrated. This Dipping also had an under-manager who reported to Lochhead's manager, but like the original Victoria pit, it was a separate coal-producing entity in it's own right, with it's own ancient steam-driven winding gear, that is fully described elsewhere within the narrative of this book.

This book also attempts to explain the 'lot' of the coalminers employed in Lochhead Colliery, some of the conditions in which they were prepared to work and parts of the underground environment in which they toiled. It also includes and describes much of the sometimes simple, primitive machinery that was abroad at the time. I have also included many confirmed stories and anecdotal situations that many miners found themselves in, both below and above ground. Most of these incidents taking place at a time when money was scarce, entertainment was fragmentary and possibly of a localised nature, with working men followed activities and pastimes, all of their own devising. Miners and their wives and families, were for the most part, law abiding citizens, who accepted the sometimes constricting social rules placed upon them, but that didn't stop them from cunning invention, inspired innovation and sometimes downright chicanery if it served their purposes.

Part of this book is also semi-autobiographical, in that I have mentioned several escapades in my early life, which may or may not be familiar to some of my contemporaries *(or co-conspirators!)*. I have described my initiation into the coal mines, how this came about and my subsequent development as a coalminer. I have gone on to described my progress though the pithead to my first underground job, and thence my fairly rapid initiation into the ranks of the better paid face-workers and on to the various types of coal-getting procedures that I was involved with.

As I progressed though the various types of coal-getting, I met and worked alongside miners from all walks of life, who had many and varying interests and whose spare time was fully utilised in the pursuit of their secondary occupations! I have also done my best to describe or delineate some of those uncommon activities, which may be of some interest to others in different walks of life!

Miners for the most part were past masters of invention and opportunism, especially if they themselves stood to benefit from any small enterprise, whether it concerned work or otherwise. Some of them were also very closed-mouthed about their activities, especially about any small nefarious pursuit that they might be involved in, just in case it didn't pay-off, or if it did, it was therefore their own ingenuity that deserved it's just reward. I must admit that some of the ex-miners whom I interviewed, were rather reluctant to admit their part in some of the escapades and anecdotes that I uncovered/discovered, whilst some miners were only too keen to share their knowledge of incidents and happenings of what they now regard as an absolute hoot! I can but only say, of how dismissive most of these miners were of events that took place all of fifty or sixty years ago, not caring in the least, of any late possible consequences and only too happy to share their knowledge and remembered experiences. I can also say that some old ex-miners remain very close-mouthed to this day about some these same events and happenings. Old habits die hard!

This narrative not only includes such coalminers and their families that were employed in Lochhead colliery, but describes people, places and events that took place outwith the limited area of the underground confines of the coalmines. The people, places and events described or referred to in this book, are all relative to the coalmining industry, firstly, as part and parcel of the Wemyss Coal Company 'extended family' and then later, the early years of the National Coal Board as operated within the East Fife coalfields.

The old mining village of the Links of Denbeath and the new 'model' mining village of the Denbeath 'Rows' *(as it was then!)* are often mentioned throughout this text, for the simple reason that they in their respective era's epitomised the expense and the lengths that the 'Company' would aspire too, to serve the 'King' that was Coal! The new village as exemplified by the Denbeath Rows was seemingly built for one reason only, and that was to accommodate the immigrant miners from other defunct coalmines in Scotland, that were being brought in to serve the needs of both Lochhead pit and the Earlseat mines, as well as that of the increasing numbers of local miners being employed by the burgeoning Wellesley colliery. *(Not to mention the overspill from the old Links of Buckhaven!)*

I particularly wish to point out that the parts of this narrative that mention and discuss the Wemyss Coal Company, or indeed, any member of the Wemyss family dynasty, is in no way meant to be an attack on the integrity of any individual or members of that illustrious family. During the course of my research, I have come across numerous examples of the consideration, thoughtfulness and imaginative planning that seemed to be the hallmark of the late Randolph Gordon Erskine Wemyss, the last named Laird o' Wemyss mentioned in this narrative. Indeed, the on-going largesse of this long established family, especially in the provision of worth-while local amenities, where the health, basic needs and recreation of the coal-miners and their families, has seemingly come a very close second to the primary aim of 'Coal-getting!'

I also feel that I waited far too long in attempting to write this book, in that most of the Lochhead men of my coalmining days are long gone, as is the coalmine in which they 'wrocht!'. The passage of fifty years is enough to dull all but the keenest of memories, even if the miners themselves are still living. I, at the time of which I mostly write, was but a lad of seventeen years when I went underground for the first time, and some of the miners who pass through the pages of this book were in their middle thirties and early forties at that time. At the very onset of my research in the greater Glenrothes area - and in mentioning to Bill and Betty Hay at the Amateur Radio emporium in Woodside my desire to contact ex-Lochhead miners. Bill directed me to a customer of his who had mentioned to him, that my voice coming over on the 'Forty metres'** band seemed familiar to him. The result was that I was fortunate enough to 'find' a coal-miner whom I had previously known, who turned out to be a very fit and able 79 years old Bob Stevens, resident in Woodside, who was instrumental in my being able to trace additional old Lochhead ex-miners. From that initial interview and Bob's increasing interest in my project, came the valuable flow of information that enabled my research to produce much more that I had ever anticipated.

To those Lochhead ex-miners whom I have not mentioned by name and who were employed in Lochhead at that time of which I write, I can only apologise if I may have got some of my facts slightly wrong, or if they feel ever so slighted at being left out. This was never my intention. I only wish that I could have found and spoken with a greater number of ex-Lochhead coalminers before this volume was published!

Readers, on browsing through the opening pages to this narrative, shall find that I have included a longish Historical Foreword to this book, that partially describes the near unbroken line of the Wemyss dynasty succession from Sir John of Wemyss in 1372, to the death of Randolph Wemyss of that Ilk in 1908. Readers may also ask as to why I have done so? The incontrovertible answer is, that the Laird's of Wemyss owned* the lands under which the coal was to be found. The Wemyss dynasty in fact, can be traced as far back, as some very early historical scribes who have attempted to trace their descent from the ancient Earls of Fife. *(A lineage not quite proven at this time!)* However, part of my intention in this narrative is to show and describe the direct connection between the successive Laird's of Wemyss - and the men and women who worked and took the coals, both above and below ground on the Wemyss-field estates.

*Sometimes co-owned, sometimes part-owned, eventually completely owned, up to Vesting day on 1ˢᵗ January 1947.

David Morris Moodie

**This Author is a licensed British Amateur Radio operator with the Call-sign GM4FOZ.

Author's recommendation: - In light of the fact that there are only a few photographs and illustrations in this book - I would prevail upon readers to beg, borrow or obtain copies of the following illustrated publications by Stenlake Publishing, in order to savour the 'spirit of the times' as described and experienced within the mining community in the Kingdom of Fife.

1. METHIL - No More! . . . . . . . . . . . . . . . by Paul Murray.
2. Old WEMYSS . . . . . . . . . . . . . . . . . . . . by Margaret Thomson.
3. Bygone LEVEN. . . . . . . . . . . . . . . . . . . by Eric Eunson.
4. The Lost Village of BUCKHAVEN . . . . by Eric Eunson.

D.M.M.

# Chapter I

## INSIGHT (Part I)
## *A Fulsome Intermination!*

In the 1950's in the county of Fife, one in five* of the male population were employed in the coal industry! In the not so distant future, this twenty per cent figure was expected to rise even higher! Indeed, the Kingdom of Fife was fully expected sometime within the next one or two decades, to become Scotland's premier source of coal. The signs were that this would no doubt soon materialise, due to the even more sophisticated, coal-cutting equipment being quickly and extensively introduced into the N.C.B's coalmines, signifying the welcome change from labour intensive hand-stripping during one shift in three, to machinery such as coal-shearers and mechanical power-loading equipment, capable of producing coal during three working shifts out of four! *(Eighteen hours production time, with six hours maintenance time in any twenty-four hours period).* Concurrent to this new mechanical equation, remained the undeniable fact that the underground coal reserves in the county of Fife, had been constantly revised and up-rated to the point where they were almost inestimable! The considered deliberations of the **Coal Commission Report of 1905** put the estimated figures at **Six Billion Tons.**

*Third Statistical Account of Scotland (1952). Vol X. (Fife) by Alexander R. Smith, M.A.

(See below!)

*

**Just how great were the reserves of coal that lay under the Kingdom of Fife ...?**

The County of Fife covers an area of approximately 505 square miles, or approximately 323,000 acres, *(in today's terminology that would equate to approximately 132,000 hectares.)* The length of the county East/West from Fifeness to Kennet-pans is approximately forty-one miles, while the breadth of the county North/South from Newburgh to Burntisland, is approximately twenty-one miles. The length of the sometimes rugged, rocky and often sandy coastline is approximately one hundred and fifteen miles long.

Within the boundaries of Fife at the start of the Nineteen-fifties, there were approximately thirty-four working coalmines, with a further eight or nine surface or drift mines projected.  The main coal-bearing strata was to be found in two distinct coal measures separated by the Millstone Grit, namely, the Limestone Coal Group and the Productive Coal Measures, euphemistically named the Lower coal measures and the Upper coal measures respectively.  The Fife coalfields could be geographically further separated into three regions, namely, West Fife, Central Fife and East Fife.  *{There were virtually no coal mines in the northern two-thirds of the County of Fife. The cut-off line (both geographically and geologically speaking) commenced from Dollar to the east and ran above the Cleish Hills, south of Loch Leven, on to Leslie and thence to Largo.  This was known as the* **Ochil Boundary Fault.**}

Both of the separate coal measures that appear *(and in some instances, surfaced)* in different areas of Fife, are to be found at depth under the sea bed of the Firth of Forth.  The upper coal-bearing strata rising from under the Forth, appears *(at a slightly irregular incline of between approximately 16 to 22 degrees)* as the 'Productive Coal Measures' in that part of East Fife, that stretches along the coast from Largo to Kirkcaldy, a distance of perhaps ten to twelve miles and to a width of approximately three miles inland from the sea-shore.  This roughly triangular-shaped area covers approximately fifteen percent of the coal-producing areas within Fife, but contained what was probably the most prolific seams of coal every to be mined anywhere in the United Kingdom.   The lower coal bearing strata, *(below the Millstone Grit)* the 'Lime-stone Coal Group', appears to have been worked mainly in that large area from Kirkcaldy westwards, through middle and west Fife and into Clackmannan.  The areas in which the Limestone Coal Group *(Lower coal measures)* were worked, amounted to perhaps the other eighty-five percent of the geographic area of the Fife coalfields.

*{The Productive Coal Measures (U.C.M.) that re-appeared in Alloa are not part of this narrative, even though for purposes of practical administration and production data, the coalfields of Alloa were lumped together with those of all of Fife after Vesting day on Jan. 1ˢᵗ 1947.  These were known as the Fife and Clackmannan Coalfields, with the regional head-quarters situated in Cowdenbeath!}*

The working coalmines that took coal from the Limestone Coal Group *(L.C.M.)*, heavily outnumbered those working in the Productive Coal Measures *(U.C.M. in the East of Fife)* by a factor of probably five to one, even though this is not reflected in the overall production figures for the Fife area in general.

### Just how many coalminers were employed in the Fife Coalfields?

Before Nationalisation in January 1947, there were probably ten separate coal-mining leaseholders producing coal in Fife.  The three largest enterprises being the

Fife Coal Company (F.C.C.), the Lochgelly Iron & Coal Company (L.I.& C.C.) and the Wemyss Coal Company (W.C.C.). These three companies were responsible for approximately eighty-five per-cent of the coal produced from under the Kingdom of Fife and in the case of some of the East Fife coalmines, from under the *Firth* of Forth. *(The remaining 15% came from privately owned, or limited leased mines or drifts.)* Out of the total amount of 34 coalmines worked, the Fife Coal Company operated perhaps fifteen or sixteen collieries, *(plus several surface dippings)* that produced approximately 50% of the 85% total coal output, whilst the other two companies, the L.I.& C.C. and the W.C.C., produced the remaining 35% from four coalmines each.

These three companies owned fourteen of the sixteen coal mines, that employed more than 500 men just prior to Nationalisation. Of the three companies mentioned above, the Fife Coal Company was by far the newest, having been formed as late as 1872. It soon became the largest of the three, by a process of vigorous expansion and fruitful amalgamations.

*{The Wemyss Coal Company (per se) was actually formed in 1894, but the Wemyss family dynasty had been mining different coal seams on their own lands of Wemyss-field since before 1428. The W.C.C. did not operate coal-mines outside of their own fiefdom, although they did at one time operate up to nine or ten coalmines within the P.C.M.}*

There is also no doubt, that the coal taken *(extracted)* from the Wemyss-field lands *(including that taken by the F.C.C.),* was initially much easier to extract and could be got to the docks and harbours *(and salt-pans at Wemyss and Methil)* at a much lower transport cost, than the coals taken from inland coalmines working to much greater depths.

In 1870, the total amount of workers employed in the coal mining industry in Fife was around 4,500, where the output from the coalmines was around 1.5 million tons annually. This was quite a small percentage of coalminers, considering that the county's total population numbered around 155,000 persons. In 1914, the number of workers employed in the coal industry had risen to almost 30,000, from a total county population of around 250,000 persons, *(approximately one in eight or nine of the population)* while the amount of coal being produced by all of the companies *(and lease-holders)* at that time from the Fife coalfields, was around 10 million tons annually, with up to twenty-five percent of it being exported from Methil Docks. In 1914, at the outbreak of the Great War, a large number of coal-miners were by one means or another, drawn into the Armed Forces and by 1918, coal production in Fife had actually fallen by approximately one-third from its 1913-14 level. After the end of the war, the export of coal from Methil fell considerably, mainly due to the disappearance of some markets and the reduced needs of others, while at the same time, the coalmines were beginning to suffer from the lack of development and the spiralling cost of new mining equipment.

In the months after the end of the Great War in 1918-19 when some of the ex-miners returned to the coalmines, the industry experienced a small boom where the output from the coalmines was hard-put to keep up with the demand. This was however, relatively short-lived and very soon the coal industry like many others, was headed for a longish period of labour unrest and social strife that severely crippled a once booming industry. The period from around 1918 to around 1931, was a very bleak one for most of the coalminers and their families, who had lived through both the National strike of 1921 and the General strike of 1926, where the state of wages and conditions had not improved to any extent, with many miners leaving the industry to seek regular work in other areas and other occupations.

This was also a time when some of the coal owners were trying to move away from some of the more labour intensive methods of coal-taking, by investing in what at that time, was more modern mechanised methods of coal-getting, that invariably involved the employment of a lesser number of miners. Between the years 1924 and 1938, the number of miners employed in the Fife Coal-fields actually dropped by about 25 percent, while production figures for those two different years *(separated by a fourteen year gap)* were similar at approximately 7.5 million tons.

In the early 1930's when the Country began to pull itself out of its depression, the number of employed miners in the Fife coalfields showed a slight increase, which may have been due in part to the re-establishment of the partial trade links with some Scandinavian Countries, where several agreements included the shipping of coal from Fife ports were concluded. Some of those agreements were bilateral in nature, which meant that North Sea shipping did not sail one way in ballast, but carried worthwhile cargo's in both directions.

*(The relative cost of shipping at that time was so much less than that of rail transport, that it was cheaper to send coal from Fife to London by boat, than it was by rail!)*

During, or around the year 1932, the Fife coal-owners were again up in arms, this time against a proposal by the Coal Re-organisation Committee to prepare a scheme for the general amalgamation of the coal industry. *(Scotland?)* However, no such scheme was ever submitted, much to the relief/delight of the hostile coal owners! *(Could this have been an early attempt at a Co-operative system or even a pre-emptive 'Nationalisation' of the coal industry?)*

The years leading up to W.W.II, did not see too much change in the coal industry in Fife, but in 1939-40, a great sea-change swept through the coal industry. A state of general war now existed between Great Britain and the Axis Powers, with all that that presaged in the form of the one intangible energy source that Britain possessed in great abundance - COAL! Coal was energy! Energy was power! and this specific raw material for the source of both energy and power, was the one mineral entity that was going to be needed in vast quantities.

Nearly every power source needed coal. Railway engines, steam ships, electricity generating stations, munitions factories, the steel industry and even the coalmines as well, needed large quantities of their own coal to produce the press-urised steam needed to drive the massive winding engines, the pithead machinery and even some of the underground haulage systems! Coal was now at a massive premium, the extraction thereof to be made a grade one Government priority, with the retention of experienced coalminers even more-so! For the last ten years, the coal industry had simply not produced enough coal to the satisfaction of heavy industry and the domestic market. This was to change somewhat, but not too drastically! During the Great War, many coalminers and with it their expertise, was lost to the industry through miners being allowed to join the Armed Forces. This did happen to some extent at the start of W.W.II. - but this was soon stopped with coalmining being imperatively and deservedly classed as a 'reserved occupation'. *(During the course of the hostilities, the many 'felons' passing through the British court system, were on conviction, given the option of either joining the Army, or entering the coalmines! I have no idea whatsoever as to what percentage did either!)*

The movement and final destination of most of the extracted coal also changed somewhat during this time, in that there were little or no exports of coal from any of the coal ports on the *Frith* of Forth. Some coal was moved by rail to ports on the west side of Scotland, but the bulk of it was transported by rail *(and then sometimes by road)* to user destinations all over the Country. However, in spite of the fact that coal was in great demand *(and by implication, coalminers also!)* during the period 1939 to 1945, many coalminers did leave the industry and overall coal production fell by almost one quarter.

*{It may have been due to the fact that the Armed forces were suffering great losses on all fronts, that men from civilian industries and experienced miners from the coal industry were being accepted to supplement the much needed replace-ments. Many coal miners were still wont (perhaps for patriotic reasons?) to join the armed forces!}*

At the cessation of hostilities in Europe in May 1945, in the two years before Public ownership, the coal-owners in Fife *(and probably elsewhere!)* found them-selves in an inevitable predicament. During the last six years of war, the country had needed and got 'quick coal', and this from just about every coalmine in the Country. The coalmines were now in a semi-perilous state. Coal had been taken at great risk and speed, without too much regard for the well-being of either the miners, nor the conditions in which they 'wrocht!' Most of the underground workings were near to being described as unsafe, and much of the underground support equipment used, had been of inferior quality, *(girders, wooden bars, wooden props, etc.)* and a great percentage of the mechanical installations were on the point of collapse due to overwork, lack of spares and disregarded maintenance. And, the miners themselves were almost at the end of their tethers with regard to conditions and wages! In other words, the coal industry was in an utter shambles!

It needed a breather!  It needed re-planning!  It needed better overall organisation!
It needed huge investment.  It needed a lifeline in the form of unequivocal and
unstinting financial help!

It got it!  It came with the recently elected Labour Government, but it also
came with chains attached!  It was called '**Nationalisation!**'

### Just how many working coalmines remained in Fife on Vesting Day?

At Nationalisation on January 1$^{st}$ 1947, the following coalmines were still in
production in the County of Fife. *{I have disregarded those ten coalmines in Alloa/
Clackmannan that were lumped together with those in Fife for purposes of
administration and statistics. They are simply not part of this narrative! However,
it is worthy of note to remember that the coalmines in Alloa (due to the vagaries of
geology) worked several coal seams within the same Productive Coal Measures
(upper coal measures), as did the coalmines in East Fife.}*

1. Isle of Canty pit.   2. Muircockhall pit.   3. Dean pit.   4. Lady Anne pit.   5. The
William pit.   6. The Alice pit.   7. Mossbeath pit.   8. Kirkford pit.   9. Cowdenbeath
No 7 pit.   10. Lumphinnans No 1 pit.   11. Lindsay pit.   12. Aitken pits No's I & II.
14. Lumphinnans pits No's 11 & 12.   16. Lochore pits No's I & II.   18. Bowhill pit.
19. Kinglassie pit.   20. Glencraig pit.   21. Nellie pit (Lochgelly).   22. Minto pits No
I & II.   24. Dundonald pit.   25. Jenny Gray pit.   26. Dora pit (Little Raith).   27.
Francis colliery (Dubbie).   28. The Randolph colliery (Dubbie).   29. Wellsgreen
colliery.   30. Balgonie pit.   31. Lochhead colliery (inc the Surface Dipping and
Victoria pit).   32. The Michael colliery.   33. The Rosie pit.   34. The Wellesley
colliery.*

In addition to these established pits and collieries, *(some of them very old!)*
there were two further pits or collieries and two further dippings or mines in
planning, or in being *(but not in production!)*  The two collieries that were to be
sunk, were Seafield pit at the west end of Kirkcaldy and the Rothes colliery at
Thornton.  The two Fife mines *(dippings or drifts)* that were planned, were Earlseat
mine and Cameron Mine.  Of the latter two, Cameron Mine was the first to bring
mined coal to the surface, by methods that involved the coal being carried all the way
from the face-line to the surface by a limited series of wide, endless rubberised
conveyor belts.  This was followed a few years later by the even more prolific output
from the Earlseat mine, a working that was also unique in its time, in that it was
substantially equipped with the new Anderson-Boyes, short-wall, coal-cutting
machines, with a 'pan-engine' and a long train of jigging pans allocated to each coal-
cutting machine.*  Most of the blown-down coal from each short-wall face, was
virtually shovelled-up by the duck-billed end pan, carried and transferred to a 36

inches wide belt system running the length of the mine, and conveyed directly into waiting coal wagons at the mine head. *(Eighty percent of the taken coal\* was untouched by miner's shovels!)*

\*See Chapter, 'Out of the East!'

*The life expectancy of a Drift mine as opposed to a Surface Dipping was usually not very long! Drift mines (i.e Cameron mine, Earlseat mine) were developed to take quick, easy 'surface' coal that required comparatively small development costs by way of 'getting' to the coal. Drift mines can be compared to the early coalheughs, in that the target seam of coal either extrudes to the ground surface, or lies immediately under the ground surface. The 'angle of attack' is usually directly into the seam at not too steep an angle, that will help facilitate the installation of a long conveyor belt to the very bottom of the mine. If the angle of descent is too steep, such an installed conveyor would 'spill' too much coal on its upward journey, so a different, slower, more expensive coal-raising system would have to be deployed. Fortunately, the angle of descent in both Earlseat and Cameron mines was such, that long conveyor belts were successfully installed in both mines to bring the extracted coal directly into coal wagons waiting at the mine-heads.*

*Cameron mine was never a very large enterprise and employed perhaps around 180 miners at maximum deployment. It produced moderate quantities of land-sale coal from the remnants of the Braxton coal seam, which was approximately 24 -30 inches thick at that location and had previously been abandoned by the W.C.C. in 1939. The mine lasted all of ten to eleven years, before it was ignominiously closed by the N.C.B. in 1959!*

*Earlseat mine was a completely different entity for several reasons:- 1. It was driven into perhaps the second thickest seam of coal in the upper coal measures,- the Lower Dysart seam. 2. The surrounding strata and 'ground' was already well known, in that the mine encroached into an area well within Lochhead's coal boundaries. 3. The coal was taken using the very latest 'hi and run' methods, utilising the now ubiquitous and recently introduced A.B. short-wall machines, working in conjunction with pan-engines and trains of jigging pans, with uninterrupted coal conveyance coming directly from the many individual short-wall coal faces, directly up to the mine head. (The loaded wagons were streamed daily onto the W.C.C. mineral railway and thence to the Wellesley washer!)*

Even though Cameron Mine and Earlseat mine were commenced around the same time in the late Forties, production totals and relative outputs cannot be readily compared for two good reasons: - 1. The difference in thickness between the Lower Dysart seam and the unsubstantial Branxton coal. 2. The great difference in the types of coal-getting machinery used in the separate mines.

Author's Note: - The Lower Dysart coal, also known as the Leven seven-foot coal, was not an fully homogeneous seam as will be made clear in the chapters dealing with the 'Lower Dysart West', even though the quality and friability of the coals contained therein were much of a sameness! There was contained within its approximate 155 inches *(3.9 metres)* thickness, *(in Lochhead colliery)* three separate, but closely bonded 'sandwiches' of coal, the top and bottom leafs of the same relative thickness, with the middle leaf being slightly thinner. All three leaves were effectively separated by layers of stone or blae, parrot coal or soft blae. The average thickness of the top leaf was between 39 and 43 inches, separated in its middle by a 1 to 2-inch band of stone. The middle leaf was approximately 31 to 33 inches thick, but separated from the top leaf by a 30 to 36-inch band of soft blae. This middle leaf of coal was absolutely clean! The bottom leaf was between 37 and 42 inches thick, but also separated by another 1 to 2-inch band of stone at its centre. This bottom leaf was separated from the middle leaf by a 4 to 5-inch band of stone, adjoined to a 5 to 6-inch band of parrot coal.

It can therefore be adjudged from the above description, that even though the different leaves of the thick seam were relatively 'clean' - the good quality coal contained within the seam taken as a whole, was termed to be 'dirty!' This then, had to be taken into account by the individual colliery managers in deciding as to how the various 'leafs' were to be extracted.* It must also be borne in mind, that at the Earlseat location, the Lower Dysart seam had thinned somewhat to between approximately 8 to 10 feet in total thickness, - but *(on a pro-rata basis)* with the same percentage of 'dirtiness!'

*See Chapter: - Lower Dysart West and Lower Dysart Dook!

**A full description of this type of extraction appears in the chapter; 'Out of the East'.

*

In the days before Nationalisation, at a time when colliery managers were responsible to the coal-owners for the quality of coal that actually came up the mine, this responsibility was sub-delegated to under-ground managers and thence down the chain, through section gaffers and Firemen *(deputies)*. In earlier times, and in the eyes of the colliery management, it was almost a sacking offence for an experienced collier to load 'dirty' coal into his 'tagged' hutch *(coal tub)*.  In Lochhead pit, under-managers of the calibre of Isaac Young and Andrew Cairns, were forever warning colliers against the temptation of filling even the tiniest amounts of redd into their hutches.  *(Many Lochhead pick-place colliers tell of instances where the under-manager's had demanded to be shown the whereabouts of the current pile of separated rédd, that had to be purpose built by the roadside to show compliance with this sometimes spiteful diktat!)*  They both had the reputation of railing hard against transgressing miners and would on a whim, cause any tagged hutch to be 'couped' at the pithead in order to inspect its contents.  Many colliers suffered this archaic, but wage-docking ignominy and were usually treated quite unsympathetic-ally by the colliery hierarchy *(and by their fellow colliers!)* if they complained about 'lost' weight!

With the introduction of modern coal-cutting machinery and coal-conveyors, came the slow demise of old-fashioned 'pick-places', 'aeroplane-braes' and tagged hutches!  Then came the mechanised long-wall faces with the indiscriminating face-conveyor belts, whereupon up to twenty 'coal-strippers' simultaneously and system-atically, hand-loaded their ten yards of blown-down coal over a seven hours period of time.  If there was any redd or stone downed with the coal, they were at least, morally bound to manually separate this from the downed coal, before throwing it over the conveyor belt and into the condies *(wastes)*.  Few, if any of the colliers *(strippers)* did so!  It was time-wasting unpaid work, where the pressing need was to shovel coal while the face-conveyor was actually moving!  Often it didn't, due to break-downs, broken belts, lack of empty hutches, etc., etc!

It can therefore be seen, and indeed was a plain fact, that the incidence of 'dirty' coal greatly increased with mechanisation - and the fact that the begetting and taking of coal had become far more impersonal.  It was also plain to see, that most of the 'cleaner' seams of coal taken at shallower depths had been exhausted and that deeper, richer, but potentially dirtier seams such as the Lower Dysart, had to be eventually tackled!

*

### Just how much mechanisation was there in the Fife coalmines?

At the turn of the nineteenth century and into the first decade of the twentieth, partial electrified mechanisation was introduced into several coalmines in Fife.  This was in the form of one of the earliest types of new coal cutting machinery available, the D.C. powered **Disc machine**.  This rotating disc coal-cutter was something of a revolution in the coal industry at that time, in that it actually worked, albeit slowly!  It was also prone to constant breakdown, but it was made to work in the environment that it was designed for, and thus, the era of coal-cutting had commenced!  These new machines were so successful in the Fife mines that by 1910, over 100 of these machines had been installed by the Fife Coal Company in its own coal mines. In the fifteen years that followed, nearly every coalmine in Fife had installed several more machines, with the total then in use, approximating to 340 machines of the type!

The next generation of coal-cutting machines saw a welcome change in the method used to undercut the coal. The old disc-type machine was by comparison, too light in construction, too easily broken or damaged and often needed replacement parts. It was also relatively slow! The new generation machines, **the Anderson-Boyes 17-inch** chain driven coal-cutters, employed a slightly different principle to undercut coal, in that it incorporated a moving endless steel-linked chain, driven by a heavy sprocket moving around the periphery of a four and a half foot long flat, but narrow jib that could be locked at right, or left angles to the body of the machine, to effect a continuous undercut in either direction. The endless 'chain' was comprised of linked 'pick-boxes' that held the individual sharpened 'picks', that when rotated at speed, had the effect of ripping an undercut to a depth of approximately four and a half feet at ground level and to the low height of approximately 6 inches above the pavement.

These machines were powered by 550 volts A.C., with both electric motor and mechanical gearing completely encased in an oblong shaped, armoured steel casing. These coal ripping machines pulled themselves along the pavement, tight to the face-line, by means of an anchored chain being slowly wound in through a mechanical sprocket chain-drive situated at the front of the machine, that 'discarded' the 'used' chain to one side as the machine passed along. This coal-cutting machine was relatively fast, powerful, dependable and left a smooth pavement in its wake. It was a thundering success!

These A.B. 17-inch machines were slowly replaced within ten years by the Mark II version, that was slightly smaller by two inches in height, was a little more powerful - and had a half-inch steel-wire rope drive wound onto a horizontally placed 18-inch drum at its bottom front end, to effect its forwards and reverse travelling motion. Its official designation was the **A.B. 15-inch** long-wall coal-cutting equipment. It was also affectionately known as the **'*Bluebird!'**, probably named after the fleetest of W. Alexander & Sons coaches? *(The bus company having the passenger road transport monopoly in the Kingdom of Fife.)*

These long-wall machines had a long cutting life, with some of them still in use in 'stable-ends' on the long-wall shearer faces in Lochhead's New Dook, in its terminal decade! (1960-1970).

*Its given name always puzzled me, the machines were always painted bright red!

\*

## Just how much coal did the different areas produce?

In 1939, there were approximately 34 collieries producing coal in Fife from three different areas:- East Fife, Central Fife and West Fife. None of these collieries were spaced very far apart from each other, and this was mainly due to the fact that many of the multiple under-lying seams of coal were easily accessible and rich in content, with many of the seams containing 'absolutely clean' coal. Indeed, many of the collieries clustered around the centre of the county were within easy walking distance of each other. Some of the seams of coal worked were very prolific and had names that were well-known within the coal industry. i.e. The Dysart main, the Chemiss, the Dunfermline splint, the Lower Dysart, the Lochgelly splint, the Diamond-Jersey group, etc, etc.! Some of these named seams had developed a thickness that almost defied belief. The Dysart Main up to a maximum of 32 to 33 feet *(10 metres)* thick, the Diamond-Jersey group at nearly 16 to 17 feet thick in places, the Lochgelly splint at approximately 8 to 9 feet thick and the Dunfermline splint at fully 5 feet thick. Other workable seams such as the Barncraig, the Coxtool, the Bowhouse and the Branxton were thinner for the most part, but were eminently suitable for extraction by long wall coal-cutting machinery of the ilk of Anderson-Boyes 'bluebirds!'

The eight major collieries that were operated and produced coal from the Upper Coal measures *(East Fife area)* were, around 1938-40, responsible for approximately 40 percent of the total output of the county. The reason for this, was the prolificacy of coal from the Dysart main, the Chemiss, and the Barncraig seams. The total reserves in the Upper Coal measures *(P.C.M.)* at this date, were reckoned to be in the region of 1.5 billion tons, with the total reserves in the West of Fife *(L.C.G.)* estimated at 2.26 billion tons. This total figure of 3.76 billion tons comprised of approximately 50 percent of the total Scottish coal reserves!

*

## Just how and where in Fife did the first 'coalliers' get their coals?

Author's Note: - A heuch or heugh : *a crag, a cliff, a rugged steep, a hollow, a deep glen, a deep cleft in rocks, a coal pit, the shaft of a coalmine.*

A heuch-man or heugh-man : *a pitman or coallier.*

A heuchster or heughster : *a pitman or coallier.*

A Heuch or Heugh-head : *the top of a crag, cliff, or precipice.*

## What is a Coal-Heugh?

A coal heugh was the name given to the working, taking, or extraction of coal, commencing at ground level by means of a basic form of 'opencast' working, where a seam of coal extruded onto the earth's surface. These extrusions of coal *(in spaced seams)* very rarely appeared as vertical entities coming directly up from the bowels of the earth, but appeared as irregular longish ridges across the landscape, at a width that belied their true thickness. In the case of the original coal heughs at Wemyss, the average slope of the rising strata was between one in three and one in four. This meant that a regular seam of coal that had a true thickness of approximately four feet, would appear at ground level to be all of 12 to 16 feet wide. An entirely false impression that was realised as soon as the heughster's started to 'take' the coal. These coal ridges or coal heughs at Wemyss, did not lie parallel to, or level with the line of the sea shore. Quite the reverse in fact, that they originally lay at nearly right-angles to the shore-line!

The multi-layered 'tongue' of the Productive Coal measures (U.C.M.) that intrude into the South-east coast of East Fife from under the *Frith* of Forth, does so in a north-westerly direction and at a irregular inclination of approx 1 in 2.5 to 1 in 4, or between 14 to 22 degrees. The intrusion of this multi-layered tongue is neither wide nor long! It is almost semi-circular in shape, having a width of approximately 9.5 miles or 15 kilometres. The point or tip of this tongue at maximum intrusion, extends inland for a distance of approximately 3 miles or 5 kilometres. The deepest, generally worked seam of coal within this tongue, *(this being the Lower Dysart, for the reason that the Lethamwell seam was rather irregular and heavily faulted!)* appeared as an outcrop *(coal heugh)* near Dysart, swung inland in a semi-circular arc passing near to Gallatown, then Thornton, then swinging clockwise to the south of Markinch, around past Windygates and back towards the shore-line between Leven and Lower Largo. All of the other seams immediately above the Lower Dysart in depth, also outcropped within this arc to a greater or lesser extent, but in ever-decreasing semi-circles. However, this was by no means a series of perfect extrusions. If it had been, it would then have been possible with hindsight, to have traced the semi-circular route of the outcropping of nearly every seam of coal within the Upper Coal Measures.

There were of course, many small and some larger fault-lines within the Wemyss-field 'metals', that were encountered as various underground seams were worked. Many of those faults were encountered in all of the W.C.C. coalmines and those belonging to the F.C.C. But, in most cases, the coal-seam was rediscovered either as a limited up-throw or downthrow, with minor redevelopment quickly regaining the initiative.

The two great hiccups *(faults)* in the Upper Coal Measures that all but separated the Wemyss-field from the Methil and Leven fields, were the Buckhaven/Earlseat fault and the Muiredge North fault. At the north-east boundary of Lochhead pit where its coal seams abuts onto the Buckhaven fault, the downthrow is of such magnitude on the other side of the Dyke, that Dysart main coals taken in Wellsgreen colliery were fully 25 fathoms deeper than in Lochhead at coincidental points. This of course, means that nearly every layered coal seams within the metals to the north side of the Buckhaven fault are effected in the same manner.

The Muiredge fault seems also to have had a much greater effect on the lie and depth of the seams on the right side and edge of the intrusive tongue of coals, in that the depths at which the relative seams were found, bore little resemblance to that on the left side and edge of the tongue. The coal seams found and worked in the Methil and Leven pits, did bear semblance to similar coals found within the Wemyss-field area, but the seams were actually given different names.

The 'main' seam of coal worked in Leven and parts of Methil, was actually the named Chemiss seam as worked in Wemyss-field collieries. *(The Wellesley, the Rosie, the Michael, Lochhead, Victoria pit and the Francis.)* Under the burghs of Leven and Methil, the Chemiss seam was fully nine feet thick and known as the Main coal. *(This Main coal must not be confused with, or mistaken for the Dysart Main seam as described in the Wemyss-field areas!).* This Main coal at Leven was found and worked at a depth of 60 fathoms, while at the Kirkland pit in Methil, the Main seam lay at 144 fathoms. The two other workable seams taken at Leven and Methil, were the Six-foot and the Eight-foot seams that were also found within the Wemyss-field coals, but named the Coxtool and the Barncraig respectively! At the Durie colliery at Leven, the Main coal was found at 60 fathoms, the Six-foot coal at 33 fathoms and the Eight-foot coal at 28 fathoms. Even though the general inclination of the Upper Coal Measures is in a north-westerly direction, the inclination is by no means uniform. There are several undulating folds within the strata/coal measures in this part of Fife, which meant that the depth and lie of the known coal seams were somewhat irregular in certain areas. This was further complicated by the now identified large fault-lines or throws, of which there were several!

To the North-east of this location, there is yet another fault known as the Scoonie fault, that reputedly had a throw of approximately 130 fathoms. At the new Scoonie pit that was located near the first milestone on the high road from Leven to St Andrews, the Chemiss seam *(Leven main coal)* lies at a depth of 99 fathoms.

Edited extract from the Second Statistical Account 1845. (Fife) re the Leven/ Scoonie coals.

Beds of coal, varying in thickness, and at different depths under the surface, pervade the whole of the Parish of Scoonie ; but none of these seams are at present worked. The coal upon the estate of Durie, which was wrought for upwards of a century, and was drained by a water engine, consisted of three seams, the two upper (seams) each four feet thick, and the lower (seam) eight feet thick.

(This probably, was the Barncraig, the two, close, thin Coxtool seams and the Chemiss seam.)

There is understood to be a fourth seam below these (three), called the craw coal, the outcrop of which comes out at about 120 yards south-east of the mansion house (Durie?). The third seam, called the *main coal* (Chemiss) was considered the best in the country. Considerable quantities of it used to be exported to Holland, where it met with a ready sale ; and it is said, even at this day that the best Scotch coals in the market go under the name of Durie coals.

{The Barncraig seam was found at a depth of 19 fathoms, the Coxtool at 24 fathoms and the Chemiss coal at 40 fathoms. The fourth seam, the 'craw coal' was likely to have been the 'Bush' coal, (*which had parrot-like qualities*) for the simple reason, that the next underlying seam was the Parrot coal, that lay a full 50 to 55 fathoms below that!}

It can therefore be realised, that the coal seams that lay within the south-west half of this intrusive tongue, were far more predictable in lie and much easier to work, than the same seams that lay within the much-faulted and altogether unpredictable north-east part of the intrusion, which, when lumped together, were known as the Productive Coal measures, or the True Coal measures, within East Fife.

Most of these outcrops were wrought many centuries* ago, within the whole of the area bounded by this intrusive multi-seamed tongue of coal-bearing strata! The greatest depth at which the Dysart main seam (*third from bottom*) was breached in the Wemyss-field area, was at the initial depth of approximately 270 fathoms at the Wellesley pit at Denbeath.

This then, was the deepest that any of the W.C.C. controlled pits was taken, to 'broach' this seam. (*The internal workings of course, were taken much deeper!*)

The seam itself was worked to much greater depths, out and under the *Frith* of Forth, but the initial intrusions were always made in solid strata (*metals*) away from the sea-shore. This broaching of the Dysart Main at the Wellesley, could be said to be the deepest point of the seam within the Productive Coal measures (U.C.M.) on the East Fife coast. (*Not under the sea!*)

The coal-bearing strata then inclines upwards in all directions landwards in a wide arc, where most of the coal seams out-crops in a series of undulating 'waves' radiating out from this general area. (*Except of course, for the faulted irregularities at Muiredge, Methil and Scoonie!*)

* The following is an edited extract from the Second Statistical Account of 1845, concerning the coals belonging to the Earls of Rosslyn, whose eastern march

lands abutted those of the Earl's and Laird's of Wemyss. *(Both of these landowner's/ coal-masters between them, worked the sixteen or seventeen coal seams that out-cropped on their respective estates.)*

> The coalmines which are on the estate of the Earl of Rosslyn consists of fourteen beds, most of which, however, are thin, and have been wrought out above the level of the sea. Three of the thickest of these beds are now working. The uppermost is five feet thick, the second (middle one) is eight feet (thick) and the third (lowest one) is five feet thick.

(These three seams can only have been the Barncraig, the Chemiss and the Bowhouse.)

> At present these beds are wrought about sixty or seventy fathoms below the surface. Dysart coal was amongst the first to be wrought in Scotland, operations having begun upwards of 350 years ago.
> *(This was written in 1845!)*

> It has a strong heat, but being rather slow in kindling, and leaving much ashes, is not so pleasant for rooms as some lighter coals. Like most minerals on the sea coast of this parish, it dips to the south-east one fathom in three near the shore ; but is flatter as it goes north. It has been repeatedly on fire, the effects of which may still be traced by the calcined rocks from the harbour (to) more than a mile up the country.

Author's Note: - This paragraph seemingly describes all of the coal taken from the Earl's coal-heughs and pits as 'Dysart coal!' It is described as having a strong heat, but rather slow in kindling! This surely did not apply to all of the coal seams worked by the Earls of Rosslyn? (And the Lord's Sinclair?) Was it merely that all shipped coals were mixed together and described as Dysart Coals? Also, the above paragraph does not describe which of the above-named three seams had the propensity to catch fire! They were not all subject to spontaneous combustion and calcination, as proved by later extractions!

★

The coal-heughs that I describe here, are those that appeared on the left edge of this tongue, where, fully twelve seams of coal were discovered as outcrops coming out of the waters of the *Frith* of Forth between East Wemyss and Dysart, probably initially worked in the 13th or 14th Century!

As already discussed near the beginning of this chapter, none of the named coal-seams were at the same regular depth along the line of the seashore at Wemyss. Those that surfaced *(outcropped)*, did so in a north, north-westerly direction, for the reason that the further east the lie of the seams, the deeper they were - and therefore surfaced farther inland!

*(This surfacing of coal seams is not to be confused with the general rise of the strata towards the northwest, as verified by vertical drillings taken at various locations on the lands of Wemyss-shire.)*

The coal-heughs/ridges, therefore lay at a shallow angle to the line of the seashore - and in most cases ran indirectly into waters of the Forth, where they disappeared under the waters of the shoreline.

A coal-heugh could be worked simultaneously over the full length of its exposure at ground surface, as indeed many of them were. The coal-heugh is also quite unlike a coal mine, where there is limited access to the seam, once a mine or dipping has been commenced. The coal-heugh can employ as many *coalliers* as there is space to allow them to function, the only limiting factor being the number of *coalliers* the coal-master wished to employ. In those days, the types and amount of various iron working tools that could be employed, would depend on the skills of the local blacksmiths. A miner's lightweight 'holing-pick' would probably be the foremost tool, along with some sort of large coal chisel or wedge and a mallet. A shovel of sorts would also be necessary, bearing in mind that there would not be a 'level' pavement as such at the base of the workings. *(Much of the larger coal would probably be removed by 'hand and bucket' or a wooden sledge after extraction by the men, with the smaller coals hand-gathered and transported by women and children)*

As the coal-heugh opened up in length, more colliers would have access to the workings, which would probably proceed as follows. As the easiest surface coals were taken first, the heugh would take the shape of a wide-mouthed drift mine being gradually deepened over its full length *(up to several hundred metres in length)*, with the wide newly exposed pavement showing the true angle of dip. The 'overhang' above the coalliers heads would become more precipitous, as the full coals were being slowly extracted along the length of the heugh. If the taken height of the coal seam was approximately four feet, then that would be the length of the wooden supports needed to 'hold' the roof in position, lest it all collapsed thus rendering the coal-heugh unworkable, through being buried. At the bottom *(lowest)* end of the heugh, the coal would be worked into the sea all the way down to the low water mark, even though this meant removing the incoming tidal silt down to every low tide. The terminating factor at this bottom-end of the coal-heugh, was when the colliers could no longer overcome the deepening silt and the rising seawater.

As the coal-heugh deepened over its long length, the number of 'angled' supports needed to hold the roof, became directly proportional to the greater over-head weight that needed to be supported. *(Readers will realise that with a long open coal-heugh, there were no solid stoops left at regular intervals to absorb the overhead weight and increasing down pressures.)* With this method of coal extraction, all of the weight of the thickening precipitous overhang bore directly down onto the wooden supports, which bore greater weights that increased with depth. This then, was one of the limiting factors on how deep a coal-heugh could be cut, before it all collapsed on itself. When that happened, the only way to take the deeper buried coal was to work it as opencast, an option not really open to those early colliers, because of the increasing difficulties in bringing the coal to ground level and the likelihood of consistent flooding. *(The beauty of the coal-heughs at Wemyss and Dysart, was that by the method that they were worked, they were usually self-draining to the sea!)* As the coal was wrought, each piece/lump of coal was labouriously 'howked' from the solid. Coal being generally multilayered, did lend itself to this type of systematic 'stripping', with the collier standing to the downside of the grain.

The colliers, using a 30-inch long-handled 'pick', would cut *(hole)* a narrow channel across the grain of the coal to give himself an open end, where he then used his pick to howk and split/strip a long ream of coal from an exposed edge. *(There would be several such long parallel reaming strips taken across the full width of the heugh.)* This would be done systematically, until the colliers had taken the full width and length of the exposed seam to a further depth of approximately 12 inches along his appointed stint. Then, the collier would begin again!

When the heugh had been surface-stripped and was wrought down to a given depth, the method of extraction would undergo a marked change. If the height of the coal seam was less than that of the collier, he therefore no longer had sufficient headroom to enable him to stand erect to work the coal. The 'face' of coal along the length of the heugh would resemble a 'long-wall' face of coal showing it's four feet height of seam, but at a direct dip following the lie of the seam.

If the coals were to be worked deeper, it would then make sense for each collier to work 'upwards' from the bottom of his place in the heugh, so that the subsequent 'bottom-holed' and downed coal would gravitate to the bottom of his place, thereby negating the collier having to work to the dip and causing additional grievous strain on an already over-worked body.

As to how deep a coal-heugh was actually taken, would depend on the coal-grieve *(superintendent)* in charge of that particular heugh. Coal-heughs were reasonably safe as long as the surface coals only were being taken. The problems came when the colliers had to go deeper to howk and extract the coals from the dipping, deepening seams. The question of how to support the overhanging precipitous roof was always uppermost in the colliers minds, along with the need to fulfil the everyday coal quota needed to satisfy the demands of the coal-grieve.

\*

Working a coal-heugh on the seashore and to the rise, is an entirely different prospect to working a coal-heugh in the middle of a field on level ground. Indeed, it must be allowed, that most of the early coal-heughs in the East Fife coal fields, were in fact, away from the seashore. It also must be realised by readers, that irrespective of how potentially rich a coal seam may be, it is virtually impossible to extract coal from any single seam that lies under a built-up area. Here, we are discussing the potentiality of coal extraction from at least 12 of the 20 named seams, some of them at six, eight, twelve and twenty feet thick, as they lie underneath. It is simply not possible to extract coal to any degree from under man made structures - and maintain the integrity of those buildings. That therefore, is a positive no-no! It is therefore not surprising, that nearly every coal mine takes coal from seams that lie under agricultural land, or in the case of the Wellesley, the Michael, the Francis and Seafield collieries, from under the sea.

It will therefore come as no surprise to readers, to discover that the bulk of the early coal-heughs were to be found in the middle of ploughed fields, on common or waste ground, or in the woods and forests. No doubt, coal outcrops were to be found in or near early habitations, indeed, it may be that small villages or dwelling houses were to be found clustered around the means to provide cheap and plentiful fuel, but no one in their right mind would dig a large hole underneath their own house!

At the early heughs at Boreland, Thornton, Earlseat and Balgonie, the coal appeared on the ground surface as long broken furrows that stretched for many hundreds of yards *(metres)*. That the coal was taken and to a reasonable depth, there is no doubt! That more than one seam of coal outcropped in one or more of these areas, is also not in doubt! That which appears to be in some doubt, is exactly how the different heughs were extracted!

Old records show that at Thornton, the coal was extracted to a depth of approximately 30 fathoms *(180 feet or 55 metres)* by a method called 'Wrought level free!' This does not of course, imply a vertical depth of 30 fathoms, but an extracted 'depth' of 30 fathoms on a slope of approximately 1 in 7 or 1 in 8. A much shallower proposition! This coal is described as being wrought *(worked)* level free! This can only mean that the 'Stoop & room' method of extraction was not used, nor were any supporting stoops of coal left in situ! It would seem that the extraction was total, down to the given depth with the coal being '*wrocht*' and in a side-ways stripping manner along the full length of the heugh. This method implied that the colliers set two, or three rows of wooden supports to the length of the sloping 'face-line' to protect themselves, but allowing the worked-out ground behind them to collapse as they constantly withdrew the earlier *(first)* line of roof supports. They would do this to ease the immediate weight above their heads! This would also have the effect of limiting a possible large reservoir for standing water, which in fact, was one of the limiting factors in how deep they could take the coal extraction in the first place.

\*

### Just what did the previous Lairds of Wemyss do with their early coals?

Once upon a time a few centuries past, there appeared on the shoreline between the villages of West Wemyss and the east-end of Dysart, the black outcrops of fully ten different seams of coal. They appeared to run directly out of the waters of the Forth in long parallel lines, kink a little to the west and then run almost due north for a considerable distance inland. These out-crops seemingly appeared long, long before the value of what they contained, was realised for the 'black diamonds' that they were later to become. Seemingly, the monks of old were amongst the first of semi-educated people to realise a little of its true worth. There is on record the fact, that the monks of Dunfermline mined shallow coals as early as the eleventh or twelfth century!

The coal seams that appeared as outcrops between West Wemyss and Dysart were as follows from east to west: - 1. Barncraig. 2. Chemiss. 3. Wemyss Parrot. 4. Wood Coal. 5. Earl David's Parrot. 6. Bowhouse. 7. Branxton. 8. More Coal. 9. Boreland. 10. Sandwell Upper and Lower.

*(These named seams of out-cropping coals plus two other even shallower seams, ran directly into the sea (or out of the sea) between East Wemyss and Dysart and continued on into the sea-waters as open coal heughs. Obviously, time, tide, weather conditions and sea currents have played their part in exposing, covering, re-exposing and re-burying these once open sea-heughs. To my mind, it is obvious that Mother nature has, over the period of many centuries, eroded and decimated the once exposed coals to that much of an extent, that all, or nearly all of what lay on the sea bed, has all but been deeply buried with silt! I would also hazard a tentative, or even reasoned guess, that the sea-coal bonanza experienced in the forties and fifties, between Buckhaven and Dysart, was the result of one such timely exposure of parts of the coal seams, lying on the bed of this part of the Firth of Forth.)*

   ***It could eventually happen again!***

There is of course, ample evidence that the out-cropping coals were worked and taken at both of the Wemyss's as early as 1428, if actually before then! The question does arise though, as to what use the extracted coal was initially put to, at both East Wemyss and West Wemyss? It would seem from past writings on local history, that the coal-owners/land-owners were more interested in the production of salt as a commodious export than coal! Salt of course, being an substance that was appreciated for its culinary qualities, effects, and its inherent need to peoples of every kind and creed! It is also a known fact, that the respective Laird's o' Wemyss, over many decades, were engaged in the making/production of salt as a local industry at East and West Wemyss, and that this same industry also flourished at Dysart. *(Perhaps at the insistence of the Earl's of Rosslyn and the Lord's Sinclair!)*

As to the amount of saltpans at the different locations, this varied with the different times and the then current Laird. There was situated at West Wemyss, up to nine or ten separate pans in operation at one time, with the numbers as low as four at other times. This produced sea-salt was not limited to local consumption and inland countrywide use only, but was also manufactured for export to various lowland countries across the North sea. The chances are that it was a recognised export before that of Dysart coal!

*(It would also seem that for many centuries, many unenlightened people just did not know what to do with coal?)*

Before I turn to the description of a typical salt-pan, I feel that it may be necessary to make a few assumptions with regard to their possible construction. Contemporary descriptions and drawings of various salt-pans, show some of them as being large evaporation tanks partially filled with clean salt water, sitting exposed in the open air, while some drawings show a stone-pillared and roofed 'Dutch-barn', with the open tanks sitting underneath. The last set of drawings that I inspected, showed a stone build roofed 'house' completely enclosed on all sides, built internally with split levels, but containing only one coal-fired iron tank.

From what I have seen, read and discussed, I feel that evaporation tanks set in the open air were at a distinct disadvantage, therefore I shall describe the indoor type, before I make comparison with those at West Wemyss.

The whole principle of sea-salt making, depends on rapid evaporation of the sea water contained in the flat tanks. If the tanks were to be constructed in the open air, they must by all accounts be exposed to the vagaries of wind and weather, which during any rain-shower would surely negate, or severely curtail most of the evaporation process to date.

If the tanks were housed in 'Dutch-barns' this then would surely obviate the untimely dilution of the sea-water in the tanks, but it would still expose the manufactured salt to the effects of damp or inclement weather.

It therefore made absolute sense to completely house each and every separate tank within a weatherproof enclosure. This would have the distinct advantages of:-

1.  No contamination of the clean sea-water in each tank.
2.  Salters would work in a warm, humid, but controlled atmosphere.
3.  The salt produced would be kept dry and clean as the last of the water in the tank was allowed to dry-out.
4.  Faster evaporation due to higher internal heat.
5.  The salt produced could be weighed and bagged simultaneously in dry conditions.

*Each salt-house contained one large iron pan approximately 18 feet long, 9 feet wide and fully 18 inches (one and a half feet) deep. This large pan would sit on a purpose-built stone furnace standing approximately three feet high and having possibly two separate, but wide fire-boxes each with its own chimney. The stonework of the furnace would fully underscore the area of the salt-pan. The large pan would be half-filled with clean sea-water with both fire-boxes set alight. The furnaces would be constantly fed with coal, being previously howked from the coal-heughs, a mere stone's throw from the salt-house. The saltwater in the pans would reach boiling point in a very short time indeed - and the boiling momentum was vigorously maintained by the furnace-man, who must not on any account allow the contents of the pan to over-boil! There was a fine line between boiling and boiling over!*

*As the water in the pan decreased due to rapid evaporation, more clean salt water was slowly added to the pan to bring the level back to half way, with this being done in such a way as maintain the boiling action within the pan. This cycle of the addition of diminishing quantities of cold sea-water, was finally terminated by the resident 'salter', when he judged that he now had what he knew, (or guessed) to be a completely saturated solution within the tank. (Crusted salt would have started to appear around the inside periphery of his tank.) The salter would then suspend the further addition of sea-water to the tank and 'order' the reduction of coal to the fire-boxes. These actions came from hard won empirical knowledge, in that the diminishing of the heat from underneath, needed to coincide with the evaporation of the last of the sea-water. If all had been timed correctly, there should be up to a 2 to 3 inch layer of caked sea-salt lying on the bottom of the pan, which needed to scraped out before it burned hard. During the latter stages of the evaporation process, several hands-full of dried 'bull's blood' may have been introduced into the boiling mixture to help purify the final product! This salt was now scooped out of the pan, inspected for cleanliness and bagged whilst it remained in its dry state!*

The ratio of coal burned to salt produced, stood at approximately four or five to one. One ton of salt produced for every five tons of coal burned! Salt probably weighs more than its volume in coal, which of course, begs the question; was the price of salt more than five times that of coal? *(And of course, salt was not taxed at one time! The 'salt tax' had yet to materialise!)*

At both East Wemyss and West Wemyss, the type of salt pan in use was neither four-sided, nor made of iron. Neither were they housed in stone buildings, or Dutch barns! The pans were hand-crafted and individually constructed from thin layers of beaten copper, they were also circular in shape, having a depth of between 15 to 18 inches *(approximately 38 to 45 cms.)*. The diameter of the pans could have been anything from 10 to 16 feet *(approximately 3 to 5 metres)*. The pans were supported all around their periphery and in the centre, by short columns of cast iron bars. The sea-water was seemingly pumped into each pans directly from the sea, which had the added advantage that additional sea-water could be added at any time. *(which it was!)*. The fires *(furnaces)* underneath the pans could be stoked and tended to from any side *(a fair precaution on a windy shore!)* which no doubt they were! Again, the sea-water was brought to the boil and this was maintained by the constant stoking of the fires and the measured addition of clean seawater, until such times as the solution became salt-rich and was ready for total evaporation.

This method of salt-panning in the open air had many disadvantages, not the least of which, was that much more heat was needed to counter the cooling effect of the outside air. If the weather became inclement, this would surely also hinder the taking of dry salt from the pans. It is recorded, that at the Wemyss's, it needed up to thirty tons of sea-water to produce one ton of clean salt. Also, the coal consumption was greatly increased to between ten to sixteen tons burned, to one ton of salt produced!

*(Just what was the value of salt at that time?)*

As far as the Lairds of Wemyss were concerned, it was obviously more profitable for them to sell and export one ton of salt, than sixteen tons of coal! The one redeeming factor in favour of the salt manufactured at both East and West Wemyss, was that it was reputed to be of the very finest quality!

In the year of 1707, just after the Union of the Crowns, the parliament of the day introduced what was to become known as the salt tax! The industry at Wemyss did not actually die a death at this juncture, but the writing was on the wall! During the time of David, the second Earl, there were seven salt-pans in operation at the Wemyss's. By the mid 1830's, there was on one pan left in operation, and that probably only for local consumption!

*

## The 'Water Equation' in the East Fife Coalmines!
### *Or, Just how much standing Water was there?*

In the Wemyss-field *(East Fife)* coalfields, there were a few coalmines that were 'pepper dry', but unfortunately, there were a greater number *(over many years)* of coalmines that were in constant danger of being severely compromised, if not entirely flooded, but for the reliability and dependability of the installed powerful steam engines and high powered electric pumping machinery. Lochhead pit and the Earl-seat mines were a typical example of 'dry' coal mines. They were not however, entirely free from water. The criteria, ostensibly being, that if there were no pumping machinery installed on the pithead, or if the water came up the pit in loaded hutches, or in the case of Lochhead, the small amounts of water gravitated to the Michael pit, then this was deemed to be a dry pit. The 'water pumped to coal gained' ratio in such pits was therefore nullified - and did arithmetically, become a negative equation!

However, in wet pits such as the Wellesley at Denbeath and the Michael at East Wemyss, there were great amounts of water to be contended with, which had to be controlled to such an extent that it presented its owner's with a large, meaningful, but positive 'water equation', that weighed heavily with the mine management and needed to be considered in conjunction with the proposed development of every new section within the coal-mine.

Here, in this pass, this was not a case of a few gallons of water that could be handled by a single permanent pump working a few hours each day. This was the reality of many hundreds of thousands of gallons of water, that warrants being described in tens and hundred's of tons, that needs must be carried to the surface every minute of every day, on a permanent and never-ending basis. This was not salt water seeping in from the sea through cracks and fissures in the strata, but land water, sometimes heavily mineralised on its journey down through the various metals. The over-riding characteristic being, that water shall always find a lower level and unfortunately, what better place than in the depths of a working coalmine!

In coal mines such as the much-hyped, but unfortunately ill-fated Rothes colliery at Thornton, at the time when both shafts were being sunk, the reported rate of flooding was confirmed at 100 gallons per minute. At the Michael Pits and especially the Wellesley colliery, the amount of ingressing water was measured in hundreds of gallons per minute, a degree of measure that was soon to become obsolete, when the increasing amounts of pumped water began to be measured in tons per minute!

To initially help overcome the 'water problem' at the Wellesley colliery, the management/planners installed two steam pumping engines each capable of raising 1,000 gallons of water to the surface per minute. *(Approximately 4.5 tons per minute!).* These two pumps were powered by super-heated 'dry' steam at 150 lbs. pressure, to ensure the necessary work-rate required. In addition to the two steam pumps, two electro-turbine pumps of similar pumping efficiency/capacity were also installed, to supplement and add to the existing installations. These electro-turbine motors were each driven by squirrel-cage motors requiring an E.M.F. of 6,600 volts A.C. The combined handling capacity of both sets of pumps were capable of raising approximately eighteen tons of water per minute to the surface, if and when the need arose! *(This would equate to approximately 4,032 gallons, or 18,150 litres per minute, during every minute of every hour of every day, etc., etc. !!)*

As an incredible and uneconomic statistic, it was expensively, but acutely realised and understood, that for every ton of coal that was brought up the Wellesley pit, there had to be pumped, a heavy five tons of water. This pumped water was carefully channelled so that it did not have the opportunity to return underground thereby negating the efficiency of the pumps, but in spite of this, the damnedable water equation at both the Wellesley and the Michael Pits was never to diminish, in the resultant longer and deeper workings developed in the chase for coal. As the underground working extended outwards and deeper under the sea, more, larger and more powerful, localised pumps in separated tandem, were brought into use to keep coal faces and in-going roads from being submerged. The deeper the workings went, meant that the water became increasingly 'heavier' - and the power needed, to water pumped ratio, became decidedly top heavy!

The one saving grace with regard to 'coal raised up the pit, to water pumped', was that the energy from one ton of coal, was more than sufficient to raise many, many tons of water to the surface - and that the profit from coal, greatly exceeded the negative cost of pumped water!

In 1951, Alexander R. Smith in his illuminating and authoritative 'Third Statistical Account' *(of Fife)*, advances three solid and expansive reasons for this coal phenomenon in the Fife coalfields.

Extract from the chapter: - Coal mining. Page 232.

1. The demand for coal increased enormously, now that the industry had to meet the needs of growing industries at home, of markets of various kinds overseas, and of the bunker trade. 2. The improvements in communications that made rapid progress during the century *(early part)*. British railways and ships, besides consuming vast quantities of coal - also enabled this same coal to become transportable to an extent hitherto uncontemplated. Ships bringing home raw materials from all over the world found coal to be a very profitable ballast on the outward voyage. 3. Coal was now available in much greater quantities. New and rich coal seams were discovered in Fife in the Cowdenbeath/Lochgelly area and in the Wemyss-field area. *(Should this read - Re-discovered?)* New technical developments in machinery and technology allowed older known seams to be re-exploited and developed to a much greater depth than previously envisaged. **Four.*** Furthermore, the types of coal found in Fife, fortunately, included all of the principal types needed and used by the main industries including that suitable for export. *(With the single exception of 'coking' coal!)*
*Author's additional No. 4.

Author's Comment: - *I can find no fault whatsoever with any of the above, which at the time of writing (in my own humble opinion), must have been absolutely true, therefore, I shall make no attempt to gainsay any of Alexander Smith's erudite and illuminatory conclusions.*

I shall however, make one comment, the exact same comment that was asked by the many perplexed and bewildered ex-coalminers in the East Fife coalfields, - **"How could it all have gone so horribly wrong - and so quickly?"** In 1950's and 1960's, the coalmines in East Fife were **spewing coal**! In **1970**, there was **near silence!**

*

# Chapter II

# Black Bloody Bings!
## Captains & the Kings and unlovely 'bings!'
### *(and Troubles & Strife's and uncommon things!)*

This narrative is mainly centred around two particular coalmines in the East Fife Coalfields, namely Lochhead Pit and its Surface Dipping near the Coaltown of Wemyss, and the earlier, but now derogated Victoria pit at West Wemyss, which only ever had an under-manager after the much later opening-up of the Lady Lilian, *(now named Lochhead)*. The under-manager at the Victoria pit was one of three under-managers who were ever responsible to the manager at Lochhead pit. *(One for each of the Pit, the Surface Dipping and the smaller Victoria Pit!)* This narrative also concerns and describes some of the coal-miners who worked in one, or two, or even all three of these underground mines, where much of this sometimes unbelievable story takes place. However, not all of the narrative takes place under-ground. It touches on places, events, happenings and people, many of whom had very little to do with coal or coalmines, but were nevertheless part and parcel of the extended W.C.C. 'family' and as such, involved in one or other aspect of the industry.

In order to give readers an overall general picture of the 'coal' topography in this part of the Kingdom of Fife, I shall attempt to name and describe the individual coalmines as they appear along this part of the coast of Fife in the middle of the century. Lochhead Pit was one of the inland Pits along with Buck'hind Pit, Cameron Mine, Wellsgreen Pit, Earlseat Mine and the Randolph Colliery. *(Some of the collieries/pits/mines in the Fife coal fields were often referred to by their nick-names. In the course of this narrative, I shall however, use the each colliery's given name!)* These coal mines stretched all the way from the 'lang toun o' Kirkcaldy' *(Long Town)* to the huge and spacious railway sidings at Methil Docks. *(A short description of these sidings appears at the end of this chapter.)* In the year of 1950, a sight-seeing tourist in a double-decker bus journeying from Leven to Burntisland, the order of Coal-mines passed *(but, not necessarily seen);* would be the remains of the Pirnie Pit *(defunct and abandoned and situated right in the middle of the Methilhill village would you believe?);* the Wellesley Colliery at Denbeath, originally known as the Denbeath Pit, *(in massive full production.);* Muiredge Pit at Buckhind (Buckhaven) *(defunct, but with some pithead remains.);* Cameron Mine on the outskirts of Buckhaven *(producing coal);* the Rosie Pit *(rumours about the demise of this pit abounded unceasingly);* Wellsgreen Pit between Buckhaven and Earlseat *(quite recently abandoned, but easily visible);* the Michael Colliery at East Wemyss *(very large and operational deep pit with two shafts almost on the sea shore);* the Lochhead Pit and Dipping to the north of Coaltown of Wemyss *(both producing coal);* Earlseat Mine *(probably the fifth or six of that name and producing coal from the outcropping Lower Dysart seam),* but hidden from view! the Francis Colliery at Dysart *(called the Dubbie for obvious reasons and producing Coal);* the Randolph colliery near the Boreland end of the Standing Stone road. *(Euphemistically named 'the Randy' and producing coal from the Dysart main seam);* the Rothes Pit at Thornton *(a great 'black' elephant?)* And finally, the Seafield Pit to the South-west of Kircaldy *(a successful late pit, but aptly named. It also wound to the surface, upwards of fifty percent of the coal 'taken' from the deep mines within the Francis\* colliery!)*

*(\*In the auld Scots tongue: -   Dub means - a puddle of water.*
*Dubbit means: - mud-stained.*
*Dubble means: - mud and dirt*

*Readers may therefore deduce as to why the Francis colliery was euphemistically nick-named 'The Dubbie!'*

Most of these Collieries were situated along the coast line and were easily visible from nearly anywhere on the Firth of Forth! *(Or at least the tall chimneys stacks were.)* They were, to be quite blunt, a great dirty blot on the coastal sky-line! These collieries situated on the coast, each on a daily basis, spewed and spilled their ever-increasing waste onto great, dirty, grey-black slag-heaps *(bings!)* These 'bings' consisted mostly of the useless stone and slag, that was the inevitable by-product of coal-getting and ever greater production! Some of these bings were so large and extensive, that they inevitably formed headlands onto the open sea. There was to my knowledge, four separate bings along the coast-line being constantly washed by every sea-tide and carrying both toxic waste and indescribable filth into the open sea! Over a relatively short space of time, this binged waste had the irreversible effect of destroying forever, the beautiful sands of the Buckhaven beaches, that lay directly under the wide green expanse of the Ness Braes! *(Recognisable from the Lothians across the Firth of Forth!)*

It is said that for every minus, that there must be a plus! This was certainly the case on Buckhyne *(Buckhaven)* beaches. There was two very distinct and lucrative industries that grew from the 'bings,' and from the coal that 'magically' appeared on the beaches, from between the lang toon o' Kirkcaldy and the bing at Denbeath. One industry could be classified as an 'individual enterprise,' whilst the other developed into jealously guarded 'big business!' On the home-front, men and women who were brave enough, or desperate enough, would travel on foot or bicycle to the braes above Buck'hyne, descend to the beach front and make their way onto the bings. Once there, usually from early morning *(they had to arrive early to stake their claim to the newly-tipped waste!)* and armed with up to half a dozen sandbags or gunny-sacks, they would spend all of the daylight hours in sifting though as much 'waste' as they could, in the sometimes over-optimistic hope that they would find enough for their immediate needs and perhaps one or two bags to sell! Many men and almost all of the women *(some with young children)* suffered from torn and partially mutilated hands. A good few actually lost fingers thro' accidents from slippage and frostbite *(black-finger)* and from continued exposure! The sloping sides of the bings were very unstable, in that the newly deposited waste was tipped from 'on high,' and had a tendency to slip and slide, especially in wet weather and regularly in the Winter ice and sleet. *(Snow did not settle for long in the heavily salt-charged atmosphere!)*

I have seen for myself, some of these pathetic souls waiting and shivering on the bings in the failing light, for a father or husband finally arriving to collect the 'fruits' of their labour. These people forced to stand guard over the days 'pickings!' Theft was rife! Many unguarded bags just disappeared! Arguments were many, which often resulted in fist-fights to a finish, some of which ended with one or more of the protagonists finishing-up in the sea. There was however, an oligarchy of sorts on the bings, the pickers tended to 'police' their own! After all, what all were doing was highly illegal! Coal-mines and bings were private property, but only sometimes guarded by the Railway police.

The most lucrative of these enterprises was the business of 'Seacoal!' In the past and over a great many years, there had been vast amounts of waste washed into the sea. The Firth of Forth is a great wide expanse of sea, is fully tidal *(anything from 12ft to 24ft)* and can be very rough! The constant wave and tide action, had taken these many tons of mixed waste far out into the seabed and very gradually separated the coal from the slag! 'The sea shall always give up her bounty' and so she did! But in this event, it was sea-coal! This however, was not the only coal that the sea was giving up! Within the Productive Coal Measures, *(upper coal measures)* lying in the strata beneath the Firth of Forth, there was sandwiched, twenty named seams of coal *(this included the upper and lower Coxtool seams and the upper and lower Sandwell seams),* plus the deepest and faulted Lethamwell seam, that were worked and taken to a greater or lesser degree by the surrounding coal-mines. Of these twenty coal seams, the topmost twelve *(commencing from the third seam down),* appeared as outcropping seams running up the beach and partially inland in a northerly direction between East Wemyss and Dysart, and were worked in centuries years past as coal-heughs.

Those same seams that surfaced on the sea-shore, ran directly into the sea and remained 'surfaced' on the sea bed for some considerable lengths underneath the waters of the Forth. Most of the upper seams that were to be found outcropping on the shore and coastal grounds between Buckhaven, the Wemyss's and Dysart, were picked clean many years ago, but those seams that 'surfaced' and outcropped on the sea bed near the coast, lay partially exposed to the effects of tide and sea currents for many years, being slowly eroded and separated by both, before being carried by wind, wave and undercurrents onto the beaches.

This naturally recurring phenomenon seemingly reached a peak in the years after WW II.- and also seemed to coincide with the sea returning the waste coal from the bings, back onto the beaches at the same time. It was also noticed by one or more of these intrepid sea-coal entrepreneurs, that during rough weather and adverse sea conditions, this new black 'bonanza' was frequent, manifest and absolutely free for the harvesting!

At first, the coal that appeared along the shoreline could be picked without undue effort. Needless to say, the early finders did not actually advertise this new find. However, as is the way of things, word did get around within a short space of time, with more people turning up on a near daily basis. *(It was very easy to spot those people who frequented the beaches, their leather footwear soon became white salt-stained!)* Very soon, such was the volume of this wave-borne coal and its ubiquitous spread along the length of the beaches, that a few of the local coal-merchants began to cast an appreciative eye to the possibility of harvesting this valuable flotsam! Picking-up by hand was out of the question, as only some of the in-coming coal was wave-borne, with most of it lying just slightly out of reach under the incoming waves, besides, it would take a hell of a long time to gather one ton of coal even using two hands! However, *'needs must when the deil drives'* and inevitably, one enterprising entrepreneur solved this little problem! He acquired a stout 8 feet long wooden pole, and on the end of it, he fabricated a wire-mesh shallow basket shaped like a scoop. He then borrowed a pair of thigh-length waders, and fully armed, he then hitched one of his heavy Clydesdale working horse to his one-ton coal cart. *{He had thoughtfully fitted large wide rubber-tyred wheels (ex R.A.F.) to his cart, so as not to sink into the black sand and shale!}* Having successfully arrived on the sea-shore, he then backed his cart into the murky sea water and began to reap the in-coming wave-borne black diamonds! His 'basket' was being nearly half-filled every time he propelled it forwards into a coal-carrying wave! He therefore managed to completely load his cart to almost over-flowing in less than an hour! *(This sea-coal did seem to be rather light in weight!)* Now it was the turn of the horse! Neither he nor his horse had tackled this wet loading before now! The coming heavy pull over the soft beach and the sustained pull up the 'Rising Sun Brae' just might well 'do' for his horse! The animal however, was well shod, well fed on oats, and came from a breed renowned for their strength! He need not have worried, the horse was more than equal to the weight of the load and the severe up-hill climb! *(This should have told him something, but that came later!)*

This coalman sensibly rotated his work-horses at lunchtime and made several more trips to the wet shore in the course of the afternoon. He, of course, did the sensible thing and stock-piled the not-so-black diamonds! For him, this was a very lucrative venture while it lasted. The coal was initially very in much demand for those who did not have miners concessionary 'firecoal' as of right - and did not wish to pay merchant's prices for land-sale coal.

This sea-coal was sold at a price much below that of the normal prices charged by the local land-sale merchants. However, this sea-coal just could not keep coming in forever and of course, it didn't. Which was just as well! This coal had some very peculiar burning qualities! It was light in weight, it was porous and it was full of salt! Its initial burning qualities were very poor, but once it was fully ignited it did give good heat. The only trouble was that it burned out very quickly! More and more extra fuel seemed to be the answer, or was it? Yes! This was the infuriating crux of the matter! The stuff was virtually useless! Even the 'bloody' horse knew it was on to a good thing! It knew by the absence of weight. It did seem to spend an extraordinary amount of time just paddling about in the water, probably thinking that it had gone to heaven and was on permanent 'light duties!' I mean, how can anyone be really serious about 'coal' that actually floats on water . . . albeit salt water!?

*(Some of these draft horses were so intelligent and well-trained, that they anticipated the in-coming tide and slowly ambled forwards whilst loading, to maintain the correct height of the cart in relation to the waves!)*

\*

\*Bings: - *(slag-heaps)* The inhabitants of the village of Methilhill in Fife, all suffered slightly from the effects of carbon-monoxide poisoning! Or at least, they should have! The village had grown outwards in all directions from the centre, which was of course, the rather unlovely and stinking bing! The village had developed and grown after the demise of the Pit and it's now redundant buildings, which had been knocked down, thereby leaving vacant building space! Hence the dwelling houses. What to do about the unlovely bing? Nothing! Just leave it! It's sure to go away! - And that's just what it did! The only trouble was that it took approximately 30 to 40 years to do so! Anyone who knows anything about unlovely bings and the contents therein, will know exactly what is bound to takes place! One of the basic laws of physics, the damned thing will continually self-ignite! It is packed with all of the necessary ingredients! Stone, shale, old wood and undiscovered discarded coal! *(Which in itself contains that many by-products, that it would take another book just to describe!)* This mixture is inherent in any self-respecting bing! - along with unlimited quantities of fresh air and plenty of rain water! The process of internal combustion and self-ignition will continue on regardless, as long as there was enough raw materials available.

The bing at Methilhill was no exception! It turned into a living, breathing, stinking, underground, slow furnace, to which the local Fire-brigade were summoned to attend with monotonous regularity! *(Some of the younger members of the local population often called them out just for a little bit of excitement!)* Even they, sometimes had to wear breathing apparatus. At the end of approximately 25 to 50 years, these bings would have slowly burned themselves out and in doing so, left a curious but sustainable legacy! Several old burned-out bings could be instantly recognised by the fact that all of the material contained within would have turned bright pink and various shades of brown and red - and was totally inert! It had been cleansed by fire! It transpired that this purified waste was one of the ideal ingredients in the making of kiln-fired house bricks! It was cheap, it was plentiful, it was readily available and of more import, it was on the doorstep! There was one at nearly every worked-out Colliery! This material was also used extensively in road-making, even though it did require compressing with steam-fired road-rollers! Inevitably, most of the 'red' bings became exhausted with the ground being reclaimed, re-levelled and returned to other uses.

Several of the unlovely blots on the Fife landscape, slowly began to be reduced in size and amazingly disappear from sight, to be replaced with new buildings, houses, factories and in some cases returned to farmland. Who knows what? Unfortunately, a few of these black bings have not disappeared from view, are not hidden, and have yet to be dealt with. They still remain an eye-catching 'blot on the landscape!' One of the remaining bings, the one at the site of the old Randolph colliery, has apparently not 'fired-up' and remains 'as is' since the day the pit closed. This is an illusion! This bing has recently been broached and fully opened, to reveal just about every colour within the Autumn rainbow. A sure sign that complete combustion has taken place and that the slag-heap in now inert!

*The site of the old Randolph colliery is at this time, (2000 - 2001) in the process of being ripped open to a given depth, to extract the unexplained, but thickened Branxton seam, lying at no great depth! This coal is being taken as a form of 'opencast strip mining', where the excavations are being re-filled as soon as the coal is extracted. The land surface is therefore being 'reclaimed' almost immediately!* The large black bing at the Francis colliery, even though it was partly levelled, remains dirty black and still touches the sea waters of the Forth. The largest bing* in the county of Fife, at the Wellesley Colliery, equates to the size of at least twenty football pitches and is fifty feet in depth. It has been for the most part flattened and levelled, but much of it still remains as a hideous black blight to the residents of the north-east end of old Buckhaven. Its black slag still reaches into the waters of the Firth of Forth and the adjacent sea water is still opaque and coloured a murky grey! Fortunately, much of King James' 'Beggar's Mantle' has all but disappeared and is now partly grassed or built over, but unfortunately, so has much of Fife's 'Fringe of Gold!' That part of the southern shore-line of the Kingdom of Fife, has been incontrovertibly changed forever . . . !

Author's Note: - *James VI, King of Scots, on a visit to Wemyss castle in the early part of the 17th century, was reputed to have described that part of Fife as 'The Beggar's Mantle fringed with Gold!' This has somehow, during the passage of time, become corrupted to 'The Black mantle with the Golden Fringe!'*

*\*Beautifully photographed, old, and rather stark black & white photographs of this smoking bing, can be seen at the Heritage Centre in Lower Methil, Fife!*

*Note: This sea-coal bonanza did last for a few years. The areas where the 'pickings' were richest, stretched from the beaches of Buckhaven to the beaches of Dysart and a little beyond. The hauliers concerned, and that amounted to usually small family groups from each of Buckhaven, Rosie, MacDuff, East Wemyss, West Wemyss and Dysart. These groups became even better organised as soon as it was realised by one individual, (I don't know who!) that a farm tractor and a low-loading trailer was a much better combination than a horse and cart!, especially when they discovered that the much-maligned, Ferguson petrol/paraffin tractor could be bought for a modest sum, second-hand! This obviated the need to 'change' horses at lunch-time and ensured a longer and smoother operation. By this time, the groups had begun stockpiling the coal in secure compounds on the beaches! Needless to say, these were guarded with a crude efficiency that would put a military operation to shame!*

*

## The Wellesley Bing and the Lost Hamlet of Buckhaven

This bing was surely the largest blot/blight to be seen anywhere on the landscape of the Kingdom of Fife. I would even go as far as to say anywhere over the whole of the coast of Scotland. It stretched from beneath the great Baum washer at Denbeath, down to the immediate shore-line, and its length finally extended westwards along the Old Links of Buckhaven to the East end of Old Buckhaven, immediately below the Rising Sun Brae. Its width encompassed the large expanse of ground between the Buckhaven to Methil railway, all the way to the Sea shore. Its height was the equivalent to that of a four or five storey building. It was comprised of stone, shale, slag, sclits, dirty coal, unburnable coal *(known locally as 'dugger' coal)* and probably steam-boiler ash and clag as well. It contained old broken wood, rotten pillar wood and discarded support wood. It also contained every sort of rubbish, garbage and waste material, that was the daily by-product of a work force of around 1500 miners and tradesmen. The great Baum Washer\* that operated on a daily basis, used literally millions of gallons of water to clean and 'polish' the coal, with the resultant thick, filthy-black, running scum, being eventually scourged into the waters of the Firth of Forth in an unending stream, for a full six days of every working week.

This bing eventually became so large, extensive and all pervading, that it overcame and completely buried the whole of the old village/hamlet, that was situated on the then Links of Buckhaven. The village itself was quite extensive and was situated on the low level 'links' between the Wellesley colliery and the East end of old Buckhaven. It lay between the Buckhaven to Methil passenger railway and the line of the sea shore, and to the north side of the original Buckhaven to Leven coast road. The village was totally sacrificed to the encroaching bing for the simple, but incontrovertible reason, that there was simply nowhere else to dump the waste. To the east, the great railway* sidings feeding Methil docks with coal for export, occupied all of that available space, and that ground was untouchable. To the north and east side, was the higher ground upon which lay the mineral railway, that brought the mined coal from most of the other East coast collieries in company wagons, and were lined up in long parallel lines to be fed through the Baum washer. The only low-level suitable space that could possibly accommodate the likelihood of hundreds of thousands of tons of colliery waste, was that occupied by the houses comprising the village of the Links of Buckhaven.

In the first decade of the twentieth century, the announcement of the beginning of the evacuation process of the families living on the Links of Buckhaven, was promulgated by the Wemyss Coal Company and delivered up by the local council. It was not to happen all at once, there was to be some breathing space for those miners and their families living in the middle and subsequent to the south-west end of the village. Colliery bings do take a considerable time to build, as will soon become clear.

The old hamlet/village of the Links of Denbeath/Buckhaven, lay on a wide flat area of ground between the passenger railway and the sea-shore. It lay to the north side of the old thorough-fare between Buckhaven to the south-west, and the Wellesley colliery to the north-east. It was a longish village rather than a cluster of houses. It covered a length of approximately one kilometre, in two, perhaps three, roughly-symmetrical, but irregular rows, and was mostly occupied by locally employed miners and their families. The village did not lie on the fore-shore, but was perhaps one hundred and fifty, to two hundred metres from the shore-line. It was comprised of stone built houses and cottages with slated and tiled roofs, with perhaps, no two houses built to the same design, but with just about every house having several chimney pots to grace their gable ends. Coal, it would seem, was used extensively to heat every room within the houses.

In the year of nineteen hundred and five, the villagers were informed that over the next few short years, they would have to abandon their houses on the Links of Buckhaven and move to other accommodation to be provided for them in the near vicinity. *(Most families were allocated houses in College Street, Buckhaven and* *some allocated houses in the \*newly completed miners rows in the new Denbeath village.)* The empty houses were to be abandoned and sacrificed to the encroaching behemoth, that would slowly, but inexorably engulf them. 'Coal was King' and nothing was to be allowed to stand in the way of this burgeoning bing! The evacuation of the houses began in 1906, with the houses nearest to the existing bing being emptied first. Each and every occupier in his turn, was required to vacate his home when the creeping bing reached the next house but one, where the newly vacated and abandoned house had all of it's downstairs windows heavily boarded-up and the doors securely locked, *(probably from the inside)*. The houses were not demolished or legitimately cannibalised in any way, but were boarded-up as they stood. Initially, the only way to distinguish these empty houses from occupied dwelling houses, was the lack of smoke from the cold chimneys pots!

The method used to initially create this great bing, was by a quartering system where the bing advanced, not on a broad front, but by a series of parallel enfilades commencing with the innermost flank, which unfortunately, encompassed the whole of the long row of houses, situated to the North side of the main thoroughfare between Buckhaven and Leven. The general plan of course, was to 'bing' the up-coming waste as far from the sea shore as possible. The existing small bing below the Washer that had been built towards the shore, now needed to change direction and the new direction re-started at a point immediately underneath the 'Washer,' where it began its drive in a south-westerly direction, along the line of the village houses. The area of ground that was the Links of Denbeath/Buckhaven, was perhaps just over one kilometre long and between four hundred to six hundred metres wide and lay in an south-west to north-east direction. The links were for the most part quite flat *(hence the name!)* and probably ideally suited, *(company-wise)* for what was about to submerge this picturesque mining and fishing village forever!

And so! The quartering began. The boarded-up houses at the north-east end of the village were slowly enveloped and smothered, in the creeping and sometimes stinking, black waste. The height of the advancing bing was greater than the height of the two storey houses and buried them in a single sweep. The bing was initially surged forward and served by a single narrow gauge railway line sitting atop its 50 feet height *(approximately 15 metres),* - and a never-ending stream of mine-cars carrying the pit waste, was constantly tipped over the leading edge of the bing, thus creating its own Massadian precipice. The long, first quartering, eventually reached its south-west boundary and in doing so, had completely enveloped and destroyed an old village, whose only crime was to lie in the path of King Coal! Over the years, the quartering continued, with the black mass creeping ever-nearer the sea-shore. By this time, the underground workings of the Wellesley pit had been greatly extended, with its tentacles now spread far and wide. In the ensuing years, the amounts of waste had greatly increased as had the build-up of the great bing. The narrow gauge rails had now been replaced with full gauge permanent rails and the waste was being delivered to the bing in trains of coal wagons, that disgorged their contents through the side trap-doors of the wagons, directly along a wide front. This process continued unabated until the middle Forties to the early Fifties, where the bing had now reached its land perimeter. The bing had now reached the sea-shore, with the trains of wagons now discharging their waste contents down the high steep slopes of the bing, with the waste materials falling into the incoming waves.

The previously clear seawater was now being stained a cloudy grey colour and contaminated with the filth of mine waste. This was being carried to and fro with every tide and gradually extending outwards and sideways in both directions, where the tidal waters of the Forth carried and deposited this filth on the sandy beaches, all the way from Leven to Burntisland. When I was a 12 year old scholar at Buckhaven Higher Grade School, I used to spend my lunch-breaks down on the rocks at the waters edge. The in-coming waves at that time were dark and murky and the seawater was certainly not fit to swim in!

The permanent legacy of this bing and the three other bings that encroached onto the sea-shores, now meant that the open beaches from Buckhaven to Dysart, which were once covered with golden sands, now look as though they had been heavily painted with black lava dust and grit, and were being washed with salt water that looked as though it came from a coalminer's bath-tub!

*

*A few miles inland from the coast, immediately to the north of the Standing Stone road between Buckhaven and 'The Gallatown,' the two rich coal seams, the Dysart Main and the Lower Dysart had only been partially extracted from underground. The reason seemed to be that these seams were rather to close to the surface over this wide area and given the heavy equipment that was now available to contractors, it was deemed to be more than feasible to extract these reserves of millions of tons by outcast extraction. It must be said that this method of extraction would take every ton of coal that was visible in any given seam, and was reckoned to be much more viable than the deep mine methods, which by necessity, leaves most of the remaining coal inaccessible. This method of extraction involved massive holes gouged deep in the countryside and worked over many years. The machinery used was simply gigantic and extremely expensive. The load carrying vehicles were bulky, large-wheeled, dumper-like trucks, capable of sustained operation and usually filled and loaded by both wheeled and tracked vehicles capable of pushing, scooping and lifting large quantities of coal at one time and deployed in their tens!

*The railway sidings that covered the wide area from the Wellesley pit to the Methil docks, were used exclusively for the transport of wagon-loads of coal coming from the Wellesley washer and the collieries beyond. The area consisted of many miles of converging steel track, with hundreds of crossovers and turnouts which served a three-fold purpose. The main purpose being, to cater for the hundreds of loaded coal wagons lined up in long trains, awaiting their turn to be unloaded into the waiting coaling ships lined-up against the great, coal wagon hoists. These hoists were capable of bodily lifting a railway wagon of 20 tons capacity and emptying it into a ship's hold in one fell motion. The wagons lying in these sidings, belonged to either the Wemyss Coal Company, or the Fife Coal Company and latterly the N.C.B. - and ranged from the small eight tons capacity wagons, thru' ten, twelve, sixteen and twenty tons capacity wagons. They were often thoroughly mixed! The area was also used as a large shunting yard, where the emptied wagons were re-grouped and re-sized into trains *(where possible)* belonging to each Coal Company, *(before Nationalisation)* before being re-routed back to the various collieries in readiness for their next load.

Before the start of W.W.II, there was approximately twenty-five miles of railway sidings between the Wellesley colliery and into Methil docks area. *(This does not include the vast array of parallel rail tracks leading to and from the great Baum washer and grading tables at the Wellesley colliery.)* This gave this network of sidings a holding capacity of around three thousand railway wagons, *(standard British gauge)* which, at an average capacity of sixteen tons per wagon, amounted to the staggering total of approximately 50,000 tons of coal, just waiting to be exported at any given time!

These huge, 100 feet *(30 metres)* tall, steel, moveable, coal hoists on the docks at Methil, were built directly onto their own heavy rail tracks, embedded into the concrete and stone-work of the dockside and tight to the edge, with full gauge, railway tracks running immediately through their open bases. This meant, that a train of loaded coal wagons was fed through each hoist where the wagons were individually uncoupled, quickly raised on high, with the entire contents being tipped into the large coal chutes, which fed directly into the coal ship's holds. This individual operation took but ten to twelve minutes, where thousands of tons could be mechanically loaded in the course of one day.

\* The miners Rows at Denbeath are described in the chapter: - The Denbeath Village.

Author's Note:- *The large Baum washer at the Wellesley colliery at Denbeath, installed in 1906, was a feature that quite dominated the local landscape. Its size was very impressive at 60 feet wide, 90 feet long and approximately 100 feet tall. Its task was to wash and grade the coal, not only from the Wellesley pit, but eventually, from all of the coalmines operated by the W.C.C. at that time. It was originally designed to handle, wash and grade around 1,000 tons of 'dirty coal' during the course of an eight hour day. The initial installation in 1906 was capable of handling 100 tons of coal in one hour. In 1910, a second unit was added that increased the handling capacity to 250 tons per hour. In or around 1950, a third unit was added by the N.C.B, that brought the handling capacity to approximately 400 tons per hour. That meant, if the whole system were to work two eight-hour shifts in every day, (which it eventually did!) It would then have the capacity to handle up to 35,000 tons of coal every week! The whole solid construction, stood on heavyweight brick pillars, under which ran several parallel tracks of railway line, to accommodate the trains of coal wagons needed to maintain its efficiency.*

*The incoming trains of loaded wagons were individually tipped into an intermediate underground reservoir between the rails, and mechanically raised to the input side of the washer by a steeply inclined, continuously moving, hopper. The washed and graded coals were then re-loaded into the same trains of recently emptied wagons within a few minutes, without them having to be re-shunted! This system seeming separated the large coals from the various named seams within the coal field, into different trains of wagons and further separated the remaining coals into five different sizes of nuts, excluding the duff, i,e. chirles, trebles, mixed trebles, doubles and nuggets. All in all, a very efficient system! With this washer working at two-thirds capacity, its output was capable of overloading the extensive wagon sidings at Methil docks within two weeks, if in-coming coal ships were late in arriving, or were delayed in loading.*

*

# Chapter III

## Lochhead Pit Bottom –
## *An Unexpected Illumination!*

The Pit bottom at Lochhead was probably not unlike any other of its kind, except that it was much smaller and used less manpower. It was one of the very few underground areas that was actually wholly lit by electric lamps using mains electricity. Not quite the voltage that would light a living room, but the same type of domestic bulb, fully enclosed in protective glass and covered by a solid wire mesh. All underground electric cabling was also enclosed in conduit tubing. In addition, all pit bottom workers were issued with electric battery cap-lamps just in case of mains failure. This was, of course, very necessary in that the normal type of miners carbide gas flame-lamp would simply be of no use, due to the consistent blast of cold ventilating air being forced down the shaft and through the pit bottom.

The number of miner's who were actually employed on the pit bottom and in the vicinity of the cages, numbered about one even dozen. The main task, in fact the over-riding task of the oncost workers employed on the bottom, was to ensure that during coal-winding, the time taken between the grounding of the two-hutch cage and its subsequent lift-off, was measured in seconds rather than minutes. The two men employed on the physical pushing of the two full hutches of coal into the cage, which in turn pushed the two empties out of the cage, was surely the most demand-ing, repetitive and consistently strenuous type of work that could be asked of an oncost worker. It must also be remembered, that this exact physical operation was happening simultaneously at the pithead with the opposing cage. The coal-winding operation started at exactly six o'clock in the morning, irrespective of whether all of the day-shift workers were down the pit or not. Coal winding was the priority and this was strictly adhered to, come what may. The remaining day-shift miners still on the surface, had to wait until six-fifteen until a short man-winding session resumed.

Meanwhile, back at the pit bottom, the loaded coal-hutches would be arriving from all of the working sections. From the New Dook, one or two long wall faces and several stoop and room workings. From the Old Dook, mostly stoop and room workings and one or more new developments. From the Lower Dysart sections, where the loaded, long races came onto the pit bottom, first, via a diesel Pug and then, through the long tunnel of the Lower Dysart endless haulage. The Lower Dysart at this time, produced coal from two places in the section, one, being the long-wall face (the slopes) of the Lower Dysart coal and the other, being the developing and sometimes wet, long-wall faces of the Lower Dysart Dook, that initially undercut the standing coal to the dip. At full production, these two sections simply spewed coal, which was built up into 50 - 60 - 70 hutch races and was transported to the pit bottom, via the Lower Dysart haulage-way. When the coal began to flow, these long races arrived at the pit bottom with timed regularity. Coal from the Old Dook arrived at the pit bottom one hutch at a time, on the fully-clipped, endless haulage, with probably 5 to 8 seconds between hutches, as did the coal-hutches from the New Dook, using the same type of endless rope haulage and with the same regularity.

The hutches going into the cages bereft of their multi-linked steel couplings, was yet another job on the pit bottom. The uncoupling of the arriving full races and the re-coupling of empties going back into the long haulage of the Lower Dysart, was a never ending task and, often involved the carrying of heavy couplings for most of the working shift. It was a very skilled coupler who could uncouple a full race and re-couple an empty one, without having to transport at least a few couplings! The pit bottom was also a very noisy place, which made the practice of shouting almost a habit, although, most people were far too busy getting on with their work and at the same time, looking to their own safety. Accidents were never very far away and the men were much concerned that they did not get in the path of constantly moving hutches. If Walter (don't call me Wattie!) Johnstone, the pit bottom Gaffer wanted to call for any man, or needed to pass-on any instructions, he had physically to confront the man in order to make his wishes known. That sort of thing always seemed to go against the grain to Wattie, he preferred to pass on his instructions through better beings. Not that Wattie would ever hold a long discourse with an oncost man, it was quite sufficient that Walter had even to speak to those horrible, dirty, uncouth miners of the Old Dook, over whom he also held full sway, being nominally the section oversman (Gaffer). *(That also included this author!)*

At the time of my employment in Lochhead pit, all of the loaded hutches of coal that converged on the pit bottom came from three distinct sources and from two given directions. The feed rails/roads to the cages, were of course, partially inter-linked, to be able to feed loaded hutches from any source onto either cage, but this mainly applied to the coal coming from both the Old Dook and the New Dook, when there was a dearth of coal from the Lower Dysart haulage. The coal from the New Dook was fed into the right hand cage whilst the coal from the Lower Dysart was fed into the left hand cage, with the up-coming hutches from the Old Dook capable of being directed to either side, by the use of switched rails. As the amounts of coal from both the New Dook and the Old Dook intensified on the productive day-shift, the coal from the Lower Dysart could quite easily be backed-up along the length of the 1000 yards long haulage-way, without any undue strain on the on the system, as could the sheer numbers of empty hutches being directed into the section. This backing-up of the Lower Dysart output was done on a daily basis for two simple reasons.

1. There was very little room on the pit bottom to allow the piling-up of coal

from the two main Dooks. 2. The long races of loaded hutches from the Lower Dysart haulage, would be used to feed the two cages at the commencement of winding the next morning, whilst coal production from the two Dooks intensified after start time. If there was a huge surfeit of coal waiting along the length of the lower Dysart haulage-way, then, a couple of hours extra winding time on the back shift, would soon reduce the excess and restore the critical balance.

At this time, coal was only wound on the extended day-shift, but, as the production from the East-side mine and the developing Lower Dysart seams in both the East side and the West side began to spew coal, this was extended to both day-shift and back-shift winding.

Of the miners and 'characters' who frequented and were employed on the pit-bottom, were some who had seemingly been there for 'yonks' and a few for only a short length of time. One such person was Laurie Gibb, one of three brothers working in Lochhead pit and all of them underground. Two of the brothers were oncost workers with particular skills and one them was a coal-stripper in the Lower Dysart West, who was also the local union representative cum delegate. Laurie Gibb was classified and employed as a rope-splicer, but he could put his hand to just about any task on the pit bottom, (except Banksman) and usually did! He was constantly on the move. He was often likened to a 'blue-arsed fly' and seemed to move just as quickly. He, at that time, seemed to be of a nervy disposition which was reflected fully in his movements, his manner and his conversation. He was difficult to talk to, in that he just would not look a man in the eye and converse for anything longer than a few seconds. His eyes seemed to be seeking and probing for anything untoward or 'out of sync', within the general hub-bub on the pit bottom.

Laurie's duties as a steel wire, rope splicer extended far beyond the pit bottom. His fiefdom covered the Lower Dysart haulage, the Old Dook haulage, the New Dook haulage and the main level No. 3 haulage down the New Dook. His duties did not extend to Nicholson's Dook, the Victoria haulage or the Lower Dysart Dook, which were covered by his older brother Bob, who did virtually the same job as Laurie, but in a different part of Lochhead Pit. The two brothers did however, find themselves working closely together on the many occasions, where the expertise of both of them was required at a particularly involved task. I am also reliably informed that there did not seem to be any great conflict of knowledge or application, when the two brothers were pressed into combined service. They both seemed to revel in their common element, when confronted with several long 'spoons' and two times 10 yards of opened-up wire, just waiting to be 'joined' together and be happily 'married' into a 20 yards long splice. *(I am reliably informed by Laurie, that the length of the splice depends wholly on the inherent diameter of the individual wire rope and the amount of strands therein!)*

An essential part of Laurie's task's which, thankfully, he was not called upon to attend too often, was when there was a 'wreck' on the Dook! The sort of wreck described here could not, and did not, take place on a level haulage way, but only on haulage-ways where, by accident, the force of gravity was allowed free rein, due either, to the frailty of man or the failure of equipment. On the great Dooks such as there were in Lochhead pit, the method of transporting coal by rail, was either by a single rail system using coal races on the end of a single roped tail, or a twin rail system using an endless-rope haulage with removable clips.

In the case of the former, the Dook would be laid from top to bottom with a single, semi-permanent rail-track. There would be a powerful electrically driven motor, coupled to a winding drum, onto which would be wound a steel rope capable of hauling a fixed number of loaded hutches up the steep incline. This large 'tugger' motor would be situated at the top of the Dook, within its own girdered motor-house and would perform endlessly during the course of any coal-shift, repeatedly raising, and lowering, full and empty hutches in an endless stream.

In the case of the latter, the length of the Dook would be laid with two separate sets of semi-permanent rails, one side being for up-coming loaded hutches, while the other side would be used for down-going empty hutches. The dook would then be fitted-out as an endless-rope haulage, designed to run continuously with both loaded hutches and empty hutches being delivered in and endless stream, so long as the haulage motor was running! The individual hutches would be coupled to the running wire rope by means of a device called a wheel clip, which, could best be described as a pair of 6-inch long, metal-lined, vice-jaws which were opened and closed by a very simple, but effective bi-directional, screw-wheel, situated at the middle of a six inch shaft that was threaded in opposite directions, which, when rotated, effected an opening and closing action on it's vice jaws. This was attached to a steel arm approximately 18 inches in length, that terminated in a closed steel ring which looped over the draw hook at either end of the hutch. *(This type of wheel-clip was actually used in the Victoria Dook and haulage and also throughout Lochhead colliery, on all of the long dooks and main level haulages. It was actually invented by Johnnie Brown who was the manager at Lochhead Pit at the time of its inception.)* The simple principle of operation was when a full hutch of coal was sitting on the rails on the up-side of the haulage, the 'clipper' would throw the end loop of the clip over the draw-bar hook of the hutch and drop the open jaws of the clip onto the moving rope. He would then spin and tighten the oiled screw-wheel on the top of the clip, thereby clamping the clips bottom jaws hard to the steel rope, whence the full hutch would jerk forwards at all of 2 to 3 m.p.h. following the rope, whereby, the clipper would then follow-on, giving the screw-wheel an extra part-turn with his hand-tool. Each and every loaded hutch would be coupled on to the haulage rope in turn, with perhaps a spacing of 20 to 25 yards between hutches. On the down side of the haulage, the amount of empty hutches appearing at any given numbered level, would roughly equate to the amount of full hutches going up, for the simple reason, that at the top of the dook, the permanent 'clipper' would have to await the arrival of a full hutch coming up, before having a fresh clip available for the next empty going down. The system was self-perpetuating and the frequency of a full hutch arriving at the top of any Dook, was regulated by the amount of clips available and in use, and the speed (work-rate) of the intermediate clippers at the various levels. *(As well as the rate at which the miners produced the mined coal!)*

Laurie was sometimes euphemistically *(and unfairly)* referred to as the Five pound-fifteen Gaffer on the Pit bottom, a full unpaid 'assistant' to Wattie *(call me Walter)* Johnstone *(oversman)*. One of the supposedly 'better' beings that frequented the pit bottom!

Of the three brothers Gibb, I have named two. Both Bob and Laurie were deemed to be oncost workers, but with particular skills in rope-splicing and road building *(with steel semi-permanent rails)*, with much of their work taking place out of shift hours and at week-ends. Tom Gibb, who was the local union delegate at that time, was also a face-worker (stripper) on the Lower Dysart slopes, during my short time as a clipper on the Lower Dysart Dook in 1951. Tom Gibb's career within the miner's union did take-off and soon manifested itself amongst the Scottish N.U.M. hierarchy as the years progressed, but that is out-with the remit of this Lochhead narrative.

*

The Onsetter on the Pit bottom, was the one man who stood firm for most of the long working shift. The blast of cold and sometimes damp air that was forced down the shaft, met it's first obstruction in the shape of Big Charlie Fleming, who stood foursquare on a small, precarious, platform position, exactly between the two cages and facing the blast. This forced air was pushed to near gale force by a huge multi-bladed, electrically operated fan, measuring 18 feet in diameter and positioned strategically on the Pit-head. It operated non-stop through twenty fours of every day. This was the life giving air, on which the very existence of the coalmine depended. If the fan stopped blowing, then the mine stopped working. It was as simple as that! Charlie's job, as pit bottom onsetter, was to oversee the efficient loading operation of the two cages and to operate the mechanical signalling system that was in use at that time. This consisted of an horizontal metal lever at approximately head height, that was affixed at it's rear and coupled to a long length of steel wire, that ran all the way up the sixty fathom shaft to a large bell in the winding engine-man's house. A similar arrangement was also used at the pit-head, albeit shorter! Part of Charlie's job, was that he had to make a snap judgement every time the cage bottomed. In the same second that the full hutches were manually rammed into the cage, he must operate the weighted lever to send the appropriate signals to the windsman on the surface. This happened approximately every 50 seconds or so whilst coal-winding was in progress, he therefore, could not leave his post for a second. Charlie was usually clad in work clothes that protected him from the worst of the blast, but that did not stop Charlie from becoming the dirtiest man on the pit bottom. The blast of air always carried with it, small and finite particles of coal (*which was not only dust*) that was duly deposited on Charlie, all over Charlie and even inside Charlie. He was covered from head to boots in the stuff and was never heard to complain. He daren't, he was liable to get a mouthful of grit! He was 'happed-up' to the neck in heavy scarf and buttoned collar and wore goggles against the unremitting gale.

The one thing that Charlie always felt highly aggrieved about was, that he consistently experienced great difficulty in trying to light his tobacco pipe whilst standing-to! Matches were no good, they would splutter and never take hold, even half a dozen at a time. That sometimes, could be made to work but, the taste was vile! A solution was needed, and with some thought and help from the engineering shop, the problem of igniting Charlie's pipe was overcome. He commissioned, through the engineering shops, a giant sized, liquid fuel, spark lighter, a full 4 inches high, 2 inches wide and a full inch thick. This was topped-of with a three-sided perforated windshield, a great three-quarter-inch wheel and a commercial gas flint. It had a large hinged top to enclose and protect the workings, and looked all the world like an oversized Zippo lighter. It must have weighed at least half a pound. To ensure perfect ignition and a fierce burning quality, the inside was stuffed full of cotton wool to absorb and sustain, the large quantities of lighter fuel that was used to fire the infernal machine and ensure consistent ignition. It was said, that when Charlie lit his pipe, it was a toss-up as to what would burn first, Charlie's pipe or the end or his great nose. Wattie referred to it as the Pit-bottom flame-thrower!

Of the two oncost men who worked as a matched pair (side by side) in conjunction with Big Charlie, they were the two men who worked unremittingly and unceasingly in pushing the next two loaded tubs into the arrived cage, thereby pushing the next two empties out of it. *(Hydraulic ram power had not reached Lochhead pit at this time!)* It might be supposed that these workers would be large powerful men with huge reserves of strength. Far from it, one in particular, was one of the slimmest and lightest men employed in Lochhead pit, yet, he had the strength and endurance to cope with the most demanding of tasks without complaint. The one perk that these men enjoyed which none would gainsay, was to be the very first in the queue to be surfaced *(on the 'first tow')* at the end of the shift, and thence to the baths. *(The sheer luxury of being able to stand naked under the soothing cascade of warm water was without peer!)*

One of the coldest and loneliest jobs pertaining to the pit-bottom, was to be a bogie clipper-man on the Lower Dysart haulage. This was a double-track semi-permanent railway that linked the pit-bottom to the Lower Dysart mine. It was approximately 1000 yards long - and was used to convoy both the long races of empty tubs into the section and the equally long races of loaded tubs out of the section. It was termed an endless haulage, in that a seven-eighths-inch, steel wire rope ran between each rail-track throughout the length of the haulage, at a speed of about two to three miles per hour. The left-hand track going into the section was for the ingress of empty races and supplies (wood and steel etc.), while the opposite track was used for the egress of loaded races, coming from the Lower Dysart West to the pit bottom. The large and powerful, electric motor system and the Becander gear, were housed in a custom built motor-house on the pit bottom level, approximately 100 metres from the pit bottom. The haulage road itself, was just wide enough to take the width of two races passing each other at the same time and any miner caught travelling this haulage when two such races met, would just have to wait in the nearest manhole until one track at least was clear. To do otherwise would court a serious risk of injury. This mine at various parts was girdered with both arc girders and flat girders, and walking the length of this mine twice every day was akin to an assault course. It was high and low in turn. It was wet and dry in places. *(But, only on the pavement!)* It was muddy *(coal mud)* and dirty and the moving air was bitterly cold. The main air stream could feel like a gale, especially in those parts that were low and the air-flow therefore increased. The men who operated the clipper bogies, both in and out of this haulage, had just to sit within the bogie and brave the chilly blast whilst travelling at a top speed of three m.p.h. The only 'excitement' to be had in doing this otherwise mundane task, was when one or more of the race of tubs became de-railed and then, it needed perhaps a half-dozen or so to do so, before the extra 'drag' would be felt by the clipper himself and even then, it sometimes took a continuously 'slipping' bogie clip to alert his attention amidst the rumbling noise. Then, of course, he was on his own, where he would probably expend a full shift's worth of energy in labouriously lifting-on all of the de-railed full tubs, one end at a time! *(Amidst many descriptive expletives and calls for help to a much higher authority!)*

The Lower Dysart haulage to the pit bottom did not run in a straight line, there was one distinct bend along its passage with several pavement pulley-wheels to help the rope negotiate the bend. Where an endless rope haulage is made to run between the rails along a perfectly straight tunnel, it is a simple matter to keep the rope from dragging on the pavement, by installing a system of flat, wide, steel rollers at regular intervals between the rails. These wide rollers also had a raised flange at either end, that played a part in helping to keep the moving steel rope centralised between the rails. When the line of the haulage-way, therefore the rails and haulage-rope, had to negotiate a bend in a tunnel, the wide flat rollers were in effect, quite useless in guiding the haulage rope, so, a different type of guide pulley was brought into use. This circular steel pulley wheel was about 12 inches in diameter and 4 inches deep, and was drilled in the centre to take a 'one and a half inch' bolt, which was firmly affixed to a strong-point on the pavement. The number of these pulleys used would depend on the degree of bend and would probably need one for every 3 degrees of turn. These pulley wheels had a deep-set groove, with the lower flange standing prouder than the top flange, which was slightly shallower with a rounded edge so as not to have a cutting effect. This was to ensure, that when the 'clip-bogie' with its vertically positioned vice-clamp, (*previously tightened onto the rope,*) did pass around the pulley wheel, it would do so, without the rope jumping out of the wheel and with the minimum of friction. Also, more importantly, it should not cause the haulage rope to momentarily stall, as it could do, due to the dead weight of a heavy race and a therefore, very tight bogie clamp. An attentive 'clipper', on approaching such a series of pulley wheels with a heavy race in tow, might just pressurise the top of the vertical clip in the direction of the turn, just to ease the jarring effect of the bottom part of the clamp meeting the pulley wheel.

Author's Note :- *I have described and emphasised this last paragraph by way of explanation, in order that the reader may visualise and understand the meaning of an 'Endless-rope haulage! It matters not as to whether the haulage-way is on level ground, or on a 1 in 3 incline. The principles are the same! The degree of potential danger however, is somewhat different!*

Within 40 metres of the shaft bottom and in the direction of the Lower Dysart haulage mine, was the top of the Old Dook. This Dook was driven down through, and followed the dip of the Dysart main seam on a map bearing of approximately 140 degrees, as did it's companion Dook to it's right. The Lower Dysart Haulage mine, again driven through the Dysart main seam, (*this straight mine was not known as the Lower Dysart haulage mine until it was re-brushed and started transporting coal from the Lower Dysart West - i.e. the Lower Dysart Slopes and Lower Dysart Dook*) was initially commenced at the pit bottom, and driven at a map bearing of approximately 180 degrees for a distance of 100 metres whence, it made a right turn of 45 degrees, and continued on another 1600 metres through the Dysart main coal at a bearing of 225 degrees, until it reached nearly to Lochhead's western boundary. At the 900 metres point along this level mine was situated, the head of Nicholsons Dook, which took a left turn and plunged into the depths of the Dysart Main coal on a map bearing of 125 degrees and to a depth of 650 metres at most. There was no companion dook to parallel Nicholson's Dook.

At the other side of the pit bottom, called the East side, but actually to nearer to the North, another indeterminate level mine, also in the Dysart Main seam, was driven to a map bearing of approximately 30 degees for a distance of about 950 metres, until it too, petered out at the edge of the Buckhaven Fault. At a point along this mine, about 65 metres from the Pit bottom, another Dook was commenced down into the Dysart Main seam, on a map bearing of 80 degrees. This dook, which also had a companion dook to its left, was ultimately driven to a depth of 950 metres and named the New Dook.

*

# Chapter IV

## A Not So Wicked Youth!
## *(In the Spirit of Free Enterprise)*

*Fur aw' that - and aw' that,-*
*a lads' a lad - fur aw' that,*
*and Faither's lad's the world ower,*
*shall laddies be fur aw' that!*

Whilst I was still a pupil at Denbeath Public School, there took place two particular incidents *(out of many I may add!)* which are thoroughly embedded in my earliest memories, for the simply reason that I had painful cause to remember both.  The first of which was when I was eleven years old and in the top class of the school, having recently passed my 'qualifying' examination and was awaiting the transfer and elevation to Buckhaven High School.  I had quite recently 'discovered' the manly, though sometimes sickly joys of cigarettes.  Old Harry's cigarettes to be more precise!  Old Harry had not as yet discovered the quiet satisfaction of 'the pipe' and usually had a small stock of 'Gold Flake', which he kept in a fifty-size cardboard box in the bottom drawer of the living room bureau.  He watched over this precious stock and I am sure that he counted them daily, but after satisfying himself that all was accounted for, that was when I struck! - only one at a time though!  Old Harry was nobody's fool!

These few cigarettes that I did manage to acquire from under everyone's watchful eyes *(Harry had quietly voiced his suspicions, he suspected Mother who sometimes enjoyed a quiet drag),* I did manage to keep hidden until the next days 'leave-time' *(mid-morning break)* at school, where a group of us bigger boys would chase the younger ones out of the toilets and practise our new-found skills at lighting and puffing S.C.W.S. cigarettes.  Wills' Capstan *(medium strength)* and Gold Flake being the most readily available.

At this time, the School Janitor was Andrew Goodwilly and he was the only full time janitor employed at Denbeath School at that time.  When 'Awnd' Goodwilly took his summer break, a replacement janitor was called in - and it just so happened that Old Harry was on constant night-shift at that time.  Mr Lawrie, the Headmaster of the School, lived in the school-house almost directly opposite our house in Barncraig Street, and he approached Harry with a view to him helping out during the absence of Awnd Goodwilly.  Old Harry accepted this daytime job on the understanding, that it would only be for a restricted period and be treated as part-time additional employment, with Harry's night-shift work at the pit taking priority.

Harry came home from the night-shift around 7.30 a.m, was fed a good breakfast and immediately went across the street, into the school grounds and thence to the boiler room, where he remade the remnants of last-night's banked fire, attended to by him on the previous evening, prior to his going to Lochhead on the night-shift. Old Harry was working a full week's work on the night-shift at Lochhead Pit, and also doing a full days work as School janitor during the day. Understandably, he did get some sleep in the afternoons and early evenings, but had to make his rounds and stoke the furnace each night, before departing to Lochhead on each successive night-shift.  It was therefore quite understandable, that Harry was somewhat unusually 'crabbit' during the day, from the lack of sleep.

One morning at 'leave-time' *(eleven o' clock break),* our usual small group of furtive smokers gathered in the outdoors, but enclosed toilet and proceeded to light up our mostly ill-gotten 'ciggies'.  Just after we had lighted-up and were manfully puffing clouds of uninhaled smoke into the atmosphere, the main outside door burst open and in strode old Harry.  We were taken completely by surprise, in that he had been waiting just outside the other door, having previously been warned by Mr. Lawrie as what was taking place.  Mr. Goodwilly had never been able to catch any of the previous culprits, he was ponderously heavy and far too slow.  Old Harry was a different kettle of fish.  He was of course, in the prime of his life and 'fast' with it!  We never stood a chance.  Harry didn't say very much.  He pointed a finger at Jimmy Gough, Bill Anderson, Pete Donaldson and Adam Laing, and jerking his thumb, abruptly said, "Headmasters Study - Upstairs!".  I stood still!  This was 'My Dad!' . . .  My initial surprise and cockiness was short-lived!  Almost in the same breath he pointed his great finger at me, jerked his thumb again and said, "You too!"

Understandably, my inclusion gave me quite a shock!  Very slowly we all made our way up to the headmaster's study, knowing quite well that 'this' was going to hurt!  Mr. Lawrie had a well deserved reputation, not for being a bully, nor for undeserved thrashings, but for the very opposite.  It was said that his 'belt' was that old and unused, that the leather was black, shrivelled and hard.  I can certainly vouch for this 'legendary' fact.  Before my turn came, he had practised this 'black art' on my four co-conspirators and as he attempted to administer the last of four on Pete Donaldson, Pete suddenly withdrew his hand.  This same wrist was then suddenly grabbed by the now incensed Mr. Lawrie, to ensure proper contact for the last 'strike', when the old belt at the top of its swing, came into full contact with the study roof lamp.  Needless to say, it simply exploded with a very loud bang, showering all of us with broken light bulb and shattered shade.  This seemed to make Mr. Lawrie just a teenie wee bit more angry and as a result, instead of feeling the effects of a tired arm, I bore the brunt of a re-invigorated swing which hurt me as I had never been hurt before and not surprisingly, I spent the rest of that day nursing a stinging left hand. I did get my own back on Old Harry though, I bravely pinched two cigarettes next time round. I had been smitten!

The second incident took place over a period of several short months and I really thought that it was going to last forever, or at least, until the almost never-ending supply of materials ran out. Old Harry was forever warning me on Saturday mornings not to go away anywhere with my 'gang', or to stray too far from the back garden. He needed my help in wheeling the two wheeled barrow 'to and from' the sawmill at Muiredge, where he seemed to be able to purchase enormous quantities of cut wood for next to nothing. The most that I ever saw Old Harry part with, was three and sixpence for a full, barrow load of cut, half-inch sarking and then, he would grin at me once clear of the sawmill and say, "What do you think of that then?" I really couldn't say anything, three and sixpence seemed like a fortune to me!

This wood of course, was used by old Harry to allow him to pursue his long time secondary job of building garden sheds, garages, coal bunkers and anything else to order. He was mainly on night shift at this time and he used the morning hours up until lunch-time in construction and took his much needed sleep during afternoons and evenings. Not everything that Harry built turned out just as he had initially visualised, he did all of his construction from memory and from empirical knowledge, never really committing anything to paper, except the overall size and shape. This method of working sometimes resulted in Harry getting one or two things wrong on occasion and on discovering this, Harry would immediately strip the offending item and salvage nearly all of the materials, including all of the wire nails. I can surely vouch for this, I often suffered from 'chackit' finger nails in my efforts in trying to straighten-out Old Harry's bent, recovered wire nails! This was how I managed to get the odd threepence for the Wednesday afternoon picture matinee.

As a result of these small misfortunes, there was in a separate pile, all of the reclaimed wood with some of it badly split and slightly warped. *(Green wood does this!)* This wood was used for smaller items such as coal bunkers and log-boxes. Old Harry never wasted any of his old wood or left-overs, as this would be hand-sawed down into seven or eight-inch lengths, and then meticulously chopped-up into perfect firewood. Harry then tied-up and sold the bundles around the local neighbourhood for threepence, fourpence and sixpence. I know this, because I was the one that carried the wood to the houses concerned and took the money. This was how Harry always managed to have enough money to pay for the next Saturday's purchases of new wood!

The fact that the 'firewood' business seemed to have a small but regular turn-over did give me an idea, but how to do it did present a small problem, after all, the wood did not belong to me. I knew that old Harry did have his disappointments, especially when someone would cancel an order halfway through it's construction. Orders were not that forth-coming, so Harry would simply strip the work down and salvage what he could. This meant, that very often there were simply stacks of used wood available and this availability was what got me started in my own little 'free enterprise'.

I carefully selected the oldest and driest of this wood and commenced my 'very own' clandestine cut-firewood business. *(I was a practised hand at chopping 'neat' firewood!)* I was most careful not to shortchange any of my customers with the size of my bundles and gave good value for money. I was also most careful not to be seen to be engaged in this nefarious enterprise, and chose my potential customers well away from Barncraig Street. This worked an absolute treat and I was at least two shillings richer every week as a result. This sum was added to the one and sixpence that I already earned on a weekly basis, by going to the local baker's shop every school-day morning and collecting the morning rolls for several pensioner neighbours. However, this simply could not last, but I was too young and naive to notice, the chinking of small silver coins in my pocket and the feeling of what I now recognise as 'independence' was my undoing. I had forgotten to ask for my regular pocket money! This had been quickly noticed, but suppressed by mother until 'enquiries' could be made. There was no need! One fine day at the Denbeath Co-op, Mrs Gibson from Bow Street remarked to my father, "Those are fine bundles of sticks that your Davids' bringing round to the house. They're fine and dry and just the right size - and what a lot I get for sixpence!" Needless to say, my immediate fate was sealed and when Old Harry finally realised the scale of my small enterprise and discovered the amount of missing wood, he was more than just rendered speechless. He was bloody angry, not so much at my ingenuity, but at the fact I had gotten away with being undiscovered for so long! He 'tanned' my backside for that, long and that hard, that I missed school the next day, my not being able to walk properly without pulling faces! I will give Old Harry his due though, that was the last time that he ever tanned my backside and the reason was, that as I was being tanned, I just shut my mouth and 'took' it, never uttering a cry and with a look of sheer determination on my face. When Harry had expended his anger, he turned to my mother and firmly said, "That's the last time that I shall ever do that, all I saw there was sheer defiance!" And to his eternal credit, he never did!

A third little episode is well embedded into my memory, but this did not happen until I had grown a little more, was a full two years older and was then a third-year pupil at Buckhaven Higher Grade school. All of the times previous to this episode, where old Harry had fulfilled orders for any wooden structure, whether it be coal bunker or large garage, Harry would always finish the construction with at least two coats of dark creosote. I should know, I grew up with the smell of it in my nostrils and on my clothes. It was not the first time that I had been accused of 'stinking of creosote' whilst sitting down to meals. I was never with Old Harry when he obtained this creosote and did not know where it came from? I was soon to find out!

One Spring evening just as it was growing dark, my father said to my mother,

"I'm going out for a while and I'm taking David with me!" The reply that I remember was, "No!, not to the sawmill Harry!" But Harry was not to be thwarted and his retort was, *(as I remember it).* "He's got a fine pair of legs on him now!" Still the penny did not drop! So, off we went in the growing twilight with old Harry remarking that I was wearing 'sand-shoes' *(plimsolls/trainers)* and not tackety boots. We made our way up to the vicinity of Muiredge Pit by Den Walk, Buckhaven, over and around the old red bing and then past the bottom side of the closed and unlit sawmill. This was when I felt that I was about to be involved in something that grown-ups did and was not sure as to how I should react. I was however, just a little excited! Advice was not long in coming!, "You sit there, crouch down in the grass and keep quiet! If anybody asks what your doing, just shout!, - then run like hell and I'll see you back at the house!" No-one came, and after about 10 minutes, Harry did appear walking at a crouch and with two, one gallon cans that were dripping with creosote. He wiped the outside of the cans with handfuls of dry grass and motioned me to follow him back the way we had come. Once on the quiet roads, he then covered each can with a brown paper carrier bag, and we proceeded home as though on an evening stroll, with Old Harry puffing at his Stonehaven pipe. It was not until several years later when old Harry was just a little more 'flush', that I actually discovered as to where that creosote had come from. He was as usual, at the sawmill one Saturday morning, haggling over the price of cut sarking, when he asked for the price of a gallon can of creosote to be included in the total cost. The sawmill foreman looked directly at Old Harry and said, "That'll be an extra tanner *(sixpence)* Harry - and **you** know where to get it!" Harry did, and I followed him to the open, but roofed creosote tank, where I discovered that this was the place where the Post Office telegraph poles were immersed in creosote for about six weeks, *(water-proofing?)* prior to drying and purchase. Now I knew! Old Harry had been raiding that tank for years and kept it to himself, until embroiling me in his nefarious activities. Just the thought of it makes me shake my head in absolute wonder! And to think, that I had had my backside tanned for pinching just a miserable few planks of old, discarded wood!

Before I left school and started to work, there was only one other incident worthy of note that took place that concerned Old Harry, and his involvement in this incident happened quite by chance without my even asking him to do anything at all. I was 15 years old at the time and still a pupil at B. H. G. S. At school, one of the lesson periods that I always enjoyed along with a few others of my ilk, were the periods devoted to physical training and box-work. I was at that time, a member of the S.A.B.C., *(Scottish Association of Boy's Clubs)* which was organised and run, under the auspices of the Miner's Welfare Institute, who employed professional trainers and organisers in each mining district. Its purpose was to physically educate, train, develop and help the young sons of mining families and keep us of the streets, and at the same time, to probably inject some sense of sporting knowledge and fair play into our heads as well! It was also possible to travel around the various village clubs as a visiting guest on almost any week night, as the same instructor did his regular rounds of the five or six clubs in his area every week. This particular instructor's name was Harry Maquire, a middle-aged man who was well liked by both members and parents - and did his job very well to the satisfaction of both club member's and his organisation.

These clubs were devoted to the physical well-being of its members, where the accent was on fitness, co-ordination, strength-building, discipline and teamwork, and catered to youngsters between the ages of 11 to 17 years old. If a young member did not wish to conform to these rules, he was tossed out on his ear with Harry Maquire paying a visit to the parents to explain why. Harry had infinite patience with the members that showed promise in any of the sporting disciplines, whether it be running, jumping, or any other of the field discipline's. Sufficient to say, that any youngster who had spent any time in the S.A.B.C. system, received a physical education that was way beyond anything that could be taught at school.

At school therefore, there were boys like myself who were way beyond the fitness, strength and endurance of most of the rest of the class and we took great delight in showing-off to all and sundry, especially at end of term, where the gym teacher was always asked to put on a display of team games and finish with a running tableau of 'box-work'. Needless to say, the boys that figured greatly in this final display were invariably those brought up, developed and exercised through the S.A.B.C. system. We would do the most daring and daredevil displays, with no though to life or limb and the 'credit' of course, went to our P.T. teacher, whom I can only now describe as being rather tolerant.

Nothing however, lasts forever, and all of this happy camaraderie amongst my school-chums suddenly came to an unhappy and abrupt end. In point of fact, it took a sudden and near disastrous nose-dive. Enter the villain of the piece, a new P.E. teacher by the name of Mr. A. Milne, ex-W.O.II. *(A sergeant-major no less!)* Army Physical Training Corps. He was one hell of a 'piece'! He was of course, a mature being, not yet reached middle age. He was upright, strong, well-built and sun-tanned. He was also a rigid disciplinarian with short back and sides, and a polite smoothie who always 'delivered' a smarmy 'good morning' to the lady teachers. He always looked down his handsome nose at all youngsters and mostly deigned not to speak to them. He only issued orders! He was also, as it turned out, a spineless bully!

During our first class session with him, he did not reveal his true colours. He was distant, but firm and seemed to know his job in putting us through a few routines that were new to us. We did not seem to get very much box-work, as he seemed to concentrate on exercises that we felt were only fit for wee lasses! *(Undeveloped school-girls!)* That didn't go down at all well, and neither did he! Very soon, as happens in schools, word began to circulate that he was being very heavy-handed with some pupils, whom he felt, were not giving of their best and he had forewarned *(threatened?)* some lads, that he believed in corporal punishment. We did not have long to wait, as it soon became public knowledge that he had made a fourth year pupil bent over in front of his P.E. class, to be given three whacks over his rump with a two-tongued leather strap. That made his future classes look to wearing thicker shorts and junior 'jock-straps!'

The next week as our class trooped into the changing rooms, Mr. A. Milne was present and standing on one of the long forms *(wooden benches)*. He informed us in no uncertain terms that there was a new broom being used and that in future, we would all 'toe the line' without complaint and from now on, 'box-work' would be an earned privilege. He added, that if anyone stepped 'out of his line', the result would be more and more disciplined exercises. He then left the changing room at an exaggerated 'slow double' with several small 'toots' on his 'Acme whimperer', *(thunderer)* with the added warning, that he was giving us exactly two minutes to be out in the hall ready for exercises and that any undue tardiness on our parts would be instantly punished!

In the changing room, there was a mad scuffle to get into shorts and plimsols before the expiry of the two minute whistle, with all thoughts of our previous neatness being cast aside. To say it was a scramble would be an understatement and inevitably, there were two pupils who did not quite make the deadline. I was one of them and I can now vaguely recall thinking in terms of stubbornness and brinkmanship, *(in those days of course, it was just utter bloody dourness with me bottom lip stuck out!)* with the though of 'what can he possibly do to me?' I had barely started to undress with only jacket, jersey and tie removed, when he reappeared, shouting an instruction that sent everyone remaining out into the hall. I was soon to find out what was to happen! On spying me in a state of near full dress, he ordered me to 'bend over', at the same time, whistling the belt through the very still air. My bottom lip went out! . . . "No!" I said defiantly, with mouth held tight. "Bend over", he shouted, in an angrier voice, to which I replied, "No! - Never!", still showing sheer defiance. He moved very quickly. I had seemingly given him all the excuse that he needed! He grabbed my left forearm with his right hand, and demanded that I stretch out my right arm with palm extended, which, like a fool I did! He drew back his belt hand and brought the heavy leather down across my hand with a terrible crack that made me almost drop. He had hurt me - and I refused to hold my hand out again. He let go again!, and this time with no hand to aim for, the blow landed across my wrist and forearm, which immediately drew a large reddening welt and copious amounts of blood! The bugger had cut me open! I ran into the changing room, grabbed my belongings and ran out of the school, with Mr. A.Milne shouting words of dire, probable consequence *(threats!)* if I did not return immediately. After a few hundred yards I slowed down to a painful walk, my arm as sore as hell with the blood still flowing, but starting to congeal. I passed the College Street bus stop, still with the blood dripping from my downed arm - and who was standing there quietly puffing at his Stonehaven pipe, but Old Harry. I walked over to the shop door where Old Harry was standing, where by now, he was alerted to the fact that all was not well. I held out my damaged arm, to which Old Harry darkly uttered only one word "Who . . ?" "Milne!", said I. Old Harry stuffed his pipe in his pocket, turned sharply and headed of in the direction of Buckhaven High School. Old Harry was angry! I could tell! He hadn't stuttered! That meant his anger was stone cold. Someone was going to get it! - and I wished that I could have been there to watch!

*

I was never actually told exactly what Old Harry did when he got to the school that lunch time, but I was later able to piece together the snippets that came to my ears. The story of how Old Harry went to the staff-room door and 'quietly' asked for Mr. A.Milne to appear makes for good telling. Old Harry was told in no uncertain terms by one of Milne's colleagues *(cronys/protector?),* that Mr. A. Milne would not come to the door of the staff room and if Mr. Moodie wished to make a formal complaint, then he must make an 'appointment' with the headmaster! Poor fool, he should have known that Old Harry didn't take kindly to being thwarted - and besides, old Harry had a shift's worth of 'brushing' awaiting him in the Lower Dysart West. Old Harry, being a man of very few words, just 'pushed' forwards through the door. He didn't need to be told as to who Mr. A. Milne was, one glance was enough! Unfortunately, old Harry did not manage to lay hands on Mr. A.Milne at that time. Mr. Milne was too fleet of foot - and besides, he was man enough of the world to read the *(his)* impending disaster in Old Harry's eyes. Harry then chased his quarry all around the school, until Mr. A. Milne obviously decided that the manly thing to do was to cease this ungentlemanly conduct and face-up to his tormentor. He called a halt when he reached the supposed safety of the raised stage in the main hall, then threw up his hands, calling for peace and calmness and an end to this 'unseemly conduct!' *(On his part that was!)* The poor fool! Old Harry caught him and grabbed him in a bear-hug, not wishing to punch his lights out in full view of the assembling school. He did not hit or strike Mr. A. Milne, but merely held him very firmly. Old Harry was still gripped by a cold anger and momentarily loosed Mr. A. Milne, merely to get a better and different grip, in an effort to seemingly, throw him bodily off the stage. Old Harry was on the point of doing so, when common sense seemed to prevail and just as quickly, old Harry's previous anger died, which was just as well for Mr. A. Milne. Old Harry just dropped him bodily to the floor, not really caring if he hurt himself or not! Old Harry then walked out of the school building and not a single person followed him. He caught the next bus to Lochhead pit and disappeared from view.

The following Monday when I went back to school, the talk was all about this big coal-miner who had appeared at the school and chased Mr. Milne onto the stage, with the supposed intention of throwing him off. The general opinion amongst the lads seemed to be, that he hadn't been thrown far enough! The police did get involved and were called to the school *(the local police station was only across the road)*, but no charges were brought against anyone.

The upshot of this event was, that Mr. A.Milne was suitably chastised by Wullie Byres, the school's diminutive Headmaster *('guid girth gangs in wee book!')* and banned from using corporal punishment on the school's pupil's. Mr. A. Milne then attempted to call me 'David' after that incident, but all he got from me was an insolent stare, a protruded bottom lip and a first-hand finger-pointed view of my still damaged forearm. I carry this belt-mark on my forearm to this day!

✻

# CROSS SECTION OF THE DYSART MAIN SEAM
## Showing a Thickness of 22 feet at Lochhead

Stone / Blaes

Upper Head Coal .... 31"

Lower Head Coal ... 16"
Stone ........................... 1½"
Sparcoal ...................... 16"
Stone ........................... 1½"
Toughs Coal .............. 21"

Clears Coal ............... 17"
Stone ............................ 3"
Myslen Splint ........... 10"
Stone ............................. 8"

Nethers Coal .............. 33"

Grounds Coal ........... 26"

Sclits ........................... 21"

Stone ........................... 13"

Coronation Coal ...... 47"

Stone / Fireclay

# Chapter V

## The Taking of the Dysart Main
### *An Incredible Extraction!*

The Lady Lilian* as Lochhead Pit was once named, *(this pit was the second of that name!)*, seemingly did not take any of the other shallower seams of coals from much nearer the surface, before the shaft bottom struck the Dysart main seam of coal. This was the targeted seam that lay at the relatively shallow depth of only sixty-three fathoms *(approximately 378 feet)*. The whole purpose of this sinking was to broach and open up, the known, but as yet untouched twenty-three* feet thickness of this 'beautiful' seam of mixed coals. At this juncture, I must make it very plain to all readers who may be just a little confused, *(just as I was!)* that this was not the original Lochhead pit! The original site of the Pit of that name *(long abandoned)* is now well hidden, ploughed over and mostly forgotten. It was situated in the field directly behind the houses that now form Anderson crescent. The old shaft was near to Lochhead farm and approximately 100 metres from the Check-Bar road, that links the Coaltown of Wemyss to the Standing stone road. *(The approximate Grid Reference of the old shaft is 3175 9605 on O.S. sheet No. 59. Land-ranger second series, (metric) one over fifty-thousand/1 : 50,000.)* I have very little information regarding this original pit, except to say that it was probably sunk around the early 1800's to take some of the shallower seams in the upper coal measures. I do know that it extracted several seams of coal, at least down to the Branxton, which was worked-out and abandoned as early as 1839. *(This Branxton seam lay five seams above the Dysart Main!)*

Author's Note:- *The Dysart Main coals was well known to both the Earl's of Rosslyn and the Lairds o' Wemyss, and to the many of the old colliers who had wrought this coal, both as 'open coal-heughs' and 'level free' workings. This thick seam of coal had outcropped from a few hundred metres inland at Dysart and could be traced cross-county to near Markinch. By the year 1900, all of the Dysart Main outcropping coal between these two points had been wrought down to water level, hence the need for the coal-owners to employ pit-sinkers! In all, the approximate amount of coal within the Dysart Main seam that lay under the Wemyss estates, could be conservatively estimated at 120 million cubic metres! This excluded Dysart Main coals taken from under the Firth of Forth - and the different coals from any of the other six or seven fully accessible and workable seams within the area of the estate! Practically speaking, any seam that has a thickness of less than twenty-four inches (approximately 0.60 metres) is both difficult to work and would probably be an uneconomical proposition, if worked as a single entity.*

In the early 1890's, Randolph Wemyss and the Wemyss Coal Company decided to originate the new Lochhead colliery *(and slightly later, the Surface Dipping)* at the new site, where the shaft was sunk down to the Dysart Main seam of coal *(deliberately by-passing all of the other seams of known coal),\** where, at a depth of Sixty-three fathoms *(approximately 115 metres),* the splendid thickness of the Dysart main was realised. It was fully twenty-three feet thick at this point and showed nine different coals, separated by mostly very thin layers of stone, with only one thickish 18-inch layer of stone between the Coronation coal at approximately 50 to 55 inches thick and the Sclits at 30 inches thick. *(The Sclits proved to be a 50/50 checkered mix of coal and stone that was naturally bonded and economically inseparable).*

*\*By virtue of the principle of opposing twin cages winding and drawing, to and from a fixed pit-bottom, that in effect, means that only one pit-bottom can be serviced at one time. The twin, but opposing steel winding cables (and cages!) are each 'set' to a given winding length, that in practice cannot be quickly altered!*

*(On 28<sup>th</sup> July 1884, **Randolph Wemyss** married the **Lady Lilian Mary Paulet**, daughter to the Most Hon. John Paulet, Marquis of Winchester. Hence the original given name to the second Lochhead colliery).*

The listed seams of coal that are contained within the strata above the Dysart Main seam in the Wemyss district, are described below, along with their approximate depth in yards and approximate thickness in feet. *(Since their discovery and subsequent identification, some of the seam names have changed. Any known change of name is shown.)*

| Name of seam | Depth of seam | Thickness of seam |
|---|---|---|
| Skipsey | Not quite a phantom seam, but not worked! | |
| Pilkembare | Outcropping in the Wemyss-field area. | |
| Wall Coal | 16 yards | 3.0 ft. |
| Barncraig *(Leven eight-foot coal)* | 32 yards | 6.0 ft. |
| Upper Coxtool *(Leven six-foot coal)* | 51 yards | 2.5 ft. |
| Lower Coxtool | 57 yards | 3.0 ft. |
| Den | 74 yards | 2.0 ft. |
| Chemiss. *(Leven main coal)* | 116 yards | 9.0 ft. |
| Bush coal | 136 yards | 3.75 ft. |
| Wemyss Parrot | 189 yards | 3.25 ft. |

| | | |
|---|---|---|
| Four Feet. *(Wood Coal)* | 208 yards | 3.75 ft. |
| Earl David's Parrot | 236 yards | 2.0 ft. |
| Bowhouse. *( Bu'hoose)* | 252 yards | 7.0 ft. |
| Branxton. *(Brankstone)* | 275 yards | 2.5 ft. |
| More Coal | 313 yards | 1.75 ft. |
| Boreland. *(Mangie/Balgonie Splint)* | 321 yards | 2.3 ft. |
| Sandwell Upper | 335 yards | 1.25 ft. |
| Sandwell Lower | 355 yards | 2.0 ft. |
| Dysart Main | 421 yards | 24 ft. |

And, at approximately 25 yards below the Dysart Main seam:-

Thickness

Lower Dysart. *(Seven-foot coal) 448 yards.*     *Top leaf = 37 to 43 inches*
*Middle leaf = 25 to 30 inches*
*Bottom leaf = 38 to 44 inches*

*(The top and middle leaves in the seam were separated by approximately 30 inches of soft blae/redd, while the middle and bottom leaves were 'separated' by a four-inch band of stone adjoined to a five-inch band of Parrot coal!)*

*The topmost seam in the P.C.M, the Skipsey, did not actually appear in the East Fife coalfield, but was found by test drillings, out and under the waters of the Forth. At the very bottom of the P.C.M. in East Fife, there lies the heavily faulted Lethamwell seam, that did not extend over the full Wemyss-field area. Therefore, it suffered only limited working at a few W.C.C. collieries. It was seam No. 21 in the P.C.M. within the East Fife Coalfields!*

Author's Note:- *The above listed depths were calculated by a drilling NOT taken at the site of Lochhead Colliery. However, the above table is perfectly accurate at the point of the test drilling. Comparisons can therefore be made on a pro-rata basis, as to the vertical distances between the named coals at the Lochhead shaft. Lochhead pit shaft struck the Dysart Main seam at the relatively shallow depth of 63 fathoms. (126 yards/378 feet)*

\*

Within the pages of the some of the books that I have read, with reference to the Dysart Main seam of coal, the seam itself seems to have been treated and described as a varying thickness of homogeneous coal, as much as thirty-two feet thick as it dipped out and under the waters of the Firth of Forth, and as thin as twelve feet in thickness, as it outcropped to the North-west side of the Standing-Stone road between Wellsgreen pit and the Randolph Colliery, near the Boreland. The variations in thickness of this great seam are quite well documented, but seemingly not so, are the many and various named coals within the seam.

I do know that it is sometimes very confusing and frustrating to *inquisitive readers* to discover, that in reading different books by different authors, who have researched different sources and all dealing with the same seam of coal, to discover that they may be discussing the same seam of coal that was being mined in several different pits, and which had varied in *thickness* and in *Name!* The Dysart Main seam of coal as worked *continuously* in Lochhead pit between 1890 and 1960, *(except for the duration of the1912 minimum wages strike, the 1921 National strike and the 1926 General Strike)*, had nine different named coals contained within it's thickness. Each one of these nine coals had different and distinguishing characteristics and friability *(and in some cases, narrowly separated by thin bands of grey stone).* These different coals when worked, were quickly recognised and easily identified by the colliers/miners who 'took' this ubiquitous and bountiful seam.

The named coals *(inc. other names if known)* within the Dysart main seam as worked in Lochhead pit were, starting at the top of the seam:-

| *Name Strata* | *Description* | *Thickness* |
|---|---|---|
| Upper Head coal ....................................*Roof coal* ...................... 3' – 0" (0.92m) | | |
| *Separated by a 1-inch band of stone* | | |
| Lower Head coal ................................*Same Roof coal* ................. 1' - 6" (0.46m) | | |
| *Separated by a 2-inch band of stone* | | |
| Sparcoal ................................ *Sometimes named the Head coal* ..... 1' - 8"(0.51m) | | |
| *Separated by a 2-inch band of stone* | | |
| Toughs .................................................. *Head coal* ...................... 2' - 0" (0.61m) | | |
| Clears .......................................... *Clean coal or Clarus coal* ......... 1' - 8" (0.51m) | | |
| *Separated by a 5-inch band of grey stone* | | |
| Myslen ....................................... .....*Splint coal* ...................... 1' - 0" (0.31m) | | |
| *Separated by a 6-inch band of hard stone* | | |
| Nethers .................................................. *Bright coal* ......................3' - 3" (0.99m) | | |
| Grounds ........................................ *Same* ...........................2' - 6" (0.77m) | | |
| Sclits :- *a 50/50 mixture of coal and stone. (Utterly useless)* ........... 2' - 6" (0.77m) | | |
| *Separated by soft blae at approximately 15 inches thick* | | |
| Coronation coal ......................................... *Thief coal* ...................... 4' - 6" (1.37m) | | |

*The above thicknesses were measured at a point slightly to the South of Lochhead pit bottom, where the seam was approximately twenty-six feet thick in total (inc. stone and blae!).*

\*

In the following chapters within this narrative, I shall endeavour to explain as to how the coal was taken from the Dysart main seam, approximately how much of the coal was taken from each section/level during that time, and the actual coals taken by name. It must be remembered that even though this seam of coal was fully twenty-three feet thick as worked at the base of the Lochhead shaft, it generally thickened as it deepened out to and under the Firth of Forth, and shallowed as it outcropped inland, and that it is just not possible to successfully fully extract such a thickness in safety, or in volume at one go! I must also point out to the reader, that to the south and east and under the 'Frith' of Forth, the Dysart main seam is almost thirty-two feet thick, whilst to the north and west the seam peters down to approximately twelve feet thick, where it outcrops to the north of the Standing-stone road in Fife. *(I use the present tense at this juncture, for the good reason that the seam is still largely intact, but completely abandoned and probably flooded, except for where it crops-out to the surface to the north-west of the Standing Stone road.)*

I shall also attempt to explain by description and delineation and with reference to named places and datum heights, the general areas from which the coal was taken. Considering also, that most of the original levels, headings, dooks and sections may have had particular names in the past, most of these names have long since fallen into disuse and are disremembered along with the miners who developed them! I shall therefore make reference to the levels and dooks that I have named and introduce new levels by number or name where possible - and further headings, dipping's and dooks by reference to a known point or datum height. I do hope that readers will understand, that in the taking of the coal and the getting of coal to any pit bottom *(where it must eventually come up the shaft!)*, the taken coal seam is usually 'worked out' in given sections *(areas)* that radiate outwards in given directions from the Pit bottom, with the nearest and 'easiest' coal being taken first. In the case of Lochhead colliery, with reference to most of the very long levels and forever extending dooks, I have given where possible, the months and years of intrusion and development, along with the datum heights and the approximate date of eventual abandonment, or 'sealing-off' from the circulating air flow, if this was actually done.

Readers should also realise, that in the taking and getting of underground coal in the 1890's, there was very little in the way of machinery or haulage systems, *(except for pit ponies and human drawers)* and the economics and practicalities of getting 'cheap' coal was the over-riding premise. Thus, the quickest and cheapest coal was to be got, in and around the areas of the pit bottom *(from out-with the area of the pit-shaft's circular, solid, supporting stoop)*. Also, considering the fact that the whole of the seam of the Dysart Main coal did follow the curvature of the metals *(strata)* in slowly rising towards the north-west, *(including a rolling fault with which I shall not confuse readers),* it made sound sense and good economics to follow the coal to the rise in the first instance, where utilisation of the forces of gravity would favour the process of wheeling the loaded hutches on their way back to the pit-bottom.

*

# First to the rise!

Later on, I wish to pose an indirect question and perhaps raise an issue that may have perplexed several generations of colliers who 'wrocht' at Lochhead pit. This question seemingly existed from the start of the Dysart Main extractions in the three main Dooks at the Lady Lilian. *(I shall now refer to this colliery as Lochhead pit, since it was re-named as such by Randolph Wemyss after his divorce from the Lady Lilian Paulet!)*

The second Lochhead pit was sunk with the explicit intention of broaching the Dysart main seam of coal, hither-to untouched in this area. The W.C.C. quite intentionally decided from the onset, this would be a one-seam extraction pit, taking the Dysart Main seam first to the rise over a broad front, and then to the dip over an even wider front! It was also a forgone conclusion, that the potential coal sections within the Dysart Main workings taken to the dip from the pit bottom level, would in fact, be the mainstay of the Lochhead workings for many decades into the future. It was also realised at the time that the workings could not be mechanised to any extent and that hand-operated hole-boring 'machines', picks, shovels and wooden hutches, along with low explosive powder and pit-ponies, would be the primary means of coal-getting! *(It would be many years hence before powerful machinery of the sort needed to tackle the Dysart Main seam would be developed!)*

It was also eventually realised after some hard lessons, that the coal-taking would probably have to be by the 'Pillar & Stall' (Stoop and Room) method of extraction and on a general widespread scale. When this decision was made is unclear, since all of the Dysart Main extractions/workings taken to the rise from the long pit bottom level, were extracted using methods other than Stoop and Room! So far, so good!

Before the reasonably extensive limits of the seam had been fully exploited to the rise and subsequently worked-out, the planners/management decided to sink first one main Dook, then a second main Dook and finally a third shorter Dook, *(in different locations),* to take the very extensive Dysart Main coals to the dip. No mean undertaking, considering that the average dip on the wide coal seam from the pit-bottom, was in the region of 1 in 3.5. This then, *(in my opinion),* was where the question of how best to take this thickening seam of near homogeneous coal, should have been carefully considered!

Suggested Probabilities and Concessions in the adoption of the 'Stoop and room' method of extraction in Lochhead pit: -

*1.    Minimum mechanisation.*

*This for two reasons: -*

*(a)    The scarcity of dependable heavy-weight coal-taking equipment!*
*(b)    The initial overall cost of whatever light-weight equipment that was available! (Also, coalminer's themselves did not 'breakdown' in the short term!)*

*2.    One of the principles of 'horizon mining' could be (and was) utilised alongside the Stoop and Room methods, in the driving of the long parallel levels that eventually ran from the Francis boundary to the Buckhaven fault boundary. (This, from a practical point of view, made the transport of coal in hutches much easier)*

*3.    This method was to be adopted and actively pursued in approximately 60 percent of the seam, because every other method of extraction had been attempted and found to be unsuitable for this thickness of seam! (Perhaps again, in view of the dearth of suitable 'coal-taking' equipment!)*

*4.    This method involved coal explorations radiating slowly out from the pit bottom, in ever-lengthening Dooks and levels, until the limits of the colliery boundaries had been reached. (This meant that coal was being taken while developments continued!)*

*5.    Limited Stoop and Room extractions as practised in Lochhead pit, ensured (resulted) that the minimum of coal was taken, for the underground distance travelled. It also necessitated the leaving of even larger stoops of coal to maintain the integrity of roof (and roadways) as the workings deepened!*

(In using this method of limited extraction in narrow roadways, as long as the Head coal and sometimes the Sparcoal was left insitu and unbroken, then, the need for the miner's/collier's to set immediate roof supports was obviated in the short term.)

I am also fairly certain, that when the Dysart Main seam was first broached at Lochhead pit bottom and part of the fulsome beauty of this consistent* thickness of seam was realised for the first time, Randolph Wemyss and the Wemyss Coal Company, must have thought that this was the greatest potential coal bonanza ever to be struck within the Wemyss-field area.

The Board must have surely wrung their hands in sheer joyful anticipation, at the hundreds of thousands of tons of coal, that Lochhead's shaft would most assuredly bring to the surface for many decades into the distant future. At Lochhead's pit

bottom, they were faced with a solid, upright wall of coal, that stood all of 23 feet *(7 metres)* high, with the prospect of this solid wall steadily increasing to a maximum height of 33 feet *(10 metres)*, under and along the Wemyss foreshore! *(A distance of approximately one mile, along an expanding two miles front!)* But, how were they going to take this wall of coal, to successfully exploit it to the maximum?
I am fairly positive that they just didn't know!

*This consistent thickness of seam remained at between 21 to 23 feet, along the full length of the pit bottom level, from the Francis Barrier (boundary) to the Buckhaven fault (the Dyke)!

Obviously, the board knew that it was a virtual impossibility to safely extract such a height of coal, over even over the smallest area without the roof caving in! It was just not feasible at that date, to leave a great empty void underground when cutting through mixed metals. *(This can be achieved however, if the void is created or situated within a single medium, as in the salt mine at Heilbronn in Bad Friedrichshall in Deutschland. In that rock-salt mine, there is an unsupported void, the size of the interior of a cathedral!)*

The decisions facing the board and the planners/mining engineers at Lochhead, was not so much how much coal to take, but how to take the coal in such a manner as to allow, either the maximum extraction at one go, or to take selected sandwiches of coal at first bite - and then re-visit the same site sometime later, with the prospect of different 'sandwiches' being safely extracted!

The Dysart Main coals to be taken first, was the 21 to 23 feet high, wide front to the rise, that lay between the Buckhaven fault and the Francis colliery boundary. These coals were taken by section areas of differing sizes and worked using various methods of extraction, *excluding* Stoop and room.

i.e:-

1.  *Long-wall pick-places of perhaps 100 to 150 yards long and at limited height, taking one or more named coals over a wide area, packing the wastes with stone, then allowing the whole section to 'weight' (or crush down) as an unbroken whole, before revisiting the same section area, to extract a second and possibly third sandwich of different coals.*

2.  *A series of parallel intrusions approximately 10 to 12 feet wide and approximately 30 feet apart, originating from a straight level roadway. Within these parallel intrusions, the collier's would each 'take' the coal to a given working height and to the full length of the driveage, perhaps up to 100 yards. When the individual roadway had reached its maximum viable length, the miner's would then exercise an about turn and start to take the roof coal to the retreat. The enlarged empty void left behind the miners, would not be supported in any way and be allowed/encouraged to fall, thereby relieving the*

*downwards pressure within the limited section area. When this happened, this method of extraction had the negative side-effect, that whatever solid coal remained within the section, was now irretrievably lost forever!*

3.      *If this didn't happen and the roof remained solid, then inevitably, there would probably take place a long, slow, sustained crush upon the whole of the section area with the certain result, that the side walls along the full lengths of each parallel roadway would burst out under the pressure, filling the voids with lumps of broken coal! When the maximum crush or weighting had taken place and the section 'settled', and as long as the Head coals remained unbroken, this fractured, broken coal could be harvested at minimum cost to the coal-owners! It was there for the taking! It was also a rather dangerous environment!*

What is apparent, was that from the start, there was a great inconsistency in the methods used to take the different coals within this seam, as it was worked to the rise from the pit bottom levels! Not only did there seem to be a lack of cohesion or planning in the size and shape of all of the different section areas, but there also seemed to be some doubt as which coals to take first! It must be remembered that as the seam was slowly worked to the rise, two opposing factors would determine how much *(layers)* of the coals could be taken!

1.      *The seam would gradually thin down to approximately 12 to 14 feet thick as the workings approached the Earlseat boundary. (With all of the named coals thinning down on a pro-rata basis)*

2.      *The Dysart Main workings were becoming shallower as the seam rose towards the north-west! (With the subsequent lessening of the overhead pressures!)*

At the commencement of working the seam along the broad front from the pit bottom levels, from datum heights of approximately 9770 feet up to datum heights of 9850 feet, much of the bottom-most Coronation coal was taken before the extraction of the Grounds & Nethers and the Myslen. *(The waste redd/blae lying immediately above the Coronation coal could have been 'downed' and used to partially pack the void left by the extracted Coronation coals!)* Where the Coronation coal had been removed, the subsequent extractions were taken only up to the Myslen splint, leaving the Toughs & Clears, the Sparcoal and the thick Head coal as a solid roof.

If the Coronation coals had been left insitu, then the Toughs & Clears and probably the Sparcoal would be taken first, with the secondary taking of the Grounds & Nethers along with the Myslen Splint at the next re-visitation. This selective choice, ensured that a specific thickness of coal was removed in two or more sandwiches from all working section areas, regardless of which coals were actually removed!

As the workings became slightly shallower and the seam thinned slightly, the miner's began to concentrate less on the extraction of the Coronation coal and more on the taking of the Grounds & Nethers and the Myslen, and then returning soon afterwards to extract the Toughs & Clears and occasionally the Sparcoal. This invariably occurred between datum heights 9850 feet, up the datum height 9950 feet. The seam was certainly thinning down a little, but the miner's were still extracting approximately the same relative thickness of coal!

Within the remaining Lochhead areas above the 9950 datum heights, *(this area was now only around 150 feet/45.7 metres below ground surface)* the Coronation coal was seemingly left severely alone, with the miners now taking the Sclits, the Grounds & Nethers and the Myslen, and then the Toughs & Clears with the additional 12 to 15 inches of the Sparcoal, in probably two or three separate sandwiches. As can be seen, this was coal extraction to near maximum, but really only possible because of the proximity of the seam to ground surface!

# And Then to the Dip!

During 1899, the first attempts were made to exploit the Dysart Main seam to the dip. This first underground dipping mine was commenced from the Pit bottom level at a point approximately 50 metres from the pit shaft. It was cut through the middle coals of the Dysart Main and was named, the Old Dook. Three or four years later *(around 1903),* a similar dipping mine was commenced on the Pit bottom level at a point approximately 100 metres from the pit shaft but, at the other side of the pit bottom. This second dipping mine was also cut through the middle coals of the seam and was named, the New Dook. A third dipping mine was also commenced from same height as the Pit bottom level, but that mine was commenced on a level, at a distance of approximately 1000 yards from the pit bottom and in a south-westerly direction *(first known as Nicholson's level and later named the Lower Dysart Haulage-way!).* This third, short, dipping mine would eventually be named for its originator, Nicholson's Dook. *(The Old Dook, the New Dook and the Lower Dysart haulage-way are described in greater detail elsewhere within this narrative!)* The upper part of Nicholson's Dook was later utilised for the transport of loaded hutches from the head of the Victoria Dook up to Nicholson's level. *(Later known as the Lower Dysart haulage-way!)*

As these three named Dooks were deepened and developed down to what is described and known as their first level, these levels were eventually extended left and right in both directions, until such times as they were interlinked at approximately the same datum heights. This No. 1 level eventually extended without break, from the Francis barrier to the Buckhaven Fault. However, the separation distances between the long Pit bottom levels and the newly developed interlinked No. 1* levels were not equidistant over their near 3.3 kilometres length, simply because of the vagaries of the rolling underground strata. At the Buckhaven fault, the levels were only 125 metres apart, whilst near the Francis barrier, they were fully 575 metres apart! This means that at these datum heights, the declivity *(downwards slope)* at the Buckhaven fault was almost 1 in 3, while the declivity at the Francis barrier was only 1 in 9.5!

*(This 3.3 kilometres long No.1 level on the Old Dook, was known as the 200 level at the New Dook. As it meandered to the south-west at an approximately datum height of 9600 feet, it cut through the mid-point of Nicholson's Dook and by-passed what was to be the head of the Victoria Sea Dook, before it terminated at the Francis barrier!)*

Between these two levels and along their full lengths, there were approximately twenty-one different section areas of different shapes and sizes, all worked in a seemingly haphazard fashion with apparently no rhyme or reason as to how the coal was approached. It was certainly a rule of thumb in those times, that if the roof held and the access roads could be kept in good repair and the coal was easy to take, then a section area would be 'milked' for all that could be taken! If the section was 'faulted', or if the roof constantly failed, or something stood in the way of coal-getting, then the section area would be swiftly abandoned. There was a great abundance of workable coal!

The taken coals were mainly and firstly, the Grounds & Nethers and the Myslen, followed by the Toughs & Clears and probably the Sparcoal at a much later date! *(In one or two of these smaller section areas, the Coronation coal was taken first, in addition to all other coals below the Head coal. This amounted to an approximate 85 percent extraction!)*

Of the twenty-one sections worked, nineteen of them took the coal in at least two separate sandwiches and at different dates, sometimes years apart. The two section areas that worked the coal as limited 'Stoop & room' extractions, lay directly under the Bowhouse farm Buildings at the south-west end of the Coaltown of Wemyss and the main road linking them both. The taken coals by Stoop & Room were mostly the Grounds & Nethers and the Myslen, followed by the Toughs & Clears almost immediately! *(The approximate size of the remaining coal stoops left in this area measured around fifteen to eighteen metres square! The extractions were not uniform!)* These two Stoop & room section areas suffered constant re-visitations from the time of their original working in 1893, up until 1949, where limited cheap easy coal, much of it downed, was still available without the need for expensive machinery!

At the same time as the section areas that lay between the No. 1 levels and the Pit bottom levels were being worked and extracted, work continued on the deepening of all three Dooks within the Dysart Main coal. The deepening *(developing)* of the

New Dook initially progressed at a much faster rate than that of the Old Dook, for the simple reason that the section areas within the New Dook between the No. 1 levels and the Pit bottom level, were much smaller in size and therefore in volume. *(Only three of the twenty-one section areas were served by the New Dook at this depth!)*

As the working sections above the long No. 1 levels slowly became exhausted, each of the three main dooks had now been further deepened and were to be developed and worked independently. The subsequent levels and section areas within each dook, would now be named or numbered and designated east or west and in such a manner, so as not to cause confusion to anyone working within the pit. Also to be considered, was the subsequent naming of the yet to be developed levels in the newer fourth dook within the Dysart Main seam, the Victoria Sea Dook! *(All of the different names and numbers pertaining to the individual dooks, are contained and described in the relevant charters dealing with these dooks, elsewhere within this narrative.)*

As is the necessity with progressive mining, new section areas are explored, opened up and developed, long before the previous section areas become exhausted! This was certainly and extensively practised in Lochhead pit. All three dooks within the Dysart Main coal were being steadily deepened as the coals were being taken from the previous levels. As the three dooks deepened down towards the 9430 feet datum height, each dook then took on its own characteristics. Nicholson's Dook was terminated in length and used as a haulage dook. *{Firstly at a length of 650 metres, then later reduced to 375 metres, (No. 1 level) to cope with some of the coal output from the Victoria Dook!}* The New Dook's *No. 1 level* was actually named the 200 level, but did not develop a set of levels to equate to the Old Dook's No. 2 level.

Within the Old Dook, at datum height of approximately 9430 feet, the No. 2 set of levels were developed only to the *(south-)* west, to cut through the bottom terminated length of Nicholson's Dook, and to cut into the Victoria Dook's 300 levels. The section areas between the Old Dook and the Victoria Dook, that lay between No. 2 levels and the No. 1 levels, were four in number. Three of these section areas were taken by methods other than Stoop & Room, where the extractions were in the region of 80 to 85 percent of maximum. *(Leaving only the Head coal.)* The one section that was taken by limited Stoop & Room extraction, lay directly underneath the village of the Coaltown of Wemyss! The one large section area to the west of the 300 level in the Victoria Dook, was taken by methods other than Stoop & room, but did not conform to a regular extraction. The Coronation coal was removed over the whole of the large area, but only small sections of the remaining coals were tackled and then in a haphazard fashion.

The No. 3 levels in the Old Dook were now developed down around the 9270 feet datum height. This No. 3 level(s) did eventually stretch from the Buckhaven fault to the Francis barrier. In the New Dook, it was known as the 400 section. At the Victoria

Dook, it was known as the 600 level. *(Nicholson's Dook was not developed down to this depth!)* Within the large near triangular area that lay between the Old Dook and the New Dook, and in the areas to the east of the New Dook and all above the No. 3 levels, the maximum of coal was taken in successive sandwiches from all of these section areas, in most cases, leaving only the Head coal as the immediate and final roof. That these extractions were successful, there is no doubt, except for many and sustained roofing problems within the 400 section, to the east side of the New Dook!

At this point, *(approximately datum height, 9270 feet)* the method of extraction changed quite dramatically within the Old Dook and the Victoria Dook. From now on, the only method of general coal taking/getting that would be operated, would be limited Stoop & Room extraction! This did not yet apply to workings or section areas within the depths of the New Dook, but for some reason, *(or several reasons)* the miners were experiencing much roofing problems in the many widespread areas of the 400 section. *(Of the three main Dooks in the Dysart Main coal, the New Dook was the farthest inland!)*

In the application of this method of coal-taking, the principle of 'pick-places' and long-wall extractions were to be temporarily abrogated, *(but not abandoned!)* The basic tenets of Pillar & Stall *(Stoop & Room)* working were to be strictly adhered to, with the intervening, supporting, coal stoops left at a uniform size! The coal extractions were to be in the form of a regular martrix, where the distance between parallel headings in each section area would be accurately measured and rigidly enforced. The spaced headings would be absolutely true to direction, but the 'line' of the successive intervening levels that bisected all of the parallel headings at regular intervals, would be determined by the working colliers in each heading. *(As each heading was advanced upwards, the collier's in each heading would, at measured distances, cut levels both to the right and left of their own headings, until such times as the interconnecting levels met! This ensured that the taken coal in each level had only the minimum distance to travel, before being transported down each individual aeroplane brae and thence along the main transport level!)*

At this underground depth, the matrixed stoops that were left, seemed to be in the region of thirty metres square! The parallel headings and bisecting levels were ostensibly cut to a width of approximately 12 to 13 feet *(four metres),* and the coals taken within these matrixed section areas, were the Coronation coal, the Grounds & Nethers, and the Myslen splint! *(This left the Head coals, the Sparcoal and the Grounds & Nethers as a solid unbroken immediate roof!)* This extraction amounted to a taken coal height of between 150 to 160 inches. *(13 feet or 3.95 metres.)* It is my considered opinion that this coal was taken in two separate leaves. The Coronation coal was separated from the Grounds & Nethers by the thickness of the useless Sclits and approximately 15 to 18 inches of stone. This would have determined as to how the coal might have been taken, for the reason that neither of these useless layers would have been loaded into hutches.

It is my asservation that the full 65 inches height of the Coronation coal was taken first, at the nominal width and to the full length of every heading, leaving the stone roof intact, until the coals' extraction. The collier's would then return to the bottom of each heading. At that juncture, they would then systematically 'down' the 3.5 to 4 feet height of the stone and Sclits into the void created by the extraction of the Coronation coal, to form a new pavement, from which they would then work to 'down and take' the Grounds & Nethers and Myslen splint. *(This is a likely description of one scenario only, I am guided by the fact that it was considered a sacrilege to load redd or stone into coal hutches!)* In any event, no matter as to how the coal was taken, the waste material would have been left!

This changed method of extraction would not have dramatically reduced the coal output to any great extent. After all, the chances are that even more collier's would have been employed at the many 'shortwall' coal places. What is evident, is that the percentage of coal taken from each section area would now be much reduced.

For example, a single, coal section area measuring 275 metres long by 225 metres wide at 8 metres high and worked by this Stoop & room method, would suffer the following extractions:- 9 parallel headings each 200 metres long, equals 1800 metres of roadway. Each roadway is 4 metres wide with approximately 4 metres taken height of coal. This amounts to 1800 times 4 times 4, for the volume extracted by the parallel headings. This equates to 28,800 cubic metres. Seven parallel, bisecting levels each 250 metres long, but each reduced by 9 times 4 metres in length, to compensate for the volume already taken in the headings. This equates to 7 times 214 metres *(250 minus 36)*, which equates to 1498 metres of roadway at 4 metres wide and 4 metres high. This amounts to 23,968 cubic metres of coal, giving an extraction total for the whole of the section area as 52,768 cubic metres, or 44,000 metric tonnes. The total reserves contained in this area would have amounted to approximately 495,000 cubic metres of coal. The amount of coal extracted from this single section area would therefore amount to perhaps 9.5 percent of the total available. This is a far cry from the 80 to 85 percent extraction taken from the No. 2 levels above!

As the workings within all three main Dooks progressed deeper, the size and volume of the remaindered stoops in the Dysart Main were to increase dramatically, whilst the height of same taken coal was to increase gradually, due to the increasing thickness of the seam! However, this did not compensate for the wider spacing of headings and levels in the new deeper section areas, with the inevitable result that the taken percentage of available coal was being reduced!

With the adoption of the Stoop & room method of coal-taking at the comparatively greater depths below the No. 3 series of levels, the law of diminishing returns was being keenly felt! The amount of coal taken from the average section area at the Dook's maximum depth, had now reached a miserable five percent! There surely could have, or must have been, a better way to tackle this behemothic volume of coal?

Author's Note: - *At the 1200 level (west) in the Victoria Sea Dook, the Coronation coal had thickened to 82 inches (2.1 metres), while the Grounds & Nethers had also slightly thickened to approximately 80 inches (2 metres). It was an unfortunate and wasteful irony, that as the seam deepened and thickened, the overall percentage of coal taken had become less!*

\*

In underground mining and development, whilst working in a seam such as the Dysart Main, it made good sense to intrude into a new section by driving levels that actually were level and followed the line of the coal, or parts thereof that were being taken. This would sometimes mean that over a given distance, a level mine in order to remain nearly level, would have to be turned slightly left or right in order to maintain the integrity of its level plane. Better a haulage-way with a few small turns along its length, than a haulage-way that undulated or had several reverse slopes. This latter made for difficult and sometimes dangerous transport systems.

*(Readers must bear in mind that very few levels, headings or dippings were initially driven for the specific purpose of getting immediately to the extremity of a section area or coal-field, except where the original plan was to extract the coal on the retreat as in the East-side mine - and on the very last power-operated coal-face extracted within Lochhead pit. Many early developments were approached on the 'suck-it-and-see' principle, the over-riding premise was to 'take coal!')*

In the Lochhead pit area and in working both the Dysart Main and the Lower Dysart seams, reference will be made to two other shafts that were pits in their own right in the past - and two shafts that was only ever used as air outlets. They were, all four, used as return air outlets to rid Lochhead workings of its stale and sometimes foul air. i.e. The Victoria pit at West Wemyss (after 1947), the stair pit at Earlseat and it's neighbouring shaft and the Duncan shaft at the North end of the Checkbar Road, where it meets the Standing Stone road. Warm, fouled air from Lochhead pit was expelled from all four shafts *(later three)*, sometimes assisted by booster fans strategically placed in distant sections to boost the underground airflow.

Lochhead pit was basically a dry pit and with the exception of some parts of the New Dook, it could be a cold pit. Many of the fit and hardened miners *(collier's and strippers)*, actually wore their woolly jerseys while they worked, or even their jackets in Winter, where the blast along the Lower Dysart haulage-way *(within the Dysart Main coal and once named Nicholson's Level)*, would sometimes freeze the puddles. Any water that gathered in the pit, especially at the bottom of developing dooks, was merely pumped up to the nearest level, where it flowed into the cut gauton *(narrow and shallow ditch)*. It then ran away and down to the next lower level or dook. *(Do bear in mind that most levels were driven to the slight rise where possible, to gravity assist the transport of loaded hutches back to the pit bottom.)*

There was never any water pumped to the surface via the shaft, during my time at Lochhead. Water from both the Dysart Main and Lower Dysart workings, eventually gravitated to the workings of the Michael Colliery at East Wemyss, via the twin levels cutting past the bottom of the Old Dook, which was named the Michael level. These twin levels ran from the Twelve Hundred level on the Victoria's Sea Dook *(well below the level of the pit bottom),* past the bottom of the extensive workings of Lochhead's Old Dook and terminated in a level, near the shaft of the Michael colliery at East Wemyss. *(The New Dook in Lochhead pit was not directly connected to the Michael levels, but through connections were made via the deeper, 540 metres long, parallel headings.)* Along the length of the Michael levels, there was a gentle downwards slope nearly all of the way from the Victoria Dook! The depth of these connecting levels, reference to M.S.L, starting at the Victoria Sea Dook were:- Victoria Dook:- 1775 feet, Lochhead Old Dook: - 1782 feet, Michael levels:- 1784 feet.

*

Author's Note : *In order that the reader may better understand the relative depths of various underground workings and the difference in relative depths, I shall now revert to the surveyors ploy of using the common 'datum' measurement with regard to all under-ground depths/heights. This makes for easier arithmetic calculations, in that all heights are at a 'plus' to the 'datum plane'. This means that readers do not have to look at a map, find the contoured height and calculate a depth from the surface.*

*To initiate a Datum plane, a mythical horizontal plane is 'floated' or 'laid' under-ground at a given depth. In this case, at a point exactly Ten Thousand Feet (10,000') below Mean Sea Level. This then, is the start point for vertical heights. (The discerning reader will immediately realise that provided the mine workings do not go below 10,000 feet, then every datum height given, shall be at a 'plus' value) The beauty of this system, is that there is no need to take into account the changing contours value of the surrounding countryside above, and all negative arithmetic calculations are summarily banished, (except for the surveyors!).*

*All underground heights are therefore given as a 'plus' value reference to this datum plane - and to find the difference height between two given places, merely requires the subtraction of the lesser from the greater.*

*Therefore, if a height (or datum level) of 10,000 feet is read from part of an under-ground plan, that simply means that the height equates to mean sea level, but below the surface of the ground at that point! If the height of the ground surface shewn on an Ordnance Survey map at such a point is 231 feet, then the mine Datum height at that point would be 10,231 feet! - but only for that coalmine!*

# Initial Intrusion at Sixty-three Fathoms.

The first developments from the new pit bottom at 63 fathoms *(from a surface height of 146 feet above M.S.L.)*, took the form of two levels, the first level going due north at Zero degrees for approximately 25 metres, then taking a right turn to continue on at bearing of 30 degrees. This level roadway was driven through the middle of the seam, and eventually to a length of approximately 975 metres from the Pit bottom, gaining only 12 feet in height over that distance! It struck up hard against the Buckhaven Fault *(The Dyke)* in April 1904.

The second level was driven exactly due south for a distance of 50 metres, where a proliferation of parallel roads were driven short distances. The long level that was eventually driven, was on a bearing of approximately 220 degrees, almost in the opposite direction from the north pit bottom level. This long level to the south would eventually reach a distance of 1625 metres from the pit bottom, stretching out to the boundary line with the Francis Colliery. This long level, driven slightly to the west of south, rose only 13 feet in height over it's full length.

For the want of a proper name, I shall describe these two levels, cut in the middle of the Dysart Main seam, as the **Pit Bottom Levels,** for the reasons that they constituted one long level from the Buckhaven Dyke in the North, to the Francis Colliery boundary to the south *(with the actual Pit bottom and the head of Nicholson's Dook in between)*, and also, so as not to confuse this **Pit Bottom Level** with the consecutively numbered levels to be systematically developed much later, in both the Old Dook, the New Dook, and later still, the Victoria Dook.

*(Part of this pit bottom level to the south-west on bearing 220 degrees, would be named **Nicholson's Level** (not to be confused with **Nicholson's Haulage-way**) and later, would become and be re-named the **Lower Dysart** haulage-way, even though it was cut through the middle part of the Dysart Main coals.)*

The first coals to be taken in the **Dysart Main,** were all to the **West side** of the **Pit Bottom Level,** where the Dysart main seam rose at a near steady incline of 1 in 5. The first section to be developed, was along a heading approximately 350 metres from the pit bottom on a bearing of 315 degrees. This section covering an area measuring 450 by 400 metres, opened up in October 1891 and the coal that was taken was probably the Myslen, the Grounds & Nethers and much later, the Sparcoal, the Toughs & Clears. Conditions at that time and the method of extraction used, would suggest that after the first 'sandwich' of coal had been taken over a wide area and, fully 'packed' using regular pillars of wood and waste stone, the whole section would have 'weighted' evenly without the roof coal breaking, and thus the return visit to take the Spars, Toughs & Clears would not involve working to any 'height!'
The Head coal was left as a strong untouched roof and the Coronation coal underneath was also left untouched. There was, just to the north of these workings, the **Earlseat, Stair-pit, air-shaft** at a depth of 140 feet from ground level, located

just 100 metres to the north of the Standing-stone road and sitting in the middle of an open field. There was also another air-shaft situated to the west of this section lying just 25 metres to the south side of the Standing Stone road at a depth of 190 feet. The linear distance between these two air-shafts was only 170 metres.

The first coals that were extracted from this section *(as mentioned)*, probably the Spar, Toughs & Clears, were exhausted by February 1896. The methods used to 'take' the coal, were probably levels and headings with long-wall pick places and wheeled braes, using hand-operated boring machines and slow, rough, blasting powder. *(This section was re-worked in 1924 and 1928, where the miners probably took the Grounds & Nethers along with the Myslen splint. It was abandoned as a collapsed, worked out section in the Summer of 1928, with no through roads remaining.)* This worked-out section was later by-passed by a circuitous roadway to the east, then north, then west to link up with the Earlseat air-shafts in 1896. This large section would have yielded approximately 375,000 tons of coal in total.

The Pit bottom level to the south, driven around September 1892, did have a rather indeterminate start in there was not just one level, but approximately four,different, nearly parallel levels for a short distance that converged into three, then two levels, before driving on to the south-west over the next few years. This was probably more to do with ventilation into the next 'intrusion', than indecision on the part of the planners, and superfluous roadways could always be 'packed' *(stowed tightly with waste/redd)* and closed if so desired.

The next section to opened-up near the pit bottom in late 1892, was by three, separate, short intrusions from two of the above levels, only 125 metres from the pit bottom and this was to open up a section having an area of approximately 65,000 square metres or 6.5 hectares *(13.27 acres)*. It was oblong shaped and measured 200 metres by 325 metres. The coal to be taken was directly to the west of the pit bottom and was probably the Grounds & Nethers and the Myslen. These named coals were extracted fully throughout this section by 1896. The area was re-visited again shortly afterwards, where only about 50 per cent of the Toughs & Clears and the Sparcoal was extracted up to 1900, when the section was officially abandoned, leaving the Head coals untouched, along with the bottom Coronations coal. The amount of coal extracted would have been in the region of 150,000 tons.

At a different point along this south level around 1893, at a point about 450 metres from the pit bottom, another pair of short intrusions to the immediate west opened up another area, this time measuring approximately 225 metres by 375 metres. Again, the coals taken were the Grounds & Nethers and the Sparcoal, followed by the Toughs & Clears and the Myslen. This section area would have yielded an output of approximately 225,000 tons of coal and was abandoned by 1904.

The practice of extending the 'taken' areas all around the base of the pit bottom *(out-with the circular area of the Pit bottom solid stoop)* into different sections, did have its advantages, in that as more sections were opened up, more and more miners/

colliers could be employed with the subsequent gain in coal production.

*(The initial development of Lochhead pit can be likened to a very slow moving explosion, emanating from the Pit bottom centre point and slowly expanding in all directions, although, not at a consistent or uniform progression.)*

One of the main disadvantages was that the further a section was from the pit bottom, the greater was the transport problem, with the inevitable rise in production costs. It must be realised of course, that none of this could have happened overnight. No matter how modern the machinery or the methods of extraction, mining is inevitably a slow on-going process, with the prospect of millions of tons of coal there for the taking. In the case of Lochhead colliery, I am positive that records will show, that there was more quick coal mined from Lochhead during the last twenty years of it's life, than there was during the previous sixty years and that most of this coal was extracted well away from the pit bottom.

At a point along the pit bottom level to the south, just short of the head of Nicholson's Dook, about 60 metres to the north, there ran a fault line that lay approximately in a north/south direction. This fault started just above No. 1 level *(the Victoria level)* between Nicholson's Dook and the head of the Victoria Dook and finished at the Buckhaven fault in the Earlseat pit workings to the north of the Earlseat stair pit. *(Return airshaft.)* In the area of the triangle formed by the pit-bottom level and the two sides formed by the two faults, all of this area was worked out taking the chosen coals in the years from 1890 to about 1910. The size of this triangular area, measured 1700 metres along its base and 1150 metres and 1050 metres respectively on it's other two sides. This would amount to an overall area of approximately 600,000 square metres *(60 hectares or 146 acres)*. The taken coals were again were the Grounds & Nethers and the Myslen, probably on short long wall runs with 'pick-places', cut or picked-out to the rise. *(The slope of the strata was fairly uniform at 1 in 5.)* These runs *(coal-faces)* would have been advanced side-ways, without the use of isolated or regular coal stoops, which if left, would have resulted in a series of broken Head coal in the roof, which was not desirable if one of the bottom coals were to be taken at a later date. The system of roof supports in use at that time, would have seen the regular building of matrixes of wood and redd *(waste stone)* pillars over the whole length of the face-line as it advanced, thereby allowing the Head coal roof to weigh and 'bend' in a controlled fashion, without actually breaking up. This, when 'weighted, would allow a later intrusion over the same ground, taking coals such as the Toughs & Clears and even the Sparcoal, if desired. This method allow for the extraction of a thickness of at least 11 to 13 feet of good coal over most of this large area. A conservative estimation of the amount of the coal extracted over this 20 years period, would have approached 1.5 million tons.

*(Within this triangle, there were eleven separate and distinct sections of various sizes, with only small stoops left between them. The reason for this, was probably to allow a single section to collapse and close completely, so that it may be forever*

*sealed, thereby helping to ensure an unrestricted and untainted air-flow into the next succeeding section. Where I have lumped a few close sections together in a given area, simply means, that the coals extracted and the mining methods used were similar.)*

At a point on the pit bottom level *(south)*, approximately 100 metres beyond the head of Nicholsons Dook at about 1100 metres from Lochhead pit bottom, there was driven to the west and to the slight rise, a straight, double heading approximately 435 metres long. It commenced at a datum height of 9780 feet. It then took a right-angled turn for approximately 200 metres, where it joined up with twin dooks coming down from the Duncan pit, which were 525 metres long, and had previously driven in a southerly direction from the Duncan pit shaft. The twin dooks had been commenced at a datum height of 9950 feet and declined to a datum height of 9865 feet, a drop of 85 feet over a length of 525 metres, where it met up with the 635 metres long, dog-leg heading from near the head of Nicholson's Dook. This twin dook/dipping from the old Duncan shaft was in effect, a rather shallow, double dook with a declination of 1 in 20.

I make mention of this connection for the simple reason, that I am not totally convinced as to how the coal from this largish area was brought to the surface. I am convinced of course, that it was brought to the surface, but I would hazard a guess that most of the coal would have been drawn up Lochhead shaft. Lochhead cage was larger and probably faster, the pit bottom probably better manned and though farther from the section, it was on a slight down-slope, always a favourable plus in any transport equation. The area in question was a regular, diamond shaped area, but lying on its side. It was bounded by the pit bottom extended level and the Francis workings on the left and the Nicholson's fault and Earlseat workings on the right. It measured approximately 1600 metres in diameter across its acute angles and approximately 750 metres in diameter across its obtuse angles. The size of the extracted area was approximately 595,000 square metres *(59.5 hectares or 145 acres)*. There were approximately fourteen distinctly separate, but adjacent sections in this area and again, there was quite a difference in the choice coals taken and the extraction methods used. The coals taken were mixed lots, in that some parts, the 55 inches of the Coronation coal was taken in addition to the Toughs & Clears and the Grounds and Nethers, but in other parts, the bottom-most Coronation coal was left untouched. The area where the Coronation coal was also taken, was to the southern part of this diamond area, where it bordered with the Francis colliery workings. It must also be remembered, that this was the area of the seam that was at it's closest to the surface within the Lochhead/Duncan pit boundaries. The depth of the Dysart Main coal in this area was at datum height 9850 feet. *(This would equate to an underground depth of approximately 52 to 53 fathoms from the surface)*

The approximate amount of coal taken from this diamond shaped area would have probably exceeded one million tons from 1901 to 1915, including that taken with the extensive re-visitation around the 1930's.

It is to be noticed that most of the coals taken from these last two large areas, included the Toughs & Clears, the Myslen, the Grounds and Nethers and very occasionally the full thickness of the Coronation coal. These coals were extracted nearly to the maximum, with the Head coals and sometimes the Sparcoal left intact to serve as a solid unbroken roof. The method of stoop and room extraction was not used to any degree at this relatively shallow depth and usually only applied where an intrusion into a specific area was initiated. *(New permanent development.)* The long, adjacent and parallel stoops were always left in situ to maintain the integrity of the 'permanent' roadways. The farmlands on the ground's surface immediately above these extensive workings *(considering their shallow depth),* were allowed to subside over a long period of time, which of course, would show no discernible damage to ploughed fields. On the ground's surface, a farmer could look out from his farmhouse windows and survey his expansive lands with an unbroken view *(and perhaps a jaundiced eye)*, and never realise that nearly the whole of the surrounding countryside had dropped 10 to 12 feet. The areas effected sank like stones, which of course, was preordained and according to plan, considering the type of extraction followed. This same farmer may just have noticed that his fields somehow undulated more than they used to, but I don't think that he could have placed his hand on his heart and sworn to it! Subsidence on the surface, did not for some reason, follow exactly the perfect area of underground extraction. Something to do with the formation of the strata, I would imagine!! This most certainly was bound to have happened in the area from Buckhaven to the Boreland and between the A955 and the A915 roads, the main road coast route and the near parallel Standing Stone Road. This wide area was the site of some of the most comprehensive underground extractions that took place within the boundaries of Lochhead pit. Only in one other part of the pit was so much coal extracted from one section, so much so in fact, that I have devoted a short exclusive chapter in this book to the after-effects of such an extraction. An area by name, that some ex-Lochhead miners think of with a great knot in their guts, an instinctive feeling of grave disquiet, and a sinking heart! The 'Baund' Section . . . !!

*

Before I go on to describe the driving of three of the main Lochhead Dooks and the extractions therein, there was one other area where numerous extractions were carried out and curiously enough, made to the dip along and below a long straight level. The level in question, was part of the long pit bottom level between Nicholson's Dook and the extremity of this level to the south and west *(the Francis Boundary),* a distance of 700 metres.

There were approximately 21 short intrusions made to the dip and into the several sections below this pit bottom level, and above the line of No. 1 level from the Old Dook *(not yet started)* to the head of the Victoria Dook. The amount of coal taken from these 21 intrusions is indeterminate, for the reason that I simply not know how the coal could have been got out in quantity from so many short dipping workings. *(Coal could not realistically have been brought uphill, without the use of much needed horse-power!)* I feel that most of the coal from this section area, would have eventually progressed downhill to the No. 1 level at a much later date, where there was eventually installed an endless rope haulage-way, between the head of the Victoria Dook and the mid-point of Nicholson's Dook. This short haulage-way later became known initially as Nicholson's haulage, and then as the Victoria haulage. The coals taken from the sections along the narrow top half of this large area, were the Toughs & Clears, the Myslen and the Grounds & Nethers. The Sparcoal, lying directly under the Head coals and the bottom-most Coronation coals, were left untouched.

# In Summary

The mineral area of the Dysart Main seam covered by the Lochhead workings, stretched from the Francis colliery in the south and west, The Randolph colliery in the west, Earseat workings in the west and north, Wellsgreen colliery workings to the North on the top side of the Buckhaven fault, *(The Dyke!)* and the Micheal colliery workings to the east and north-east. At the turn of the last Century, *(1900)* the Dysart Main seam was worked fully to the north and west and to the rise, by methods that took the coal in one, but more often, two or three separate leafs. Very occasionally, the Head coal was taken just after the Sparcoal and the Toughs & Clears, but in this case, the Grounds & Nethers would be left, with every likelihood of the Coronation coal being taken at a later date. The extractions from the Dysart main seam was the most comprehensive working ever to have taken place in Lochhead pit, with something of the order of 90* per cent of the above-mentioned areas visited and re-visited by one or more types of extraction!

Where limited underground workings were permitted under villages or farm buildings, the method of extraction was always Stoop and room *(Pillar & Stall)*. In the pursuit of the extractions from under the whole of the area covered by the village of Coaltown of Wemyss, the method employed was very regular Stoop & room with only limited coals taken. It is interesting to note, that the limiting factor in these extractions saw only the Grounds & Nethers, the Myslen and the Coronation coal taken from under the village, and then only by regular intrusions in the form of a regular matrix, where the parallel levels and headings were only 12 to 14 feet wide with regular stoops at least fully 30 metres square. This matrix would have suffered only one intrusion/extraction, never to be visited again.

A point to note, is that additional strength was left in the roof, with the leaving of the Head coals, the Sparcoal and the Toughs & Clears, giving a thickness approaching 10 to 12 feet of solid coal. This area would not suffer to any extent from subsidence, as the large remaining stoops directly underneath the village buildings would more than likely absorb the additional weighting without crushing. *This does not mean that 90% of the available coal was extracted!*

The one area that could not possibly suffer from any form of subsidence, was the large circular stoop left untouched under Wemyss Castle and to a lesser extent, the Home farm. The Castle and it's grounds were protected from 'possible underground movement' by a whacking great exclusion area in the form of two interlocking circles, each measuring 400 metres and 300 metres respectively in diameter. There was absolutely no coal whatsoever, mined from under Wemyss Castle within this exclusion area. This did not fully apply to the Home farm, where a limited amount of coal was taken, again in the form of a regular matrix and again leaving a solid coal roof of approximately 9 to 11 feet in thickness. The one overall difference being, that the retaining stoops measured approximately 30 metres by 60 metres. As a point of interest to those old ex-miners who can recall working on No. 4 level *(west)* on the Old Dook, the three parallel levels that ran from the Old Dook and through the Victoria Sea Dook and on to the Francis boundary *(aka the Michael levels)*. These parallel levels ran directly underneath the Home Farm buildings.

In general, the further inland was the seam, the nearer it was to the surface. This in theory, meant that there was less of the weighting factor to take into consideration and subsidence meant little or nothing, where the surface area was 99.5 per cent farmland. In these areas, the maximum amount of coal seemed to have been extracted, with little regard being paid to the possible depth of subsidence in the surrounding countryside, but hopefully, with due regard to the miners safety. These extensive workings to the west of the pit bottom level, were far enough away from the sea to obviate the need to think of broken strata above the weighted and crushed workings, and irregular subsidence was wholly accepted in the wide areas of farmlands surrounding Lochhead colliery.

(*One only needs to travel along the old Roman Standing Stone road (A915) from the Boreland (near Gallatown) to Cameron Mine (Buckhaven), to fully appreciate the regular undulations over it's straight length!*)

As the workings and extractions neared the coast line and indeed, ventured out under the sea, much thought must have been given to the type of extraction that would maintain the integrity of the roof and not cause undue weighting and crushing, with the increasing possibility of damaging fissures extending upwards towards the sea-bed. As the Dysart Main seam extended out and under the Firth of Forth and dipped as it did so *(as did all of the other seams),* so did the sea get deeper. The seam did flatten out, as did the sea bed, so therefore, the thickness of the intervening metals/strata did remain fairly constant. The one imponderable that had to be taken into account, was that in comparing strata to sea-water, strata when intruded into, provided that sufficient width of stoop was left, could be depended on to support a narrow intrusion, *(tunnel)* due to it's inherent strength. No such thing could be said about water. It is heavy, it is fluid, it is moveable and it is under unrelenting pressure! There is no support in sea water! Water under pressure, will find the least crack or fissure and once having found it's way into a tunnel or mine, could well prove to be the beginning of the end. The planners, in thinking up ways to take this coal *(Dysart Main)*, decided that in the Victoria pit, as the sea dook plunged even deeper, the method to be used was to strictly conform to the accepted premise of the regular matrix, with larger stoops to be left in situ, somewhere in the region of 40 metres wide and 60 metres long. The coals to be taken, were somewhat different to that in other parts of the 'field', in that the seam had thickened as it deepened out under the Forth. The named coals taken, were usually the Grounds & Nethers and the Myslen, but additional thin layers of coals had appeared between the Grounds & Nethers, measuring 10 inches and 12 inches respectively, making a total extraction thickness of approximately 9 feet of coal, *(containing about 10 per cent stone)*. This left a good strong thickness of approximately 10 to 12 feet of unbroken coal overhead, with the useless Sclits and the increased thickness of the Coronation coal underneath. This limited method of extraction was rigidly adhered to, in that any deviation would cause unnecessary weighting and unwanted crushing, with probable cracking of the overhead strata, which in turn, might bring in the sea!

If any reader would care to make the necessary calculations as to just how much coal was actually extracted by this method from the deepest parts of the Victoria Dook and exactly how much coal was actually left, that reader would then probably arrive at the same conclusion as myself? *(Would 8 to 10 percent be an exaggeration?)*

# Conclusion!

This coal is now lost forever! The great stoops are probably still standing, an ever-lasting tribute to the cannie planners and the wary mine managers, but the dark waters of the River Forth just may have pervaded the exploratory mines and levels, headings and dippings, that took relatively small amounts of coal from perhaps the greatest seam of coal in the country! I estimate, that the amount of coal taken from the Dysart Main seam as worked from the Victoria pit via it's sea Dook and from the more extensive workings in Lochhead pit, could not have amounted to more that 12 to 15 percent, even by the most optimistic of estimations. I would not be in the least surprised if it only amounted to perhaps 10 or 12 percent!

This colossal seam of beautiful coals is lost forever in the general area of the East Fife coalmines and the miners who worked the coals are all but gone. Those old collier's and coalminer's that are still alive, are well into their seventies and eighties and in some cases, can not even remember the names of the taken coals, even though, there is still the odd old collier, who can look at a piece of coal, feel it, and correctly state that it came from part of the Dysart Main!

Tailpiece:- At this date in the year 2001, there are very few coalminers left who are still capable of remembering their experiences at the coal of a 'pick-place', or on an 'aeroplane brae, or an old Stoop & room working within the Dysart Main at Lochhead pit. Those miners are now very thin on the ground.

I am however, pleased to report, that most of the old miners *(several in their mid and late eighties!)* whom I have interviewed to date, appear to enjoy a reasonable health with adequate strength, but a general weakening of the body *(but, not the spirit!)* - and perhaps a quality of life that just may have stemmed from the very active and demanding work that was asked of them during their years in the coalmines. However, I must make mention here of the large numbers of ex-coalminers who, having spent most of their working lives in the 'pits', have developed incurable pulmonary diseases and additionally, some of those who enjoy a somewhat less than joyful existence, being almost permanently attached to either a nebuliser, or an oxygen machine.

I can therefore say, that old 'pick-place' collier's and their ilk, who still retain good health, a small degree of strength and a measure of mobility, can be likened to the old Scottish heroes of times long past - 'Gey few an' maistly deid!' *(Limited numbers and mostly dead!)*

\*

Author's Note:- *Since the writing of this chapter, it has been suggested to me that the salt waters of the Firth of Forth never did permeate the workings of Lochhead Pit or the Victoria Sea Dook at any time during their operational life. Having worked down in the depths of two of the three Dooks in Lochhead pit, I can confirm this, at least until 1960. I cannot speak for the Victoria pit, even though from all reports that I have heard to date, this also would seem to be the case. Also, I can find no evidence whatsoever, that sea water ever ingressed into the lower workings of Lochhead pit (below sea level), even though all of Lochhead's relatively small amounts of land water, was allowed to gravitate to the Michael colliery via the Michael levels.*

*Sea water from the Firth of Forth did find it's way into the deep workings of the Micheal colliery, the Francis colliery and the Seafield colliery. (And possibly the Victoria pit via the Francis colliery.) It has also been suggested to me, that since the closure of the Micheal pit in East Wemyss, (and hence it's pumping operations) that the egress of acid/alkaline tainted water from the old workings of Lochhead pit into the Micheal and Victoria workings, has now mixed with the rising salt water and has reached near maximum proportions over the last 30 years, with very little underground trapped air being allowed.*

*The land water that seeped into some of the workings of Lochhead pit would have slowly accumulated over the years since it's closure, and then gravitated into some of the upper workings of the Micheal pit and into depths the Victoria Pit Sea Dook, before mixing with the incoming sea water, with the inevitable flooding of Lochhead as a final resort. (The Victoria Sea Dook would have commenced to flood as soon as the waters reached the Micheal levels.) I also feel that the relatively small amounts of water that was present in Lochhead pit, would be far outweighed and overcome by the substantially greater volume of salt water now pervading the workings of Lochhead from the now defunct Micheal Colliery.*

*The inherent, but slightly convex slope on both the Lower Dysart and the Dysart main seams throughout the coal-field, would have exacerbated and ensured a perfect flooding, with the higher levels of the abandoned workings of Lochhead pit being the last to flood, up to M.S.L. at least.*

*

Author's Tailpiece:- *Strictly speaking, the Laird's coalmines/dippings at Earlseat have little bearing on this narrative, even though one of Lochhead's escape shafts could be said to be in the Earlseat area. This under-mentioned information is of great interest for the reason that in the whole of Scotland, there does not seem to be any other location where this rare geographical occurrence was realised and then profitably exploited! The following extract comes directly from Andrew S. Cunningham's book on R. G. E. Wemyss, published in 1909.*

*'At the beginning of 1902, developments were pressed forward at the Michael Colliery, the same course was pursued at Lochhead and in 1904, the opening up of the northern section of the Wemyss coalfield was commenced at Earlseat, near Thornton.*

*Earlseat may be said to be the apex of the Wemyss, Dysart and Balgonie (coal) fields, and five seams of coal, including the Dyasrt Main converge within a comparatively small area. From the top of the ridge, the Dysart Main (seam) dips in different directions.*

*This situation gave the Laird of Wemyss (Randolph Wemyss) the opportunity of founding a colliery unique in the (mining) history of Scotland. Full advantage was taken of the peculiarity, and in place of (vertical) shafts, five 'day-mines' were started, and are being run through the fields for distances of from a mile to a mile and a half. The results quite justified the rare innovation.'*

In addition, drift mines/dippings continued to be the best method of taking the coals near this location, within this abnormal confluence of the rich coal seams. With the coming of the A.B. Short-wall, coal-cutting machines in the early Fifties, this opened up a whole new vista of coal-getting within the outcropping Lower Dysart seam *(twenty metres below the Dysart Main at this location),* in that the final dipping mine at Earlseat under the managership of Jack Kennedy, was operated using only A.B. short-wall machines, jigging-pans and a 30-inch wide endless conveyor belt to bring the coal up the mine, tumbling the taken coals directly into 20 tons capacity, coal wagons waiting in-line at the mine head.

The mine was so designed and operated, so that there was a continual series of parallel levels up to 150 metres long, being driven to the left and right of the main mine, where all the levels were advanced *(lengthened)* using short-wall machinery to the maximum length allowable, by the jigging power of a strong pan-engine train. *(These levels, to the right and left of the main dipping mine, were commenced and driven so that the 'levels' followed a slight inclination, thereby favouring the 'jigging' action of the pan-engine trains!)*

At the event, where the pan-engine could not cope with the length and weight of the coal load, the machinery was withdrawn and subsequently re-installed in the next new level to be commenced, at right or left angles from the straight, main mine that followed the downwards slope of the seam! This was very fast and modern 'Stoop & room' type of extraction, with only the minimum distance *(solid coal stoops)* being allowed *((left in situ)* between successive levels. *(This was probably the nearest approach to the 50 percent extraction that could be maximised with Stoop & room mining!)*

See chapter: - 'Out of the East!' for comparative extractions.

*

Author's Tailpiece II :- *I was employed in this mine for several months before I joined H. M. Forces. It was at this period and using this machinery, that some miners wages grew to be somewhat respectable, with especially fit and strong colliers able to extract the maximum yardage on a weekly basis for many, many months, before their overall endurance began to suffer!*

*Earlseat Mine, in its relatively short life, probably had one of the best output records of any colliery within the Fife Coalfields, in that the coals extracted must have been the highest ratio anywhere, of 'Tons produced to number of men employed!'*

*Literally: - "Multum in Parvo!"*

*

# Chapter VI

## "T-a-l-l-y C-a-r-b-i-d-e !!"

## Pete Dickson - *and the 'Killing Ground!'* *(From Horse & Cart to Bantam Karrier!)*

On his return to civilian life after war-time service in the Army, Peter Dickson immediately picked-up his previous self-employed occupation almost immediately his great store of trade-goods remained almost completely intact and securely locked away for the 'duration' in his heavily padlocked, thick, stone-walled warehouse within a high walled compound known as the 'dairy'. His good wife had kept the business alive in that she 'served' anyone who came to her door looking for goods. She would make herself available to needy customers requiring goods or 'tally-carbide' almost at any time outside of sleeping hours. Mrs. Dickson did keep Pete's business alive during the war, for the simple reason that he had had the foresight to stock the items that was needed to sustain the miner's families for a long period of time! To be allowed to visit and view this warehouse, was like browsing in an over-stocked Aladdins cave! His main stock was reckoned to be at least ten times that of his business cart's capacity.

Peter Dickson was a man of 'letters'. Unfortunately, most of these letters probably went before his name and formed the type of adjective that is wholly unprintable. He was a seller, a tout, a tradesman, a merchant, a salesman and a hawker extra-ordinaire, who did hold an official 'Hawkers' license. He was clean, tidy, daily-shaven and close-cropped. He stood 5ft -3ins tall with a strong stocky build. He was smartly turned-out in highly polished, brown, leather boots, polished leather leggings, whipcord breaches, a khaki shirt and tie and a brown ex-army leather jerkin. He was very smart, quick of movement and bright of eye! He was a veritable 'pocket dynamo!' He was also, rightly or wrongly, sometimes accused of being a mischievous rogue by his many customers. As to whether he practised outright chicanery in his dealings with people, I am simply guided by the opinions of some of his late customers. Sufficient to say - and to quote an old Fife expression: "ye wad need a gey lang spin tae sup wi' Pete Dickson!" *(A person would need a long spoon to dine with Pete Dickson!)*

He was one of the very original 'Tally-Carbide' men. His raucous, but resonant hawking cry was indeed:- **'T-a-l-l-y  C-a-r-b-i-d-e ... ! ! !'**, which he used to good effect, to announce his once-weekly presence to one and all in every street that he visited. *(Along with his earlier horse and cart of course!)*

Part of his home 'patch', which he guarded jealously and robustly, covered an area of approximately one square mile and roughly enclosed the residential area from Denbeath bridge to the Buckhaven Co-op, and Wellesley road to Den Walk, Buckhaven. This was an heavily populated area consisting of seven double streets of double-storeyed, tenement houses and, many more streets of semi-detached blocks of houses, quite closely run together towards Buckhaven. This area was known as the Denbeath Rows, originally built to accommodate the influx of coalminers from Lanarkshire and Ayrshire, serving the Earlseat and Wellsgreen collieries. Most of these houses were occupied by miners and their wives and often, large families *(miners were always lusty men),* and their daily needs were many and varied.

Peter Dickson would trade in anything and everything that a mining family would and did need. It was not foodstuffs or clothing *(that were still in short supply and, on coupons)* that Pete traded in, but every sort of kitchen ware, kitchen supplies, hardware, tableware, candles, paraffin, tallow, carbide *(for miners flame lamps)* and even limited quantities of electrical goods. Give Pete a single week, and he could probably lay his hands on anything that was available in the country at that time. *(Such were his connections?)*

How Pete Dickson was paid for most of what he sold, was always one of the great mysteries of his profession. He was very rarely paid in cash, the full value of individual items, very few wives could afford to do so, the last items purchased were still being paid for on a regular basis. As was the way of his business, everything was bought on 'tick'. This system worked a treat for those miners' wives whose husbands were in steady work, as most men were. Each wife's 'worth' was carefully assessed by Pete, and a customer would be given an unofficial credit rating according to Pete's deliberate calculation of her husband's weekly earnings. *(Pete knew this to the last penny! He could reel-of from memory, the basic weekly pay of strippers, brushers, machine men and preparatory workers. He even got to know the amount of overtime that some oncost workers were paid in a given week.)* No matter how much a wife might plead for an additional item over and above her 'credit', Pete's benevolent heart would turn to stone! A typical example of Pete's 'benevolence', was exemplified during one unsettling incident that took place in one of the Denbeath Rows, where a miner's wife had purchased an 120 volt H.T. dry battery for the family wireless set from Pete, on one Friday evening. On the following Friday evening at his next visit, she told him that she could pay her 'weekly' dues that week! Pete immediately left her at his cart, stormed into her home, made a bee-line for the wireless and pulled out the H.T. leads from 'his' battery, and then returned it to his cart with the words, "You can have it back next week after you've payed me!"
*(Pete was sometimes a mite high-handed!)*

Pete's 'patch' also included the Rosie, Macduff, East Wemyss, the Coaltown of Wemyss and West Wemyss, all of which he visited on a weekly basis. This was the western limit of Pete's round and regular as clockwork were his 'welcome' visits. Pete Dickson was up at the crack of dawn every morning except Sundays and could be seen on the open road returning home at dusk every evening, holding a tilley lamp in one hand and leading his tired horse by the other.

*

Within two or three years after the end of W.W.II, Pete's business had flourished so well, that he could afford to have custom-built - a specially adapted 'Bantam-Karrier' motorised vehicle, so that he could dispense with his overloaded horse and cart! This vehicle was a two-axled, six wheeled, load carrier with an enclosed body sitting on small road wheels, that was fitted-out inside like the interior of a multi-shelved mini-market! Entry to this vehicle was by a narrow side door at the rear of the vehicle, where potential customers could view all of its contained wares, but were not in a position to touch! (Pete Dickson was not altogether, a very trusting soul!) This Bantam Karrier, was, to say the least, so grossly overloaded that travelling at speeds of over 10 m.p.h. was entirely out of the question! Any faster and Pete found that some of his stacked ware were continually ending up in the middle passageway of the vehicle. This did not worry Pete unduly, it was usually very rare for him to use anything other than first and second gear in any case, with his customers being in such close proximity to each other! This vehicle was a godsend to Pete, in that he did not have a horse to feed and tend to, after a tiring day at the 'hawking' - and he absolutely hated 'mucking-out' at the weekends! After all, what was the point of all of his military training in the R.A.S.C., if he couldn't put it to good use in civilian life!

After a few months with the new vehicle, Pete found that its use had not really speeded-up his daily rounds, merely that the actuality having to get into it and drive it, was something of an inconvenience. With the horse and cart, he had merely walked on to the next customer's house, with the intelligence of the well-trained horse prompting it to amble forwards without being told! This 'Bantam-Karrier' had no such 'horse-sense'! Where the vehicle did come into its own, was when he was returning home from either East Wemyss, the Coaltown of Wemyss or West Wemyss. He rather enjoyed the slow, relaxing five miles drive with the engine ticking over at low revs and in second gear! The only problem experienced on his rounds, was the time-consuming 'first-gear' haul, up and out of the mining village of West Wemyss. He was fortunate that his new vehicle had to make that journey only once per week.

Peter Dickson even managed to include the village of Methilhill in his weekly rounds, after the purchase of the Bantam-Karrier, where no doubt, he made just as many 'friends' amongst the population of that small village, as he did elsewhere! Many of the local miners resident in Methilhill worked at the nearby Wellesley colliery, where, just adjacent to the Wellesley colliery, at the pit-head baths gate, was situated the extensive warehouse of goods belonging to Pete Dickson. It was therefore of great convenience to some miners, whose wives needed something or other from the great emporium at Denbeath!

I have no idea just how long the 'empire' of Pete Dickson lasted, or indeed, if it still flourished in the light of increased competition from the likes of Carr's ironmongery at Buckhaven, Cassel's ironmongery and hardware shop at Denbeath and Cairns' hardware shop at Lower Methil. Sufficient to say, that Pete Dickson was an entrepreneur of the first grade, who made his 'killing' in the early post-war years of W.W.II - and that he was eminently successful in his endeavours, with of course, the added incentive to his many customers, that the principle of 'Live now - Pay *(grovel)* later' seemingly worked well for all concerned!

*

Author's Note: - *Just when Pete Dickson retired from his lucrative business of 'street hawking' - I have no idea! As to what Pete actually did when he stopped trekking the local streets, I equally have no idea! As to whether Pete Dickson and his lady wife ever had any family, I cannot recall ever seeing, or hearing of any off-spring! That which I do know, is that after the deployment and undoubted success of the Bantam Karrier, he actually bought another horse, but this horse was meant for riding, which Pete seemed to have a particular yen for at the time! What the general public didn't see of course, was the mini-performance that preceded Pete's attempts to mount the tall horse. (It was maliciously rumoured that Pete used a household stepladder to gain the necessary height - after all, he was only five feet - three in his gaiters.) This 'union' didn't actually turn out too well for either man nor horse, in that after a while, both of them experienced an 'estrangement' bordering on an official 'legal separation'- due to several misunderstandings between horse and rider! Sufficient to say that the 'rapport' between man and beast was short-lived!*

*As to what eventually happened to the Bantam Karrier, I did hear tell that he actually drove the vehicle into the ground with much and prolonged hard usage. But it was also jokingly rumoured, that when he eventually sold the vehicle, he seemingly had the temerity to advertise the gearbox as having third and fourth gears as relatively new and barely used!*

*

# Chapter VII

## Dauvit's Progress I.  -  Initiation!
### Young Dauvit goes Forth!  *or, Was he Pushed ?*

At the very tender age of twelve when young Dauvit was newly installed at high school *(and still in short trousers),* - his Mother was reported to have uttered to her spouse 'Old Harry', "Has someone hit that wee bugger over the head with a hammer?  He's certainly not one of us!  For God's sake, look at his older brother, he's only 15 and he is about six inches taller!  And you, Harry, are a big braw man!" *(Dauvit's mother, 'Maidie' by name, stood only 5 feet 1 inch in height.)*  Harry stood a full 5 feet 10 inches!  From the age of 13 years, Dauvit started to grow, and by the time he had reached his sixteenth birthday, he stood one inch taller than his big brother and looked 'Old Harry' right in the eye!  *(Metaphorically speaking, that was!)*  When David *(as he was now called, since he started wearing long trousers)* left High School at the age of sixteen, armed with what are described as Junior 'Highers'.  *(I did not realise that at the time!)*  The question of 'what to do' then arose!  Neither David's mother nor he, had given much thought to this matter *(so David had thought!)* and nothing seemed to be in the offing!  David wasted approximately three weeks at this time, ostensibly looking for work, but in reality hovering around Buckhaven High School on his brand new Raleigh 'all-steel' bicycle *(with the new cable-brakes),* especially around break -time and lunch hours. He was hoping to catch a glimpse of a certain well-built young lady *(blond)* of 15 years who had taken his fancy.  Her name was May M . . . . . . n!

    The first job interview that David attended, was prospective employment as an apprentice draughtsman in the Durie Foundry in Leven.  *(This factory was a very successful organisation, in that they had produced up to one pair of 'Spitfire' wings every week during the latter part of W.W.II.  Although, it was rumoured, that they had seemingly taken all of two years to produce the first pair!)*  David duly turned up for interview one bright Monday morning in the month of May 1950.  He reported at the appointed hour, was given a quick guided tour around the immediate offices, introduced to a few loungers/loafers and was rapidly introduced into some of the peculiarities of the job!  David then found out all about 'pencils', - pencils hard, pencils soft and pencils plentiful!  He also learned all about the umpteen ways to sharpen pencils!  He also learned very quickly that no two draughtsmen wished to have their pencils sharpened in the same way!  And, that pencils under a certain length were no good whatsoever, and don't sharpen away too much wood at any one time! Pencils cost money! By the end of the longish interview, David decided that being an apprentice draughtsman was not all that it was cracked up to be and when he learned exactly how much of a 'monthly' salary he would receive, that put paid to David's aspirations as a potential apprentice draughtsman.

David then did as all sixteen year old youths did when job-hunting, he got on his bike again! Before a week had past, David then got himself fixed up with a small firm of Marine engineers and Welders in Lower Methil. This small business was run by three working brothers who each specialised in different aspects of the business, i.e. welding and engineering, metal turning and boiler-making and pre-fabricating on a middling scale. David was to be employed on the welding side initially. The brother to whom David was to serve under, was named H - - t and he proved to be one of the most dour and uncommunicative of men. He would never speak if a nod would suffice and would never take the trouble to explain anything. If a welding task was in the offing, he would merely grunt and nod in a given direction. From that, David was supposed to know what the job was, where it was to be done and what machine was to be used, and whether it was to be done using either electric rod welding or acetylene and oxygen bottles. Either way David usually got it wrong!

*

The one redeeming factor that always helped to overcome my frustrations with the non-communication factor, was when H - - t was called out to work on a ship berthed in Methil docks. Ship repairs were usually carried out on board the vessels concerned and H - - t would usually go directly to the scene of the repairs accompanied by an apprentice or a helper - in this event, myself. The innards of big ships were quite large, imposing and utterly fascinating to me and quite often my attention would wander and I would get purposely lost! On one such occasion, one Friday afternoon after H - - t had been called onto quite a large ship, we descended into the very depths of the boiler-room, where one of the massive boilers had been taken off-line because of a possible collapse of the bottom fire grating. The boiler was of course, disconnected, the fire drawn, with the fire-box completely cleaned out and cooled. H - - t immediately crawled into the large fire-box through what seemed to me to be a rather small steel doorway and disappeared inside. After about two minutes he called for certain items of equipment to be pushed through the doorway and for me to follow. I was wearing a boiler suit at the time and took the precaution of buttoning it up to the neck and pulled my cloth bonnet well down over my head. I managed to wriggle in through the smallish door and into the 6 feet wide fire-box on my belly and crawled up beside H - - t. He pointed to several of the cracked and broken fire-bars and then actually described to me what 'we' *(him and me!)* were going to do about it! I, for my part was quite chuffed! He had actually spoken to me in words that I could fully understand and needless to say, I was quite delighted.

Whilst I was in this semi-dazed state, I gradually became aware of a rising commotion that seemed to be taking place immediately outside of the fire-box door. H - - t's attention was quickly diverted to the cause of this commotion and with an equally hurried decision, ordered me out of the fire-box with all haste, then commenced to push and pull at my clothing in an attempt to hurry me. We were too late! Suddenly our ears were assaulted by a great whooshing noise and the fire-box was immediately filled with hot soot-filled steam, that scalded and choked and completely blinded both of us. We were caught within this fire-box with both of us spluttering, choking and retching horribly. We made hurried attempts to get out of this cage, but were soon both reduced to gasping wrecks, unable to speak and gasping for air. We were also completely disoriented! Just as quickly as the horror had started, it suddenly stopped, but still leaving the air in the fire-box thick with the choking soot. Within seconds, a different sound was heard, and this had the effect of drawing the thick leaden atmosphere up and away though the flue system, with relatively clean fresh air rushing in through the fire-door.

It was quite a while before either of us could move, as we were still coughing miserably through tight chests and pain-racked throats. We were then forcibly dragged out through the fire-door, where we just continued coughing, spewing and retching! As we both started to recover, H - - t, with eyes still streaming, began to look around him for the possible cause of our near choking to death. It did not take him long to discover the culprit, he was after all, a marine engineer who knew his Port from Starboard. H - - t's eyes grew wide as he identified the cause of our near fatal experience and without saying anything, slowly walked forwards to a tall 'sweaty' fat man who was holding what appeared to be a long tube with a nozzle on the end, connected by a heavy hose to an outlet on one of the live boilers. This big, dirty, sweaty, 'blubbery' man, was stripped down to dirty white vest and stood smirking and posing, with the steel lance held across his chest like a long-staff. H - - t slowly approached this man and from about five feet away, delivered one of the quickest and most devastating kicks to the goolies that I have ever had the pleasure of witnessing. *(The appreciation came later!)* The blow was bang on target and the man literally crumpled. He threw the high-pressure steam lance away from him and hurriedly clasped his nether regions, falling to his knees at the same time. I looked at H - - t, and he looked at me, and he quickly motioned me to help him pick up our work tools where we quickly departed the ship's boiler-room. I certainly revised my opinion of H - - t after that incident. I had though of him as a dry, dour, acerbic and uncommunicative man, and there he, was picking on blokes nearly twice his size and weight, and all because of a little bit of hot, dirty soot!

This 'apprenticeship' had actually lasted for two full weeks. All apprentice boys were paid by the week, but considering the fact that the firm operated a 'lie-week' system - I had to wait two weeks before I received a 'pay-packet!' That

Friday evening, which was the same day as the 'Fire-box' incident, saw me travelling home in a terribly dirty state. I had made to get on the bus at Bayview, Lower Methil, but the conductress took one look at my filthy state and literally pushed me back onto the pavement with the words, "Not on my bus, you don't!" The long walk home was all that I needed!

When I did eventually arrive home, I was very proud of the fact that I had finally had produced some 'earnings' to take home, which I duly did, unopened of course! I proudly handed-over the gummed weeks wages and eagerly awaited my 'pocket-money' in return!

*(What follows next is very difficult to describe, in that it surely marked a 'milestone' in David's relatively short life! David's mothers' reaction was not as he had anticipated. She took one long look at the contents of the 'wage-packet', looked at the dirt and filth that covered David and literally threw the contents therein, at him! David gather up the coins that had scattered over the kitchen floor and rather slowly realised the cause of his mother's opprobrium!)* The total wages for one weeks work amounted to the princely sum of twenty-one shillings and sixpence, which was exactly one half of what my father earned in one day!

This was 'disaster', as I was soon to find out! This miserable wage, coupled with the fact that I was immeasurably dirty, seemingly put paid to my marine engineering career.

\*

My mother said nothing more at that time! She was impatiently waiting on 'Old Harry' coming home from the 'back-shift'. He duly arrived home around 10.45p.m. *(in scrubbed 'life-buoy soap' pristine condition),* to a hot supper of 'soused herring,' and an atmosphere that could be cut with his dinner knife! *(I do remember!)* Old Harry was no fool, he knew something was in the air the moment he came through the back door. Nothing was said, but Harry had to be fed first! Harry finished his supper, looked at my mother, told me to stay in the living room, and disappeared into the kitchen with my mother! The doors were closed! Two minutes later my fate was sealed! My future decided! Twenty-one and sixpence a week indeed! *(I was accused at this time of eating that much food in one day! I was getting to be a 'big lad!')* My father returned first, looked at me, but said nothing! My mother followed, a grim determined look on her face, a pointed finger levelled at my nose, and uttered the doom-laden words:- "Lochhead Pit for you on Monday Morning !" I was totally bereft of suitable words!

Saturday morning arrived, with the same grim atmosphere seemingly affecting all of the members of the household. 'David's going to the Pit' was whispered between my younger sister Marlene and even younger brother, Junior. He couldn't care less, he was only two years old! My older brother Bill was away doing his two years national service in the Royal navy at this time, so, there was no-one that I could 'talk' *(or complain)* to! Breakfast was taken in sober silence, I was totally ignored! Harry having finished his, then merely jerked a thumb in the general direction of the back door and grunted; "Git yer jeckit on son, wir gaun tae Lochhead Pit!". After what seemed to be an altogether swift bus-ride, the Coaltown of Wemyss came into view. A quick exit and a hurried walk down the Pit road ensued, with nary a word from 'Old Hairy' *(as he was sometimes known).* On arrival at what I later found out was known as the 'Checkbox', old Harry had a quick word with Andrew Pride *(one of the male clerks)* and I was deposited inside with a grunted 'wait!' Old Harry then disappeared into the manager's office. He reappeared several minutes later with William Hampson in tow. Without even being looked at, I was signed-up in minutes and given the order to appear at the Pit-head baths before six o'clock on Monday morning, to collect a sixpenny bar of concessionary soap and be allocated a clean and 'dirty' locker. *(No. 346. I remember it well!)* My mining career was about to commence!

On arriving back home, Harry was greeted with an inquiring look that didn't need translating and was completely answered with a silent nod. The look was then transferred to me with the words, 'Your working clothes are laid-out on the bed, try them on!' It seemed to me at that time, that my early fate had been decided on, even before I had left Buckhaven High School.

\*

# Chapter VIII

## The Denbeath Rows
### *'Garden City, or Model Village?'*

The new village of Denbeath was bounded to the North by a long slightly curved, but broken row of terraced cottages, that stretched from Denbeath 'brig' *(bridge)* to the Wemyss Tile & Brick Company, that nearly touched the edge of Stark Street. *(The break in the middle of this long row is 'closed' by a seven feet high brick wall, which stretches for approximately eighty metres. Why there was this break, I simply have not been able to discover!)* This row of cottages was probably built around the mid-1870's, to house the miners who were employed in the original Denbeath pit, leased from the Wemyss Estates *(later re-named the W.C.C.)* and sunk by Bowman & Coy around 1875. *(This pit reverted back to, and was reclaimed by the W.C.C. in 1905 when the lease expired. Two new shafts were then sunk and renamed the Wellesley Pit.)* This curved row of pan-tiled roofed and harling-walled cottages was named* and is still known as 'Cowley Street!' The cottages in Cowley Street did not have any front gardens, as the main doors fronted directly onto the five feet wide, foot way adjacent to the carriage road. The cottages however, did have large back gardens that sloped gently down towards the mineral railway line, that later ran from the Wellesley pit to Thornton. That mineral railway roughly paralleled the back of Cowley street at a distance of approximately 25 metres. To the south-west side of the village, the parallel rows of the Denbeath streets are bounded by another straight row of different type of terraced cottages, with the exception, that this straight row is broken near the middle to allow access to the length of Barncraig Street *(where this author was brought-up!)* Bow Street, Den Street and Wall Street.

*(It may be of interest to readers to note, that three of these street names are directly related to individual coal seams contained in the East Fife coalfield - and, that the fourth name: Bow Street, could be named after the Bowhouse (bu'hoose) seam of coal!)*

This straight, but interrupted row of terraced cottages were of a slightly more modern design *(than Cowley Street),* in that there was a small built-in scullery cum wash-house and inside toilet incorporated into the design. *(The flushing toilets were built into the structure of the houses, but had their own separate outside access doors.)* This slightly shorter row of cottages was named Wall Street. *(Again, another named seam of coal!)* A most peculiar and possibly strange aspect of this second row of cottages, was that they were only 'one-sided!' That is to say, that both the front doors and the 'back' doors and all of the cottage windows, were actually on the same facing wall! The rear walls of all of these terraced cottages, was one long uninterrupted harl'ed brick wall along their complete length. *(Roughcast outside cladding comprising of small chips or pebbles affixed to the brick exterior with 'strong' cement).*

The larger type of 'four in a block' semi-detached and terraced cottages in Barncraig Street, Bow Street, Den Street and Wall Street, had a conventional front door and a oppositely placed back door, which looked out over an equally large back garden. They also has large double windows for each room, allowing much greater natural light to infiltrate into the rooms. These cottages contained living rooms that could be described as large, for the reason that they were a combined living room, dining room and contained two recessed spaces for double-beds, which indeed, most of them did contain! There was one additional double bedroom only, along with one separate toilet *(no bath!)* and a concrete floored scullery with the inevitable coal-fired 15-20 gallons capacity zinc boiler, but with cold mains water only.

At the time of which I describe, from the first occupancy of these houses, cooking was accomplished on a blackened, cast iron and steel fireplace cum cooking range, which was in fact, installed into the wall in the main family room. The coal-fired range stood about 32 inches high and consisted of an upright, off-centre, open fire grate with a large, square, ash-box directly underneath. There was a large capacity oven built in to the left of the fire-box, which was covered by a flat cooking surface, taking it's heat from the oven directly underneath and the open fire at the off-centre. The right side of the range also had a cooking surface which was approximately half the surface area of the left side. This was the fire range that supplied the main living/family room with heat and cooking facilities during the day - and some of the room's illumination during the early hours of darkness.

*(Many coal miners and their wives had developed and practised the very pleasant and peaceful habit of sitting and chatting together in the darkened living room during the long winter's evening, where their only entertainment was nothing more than gazing introspectively into the bright, flickering flames of a warm coal fire - and perhaps listening to the wireless, if they were fortunate enough to possess one!)*

This combined arrangement also had the tremendous advantage of keeping the house at a comfortable warm temperature, even in the depths of winter, with the distinct advantage that this coal fire also made the most beautiful thick slices of 'Po-toast' imaginable. Its main disadvantage, was that it was ugly, dirty, needed constant cleaning and to the discerning housewife, needed to be left blacked out and cold at least one day in the week, to be 'Zebo'ed black' to restore its pristine condition!

*(Zebo:- A jet-black, metallic, bare metal polish that came in a black and gold striped tin with a screw-top seal. It was applied either with a rag or a paint brush and left impossible stains on exposed flesh!)*

When the 'old, ugly, dirty ranges' were finally ripped out of the miner's houses, the miners wive's almost wept with joy at it's removal, but bitterly rued the great loss of the inherent heat that it had produced. The old ranges were quickly replaced with the 'modern' four burner, gas cooker installed in the sculleries and supplied with piped mains coal-gas, of which there was a plentiful supply. *(The coal gas was produced at Leven, only two miles away)* Heating in all rooms was still by open, coal fire-place *(which, in the case of the main family room needed to be re-built and replaced by a more 'modern' fireplace, to occupy the space vacated by the large range, again coal-fired, but in plentiful supply and reasonably cheap to employed coal-miners!)* Internal illumination within the houses was something of a problem, in that mains electricity had yet to materialise. Gas lighting *(gas mantles)* in the houses was by no means common, but was later installed and used extensively before electricity. The main lighting, if it could be described as such, was either by firelight and candles, or by oil lamps. Oil table lamps *(paraffin)* were very much to the fore, with the slightly more 'affluent' families able to afford the larger ceiling mounted 'branded' lamps which, I now recall, as being one of the greatest fire hazards ever encountered! *(In my opinion!)*

*(I distinctly remember my mother refusing to have that branded lamp lit, if old Harry was not at home. I can also remember two separate occasions, where small house fires were caused in our living room by the one of these ceiling mounted lamps overheating. Fortunately, old Harry was on hand on both occasions to suppress the fires and limit the damage.*

*On the occasion of the second flaming eruption, whilst I was still a very young, but obviously impressionable lad, I witnessed old Harry bare-handedly grab the flaming lamp, tear it from the roof and run though the house, to throw it boldly into the back garden. Old Harry suffered burn injuries to both hands on that occasion and lost several working days at Wellsgreen pit, through large fluid burn blisters. I can also remember my mother vowing never to use oil lamps again and that a three mantled 'gas-light' was to be installed soonest, even if it took all of Harry's next weeks wages!)*

One form of limited household lighting that was used by more miners than would care to admit, was that some miners brought home their carbide lamps on a daily basis, to supply a source of lighting for parts of the house. Carbide lamps work very well in the confines of a dwelling house, in that there was more than enough light reflection from the white, or light tinted, distempered or shami-texted* walls within the house.

*A war-time utility measure, whereby coloured distemper was daubed and textured onto prepared wall, by the judicious use of a natural sponge or similar implement.*

## Anecdote or Myth?

In one such miner's home in Clyde Street in the stone-floored scullery and, at a time while I was still a pupil at Denbeath primary school, I personally witnessed, from outside the scullery door, a sight that remains vividly in my memory to this very day. One miner's wife, a large and energetic red-haired woman, wore her own brown cloth 'yankee cap' upon which she carried her own carbide lamp and went about her household chores in the dark evenings, carrying her own light source with her on her head. *(Seemingly, she was just as adept at controlling this lamp as her miner husband, who worked in the Wellesley pit!)* She did this on a regular basis in the dark evenings during preparation of the family meal, whilst the remainder of the family sat waiting patiently at the supper table, on their mother finally re-appearing at the table to sit down, in order to see their own plates! I was told *(by the second son)* that when she reappeared at the dining table with the food, the carbide lamp was placed on the high overhead mantel-piece to illuminate the dinner table. This revelation is not an exaggeration and but for the possible embarrassment that it could cause her still living family *(of which there was four, three sons, the second of whom was my age, and one daughter),* I would have no hesitation in naming such a proud, practical and industrious woman!

*(On second thoughts, I would not dare to name, expose or ridicule such a capable woman, for I have no doubt that these same family members still remember their own mother with great affection, even though, I wonder, do they still remember the carbide lamp?)*

\*

The main part of Denbeath village consisted of parallel rows of double storeyed terraced houses, that lay at right-angles to Wall street and were further split by Swan Street, that bisected all of the rows except the last two. The lengths of the seven rows of houses grew progressively shorter as they developed Northwards, for the reason that Cowley street was not parallel to Wall Street, but curved inwards over it's length, to almost converge with Stark Street and Den walk at their eastern ends. The area was therefore shaped like a triangular wedge, having a long curved hypotenuse, a long opposite side, with the half-length adjacent side at Wellesley road. The street names that were eventually given to the separate rows were quite unique, in that they were all named after Scottish rivers:- i.e. Forth Street, Clyde Street, Tay Street, Tweed Street, Dee Street, Don Street and Spey Street. Swan Street cleanly bisected the first five named streets to Wellesley Road. Don Street and Spey Street were but short single rows.

The gardens arrangement in Denbeath village *(this village was at one time named 'The Garden City')* were also quite unique, in that every house did have a garden area, but they varied in size and shape. This was entirely dependent on what part of the street a house-holder lived! The rows of houses may have been parallel, but the roads between were not parallel to the rows of houses. *(With the exception of the south half of Tweed Street that formed the extension to Barncraig Street!)* The streets/roads were offset to the extent, that the houses at one end of the road had extremely short and small gardens, whilst the houses at the other end of the road had rather long gardens. This peculiar arrangement was apparent between every row of houses from Forth Street to Spey Street! *(Barring Tweed Street!)*

Some of the more horticulturally minded of the miners living in the village were quite successful in what they managed to grow in their gardens, mostly because they managed to import cart-loads of horse dung, which was quite prolific at the time, because most of the local transport was horse-drawn. *(Self-employed carters used this means to transport coalminers concessionary\* coal!)*

Even the 'Scaffies, *(actually, a solitary, ambulatory, weather-beaten individual, who classified himself as a street cleanser and was thus 'armed' with a two-wheeled, twin-dustbins contraption, complete with shovels and long heavy bass brooms. Who can forget the redoubtable Archie Cooke, who was a better story-teller than his namesake Alistair Cooke?!)* Even the *'Scaffies'* had their favourites, in that they selected the particular gardens of miners into which they shovelled the newly produced 'hot' dung.

*{Archie Cooke often carried fickleness to the point of pawkiness in the distribution of his gardening favours! (In the considered opinion of my mother!)}*

It was also quite a common sight in the streets around Denbeath, to see children *(and adults)* armed with a bucket and a hand shovel, trawling the streets one by one - on the lookout for some fresh steaming manure. Some of the more enterprising youngsters actually followed the horse-carts and attempted to 'sell' pails-full of the high-smelling produce to willing takers, for tuppence or even a thripenny bit! *(Two pennies or a three-penny piece!)*

This natural feeding of the soil was almost a necessity in the gardens of Denbeath, in that the two tall chimneys belonging to the Wellesley pit constantly spewed copious quantities of hard black soot particles, which was naturally 'dumped' on the local gardens as soon as it cooled. Mother nature and the vagaries of wind and weather, ensured that all of the gardens in Denbeath received their 'fair share' of this black carbonised grime!

Author's Note:- *The New model village of Denbeath was commissioned directly by Randolph Wemyss around the year 1903-1904. Its design was in the capable hands of Alexander Tod, an architect in the resident employ of the Wemyss family estate, whose brief it was, to design and build a large quantity of good quality double-brick houses having an harl'ed exterior, to house some of the families who were forced to quit the soon to be overwhelmed hamlet of Denbeath, but mostly, the expected influx of migrant miners, who were to be brought in from the now defunct coalfields of parts of Lanarkshire and Ayrshire. The model village was to be designed and built to a high standard (of the times), with spacious individual drying greens/gardens allocated to each and every house. The plan was to build the terraced houses in serried ranks, to completely fill the immediate area to the south of Cowley Street* *(already in existence!) and bounded by Ward Street.  Each house was to have a large family living room (named:- the kitchen) with an in-built cooking range, two large double bedrooms, or possibly one in a small percentage of houses - and a scullery with running mains cold water, a coal-fired water boiler and inside lavatories. (Only the best was good enough for the miners!)  The only facility that was not catered for, was the internal illumination of the houses!  The houses were built in record time with the first terraced row (and the longest), built to face onto Wellesley road.  That, which was a little unusual with the design of these houses, was not patently obvious at first and could easily be missed by the casual observer.  The upstairs houses each had a concrete, dog-legged shaped outside access stairway, which meant that there was only entrance/exit to the houses - also, the doors and windows all faced in the same direction.  (The reason for which will become clear as this paragraph moves on!)  The downstairs houses, though of the same size and shape had two\* different entrances/exits, of which the main doors were directly under the outside stairs and were therefore quite hidden.  This first row of houses were built approximately twenty yards (18.28 metres) from the pavements' edge which overlooked the tramcar\*\* lines that traversed the length of Wellesley Road and beyond.  (The main means of transport to and from Denbeath, for those miners employed in Lochhead pit and Earlseat mine, even though some hardy miners cycled this distance on a daily basis for six days of every week!)*

*Immediately adjacent to Wellesley Road on the north-west side, was the terraced row of houses that comprised the south side of Forth Street and simultaneously, the front row in Wellesley Road.  The interesting point about these terraced buildings, was that each and every row of terraced houses in Denbeath were double-sided! The row of houses that comprised the south side of Forth Street, was exactly the same 'block' of buildings that comprised the front row of Wellesley Road.  These houses on the south side of Forth Street faced in exactly the opposite direction to those in Wellesley Road but, they were in fact, the same solid block of buildings. They had been designed and built back to back, for reasons of economy, common plumbing, mutual strength and shared internal heat!  And, strangely enough, each individual house had one of its double bedrooms looking out (windowed) onto a different street.  (Clever fellow this Alexander Tod!) i.e. All of the houses in Wellesley Road had the living room (kitchen), one bedroom and scullery windows and their doors facing into Wellesley Road, however, the second and 'back bedroom' window, faced and looked 'over and into' Forth Street.  The same applied to the houses on the south side of Forth Street.  Those houses had their back bedroom windows facing onto Wellesley Road!  The remaining rows of houses were built in exactly the same manner (except for the single-storeyed semi-detached cottages of Swan Street, half of Wall Street and half of Cowley Street), with the two long halves of each successive block facing in opposite directions and therefore, in different streets.  To accommodate miners and their wives who had only one child or none at* all, the whole of Swan Street and the two opposite halves of Wall Street and Cowley Street, were interspersed with semi-detached cottages of the type to be built in Barncraig Street and Bow Street.

In the years following Nationalisation of the Coal industry in 1947, the whole of the structure of the village of Denbeath began to suffer neglect and partial decay by the new State owners, and the houses were gradually allowed to deteriorate and suffer depopulation. Many of the houses at the top end of the village have since been pulled down, to be replaced by modern brick-built cottages - and the original Spey Street has all but gone! However, the first row of houses that comprised Wellesley Road are still standing, as are many of the original house in the other named rows. Much has been done in the last twenty years to refurbish and renovate these solidly built houses, which to my mind, must stand as a lasting tribute to the competence of Alexander Tod and, the foresight and benevolence of Randolph Wemyss and the Wemyss Coal Company, in that first decade of the Twentieth Century.

*Almost all of the shorter Southern halves of all of the streets of the Denbeath Rows, have been pulled down to make way for more modern detached and semi-detached cottages, but many of the original Rows still remain. Extensive refurbishment has taken place on these 95 year-old houses, even to the extent of removing the old outside 'scullery' doors and replacing them with an extra window. The only clue to the original existence of this extra downstairs door, is that on inspection of the recessed kitchen walls along the full length of the block, between the individual stepped gable ends, it can be seen that the upstairs walls has TWO small windows, whilst the downstairs walls has THREE small windows.

Author's Note:- *In an effort to determine just how many houses were built in Denbeath by Randolph Wemyss and, after some rather detailed inspection of such plans that still remain in being, I finally and severally, walked around the whole of the village counting houses and compared old house numbers with new house numbers. I have come to the following conclusions: -*

There were a total of 299 dwelling houses in the village of Denbeath bounded by Wellesley Road, Cowley Street and Ward Street. There were 216 double-storeyed tenemented houses in four, double, broken rows, and three, single rows from Wellesley Road to Spey Street. In Ward Street, there was 20 terraced cottages on one side and 10, two bed-roomed, semi-detached cottages on the other side. In Swan Street, there was 10 two bed-roomed, semi-detached cottages on both sides of the street, interspersed between the main parallel rows. In Cowley Street, there are 25 terraced cottages in two groups down one side of the street, and 8 two bed-roomed semi-detached cottages on the other side of the road, also interspersed between the main rows.

At the time of completion of the two-storeyed, tenemented houses and semi-detached cottages, it seems that street names were not to be allocated to the different *rows in the village, individual house numbers being thought sufficient. The original row of terraced cottages in both Cowley Street and Ward Street already had house numbers, which were seemingly changed to accommodate a new numbering system within the village. The numbering system adopted was thought to be relatively simple, but as it turned out, a veritable postman's nightmare!*

*Commencing with the number 1 at the downstairs house at the bottom right of the village (Wellesley Road), the numbering commenced right to left and then followed a system emulating the numbering system as that on a 'Snakes and Ladder' board, until the last house in Spey Street at the top of the village received the house number 216. Cowley Street, Swan Street and Ward Street were individually numbered in there own right, with odd numbers on one side and even numbers on the other. The relative, consecutive numbers in each street were not adjacent to each other, except in Swan Street!*

The individual address for each house-holder was therefore:- Mr. R.U.Collier. No ?, Parish of Denbeath. Fife. ALBA.

The individual street names each emulating a Scottish river were allocated much later!

N.B. The 18, double storeyed, tenement houses that comprised and faced into Wellesley Road, were later re-numbered to coincide with the consecutive numbering system adopted for the whole length of the long Wellesley Road, from the Tower Bar at Aberhill to Muiredge roundabout, Buchaven. *(At Muiredge S.C.W.S.)* i.e. Odd numbers 625 to 659 inc.

*

\* Every coalminer, if he was a married man and a householder, was entitled to an approximate seven to nine tons of workmen's fire-coal, every year of his working life in the coal mines. *(This concession still applies in part to retired coalminer pensioners, who are house-holders.)* If the coalminer was unmarried and in lodgings, he would still qualify for the reduced concession of approximately two tons per annum. This concessionary coal which, in the late Forties and early Fifties, cost around nineteen and sixpence *(97.5 pence)* per ton *(tonne)* was usually requested by the miner every eight weeks or so in the summer months and approximately every six weeks in the winter months. The recipient miner would place a written request into the 'check-box,' asking to be supplied with up to 20 cwts., *(20 times 112 lbs. Imp. equals one ton, or approximately 1000 Kg.)* of workmen's 'fire-coal!' *(The pithead office where the general clerking is done and where miners uplift and deposit their underground, individually numbered, identity disks.)* This authorisation chit was the written authority that any coal-carter needed, to enable him to approach the 'Company check-weight-bridge' at

Denbeath, and demand access into the 'Depot' to obtain fire-coal to the weight described on the authorisation. *(This numbered signed chit alone, was worth the gross weight of the coal described therein. It was pre-paid!)*

The procedure was simple. At the onset, the *(cairter)* carter, whether he had motor vehicle or horse and cart, drove/walked on to the large steel weigh-bridge at the coal-depot, where the all-up empty weight of the transport means would be weighted to the nearest one pound. *(1 lb. imperial / 454 gms.)* That empty weight being recorded against his 'vehicle' with the check weight-man. The carter would then proceed down to the single row of company railway wagons that contained the clean mixed *(small chirles, trebles, mixed doubles and broken lumps)* work-men's fire-coal, which had been shunted into the special siding for this very purpose. There was no means of mechanical loading, so therefore, the carter wasted no time. He carefully backed his horse and cart onto the side-door of the nearest available wagon and proceeded to fill his cart by the only sustainable method open to him. He used the physical power of his body! Wielding a grape,* he would heave all of 20 cwts of loose coal into his own cart whilst his horse stood patiently waiting. There was little respite in this task! He was self-employed! Time was money and little time was wasted on such mundane things as having a rest, that would come later when the horse did the pulling work during the delivery stage.

*\*(A 'T'-pieced, long handled, shovelling tool shaped like a shovel, except that it had approximately ten or twelve 15-inch long tines in place of a blade. This was to ensure that the 'dross' remained on the floor of the coal wagon - and not into the cart! Recipient miners would not, with justification, accept 'dross' with their load of fire-coals!)*

When the carter reckoned that he had loaded an imperial ton of coal *(20 cwts.)* into his cart, he then led the horse and cart the short distance uphill to the weight-bridge. If the bridge was free, he would move straight on to the large steel 'live' platform, where the total weight of horse, cart and coal load was immediately registered on the large-faced, three feet diameter, precision scale, to the nearest one pound. *(Imp.)* If his weight estimation was good *(very few carter's ever got it precisely correct)* and the scales showed a slight over-weight, all that needed to be done, was to throw off one or two of the larger lumps into the excess coal reservoir, which was custom made for such an event! *(Estimations could be underweight as happened many times. This coal reservoir also precluded a carter from having to return to the coal-wagons if the load was too light!).*

Then, and only when the load was correct, the carter would hand over the miner-recipients 'coal-line' to the check-weight man, who in turn, would 'spike' the 'coal-line' thereby cancelling it's validity. The carter would then begin the delivery journey to the home of the recipient miner, where the load would be literally 'dumped' onto the street/ pavement at the nearest access to the miner's home coal cellar. For this service, the carter would be paid the standard fee of six and sixpence,

*(32.5 pence)* irrespective of the distance travelled. A hard working and conscientious carter would repeat this process, six to eight times during the course of a working day, depending on how far he had to travel to each miner's home address and providing that he did not have to wait on further stocks of miners' coal arriving at the depot. A carter whose business was won using a horse and cart, was only half finished when he returned home to his stables. The horse usually had to be washed with warm water, dried and groomed before being fed and watered *(or, at the same time)* and then the heavy-weight, leather 'tackle/harness' had to be checked and repaired where necessary. If the carter were fortunate enough to own more than one 'Clydesdale,' then the grooming work would be doubled, but then, he would be able to work from seven in the morning to six at night continuously, with a change of horse around midday!

Inevitably, creeping mechanisation began to slowly oust the carter with only a horse and cart, in that firstly, a single 30 cwt. lorry made an appearance at the depot, which, in spite of the fact that the coal still had to be hand-loaded, the faster turn-around time for the light lorry, meant that the driver-owner could manage around two to three more loads in any one day, which of course, was needed to cover his extra running costs. The mechanised carters *(of whom some were in the sea-coal business on the sea-shore at Buckhaven)* gradually began to take most of the business from the original horse transporters, when vehicles such as the ex-W.D. five ton, flatbed lorries became available for resale, along with the partial de-restriction of petrol. One particular 'family business' in Buckhaven acquired such a vehicle, then quickly fabricated 30-inch high wooden side-wings and a dropping tail-gate. They then sub-divided the flat-bed crosswise, to accommodate up to three or four, separate, one-ton capacity compartments. This meant that the loading and weighing arrangements were a little more complex, with a separate short journey to the weigh-bridge for each successive accumulative load, but with four or five men hand-loading the coals, they could more than outdo a 'one-man' operation. This inventiveness, allowed several different loads to be carried simultaneously to different locations, all for the running costs of the longest journey! Fortunately, there was enough business for all concerned with the added side-effect, that the waiting time for horse and cart delivery was speeded-up due to the increased competition.

Author's note:- *At the time of which I describe, there was an illegal business practice being conducted in some of the mining areas, whereby one or two of the mechanised carters would 'purchase' a 'coal-line' from a hard-up miner. The 'coal-line' as requested by the recipient miner cost him 19/6d (97.5 pence) deducted from his wages, therefore, the pre-paid 'coal-line' itself, being undelivered coal on demand, was worth up to Five pounds (sterling) in land-sale coal to the purchaser. The acceptable price therefore to the vendor, was the split difference between the initial cost and the selling price, namely Three pounds (sterling). To an oncost miner, this was two days wages. To a contractor dealing in illegal 'coal-lines'? He was doing the miners a 'service'!*

# A Street-Wise Endeavour

To the residents and householders of Denbeath and parts of Buckhaven and Methil, where it concerns the legitimate dumping of a load of coal at any given address, there was and still is, one young man's name that is remembered with great affection and for many tired miner's, a great sense of relief. That young man's name was and still is, I am pleased to report; Bertie Christie!

To the tired married coalminer coming home from any one of the three shifts, the sight of one ton of fire-coal sitting on the road/pavement awaiting to be transported in to the recipient's coal-house, was almost too much to countenance. The last thing a newly-showered, tired miner needed was to be faced with, was the thought of another one ton of coal to load, carry and deposit in the coal-house. *(Some of the garden paths in the Denbeath rows were up to 25 metres long!)*

When my brother and I were in the coalmines and still living at home, we often came home from the day-shift to the sight of a great load of coal lying on the pavement, waiting to be uplifted to the coal-house. Old Harry, cannie man that he was, would not even think of taking in his own coal, when he had two, young, strong, miner sons living at home. *(Bertie Christie was not part of the coal equation at this time!)*

I became aware of Bertie Christie after I had left High School and soon after I had commenced at Lochhead pit. Bert Christie had attended Denbeath public school, but he had not received an education! My mother told me, that he had spent more time outside of the classroom than he had inside. He was seemingly not a disruptive youngster, nor was he violent in any way! He was apparently intelligent, but not clever. He was attentive, but slow on uptake! He was, in a later time and place simply dyslexic! - a condition not readily recognised in the 'forties'. This was a phenomena that was not properly understood at this time, with few teachers prepared to devote the time and patience needed, when faced with this unfortunate affliction. Most teachers simply gave up on him, with the result that he left school with their blessing and no qualifications whatsoever!

Bertie Christie was no fool! Even without the benefit of formal education, he was 'street-wise' before the term had been invented. He knew the important things in life were his mother, the need to be clothed and fed, and most important, the need to earn money. Bertie Christie was all but unemployable! That much was apparent. But what to do? That which Bert had going for him was the fact that he was fit, healthy, quite strong with a willingness to do any task whatsoever! Bertie Christie therefore simply created his own employment. He saw the likelihood of continuous employment in the streets of Denbeath on a daily basis, whether it be rain, hail or sunshine! *(Better if it was a dark, miserable wet Winter's day, the wetter the better!)* Bert armed himself with a builder's rubber-tyred, steel bodied, wheel-barrow, a long-shafted No. 6 sized shovel and a bass broom. He then walked the streets of Denbeath until he spotted a newly dumped load of miner's coal, where he then approached the housewife and offered to load, transport and stack the coal in the recipients coal cellar and clean up the pavement afterwards. All for the sum of half-a-crown. *(2/6d or 12.5 pence).*

Housewives may have been initially reluctant to part with two and sixpence, but their coalminer husbands were not! So, Bertie Christie the entrepreneur, was born. Bertie Christie soon hardened to this work, so much so, that he was to be seen at all hours of the day, both summer and winter, either going to his next 'appointment', or trawling the streets for the next likely dumped load. As word of Bert's depend-ability and 'tidiness' spread, more and more housewives *(and miners)* came to depend on Bert's willing expertise with his wheel-barrow and brush. Bertie Christie built up such a following in Denbeath, that within a short space of time, his mother, with whom he still lived, was hard put to keep up with the callers who required Bert's services. Not only that, but Bert had to depend on his mother to run his booking book up to three days hence, for the reason that Bert's services were now so much in demand, that he sometimes regretfully had to disappoint non-regular customers. *(I do believe that one of the local bye-laws stipulated, that dumped loads of miner's fire-coal was not allowed to lie on the pavements overnight, probably something to do with pedestrians right of way and safety, during the hours of darkness!)*

Bertie Christie's familiarity with the needs of his regular customers, soon grew to the tacit acknowledgement that Bert no longer informed the housewives that he had arrived. He merely commenced the work, whistled whilst he did so and after the final clean-up, reported to the back-door of the house in anticipation of a welcome drink of cold water or lemonade, along with the grateful thanks of his host and his hard-earned obligatory half-crown!

Bertie Christie, through his diligence, hard work and genuine affability, became a quiet institution, even a local legend, around the streets of Denbeath and Buckhaven. He was dependable, tireless and ever pleasant. Some ignorantly stupid and pusillanimous people made fun of Bert, but nothing very much fazed him. His lack of formal education didn't seem to worry him very much either. It certainly did not stop him from earning more than enough to support him and his widowed mother during her lifetime. During his working life and after many countless hundreds of tons of coal removed, many of the Denbeath, Methil and Buckhaven housewives were proud to say, that their 'coals' had been 'taken-in' by Bertie Christie!

Author's Note:- *On Saturday the 22nd of May, 2000, my wife and I were slowly walking up the main street in Leven, when I spotted an instantly familiar gait walking smartly down the middle of the road. I recognised the slight figure instantly! It was Bertie Christie! His gait had lost none of its spring and his figure seemed as slim as ever. The only noticeable change to his outwards appearance was that his always slicked-back hair was now a perfect grey. I called out twice to the hurriedly departing figure, but to no avail! He hadn't heard me or seen my raised hand. But, then I remembered, Bert always suffered from slight deafness, which no doubt, with the passage of time, is much exacerbated!*

The passage of fifty years does much to a man's appearance. Not so, for Bertie Christie! *"If I don't meet up with you again Bert, before one or both of us depart this world, I hope that you are given the chance to be told of the contents of this chapter. Your presence at the coal, warmed the heart of many a tired coalminer, after a hard shift's work!"*

*

*In the downstairs end house on the north side of Tay Street at its junction with Ward Street, there lived the family of Donald Findlay Sr. Mr. Findlay was a coalminer in the Wellesley pit and he and Mrs. Findlay produced and raised four offspring. Three boys and one girl.*

*Of some mining families that occupied the Denbeath rows and where there was a proliferation of siblings, it was often knowing said, (with pursed lips, a wink and a common nodding of heads) that some children 'suffered' the consequence of being 'dragged-up', as opposed to being 'brought-up' or raised. Let me hasten to say that this was not the case with the family Findlay! This family was well fed, well clothed, well cared for and scrubbed clean on a nightly basis. Discipline was exercised in the usual fashion with the ever-present (but rarely used) threat of the father's pit-belt.*

*There was nothing unusual or strange about this family, they were well known, well liked and used the Denbeath Co-op on a daily basis. In fact, my older brother Bill used to chum with Donald the eldest sibling, whilst I, on occasion, used to chum about with the second oldest, Dod (George). In fact, we were light drinking chums for a while after he got married and lived in Leven. However, that has absolutely nothing to do with this particular part of my narrative. That which has to do with this part of my story, concerns a ritual that took place nearly every night, just as the light was fading on the rows of Denbeath. At this time of the evening and probably slightly before-hand, a few wives in Tay Street would appear at their front doors or garden gates to witness a most unusual happening, that invariably shattered the relative peace and quiet of a long summer's evening. Mrs. Findlay's actions were utterly predictable, if not unfemininely strident! She would come out of her scullery door, take a few steps into her back garden (there were no front gardens) and take a deep breath. From her strong lungs would emanate a longish cry that would resonate, and reverberate around the surrounding rows of houses: -*

" Donald,Geordie,Betty, Johnny-y-y-y!"

*To be followed by the second deeper intake of breath:-*"**DONALD-GEORDIE-BETTY - J-O-O-H-N-N-N-Y-Y-Y-Y ! ! !**"

Mrs. Findlay was calling her brood to the homestead and to the bathtub, to be scrubbed in ascending order and then suppered. This was a nightly ritual that was practised with loud rhythmic regularity and was a joy to behold *(As some may remember!)* A very few, mean-spirited wives in the Denbeath rows attempted to mimic and ridicule this simplistic ritual, and this was known to the Finlay family, but nary a one could hold a candle to this caring and industrious woman, to whom her family meant everything!

\*

Author's Note:- *In life, the Laird, Randolph Wemyss had expressed a desire to see the provision of an 'Accident Hospital' for coalminers in the Wemyss area. He had, in 1907 said, that he had been thinking for long enough that 'a great many of the accident cases that occur in the course of any year could be treated at home!' (He was of course referring to the fact that Edinburgh Royal Infirmary was a considerable distance away in the Lothians.)* He had stated:

"I hope to give the erection of a hospital for accident cases a start next summer!"
Unfortunately, he did not live long enough to see this hope materialise!

In death, however, his aspirations were realised. His widow, the Lady Eva Wemyss, though deep in grief, determined that her late husbands' wish to provide such a hospital for the miners and their families should bear fruition. The Lady Eva accordingly stipulated, that she would cause this hospital to be built entirely at her own expense, without the need for public donations. This was to be her tribute to her late beloved husband. She did however, acquiesce to the gifts of articles of hospital furnishings, which she gladly accepted.

The site that was chosen was on an open stretch of level ground approximately one quarter mile from the Wellesley colliery and approximately one hundred metres in front of the still to be built Denbeath Primary School. The site enjoyed a magnificent, uninterrupted view of the waters of the *'Frith'* of Forth. The foundation stone was laid on the 30th of March 1909 by the Lady Eva Wemyss and the hospital was formally opened on August 28th of the same year, to a tumultuous gathering of dignitaries and local people. Amongst the many contributors to the furniture of the building, was Mr. Charles Carlow, chairman of the Fife Coal Company, who presented Lady Eva and the hospital with the gift of the four-faced striking clock with bells. This clock is unique in its perception, in that it's four pairs of 'hands' are depicted by the stark representation of coal-miner's picks and shovels! This hospital was dedicated to the memory of Randolph Gordon Erskine Wemyss, whose dream it originally was and fittingly named:- 'The Randolph Wemyss Memorial Hospital'.

*

## **The Garden City Express!
*or, the first Denbeath Rocket!*

In the first decade of the new century, when the planning of the new Denbeath Village was still on the drawing board, Randolph Wemyss also came up with the suggestion that there ought to be a cheap and regular form of transport covering all of the small coastal towns, villages and hamlets that served the coalmines from Leven to Dysart, including the three or four inland collieries! This would take the form of a system of electric-powered tram-cars running the full length of the route, probably starting early in the morning before the day-shift start-times and finishing late at night, to coincide with the miners coming home from the back-shift. This would mean a non-stop regular service for approximately eighteen hours of every working day, Sundays excluded. *(Limited schedule!)*

To achieve these ends, Randolph Wemyss seemingly first approached the North British Railway Company with a view to a possible partnership, but they were not interested. They had their own railway system, which might suffer from the potential competition! Randolph Wemyss then approached Kirkcaldy Town Council. He did this, because Kirkcaldy local council had recently opened their own tramcar system within the town limits, which so far, had not shown any profit. They too, were not interested! Randolph Wemyss then re-approached the K.T.C. with the offer of his laying the tracks at his own expense, over his own land and have them run the schedule - but that too, was rejected!

Not to be thwarted, he then approached Parliament for permission to build his own private tramway system, 'ostensibly' at his own expense and, having obtained the necessary permission, then proceeded *(with financial backing!)* to organise his new Company. It was to be named the Wemyss & District Tramways Co. Ltd.

The lines were laid over the next two to three years and stretched from the Carberry House Gates in Leven, to Rosslyn Street in the Gallatown. Some stretches were double tracked, whilst some small stretches were laid with single track only, which was very quickly up-graded to double-tracks as the service needs quickly increased. At the opening of the line in August 1906, the route could be quite easily followed by the appearance of a long, double row of high pavement-edged 'telegraph' poles, that lined both sides of the route. *(The high poles were needed to lift the 'power' cables to a height greater than a double-decker tram-car, with the addition of it's trailing conductor pole.)* The line was initially populated with nine, single-decked tram-cars, finished in a buff-coloured livery that very quickly earned them the sobriquet, *'The Mustard boxes!'* This transport system was an immediate success and served nearly every coal-mine in the East Fife area, with the exception of the Francis at Dysart, but a quick connection could be made with the Kirkcaldy town service at the top of the Boreland, to connect to the Francis Colliery at Dysart.

Those miners who worked in the Isabella *(Cameron mine),* Wellsgreen pit, Earlseat mine and the Randolph pit *(the Randy),* did have extra distances to travel by foot, in order to reach their respective coal-mines. *(For those miners employed at Wellsgreen pit, they alighted at the Rosie village and trudged over the fields to Wellsgreen. For those miners employed at Earlseat mine, they probably alighted at the Coaltown of Wemyss, followed the miners down the Lochhead pit road and continued across the fields to Earlseat!)*

This transport system was seemingly an instant success, in that within a few months in 1907, another four, larger capacity, single-decker 'cars', all of forty feet long were added to the fleet, now making a total of thirteen cars. In the years 1912 to 1913, a new livery was adopted by the Company, changing the 'mustard boxes to a smart 'Burgundy' coloured design, which I understand was retained until the later, but subsequent collapse of the tram-car system. However, the tram-car system continued to flourish, in spite of the fact that the miner's 'fares' were subsidised by the Company. Notwithstanding, the small 'hikes' on this still subsidised fare seemed to infuriate the miners, in that at every small 'hike,' they seemingly were on the point of 'striking' at the minimal fares increases. *(These East Fife Miners learned very early, the value of 'voting' with their feet!)* However, before very long, it soon became apparent that there was not a sufficiency of tram-cars to maintain the service that was now needed and another four tram-cars were this time hired from Kirkcaldy town council. These cars were of a different type to those already in use, in that they were double-decker cars. Or maybe, they should they be described as promenade cars, in that the top decks were open-topped and completely exposed to the vagaries of the coastal, winter weather. *(They also had an open front and back on the lower deck, which meant that the drivers/conductors were also exposed to the elements!)* This did not go down at all well with the tired, dirty coal-miners who were returning to Denbeath from Lochhead pit *(or, any pit!)* on a dirty Winter's evening. I guess that they thought that the Company should 'pay them' to endure the dark, foggy, wet and miserable exposure to the elements!

*{Initially, the electric power required to operate the tramways system was supplied by the Wemyss Coal Company from a power source situated at Denbeath, probably the same generating source that supplied the power needs of the Wellesley colliery. However, in 1912, this system was changed, updated and improved upon, in that the power source was changed to that supplied by the Fife Tramway Light & Power Company and from a large, transformer station situated at East Wemyss. (This also, may have been the same transformer station that supplied Lochhead*

*colliery with its electric power, before it was coupled to the National Grid system!)}*

There were, along the length of the route, three separate sets of 'car-sheds' which acted as garages. One at Aberhill, one at East Wemyss and one at the Gallatown, with the main company offices being situated at Aberhill. I am also led to believe, that each of these 'car-sheds' had their own maintenance facilities, but that most of the major repairs were carried out at only one of them. At Leven, where the line terminated at Carberry Gates, the tram-cars did not turn around, but merely stopped, with the driver moving the 'points' and then commenced the return journey on the other tracks! *(Tramcars were capable of being driven equally well in both directions!)*

At the Coaltown of Wemyss, coming from the direction of East Wemyss, the tracks passed in front *(to front and south)* of the Earl David Hotel at the north-east end of the village and continued on through the main thoroughfare, past Lancer Terrace, the S.C.W.S., Barn's Row and on to the end of the Check-bar road at the south-west end of the Coaltown. The tracks then by-passed The Bowhouse *(Bu'hoose)* farm, before cutting through the Boreland woods to emerge at the North Lodge Gates *(Clay dales)* at the top of the Boreland. *(Yes! The same woods as described in the chapter - 'Pitch & Toss'.)*

*Not too long after the inauguration of the tramway between Leven and the Gallatown, there occurred an isolated, but terrible tragedy on the main street of the Coaltown of Wemyss. A small six-year old schoolboy by the name of Andrew Bell was struck down by one of the moving 'mustard boxes' whilst he was crossing the main road and killed instantly! In those days, there were no traffic lights, pedestrian crossings or 'lollipop ladies'. The tramways ran directly through the middle of the village, effectively cutting it in halves and, seemingly without any form of 'guard-rails' to prevent the good folk of the village intruding onto the double tracks. The villager's simply had to go about their daily business and probably crossed the lines at the most convenient point and children being children, probably got far too close to these tram-cars as they trundled past through the village. With hindsight, this probably meant that the ever-present danger of unregulated crossings was always an 'accident waiting to happen!' It must of course, be realised, that at the time, very few people were really aware of the potential dangers involving road safety, for the simple reason that the sad experience was yet to be fully realised! It must be said, that the one person who was quickly made aware of the potential risks and imminent danger that the 'middle of the road' tram-cars represented, was Randolph Wemyss himself.*

*Randolph Wemyss responded quickly! I have been unable to discover whether it was from village complainants (his tenants) and possible pressure by the 'professional residents' of the village, or whether he acted on his own cognition, with realisation of the possible dangers represented by the 'open' tram-lines that split the village in halves. Probably a little of one and a lot of the other! In any event, the Laird initiated immediate plans to re-route the double tram-lines away from the centre of the village and out of harm's way! The new modified route commenced at the Earl David Hotel, where a double row of permanent rails were laid to the north side of the hotel, and advanced up behind the full length of Lochhead Crescent at a distance of approximately twenty metres from the long row of semi-detached cottages. This new route rejoined the original tracks at the south-west end of the Coaltown, before the Bu'hoose (Bowhouse) farm, thereby still actively serving, but effectively by-passing the congestion of the Main Street! This new route was now also effectively 'screened' from most of the village inhabitants, except for access points at the tram-halts!*

The service reached its peak with this addition of the extra four 'promenade' cars hired from Kirkcaldy Town, in that the regular service was seemingly maintained at one car every 15 to 16 minutes and in both directions. I cannot say if this regularity was maintained 'between' shift peaks, but to have organised and run such a service for his miners in the first decade of the Century, must have shown everyone, the lengths that Randolph Wemyss would aspire too, in his attempts to help the miners maintain a uniform and regular attendance at his company's coalmines!

By the start of the 1920's, with the advent of the omnibus and their increasing reliability, the miners from the coalmines untouched by the tram-lines, began to change their mode of transport, especially when it was realised that the omnibuses could and would, transport them directly to and from their respective coal-mines, without them having to walk any distance whatsoever. This meant that the miners in the out-lying inland pits now needed less travelling time, which suited them much better! The competition was perhaps quite evenly matched at first, with not too much loss being felt by the Wemyss & District Tramway, due to the steady increase in the numbers of miners being employed at all of the Company's coalmines. The competition did last for about twelve years, until the owners of the W. & D. Tramway realised that the omnibus was here to stay and was improving in reliability! The Company then decided that annihilation was the better part of 'greater competition' and closed down the 'wonder railway' in 1932. It has never been replaced or revived!

Author's Note:- *In order to maintain a 15-minute regular service in both directions, over a total track(s) length of approximately seventeen miles, where a one-way journey from terminal to terminal lasted approximately 65 minutes. This meant that there must have been at least ten tram-cars on the lines at any one time, including those waiting at the terminals to maintain this schedule. If the total fleet amounted to nine + four + four, - a total of seventeen cars, then it is entirely reasonable to suppose that an operational number of ten cars is entirely feasible, taking into account wear and tear, servicing, repairs, cleaning and probably limited running hours for each car, during any twenty-four hours period. Eighteen hours running for one car, each and every day, would cause massive wear and tear, much replacement of worn parts and long maintenance periods!*

*Also, I would question the frequency of the 15 minutes service between the one and a half hours prior to the onset of both day-shift and back-shift, and the one and a half hours periods after these shifts had finished. There were literally hundreds of miners on the move, both prior too and immediately after shift-change periods. I estimate, that a frequency nearer to a five minute schedule would have been needed to transport the greater movement of miners at these times! Miners did not all work just at their local coalmines! Many, as lived in Denbeath, travelled as far as the Coaltown of Wemyss and beyond on a daily basis.*

During 1951, whilst I was on constant day-shift at Lochhead pit, I travelled from College Street, Buckhaven to the Coaltown of Wemyss on a daily basis. I do seem to remember that the frequency of the bus service run by W. Alexander & Co. at that time, was either a single or a double-decker bus every five minutes, from 4.45 a.m. to 5.20 a.m. every morning except Sundays. The 'service' commenced at Leven and terminated at either East Wemyss, Kirkcaldy, Dunfermline or Rosyth. This frequency was still needed then to cope with the large amounts of miners travelling to the various coalmines and day-shift workers to Rosyth Dockyard.

\*

The following statistics should give readers an small insight as to just how much of an phenomenal success this tramway turned out to be:-

From August 1906 to August 1907, the 'Company' had carried approximately 1,580,000 passengers, over an approximate total of 360,300 miles travelled. The income for the first full calendar year of operation in 1907, amounted to approximately £13,110 sterling, giving an average monthly total of approximately £1,092. This compares greatly with the average monthly income over its first four months of operation, where the average monthly income was approximately £1,022 sterling. In 1908, the average monthly receipts rose to approximately £1,184 sterling.

Such was the success of this venture, that in the year 1908, the Laird commenced upon a planned undertaking to link up the towns of Dunfermline, Rosyth, Cowdenbeath, Bowhill, Lochgelly and Kelty. Unfortunately, Randolph Wemyss suffered a further deterioration in his already troubled health, and was forced into a sojourn to better climes on the European continent in search of improved health. The Laird of Wemyss did not live to see the fruition of his 'Kingdom' tramways, nor his dream of a miner's hospital in the Wemyss district. Whilst returning from the Continent, the Laird and his lady wife stopped over at their

London residence in Chesterfield Street, where an enforced stay necessitated, due to the deteriorating condition of the Laird's health. The Laird had been anxious to return to his home on the coast of his beloved Fife, longing for the relative peace and quiet of its familiar countryside.

This was not meant to be! The Lairds' condition had become very grave and his limited strength was ebbing. He was therefore confined to bed rest, but with no apparent improvement to his already weakened constitution.

In life, the Laird of Wemyss did not complete the long journey northwards to his beloved Wemyss Estates, but succumbed to his illness and slipped quietly away in his sleep, in the presence of his lady wife Eva, on Friday the 17th of July, in the year of 1908 A.D. He was exactly fifty years of age!

*He was mourned long, lovingly and with great affection by all who knew him!*

*Mr. Wemyss' remains were conveyed to Wemyss Castle in Fife, where, on 22nd July 1908, he was laid to rest in the very Chapel Garden that he himself had converted into a sepulture for the Wemyss family, only fourteen years previously!*

*

# Chapter IX

## The Victoria Pit!
### *The Long Dook Coal-Mine!*

The Victoria Pit as it was named, was one of several pits belonging to the Wemyss Family in the Forties and under the named ownership of the Wemyss Coal Company, before Nationalisation in Jan. 1947. The Wellesley Pit at Denbeath, the Rosie pit at the East of Wemyss, the Michael Pit at the west of Wemyss, the Randolph colliery near to the Boreland, the Earlseat pit, the Lochhead Pit and Surface Dipping at the Coaltown of Wemyss and the Victoria Pit that surfaced near the beach at West Wemyss to name but seven. The Victoria Pit was not a deep pit nor was it a large pit. It did have a narrow vertical shaft with two cages and a winding house, so therefore, could be described as a conventional pit.

The Victoria Pit was quite an old pit that had initially been sunk in 1824 *(during the Lairdship of General James Erskine Wemyss 1822-54)* at West Wemyss in the Kingdom of Fife and named the No. 7 pit, later named for Queen Victoria who succeeded to the British throne in 1837.

In 1841, Mr Thomas Byewater the manager of the Wemyss pit, is on record as having stated that of the 269 persons that were employed underground, 25 of them were females, and of the boys employed, 25 of them were under 13 years of age! The New *(second)* Statistical Account of Scotland 1845, describes this Wemyss pit being sunk upon the 'main seam', which is all of nine feet thick and has been 'wrocht' to a depth of 100 yards *(approximately 91.5 metres)* below sea level. The 'Account' goes on to say, that the coal-pit employs 140 men, 24 boys and quite surprisingly, 42 girls! *(It must also be correctly assumed that a goodly proportion of both boys and girls were under 13 years of age! I also feel that the disparity between the two sources in the number of persons employed in this colliery, may have been due to the lack of comprehensive knowledge amongst different officials within the parish. Statistical Accounts were often and usually written by the one 'educated or erudite' person in the parish! (i.e. the reverend minister or the resident doctor!)*

The 'Account' also mentions the coal that was being 'wrocht' was the nine-foot seam! This could only have been the Chemiss seam of coal, for the simple reason, that it was the only seam in the Productive Coal Measures other than the Lower Dysart of comparable thickness in this area!

*(The Lower Dysart seam lay much deeper and was completely untouched at this time. It was never broached within the Victoria pit workings!)*

There were those in the mining industry *(in the forties!)* mostly employed in the larger and more productive coalmines, who rated the Victoria Pit as nothing more than a 'flea-pit' and spoke of it with unfettered humour and mild contempt. How wrong could they be, as the ageing gravestones in the village cemetery will surely testify. It shall surely strike the casual visitor to this 'out of the way' small, quaint, mining and fishing village on the coast of East Fife, that some of the gravestones in the small cemetery are dedicated to young boys of fifteen and sixteen years of age, who met their fate in an murderous environment that would seem barbarous and heart-breaking, in light of the health and safety acts of today's society.

*{With the sinking of this No. 7 shaft near the sea-shore (and below the cliff face) to take the Chemiss coal, the decision was taken prior to 1835, to develop the workings out and under the sea in ever-deepening dooks. This was known to be a risky venture in light of previous knowledge, with much hard won experience gained in dealing with the expected ingress of water, whether it be salt or mineralised (land water). To cope with the expected increase of water as the workings became deeper, the resident engineer, Mr. David Landale reported, that to cope with the potential flood water, four large steam-engined water-pumps were installed and brought into play (and having the combined power of 172 horses!).*

*Mr. Landale also reported that in working the main seam (the Chemiss?) out under the Forth, the firemen (later named deputies) had to issue orders for the collier's to drill long holes at regular intervals along the line of - and into the roof, to allow the pressurised hydrogen gas to escape. The 'fireman' then had the unenviable and dangerous task of igniting the gas at the mouth of the holes! Many firemen of the era were often recorded as having their hair, eyebrows and beards gloriously singed, as the released gas burned for several minutes!*

*With the empirical knowledge gained from the tribulations and near disasters of the taking the Chemiss coal directly from the bottom of the shaft, then working to the dip, it seemed that the planners were not about to repeat the same mistakes when the eventual prize was to be the lasting bounty of the Dysart main coals! Hence, the abrupt change in 'Modus Operandi!', and the expected conversion from a wet pit to a relatively 'dry pit!'}*

The Victoria pit shaft was only one of six or seven attempts to get to the deeper coals that lay in abundance under the shores of the small village of West Wemyss. Many seams of coal had been previously discovered and worked within the Wemyss area many centuries beforehand, and had been identified and named down to a given depth. In the middle of the first decade of the 1900's, the decision was made to sink the existing vertical shaft down much deeper to the **'More coal'**, to gain access to the greater bounty of the Dysart Main seam. This was a known seam of supposedly homogeneous coals, a full 30 to 32 feet thick, *(with some thin bands of stone interleaved)* that lay under the *nearly the whole* of the area of Wemyss-field, but in the West Wemyss area, it lay deep under the foreshore

The shaft at the existing Victoria pit was then sunk down to a depth of 72.5

fathoms (145 yards or 435 feet). This meant that the **new bottom** of the shaft was at a datum height of 9589 feet, the surface datum height being 10024 feet. (24 feet above mean sea level.) This deepened Victoria shaft **did not** strike the **Dysart Main** seam of coal at this **bottom.** The Dysart Main seam was fully 58 fathoms **below** this level at a datum height of 9240 feet, a vertical difference height of some 58 fathoms or 348 feet. (*Some previous local publications that superficially describe the Victoria pit and include the depth at which the Dysart Main seam was **not struck with this shaft,** have not made this very clear!*) This new pit bottom **did not** terminate in any other seam of coal, but in solid grey stone. The adopted plan was to cut through the metals (*strata*) at this depth (*new bottom*) with a **level crosscut stone mine,** and *drive* in the likeliest geological direction that would strike and cut into the **rising Dysart Main seam,** over the shortest possible horizontal distance.

To achieve this objective, a **level stone mine** measuring approximately 12 feet wide and 9 feet high was commenced from the bottom of the vertical shaft, (*at 72.5 fathoms*) and driven directly into the metals from the 9588 feet datum height (*shaft bottom*), on a grid bearing of 298 degrees. This mine was quickly paralleled by a companion mine measuring approximately 8 feet wide by 6 feet high and commenced exactly 20 metres to the right. These **two, level, parallel mines** were driven forwards (*driveage*), cutting through the rising strata at it's steepest inclination, this being the quickest route to the target seam. These two mines were driven truly level and straight, with the wider mine (*later, to be utilised as a coal transport mine*) being regularly steel girdered at 12' by 9', with the smaller travelling mine only being girdered where needed! (*If the metals are absolutely solid, supports are neither needed nor necessary!*)

At a distance of 135 metres from the start point, the level, cross-cut mines cut into the rising **Sandwell Lower** seam of coal, which first appeared coming up through the pavement, progressively rising-up through the 'face' of both mines and disappeared up through the roof as the level mines progressed. This then meant that the **Dysart Main** seam would be the next coal seam to 'rise' up through the pavement. The level mines then progressed another 430 metres at the same level and in the same direction - before the 'Head coal' of the Dysart Main seam was struck, coming up through the pavement (*at a gradient of approximately 1 in 3.5*) and showing more and more of it's thickness as the level mines were driven into its middle coals, exposing the Grounds & Nethers, the Myslen and the Toughs & Clears coals.

*{This would have been the named coals that were to be wrought. The plan being (in this area only?) to keep the full thickness of the Head coal and occasionally the Sparcoal as a solid roof, with the all but useless Sclits as a solid pavement with the full under-lying thickness of the Coronation coal left intact! (See section on* **Dysart Main** *coals and thicknesses!)}* The total driveage of the **level cross-cut** mines was now fully **565 metres** long and cut directly through the steepest incline to the target seam. (*Datum height 9590 feet.*)

*{A point worthy of note, is that these 'level' mines only gained two feet in*

*height over their 565 metres length.* A *true testament to the quality and skill of the* **stone miners** *of the time.* This *also meant that any ingressing or encountered* *standing water would gravitate back to the Victoria pit bottom!}*

These two, cross-cut mines had been driven through the **solid metals** without benefit of electric power, where the necessary bore-holes were hand-drilled using nothing else, but a simple mechanical hand rachet implement, that usually needed 20 to 30 minutes to bore one hole. *(Poker drills as used with coal, were virtually useless for this type of drilling in stone!)* The level cross-cut mines did not actually stop at this point. They had in fact, been driven into the Lochhead's Old Dook's No. 1 companion level, that ran from Nicholson's Dook to the Francis barrier. *(Solid stoops of coal in each seam, left in situ, to delineate the boundaries between closely located, but separate coalmines.)* This No. 1 companion level had been driven a little earlier around 1900 (in Lochhead pit) and within the Dysart main seam. The two, parallel mines were then driven upwards for a short distance of 20 metres on the same straight line, following the middle coals in the seam, and now cut into No. 1* level proper. These two mines now joined-up with a pair of parallel **headings** that were directly in line with the new cross-cut mines and, could now therefore be described as extended (upwards) headings to the cross-cut mines.

These **headings** followed the slope and rise of the seam for another 350 metres, where the left hand heading drove in to the start of a new developing section, *(datum height 9724 feet)* while, during October 1904, the right hand heading turned slightly clockwise onto a bearing of 332 degrees, and went on for another 310 metres, where it joined up with the extended **pit bottom level**, *(west)* at 1600 metres *(1750 yards)* distance from the **Lochhead pit bottom**, (datum height 9780). The connection was completed in October 1905. The Lochhead pit and the Victoria pit were now well and truly joined, which served two practical purposes! The Victoria's cross-cut mines and its pit shaft, would serve as a much needed return airway for nearly the whole of the east side of Lochhead's Old Dook and of course, coal mined in either pit now had two exit routes to the surface! However, the Victoria pit **Sea Dook** was *not yet* commenced in the Dysart Main seam of coal, even though this underground area was to be the designated head *(or top)* of that great Sea Dook!

* *This was the endless-rope haulage level, between the head of the Victoria Dook and the working bottom of Nicholson's Dook, that is described in the introduction. It was brought into full use for the transportation of coal, after the abandonment and subsequent sealing of the 'over-ground' Hugo Tunnel between West Wemyss and the Coaltown of Wemyss.*

*

The **driving/sinking** of the Victoria Sea Dook commenced in January 1907 *(datum height 9590 feet)*, with the new left-side **coal** dook exactly 15 metres to the left (north) of the left-hand cross-cut mine. *(Now looking down!)* The companion dook *(also in the middle coals of the seam)* was commenced in October 1907, approximately 10 metres to the right of the right-hand cross-cut mine. That meant that the **Victoria twin Dooks** were to be initially approximately 45 - 50 metres apart at the commencement of their driveage. This would change at the **300 level** *(length)*, where the right-hand companion dook was moved approximately 25 metres to the right, giving a separation of fully **70 metres** between the two downwards, developing dooks. At this point on the **300 level,** the new start point for the companion dook was to be later extended back upwards as well as downwards, thus creating a new parallel heading, with a spacing of 70 metres from the main coal dook on the left. *(The 70 metres, spaced, parallel heading from the 300 level upwards, was not actually commenced until 1934.)* This now meant that there would be a 70 metres, parallel spacing between the two dooks over their full driven lengths to the dip and, following the middle coals of the seam all the way down to their maximum length/depth.

*{The miner's commenced the driving of these dook's by taking only the Grounds & Nethers along with the Myslen coals, at a width of approximately 12 feet. This would have produced a more than sufficient height in which to work and walk. The seam thickened as it progressed deeper and this would have resulted in the miner's having to 'take' less of one of the named coals in order to maintain a fixed height. Also, considering the solid environment that the miners were driving through (i.e. solid coal), the premise would have been to leave as much thickness of coal above their heads, as was conducive to maximum safety. An overall height of between six to seven feet would have been more than sufficient in which to operate a haulage system.}*

The twin, deepening dooks were developed critically true to direction and matched the parallel lines of the cross-cut mines exactly. They were therefore, driven down on a map bearing of 118 degrees, *(in the exact opposite direction by 180 degrees from the cross-cut mines, but taken to the dip to follow the grain of the coal seam)* and the coals taken as the miners progressed deeper, were still the Grounds & Nethers and probably the Myslen. A total thickness of approximately 98 inches *(2.5 metres)* at this depth. *(The vertical height was of course, somewhat greater!)* This left a good, solid and unbroken, and more than sufficient thickness of coal in the roof, to negate the need for steel girders. The immediate roof consisted of the Head coals, the Sparcoal and the Toughs & Clears coal. The Myslen coal may also have been left, depending the taken height at any given place. *(The immediate coal roof was therefore comprised of approximately 10 to 12 feet of unbroken coal. More than enough to be completely self-supporting!)* The unusable Sclits, the intervening redd *(stone)* and the thick Coronation coal, to sit underneath as an unbroken pavement. One of the governing safety factors in the driving of these parallel dooks, was that the stoops of solid coal on both sides of the driven dooks, needs must be left untouched and unworked, and to be effective, needed to be in the region of 40 to 60 metres wide to maintain the integrity of both dooks. *(The weighting factor would increase with greater depth!)*

By 1909, the dooks had been driven down to 285 metres in length *(datum height of 9425 feet)*, with a drop of 169 feet in depth. The angle of depression *(declivity)* was therefore 1 in 5.6. Over this 285 metres dipping length, there were eight connecting levels *(approximately 4 metres wide)* between the two dooks. The bottom two of the eight were actually **No. 2 level** and it's companion level coming from the Old Dook, that bisected the Victoria Dook at right angles and continued on to the Francis barrier. *(Over the dipping length of the Victoria Sea Dooks, each of the new sections/levels had their own successive, distinctive, identification **number**, which **did not** tie-in with the through connecting numbered levels from the Old Dook to the Francis barrier).* As the dooks developed downwards, there was by necessity, several intrusions made into the solid coals on either side of the dooks. These intrusions were cut at right-angles to the line of the dooks, so as not to damage the integrity of the statutory wide stoops. Therefore, they were actually horizontal levels. However, these short levels had to be driven, to enable the miners to get into the coal and establish new working sections on either side of the two dooks. These two sections, one either side of the dooks and the coals taken, have already been fully described in the chapter relating to the **Old Dook**. These two sections are the areas immediately to the left and right of the Victoria Dooks and lie between No.1 and No. 2 levels as named from the Old Dook. As mentioned earlier, the numbered levels from the Old Dook that eventually ran through *(bisected)* the Victoria Dooks, are known to the Victoria Miners by a different numbering system. Within the Old Dook, the levels are numbered consecutively from the top down as levels No. 1 to No. 7. *(The No. 7 level at the bottom of the Old Dook is actually the long, twin levels that connect the 1400 level at the bottom of the Victoria Dook to the Michael pit shaft, at datum height 8250 feet! This is known as the Michael Level!)* On the Victoria Dook, the different levels were numbered according to their distance in metres from the head of the Victoria Dooks. *i.e. The 150 level, the 300 level, the 600 level, the 900 level and the 1200 level, and all the way down to the 1400 level at the termination depth!*

The point at which the Victoria Dook was now driven to *(datum 9425 feet)*, was known as the **Three Hundred** level. The half-way point above that, at the middle level intrusions was known simply as the **One-fifty** level. *(This **300** level on the Victoria Dook equates to, and is synonymous with **No. 2 level** coming in from the Old Dook.)*

The section of coal *(area)* to the right of the Victoria Dook and above the No.

2 level, was wholly taken by Victoria miners and measured approximately 675 metres by 275 metres. The coals taken from 1921 to 1923, were the whole of the Coronation at 54 to 64 inches, possibly worked as long wall faces (*no other method would have been feasible*). (*This was the first time that this Coronation coal had been taken to any great extent - and I just do not know why?*) The section was re-visited again several times, with different parts of the remaining coals taken from different areas. This happened in 1929, 1930, 1931, 1932 and 1935. This area seemed to have been the test bed for several different types of extraction, as the method of close Stoop and room was used on the Myslen, Grounds & Nethers on the re-visitation, and then the Toughs and Clears in addition a while later. Mining history does not appear to record the troubles encountered, but these methods were not attempted again in any other section until 1944-45, and in a small Victoria section under the sea.

*

During 1910 and up to January 1911, the coal dook (left-hand dook) was extended a further 250 metres down to the newly named **600 level**, at a datum height of 9265 feet. The companion dook was a full six months behind this and did not reach this 600 level until the middle of 1911, at the same 9265 datum height. This **600 level** was almost on the same datum plane as the **No. 3 level** from the Old Dook and was the lowest of three levels that ran from the **Old Dook** to the **Francis barrier,** bisecting the Victoria Dook. (*These three parallel levels eventually stretched all the way from the Buckhaven fault to the Francis barrier, a distance of approximately 3750 metres or 3.75 kilometres!*) These parallel levels interconnected the Victoria Dook the Old Dook and the New Dook

To the right (*south and west*) of the Victoria Dook at the **600 level** (*adjacent the Old Dook's No. 3 level,* three new parallel levels approximately 20 metres apart, were commenced during the month of January 1924. (*At datum heights 9265 feet, 9279 feet and 9284 feet!*) These three levels followed the level plane of the coal and were driven eventually approximately 650 metres to the Francis barrier, where the arrival date was January 1928. The distance (*solid stoops*) between the three parallel levels along their full length, was as stated, approximately 20 metres, with the three levels being interconnected approximately every 75 metres. The solid stoops of coal at this depth measured 20 metres by 75 metres, hopefully giving extra stability to the roof, bearing in mind, that one or more of these levels was bound to have been used as a coal haulage-way of some sort!

This coal section therefore measured 650 metres by 225 metres and the method of extraction was again by Stoop and room. The area immediately above the parallel three levels was much intruded into, in that the solid stoops being left after extraction were reduced to approximately 25 metres square, with the additional driving of five parallel levels, running the full 650 metre length from the dook to the Francis barrier. The driven parallel headings that bisected the top five parallel levels, were 25 metres apart, so that meant, that slightly more of the total reserves available were taken than was usual at this depth. The named coals that were extracted, were initially the Coronation and the Sclits, a height of approximately 81 to 87 inches plus 9 to 12 inches. *(I rather suspect that the Sclits were taken to give added height to the extraction - and so that the next 'sandwich' of coal, the Grounds, would form a much more solid roof. The Sclits were utterly useless as a form of fossil fuel!)*

Within this working area interspersed by the 25 metre solid stoops, these two bottom coals would have been taken first, until all of the levels and headings had been driven, and a re-visitation was made to take the Grounds & Nethers and probably the Myslen coals at a later date. This would have then amounted to an additional 110 to 120 inches of coal, as the seam thickened to the south and west. This additional coal would have been much easier to take, considering that it amounted to 'dropped coal', which only needed 'blowing down! The only problem with this downing of the coal, was the colliers getting up high enough to drill the holes. The percentage of coal taken in this section would have been slightly more than average at approximately 12 to 15 percent, in that even though more of the coal was taken, the seam was now considerably thicker!

This extracted coal would have been transported along the **600 level** to the Victoria coal Dook, up the dook via the endless haulage onto the Victoria Haulage-way, along the Victoria haulage, to the bottom of Nicholson's Dook, up Nicholson's Dook via another endless haulage to the inside end of the pit bottom level, and then along the length of the Lower Dysart, endless haulage-way and thence to the pit bottom. It can be seen that each and every hutch of coal from this section, was being handled by five different groups of oncost miners, and five different means of mechanical haulage, on it's way to the pit bottom. *(Such transport costs did not bode well where the 'pick and shovel' miner's were fighting for a decent tonnage rate!)* It was entirely possible however, that some of the loaded hutches of coal could have been transported up to the head of the Victoria Dook and taken back along the cross-cut mine to the Victoria pit bottom, even though that winding capacity was only half that of Lochhead, and slightly slower as well.

To the left of the Victoria Dook between the **300 level** and the **600 level,** *(where the Old Dook's **No. 3 level** eventually connected),* the inroads to the new developing section was also made in 1924, with a first level to the left at datum height 9380 feet, and a second level at the 9265 datum weight. *(These levels were 185 metres apart!)* The section to be opened up would cover a area measuring 325 metres by 175 metres, and comprised of three new levels, approximately 20 metres apart at the 600 level, with five interconnecting headings approximately 80 metres apart. The remainder of the section above the levels was taken out by Stoop and room working, where the stoops were approximately 20 metres square and matrixed by a further 5 levels and 11 headings. The coals taken were the Coronation coal at 65 inches, the Grounds & Nethers at 68 inches, and the thinner Myslen at 10 inches. The Toughs & Clears, the Sparcoal and the Head coals were all left as a thick, solid roof, probably 11 to 13 feet thick. This section was worked between 1924 and 1929 and then left abandoned, except for the three through levels to the Old Dook.

The next section to be worked along this **600 level,** *(by the Victoria pit miners),* was in fact, merely a continuation from the last section, but with a solid 50 metres wide stoop of unworked coal between this new section and the preceding section. The three bottom-most parallel levels towards the Old Dook, were continued on until the half-way point *(between the Old Dook and the Victoria Dook)* where the bottom level was discontinued at a distance of 675 metres from the Victoria Dook because of a fault line that had appeared from the Old Dook direction of this bottom level. There was therefore a break of approximately 135 metres length along the line of this bottom level. This made no difference to the connection between the Old Dook and the Victoria Dooks, as there was two unbroken parallel levels still interconnecting the two Dooks.

This middle section between the Old Dook and the Victoria Dook was fully 660 metres long and 190 metes wide *(and on a gradient of approximately 1 in 5).* Again, it was worked by the Stoop and room method of extraction, with five parallel levels running the whole length of the 660 metres and 22 parallel headings across its length. The coal stoops left standing between the three, bottom, parallel levels measured 20 metres by 60 metres and the solid coal stoops within the taken matrix measured approximately 20 metres square! The numbered amount of solid coal Stoops left in this section alone, amounted to a staggering total of 95. This section saw the drawing together of miner's from both the Old Dook and the Victoria Dook. Indeed, they did meet up around the centre point of this large working section, in the years 1928 and 1929! The coals taken were again the Coronation coal, the sclits, the Grounds & Nethers and the Myslen.

The one decidedly strange operation in this section, happened prior to 1949, where three dipping intrusions were made from No. 2 level above and into a small strip of this section, measuring 275 by 35 metres. All along this small strip, the same coals was removed, but this time, by a long-wall face using wood and redd *(inert stone)* packs to fill the space vacated and sustain the roof without breaking. This happened only once on this long level and the reason for the extraction does remain a mystery to me! *(I rather suspect it was a form of 'day-light robbery' in the ever-lasting quest for cheap easy coal.)*

The coal from this shared section extracted by the Victoria miners, probably went back along the 600 level to the Victoria Dook and thence upwards, whilst the coal extracted by the Lochhead miners would travel along **No. 3 level** *(opposite to the Victoria's 600 level)* and be carried away up the Old Dook - direct to the Lochhead pit bottom. I can not even make a guess at this stage exactly how much coal was transported in either direction, but it does seem that the Old Dook haulage-way offered the shortest route, even though it would have been utilised to capacity. The coals in this large section were worked out by 1930 at the latest, with this section and the sections immediately to either side, being extracted by an almost perfect matrix system, which by it's very regularity, suggested accurate and diligent survey work!  Simple arithmetic would put the amount of coals taken, at approximately 11 percent of the total reserves available in this large section!

*{This section was somewhat plagued by a single downthrow fault (from the Victoria side) that showed a full seven feet of throw and also appeared to run under the Coaltown of Wemyss!}*

Around 1926, both of the Victoria Dooks were again in the process of being extended deeper.  The same strict bearing line was adhered too, with the spacing between the Dooks remaining constant at 70 metres.  By the beginning of 1929, the dooks had been extended a further 325 metres down, *(datum height 8935 feet)* and the dooks now lay directly below the low tide mark on West Wemyss beach. *(And 1065 feet underneath the shale!)*  The running lengths of the twin Dooks was now approximately 890 metres, and so this new section was invariably named the **900 level.**  The difference height between the **600 level** and the **900 level** was 330 feet, therefore the angle of descent *(declivity)* was approximately 1 in 3.  The dook was obviously getting **steeper** as it progressed downwards!  This would not cause too much of a problem as long as the slope remained constant.  The problem would arise if the angle of descent grew shallower, in that the haulage-rope would be pulled tight on its upwards journey.  The wire ropes would then cut into the line of the coal roof and perhaps gouge a long channel along its path. *(Which it did within a short space of time!  This however, was quickly alleviated by the fitting of metal roof-rollers at appropriate places on the dooks.)*  The wire ropes under tension, would also tend to 'lift and overturn' any loaded hutches of coal on the way up, if sufficient weight *(a greater number of loaded hutches)* were not clipped to the up-side of the haulage-rope and evenly spaced.

At this new **900 level,** the miners again commenced to drive levels at right angles to the right and left of the dooks.  The now customary three parallel levels spaced approximately 20 metres apart were now *de rigruer,* and interconnections between the three levels were made every 75 metres at this depth.  An increased length of approximately 15 metres, taking into account the added weight and extra depth.  However, the interconnections between the first and second levels were only 37.5 metres apart. *(I can but guess, that this was something to do with the initial ventilation as the levels were being developed!)*  The parallel levels to the right of the main dook, were driven approximately 775 metres in length and driven to come up against the Francis Barrier.  The most noticeable characteristic about these levels, was that they were driven in an almost straight line. *(The lie of the sloping strata was almost completely uniform in this area!)*  The difference datum height between the dook and the Francis barrier was plus 37 feet, therefore a slight downwards slope of 1 in 68, would no doubt help the loaded hutches on their way to the main dook!

This section was entirely unique within the Victoria workings, in that it was by far, the largest in overall size and area, and for the problems that were confronted and overcome. There were fully, five, separate, long, strata faults encountered in its enclosed area, some of which were double faults. *(An up-jump followed by a down-jump, or vice-versa.)* Most of these faults/jumps were not very great, probably between one foot to six feet and either up or down, but they did stretch from the **600 levels** down to the **900 levels** in most instances! Within the large area of the section above the three parallel levels at the **900**, the full size of the section was opened up by 19 separate headings each 30 metres apart and to a length/height of 275 metres.

These parallel headings were cross-cut by 9 parallel levels, again 30 metres apart and driven to full length, up to the Francis barrier. The remaining, solid, coal stoops amounted to 175 in number, which made this the largest single section in overall area to be worked within the Victoria pit. The coals taken at the following thicknesses were:- The Coronation coal at 70 to 75 inches thick, the Sclits at only 5 to 10 inches thick, which had thinned and lost some of it's stone content, followed by the Grounds & Nethers at approximately 85 to 90 inches thick, and the Myslen at 15 to 18 inches. The Dysart Main seam had thickened considerably at this depth to a full 30 feet, which meant that the Toughs & Clears, the Sparcoal and the Head coals were left untouched. The thickness of coal extracted in this section was approximately 16 feet, which meant that a full 14 feet thickness of unbroken coal remained overhead. The amount of coal extracted from this section alone, would have amounted to an amount in excess of 300,000 cubic metres of coal. Again, by this method of extraction in this large Stoop and room area, only about 15 per cent of the total reserves could have been taken. And unfortunately, this section was now classified as **worked out!** The section was therefore left as it stood, classed as abandoned in 1936, having been worked, albeit successfully, by the pick and shovel methods of the time.

At this same **900 level,** the customary three parallel levels were driven to the left of the Victoria Dooks and these were interconnected to the same levels coming in from the Old Dook. These were known as **No. 4 levels** on the **Old Dook.** *(The name description given to the level, depended on whether the collier descended the Victoria pit or Lochhead shaft.)* The system already in use was rigidly adhered too, in that three parallel levels were driven from the Victoria Dook towards the Old Dook in 1929, and three parallel levels were commenced from the Old Dook, to link up with those from the Victoria Dook in 1931. These three parallel levels did take two years to link up, for the reason that the miners from the Victoria did open up a coal section immediately to the left of the dook at the 900 level, leaving an 80 metres wide stoop between the first heading and the left dook. This fair-sized section area was almost square in size, measuring 350 metres long by 315 metres wide.

The section was developed and opened-up by the now customary matrix extraction system, with 11 parallel levels, *(including the original three parallel levels)* and 11 spaced, parallel headings. There were 100 solid stoops developed, each approximately 30 metres square, and the coals taken from between them, were the Coronation, the Sclits, the Grounds & Nethers and the Myslen. In this area, the coal thickness extracted was slightly less at 14-15 feet, with approximately the same thickness being left as a solid roof. This section was worked between 1929 and 1934 and abandoned with no further intrusions recorded.

There had been exactly six level intrusions made into this section, through the 80-metre wide, solid coal stoop that lay parallel to the line of the main dook, three at the 900 level, and three more evenly spaced between the 600 and 900 levels. Exactly the same amount of intrusive levels had been made into it's sister section on the other side of the dook, *(the large section)* between the 600 and 900 levels.

Between the Old Dook and the Victoria Dook on this **900/No. 4 levels,** there were three separate wrought sections of approximately the same area. The middle section of the three, was where the miners of both collieries met and worked together. *(Readers are reminded that both Lochhead and the Victoria pits came under the same manager, who was based at Lochhead. The Victoria pit had its own under-manager.)* The system of extraction was precisely the same as before with the exception that the extraction matrix was on the slew. *(This meant that the remaining solid stoops were slightly rhombic or quadrilateral in shape.)* The eleven driven parallel levels all roughly followed the grain of the coal, but they were all slewed slightly to the left, following the changed steepest rise of the seam. *(The strata had somehow developed a slight north/south roll in this underground area!)* The bisecting, parallel headings were at approximately 80 degrees to the levels. This made no difference to the taking of the coal, merely that the three, parallel interconnecting levels between the two dooks were not in a straight line, but curved gently left in direction towards the Old Dook in order to maintain a level datum plane, hence the need to slew the parallel headings through this section.

This bottom level of the three that connected the Victoria Dook to the Old Dook, differed slightly from its two companion levels, in that this bottom level ran directly under the edge of the buildings of the Wemyss Castle home farm. The stability of these farm's buildings, was primarily protected from underneath by a 300 metres diameter stoop left mainly untouched, so as not to effect the integrity of the farm buildings. So, as this bottom level passed underneath the farm area, the only coal to be extracted was the 65 inches height of the Coronation coal, to effect a low through roadway. The two companion levels above the bottom level had the bottom fifty percent of the seam extracted all the way through the circular stoop. *(These extractions were carefully monitored by the resident surveyor at the time of the 'diggings', but no reports of subsidence were received. The colliery manager's head would have probably been served for supper if structural damage had been caused to the home farm.)* The north-east quarter of this middle section, *(bottom right just above the levels)* was actually worked under the fields of the home farm, where the Coronation coal, the Sclits, the Grounds & Nethers and the Myslen coals were extracted without concern, on the part of the owners. *(Who cared if the level of the ploughed fields drop a few metres? It didn't make any difference to the cattle-beasts, or to the ploughing!)* This middle section was worked out and abandoned by 1934, with no return visit ever logged.

In or around 1930, the Victoria twin Dooks were further extended down to reach a datum depth of 8480 feet. This was completed by October 1932, which meant that the dooks were now approximately 1160 metres long and at a depth of 1520 feet (463 metres) under mean sea level. *(This was named the **1200 level**!)* Not only that, but at this length the Victoria Dook was now under the open sea bed. It is now also worthy of note to mention, that the only coals extracted from the thick seam in the deepening of the dooks, was the 72 inches of the Grounds & Nethers coal plus 6 inches of redd. The incline on the dook had now increased to a gradient of 1 in 3.3, an angle of 18 degrees depression. This increased dip on the dooks, therefore made the 78 inches of extracted coal appear rather more to a miner standing upright.

This added thickness of coal left in the roof, which amounted something like 15-16 feet, was deliberate, in that it gave added strength to the roof at this new greater depth thus ensuring a larger safety margin, with little or no sustained crushing-down in evidence.

The first horizontal levels that were driven from this **1200 level** to the right of the dook, at datum height 8470 feet, were a parallel pair of levels approximately 25 metres apart. They were driven through the solid to the Francis barrier, ostensibly following the grain of the seam, but with a rise of 40 feet over their 775 metres length. They were interconnected every 50 metres or so, for approximately three-fifths of their length. *(At that point the method of extraction changed!)* This section above the **1200 level** and below the **900 level,** measured approximately 725 metres by 275 metres and could be separated into three distinctly different section areas, each with different methods of extraction.

The first and largest section, covering approximately three-fifths of the total area, was worked using standard Stoop and room methods. There were five, evenly-spaced, parallel levels driven-in from the dook, at regular intervals between the 900 levels and the 1200 levels, and these levels were later bisected by eleven, evenly-spaced parallel headings running up from the two original levels driven from the 1200 levels. There was also within this section, a positive change in the size of the solid stoops to be left standing. The large stoops that were formed, *(left in situ)* measured approximately 35 metres by 60 metres and were about 55 in number. However, in the year 1943, within the roadways running through fourteen of these adjoining stoops, it was decided to strip both sides of these stoops to a further six to seven feet, giving an additional 12 to 14 feet width of coal. This experiment extended to an area measuring 400 metres long by 100 metres wide and covering eighteen large stoops. I do no know the outcome of this experiment, but, as the policy was not extended to the remaining stoops, I can but surmise that it must have showed some negative results. *(Undue or excessive crushing down of the roof!)* It has also been mentioned to me *(by colliers who were there!)* that the extra space vacated by the stripped stoops had to be stowed with redd and sealed. This does make good sense to me, in that the 'robbed' stoops were in the centre part of this part of the section and could be safely by-passed. The coals taken from this section were the thickest and richest yet within the Dysart main seam, in that the Coronation coal was 85 inches thick, the Grounds & Nethers were 84 inches thick and the Myslen *(splint)* had developed to 18 inches in thickness. This was the area where the Dysart Main seam was at it's thickest and richest! This section was originally worked between the years 1934 and 1936 and I am positive, that the intentional stripping *(daylight robbery!)* of the fourteen stoops in the middle of this section was perpetrated during the middle part of W.W.II, (1943) when the need for quality coal was at its greatest!

The innermost third part of this **1200** section *(still to the right side of the main dook)* was commenced in January 1936 and was slightly separated from the original Stoop and room workings by a 275 metres long heading, from the bottom 1200 level up to the 900 level. That 'installed' a 30 metres-wide, barrier stoop between the new developing section and the earlier Stoop and room workings. Two, separate, medium-sized sections were then developed inwards from this long stoop, one above the other, *(to the rise).* The top section measuring approximately 300 metres by 125 metres and the bottom one 300 metres by 150 metres. These two, adjacent, developing sections would terminate at the Francis boundary.

The top section was commenced in April 1936, where the Coronation coal was fully extracted over the next year, by the long wall method of pick-places working to the rise, on a face-line of 125 metres length. The space vacated by the extractions, must have been partially packed to allow the seam to settle to the pavement without breaking - for the reason that after the Coronation coal had been completely removed, the colliers went back to the start point *(the 30 metres wide barrier stoop!)* and ventured to commence another series of short levels in the Grounds & Nethers Coal. These levels were taken in to the full length of the section and when the Francis barrier was reached, the miner's then did an about turn and commenced to rob the adjacent stoops to the maximum, whilst working to the retreat. They extracted the full height of the Grounds & Nethers and the Myslen splint over this area, which measured 125 metres by 150 metres. {*It was in 1940, when the miner's drove to the extremity of this section, successfully driving fourteen parallel levels the short 150 metres distance to the Barrier. (In the Coronation coal.) During 1944, the miners also managed to extract the Grounds & Nethers and the Myslen in this small area, without too much trouble! The Victoria pit under-manager must have been very desperate to take this meagre amount of coal from such a distant, small, deep section!*}

The bottom, adjacent, lower section to the left at the 1200 level, was slightly larger that its co-section and was initially tackled in exactly the way, with the Coronation coal being taken first and by the same method. This extraction took place finishing in January 1936, with the bottom Coronation coal being removed from the whole area of the section. The followed a complete mish-mash of different extractions at different times, with attempts being made to take first the Grounds & Nethers, but leaving the Myslen splint! This was followed by a second re-visitation, where attempts were made to take the Toughs & Clears, but leaving the Sparcoal and the Head coal to support the immediate roof! *(This section was re-visited at intervals between 1942 and 1945, to extract any coal that could be safely taken without the roof coming down. I can only surmise, that considering this was the very last section to the left area of the 1200 level to be worked, that the manager took the chance to extract every last ton of coal while the roof held firm. After all, it did not really matter if the section was bedevilled with roof falls. It was to be abandoned in any case!*

Author's Comment:- *If all of the Stoop and room section areas in the Victoria pit were to have been extracted to the same degree that this last section suffered, I would estimate that the life of the Victoria pit could have been extended two-fold, with the total output in coal more than doubled! The extractions took place at a time when the country was in dire need and many chances were taken both by miner's in the pursuit of better wages, and by the management in the struggle to produce greater quantities of cheap coal!*

*

From October 1935 to April 1936, the two dooks were further extended down to what was to be the final sunk depth of the **Victoria Dooks**. The dooks were taken down a further 200 metres in length, *(datum height 8200 feet)*. This meant that the dooks were now nearly 1325 metres in horizontal length, *(the slope distance was greater)*. This new level(s) was named the **1400 level** and this was to be the final depth *(length)* of the two dooks. It is of note to record, that the angle of depression had now slightly increased and the gradient was now approximately 1 in 3. The man-haulage was not extended down to this level and the miners working below the **1200 level,** had to scramble up and down this distance on a daily basis, to and from their place of work.

To the right of the dook between the **1200 level** and the **1400 level**, a very small section was developed. *(Economically?)* By virtue of its small size, it could never have produced a sustainable output. It measured only 100 metres square and the coals taken were the full height of the Coronation coal, followed by the partial extraction of the Grounds & Nethers over 30 percent of the section area. The system used to extract the Coronation coal, was by a system of closely-spaced parallel levels taken to the advance, leaving the minimum stoops between extractions, *(probably*

*50/50 pillar and stall).* When the miner's reached the maximum boundary, they again did an about turn and commenced to strip the stoops on both sides to the retreat, leaving the wastes to regularly bend and totally collapse! This small section was commenced in 1937 and could only have lasted a few months. The Coronation coal had now suddenly thinned to only 43 inches in height. A natural phenomenon that was bound to happen sometime!

The one claim to fame that this **1400** level provided, was that from this depth, *(datum height 8250 feet).* The twin parallel levels to connect to the **Micheal Colliery** were commenced in April 1937. They were driven to strike and connect to the bottom of the **Old Dook** in **Lochhead pit,** which occurred in 1938. That then completed the interconnecting parallel levels to the **Micheal pit shaft** area, where a previous connection from the bottom of the **Old Dook** had been made in 1931.

*

The horizontal levels that were driven to the left of the Victoria Dook between the **900 level** and the **1200 level,** were never going to be of any great length, nor produce very much coal. There were six level parallel intrusions from the left side of the dook between the **900 level** and the **1200 level,** but these were merely extensions of the interconnecting levels between the two dooks, *(the main dook and its companion).* These parallel levels were at regular intervals of approximately 50 metres apart, except for the bottom two, which were only 25 metres apart, and these levels were terminated by a long heading driven between the **900 level** and the **1200 level**, parallel to the main dook, but approximately 65 metres from it. These short levels were therefore only taken in to a length of approximately 65 metres, where they were all linked up with this **single heading,** that lay parallel to the line of the coal dook. *(This heading was actually **575 metres** long and ran from the bottom of the **1200 metres level** all the way up to the top of the **600 metres level.**)* There was **no fully developed coal section** to the left of the dook between the **900** and **1200** levels. The bottom **275 metres** of this 575 metres long parallel heading was meant to be the **extreme boundary line,** so that the miners could not extract any of the coals near to the solid stoop underneath the Wemyss Castle. Later on in October 1943, when coal was at more of a premium, this boundary heading was breached by two narrow intrusions in the direction of the **Wemyss Castle Stoop,** but stopped short of actual intrusion into the circular stoops. *(This was sacred ground!)*

The first limited intrusion to the left of the dook at the **1200 level,** was at a datum height of 8547 feet and consisted of a pair of levels at 50 metres apart and driven in to a distance of 80 metres. This commenced in 1935. The narrow section was 375 metres long and between 40 to 50 metres wide. This was certainly a long narrow section worked to the advance, where the miners stripped out the 45 inches height of the Coronation coal, using wood pillars and redd as temporary roof supports. This section was finished in October 1939. The amount of coal extracted at this time, would have amounted to 16,500 cubic metres. This narrow section was re-visited in the summer of 1941, where the miners removed approximately 5,000 cubic metres of the Grounds & Nethers coal over the inside 100 metres length of the 40 metres wide intrusion. This small section was abandoned in July 1941, and never intruded into again. This small section may just be remembered for the intrusion of a close double fault that showed up as the miners approached the 200 metres length of the initial intrusions. It seemingly did not hinder the subsequent extractions!

The second and final of these narrow intrusions to the left of the dook, commenced at the approximately halfway point between the **900 level** and the **1200 level** and took the form of twin levels driven 25 metres apart, to a length of 375 metres from the left of the 575 metres long parallel heading. These parallel pair of levels were interconnected along their length by short headings spaced approximately 40 metres apart. The coal taken at the final 100 metres, was the Grounds & Nethers, the Myslen, the Toughs & Clears and the Sparcoal. This left only the Head coal as a roof and the Sclits as a pavement, with the untouched Coronation coal lying below that. This was the very last of the extractions taken from the Victoria Dooks and I feel that this was a last ditch attempt to take some extra coal from an area that could have provided much more coal, even though it did border on the Wemyss Castle stoop.

This area which lay between the Victoria Dook and the Wemyss Castle stoop, was approximately 450 metres in length and 375 metres in breadth. There had been no prior workings in the area between the **900 level** and the **1400 level/twin Michael levels** and the area remained virtually untouched before these two limited intrusions. Perhaps the 'writing was on the wall' with regard to the Victoria pit at this time, but considering the planned and systematic extractions that had taken place in every other section within the Victoria pit, I find it singularly strange that this very large and potentially rich area *(measuring 450 metres by 375 metres)* was merely tickled, and not severely scratched! After all! This was 1945 !

*

In the Victoria pit of yesteryear, there was very little in the way of modern machinery. The hutches were made from wood and iron, the winding gear was barely adequate, and each cage took only one eight hundred-weight hutch at each winding. That meant of course, that when men were carried in the cages, only four men at a time were allowed to travel inside. By today's standards, one hutch full of coal weighing all of eight hundredweights at one winding, would hardly seem to be economical, but needs must, especially at that time when coal was at great premium.

There was never a large number of men and boys *(and girls!)* employed in this pit, but those that worked there, did make it a viable concern and it produced coal in a steady stream. So much so in fact, that at times there was an excess of coal that had to be transported elsewhere, other than into the small colliers and coasters that docked and loaded coal at West Wemyss harbour. This was a very simplified affair, in that the coal as it came up the shaft, was wheeled directly around in a semi-circle onto the harbour wall, tipped into small chutes and loaded directly into the ship's holds. Such was the production from this small pit and the fact that the extra coal could not realistically be binged, that the need to get the coal out of West Wemyss loomed large. There was no railway out of West Wemyss, it was a rail-locked village on the sea-shore. There was no way out along the sea-shore, too many rocky outcrops and semi-sheer cliffs. *(The original coast road from Dysart to Aberhill had all but disappeared, with the submergence and total engulfment of the Links of Buckhaven village and the subsequent construction of the new Wellesley Road.)* Transport by lorry was unrealistic, in that the torturous climb out of the village was cost prohibitive and how far to travel? After much deliberation on the part of the Wemyss Coal Company, it was decided that the only feasible solution at this time, was to drive a below ground, tunnel straight from the beach, up through the metals to the west end of the Coaltown of Wemyss. A straight-line distance of exactly 750 metres, this mine when completed, would take the excess coal directly from the pithead of the Victoria, up through the mine and be directly loaded into the waiting coal waggons at the mine head. *(After passing through a small screening, but not grading plant).* This was a very practical though expensive solution, but would result in the excess coal getting into the main stream outlets very quickly. This tunnel came to be known as the Lochhead or Hugo Mine. *(This mine is described in a different chapter of this book).*

*

Author's Note:- 1. *The only trace of the Victoria Pit at West Wemyss that can be seen today, is a bricked-up wall set into the cliff face, where the entrance to the small twin cages is still visible! All traces of the second 'modernised' winding house have gone, with bushy trees and natural vegetation now serving to hide the original location of the old Victoria shaft.*

*The village of West Wemyss was the very epitome of a typical mining village, where the collier's spilled out of their small house and cottages dressed for underground work, with only five minutes in time separating them from the pit cage. But of course, they could also be seen spilling from these same houses and cottages, differently dressed for their secondary occupation, which again, was only five minutes from home, - Sea Fishing!*

Author's Note:- 2. *During my initial early interpretation of the underground plans for the Victoria pit, I seemingly and unwittingly misread the original diagrams, in that I almost failed to see the direct connection between the level cross-cut mines and the subsequent broaching of the Dysart Main seam in the Victoria pit! I had originally supposed that the Victoria shaft was struck directly into this seam of coal. Not so! I was shewn the errors in my research by David Rodger of the Coaltown of Wemyss, coalminer, later deputy at Lochhead Colliery, who corrected my misreading of one of the underground plans and subsequent misconceptions, and directed my attentions to the difference in datum heights between the shaft bottom and the depth of the Dysart Main seam at that point. Sufficient to say, that upon further detailed examination of the drawings, his proper interjection and interpretation corrected a few otherwise erroneous suppositions on my part!*

\*

David Rodger, ex-coalminer, erstwhile, lamp carrying deputy, is well retired and still lives in the Coal-town of Wemyss with his lady wife. Unfortunately, David's state of health does give them both cause for concern, in that he, having spent all of his life in one coalmine or another, is a victim of the coalminer's lung disease, emphysema! The probable and inevitable result of having spent many years in a fouled and dust-laden, underground environment

After qualifying as a Mines Deputy, one of David's main specialist tasks, was to walk and travel the lengths of the Victoria Sea Dook, the Old Dook and sometimes into the far reaches of the New Dook, in the daily ritual of testing for fouled air and mine gasses. This onerous, but highly crucial task, ensured that he covered many underground kilometres in his daily travels, using his glennie at regular intervals, within the confines of the many dark and almost inaccessible, disused workings within the deep Dooks. All to well can he remember the tortuous and the lung-constricting scramble, in descending and then ascending the 1 in 2 gradient, to and from the 1400 level at the bottom of the Victoria Sea Dook. *(The man-haulage did not extend below the 1200 level on the Victoria Dook!)* The traversing of this 200 metres long gradient was a nightmare journey and required the use of two free arms on both the downwards and return journeys. On the uphill climb, much use was made of both steel girders and ground rails to effect an upright position, whilst, on the downwards journey, many unwanted yards were sometimes gained by an undignified and some-times painful slide on his backside! *(Many old colliers saw no loss of dignity, in having to scramble up-wards on hands and knees, knowing that the man-haulage bogies awaited them at the twelve-hundred level!)*

This itinerant deputy knew where to find the localised areas of Firedamp *(Vapours of Methane gas)* within the depths of the hot and sticky New Dook - and how to gauge exactly what percentage was present! *(Please refer to chapter on mine gasses!)* He also knew of the various trouble spots in the long levels of the Old Dook, where the incidence of Blackdamp *(vapours of carbon dioxide)* was likely to occur. He was one deputy who did pay close attention to the surface weather conditions and where the wind was coming from! He was also a regular visitor to the aneroid barometer at the manager's office. By the time David descended the cage at the start of his shift, he had already made up his mind as to the worked-out mine areas he would especially visit, during his long peregrination. Long experience and deep, empirical knowledge had honed his perception and produced an uncanny insight, that almost guided him to those trouble spots in the mine, where the egress of Blackdamp in particular would be at its most prevalent! This was an equation where the vagaries of weather, wind, air temperature and barometric pressure, were resolved by the unerring instincts of a competent and conscientious coal mine's deputy! In Lochhead Colliery at least!

\*

# Chapter X

## The Getting of the Coal!
### *(or, How it was done in Lochhead Pit!)*

## Methods of Extraction

Coal-getting, coal-taking, or coal extractions, are probably the collective terms for any type of coal mining whether it be from deep sunken pit shaft, surface mine or open-cast mining. All involve years of planning, many complicated underground surveys, lots of test borings, and a vast outlay in terms of money, machinery and men. *(And, I may add, an awful lot of luck!)*

In this chapter, I shall attempt to describe in detail some of the more direct methods of mining coal at the coal face as practised before, during and after my time as a coal-miner *(1950-58)*, where the underground miner came in direct contact with the coal in it's original solid state. I shall try to describe fully, the types of extraction where the tools used were of the most primitive and decidedly, old fashioned kind, and where the hardened collier himself was the 'machine!' I shall also attempt to describe in detail, some of the equipment used up to and including, the most highly mechanised and sustained methods that were in use during my years as a coalminer and a little beyond, all utilised within two of the most prolific seams of coal ever to be found in Lochhead colliery - and indeed, the whole of Scotland.

## 'Long-Wall Pick-places' (Traditional)

This was a method whereby the coal was extracted all along the length of a complete face and to a given height, the taken height dependent on the overall height of the seam worked and the amount of coal that needed to be left intact, to serve as a good 'roof' or pavement. *(Either, or both!)*

If the coal to be worked and extracted was on a slope, then the very slope itself would be utilised, using a fixed wheeled system with an endless rope haulage. *(Often described as a 'self-acting' incline, but not a true 'aeroplane' brae.)* To initiate this sort of operation, a bottom level would be driven through the coal seam to be worked, so that the level would be near horizontal, giving inertia and stability to stationary hutches. From this level, a heading would be developed in an upwards direction *(following and wrought through the grain of the seam)*, to a distance of approximately 100 yards, and wide enough to accommodate two sets of narrow-gauge rails, side by side. At the top of this heading, there would be a braked pulley-wheel firmly 'stelled' to the pavement by a stout wooden pole, that was held firm by being angled and jammed tight to the roof. Around this pulley-wheel there would be several turns of half-inch diameter, steel cable, which ran from the pulley wheel down between one set of rails, and onto one side of a larger five or six foot diameter, horizontal return wheel, which was housed in a short dook *(dipping mine)* on the low side of the level. This cable would then be run up between the other set of rails to return to the braked pulley-wheel, thus describing an unbroken spliced cable, which it now was! This 'endless' rope haulage, tensioned by a floating bogie-weight at the bottom end, was made to move in one direction only, with the weight of the loaded hutches from any of the manned faces on the down side of the brae, constantly supplying the gravity force needed to 'raise' the empty hutches on the other rail track. The movement of hutches on the brae, was tightly controlled by the oncost man on the pulley-wheel brake, who was named the 'Wheeler!' Once this self-acting brae was set-up, it would stay in place until each of 10 levels cut into either side of the heading had reached a full 50 yards depth, and taking the full width of 8 to 10 yards each.

*(This meant that an area of coal measuring approximately 100 yards by 100 yards at a given height, would have been extracted and left to 'weight' before the section was abandoned. The next successive 'self-acting incline' would then be commenced approximately 110 yards further along the level, thereby leaving a 10 yards wide solid stoop of coal between each successive abandoned section.)*

Author's Note:- *This description is but one of many combinations of wire rope and chains, that may or may not, have been used in Lochhead and the Victoria pits. Every other incline or brae seemed to adopt a different variation of the same theme. The main consideration was that whatever combination was chosen, it was made to work to serve every coalminer in the section with a fair distribution of empty hutches, and remove the filled ones.*

To operate this self-acting incline at its optimum efficiency, would require an amount of hutches equal to twice the number of miners 'places' at the coal. If there were 10 miners on the brae, then 20 hutches would be the optimum number used. To start the procedure, a series of single empty hutches would be lashed by a short length of chain onto the up-going cable and manhandled up the heading so that each man in turn had an empty hutch to start with. This hutch would be 'binched' *(manhandled)* off the steel rails and run into each 'place', where it would stand ready to be filled up by the miner/s concerned. When this was done and these hutches were filled with coal at each mans place, the 'working' brae would come into being. The wheeler would be in position and the brake held on. At this point, the 'boy' would call or signal the wheeler, that he required an empty hutch to be run-up parallel to 'his' miners place, so that it could be manhandled across the rails and 'couped' onto its side, clear of the full hutch coming out of the miner's level. As each separate hutch was filled, the 'boy' would slew *(binch)* the 'snibbled' full hutch sideways on to the down side rails and firmly lash the hutch to the haulage cable using a special non-slip hitch, *(this had to be learned very quickly!)* and would signify 'Clear!' The wheeler would not operate the brake at this time, he would only do so when he was next signalled that someone else needed an empty hutch, and then, only with enough movement on the wheelbrake to bring the next empty hutch level with the recipients workplace. The golden rule of this whole operation, was that the permanent 'drawer' on the bottom of the brae, could only attach an empty hutch to the up-side rails, after he had undone the lashing chain on the last full hutch to land at the bottom of this heading. This meant of course, that there were only just enough empty hutches lashed to the up-side rope of the brae to satisfy the immediate needs of everyone on the brae, with no empty hutches ever reaching the end of the up-rails. If it so happened that there was an excess of 'empties' on the brae, they could easily be binched off and couped to one side before they were actually needed, in order to balance the requirements of the colliers. No two colliers worked at the same speed, although daily competition was rife!

# Working the Coal

To work the coal, the following procedure would then begin. One side of this heading would be selected and depending on the height of the coal, a measured length would be allocated to each man and boy, perhaps 8 to 10 yards. *(The 'length' of a pick-place would depend on the taken height of the coal. The colliers were paid by weight from volume, therefore, more tonnage would accrue from a taller stand of coal!)* This mark was heavily chalked on the roof so that each miner could identify his ground. Work would instantly commence with the skilled miner starting at the middle of his allotted ground *(stint)* to under-hole the stand of coal, with the boy filling the chippings. *(This would approximate to the size of small chirles.)* The face of coal to be taken up and along the full length of the heading, would stand near vertical from top to bottom, and his first task would be to hole-out the bottom of the coal. *(Coal blown from the 'solid,' would not produce as much coal, as that blown down into an open end! Besides, the collier would have to buy more explosives!)* When a length of about six feet was 'holed' to a depth of three feet, the miner would then commence to set up his poker drill, to position this first hole in such a way as to maximize the amount of coal obtained from this first blast! *(I exaggerate! With the limited quality of the 'gradely' powder used, it would seem more like a muted rumble!)* This first hole would be laboriously drilled into the coal near the roof and

parallel to it, to a depth commensurate with that of the holed under-cut. This hole, when charged with explosives and blown down, and with skilful use of the hand-pick, would probably result in enough downed coal to fill three or four full hutches, of eight cwts.* each. This process of holing, drilling and blowing, was a continuous process, interspersed with the need to build and set, wood and stone *(downed redd)* pillars between pavement and roof. *(This was a very necessary procedure, in that it was just not possible to leave great empty spaces behind as the coal is extracted. These packs served to control and ensure an even distribution of the 'weighting' factor, as the roof slowly 'bent' to fill the void!)*
*One Hundredweight equals 112 lbs (imp.) or 51 kg.

Considering also, the fact that the complete length of the 100 yards face proceeds forwards at a rate of about three feet per working shift, it is important that the miners build these pillars *(packs)* completely symmetrical, and packs them to the utmost with waste materials. Their very safety depends on it! The building of pillar-wood and stone 'packs', is in itself a work of art, much to be admired in its artistry and woe betide the miner who builds a crooked pack. Not only does he face the ridicule of his fellow miners, he is usually made to strip the pillar and pack and re-build it from scratch 'all in his own time!' *{Any deputy (fireman) could decide that a pack needed to be demolished then re-built!}*

# 'Stoop and Room' (*Traditional*)

This method of extraction was also known as Pillar and Stall in other parts of this country, and was an apt euphemism that describes a method that was original in its conception, simple in its planning and economic in its operation. The basic tools were, a 'holing' pick, a hand boring machine called a 'Poker' drill which required no separate 'sharpened bits', a heavy hammer and a shovel, a supply of empty tubs, *(which were originally 'drawn' by the miners themselves until such times as the levels became longer in length)* and a strong and willing back! The only concession to semi-modern mining methods was the use of explosives in the boreholes, for the simple reason that this coal was 'blasted' from the solid! *(The previously well-known and practised method of 'holing' the bottom part of the coals, was now obsolescent. It had always been very time consuming!)* This method of extraction was originated by the driving-in of near horizontal levels from a main tunnel or dook, to a given distance through the solid coal. The noticeable thing about these levels, *(they were usually commenced from dooks or headings!)* was that they all *(or, so it seemed at the time!)* seemed to be driven horizontally across the grain of the coal, thereby resulting in each heading having both a high side and a low side. There was of course, good advantage to this, which I will describe further on! When I worked in this environment, my partner was one Jackie Dryborough from 'the

Wemyss' who stood exactly 5ft 7ins in his bare feet. I was a full 6ft 1in at this time. I mention this for the reason, that this particular level was within and down the section known as the Old Dook.

We had commenced the driving of this new level adjacent to the existing No. 3 water level, *(progressively numbered downwards from the top of the Dook)* and to the left of the main haulage-way, therefore, the low side was on the right! As the level progressed in-wards, we were required to 'lay' light rails and sleepers - on which to run the hutches of coal from the coal-face to the main haulage. To my mind, it made perfect sense for me to work on the low side an 'fill' right-handed, and Jackie to take the high side and fill left-handed. Jackie, of course, being the senior miner, was the one who decided the bore-hole locations, and operated the poker drill, whilst I continued filling! This made good sense to me, in that I was younger, possibly stronger, but as yet inexperienced! This was not to be! Jackie insisted on working the low side where most of the coal fell to, and left me on the high side shovelling overhand into the hutch. He did however, acquiesce to my working right-handed on the low right side whilst he bored the necessary holes in the coalface, in that it made good sense to keep the coal flowing - and the drawers busy.

As a result of this arrangement, I was sometimes left scrambling for coal whilst Jackie was up to his waist in the stuff and could hardly see over the rim of the tub. This arrangement never changed during our partnership, and therefore, the only time that I was allowed to give free rein to my heaving coal on the right-hand side, was when Jackie deemed it necessary to drill the bore-holes for the next round of 'shots!' It may seem hard to believe, but I revelled and 'honked,' with the unaccustomed 'freedom!' *(During the driving of a horizontal 'level', the fully exposed coalface measuring approximately 12 feet wide by approximately 8 - 9 feet high, was taken all at once!)* It was not until many years later that Jackie admitted to me, that he could not operate or shovel coal in a left-handed fashion! So much for my common sense!

In this method of operation, these levels would be about 100 metres apart down each side of the Dook, and driven-in through the solid coal to a distance of at least 100 to 120 metres, or even much longer in the case of permanent roadways! *(These levels would be driven simultaneously by different pairs of miners working on consecutive, but weekly rotating shifts. This was of course, dependent on whether the Pit wound coal on more than one daily shift, or notwithstanding, whether there was an adequate supply of empty hutches available to the miners!)* When the level had reached a maximum planned distance, *(approximately 100 to 120 metres)* work then stopped. It was simply not economically viable for the colliers to hand-draw loaded hutches over this distance, and fill the daily quota of coal. The tacit understanding between collier's and management, was that working pairs of colliers only drew their own coal over the first 20 metres of any new level, after that, they were entitled to first one, then two oncost drawers as the heading progressed. I make mention here, that levels between 100 to 120 metres in horizontal distance were sufficient unto the methods of this type of operation. *(If the levels were any longer, this would then mean that the drawers would have the next to impossible task, of trying to keep the colliers supplied with empty hutches, while struggling\* valiantly to remove and 'putt' the loaded hutches!)*

*\* A true anecdote on this very subject can be found in one of the later chapters in this book!*

## Aeroplane Braes!  (*The Natural Follow-on!*)

The perception of an 'aeroplane brae' shall become very obvious to the reader as this explanation moves on!  Imagine if you will, two separate horizontal levels being driven from the left side of a dook, *(looking down!)* in the same direction, one above the other, but fully 110 yds *(100 metres)* apart and to the same horizontal distance.  One level is obviously on a higher plane than the other, but parallel to it. This leaves a great block of solid coal which, in the Dysart Main seam in this area is approximately 26 feet thick.  Mining *(driveage)* then commences with a heading being started on the lower level, but on the high side of the level, at left-angles to the level and at a distance of approximately 20 to 25 yards *(metres)* from the main dook. This block of coal will remain completely untouched as a supporting stoop, and is there to maintain the integrity of the main dook.  The miners then open up a new limited face, *(heading)* approximately 12 feet wide, and 7 to 8 feet high, *(this leaves a 5 to 6 feet thickness of the Sparcoal and the Head coal to act as an immediate solid roof support)* and following the grain/seam of coal, which is now inclined upwards at an angle of approximately fifteen to twenty degrees.  The colliers extract the full width and height of this heading for at least the first 10 yards *(metres)* of forward travel, where 'blown' coal will still land at the 'flat' base of the level.  After that, there comes a slight alteration in the 'blowing' of the face and the taking of the coal!

The taken coal must now be hand-loaded into hutches that are brought to near the face of the advancing heading.  This is achieved by the laying of two separate, but parallel sets of six feet lengths of portable metal rails which are extended on an as-needs basis to keep-up with the advancing face. *(Colliers have no wish to cast/ shovel/coals over a long throw!)*

*\* If a Long Dook (dipping) was newly driven through virgin coal and was intended for use as a coal haulage-way or coal transportation by any means, the standard policy was to leave a solid stoop of coal running the full length of the new dook to a width of approximately 40 -50 metres on both sides.  This was to prevent undue roof 'weighting' on the said dook or haulage-way.  In the case of the Old Dook*

*driven down through the Dysart Main Coal seam, roof supports were few and far between in the main Dook itself, for the simple reason that they were barely needed. Precious 'filling time' was therefore rarely lost in the setting roof supports.*

*If a long 'level' was to be developed for the precise purpose of opening up a new series of 'sections,' from which coal was to be taken in great volume, then, the level would be commenced in exactly the same way. As the level grew in length, a team of 'brushers' would be contracted to 'dig-out' a potential motor-house from the '**Solid**' which would probably then be steel-girdered and 'lined!' The length of the haulage-way would be laid with a double track of semi-permanent rails. Some form of endless rope haulage would then be installed in the lengthening level, enabling hutches to be transported in both directions. (Diesel pugs were not yet a feature of underground haulage systems!)*

At this juncture in the coal taking, two new elements are added to this equation, one of which might be regarded as creeping mechanisation. The first is that the face of coal is now taken at two separate levels, the top five to six feet of coal is blown and taken out something like four to five feet in advance of the bottom 2 to 3 feet high platform, which in turn, is used to accommodate the 'aeroplane pulley wheel'. This wheel is made from metal, is 15 inches in diameter, is 4 inches in width and has a 'u' shaped profile. It is drilled in the centre and captured by a deep forked clamp, from which a short length of heavy chain is attached. It is a pulley wheel, but with a difference. On the up side of the centre pin there is a deep thread on which sits a heavy nut, having welded to it, two extended curved lugs shaped like the steering wheel of an aircraft. The sole purpose of this manual control is to either lock or free the turning motion of the pulley wheel, by the twisting motion of the 'handlebars'(*sometimes described as a 'Jig Wheel'*). On this device rests the completely successful (*or otherwise*) operation of an 'aeroplane brae!' A single empty hutch is man-handled up to the top of the short heading on one of either set of rails, and tipped forwards so that the front pair of wheels is over the top edge of the rails. The second empty hutch rests on the other set of rails at the bottom of the heading. A short length of chain is coupled to a coiled 100 yards length of three-eighths inch steel cable (*the excess of which hangs on a removable hook affixed to the front of the hutch*), which runs up one set of rails to the pulley wheel, which is now chained to sloping stell on the coal platform. The cable is given two or three booked turns on the pulley wheel and then coupled by another short length of chain to the waiting empty hutch at the top of the rails. The cable is tensioned at the bottom of the heading with a bolted clamp affixed to the 'running' cable and the bottom tub coupled to the loosely tensioned chain, with the remaining cable firmly coiled, secured and hung on the front of the readied hutch. (*The set-up is complete. Let the law of gravity take the strain!*)

The hutch at the top of the brae is quickly loaded up to 10 or 12* hundred-weights, Jackie moves to the 'aeroplane' and gently releases the brake. I move to the tilted hutch, grasp the lifting rings and heave the rear wheels of the hutch carefully onto the rails. The weight of the full hutch then tensions the steel cable. The brake *(jig-wheel)* is further released and the loaded hutch rolls downhill under controlled braking, thus drawing the empty hutch up the other set of rails until it too, reaches the top of the incline, with it's front wheels just tipping over the edge of the rails! At this juncture, the loaded hutch has reached the bottom of the incline and onto the flat, where it can be uncoupled and sent on its way. Then, the empty 'tail' is now coupled onto another waiting empty hutch.

This simple gravity system works well and with minimum maintenance, up to the full extent of the 100 yards long coiled cable, which is of course, *(in this environment)* the maximum length of the *(inclined)* heading! The term 'aeroplane brae' was used generally throughout the Fife Coalfields, and probably beyond if the truth be known. I have no idea with whom the term originated, but to my mind, it could only have been an old taciturn Stoop and room coal-miner with a jaundiced eye to a few more hutches of hard-fought coal, who felt that time wasted pushing empty hutches up a coal-brae, was better spent on actual coal-getting! The descriptive phrase itself is totally apt, and to actually 'see' or hear an empty hutch 'launch' itself from a standing start and run rapidly uphill with no apparent means of propulsion, is explanation enough, especially when all of this takes place in total darkness!

*\* With the steel tubs (hutches) then in use in Lochhead colliery, a level tub of coal will, or should weigh-in, at an even 10 cwts. If the coal is built-up with largish lumps to approximately 6 to 10 inches above the rim and levelled, the coal should weigh-in at approximately 11 to 12 hundred-weights. (One hundred-weight equates to approximately 51 Kilograms.)*
(For 'tub' read 'hutch'. The terms are interchangeable.)

Author's Note:- *Readers will realise that in reading the above description of an aeroplane brae, that the principle used and somewhat modified, is not unlike that of a funicular railway as used on mountainsides in many countries of the world. The differences here, of course, being:*

*1. The successful operation of an aeroplane brae, (slope) depends on the permanent imbalance between the weight of the loaded hutch being lowered under control, which produces the necessary 'power' to raise the empty hutch uphill to 'stops!'*

*2. There are no switched rails within this system. There are two separate, but parallel light rail tracks, with the need to extend the six-foot long, metal rails on a daily basis. (To maintain contact with the working, moving coalface!)*

*3. The aeroplane brae is totally unpowered. It is purely a gravity system!*

*4. The down-side alternates with the running down of each respective loaded hutch!*

*5. This system can still be correctly described as a 'self-acting' incline!*

These headings were driven upwards from one level *(following the grain of the coal seam)*, until breaking through to the next parallel level above, a distance of approximately 100 to 120 yards. At this point, after all of the loose coal has been taken, the light rails and sleepers are removed and transported along the bottom level, to be stored and subsequently re-used in the next heading. *(Most pairs of colliers usually stored and hid their 'sets of rails,' wooden sleepers and 'dog-nails,' until such times as they needed them!)* The worked-out extracted heading then became just another airway, *(or not!)* with any further subsequent 'falls' of coal being left severely alone! *(Hallelujah!)* This worked-out heading was now left open and abandoned except for a daily/weekly inspection by the section Fireman. The next heading to be commenced on the same lower level, was at a measured distance from the first heading, leaving a block of approximately 8 to 10 yards of solid coal and commenced upwards, exactly parallel* to the first heading. Successive headings were then developed inwards, *(or outwards)* with the same predetermined spacing between them, until the innermost end of the original upper and lower levels was reached. These headings would not be cross-matrixed with parallel levels running through them. The headings were far too close together to make matrixing a viable proposition. Besides, this added extraction might just compromise the integrity of the immediate roof, if too much of the coals were extracted! *(These levels could be many hundreds of metres long. All depended on the economic means of powered transport used, to convey the loaded hutches to the fixed haulage system on the main coal dook! Manual drawing (putting) could only be used over short distances.)*

By this method of Stoop and room and leaving 8 to 10 yards *(approximately 7 to 9 metres)* wide stoops between each successive heading, only about 20% of the available coal could be taken. *(Bearing in mind that a given thickness of coal is left at both top and bottom of the headings/levels).* The great advantage of this method of working, being the lack of expensive machinery and the minimum cost to the owners. The going rate at this time for the coal-getters, was six and sixpence per ton! *(32.5 pence.)* The collier's had to 'blast, howk and fill' upwards of fourteen tons of coal per shift between them, to make a decent weekly wage. No coal! No wages! This was piece-work at its worst! *(Except for the bare minimum!)*

*Author's Note:- All driven mines, levels and headings were actually 'surveyed-in' by small teams of underground surveyors. Underground survey is not so different from topographical/trigonometrical survey, except for a few addition limiting factors. The basic principles are much the same with similar types of instruments are used for both. Instruments that are to be used underground, needs must be capable of being internally illuminated, either that, or be of the 'open' scale variety, so the engraved and graduated scales can be illuminated and accurately read. Linear measurement, if it is to be accurate, is usually done by 100 metre long steel tapes or chains. The underground limiting factors also include not being able to 'carry survey forwards' whilst men and machinery are in full operation. Great activity and heavily vibrating machinery are simply not conducive to accurate survey.*

*The inherent darkness did present some small inconveniences, but could be easily overcome. As a result, the survey teams could be seen to operate sometimes between shifts, sometimes during the quiet of a night-shift, but mostly at week-ends. The outward signs of a surveyors visit that appears to a miner, is when he arrives at his workplace (i.e. a new heading) to discover fresh, white, bold chalk marks on the coal, showing the location of the 'plumb-lines' now hanging from the roof! These plumb-lines are hung from wooden plugs which have been sunk into holes drilled into the roof under the direction of the surveyor. These plugs are approximately 4ft apart, the lines about six feet long and the ends weighted. The lines/cords are sometimes heavily chalked down their length to make them more visible.*

*The principle of 'sighting' the direction of the heading is basic, simple and very effective. The plumb-lines are dropped, chalked and steadied. One miner then positions himself on the low retreat side of the plumb-lines, removes his lamp from his head and, at a distance of about one to two feet, 'sights' his eye along the line of the two plumbs so that he only sees one! This is the surveyed 'sight-line' in which the direction of the heading or level is to be driven.*

*It is a simple matter for the another miner to hold his lamp at the heading face, and be given 'light-signals' to move it right, or left, to find the dead centre line of the heading. This is then heavily chalked, with the width to be mined (extracted), accurately measured left and right from this centre point.*

*This action is performed on a daily basis by all developing colliers, miners and brushers in this type of mining/development, and results in miners being perfectly able to drive mines that are both straight and true! A source of great pride to any coal-miner/developer.*

*Where miners are working to contract, in that they are paid an agreed price (*contracted with the mine manager*) for every ton of coal that they produce (*mined from the solid as in Stoop and room*), there is always the incentive to work and 'fill' to a daily quota, in order to earn the best wage possible in the given shift hours. Sometimes, through no fault of their own, or simply through circumstances, the coal blown from the face is insufficient to meet this self-imposed quota and if there is no shot-firer in sight, then 'needs must when the devil drives!' They simply go and 'search' for loose coal! This is not easy to find, simply because this situation can happen quite regularly, and 'unclaimed' coal is progressively more difficult to find. The temptation to load the tub with any old rubbish is pressing and of course, the top 10 per cent can be topped-of with real coal to hide the deception. Needless to say, I did succumb to this subterfuge at one time and actually filled two such tubs of rubbish, ostensibly, but not cleverly disguised as coal! I attached my personal numbered tag to the ring of the tubs, (No. 108) so that the weight would be credited to my rising total and smugly send the tubs on their way. *(There was a humiliating aftermath to this action, which is described in the next paragraph!)*

I was quite happy with that day's work, in that we had filled a total of 28 hutches during the course of the shift, giving each of us 14 filled hutches which would be added to our respective weekly totals. An addition of approximately 8 tons of coal at an agreed price of six and sixpence per ton. The next morning as I approached the winding cage ready to descend underground, my workmate Jackie, drew my attention to two 'couped' objects of scorn, that was wholly regarded as disgraceful amongst colliers. *(But only if you get caught!)* My two hutches of rubbish had been 'couped' *(tipped)* on the hard standing in full view of all miners and chalked with my name! This was the ultimate disgrace! I had been rumbled and everyone at Lochhead pit knew it! The 'check-weight-man' had taken one look at the 'level' tub, one look at its weight on the scales, and said nothing! He merely caught the attention of the snibbler, pointed to the two offending tubs, and then to the hard standing. He then raised both arms up-wards indicating his decision to have them tipped! A filled and levelled hutch that weighed 12 to 13 hundredweights *(610 to 662 kgs.)* did not contain coal! The final ignominy was of course, that we did not get paid for 'couped' tubs!

## Stoop & Room (*Pillar & Stall*) Modern Method

The modern method of Stoop and room *(more akin to Pillar & Stall!)* as practised in Lochhead pit, involved the use of the new A.B. Short-wall, coal cutting machines, introduced into Lochhead pit around the early Fifties. To the best of my knowledge, the above-named method of extraction using these machines was generally practised only in one large section, and two other relatively small areas in Lochhead pit. The large new section where six of these machines were in multiple operation, was in one of the very few sections where the coal was taken to the 'retreat'. It was euphemistically called 'Total Extraction!' There were other smaller sections in Lochhead pit where short-wall machines were used as the primary means of taking coal, but, never to the same extent as in the East Side mine. *(The deployment of these machines is fully described in the chapter: - 'Out of the East' - Lower Dysart Delights.)*

# The Long-Wall Machine Face

At the time of which I write, this method of extraction was widely used and could be very prolific in its output, provided that all of the machinery worked well and was carefully maintained. The basic premise was to have a coal transport system as close to the working face as possible and, extend it closer to the face on a daily basis during a non coal-producing shift. This method gave rise to the three-shift system, which I will endeavour to explain in detail and covers a 24 hours working period. *(The coal getting cycle!)*

As a long-wall face is being developed from solid coal, the near horizontal approach road across the grain of the coal seam is carefully surveyed-in, so that this main level is usually at the low side of the new area/face to be opened up. This level would probably be driven in a near straight line using 12' by 9' girders *(being approximately 13 feet wide, approximately 9 feet high and now named the Main Gate, or Mother Gate)*, and used as the transport route for the extracted out-coming coal. This width of road also allowed for the provision of a light, single-line, narrow gauge, supply railway that is extended on a daily basis to maintain parity with the 'barrel-end' *(extending end)* of the conveyor system. A second, but smaller diameter horizontal level is also driven parallel to the main level, *(also laid with light rails and used as the main supply route to the advancing face-line, to take advantage of the downwards moving face conveyor, with which to supply the individual miners on the face with straps and trees!)*, but separated from it, by the determined length of the proposed coal face. *(Long-wall coal face.)* This coalface can be anything from 100 to 200 yards *(metres)* long. This smaller sized level is usually on a higher plane, with the intervening 'coalface area' between the two levels showing a slight rise towards the smaller parallel level. *(It is extremely rare to find a seam of coal that lies exactly on a level plane!)* The main lower level *(Main Gate)* is usually also the main airway, with the higher, smaller level *(trailing-gate, or tail-gate)* being the return airway, with the daily advancing face-line being the air conduit between them. These two levels will be advanced at the rate of approximately four and a half feet on a daily basis - and always interconnected via the length of the forwards moving coal-face, over its full length.

# To develop a long-wall face in a new section!

As both of the developing parallel levels *(the Driveage)* approach the proposed *(surveyed)* start-line of the proposed new coal-face, work is stopped on both levels. The bottom level *(main-gate)* having been driven perforce, a full five yards in advance of the 'top' level. This apparent 'oddity' will usually remain so, throughout the life of the advancing coal-face! At this point, the developers on the Main-gate turn their attention to the actual development of the new coal-face and its proposed direction. The face-line is usually developed at right *(or left)* angles to the main level *(though not always!)* and usually to the rise, following exactly the grain of the coal seam. This low heading/development is driven straight and true, roughly 10 - 12 feet wide and at the 'to-be-taken' height of the coal seam, usually 3ft - 3ins to 3ft - 6ins where possible. *(Lower Dysart seam, upper or lower leafs.)* The low, embryo face-line is developed upwards *(inclined slope)* in a straight line, with the extracted coal being taken away by a conveyor system, which will be extended on a daily basis, to keep up with the developing heading. This heading when completed, and cut through to the top parallel road, is ideally the approximate size and length of a average production coal-face, the output of which, should determine the size and versatility of the conveyor system to be installed on that coalface! With determined work and good survey, this absolutely straight development probe will strike the top road *(parallel level)* within a few yards of its end, thereby 'squaring' the development. Now comes the turn of the preparatory miners and the initial installation of the coal-face machinery.

The first task is to 'line-up' all of the roof supports, so that the coal conveyance means can have a long, unobstructed and straight run from the top trailing-gate to the bottom main-gate. This conveyor can be a 'Cowley-Shaw' scraper-chain, a 'Huwood' endless underload scraper belt, or a joined set of 'jigging pans.' *(These conveyor belts or pans can be extended or shortened within 15 to 20 minutes).* At this point, after all of the machinery has been installed and after 'power' has been directed to the machinery, the three-shift manning system comes into effect to commence the daily production of coal.

# Back Shift: - 2pm to 9.30pm

The back-shift miners on a long-wall section shall most certainly include pan-shifters, *(who might have to strip coal, set steels and clean the condies)* brushers, packers, *(also on night-shift)* and perhaps steel-drawers. *(Usually night-shift.)*

The pan-shifters *(the name still applies whether the conveyor system be belt, pans or scraper-chain),* come on to the face-line at approximately 1345 to 1400 hrs. They would have had a report as to the condition of the face; i.e. whether there was standing coal to be fired and stripped, or maybe lots of loose coal lying along the face-line, which would inhibit their pan-shifting routine. Either way, the idea was to use the running face conveyor to divest the run of all standing and loose coals. Their 'breaking apart' and subsequent re-building of the conveyor system, could not commence until the last vestiges of coal had been transported down the pans. *(This was hopefully, a remunerative daily occurrence for the back-shift preparatory miners.)*

When and if, any standing or remaining loose coal had been carried away on the face conveyor, the machinery would then be stopped and the electric power to the 'pan-engine/ belt-motor' cut off at the power breaker. Depending on the type of conveyor in use on the run *(coal-face),* this would determine the procedure used to dismantle and re-build the system along its new path, which was approximately 4 feet 6 inches nearer the face-line, and at the other (face) side, of a long line of vertical roof supports. One or more of these supports could and would be temporarily removed and almost immediately replaced, but only on an individual basis. More than one or two at any one time, could result in disastrous consequences. *(Readers will no doubt realise, that were it not for the long inside line of face supports at four-foot intervals, that stood between the long conveyor and the face-line, then the need for the conveyor column to be systematically dismantled and rebuilt would be obviated!)*

# Face Conveyor : Jigging Pans

The 'pan-engine' could be situated either at the top or the bottom of the face-line. If the pan-engine were situated at the bottom of the face-line, this meant, that the train of pans *(and therefore the coal!),* would have to be carried over the top of the pan-engine, with the 'driving arms' of the engine in an upwards, vertical configuration, and extra height at the bottom end of the run would be needed to accommodate the larger profile. *(The main disadvantage of having the Pan-engine at the bottom of the column, was that if the coalface height was comparatively low, the large lumps of coal pushed onto the conveyor by the feet of the strippers, stood every chance of jamming against the roof and being the cause of untold delays. Lumps of stripped coal did not bear the 'individual stripper's signature!')* The only real advantage to this configuration, was that there was less likelihood of the pan-train shaking loose the connecting bolts that joined each set of pans, and that there was less strain on the engine itself, in that the 'gravity-stroke' was also the load stroke. *(In this configuration, the pan-train was 'pushed' back uphill, thereby helping the 'pan-doctor' in his never-ending efforts to keep the 10-inch connecting bolts tightened up, all done whilst the pan-train was in motion!)*

If the pan-engine were situated at the top of the run, *(deep into the stable-end)* this then meant that the train of pans commenced immediately at the forwards end of the engine, with the activating/driving arms oscillating like a pendulum and attached onto a large metal 'U'-shaped spur, which in turn, was bolted to the first ten-foot pan. The train of pans would then slowly snake downhill between two lines of vertical roof supports, following the line of the coalface, but approximately four feet from it. There would be no need for the pan-train to be built over the top of the engine, as the top, stable-end, coal-stripper would have an intermediate filler, or be paid 'wide pans' or, accept shortened ground! *(The top stable-end stripper was the tail-end Charlie, the length of whose 'stint' would change every day.)* With regard to the working of the pan-engine, the return stroke becomes the 'load' stroke with the force of gravity assisting the forwards down-stroke. This configuration method does throw more of a strain on the pan-train, in that the whole weight of the pan column is felt on the upwards return stroke, and this pulling action is more likely to loosen, and slacken the connecting bolts between each ten-foot pan section. One characteristic that is peculiar to a oscillating pan train, is that its rocking movements become quieter as the load increases. This is not the first time that I have been alerted to this rare phenomenon as the weight of the increasing 'gum' load, deadens the metallic noises within the supporting rocker boxes.

Author's Note 1:- *I do not know the name of the 'inventor' of the jigging (shaking) pans system as used in the coal-mines, but the principle is probably as old as time itself. A pan-engine and a column of pans is a diabolically simple machine, yet one of the most effective and awesome to behold, if only for the sheer power of its electric-powered mechanism!*

*A set of Jigging-pans, or a pan-train, can really only be used to transport coal along its length, if there is a slight slope or even the suggestion of a slope over the ground where system is to be used. Loaded coal contained within the 'pans' can even be made to be 'jig' (travel) over horizontal ground, because of the inherent principle used in the 'driving' of a pan-engine, but horizontal progress would be slow and probably uneconomical. A pan-train can even be made to work on a concave, sloping, face-line even where there is a dip in the middle of the run. The criteria being that the top end of the pan-train is substantially higher than the bottom end. In this event, the pan-train must always be full of coal from top to bottom, thereby building up a considerable, kinetically energised, down-force to the coal load. After all, this is primarily a powered gravity device. The basic premise to the efficient working of a column of pans, is that it is a mechanically powered, forwards and reverse reciprocating cycle, which is controlled by a continuous start/stop momentum.*

*If a column of 'trough' pans were to be laid end to end, along the length of a face-line having a slope greater than a given angle, the coal would them merely slide down the resultant 'chute' at speed and crash out at the bottom. (This type of conveyor has been utilised in the past! See chapter on Surface Dipping!) If the same pan-train were to be laid on a completely horizontal face line, then just a small amount of perceptible forwards movement would be instilled onto the load by the powered movement from the pan-engine. The 'load' might even stand still!*

*The criteria therefore, the ideal plane for the use of jigging pans, is on a slope where water will positively run, but where coal will not slide when placed on these pans. (The inside steel skin of these pans gets perfectly 'polished' and 'silverised', by the constant abrasive action of the moving coals, to a continuous smooth surface that is rendered almost frictionless.) The pan-train is inter-connected every 10 feet length and under each individual 10 feet length there is a 'rocker-box'. These flattish rocker-boxes (cradles)were fabricated in two opposite halves, separated by one inch ball-bearings and designed to 'roll' against one another. They measured approximately 15 inches by 12 inches and were constructed from angled iron. There was probably 12 inches of movement within each cradle. The cradles were so designed that the top half was an integral part of the under-carriage of the pan, whilst the bottom part dug itself into the pavement. The short, curved pathway within the cradle, ensured that on the slower 12-inch down-stroke, the column of pans was lifted approximately one inch in the vertical plane, whilst, on the slightly faster 12-inch return stroke, the whole column dropped down one inch in the vertical plane, thereby momentarily easing the load on the pan-engine and slightly lessening the effects of the force of gravity for a split second on every up-stroke. The forwards energy (kinetic) imparted to the downwards moving load (coal), was exaggerated by the single half-second delay imposed at the bottom of each power stroke, by the machinations of the gearing on the motor. The constantly, downwards, moving, coal load was therefore initiated, and consistently maintained by the forwards induced momentum of the 'jig' or 'shake!' The cyclic work rate of the pan-engine and train was probably around 20 to 24 strokes per minute.*

*It has often been the case, that the overwhelming load on the pan-train has caused the pan-engine to 'shake itself loose'. This was when the angled supports to the roof, or the chain sylvesters securing the engine had given way, and there happened the classic paradox of the 'tail' wagging the dog! The load and weight on the length of the pan-train (column) was actually heavier than that of the pan-engine, and was sufficient to 'still' the train of pans, thereby causing the pan-engine itself to do the 'rocking!' This was known as a 'run-away' engine! Fortunately, providing that the electric power to the motor was cut immediately, the 'engine' would not travel very far, even though a few roof supports might be displaced. Usually, this didn't not take too long to rectify with all of the nearby colliers pitching-in to help stabilise the engine! As many strippers as was needed would make short work of it's re-anchoring.*

*Author's Note 2:- It has come to my notice after reading Augustus Muir's, 'The Fife Coal Company,' that the Central Workshops at Cowdenbeath, were in the business of manufacturing those items of Mining machinery that were otherwise both costly and difficult to obtain. Amongst the named items of machinery that was produced at Cowdenbeath, was part of a face conveyor system described as 'Shaker Conveyor Troughs!' i.e. whole sections of Jigging or Shaking Pans!*

# Night Shift: - 10pm to 5.30am

On the night-shift, the machine-men, *(a coal-cutting team of two or three men)* one or two pairs of hole-borers, several pairs of steel-drawers *(expired roof-support extractors)* and several 'packers', all descend on the coalface, each group having separate tasks to perform and limited time in which to complete their work! The tasks were so arranged that no two groups should ever interfere with each other and therefore, no conflict should arise with tasked miners operating over the same piece of ground at the same time.

The steel-drawers are the miners who are charged with the task of slowly and carefully, extracting the long row of straps and 'trees' made redundant, by the ever-advancing line of the coal-face and the subsequent shift of the face-conveyor, sideways into its new position, within the next line/row of roof supports. These steel supports, now in the 'condies,' *(wastes)* are withdrawn along the whole length of the previous *(thrice -removed)* face-line, so that the length of the immediate roof line in the wastes is allowed to collapse under control and make available, the subsequent downed redd *(broken stone)* needed to build the nightly, advancing, stone-packs.

This job can be a rather dangerous in its application and definitely not for the faint-hearted! It requires patience and care, and the need to be 'quick' of thought and fleet of foot! *(It also requires a deal of prescience and a modicum of the 'second sight!')* Not for nothing is the job described as "This is Your Life!" The withdrawn steels *(provided that they are still straight and undamaged),* are needed by the coal-strippers on a daily basis, to set to the roof on the exposed face-line as the newly cut coal is extracted. Any badly distorted steel straps are usually sent to the surface to be re-straightened by the blacksmith's steam hammer and then, again, returned underground and re-used on the coal face.

The packers also work on the 'waste' side of the coal conveyor in the named condies. They build pillars of broken stone and wood into roof-supporting 'packs', that are approximately 8 feet 6 inches wide and to the depth of the previous days cut, and separated by a interval *(width)* of 15 to 18 feet. These 8 feet 6 inches *(approximately 2.6 metres)* wide pillars are extended forwards on a nightly basis to keep up with the daily-cut face. *(Approximately 4 feet 6 inches!)*

The hole-borers usually commence their work at the bottom end of the uncut coalface, for the very good reason their work must remain in front of and ahead of the coal cutting operation, else their progress be compromised or impeded by the mountains of gum* produced by the coal-cutter. They usually start their shift by dragging the 'Ram's head' and all of the long trailing cable *(approximately one inch in diameter)* to the bottom end of the coalface *(the main gate)* so that they need only look after the slack cable as they advance before the machine! Holes, approximately two inches (5 cms) in diameter are drilled into the coal face, roughly two inches from the roof and parallel to it, and to an approximate depth of 4ft - 6ins *(approximately*

*1.4 metres),* virtually the same depth as the machine undercut. These holes are drilled along the length of the full coalface at intervals of approximately 5 feet *(1.5 metres)* and without exception. This will approximate to a total of 5 or 6 drilled holes per 10 yard *(approximately 9 metres)* stint, which is the usual length of a coal-stripper's yardage. *(Any hole-boring pair who skimp on the amount of holes drilled, are usually given very short shrift by the subsequently, overworked strippers concerned.)* Pairs of hole-borers were usually contracted to deliver a fixed amount of bored holes to a pre-determined depth, over a given length of the coal face!

Author's Note: - *The term 'Ram's Head' was the colloquial name given to the electrically-powered portable drilling machine that was in use in the coalmines in the 'Forties, Fifties and Sixties!' Its name was probably derived from the fact that it sported two large fully curved operating handles (horns) from its aluminium finned 'head!' It did resemble a ram's head in overall shape and was probably the same approximately size!*

*The panel boxes that supplied the electric power for the Ram's Head could have been situated either in the Main-gate or the Trailing-gate. However, irrespective of where the trailing-cable originated from, the main responsibility for the hole-boring pair of miners, was to ensure that the borer's 'live' trailing-cable did not foul, or get fouled by the revolving chain of the imminently, progressing coal-cutting machine deafeningly, but invariably, inching its way up the long coalface!*

# The Machine-men.

When an in-depth cut of the full length of the coal-face has been achieved on the previous night shift, the machine is left parked at the top of the run in the 'stable-end', which in fact, usually extends a little beyond the Top Road. *(Trailing-gate cum return air way.)* The coal-cutting machine therefore needs to be coasted *(but still pulled!)* down the face-line to its start-point at the bottom Main gate. This is easily accomplished in that there should be no obstructions to its passage as the stilled cutting jib is locked in the straight forwards position. A skilled machine-man should be able to do it blindfolded. *(As he virtually does!)*

At this juncture, I feel that I ought to describe this machine, delineate its approximately size and explain exactly what it does! The machine is a complex assortment of powerful electric motor and precision made gearings, housed in an oblong-shaped, armoured, steel, thick skin. It is approximately 90 inches long, is 24 inches wide and is 15 inches high. It is a powerful beast of a machine powered by three-phase alternating current at 550 volts and consumes high amperage. It is fed by a long, flexible, lightly armoured cable, a full two inches *(5 cms)* in diameter. At one end of the machine at ground level, there is a flat, four and a half feet long extending jib, which is 15 inches (38 cms) in diameter and approximately 4 inches in depth. This jib can be positioned and locked at right or left angles, or in-line ahead. Around this jib, there rides a strong, flexible, continuously jointed, steel 'chain', on which there are numerous 'pick-boxes'. The chain is wrapped around and driven by a horizontal, multi-toothed sprocket, capable of being driven in both directions. The pick-boxes are slotted to a depth of 3 inches, *(approximately 7.5 cms)* and are side-drilled and 'tapped' to take a half-inch *(13mm)* bolt, which are used to tighten the small 'picks' as they are slid into their housing. The small, steel picks are four inches long, one inch in width and half an inch thick. They have an off-set head shaped to a point, onto which is welded a sizeable dollop of 'hard-welding' to give it a durable cutting edge. The total amount of individual picks available for any given machine, would usually add up to two or three times the number of 'pick-boxes' on the revolving chain, the sharpening rate of which, depended on the hardness of the coal being cut! These picks were sharpened on a daily basis by the pithead blacksmith using an powerful electric grinder, either that, or they were reshaped by the same blacksmith at his forge and re-spotted with another dollop of hard-welding! *(And God help the machine-man who went underground without his pick-box.)*

Before coal-cutting commences, the machine-man will inspect each individual pick-box to check on the cutting edge of each pick, when satisfied, he will also do a quick check to ensure that all of the picks are facing forwards, with the cutting edge in the direction of chain rotation, whether it be clockwise or anti-clockwise. *(The machine can be set-up to cut in either direction with equal power, it merely requires a reversal of the 'pick' direction.)* At the bottom end of the run, just above the lower stable-end, there is no need for the machine-man to initiate a 'swinging' entry cut into the coal. The bottom end has been 'opened and squared' by the advancing main-gate strippers and brushers. He will rotate the cutting jib to the left *(or right)* at an angle of 90 degrees and lock the jib in position. Whilst this is taking place, the No. 2 man on the machine will crawl up ahead of the machine's cutting path, pulling behind him the end of the half-inch steel cable that is being unwound from the machine's main driving drum. This end of the wire rope will be stretched-out to a length of approximately 30 yards. He will then position the chain end of this cable close to the coal face and, using a prepared wooden stell or rance, will jam the base of this chain against pavement and the stell against the roof, so that the greater the pull on the wire rope *(and chain),* the tighter the lock becomes. *(In theory anyway!)* The machine-man will check that the machine itself is suitably positioned to cut at the correct level, and he will then direct power to the cutting jib. *(Engage the driving clutch!)*

Once the machine-chain starts ripping and cutting, several things happen simultaneously. The forwards movement of the machine and cutting torque *(direction)* of the chain, will pull the machine's body tight to the face of coal and the cutting spillage - called 'gum', will spit out in great profusion and gush like hot lava, spreading itself behind and half-filling the height of the empty space vacated by the now slowly, inching machine. The speed of the cutting movement is controlled by the machine-man, who is now trying hard to balance the perfect cut that he always strives to maintain, against a forward propulsion of about 15 to 18 inches per minute. This would be a fairly average cutting speed, that will take all of five to six hours in total, without over-straining or over-heating the electric motor. In addition, the machine-man must be constantly on the move himself, in that the machine is constantly edging towards him. He, by necessity, must move sideways and backwards, so that he maintains complete control of this inching juggernaut, and at the same time, must ensure that the approaching pavement is clean and free from objects and debris, anything in fact, that could upset the horizontal balance of the cutting jib. It has been known for certain machine-men to carry a small hard-bristled hand-brush on their person or machine, to ensure a clean smooth passage for their coal-cutter! They, like most contracted miners, take a great pride in their finished work.

At the same time as the cutting progresses, the No. 2 man of the team is behind the machine and is forced to crawl over the newly produced 'gum'. His immediate task *(at a distance of approximately 4 to 5 feet behind the moving machine)* is to clear away the gum from the lip of the newly under-cut coal at given points, so that he can insert a heavy hardwood 'choke' *(shaped hardwood wedge)* into the mouth, or lip of the cut and hammer it tightly home. He does this, so that the weight of the newly undercut coal will not crush the remaining gum down onto the pavement, thereby negating the very objective of the undercut. These chokes are inserted every five feet or so and are judiciously hammered home. These chokes are kept permanently on the line of the coal face, and are usually evenly spaced out over the length of the coal face by the preceding preparatory shift.

Next morning on the dayshift when the stripper has shovelled away his gum, the chokes are re-deployed by the stripper, *(only ever one at a time!)* and strategically placed under the lip of the standing coal between the drilled, high bore-holes, to ensure the maximum effect to a prepared section of blown-down solid coal.

*It must also be remembered, that air movement and continued ventilation along any coal face, was an on-going occurrence and indeed, it usually transpired that the direction of air flow was synonymous with the direction of the cut (though, not always so!) When a machine was cutting at optimum speed, in addition to the great quantities of gum produced, there was also a prodigiously, thick cloud of choking coal dust that enveloped either the machine-man or his No. 2. In most cases, it was usually the No. 1 machine-man who had to suffer the on-going effects of the choking hazard and near blindness, and even the wearing of eye-goggles and dust masks did little to diminish the cumulative effects of these hellish conditions. A few years of breathing this sort of atmosphere, would convert the healthiest pair of lungs into a gasping entity before middle age, and render an otherwise healthy miner into a skeletal pain-racked shell of a man, condemned to an early grave!*

Author's Note:- *If there was ever a more likely set of candidates who were constantly exposed to the undeniable risks of the dreaded miners lung disease, pneumoconiosis, it was the miners who operated the long-wall face coal-cutting machines. They were exposed to this hazard for nearly the whole length of each and every coal-cutting shift. It was utterly impossible to see through these enveloping clouds, which were carried on the air currents, until the coal dust was finally deposited on every vestige of the underground environment. (In my days on the coal face, the use of pressurised, dampening, sprays of water to quell the copious amounts of coal dust produced, had not been forcibly adopted!).*

# Dayshift (strippers and shot-firers)
## 6 am to 1.30 pm

For various reasons, the day-shift was the coal-getting shift, where the number of miner's underground would be at it's greatest. There was just as many oncost workers as there was face-miners on this shift, for the simple reason that the extracted coal would all have to be eventually transported by various manners and means to the pit bottom, and thence to the pit-head. The greater the distance a section was from the pit bottom, usually meant that more men and machinery were involved in its transportation.

When the day-shift strippers arrived at the bottom (or top) of a run, they were, first of all, faced with the daunting scramble over the top of the 'gum,' which was piled high and spread wide, sometimes to a depth of 18 to 22 inches over the full length of the run.

The on-going coal-stripper's would usually wait until all of their number had arrived at the bottom of the 'pans' *(whether it was a pan-train or a endless belt),* and then, in strict rotation depending on their place number on the run, would they then proceed to scramble up the coalface, through the gum, making sure that all of their buttons were done-up! No-one liked getting a shirt-full of dusty itching coal-gum! On arriving at their familiar places *(if they could identify same),* the miners would deposit their haversacks, then look for their own graith, *(pick, shovel, mash, saw and pinch-bar)* then unlock the collector/shackle/ personal padlock to free their shovels.

The men would strip down to working rig, and starting at the estimated bottom of their places, they would dig steadily into the coal gum, throwing or shovelling it onto the moving conveyor, to quickly get down to hard, previous-cut pavement. There was usually a flurry of activity at this stage, so that the long-handled, flat gumming shovels *(no angle on the 'blade')* could be got to first, and individually used to thoroughly clean out the 'under-cut,' thus creating a long, six-inch high void, into which the 'blown' coal was blasted down into.

There was always a deal of competition amongst the strippers, as to who would be the first to call for the shot-firer, and get him!  The shot-firer of course, would have his own agenda, and much depended on where he actually was on the 'run' when the call came, and whether he actually had any explosives with him at that time.  (*One in every two or three strippers on any given coalface, were certified, authorised and delegated to collect and carry away from the pit-head magazine, a metal box containing up to 5 lbs (imp) of explosives.  He would carry this box of explosives, by hand, to his place of work, where he would personally deliver this box to his nominated shot-firer!*)  If a coal-stripper declined to carry such a box on the shot-firer's behalf (*or his own!*), he would find himself rather low on the shot-firer's prioritised 'hit-list!'

When most of the coal gum was cleared from the long coalface, and usually before any shots were fired, the deputy would start measuring each stripper's allotted 'stint', starting from the bottom stable-end.  He would direct each stripper/miner to hold the bottom end of the tape against the heavily chalked roof mark, whilst he the deputy, would measure of, to each and every coalminer, a ten linear yards length of standing coal.  These chalk marks would be heavily and vividly marked and indisputable, and warily behold the stripper who did not have a firm grasp of his end of the tape.  It would be powerfully yanked out of his fingers.  These deputies did not 'give an inch!' (*'Chappie' Cunningham did this to me on two separate occasions when I was an eighteen-year old, inexperienced stripper.  He didn't do it a third time after I 'accidentally' fell on top of him.  I did weight 14 stones at that time.*)

To allow the face-line to be initially broached by each stripper, the custom was to 'fire' two adjoining starting shots in each separate miner's place, as this made for an easier intrusion and allowed each stripper, when the blown coal has been stripped away, (*by pick and shovel*) to set one, or sometimes two sets of 'steel' (straps and trees) to his immediate roof.  This would be the commencement of a new regular advanced line of supports that would eventually stretch the length of the newly-stripped face. (*This would obviously be taking place simultaneously within each miner's stint!*)  The amount of coal taken by the first two shots should measure approximately 8 to 10 feet long, up to four feet six inches deep and the full height of the coal taken, probably 36 to 42 inches.  The weight of coal extracted from these opening shots would be in the region of 3 to 4 tons.  The next successive bore-hole in each miners place, would be primed in turn, but would not be fired until such times as the miner has cleared away nearly all of the coal from the last 'shot', and 'set' new steel supports at intervals and in line, to replace the removed downed coal.

As each miners shots are fired at different times, this means that the flow of coal being carried away on the conveyor is more or less uniform, as long as the machinery keeps going.  Any breakdowns or stoppages, usually results in prodigious quantities of coal piling up on a dead conveyor, which might just inhibit it's re-start when the power is re-applied. (*Strippers were forever being warned by deputies and gaffers, not to overload a stopped conveyor or throw good production coal into the condies!*)

Any coal-stripper worthy of his description, will endeavour to have his third shot fired and cleared, and the next steels set before piece-time, so that he can take a full 20 minute break and perhaps, have a 10 minute 'closure of the eyelids,' or relaxed smoke before re-starting. This means that he should only have two shots, or approximately four yards of standing coal to cope with in the second half of the shift. Any stripper who has not reached this stage, can only look forward to an even more strenuous second half, with the uncertain knowledge and hope that there will be no break-downs, or unforeseen delays with any of the in-line conveyor systems. Experience invariably dictated, that as long as the conveyor was running, the priority was to shovel coal, the roof-strapping and steel-setting usually taking second place to yardage taken. With any luck and with a conveyor system that did not break down, the average stripper should manage to strip his allotted stint, clean away all of the loose coal and erect seven or eight steel straps, each with two supporting trees. Each steel strap would have one tree at the back end and another, approximately four feet forwards, leaving approximately two and a half feet of steel strap protruding forwards as cantilevered, roof protection. All of this strenuous and dangerous work to be completed within six to seven hours!

When all of the coal has been stripped to the depth of the cut, along the full length of the run, a new solid face-line emerges and hopefully shall be cut anew in a near straight line. The cycle thus begins again, and so continues on, . . . ad infinitum!

*(As was often the case, the first preparatory shift, the 'back' shift, (pan-shifters) might come onto the run to find that the coal face has not been entirely stripped, where, due to delays and break-downs, the strippers had been unable to complete their individual ten yards. This then, was the pan-shifters delight! These men were in the position to earn much extra money, for the simple reason that they (quite unlike the strippers) were on fixed preparatory wages for a set task, and any stripping work that they were forced to undertake before their own work could commence, would be paid by the collective yard and the monies earned shared equally between the full face shift. If this happened more than once in any given week, their wages would surpass that of the coal stripper.)*

# The Coming of the Shearer *(Stripper's Beware!)*

It started down the Lower Dysart Dook in the west side of the pit around 1955, and from the onset it did not look like the start of a coal-cutting revolution. This first 'shearer' machine as introduced into Lochhead pit in the mid-fifties, was actually a 15-inch Bluebird machine without it's familiar 4 feet 6 inches long, flat cutting jib. This bluebird had a 30-inch diameter vertical steel drum affixed to its side rear, that measured approximately 18 inches in cutting width. Across the width of the drum, were four rows of evenly spaced 'pick-boxes' that were circumambient around the drum. Each pick-box secured a shaped, hardened, steel pick that was designed to rip the face of coal just as had the picks on the flat type jib. *(The 'picks' were one and the same at this time!)* The machine had a normal armoured base-plate, and was designed to sit and be dragged along the pavement under it's own power, just like it's predecessor. The drum was designed to cut in one direction only, in this case, anti-clockwise. *(This machine was not custom-designed to 'cut' or 'shear' coal on any particular coal face, but was an experimental Mark I adaptation of the 15-inch 'Bluebird'. It was 'fixed' to cut in one direction only!)*

This machine, quite unlike a jibbed machine, needed an open end on the coal face to begin it's cutting and once started, the ripped coal was forced down and under the ripping drum to the rear, where it lay piled-up behind the moving drum until it was quickly and cleanly lifted, turned, and poured onto the moving conveyor chains, by a purpose built plough that trailed behind the machine at a given distance. The trailing plough was an integral part of this shearer system. As the machine sheared the coal from the coal face, the 'cut' left a near perfect vertical smooth face, that was entirely conducive to the machine being able to 'hug' the coalface on it's next pass. The all-steel, coal conveyor was approximately 18 inches wide and six inches deep, and was chain driven, with spacer-bars every 18 inches or so, to 'carry' or even 'retard' the coal down the length of the face-line. The conveyor system which sat fairly close to the moving shearer, was an independent entity in it's own right and played no part in the movement of the shearer machine. After the machine had made one complete pass along the length of the face, ripping the entire length to a depth of 18 inches, the 'face-line' needed to be prepared for the next ripping cut. *(This Mark I shearer was limited to 'rip' in one direction only and therefore, had to be returned to it's initial starting point before the next long rip!)*

Before the next cut could even be contemplated, three stages of operations had to be completed and in sequence. The machine had to be 'flitted' back down the newly-cut coal-face, back along the same track to it's original start point, whence it would be man-handled to sit tight against the smooth face of coal, with its cutting drum placed to rip into the next 18-inch width of solid coal. Meanwhile, back to the point where the machine started its downwards flit. The pan-shifters would now commence the dismantling of the conveyor system, which disassembled like a large Meccano set. The 'pan-shifters' would 'break' the chain at a convenient point and using the power of it's driving motor, would 'run' the chain down into a controlled heap on the bottom, main-gate level. When the chain was clear of the pans and, starting from the top end, the pans would then be dismantled one by one, into their separate six feet sections, and immediately re-built onto their new track exactly 18 inches nearer the face-line. *(Or to within approximately 30 inches of the coalface!)* This would continue on down the full face-line, until the new installation was completed and tightened up. At this point, the broken end of the conveyor chain would be re-introduced onto the bottom end of the readied conveyor and underneath, and pulled back up inside the pan train, until it reached the power sprockets at the top end, whence the driving sprocket power would be used to drive the chain down to the bottom end, where it would be re-coupled.

*(I am reliably informed that a portable, electric tugger motor was permanently installed in the top stable end, and used to gently pull the heavy chain, up and under the 'pan-train', until it could be engaged in the recessed driving sprockets of the electric motor, with the open end being then being 'powered' and pulled down the sloping 'pans!')*

The last operation in the cycle was actually a two-fold operation, in that the face supports had to be advanced along the full length of the face as had the support packs on the waste side. This was probably an alternate 'bi-cut' operation, in that the supports needed only to be advanced after every two cuts as had the packs, so that steel-drawing and packing need only be done every second cut, with the same coalminers probably engaged in both operations.

Author's Note:- *This experimental system was first operated within the depths of the Lower Dysart Dook around 1952-53. It was used on a small coalface approximately 40 metres in length, that opened out to 70 metres length within its first 50 metres of advance. The coal taken was the full 42 inches height of the top leaf of the Lower Dysart seam, and considering the fact that the cutting drum was only 30 inches in height, it begs the question as to how the remaining 12 inches of roof coal was taken, or downed, after the machine had passed? Did it fall, was it 'pinched' down, or was it blown down? Or, quite likely, was it taken in the condies, (wastes)?*

*Readers will no doubt realise, that in order to extract a given height of coal, the coal-cutting machine, therefore the ripping drum, must be custom-designed to rip the coal at a pre-determined height. Also, considering the fact that the height of a given seam of coal will vary from pit to pit, it therefore follows, that a machine designed for use within one designated coal seam in one particular colliery, will not necessarily be of use in any other seam of coal.*

In the coal-shearing stakes, lessons were learnt very quickly by both management and designers. Under the N.C.B. administration, the mine managers were very quick to realise that the machinery was there almost for the asking, and that coal-shearing equipment was becoming bigger, more powerful and most of all, much more versatile. In the planning stages, long before any developing coal-face was opened up, a mine manager could stipulate the type and size of shearer that he needed and wanted, and a shearer could probably be custom built, to rip the coal to order. All he had to do then to justify the expenditure, was to ensure that the coal did flow!

The next generation of shearing equipment being made available, had cutting drums that were designed to rip and extract a previously determined height of coal. The depth of the shear or cut, was still much the same at approximately 20 inches, but with the height *(diameter)* of the ripping drum now previously specificd, it was now possible to take nearly the full height of the targeted seam in one pass, with no additional head coal to be downed.

*(In some cases where the need was desirable or essential, a measure of 'head' coal was left in situ to maintain a reasonably secure roof. For some strange, inexplicable reason, a coal-face miner usually felt more secure with a coal roof above his head, than a stone or 'blae' roof. I rather feel that this was due to the fact that the 'movement' of coal invariably gave more audible warning than that of soft stone.)*

The next experiment that was tried, was that the coal conveyor was widened, strengthened and custom built, so that the coal-cutting *(ripping)* machine actually sat atop the conveyor, and travelled the length of the face-line riding on top of the two vertical sides of the conveyor train *(pans)*. This ensured a steady and controlled passage for the cutting machine, even though there was some initial problems in maintaining a 'level' even cut. This was quickly overcome by having a miner precede the passage of the machine, who's sole task on the coal-face, was to 'level' the pan-train 'fore and aft' and sideways, using a spirit-level and a small sackful of assorted hardwood 'paldies'. {*Three to four inch (10 cms) square pads of redwood, having assorted thicknesses ranging from 3/8th inch to 2 inches (1cm to 5cms).*}

Considering the fact that this method of coal-getting involved the coal-cutting machine and by implication, the pan train, were now to be operated tight against the face of coal, this meant that a radical rethink of the face support system had to be instigated. Previously, where coal had been undercut and stripped by hand, there had always been a continuous line of vertical supports between the conveyor and the undercut face-line, which was extended into another long even line of supports, as the strippers advanced the full face-line by another four and a half feet. At the completion of a daily strip, the new line of the coalface would be approximately 8 - 9 feet from the conveyor, along the whole length of the face. This was about to change, and drastically so, and to the increased safety of the face-miners.

At the onset of the shearer runs and the demise of hand-stripping along the main faces, a quiet revolution had taken place with the advent of the 'Dowty' prop and the 'link-bar'. Gone, were the sticky weeping pine props and the unwieldy 7 foot wooden straps, and the crazily bent, corrugated, steel straps that forever needed to be sent to the pithead to be straightened. And almost gone, was the need for every face-miner to posses a sharp saw. The new breed of 'straps' were called 'link-bars', and it was their innovation that was their inherent strength. They were designed to link together by 'eye and socket', to form a endless progression, that could be just as effortlessly disconnected at the 'waste' side when required. *(To help 'down' the wastes!)* The dowty prop was also one of the most inspired innovations ever to be given to the coal-miner. It was an inherently powerful vertical support, in that it functioned and was extended roof-wise by hand-pumped oil pressure from its own internal reservoir. A miner merely set it to position, and quickly, using a small personal tool, could pressure it to the roof in a few seconds. *(A few minutes actually!)* Its subsequent release was effected, merely by using the same tool to release a small one-way valve. Dowty props eventually came in different limited lengths and strengths, and were purpose designed to be used in the differing heights of coal seams. *(In a very short space of time, it became 'de rigueur' for some miners to wear a 'key-holster' on their belts. An outward sign to compliment or signify their supposedly 'enhanced' status?)*

On a modern shearer face where the shearing machine sat atop the conveyor train, the space between the forward line of supports and the actual face of coal, would probably be about three to three and a half feet, just enough to allow the passage of the machine up the face-line without up-rooting any of the vertical supports. The roof between the line of vertical supports and the coal-face, would be interspersed by the forward projection of link-bars, sustained by a two 'dowties' placed together as an added precaution. As the machine passed any given four feet length of face-line on it's ripping cut, the inside dowty prop would be released and re-set a full 20 inches nearer the face line *(uncut standing coal)*, thereby helping to maintain the immediate roof's integrity.

No doubt, some of the more discerning readers of this narrative, especially those who have been following the transition from hand-stripping to shearer cutting, will realise, that there was no longer any impediment to the forwards *(sideways)* re-positioning of the 'pan-train' *(face coal conveyor)*, as opposed to it's earlier dismantling and rebuilding. There were no vertical props set between the pan-train and the coal-face, merely the protruding link-bars. This meant, that provided that the pavement was clean, there was nothing to stop the whole line of the train of pans being physically pushed over to their new station. After all, it was only a matter of approximately 20 inches. However, not just yet! The main reason for the delay in implementation, was that the shearer could still only cut in one direction, and before the face-conveyor could be moved forwards *(sideways)*, the 'dead' shearing machine had to be 'flitted' back down over the pans whilst they were still in situ.

As the shearer machine departed the top end of the run *(coalface)* on its downwards journey, the top end of the pan train was forcibly 'snaked' over to its new location, by miners using portable pneumatic hand rams, which were held tight against the roof to effect horizontal purchase. This snaking effect on the pan-train, closely followed the downwards passage of the shearer, until the shearer was re-engaged into the bottom open end of the coal face. At this juncture, the upwards ripping process would recommence. *(The pan-train was now so designed, that it could be made to 'snake' left and right without any uncoupling, or separating of their connecting bolts. Coalfaces were not always cut in a true straight true line, hence the need for this flexibility.)*

\*

Author's Note:- *During the period of time when long-wall faces were undercut with an A.B. 17-inch chain-machine or a 15-inch 'Bluebird', the night-shift machine-man usually did his best to maintain an maximum depth undercut on every weekday overnight shift. This was not always possible, due to unseen circumstances, unusual occurrences and small faults within the strata. The section gaffer was usually well aware of this \*irregularity over the length of the face-line and took steps on the last cut of the week to have the face-line 'straightened-out!'*
*\*This could easily be 'substantially proved' by a deputy or gaffer having someone shine a 'spotlight' down and along the empty 'pan-train,' so that the concentrated beam was fully reflected along its continuous silvery surface. The degree of curve(s) or bend(s) was plain to see! (The same principle could be applied to any form of conveyor system!)*

*If it were the case, that a particular strippers 'place' had suffered to be cut to the maximum depth over a five day period of week-days, then, it was always worth his trouble to turn out for a Saturday day-shift, for the reason that the machine-man might have only cut his 'ground' to a reduced depth of only one or two feet (30 cms to 60 cms). In other words, and in mining parlance - a light cut! This meant, that over certain selected lengths of the 'working ground,' there would only be approximately 25 to 50 percent of the coal taken, for which the stripper would be paid the full yardage value! (2/11½d or 14.8 pence per linear yard!) This was euphemistically called: - 'The Saturday Cut'!*

\*

# Chapter XI

## Band Aid - *or Going for GOLD*!

'Smithie' Anderson, formerly of West Wemyss now living in Kennoway. was born in the cold month of November in the year of 1919. He is a bright and cheerful old man of 80 years of age as I write this narrative and still enjoys the company of his lady wife.

When I interviewed him at his home in Kennoway, I was delighted to discover that he vividly remembered me well as the young 18 years old stripper, who had a portable radio and an antenna affixed to his green, all-steel, Raleigh bicycle. *(In good weather I cycled the distance Leven to the Coaltown of Wemyss on a daily basis whilst on day-shift.)*

Smithie's mining career started exactly two months after his fourteenth birthday, in that he was taken on as a boy filler and drawer, to an experienced miner working the Dysart Main coal down the Victoria pit. In this part of the coalfield and in this pit, the Dysart Main seam of coal was fully 30-32 feet thick and, although the policy was not to take the full height of the seam at the first bite, the height extracted was sufficient to give both man and boy good standing height. Smithie remembers being paid the princely sum of 3 shillings per day at that time, which, with hindsight and recent later information, would have made him a well-paid boy! Smithie also recalls that after his miner had bottom-holed his standing coal, and used the poker-machine to drive the boreholes to the proper depth, they then had to prepare the explosive charges in a certain way. At that time, the explosives were in the form of a loose preparation called Gradely Powder and, there was also, no standard, or issued type of initiating detonator in use. Before the powder could be 'stemmed' *(introduced into the hole and packed with fireclay),* it first had to be carefully poured into specially prepared, home-made, paper cartridge tubes, which were 'twist-sealed' at both ends to prevent spillage in the hole. In order to 'initiate' *(explode)* the powder, the 'strum' *(safety fuse)* would be folded over approximately one inch from the end and nicked on the outside of the bight. It would then be folded over once more, with a second small nick on the outside, and be tied closely together. *(A 'single loop knot' would often suffice!)* The thickened end of the safety fuse would be carefully inserted into the Gradely powder tube, before the estimated amount of explosives was placed in the hole and stemmed. The basic premise of this double fold *(and loop knot)* and twin nicks, was to give added impetus to the end 'blow-out' of the strum, thereby hopefully ensuring a first time initiation. *(Explosion or 'soft' detonation!)* This type of initiation, used trailing lengths of safety fuse cut to a pre-determined length, allowed the miner to 'light' the trailing end and walk away to a place of safety *(manhole, or bend in road, or even the inside of an empty tub))* before the explosion.

If two or more 'shots' were to be fired at intervals, the cut-length of each safety fuse would determine the interval of burning time between successive shots. Safety fuse was manufactured to burn at a specific rate. To the best of my knowledge, there was only one specific type available in the coal mines at that time, which was metal blue in outside colour and burned at a rate one foot per minute. *(I did later become aware of a second different colour of safety fuse (strum) that burnt at a rate of one foot per 30 seconds, I believe it was white in colour.)*

Smithie can also remember quite clearly, even at the early age of fourteen years old, that his miner always complained bitterly, when, after advancing a level or a heading to a given length, they were sometimes required to double back on their workings to extract the head coal, which, to the miner's consternation, was paid at a reduced rate per ton, for the simple reason that the coal was blow-down into free space, thereby producing a better yield than being blown from the solid. *(In my conversations with Smithie, I was amazed to discover that the methods of localised extraction, and the simplicity of the basic tools used at that time, was no different to the extraction methods and basic tools used by Jackie Dryburgh and myself, thirty years later. The only difference being the rate per ton paid!)*

In 1945, when Smithie was 26 years old, he transferred to Lochhead pit and into one of the great Dooks. He was employed in filling and drawing, stripping and filling and on development work. Before long, he began to become interested in 'Ambulance', *(the name given to basic and advanced First-aid down the coal mines)*. His interest grew and developed to the extent, that he soon qualified as a 'ticket-carrying' or certificated miner, which was a standard that was 'tested' every year at a given time, usually against other like, qualified miners from other collieries in the large Fife coalfield area. Indeed, these were termed area competitions and 'Ambulance' began to gain much needed credibility with the mining officials and hierarchy, and even the miner's themselves. These area competitions, involving ordinary participating miners from most of the Fife coalmines, did serve to inspire confidence in the working miners themselves, and to the extent that miners would quite happily place their faith and trust in these men in times of need, confidently knowing that they would not faint at the sight of wealing blood. They were the next best thing to a paramedic!

During Smithie's long service in the small Ambulance brigade, he faithfully and conscientiously maintained his yearly 'ticket' for a total of fourteen unbroken years. The culmination of Smithie's long service, was when he reached the finals of the area competitions, which was of an oral, practical and instructional nature. He was pitted against a very knowledgeable miner from the Michael Colliery, who had an impeccable record of annual 'wins' and was reckoned to be the odds-on favourite to secure another win for the Michael colliery. Undeterred, Smithie soldiered on, and when the results were announced, Smithie was declared to be the winner. Needless to say, Smithie was delighted! Smithie's mates were delighted! Smithie's lady wife was delighted and exceeding proud of her fastidious husband, and Wullie Hampson *(Lochhead's manager)* was as pleased as punch. This was the first time that anyone from Lochhead pit had figured in the area finals and Wullie Hampson was pleased that the honours had finally come to Lochead pit. The first thought that passed through Smithie Andersons head was, would Wullie Hampson open up the coffers and extend the Lochhead ambulance team? Smithie need not have worried! Wullie Hampson did the honourable thing!

The Coaltown of Wemyss Social Club also did Smithie Anderson proud. They organised a social gathering for the local members and miners families, invited a few very special guests, and presented little Smithie with the beautiful gold watch and chain, that he so richly deserved for coming first amongst equals, in a fierce competition involving fourteen teams from the Fife area and finally, placing Lochhead pit firmly in the 'Ambulance' stakes. This presentation took place in the Buck & Hind Pub in Buckhaven in the year of 1960, and to this day, Smithie is still the proud possessor of his rightly inscribed gold watch and chain.

Smithie Anderson spent a total of Forty-six years of his life in the coal mines. He spent eleven years in the Victoria pit, was transferred, then wrocht for another twenty-five years in Lochhead pit, before spending his last ten years in the Francis colliery at Dysart. He took the 'Lamp' *(Fireman/Deputy)* when he was 51 years of age, and successfully carried on until he was retired at the age of sixty. The N.C.B. in its wisdom, awarded Smithie the beggarly sum of Five Hundred Pounds sterling in recognition of his lifetime service in the Fife coalfields. Just about enough to buy him and his wife a comfortable bedroom suite in which to enjoy his long-overdue Lie-in!

✳

# Chapter XII

## OLD DOOK EXTRACTIONS!

(De Profundis Magnum –

*Out of the Great Depths)*

The Old Dook as it was named, could surely be described as one of the Great Dooks in Lochhead pit. It was driven down through the Dysart Main seam of coal at a width of approximately 14 feet and an initial probable height of 8 feet. It was commenced in the year of 1899, and was gradually lengthened over a thirty-two year period to its full depth by January 1931. It was driven slightly across the dipping grain of the coal to lessen the effect of the slope, and its direction from top to bottom was at a map bearing of approximately 140 degrees. The named coals that it initially went through, was probably the Toughs & Clears, the Myslen with 6 inches of stone above and below, and the Grounds & Nethers. This left the solid Sparcoal at 18 inches thick and above that, the even more solid 50 inches of Head coal, an altogether more than dependable roof that needed very little support. The coals that were left underneath were the Sclits, *(a 50/50 closely interleaved mixture of coal and stone, and utterly useless!),* a 15 to 18-inch band of stone/blae and the bottom most Coronation coal at approximately 55 inches thick (At this place!)

The Old Dook at termination, was approximately 1,550 metres in length, and its difference in depth between top and bottom was 1,510 feet (460 metres). This produced a declivity with a slope of approximately 1 in 3.7 or, a declination of roughly 16 degrees. *{As a point of pure interest, the line of the Old Dook placed it directly under the Earl David pub (the old Gothenburg) in the Coaltown of Wemyss, at a depth of 1000 feet below. The termination point of the Old Dook was directly under the high tide mark on the east to West Wemyss shore, but 1750 feet underneath the surface of the water.}* Adjacent to the main dook, there was also a parallel companion dook *(dipping or declivity),* approximately 25 to 30 metres to the right, and driven to the same overall length, but slightly smaller in size, at approximately 12 feet wide by 7 feet high, and also cut through the centre part of the Dysart Main coals. This was used as a single track man-haulage, as opposed to the main dook, which was used specifically for the haulage of coal and down-going materials. This man-haulage was laid with a semi-permanent single railway, and carried two by twenty angled seat bogies, permanently attached to a single, winding rope, coupled to a haulage motor situated higher up on the other side of the horizontal, Lower Dysart haulage-way. This parallel companion dook matched the main dook for length at the bottom, but was perhaps 30 metres shorter at the top end. These two dooks were interconnected by driven levels at approximately 30 to 35 metres apart, and 48 of these levels were driven between their parallel lengths. *(A plan of the length of this Dook showing the regular interconnecting levels, should conjure up a picture of a long, slim ladder having forty-eight rungs!)* Both dooks were started *(driven down)* from the same height as the pit bottom level, and at a tangent of 45 degrees from the start of the Lower Dysart haulage-way *(datum height 9765 feet in the Dysart main seam)*. They were approximately 35 to 40 metres apart on the tangent.

As one travelled into the depths of the Old Dook, it soon became apparent that very little support was needed to hold the roof in position. Supports were few and far between, and mostly used at intersections, if at all! The coal was not extracted to its full height at any point, which meant that in most cases, there was a good four to five feet of good, solid, hard coal, to form a dependable roof over the comparatively narrow extractions of coal. This was not a area where modern machinery had made any in-roads, in fact, the methods of coal extraction were rather basic, and the coal had nearly always been extracted by the methods known as 'stoop and room' *(pillar & stall)* and 'pick-place', long wall faces. By these methods, a good and sufficient amount of coal could be produced on a three shift system, without the need for preparatory work or falling wastes. A close study of the dook's history, reveals a basic premise to drive a series of long levels to the left of the main dook, and to the right of the companion dook. *{The solid stoops of coal (unbroken) running down both sides of the two dooks along their full length, would not be extracted for a distance of 40 metres on either side, except for these bisecting levels, so as to maintain the roof integrity of the two dooks. These were known as the solid stoops.}* These levels would be one, two or three hundred metres apart and initially be driven at a given angle to the Dooks. They would be driven to indeterminate* lengths and roughly parallel to each other, but following the level of the possibly undulating coal seam. *(This was most important, in that the driven levels were sometimes directed slightly left or right of 'straight', so that there would be a slight descent in favour of loaded hutches on their way back to the clipping point on the main coal dook.)* The new levels were numbered according to their initial intrusion, and subsequent close intrusions *(levels)* would be co-numbered, *(i.e. No. 3 level, No. 3 companion level and No. 3 water level)*.

* *The advantages of 'Horizon' mining were utilised in both Lochhead and the Victoria pits, long before the term came into vogue in the new 'modern' pits!*

The first level to be developed to the left of the main dook in 1904, *(datum height 9608 feet)* was commenced at the 180 metre depth point, where it was continued on to break through to the New Dook (coal dook) at its 190 metres depth point, a horizontal distance of approximately 285 metres but, with a gentle rise of 16 feet over this distance. *(The filled coal was obviously intended to be transported up the Old Dook!)* This was quickly followed by a companion level from the Old Dook at just 20 metres below that, at the slightly longer length of 300 metres. *(Readers will surely realise that these two, separate, long dooks were on anti-divergent courses. There was in fact, an interior angle of approximately 63 degrees between their respective directions.)* These two levels were interconnected by a series of nine short headings, placed roughly 30 metres apart, that produced a steady flow of good coal, as well as providing convenient airways as the levels advanced. These first connections between these two dooks* were completed in 1905.

*\*Readers are advised that there were three main dooks in Lochhead pit during the first decade of the century, and one main dook in the Victoria pit.*
      *The Lochhead Dooks were: - The Old Dook, commenced in 1899; the New Dook, commenced in 1904-5; and Nicholson's Dook, commenced around 1900, but not completed until around 1917. All three of these Dooks were struck through the Dysart Main seam. The Victoria Dook was commenced around January 1907, and also struck down through the Dysart Main seam, but was administered through a different under-manager working with miners employed at the Victoria pit. The Lower Dysart Dook (Lochhead) was struck through the Lower Dysart seam and commenced around 1948, but in the section known as the Lower Dysart West! Later on in this narrative, mention is made and description is given, of another two long Dooks within Lochhead pit, but these were driven much later on in the 1960's - and driven to take great chunks of the Lower Dysart seam of coal, which lay approximately 25 metres below the Dysart Main seam.*
      The very first named coals that were taken from the Dysart Main in the Old Dook, were the bottom Coronation coal *(approximately 55 inches thick)* from the base of the seam. This was taken from the area of the triangle formed between the lines of the two Dooks, *(Old and New)* and the line above this new No. 1 level. This seemingly, was nothing more than a tentative exploration into the Dysart Main seam, with no other coal ever taken from this small triangle. This limited, extractive exploration commenced in 1904, and finished in April 1908. The probable explanation as to why there no more coal ever taken from this triangular area, was that it stood too close to the pit bottom, and the main pit shaft!
      The No. 1 companion level *(at a datum height of 9619 feet)* between the Old Dook and the New Dook, was in fact, over a few short years, extended in both directions following the line *(grain)* of the seam of coal, until it reached the colliery's extreme boundaries. To the left *(facing downhill)*, until it reached the Buckhaven fault at 1250 metres long, and to the right, where it cut through

Nicholson's Dook at a distance of 700 metres, at a datum height of 9600 feet. This level to the right then carried on till it struck the head of the Victoria sea Dook at a distance of 1575 metres, and at a datum height of 9594 feet. This long level to the right, (*south and slightly west*) eventually reached 2,175 metres long, and was commenced from the Old Dook in August 1902, to meet up with another level at roughly the same datum height coming north and east, from near the head of the Victoria Dook *(not yet commenced)* as early as 1900, which would suggest that it was initiated by miners from Lochhead pit. The two converging levels met up, but not quite in line, around April 1905, immediately under the 'Baund' section, (*exactly between the Old Dook and Nicholson's Dook*) where a re-alignment of levels was carried out.

*(There had been a slight misalignment in direction of the levels coming in from opposite sides. Another interesting point that emerges from this mis-meeting of levels, is that nine separate intrusions were made upwards into 'Baund'\* section at this point, with the resultant, very confusing air-flow around the lower part of this large section).*

\* Described at length elsewhere in this narrative.

The coal extractions in this area now described, were all to the rise above this long No. 1 level, but to the dip, below the long pit bottom level. I mention this for the simple reason, that most of the coals along and above this No. 1 level, were mostly at the same datum height, which, in fact means, that the extracted coals were nearly all transported in the same direction, and the coal extractions were typical to the depth. This area includes the infamous 'Baund' section, which will not be described here, but has a small chapter all to itself elsewhere in this book. This large area between the Old Dook and the head of the Victoria Dook, and above the No. 1 level, was divided up into many smaller sections that were worked variously from 1897 onwards. It is also obvious from the underground plans, that several sections were re-visited, with further extractions as late as 1932, but the bulk of the coal would have been removed by 1915 and up to1920. The coals taken at various times, were initially the Grounds & Nethers at 72 inches thick, followed at a later date when the extensive pillar-wood and stone packs would have been crushed, by the extraction of the Toughs & Clears and possibly the Myslen, at 48 inches and 10 inches respectively. These coals would have been extracted by long wall pick-places, using wheeled braes taken at the slope. The slope distance from the Pit bottom level down to No. 1 level, was a difference height of 150 feet over a 200 metres distance. *(Approximately 1 in 4.5.)* At the head of the Victoria Dook, (*at No. 1 level*) the distance up to the pit bottom level was 525 metres, with a difference height of 190 feet. Therefore, the slope at the Old Dook was 1 : 4.5, whilst the slope to the level above the Victoria Dook was only 1 : 9.25. Obviously, the lie of the strata shallowed to the south-west.

A point of interest creeps in at this stage, in that the workings nearest to the head of the Victoria Dook, and to the left of the Dook (*directions right and left are determined looking* **down** *the dook*) above No. 1 level, were the first sign of single extraction 'Stoop and room' workings in the Dysart Main coal. The largish stoops being left in this section above No. 1 level, were rather indeterminate in shape and irregular and looked all the world like 'crazy pavement'. There didn't seem to be any rhyme, or reason to this crazy matrix, that measured about 525 metres long by 250 metres wide/deep, with about 20 per cent of the available coal taken, to include the Sparcoal, the Toughs & Clears, the Myslen and the Grounds & Nethers. This coal must have been taken by Lochhead pit miners, in that the extraction date does not fit in with the given date for the entry of the Victoria miners into the Dysart Main seam. The extraction date for this crazy matrix was around 1891-92-93, followed by what was a most unusual action, in that the miners went back into the section and 'downed' all of the Head coal in the headings and levels,as an 'after-thought ?' This was completed by 1893-94, and was notable in that this was the first inkling of this type of extraction in Victoria and Lochhead collieries. This then meant, that a total height of approximately 15 to 16 feet of coal had been taken in this area, with the Sclits and the bottom Coronation coal left untouched.

Another point of some significance with regards to the linearity of this level, was that due to the amount of small sharp twists and turns along its length, it seems very unlikely that anything else rather than pit ponies or indeed human drawers, could have been used along this level at this time. The first intrusions above this No.1 level, were of course, made by miners working from the Lochhead pit, as the first extraction dates were again, before the Victoria's cross-cut mines were struck into the Dysart main coal. It also seems more than likely that the extracted coal could have taken out, by one of two routes. 1. Along to the middle part of No. 1 level and up Nicholson's Dook to the pit bottom level. 2. Up the two new extended short headings above the head of the Victoria Dook, to meet the southern-most extension of the pit bottom level near the Francis colliery boundary. *{The twin headings that lie above the head of Nicholson's Dook (and under the Bowhouse farm near the Coal-town of Wemyss) were driven before the Victoria cross-cut mines had reached No. 1 level. The upper extensions to the Victoria Dook were incorporated into, and connected to these pair of headings, and in the same straight line.}*

The small sections to the right (*south*) of the head of the now 'extended' Victoria Dook, and above this No. 1 level, were in the shape of a small triangle and were worked hard against the Francis boundary. This No. 1 level was extended south, and slightly west commencing in February 1900, and the coals taken were nearly the full thickness of the seam with the exception of the head coal.   The bottom coals were taken first by long-wall pick places, where the whole sections were allowed to 'weight' and crush under the influence of wood and redd pillars, and when crushed, the uppermost coals would then be taken by a re-visitation.  This section would have suffered at least two, if not three visitations, and was worked out and abandoned by 1907, having been fully extracted.  The coal would have gone out by the No. 1 level, and up Nicholson's Dook, then along Nicholson's level to the Lochhead pit bottom.

By January 1916, the Old Dook *(and it's companion dook)* was extended down another 235 metres to what was to become the new No. 2 level. *(This Main Dook was now down to a length of 475 metres at datum height 9424 feet).*  This new level, commenced in mid-1918, was developed only to the right *(south and west)* side of the companion dook, and linked up quite precisely with another level being driven (north and east) from the Victoria Dook, that been commenced around 1911.  The two opposing levels, 1510 metres apart at their inception and driven at approximately the same datum height, met quite precisely around the mid-point in 1922. There was no hurry to connect these converging levels, for the reason that miners from each of the two pits were taking coal from new developed sections above this No. 2 level, as they drove inwards.  As a matter of fact, the colliers from the Victoria side took much the greater amount of coal from a section that measured 800 metres long by 180 metres deep/wide. *(All coals seemed to have been taken to the rise.)*  The coal taken in the years from 1914 to 1919,*(again by long-wall pick-places),* were the Grounds & Nethers and the Coronation coals, with a later visitation between 1929 and 1937 to take the Toughs & Clears and the Spar coals.  The area extracted by the Victoria miners, stretched from the Victoria Dook to Nicholson's Dook, and above No. 2 level up to, but short of the line below No. 1 level. *(With this form or type of extraction, there was usually a solid stoop of coal left between the long levels and the section areas worked.)* The Lochhead miners working in the area between this No. 2 level and back up to No. 1 level, were working to the rise, with the inside half of their part of this large section being directly underneath the village of the Coaltown of Wemyss. The only coal that was allowed to be taken, was a tightly controlled regular matrix pattern *(Stoop and room)* which was fully wrought from January 1929 to January 1930.
 The small rectangular section above No. 2 level nearest the Old Dook, which measured 325 metres by 150 metres, was where Grounds & Nethers and the Coronation coal was taken from 1929 to 1931.  This section was re-visited around 1943 to 1944, where the Toughs & Clears and the Sparcoal was taken, and again in 1954 to finish of any residual remains.  This small section seemingly suffered the same 'ignition faults' as the infamous 'Baund' section, but being more isolated in terms of working intrusions, it was more easily and

permanently, sealed-off!

Author's Note:- *The above described sections that lay lengthwise between the Old Dook and the Victoria Dook, and breadthwise, between No. 1 level and No. 2 level, were wrought as four separate working sections of approximately the same size. Two of these sections were worked by colliers from the Victoria pit, whilst the other two sections were worked by Colliers from Lochhead pit. The coal produced was probably transported to the respective pit-bottoms.*

The methods of extraction as practised in the Victoria pit and to a lesser extent in Lochhead pit at this depth, were about to change. The Dysart Main seam was generally dipping down, to well under the depths of the adjacent sea bed, and, as the main levels were becoming progressively deeper, the need for a more conservative type of extraction was becoming apparent. The increasing overhead weighting factor had to be taken into greater consideration, as had the fact that the underground workings were nearing the sea shore, but well below mean sea level.

This part of No. 2 level did stretch from the **Old Dook,** along past the **unused bottom** part of **Nicholson's Dook**, through the twin dooks of the **Victoria pit** and on to the **Francis** colliery boundary. It was approximately 2,200 metres long with the following datum heights.

At the Old Dook; 9424 feet, at Nicholson's Dook; 9440 feet, at the Victoria Dook; 9423 feet and, at the Francis boundary; 9442 feet.

As the Old Dook deepened a further 250 metres *(datum height of 9260 feet)* in July 1919, the No. 3 level was commenced to the right side *(south and west)* in the direction of the Victoria Dook. The new **No. 3 level** as stated, commenced at a datum height of 9260 feet from the right companion side *(south-west)* of the Old Dook which had been extended down to this depth by July 1919. The Old Dook and it's companion dook, now being approximately 685 metres in length.

*(It is of note to record at this time, that the other main twin dook, named the New Dook, had already reached this datum depth by April 1908, where, horizontal levels to both the right and to the left had been commenced, one of which, later joined-up with this No. 3 level (north) on the Old Dook. This long joined-up level stretched from the Buckhaven fault to the Francis boundary, a total length of approximately 4,000 metres.)*

The long coal section between the Old Dook and the Victoria Dook, and above this No. 3 level *(but, below No. 2 level)* measured 1400 metres long by 175 to 200 metres wide. *(There were slight variations along its length.)* The method used in the coal extraction of this extended section, was a regular matrix of Stoop and room where the coals taken, were the Myslen, the Grounds & Nethers and the Coronation coals. The Myslen varied from 12 to 16 inches in thickness, the Grounds & Nethers 58 to 73 inches, and the Coronation coal from

56 inches at the Old Dook, to 68 inches at the Victoria Dook. The full thickness of the Head coals, the Sparcoal and the Toughs & Clears were left untouched throughout this section, to maintain the integrity of the roof, at about 10 to 12 feet in thickness*. The regular pattern of headings and levels were driven approximately 12 to 14 feet wide, leaving solid stoops that were a full 30 metres wide, including the inevitable 'downed' coal which should have been left severely alone. The workings from the Victoria side commenced in 1924 and lasted to 1929, where they met up with the Old Dook miners who had commenced from the Old Dook side in 1928. The Victoria miners seeming took most of the coal from this section, for the reason that the Lochhead miners were more concerned in taking the coal from the triangle between the Old and New Dooks and above this No. 3 level, including all of the coals up to the No. 2 level above.

{*It may be of interest to readers to note, that the amount of coal extracted by the Stoop and room method in a single part of this section, measuring 250 by 250 metres and given as a percentage of the possible reserves available, worked out to approximately 12 per cent of the total. Out of a reserve of 500,000 cubic metres, only about 56,000 cubic metres were, or could be taken. This section was now considered 'worked out', unless more coal could be taken by systematically stripping the sides of the large stoops on the **retreat,** as had happened with some of the more distant sections in other parts of the colliery. There was very little evidence of general 'robbing and stripping' of standing stoops in any of the Old Dook's abandoned sections, especially in the latter stages of the extraction of the remnants of the Dysart Main seam. (What did take place in the latter half of the Fifties, was that the wide, untouched Stoops adjacent to the Main coal Dook at the No's. 3, 4 and 6 levels, were all broached and worked as limited Stoop and room extractions, to gain quick, clean coal at minimum cost.) It may also have been thought impractical, in that these deeper workings were relatively close to the sea, and that **all of the sections** were well below mean sea level. The integrity of the roof meant **everything** in the depths of both the Old Dook and the Victoria Dook, and all extractions seemed to err on the side of caution!*}

Author's Note:- *Readers will be aware, that when a 'roof' is described as being 12 to 14 feet in thickness, this does not mean that there is only that thickness of 'metals' above the miner's heads. This refers to the known thickness of solid coal remaining intact above their heads, left in situ to help support their immediate roof, as opposed to soft blae and broken stone above that, which, even at the best of times, is highly unpredictable and difficult to control, once broken!*

In the near triangular-shaped section formed by on two sides by the Old Dook and the New Dook, and above No. 3 level, this area deserves special mention, in that it seems to have been visited on four separate occasions. The size of the section area was approximately 525 metres along the bottom and 225 metres along its top and approximately 350 metres along its two sides. *(The uppermost tip of this triangle was cut through by No. 1 level between the two Dooks. The No. 2 Level on the Old Dook was not driven to the left!)* There were five level intrusions into this section from the Old Dook side, and only two from the New Dook side. The long, solid stoops that lay between the sides of this triangle and running alongside both Dooks, was fully 60 metres wide, which creates the impression that the maximum extractions taken from this section were entirely planned, with the knowledge that there could have been a great weighting and subsidence, with these extra wide stoops being left to ensure the integrity of both Dooks. The first intrusions took place around 1919 to 1920, even though intrusions from the New Dook side had taken place as early as 1916 to 1917. The first coals that were extracted were the whole of the Coronation coal, all over the area of the 'triangle'. This was completed by January 1921, with no visible internal Stoops left. It can therefore be surmised, that the method used was long-wall places, with the regular building of wood and stone pillars to ensure a slow and even crush! The Myslen, and the Grounds & Nethers coals were then extracted over approximately 80 per cent of the section area leaving a matrix of regular narrow stoops. When that was completed and allowed to settle, a third visitation was made that removed approximately 60 per cent of the Sparcoal and the Toughs & Clears. This triangle was not as fully extracted as the 'Baund' section, but the limited, well-spaced out intrusions and the wide remaining stoops, would have ensured a more than adequate sealing-of process, in greater depth, within the length of the longer headings and levels leading into the section, had that been found necessary.

*(This section alongside the Old Dook, and at regular intervals along this **No. 3 Level** to the **New Dook** (and it's companion level), were re-visited from 1954 to 1958, where several pairs of 'Stoop and room' miners, armed only with Poker-drills and explosives, were instructed to break-in to these wide stoops with Headings and levels, to take as much of the Toughs & Clears, the Myslen and the Grounds & Nethers as was possible, without compromising the surrounding roof areas. At this juncture, I would say without hesitation, that approximately **98 percent** of the coal extractions in the **Dysart Main Seam** had been completed by this late stage and, we were merely scrambling about various parts of the **Old Dook**, breaking into wide, and not so wide old stoops, in an effort to produce any amounts of easy clean coal. The management grudgingly paid the miners an agreed sum of Six and sixpence per ton for coal taken by this method. These miners were, at this time, the only miners in Lochhead pit still 'howking' coal by this now very outdated method. To howk coal and having to 'draw it' to the Dook*

*haulage-way, was something that I had only seen in mining movies' such as* **'The Proud Valley'** *and* **'How green was my Valley'**, *and now, in 1954, we were having to do it again! (When I mentioned to old Harry that Jackie Dryburgh and I were working at 'Stoop and room' down the Old Dook, he merely grunted and said, "Mind ye dinnae fill ony staines!" - a warning, barely heeded at that time!) Fortunately, as soon as a level got to beyond a certain point, we were given oncost 'drawers' to wheel away the full hutches and supply us with empties! I must admit that these men worked like 'Trojans', as Jackie and I got hardened into this form of mining. We could earn more than a long-wall stripper working on a 'cut' face-line, but, that meant working our 'drawers' to the limit. (To which they later went 'on strike' against, or, more properly, 'worked to rule'.) The names of the few men whom I can recall working in this environment were Smithie Anderson and Davie Anderson, Jimmy Higgins and Piper Lawson, Colin Allen and Jimsey Dryburgh, Wull and Jim Allen, Jackie Dryburgh, and of course, myself! Old Harry still chuckled to himself at the thought of 'his' lad working in a primitive environment that he, himself, had abandoned fully twenty years ago!*

During 1928 and 1929, the main coal dook had been extended down another 375 metres to a new length of approximately 1125 metres, down past datum height 8990 feet at the start the new developing No. 4 level to the Victoria Dook, *(there was never a No. 4 level to the left of the Main Dook!)* and further down to datum height 8795, for the start of the new No. 5 level to the left of the Dook. This new No. 5 level to the left of the Dook, was eventually to become the longest, level, haulage-way in Lochhead pit. At the new No. 4 level, two parallel levels were commenced to the right of the main dook in the general direction of the **600 levels** from the Victoria Dook. *{Strangely enough, the companion dook (man-haulage) had not been extended down at the same time as the main coal dook. This was not done until ten years later in 1938-39.}* These two levels were started in 1929, and met up with the two levels coming in from the **600 levels,** from the left side of the Victoria Dook. They were co-joined in 1931. The coals taken by the Lochhead miners along this No. 4 level and to the rise, up to just below No. 3 level, was again by the Stoop and room matrix method, from a section that measured approximately 425 metres long by 250 metres wide. The taken coals were the Coronation coal at 66 inches, the Sclits at 18 inches, the Grounds & Nethers at 64 inches, and the Myslen at 12 inches. *(I rather suspect that the Sclits were downed onto the pavement!)* The height of coal taken with it's intervening stone, was therefore in the region of 178 inches. A near 15 feet height which could only have been taken, in two or three separate leafs, and, at different times. This section was worked variously between July 1929 and July 1933. The stoops left in situ along the length and between the close parallel levels, measured 20 metres by 75 metres, while the stoops between the 12 headings and the subsequent 8 levels in the section, were a standard 30 metres square. A third parallel level to the right at this No. 4 intrusion, was extracted above the other two levels, and was connected to the main companion dook sometime after 1938, when the companion dook was extended down to datum 8990 feet. *(To parallel the main coal dook.)* The middle section along this **No. 4 level/600 level** was worked variously between July 1930 and July 1934 and involved miners from both the Victoria pit and Lochhead pit. The three parallel levels that comprised No. 4 /600 levels between the Victoria Dook and the Old Dook, were fully inter-connected and all three ran directly under the Wemyss Castle home farm, at a point 500 metres from the Old Dook and 850 metres from the Victoria Dook. No further extractions of any description were permitted from under the home farm, within the policies of Wemyss Castle grounds.

The No. 4 parallel levels *(three in number at 20 metres apart)* to the right of the Old Dook, did not extend to the left of the dook. The area immediately to the left of the Old Dook below No. 3 level and adjacent to No. 4 level, was left unworked and remained a solid stoop at exactly 100 metres wide for many years to come. *(See N.B. and chapter on Stoop and room Miners called: - Dauvit's Progress, back to the pit!)* The large, new area between the Old Dook and the New Dook, and below No. 3 level, was bounded at its bottom side by the New No. 5 level, which had been commenced in July 1930, and consisted of three parallel levels, 20 to 25 metres apart, that were driven all the way to the Buckhaven Fault, a distance of approximately 2,750 metres. The longest, double-curved level in Lochhead pit. It was shaped like a shallow, lazy letter 'S', and gradually rose in height along its long passage to termination at the fault. Its start point datum height was 8797 feet, and its termination point was at 8893 feet, a difference height of only 96 feet. I can't but surmise, that this gentle rise along its double, curving length was absolutely deliberate, in that the planners obviously knew exactly where they were heading - and were at great pains to ensure a gentle descent for the long, loaded races that would eventually be hauled to the Old Dook. This long large area and sequence of divers sections, would take fully 18 years to develop, so, the investment in a substantial, permanent haulage-way was bound to prove both beneficial and economically sound.

N.B. *This stand of coal needs be remembered, in that it remained a solid stoop of untouched coal until the middle of the 1950's. It was situated on the left side of the Old Dook between No. 3 level and No. 5 levels. This solid stoop of coal remained fully 100 metres wide and 325 metres over its dipping length. Mention will be made later on in this narrative, with regard to this untouched area of coal, that actually lay directly under the north east end of the Coaltown of Wemyss, but approximately 900 feet underneath.*

Also at datum height 8797 feet, at a Dook length of 1115 metres, the commencement point of the long No. 5 levels to the left, *(to the Buckhaven fault)* there was an short intrusion made to the right side of the dooks to the south and west. I can find no positive date for this intrusion, but, I shall refer to these particular intrusions a little later in this chapter.

From 1929, through 1930 and into 1931, the dooks were finally extended to their maximum length of approximately 1540 metres, *(horizontal length as from plans),* where the long term plan to link the Victoria Dook, the Old Dook and perhaps the New Dook, all to the Michael colliery, became a reality. *(If a near equilateral triangle were to be formed having the line of the Old Dook as its left side, and the line of the New Dook as its right side, then, the Michael colliery would lie exactly halfway along the line of its joined-up base! From the middle of the base of this triangle to its apex, the map direction would be approximately 285 degrees grid - and on a slope of approximately 1 in 3.5.)*

Meanwhile, back to datum level 8488 feet on the Old Dook, where a further two intrusions were made from the right side of the Dook, in July 1936. These intrusions took the form of two levels exactly 30 metres apart and were driven only 75 metres and 50 metres respectively. The top 75 metres level was taken into what was to become a very small section of irregular Stoop and room workings, in the shape of a triangle measuring 120 and 130 metres along its two sides and approximately 160 metres along its hypotenuse. This was the very last section at the bottom right of the Old Dook, and it could not be developed any deeper or, to any greater extent, for the reason that its workings stood hard against the Wemyss Castle stoops. The top four-fifths of this triangle was worked using very irregular Stoop and room extractions, in that there were three different sized stoops in the area, just as though the planner could not altogether make up his mind as to how much of a stoop should be left. The coals taken, were the Grounds & Nethers and the Myslen together, at approximately 80 inches thick. The dateless intrusion mentioned earlier, was in fact, used as the top road *(return airway)* out of this section, which was variously worked in 1936, 1941, 1942, 1943 and 1944, where a short heading was driven from the top of this section into the No. 4 levels just above, as another through airway cum escape route!

The section was again re-visited in 1957, when the final 2 to 3 percent of Old Dook/Dysart main extractions took place, with the latter-day Stoop and room miners. Just to crown and finalise the dying throes of this last bottom section, the planners actually advised the mine manager, who, in turn, instructed the colliers to attack the bottom corner of this triangular shaped area, measuring 75 metres by 50 metres, and make several layered extractions. The first extraction was the full thickness of the Coronation coal at 45 inches, followed by the Grounds & Nethers at 55 inches, then followed by the Toughs & Clears and the Sparcoal, leaving the Myslen splint coal as a thinnish interleaf. This was surely desperation mining, in that the last time they had attempted this degree of extraction, was within the infamous 'Baund' section, which shall remain forever as Lochhead's 'blot on the stratascape!'

Back to datum level 8797 feet. The No. 5 levels to the left of the dook were commenced in July 1930. (*See Willie's Pirouette!*) There was started, three parallel levels, each about 30 metres apart, with the top two levels commencing from the left *(coal-haulage)* dook, each at a point where the twin dooks were interconnected. The bottom of the three levels had a strange origin, in that it also commenced from the main coal-haulage dook, but it was actually later connected to the left dook by a semi-circular level tunnel that was developed over the top of the Main Dook.

Author's Note:- *This semi-circular mine was developed, so that miners travelling to and from the No. 5 level workings, would not encroach onto the coal transportation dook, whilst it was in operation. The obvious reason for this 'flyover', was for the convenience and safety of the continuously travelling miners, and the many visitors to the section, to keep them away from the inherent dangers of constantly moving rail traffic.*

As this large area between the Old Dook and the New Dook, and on to the Buckhaven fault, was in fact, the largest group of sections in Lochhead pit, and had perhaps the most diverse methods of extraction. I shall deal with the sections one at a time, starting with the one nearest to the Old Dook, and covering each one in turn as the sections lengthened and widened. This level grew to be the longest haulage-way in Lochhead pit and for many years was certainly the busiest.

This first section above No. 5 level was commenced around 1932 and to the left of the Old Dook. It measured approximately 350 metres by 210 metres and was roughly oblong shaped. There were 5 intrusions *(short headings)* up into the coal from No. 5 level, at approximately 68 metres apart and at regular intervals along a 280 metres distance. There was a long, 30 metres, wide stoop left above the top-most level, before the coal taking commenced, and the same width of stoop left below the length of No. 3 level. This meant that there was two long horizontal stoops between the two sets of levels, one below, and one above the new section. (*There was no No. 4 levels developed (connecting) between the Old Dook and New Dook.*) The southern most quarter of the section was extracted in 1935, by Stoop and room mining, where the coal extracted was the Sparcoal, the Toughs & Clears, the Myslen and the Grounds & Nethers, a total thickness of 120 inches, plus approximately 10 inches of intervening stone. This changed somewhat to long-wall face working, where the first coal taken in 1936 was the 57 inches of the Coronation coal and 19 inches of Sclits, with an new intercalated, unnamed, coal seam coming in at 11 inches thick, and sitting between them. (*This was the first appearance of this thin seam of coal, which, to the best of my knowledge was never named!*) In July 1940, a small part of this section was revisited, to take an area measuring 130 metres by 80 metres in the North-east corner, where additionally, the Myslen and the Grounds & Nether were further extracted. This never did extend to remainder of this worked section. Thus, we have seen three different methods of extraction in this one section.

The next small section to be commenced in or around 1933, lay almost half-way between the Old Dook and the New Dook. (*The length of the New Dook did not actually run down to this No. 5 level, except, by virtue of its lower, non-aligned, extended workings*)  The second section measured approximately 250 by 250 metres, and was quite square in shape.  There was 4 intrusions into its lower parts from No. 5 level, and only two intrusions into its upper parts from No 3 level above. (*These upper headings/intrusions would have been initiated to serve as an additional return airway, and, as a second escape route should it be needed!*)  The extracted coal was all brought down to No. 5 level, where it was transported to the Old Dook and thence up to the pit bottom, as it had been from the first section along this No. 5 level.  The method of extraction in this middle section was rather systematic, in that the whole of the 56 inches of the Coronation coal was removed by the **long-wall face** method, partially packed, then allowed to bend and crush, and when this had happened and had settled, the section could be re-visited at a later date to extract another sandwich of coal.

This section area was revisited in 1957-58, and then systematically divided into nine equally sized squares, in a three by three matrix of approximately 65 metres square, where the miner's then proceeded to extract the Sclits, the Grounds & Nethers and the Myslen, and then the Toughs & Clears and the Sparcoal.  A grand total thickness of 205 inches (*17 feet*) of extracted coal from nine squares, leaving a 25 metres stoop all around, and in between each small area of further extraction.  This was as good an extraction as was possible without causing major subsidence and in a controlled fashion.  The output from this small section was prodigious, and far exceeded the Stoop and room output from similar sized sections elsewhere within the Dysart Main.

By the month of March 1933, the No. 5 levels (*north and west*) had reached a length of approximately 1050 metres from the Old Dook and were at a datum height of 8843 feet.  At that point, there was commenced two headings, approximately 40 metres apart, and driven at left-angles to the levels, on a map bearing of 315 degrees.  These parallel headings to cut through the **1000 yards level** from the bottom left of the New Dook, and the **600 yards levels**, also to the left of the New Dook. (*Both to the left of the New Dook - north and east.*)  These headings were approximately **540 metres** long and served to connect the long No. 5 level, to the bottom of the New Dook, which, was never extended as a dook beyond the 1000 yards (*950 metres*) length.

The one, remaining, unworked section above No. 5 level and immediately adjacent to the right of the bottom of the New Dook is worthy of mention, for the reason that it was not worked at this time (1933-34).  This particular area measured approximately 70 metres by 120 metres and was divided into two parts horizontally.  The method used in the smaller top part was wide Stoop and room, with four short headings and four short levels and having 40 metres square stoops.  The extracted coal from this very small working was the Coronation, the Sclits, the Grounds & Nethers and the Myslen and, considering the size of the remaining stoops, I would hazard a guess that this was done to maintain the integrity of the bottom part of the New Dook. The coals taken from this part of the section were taken out too, and up the New Dook. The extraction date was October 1954. No doubt, part of the planned final extractions of the Dysart Main coal.

The bottom part of this small section consisted of three headings and six cross levels, only this time, the stoops left measured 30 metres by 60 metres and even though the same coals were extracted, they were taken out by the bottom roads onto the No. 5 level and on to, and up the Old Dook. The dates of extraction from these stoops is significant, in that the bottom, three, long stoops were re-visited and robbed of a further 5 metres strip of coal along their full length. The dates were 1942 to 1943. Coal was at a premium in the dark days of W.W.II!

At the very bottom of the **New Dook** and 50 metres below it, just above the long No. 5 level, there was a small triangular shaped section measuring approximately 110 by 110 by 160 metres. The access to this section was by two short headings from No. 5 level - and two short levels from the twin, **540 metres headings** going up past the **1000 yards level**. The coals taken, were the full height of the bottom Coronation coal by the long-wall face method, with later extraction of the Grounds & Nethers and the Myslen, followed by the even later extraction of the Toughs & Clears and the Sparcoal. An almost complete extraction of the height of the seam, with the exception of the Head coals and, in a very small section. This section was abandoned and probably sealed in Jan. 1939. The surrounding, untouched stoops were fully 45 to 50 metres wide, all around this triangular section.

At this point in the narrative and in dealing with the coal extractions pertaining to the Old Dook, I must now surely make the point that the further extractions along this level must be properly described, as coming under the jurisdiction of the New Dook and it's miners. Each separate section, or group of sections, had their own shot-firers, deputies and oversmen. It was so, that both the Old Dook and the New Dook came under the same oversman, one Walter (Wattie) Johnston, but the miners themselves were described by the section in which they worked, and by their means of egress into that section. Therefore, the miners that travelled down the New Dook man-haulage, and were employed in the sections to the left (north and east) of the New Dook. These miners, could rightly describe themselves as New Dook miners, even though, most of the coal they mined and produced, was transported along this No. 5 level haulage-way, all the way to the Old Dook, where it was carried upwards by hutches coupled by 'wheel-clips' to an endless-rope haulage, and thence to the pit bottom.

The lowest working levels on the Old Dook were numbered the No. 6 levels, and were initiated as parallel levels only, to open up the sections. The first intrusions were commenced to the right side of the dook at datum height 8492 feet, at a dook length of approximately 1310 metres. This level intrusion was dated around January 1938, and was driven only 75 metres in length where it stopped, then opened out into the start of a small working section. At the same time, another level intrusion had been commenced at a point 130 metres further down the Dook, and in the same direction. It too, was stopped at 75 metres length. These two parallel intrusions were to be the top and bottom roads into a small section, which was approximately 100 metres wide by 60 metres in breadth and on a slope of 1 in 3. The coals taken by long-wall extraction, were initially the bottom Coronation coal at approximately 65 inches thick, which was removed in its entirety during March and April 1938. Before the roof even had time to settle, the miners then went back to the same area, and opened up another sandwich within the seam, this time to take the Grounds & Nethers and the Spar coal at 80 inches thick. The area taken on this occasion was only about half the size of the original area covered by the Coronation coal extraction, and well within its original perimeters. *(The miners knew that they would only encounter broken ground if they worked too close to the peripheries of the original extractions.)* The planners however, were not quite satisfied as yet, and saw the possibility of taking the Toughs & Clears along with the Sparcoal. They left the section to weight for one full year, and then sent the miners back in to take the remaining coals. This was accomplished in July 1939, over an area slightly larger than the second extraction, but smaller than the initial extraction. In all, the planners had been very fortunate to have managed an 80 percent height extraction of the 32 feet thickness of the seam at this depth. Several factors could have accounted for this, in that the section was fully 75 metres from the line of the Dook, with no other extractions having taken place within a large radius of the section. This small section lay directly between the last leg of the Old Dook and the Wemyss Castle edifice. It actually bordered onto the 'sacred', solid stoop that protected Wemyss Castle itself from possible irreparable damage. This was the lowest section on the Old Dook and certainly the deepest worked. There would be no more coals taken beyond this depth on this Dook (except for the Michael levels)!

The coals within this last section were not however, the only coals to be taken on this side of the Dook below the No. 4 levels. There was still an untouched area below No. 4 levels, to the right of the Dooks, and to the left of the Castle solid stoop. This area measured approximately 250 long and 200 metres wide and was roughly rectangular in shape. It was also on a awkward slope of 1 in 2.7 *(approximately 22 degrees).* This new section was initially intruded into by one single level, that had been driven to the right of the Dook at a depth distance of 1290 metres, at datum height 8530 feet. The horizontal level was driven to a length of approximately 70 metres where it was temporarily stopped. This happened in July 1936.

This level then lay unworked and untouched until January 1941, where it was further extended in the same direction for another 120 metres. Here, was the termination point for this level, the miners were now up hard against the Wemyss Castle stoop again. At this juncture, the miners went back to first 70 metres point where the level had been temporarily stopped in 1936, turned to the right and commenced to drive a heading up through the 1 in 2.7 slope towards the bottom No. 4 level, which was exactly 250 metres away. This they did, during April and May of 1942, which meant that there was now a through airway from No. 6 level back to No. 4 level, on this right side of the Dook.

Here again, I must remind readers, that this was a very desperate period of time in the coalmines. The period of the extraction of this section was from April 1941 until October 1944. *(Probably the darkest period of W.W.II.)* The extractions here were by no means uniform or consistent! The area that was developed measured approximately 150 metres wide by 225 metres long and was taken by a slow series of supposedly, parallel levels and parallel, bisecting headings. There was very little uniformity in the size of the remaining stoops, which suggests that the extractions were taken in short, irregular bursts, only when spare colliers were available. The inconsistency in the size of the stoops left standing, makes it almost impossible to determine just how much coal was extracted from this section, but I am prepared to guesstimate, that the previous small section immediately below this on the Dook, produced much more coal from an area, that was perhaps only one-sixth of the latter's overall area!

This was the last section area to be mined on the bottom right side of the Old Dook!

Author's Note:- *This last section was re-visited again in 1957, for the reason that a solid stoop of coal, measuring approximately 300 metres long by 70 metres wide, remained at the left side of the dook (part of the original solid stoop) and had been left untouched by the management up to that date. The area was tackled by the latter-day Stoop and room miners /colliers looking for easy coal, near to the date of the subsequent abandonment of the Dysart Main seam in Lochhead pit. The colliers subsequently, took only about 5 percent of what was available in that area. The Dysart Main was dying!*

The last pair of levels that were driven from the left side of the Old Dook, were also at the 8488 feet and the 8540 feet datum heights. These new No. 6 levels corresponded with the other two levels that had started from the right side of the Dook. These levels to the left had actually been commenced at the same time as their counterparts in 1936, and were to be driven to a total length of approximately 1050 metres - and be interconnected every 40 or 50 metres. The coals that were taken during the driving of the full lengths of the levels, were only the middle coals of the seam. The listed extractions were the Grounds & Nethers at 68 inches and the

Myslen splint at 12 inches along with the myslen stone in between at 5 inches. This meant that the height of the roadways was 95 inches or 2.42 metres. *(This was where the G & N were at their thickest!)* There were five separate sections along the length of this No. 6 level and every single one was tackled in a different manner. The extracted coal would be transported along the level in hutches, towed by an endless-rope haulage to the main coal Dook, where they would be individually clipped and trundled up the near full length of the Old Dook to the pit bottom.

The first section area to be developed was commenced by a heading, started from 100 metres inside the level, driven upwards and parallel to the Main Dook. It was commenced in the first week of April 1937 and cut into No. 5 level above within five weeks. This heading was 210 metres long and would serve as the return airway for the new section. The section area would be 175 metres long and 80 metres wide and lying on a slope of approximately 1 in 3.

The initial coals extracted were the full height of the Coronation coal which had thinned dramatically to only 36 inches in thickness, so, I can but feel, that the system used was a long-wall face, cut by machine, and to the slope of 1 in 3. The extraction was total with no stoops being left, which suggests a conventional long-wall face with regular stone packs at intervals, to encourage a slow weighting process. This makes good sense in that the area was revisited five years later, in order to extract the Grounds & Nether and the Myslen splint, over a full 80 per cent of the original extraction. It also seems that the second extractions were also achieved by long-wall extraction methods.

The second section along this No. 6 level to be developed, was commenced at a point exactly 400 metres in from the Main Dook. This new area was roughly square-shaped and measured approximately 175 metres by 175 metres. It also was divided into four equal squares. The initial intrusion *(datum height 8500 feet)* from the bottom level, headed directly into the slope of 1 in 3, and was driven exactly 225 metres in a true, straight line. The road then turned 80 degrees to the right, and was driven on for another 90 metres in a straight line, that actually parallelled the bottom No. 5 level, but was approximately 35 metres from it. At this point, the road took a turn to the left and, was driven on for another 60 metres where it bisected the bottom No. 5 level, but terminated in the middle No. 5 level. This meant that a through airway from No. 6 level up to No. 5 level had been completed. The new section could now be developed.

The whole of this new section area suffered the total extraction of the Coronation coal by the long-wall method of extraction, probably by an A.B. 17-inch chain machine. The taken height of the coal was a full 52 inches throughout the section. The slope on the coalface was still at 1 in 3, with the extracted coals being conveyed to the bottom of the run by moving conveyor, or engine-less pans. *(Coal placed inside trough pans could only slide down such an inclined face-line.)* This extraction was completed by 1945, and this No. 2 section was left to weight! This section was re-visited in 1950-51, where the area was divided up into four separate smaller sections each about 75 metres square. The next sandwich of coals to be taken were the Grounds & Nethers at 56 inches and the Myslen at 10 inches, a total height of approximately 70 inches, including the 4 inches of the Myslen stone.

In 1953, this section suffered yet another re-visitation, in that the section was re-opened, this time for the miners to take the remaining Toughs & Clears at 42 inches and also the Sparcoal at 18 inches. This, they successfully accomplished with no adverse effects, in that the Head coal remained unbroken even after weighing. *(See Author's note!)*

The commencement date of the opening up of the third section to be developed along this No. 6 level *(north)* seems to have taken place during the summer of 1939. The parallel levels had reached this point much earlier in January 1938, but this potential section area seems to have been by-passed at this time. However, the later extractions were as follows:- At the 610 metres point from the Main Dook, at datum height 8512 feet on the bottom level, a single heading was commenced directly into the rise of the seam. The inclined slope was approximately 1 in 3.7. This heading was probably worked as a machine-less aeroplane brae, without the benefit of a conveyor system. *(The output would be fairly slow with only two to three men working in each heading!)* There was also the probability that they did have electrically powered borers, in that the levels were electrified. A second parallel heading was soon started, exactly 30 metres inwards of the first heading, to be followed immediately by a third heading, parallel to the first and second headings. There were now three headings in progression, working with the same basic equipment and manned by 2 to 3 colliers in each. The coals taken from all three workings were the Grounds & Nethers at 45 inches, the Myslen splint at 14 inches, and the Toughs & Clears at 44 inches. In addition to this, was of course, the 6 inches of stone below the Myslen, and the 4 inches above it. The taken height was therefore in the region of 113 inches or approximately 9.4 feet *(nearly three metres)*. This height of coal could have been taken in one go with the probability, that the colliers could have stood on a loaded hutch to reach the necessary height, to drill the high holes. The taken width of these three headings would have been at the standard 12 feet. To effect a through airflow, the middle heading was continued upwards for another 30 metres, to connect to the bottom No. 5 level. This ensured a return airway for the whole section.

These three, parallel headings were each taken to an inclined length of approximately 320 metres, where they were terminated by a single level joining all three, at approximately 70 metres long. As the colliers retreated back down the headings, they cut a series of horizontal levels at intervals, thereby matrixing the whole section. This was completed by July 1940 and the section was abandoned! No further visitations to strip or rob the standing stoops were ever recorded! *(Readers may realise that after a few months had passed, both sides of the roadways, whether they be headings or levels, would be piled high with crushed down and broken coals from the side walls of the stoops. This 'free' coal is a constant temptation to both management and colliers, but downed coal needs must be left in situ, as any attempts to remove this coal will result in further unwanted crushing with additional downed coals reducing the size, stability and effectiveness of the potentially weakened stoops!)*

The penultimate and fourth section along this No. 6 level *(north),* was a reversion to the extraction methods used in the No. 2 section. The extracted section area measured approximately 150 metres in width, by 210 metres in length and was worked for every cubic metre of coal that it would give up! The first intrusion into the area was by a heading commenced from the bottom No. 6 level, at a point exactly 775 metres from the main coal Dook. This was quickly followed by a second heading at the 810 metres point. The first heading was advanced a total distance of 325 into the rise of the seam, that now inclined at approximately 1 in 4. At the 275 metres distance on this heading, the miners turned sharp left with a level roadway, to cut back into the third heading in the No. 3 section. This connecting level gave the miners the needed return airway back up to the No. 5 parallel levels. Extractions could now begin!

Over the whole of the area, the bottom Coronation coal at 53 inches thick was taken, using long-wall methods of extraction, with no coal stoops remaining. The whole area was obviously partially packed, using whatever redd and stone that was available. This first extraction would have taken place around 1940 and allowed to settle. Within two to three years, the section was re-visited with the area being sub-divided into six separate, but equally-sized, smaller sections, from which was extracted the 57 inches of the Grounds & Nethers and the 12 inches of the Myslen Splint. The 4 inches and 5 inches of stone from above and below the Myslen coal would have been carefully separated and utilised in the building of stone packs, all over the new extracted areas. After these extractions, this section was again allowed to settle for a longer period of time, probably as much as six to seven years. The last re-visitation occurred around 1951, where the miners re-opened the six smaller areas within the section, to further extract the 43 inches of the Toughs & Clears, along with the 20 inches of the Sparcoal. Within these coals there was approximately 8 inches of stone and redd, which was again separated, to provide the colliers with an adequate supply of waste material, thus enabling them to build stone packs and pillars as the coal was being extracted! This section was officially abandoned in January 1952, having suffered two successful re-visitations and a probable extraction of 80 percent* of its coals.

*\* In comparing the extraction methods as used in the Stoop and room approach in No. 3 section, as opposed to the total extraction methods as used in the No. 4 section, an estimation of the volume of coal produced by the different methods,*

*makes for almost impossible statistics. The total volume of coal produced by No. 3 section amounted to approximately 13,800 cubic metres, whilst the volume of coal produced by the No. 4 section is conservatively estimated as 147,900 cubic metres.* **This ratio of approximately 10 to 1, is a typical example of how wasteful the Stoop and room method of extraction was!**

The last section to be developed and worked on this No. 6 level (north), was by the now ubiquitous Stoop and room method of extraction. The section was fairly large in that it covered an area measuring approximately 190 metres wide by 300 metres long, with its length at an incline of 1 in 4. The first of three main headings was commenced at a point 890 metres from the Main Dook, at a datum height of 8522 feet. The other two main headings were at 985 metres and 1080 metres respectively, at datum heights 8529 feet and 8524 feet. Needless to say, these three headings were parallel and were driven from the bottom No. 6 level for a distance of 300 metres. By this time, **a third parallel level** had been added to the parallel pair that constituted the No. 6 levels, at approximately 30 metres to the left and higher in the strata. The three new main headings were now interspersed with another two parallel headings, commenced from the third No. 6 level, which meant that there were now five parallel headings each at 300 metres long and approximately 43 metres apart. At the termination point of these five headings, a horizontal level was cut to join these headings together. This was completed by July 1942. The miners then commenced to cut a series of parallel levels between the first heading and the fifth heading, thereby bisecting the headings every 35 metres or so, until a block of solid stoops had been established. The end result, was that this section area was completely matrixed by September 1943, with only the minimum amount of coals extracted. This section slowly produced coals for a period of 26 months, which, considering the limited amount of coals that were finally extracted, could only have from a limited number of colliers, working without machinery.

Author's Note:- *I have based this conclusion on the following evidence. 1. The length of time taken for the extractions. 2. The limited working space in the headings and levels for the working colliers. 3. The limited amount of coal taken. The Dysart Main seam was between 24 to 26 feet in thickness in this area and at this depth, with 2 to 3 thinner unnamed seams of coal, intercalated in-between various named coals. The total coal reserves within this section amounted to approximately 430,000 cubic metres. The amount of coal actually taken from approximately 2,830 metres of roadways, at 4 metres wide and 2 metres high, amounted to approximately 22,640 cubic metres of coal. (The height of coal taken was the Grounds & Nethers at 66 inches and the Myslen at 12 inches, or approximately 2 metres.) If what was actually taken, is compared with the total reserves available within the area , then that ratio is very poor indeed!*

*Using the Stoop and room approach in headings and levels, does curtail the amount of miners actually taking the coal and severely limits output. An analysis of days worked to coal produced, shows that the rate of production in this section averaged out at approximately 4.175 cubic metres per day. In light of the above statistic, I would hazard a very strong guess that this section was only ever worked very sporadically, when and if, excess colliers from other sections were available and needed to be accommodated!*

This No. 5 section was then was the very last section area to be worked in the depths of the Old Dook. The planned excavation areas for the Dysart Main coals on both sides of the long Dook had been visited, and coals had been extracted by divers means from the huge catchment area covered by the Old Dook. Some of the extractions had been trouble-free, others had not! Most of the extractions had been by the wasteful Stoop and room method, but some had been accomplished by the sandwich system, with sometimes a third visitation. These extractions had been especially bountiful, with between 75 to 80 percent of the individual section reserves being taken.

This section was re-visited again, fully 15 years later, for reasons that I have not been able to fathom! In April 1959 *(the top end of No. 5 section was still standing open at this time!),* the two highest levels in this section were re-commenced in a northerly direction *(parallel to No. 5 levels above but 50 metres from them)* for a distance of 175 metres, where they were stopped and terminated! Why? I simply don't know! I do however, know, that these two levels had encroached into territory that was officially designated as belonging to the Michael colliery catchment area. Indeed, the Michael colliery workings were now within 40 metres of Lochhead's workings in the same seam of coal. The Dysart Main!

Author's Note:- *To the layman, the extractions of three different leaves of coal at different times, and all from this same area within the thick Dysart Main seam, may seem to be pretty straight forward. This may also seem like a simple technical exercise to the uninitiated, who have never been down a coalmine, to experience just what happens within the strata when coal is extracted from the solid metals, especially, when it is taken a great depths. To be able to extract just one layer of coal, usually results in the breaking and closing of the roof space behind the colliers, as the extractions are advanced. To be able to go into a previously extracted section, to extract a second layer from above the previous layer, is considered a very risky proposition indeed, with the likelihood that roofing problems shall be encountered! To actually go back for an unprecedented third time is almost bordering on folly! To my mind, it is only the bravest of miners who would countenance such a expedition! It therefore speaks volumes, for the inherent strength of the remaining Head-coal, that lay topmost in the Dysart Main seam of coal and at that depth! Coalminer's are very long on memory, especially after the debacle of the infamous 'baund' section!*

To many old Lochhead miners, this bountiful seam of coal was the only seam that they ever 'wrocht', and by many, it is remembered with something akin to affection. What 'Stoop and room' miner can forget the 22 to 32 feet of Dysart Main coal in Lochhead and the Victoria pits, and the long hours spent with a 'holing' pick, a poker drill, and a pocketful of personally numbered tags?

Author's tailpiece:- *Even though the Old Dook was unofficially worked-out by about 1945, coal in fair volume still continued to flow from divers places all over this dook's large catchment area. Such had been the extent of the numerous levels and headings, that sufficient nooks and crannies were constantly being found, where two and three man teams of colliers still found pickings rich enough, to maintain good weekly wages on a long term basis. Indeed, another fifteen years were to pass before the Dysart Main seam was officially abandoned!*

\*

# *Old Collier's damn-fool Questions!*

*Question:-* *What length of cord is required to tie 'bow-yanks'?*
*Answer:-* *Twice! Once for each knee!*

*Question:-* *How much coal does a 'bow-yank' contain?*
*Answer:-* *Exactly the same as doesn't fill your boot!*

*Question:-* *Why are 'bow-yanks' needed?*
*Answer:-* *So that a miner's wife need not shorten long trousers!*

*Question:-* *Who invented 'bow yanks'?*
*Answer:-* *Old Colliers who were too mean to buy coal!*

*Question:-* *Why are 'bow-yanks' so-called?*
*Answer:-* *Natural progression from a double-grannie to a double-bow!*

*Question:-*   *When bow-yanks are untied, does the small coals fall into the collier's boots?*

*Answer:-*   *Only if they're still wearing them!*

*Question:-*   *Is the measured length of cord, tied exactly one inch below each knee, with the middle of the cord first placed against the front of each leg, with each end then being wound in opposite directions, until such times as there is only sufficient cord left to effect the tying of a double bow, situated in a accessible position to the outside of both left and right legs?*

*Answer:-*   *Eh!*

\*

# Chapter XIII

## Dauvit's Progress II

## Dauvit is Taken Forth ! - *into Initiation!*

I remember climbing into bed on that Sunday evening with a great weight pressing on my mind. I had to get up at half past four the following morning to start my very first day at Lochhead! I am sure that I nursed a strong wish that everyone would sleep-in next morning so that the evil day would be postponed for at least another day. No such luck! As soon as my head touched the pillow, I was being rudely dragged from a very deep sleep by my mother who, threatening all sorts of dire consequences if I didn't get out of bed, now! It felt as though I had just fallen asleep. I managed to dress quite quickly, was sat before a lump of hot toast and a mug of tea too hot to drink, and warned to be ready to leave in five minutes. I must not miss the five past five bus. Needless to say, I was there in plenty of time, yawning my head off at the bus-stop and being given pitying glances by miners who could spot a greenhorn just by the freshness of his face. At twenty-five minutes past five the loaded bus trundled into the Coaltown, and I joined the crowd of miners already stringing out and heading down Lochhead pit road.

On arrival at the pit-head baths, I went straight to my allocated clean locker in preparation to stripping of, and collecting my new bar of soap to transfer to my dirty locker along with my new clean towel. I already had my new working clothes under my arm and proceeded 'au naturale' along the very warm corridor to the 'dirty end'. Much to my surprise, this dirty end was extremely busy and quite full of dirty miners, of whom, I rather slowly realised, had just come up the pit. God!, what dirty they were!, and all over, except for a narrow band around crotch level. There was young men, old men, men like my father, and bony old men with taut belly's and 'white' hair. I could hardly take all of this in. The one thing that did stand out, even though I did not realise it until much later was, that in spite of all the activity and bustle, the men did not have very much to say to each other. It was not until many years later that the answer quietly crashed into my mind.

A few minutes later when suitably dressed, I was taken in tow by a middle-aged man whom, it later transpired, was detailed to look out for me. "That was easy!" he said, "you look young and fresh-faced and fairly gob-smacked!" I was then taken in tow, over to the check-box, where I was given a token with my 'allocated number' stamped on it and instructed to find the location of the pit-head pegboard, hang my token on a vacant 'day-shift' hook, and wait for the pithead gaffer. This I did with due solemnity, I couldn't think of any other method. I already felt quite lost.

The pithead gaffer's name was George Peggie and, he was truly a fine figure of a well built man, neatly attired in washed-out blue dungarees with pocketed bib, and an equally washed-out light blue working shirt. The first thing that struck me about this man was that he had a very pleasant manner, and a very soft voice and, that first impression did not change during the relatively short time that I worked on the pithead. I was then directed *(introduced would be an over-statement)* to follow a smallish older man, who was in fact, the leader of a small group of 'wood-men' who were responsible for fulfilling the orders for the differing types and cuts of seasoned wood required by the underground miners. I could easily relate to this, as I had listened to many conversations between my father and mother on the subject of 'packs and pillars' that needed to be built to the roof when coal was taken out. The one thing that did strike me as rather odd, was that the 'hutches' (tubs) that were filled with pillar-wood were being addressed to a place called 'Lower Dysart', I knew the old song - 'Coal for Dysart - Dysart coal! C-o-a-l for Dysart, Dysart coal!' But, now wondered as to why they also needed this wood in the township of Dysart!, and from a coalmine! *(I just did not realise at that time that underground locations also had place names.)*

Quite apart from having to get up at an unearthly hour in the mornings, I found the light work very easy, though sometimes pressing, and also enjoyed being out in the pleasant summer sunshine in the wood yard where, often to the frowning looks of the old man, I would strip down to the waist and enjoy the heat of the sun. I spent roughly four months in the wood yard, doing various tasks and getting to know all about most of the types of wood, larch bars, steel channels, steel girder halves and, how to discover their sizes. One thing that I did get to know very quickly was that some old men guard their positions of small responsibility very jealously. I was certainly not allowed on pain of reprimand, to chalk the hutch's destination on the down-going hutches. This was the sole prerogative of the senior old man! I also discovered that there was another world, a noisy and dirty world, up on the top landing of the pithead. A place that I had not yet seen and, tried hard to stay clear of, simply because of the dust and the noise. I was, however, due for a very rude awakening, and it came quickly and quietly, and on a dirty Monday morning. This was all part of old Harry's diabolical plan, even though I was never to find out until some time later. That fateful Monday morning arrived, and I was meaningfully held back from the wood yard by the old man and told to see the pithead gaffer on the 'pit-head tables'. I went up the steel stairs hoping that he would not be there, I instinctively knew that I had seen the last of the wood yard. *(It had crossed my mind that I was the only youngster in the yard.)* The pithead gaffer was there and looking at his watch, the tables were due to start rolling in minutes and I had to be positioned and instructed before the noise made it impossible. I didn't get very much instruction, he pointed to an long-handled impossibly large shovel, then to a pile of redd and then, to a long moving table that disappeared over a lip, where the tables' contents emptied into a great 20 tons capacity wagon sitting on a railway. This was really the Pit's! Sufficient to say that I suffered countless agonies over the next two months. I suffered blisters that would never heal, aches that never stopped, cramps and spasms that made me bite my tongue, and absolute miseries that wrenched my aching heart almost to the point of tears. I finally summoned the courage to complain bitterly and loudly, about being only a young boy doing a man's work, and still, that pithead gaffer would not give me another job! Within days, I got to the stage where I was on the point of refusing even to go to work, because of the unremitting agonies involved when, out of the blue, I was told by the pithead gaffer that I was going back to the wood yard for a week, before reporting to the Mining School at Muiredge in Buckhaven for underground training. Now, that was a turn-up for the books and I hadn't even asked . . !? *(Old Harry again?)*

Mining school was a dawdle, the academic side even more so, as I was still reasonably fresh from High school, and found my considerable mathematical skills hardly challenged by square measure exercises dealing with various sizes of girders within closed areas and laboriously posed by the two retired miner/instructors. I do remember causing all sorts of mayhem in asking them supposedly stupid questions about unrelated subjects just to confuse them, which never seemed to fail. The most interesting aspect of this six weeks course was being taken down the Wellesley Pit in one of the large empty cages for the first time. It was an exhilarating and belly-lifting ordeal, which left us all speechless for a few minutes. We were all equipped with an issued safety helmet *(a present from the Coal Board)*, and a battery-powered electric lamp that fitted onto our headgear, and was connected by cable to the encapsulated wet-cell accumulator battery pack strapped onto a belt, around our waist. This lamp had a duration of approximately 12 hours before recharging, and would provide the wearer with sufficient illumination, in his immediate vicinity, to work and travel with a good degree of vision in the otherwise pitch darkness of roadways and workings.

We saw many wondrous and almost unbelievable things whilst down the Wellesley pit, which happened on alternate weekdays, and we were mostly impressed by the genuine sense of urgency and activity that seemed to prevail at every level of operation. Very few of the miners appeared to have very much free time, and most of the introductions and explanations were conducted amidst the noise and bustle of moving machinery. Thankfully, we had been previously warned to keep our eyes and ears open, and take note of anything that required further explanation. *I cannot remember anyone doing so!* The pace of the course was very slow, and I got the distinct impression that this was a fairly new venture with the 'new' instructors tentatively feeling their way. The course did come to an end without anything untoward happening, and we all 'graduated' with a single signed piece of paper explaining to all and sundry *(and especially mine managers),* that we were fit and able, to be employed underground. I did get the impression that the whole point of the course, was to learn how to tie leather boot laces, to fit knee-pads so that the buckles faced outwards and, most important of all, not to gob into the bottom of your carbide lamp, the originating spit must be wholesome and clean! We were all told of course, to report back to our own place of employment forthwith and, without delay, otherwise we would lose out on our last week's wages.

That was really the last of the fully paid 'easy working' that I was to experience in the coal mines, and the last time that I was ever on the 'piece-time' shift under the N.C.B. *(8-30 a.m. to 4-30 p.m.)*

\*

# Chapter XIV

## Banana Boy!

## Banana-ratta! –
## Or, *The Miner's Four-legged 'Friend!'*

Young George (Dod) Salmond lived and played in the small mining village of West Wemyss. He lived with an aunt, who was responsible for the care and well-being of young George whilst he was still at school, and under the age of fourteen. The age of fourteen, was when young boys who did not have either the where-with-all, or the ability to go on to further education, inevitably ended up 'going down the pit!' Young Dod as he was to become known, had for a few years watched the full hutches of coal being brought-up from the pit bottom of the Victoria pit, run around the narrow gauge, semi-permanent railway on to dock-side of the inner harbour, and tipped into the steel coal-chutes on the dock, which fed directly into the coal holds of the small colliers that plied the North sea to Norway, Sweden, Denmark and even Russia at that time. He also knew that these same boats brought back cargoes of cut trees, wooden straps and pillar wood. All of which was used underground in both the Lochhead pit and the Victoria pit. This wood, which was sometimes of a rather inferior quality and dubious strength, was used extensively for all sorts underground supports between pavement and roof, where the coal had been extracted. It could be used as a direct support to the roof, as a cut length of tree hammered vertical or, in the case of the pillar-wood, used to build three-foot, square packs filled with waste (redd), which could be packed tight to form substantial pillars. This freshly cut, resinous wood which was mostly Norwegian Pine and such-like, gave of a very distinctive pleasant odour, which was like a breath of fresh air to men, much more used to the dank and dusty environment of mine air.

Unfortunately, it was not only the miners that enjoyed the smell of the resinous wood. Some of the native Scandinavian rats, deprived of either their homes in the forests, or being chased out of the holds of the small coasters, did find their way into the tubfuls of wood being lowered underground. Not only did they find a home, they simply interbred with the smaller homogenous 'rattus-rattus', and developed into brave, inquisitive creatures who were both daring, bold and forever hungry. No miner, but nobody, on pain of a severe reprimand from the fireman, was allowed to throw any or part of their piece *(boxed sandwiches)* to these creatures on any pretext. Weil's disease was rife and there was simply no cure. Once contaminated, the victim was doomed to a slow and agonising death, through liver and kidney failure, bowel rot and heart disease. Rats were hated and reviled by most miners to the point of disgust, even though there were a small few miners who looked on them as part-time pets, and did sneakily feed them with a few tit-bits at piece-times.

When young Dod turned fourteen, his place in the Victoria pit was almost assured, so down the mine he went. In those days before W.W.II., there were very few coal mines where pit-head baths were installed, so, the normal procedure was to obtain work in a local mine when this was possible, to obviate the need to travel by public transport. *(Many mining villages grew alongside a new colliery probably because the cottages were company owned!)* All too often, the bus conductress would refuse boarding permission to a dirty miner with a pressing need to get home. In some cases, the dirty miner was made to stand in the rear well of the double-decker bus all the way to his destination. Most villages, or large chunks thereof, were often owned by the Mine-owners themselves, in that the village would probably originate and grow within a short walking distance of the pit-head, where the developing families would serve the same pit for some generations. This meant that the miners would only have a relatively short distance to walk, and therefore could and did, leave their homes already dressed in 'pit clothes' and carrying their flasks and piece-boxes on their person. Many miners wives would thoughtfully sew extra large, specially strengthened pockets onto their husbands and sons jackets, to accommodate the bulky piece-boxes and tin flasks. In this practice, young Dod's aunt was no exception, she had sewn, not only two outside pockets onto Dod's work jacket, but also, one large extra pocket on the inside, so that he could secrete any suitable piece-time 'goodie' that his aunt had provided, that he didn't want to share with his workmates. Every morning on the day-shift, Dod's routine would be exactly the same.

He would be called out of bed, made to have a quick wash, get into his pre-dusted work clothes - minus his jacket and sit down to either a bowl of porridge or a hunk of bread with jam. His piece was already made-up and packed and his flask filled with tea - and Dod instinctively knew when a 'goodie' had been placed in his inside pocket - he knew by his aunt's little smile. They had a secret pact between them, Dod would not ask as to the hidden 'goodie,' nor would he open it up until piece-time. This was just one of the shared joys of his young, simple life.

At a quarter to six, Dod left his home and made his way to the pit head - a mere 600 yards or so, which he covered in five minutes, the bulge of secret goodie tucked deep in it's pocket and pressing against his body. He made the check-box in good time, collected his numbered token and made his way onto the cage, being careful not to squeeze his unknown bounty. Once on the pit-bottom after a quick drop, Dod made his way along the haulage way to his place of work, where, with little ceremony, jackets were removed and hung up on prepared pegs against a flat wall. The reason for this was to deter and make it awkward for any of the resident rats who, forever hungry, would perform all sort of acrobatics in search of a crust. *(Most of these rats by this time were completely blind, but had an exaggerated sense of smell).* Nothing was safe from them and given time, they would suss-out almost any source of food.

By all accounts this was a normal working day - with hutches of coal moving steadily to the pit bottom, with most of Dod's work being very physical, but repetitious. This gave young Dod time to think and fantasise as to what today's 'goodie' could be? He just didn't know; often he could guess, his aunt could not hide the smell of newly baked buns or tea-bread and most often, his surmises were spot-on! At times, when there was nothing moving on the haulage - Dod's tallow lamp would sometimes pick-up the quick glint and reflection of a small pair of steady, unblinking, beady eyes within the wastes, and this gave him an involuntary shudder - he just did not like rats!

At last, piece-time came and with it, the welcome sound of silence as the haulage rope slowed to a stop. The like-minded workmates of Dod began to assemble at the bottom of the heading, in preparation of sitting together in a close group *(for warmth)* and a quick chat. Just then, as young Dod approached his own jacket - it was seen to move with a small jerk and then another. Dod stopped in his tracks. Someone shouted: - "Rat! Rat! It's eating Dod's Piece!"

Dod's heart went to his throat, this was not his piece! This was his 'goodie' *(which all of the other lads knew about).* His piece was inside his tin piece-box in his outside pocket. What to do? Dod had never faced-up a rat before, and never so close! Help was to hand! Just then the section fireman came upon the scene and without so much as a verbal instruction, handed Dod a 30-inch length of 2" by 2" pillar-wood. Dod did not need to be told what to do! With a firm two-handed grasp on the end of the stick, he swung at the offending jacket with several heavy blows until the very body fluids of the intruder began to leak through and out of the fabric of Dod's jacket. "Is it dead ?" cried Dod, - "Or shall I hammer it some more?" To which the fireman replied, "I think that you have killed it Dod. Its brains are running down your jacket!" Dod laid the length of pillarwood to a gentle rest - he did not altogether believe the fireman, who was in no great hurry to extract the offending rodent from his now - oozing, jacket pocket. He was ready to re-administer any further needed death blows. After some not so gentle goading, the section fireman was persuaded that his was the responsibility to extract the now stilled object from within the gory depths of Dod's jacket - which was what he now proceeded to do with a great deal of apparently, nervous trepidation. He eased the flap of Dod's jacket with his right hand and with his left, tentatively reached into the offending pocket - a look of sheer disgust creeping over his face. Suddenly he stopped, - his expression changing to one of puzzlement and then incredulity, then he slowly turned around to face the small stilled waiting group of men and boys. He warily withdrew his hand which he held high, to show what was the severely mangled remains of a ripe, yellow banana, which had softened and wilted in the warmth of Dod's jacket pocket. Needless to say - young Dod not enjoy the benefit of his aunt's goodie on that day, but he did enjoy the benefit of the section fireman's opprobrium for many days afterwards. *(That which the fireman had never mentioned either to young Dod or any of the gathered men that day, was that he had swiftly and surely removed the long length of black cotton that had been tied to the button of young Dod's jacket.)* Young Dod eventually became Old Dod in time and even after he transferred to Lochhead Pit at the Coaltown of Wemyss, he was still remembered as the boy who beat a banana to death with a stick of pillar-wood!

*

# "EXTERMINATE! EXTERMINATE!"
## *(RATTUS, RATTUS !!)*

In between the two World Wars and especially in the Victoria pit, as the wood from the Nordic countries was being imported and used underground, the proliferation of foreign rodents, namely Norwegian rats, was on the increase and was evident by the amounts of scurryings and visual sightings that were being reported to the firemen. Every so often, the mine owners would organise a blitz with the laying of poisons and traps in known habitats, but with only limited success. Most of the rats that were taken, were done so by the miners themselves and with the most simple of traps. Many devices were invented by miners, who, with some trepidation, always went about this task with a certain degree of secrecy, so that they might not be ridiculed if the device failed. One such device which would always be guaranteed to produce a good 'catch', was simple in its conception, economical in its materials and ghastly in its conclusion!

At most underground junctures, where hutches came out of any side road and onto a 'railway', there was no need for a set of points! Hutches were merely 'slewed' across near frictionless, smooth, steel plates and guided thereon in. Flat steel plates of a manageable size and weight, were therefore readily available. The method described here, merely involved one, large, steel plate measuring approximately 5 feet by 4 feet, placed in the middle of a juncture, with one end placed on a 2" by 2" piece of pillarwood and the other end held up by a 6-inch vertical stell, to which had been tied a longish length of light rope. The men involved in this 'kill' would sacrifice their crusts of bread, which would be tied together in a loose circle and pegged and, carefully placed beneath the steel plate, so that no single predatory rat could run of with its single prize. When the trap had been baited and an 'executioner' appointed, the men would retreat to a given distance and kneel to the pavement, so that they were in a 'starter' position ready for the command of 'Go'!

This 'order' was the prerogative of the executioner! They would all then douse their lamps, except for one miner, whose lamp would be used to illuminate the 'killing' ground, close enough to cast sufficient light, yet out of harm's way! The stage was set. Let the killing begin!

Author's Note:- *Readers of a nervous or gentile disposition should read no further! This part of the narrative develops into a very grisly and bloody story with explicit descriptions, which just might impinge into your otherwise very pleasant dreams. This following narrative is gory, to say the least! You have been warned!*

One of the very necessary rules of this 'engagement', is that all of the participants must, on pain of a good bullocking, be absolutely quiet, even to the extent of clearing their throats beforehand and settling comfortably, but readily into position. The wait could be all of ten minutes, just time enough before the haulage way and the subsequent noise would negate all of their preparations. No one moved! Not even a whisper! The men breathing lightly, with all eyes intent on the meagre flame of one flickering tallow lamp. If the rats were coming, it would be in the next five minutes, or not at all during that shift.

Silence still prevailed, the miners hardly daring to breathe lest it cover the small sounds of an inquisitive rodent. Slowly but surely, the small but distinctive sounds of the approaching vermin could be identified. They came, not in a straight line, but in hesitant, searching, arcs sniffing and moving left and right, trying to locate the source of the impending feast. All it needed, was for just one of their number to locate the jam and marmalade and their primeval instincts would ensure a fierce free-for-all. This happened with frightening speed and since all of the crusts were tied together, no single rat could escape with its new-found bounty. Rats being the primitive creatures that they were and being thwarted in their ecstatic frenzy, simply turned and bit anything and everything at hand. This was 'crunch-time' and with a deliberate pull, the small stell which was linked to the rope was tugged clear, and the heavy plate fell onto the unsuspecting rats.

The squeals of the trapped rodents became screams, as the frenzied rats sought to deliver themselves from the crushing weight of the steel plate, which, heavy as it was, was not heavy enough to kill them outright. On the shouted command of the 'executioner', the miners darted forwards and launched themselves on to the steel plate, and holding onto each other, began to dance up and down on the plate, sometimes individually, but somehow together, to enhance the down pressure needed to completely crush the still squealing rats. The plate slowly sank to the pavement with the combined weight of about six or seven men, until the squealing cries of the crushed rats had all but terminated. The miners by this time, having re-lit their tallow lamps, could now see the full effects of the carnage, which was beginning to seep from under the edges of the steel plate in the form of blood, gore and small intestines!  (In mining parlance:- puddins! - (*Scots word*).)

The 'executioner' who had experienced this method of extermination in the past, had the good sense to know, that in spite of the jubilation felt by the younger members of the 'kill', they would not be able to stomach the very necessary cleaning-up operation.  So, the youngster's were sent out of sight, so that the more hardened miners could safely dispose of the dead and bloody rats.  This was accomplished, by either placing the dead bodies singly in full hutches going to the pit head, or more likely because of the numbers involved, placed in an empty tub, soaked in coal-oil (or paraffin) and incinerated.  Dead rats would not be handled with bare flesh  (*One just never knew with rats, 'twas was not the first time that a miner that had been bitten by a 'dead' rat!*)  The bloody remains would be lifted between the extended ends of two pieces of pillar-wood, by a miner wearing heavy industrial gloves.  The killing ground would then be saturated with paraffin or such-likes and made-over with small coal or redd, with all evidence of the blood-some encounter being hidden and hopefully expurgated from memory.  This was not a method of extermination that rested easily with any miner, but a miner had only to contemplate the dreadful effects of Weil's disease to quickly harden any wilting resolve on the part of the chicken hearted!

Author's Note:-  *Andrew (Awnd) Moodie, eldest brother to old Harry, suffered either to be bitten by a rat, or was exposed to rat's urine, when he worked in the Michael Colliery at East Wemyss.  He eventually suffered the inevitable fate of like sufferers, in that his heart, liver and kidneys began to rot as the disease developed!  He also developed cancer of the stomach, which proved to be inoperable when opened up by surgeons at Kirkcaldy's Victoria Hospital!  His abdomen was then closed up and he was sent home to die!*

*Awnd Moodie did not wait for the 'Grim reaper,' nor did he wish to experience the horrible, painful and slow lingering death that accompanies sufferers of advanced Weil's disease.  He walked into an empty miner's wash-house in the Galatown, laid down on the cold stone floor - and turned on the gas-tap!*

✳

# Chapter XV

## The Hugo Mine *or, The Lochhead Mine*
## *(An Uplifting Entity!)*

The mine described in the title of this short chapter is not to be confused with the Surface mine (The Dipping) at Lochhead Pit. This is an entirely different mine that surfaced in a field, approximately 100 metres from the main road and directly opposite Hugo Avenue at the south-west end of the Coaltown o' Wemyss. The mine bottomed on the sea-shore at West Wemyss, just about 110 yards *(100 metres)* to the north-east from the top of the narrow shaft of the Victoria Pit. To the best of my knowledge, it was driven through the metals in a straight line, constantly corrected by frequent simple 'sightings', which would be 'advanced' as the mine progressed. It was not developed as a coal-mine, but was driven for two specific purposes that will become clear as this narrative unfolds.

When work commenced on the driving of this mine *(tunnel)* from the West Wemyss beach, the instructions were very clear. The given line *(bearing)* was fixed, as was the angle of elevation. This was helped by the fact, that the miners could use the angle of the rising strata to guide them as they proceeded upwards. *(The geographical lie of the Fife coal seams and therefore the strata formations, did slope upwards from under the Firth of Forth, to surface just a relatively short distance inland to the north and west, therefore, it is entirely feasible that a particular rock formation could have been identified on the low cliffs at West Wemyss and traced to outcrop somewhere at the Coaltown of Wemyss.)* As the mine started and the initial work commenced, it is of note to remember that this was before the days of the N.C.B. and mechanisation. The basic mining tools were picks and shovels, hand boring machines, long and short metal levers and loose, powder explosives. As the mine progressed upwards, some sort of short, light rails would be laid as a single track, to enable a wooden, eight hundred-weight hutch to be winched up to the driven face, at least until such times or the increasing length of the mine would facilitate something more practical, such as semi-permanent rails and double tracks.

It is entirely feasible, that some sort of aeroplane brae would be in operation during the development of this mine, in that the stone-miners/brushers would indeed, have been very foolish not to have used the forces of nature in this venture. The tunnel was to be driven to an approximately length of 700 metres through mixed metals, rising at an even rate, to gain 140 feet in height over this projected distance. The mine *(tunnel)* would need to be driven straight and true to the slope, so as not to compromise the subsequent winding/drawing rope on the sides or roof of the tunnel.  A slight convex slope would be more propitious than a concave slope, in that the haulage rope could be kept grounded and guided at all times.

The direction of the tunnel according to the seventh series, one mile to one inch O.S. maps of the period, shows that the tunnel was driven at a map *(grid)* bearing of 355 degrees, with the upwards slope of the tunnel at a rise of approximately 1 in 16, or an angle of elevation of 4 degrees.  It can be therefore assumed from this information, that there was enough of a slope to ensure the functioning of a gravity brae *(aeroplane brae)*, over at least part of this heading, with perhaps, more than one such brae within the mine, to ensure a constant supply of empty hutches at the working face.  This assumption is made on the basis of my own mining experience and the supposition, that, in a mine driven upwards through the solid, there would not be any powered means of waste transportation, back down to the level of the beach.  As to the disposal of the waste from the mine, I can only make the general assumption, that the mine waste was deposited and used to substantiate the upper beach area, either that, or it was taken down the Victoria shaft and used to fill or stow the excavated under-ground wastes.  Also, considering the use that this mine would have been put to, it would be a safe bet to assume that steel girders or iron supports would have been used extensively in it's construction, with much inter-packing between the girders to render the tunnel safe for the developing miners, whilst working and travelling.

At the small mine-head, there was a small screening plant constructed from brick and stone, and the hutches of coal that were drawn up the mine were emptied onto these tables, where the coal was screened *(dirt, stone and redd removed),* before being emptied directly into the waiting coal wagons below, which sat on full gauge permanent rails.  The wagons when full, were coupled together into a small train and pulled by a small steam locomotive.  They were pulled *(shunted)* the short way from the mine-head, across the main road that runs through the Coaltown, and down the first part of what is now Hugo Avenue.  From there, they were shunted across the area between the Coaltown and Lochhead pit, where the railway line passed the south side of the pit at the Surface Dipping, then went on to join the main mineral line, that took all of the Victoria and Lochhead pit coal on to either, the huge washer at the Wellesley Pit at Denbeath, or directly onto Methil Docks.

*(In the village of the Coaltown of Wemyss, directly opposite the junction of Hugo Avenue with the main road, it is still possible to see the remains of one wooden gate which was used to block the road when the trains ran, but is now an integral part of the roadside fencing on the south side.)*

It is still possible for any interested party to walk the 100 metres overgrown pathway from the main road, and see the dilapidated remains of this small plant.  The two entrances to the Hugo Tunnel have long since been filled in!

Author's Note:-  *When the Hugo mine fell into disuse, it was stopped-up at the top end to keep out unwanted visitors and of course, the local children, who probably were always looking for 'caves' to explore. The bottom end of the tunnel was put to a more practical use, in that there was a solid stopping put in place at a distance of approximately 50 metres from the entrance and, the tunnel opening was fitted just inside, with a pair of strong, steel covered, lockable, double doors. The bottom fifty metres part of the tunnel was converted into a dry, secure, powder magazine and used to store the explosives needed, to serve the miners working down the Victoria pit.*

This Hugo Tunnel *(or, the Lochhead mine as it was sometimes called)* was driven for the specific purpose of delivering the coal output of the Victoria pit onto the screening plant at the Coaltown of Wemyss, and thence onto the surface mineral railway line that bypassed the Lochhead colliery's Surface Dipping. The tunnel's semi-permanent railway transported the loaded hutches in five, six and seven hutch races, pulled by an largish electric tugger motor at the tunnel head, where there was built a small custom-built screening plant to 'clean' the coal before its conveyance to the Methil Docks area via the Wellesley Colliery.

This tunnel was mainly used at the beginning of the Century, between the collapse and silting-up of the harbour at West Wemyss and its use as a Coal port, and the installation of the underground Victoria Haulage, that connected the head of the Victoria Sea Dook with the mid-point of Nicholson's Dook. *(In Lochhead pit.)* Whence the Victoria pit's extracted coal was subsequently wound up the Lochhead pit shaft. I have also read *(without description)* that at one time, the traffic in the Hugo tunnel was reversed, in that coal from Lochhead pit was transported back down through the tunnel in hutches, manhandled around the dock wall and loaded into ships. At this time I can find no firm evidence that this ever happened!

# Chapter XVI

## A Day in the Life of the Dipping 'Tailsman!' *(A Local Legend?)*

Lochhead Pit (Colliery) lies approximately half a mile along a completely exposed and windy road to the north of the village of Coaltown of Wemyss, halfway between East Wemyss and West Wemyss in the "Kingdom of Fife". *{That small peninsular County that lies between the Firth of Forth and the Firth of Tay, so named by James IV. (King of Scots.) As the "Beggar's Mantle with the Fringe of Gold!"}*

Lochhead Pit *(second of that name!)* was established in the middle of a wide expanse of farmland, and was one of a long string of eleven coalmines that stretched along the Fife coast, from the lang toun o' Kirkcaldy, to the great coal sidings and coal Docks at Methil, that itself, stretched from Denbeath to Dubbieside (Inner Leven). Lochhead Pit was termed an inland colliery and was unique, in that it had both a Pit shaft and a dipping Mine. As a Pit, it had a main shaft sunk to a depth of approximately sixty-three fathoms, with conventional winding gear consisting of two opposing 'cages', which operated on the standard principle of double hawsers, *(one attached to each cage)* in that one cage rested on the pit-bottom, whilst the other cage rested at the Pit-head. These hawsers were individually wound on a huge winding drum, in opposition to each other, so that as one cage was ascending, the other one was descending! *(This was a continuous process during the course of any one working shift, in order to raise the two, half-ton hutches of mined coal in each cage to the surface. This operation was carried out at high speeds!)* At the end/start of each new shift, this coal-winding procedure was suspended in order to bring up the of-going shift and send down the on-going shift. This procedure took place three times during the course of a twenty-four day and lasted exactly 30 minutes. (Time was money!) The winding speed was of course, slowed down somewhat, men being less resilient than lumps of coal!

All of the foregoing, I mention for the reason that in its uniqueness, Lochhead Pit also had a surface 'Dipping!' A Dipping is the name given to a drift mine, or a sloping tunnel, sunk or driven into the ground at a given or fixed angle. The Lochhead Dipping was such a Mine, situated about 300 yards from the Pit, and sunk directly from the surface of the ground at an angle of about 20 degrees to the horizontal and in a perfectly straight line.

*(This line is strictly adhered to, in that the mine direction is surveyed in! The angle of descent declivity, may, or may not be uniform, depending on whether the mine was driven to follow the possible undulations of the targeted seam of coal.)*

This Dipping mine was initially driven down until it struck the first coal seam that they intended taking at the 240 level, then down to the 550 level, and finally, retreating back to the 350 level, where they extracted coal from the appropriately named 'Four-Feet' seam. *(Which was, strangely enough, approximately 45 inches thick at that depth!)*

 The extracted coal from this dipping, was brought to the surface by one of the most primitive, yet most effective devices imaginable.  It was a magnificently large and powerful, two cylinder, steam, 'donkey' engine, with a massive 10ft diameter drum, on which was wound half a linear mile of one and one-eighth-inch steel hawser. Running the full length of this Mine was a single semi-permanent twenty-five and a half-inch wide railway, over which ran races, or rakes of hutches, each with a carrying capacity of half-a-ton.  These hutches had two, fixed axles with flanged wheels and were constructed from steel, with double lifting handles at each end.  Each race, or rake, comprised of fourteen of these tubs connected together by means of multi-linked couplings and joined to the main drawing cable by means of a very special quick-release coupling. *(I shall describe this coupling in greater detail further on in this narrative.)*

 Imagine if you will, a race of fourteen hutches coupled together and standing on the mine-head level, already attached to the heavy draw-cable.  This is only one of a continual stream of rakes, that ply between mine-bottom and surface during the course of a shift!  The rake of tubs is quickly checked for continuity, the main draw cable is checked and double-checked, and the race is pitched over the edge . . .!  The first tub 'feels' the slope and the coupling tightens!, the second tub in turn feels the slope and the next coupling tightens.  As more and more tubs run over the edge, the down-weight increases as does the effect of the force of gravity and the speed of the rake accelerates.  The increasing weight of the downwards plunge, pulls the remaining empties over the edge and the increasing speed and weight of the full race galloping downhill, now tensions the heavy draw-cable, as it is allowed to pick up high speed under the direct control of the "engine windsman". *(This downhill plunge is under carefully controlled free-wheeling momentum and is continuously monitored!)*  The engine-man knows exactly when the down race has bottomed!  His cable markings tell him so!  At the mine-bottom, the underground onsetter transfers the draw-cable *(known as 'The tail!')* to the loaded full race. *(Fourteen loaded tubs of coal all coupled together.)*  This 'tail' must be attached to the race or, to be more precise, to the first tub of the race and in a very special manner! *(Indeed!, the whole outcome of it's intended successful 'landing' depends on this meticulous drill.)*

 When the new loaded race is briefly lying on the mine-bottom, the draw cable lies between the twin rails, and is supported and guided by steel rollers, strategically placed between the rails and laid-out in a uniform pattern every five metres or so. This means that the draw-cable is mechanically guided and hopefully, will not get caught-up on any projections along the length of the Mine.  When the engineman begins to wind the race, that is when the 'fun' starts! The starting speed is relatively slow so that all couplings are tensioned without being jerked, and the race is brought on to the main line and slope! Then, the engineman starts to wind-up the steam donkey! All 'hell' now breaks loose in the steel-girdered mine! The race is brought up to draw-speed very quickly. Ten ton's of coal, seven ton's of steel tubs, and a speed approaching forty miles per hour! The steel draw-cable is screaming and screeching, it is fully tensioned and oscillating wildly from side to side, and from pavement to roof. It will rip, tear, break and wrench anything and everything in its path, if the race is de-railed on its upwards passage. *(The engineman knows that he must take the utmost precautions to control this accelerating juggernaut!)* The noise generated is echoingly amplified within the confines of the steel-enclosed mine. The sound is simply deafening! It is a terrifying thing to behold! *(This, of course, takes place in total darkness! No person, whoever he may be, is allowed access to this Mineshaft whilst Speed-winding is in progress. To be 'caught-up' in this viciously whipping cable is to invite a certain bloody end!)*

As this heavy race is rapidly drawn to the surface of the mine, it may be of great interest to readers to discover, that this race of tubs is NOT slowly brought to a halt as it comes over the edge at the surface. Indeed, the whole concept of its arrival at the surface *(mine-head),* is the subject of one of the most daring, thrilling, courageous and mind-blowing exploits that has ever been my privilege to witness! *(And everyone else!)* The donkey engine and the speed winding are barely curtailed! And the race of tubs hurtles on . . . UNBRAKED . . . ! !

As I mentioned earlier, special attention is given-to, and time taken, to 'set-up' the tail at the Mine bottom. On every coal tub there is a curved steel hook at either end, on which a multi-linked steel coupling is used to build a race. This male part of the tail, is therefore a very simple device that locks over the draw-bar hook of any tub and presents a 'tractor -coupling', a flat shard of metal plate with a one-and-a-quarter inch hole in the centre.

The female part of the tail, *(on the end of the long steel 'draw-cable'),* consists of double, flat, steel plates, also with two, one-and-a-quarter inch, aligned holes, which slides over the male part of the tail! When one part is slid over the other, and the holes are aligned, a stout steel, greased pin is slid into the aligned holes. On the top of this pin there is an eyelet, from which a stout chain is secured. The tail is now firmly attached! The length of chain from the pin, is attached to a bracket that stands up vertically from the front of the first tub at a height of approximately 45" from the ground. There is no loose chain from the bracket to the pin. It falls true and straight, and is unobstructed!

At the Mine-head, all is in readiness! The draw-cable is winding-in at speed and the noise of the rapidly approaching race grows in volume! The draw-cable is running smoothly over it's rollers between the tracks and disappears under the decking towards the great 10ft drum, where its 'booked' coils are being watched and precisely counted by the apprehensive, winding engineman. He waits with bated breath, the thundering sound of the speeding race muted by the solid walls of the winding-house. He is poised, waiting on the one, single, strident sound that will galvanise him to explosive action, which, will hopefully result in the culmination of a successful landing!

On the empty landing, the Tail-puller stands absolutely still, his nerve ends jangling! . . He knows full well the inevitable consequence of a missed-tail! The noise! The dust! The 'flying' tubs! The avalanche of thrown coal, and above all, the carnage of the permanently wrecked tubs and severed draw-cable which takes many hours to repair. The inch-thick, hardened, steel plate that covers the vulnerable wall of the winding-house, must stand the colossal impact of a thrown race if the tail is lost! The tail-puller casts a wistful look at his refuge, a small, solid, steel cage, sitting bolted to the deck and having one open side. He knows that he has got exactly 2 short seconds to make a single life-saving dive to safety, if he misses the Tail!

And still he stands, balanced carefully on two feet, side-on to the up-coming juggernaut. His left hand firmly, but gingerly, holding a hanging rope that is coupled to the 'make-switch' of a strident bell, that will sound alarmingly in the locked confines of the winding-house. *(By Act of Parliament; no person can be admitted to the winding-house whilst winding is in progress!)* His right hand is half-way outstretched and slightly raised, his thumb is well tucked-in and parallel with his slightly clawed hand. *(He has already suffered the pain and indignity of one broken thumb.)* His hand is positioned exactly! He leans fully forwards to meet the on-coming juggernaut, being extremely careful not to wrench the 'bell-pull'. *(A premature pull and ring, will halt the race prematurely and reverse the forward motion of the race, thus sending it plunging back down the mine! At the moment of the 'pull', the race is not yet fully over the lip of the edge!)* And still the tailsman remains poised, knowing, that as the race appears over the edge, it's speed will be in the region of 25 m.p.h.

*

In the winding-house, the engineman knows his time is almost here! He too, waits in eager anticipation, and with a deal of anxious trepidation. All too well, he can remember the thunderous clanging din, of seventeen tons of steel tub and coal crashing against the armour-clad safety wall of his engine-house! He knows his task is to respond with lightning speed to the clang of the alarm bell, instantly cut the steam pressure to the driving pistons and jam-on the Steam brake!' The tail-puller needs only two inches of slack to make the 'Pull!'

*

*'Stuartie' Honeyman, the Dipping Tailsman, to give him his proper name and designation, was a rock of a man! He was 37 years of age, (or so he thought!) He stood 5ft - 9ins tall and weighed thirteen and a half stones 'bullock naked'! He was also as bald as a coot, and was built like the proverbial, square brick 'Kazie!' He was in the prime of his life! He was a superb specimen of manhood! He worked stripped to the waist in all weathers, especially in the summer months, and the only concession he made to the cold of winter was to wear a 'semmit'. His muscular build looked as though it were 'bolted' on!*

*Stuartie Honeyman was also unschooled and partly illiterate! He was totally bereft of any formal education. Stuartie very rarely went to School! He had difficulty reading even the simplest of words, and could barely write his name! His regular weekly wage packet was always in the form of single pound notes, one ten shilling note, two half-crowns and some copper coins! He had learned to 'count' the 'blue' notes from memory and knew that there had to be a 'red' coloured note as well, he sometimes also had some trouble counting the silver and copper, but always took the time to ensure that the amount was correct! There was no fooling Stuartie!'*

*Stuartie was also a proud man. He instinctively knew that he was something special! Stories of his skill and daring and undisputed dexterity, had filtered out throughout the coalfield. He thought that he was the only Tailsman in the whole of the East Fife Coalfields. He was a legend in his own time! He could not comprehend the accolades, could not understand the adulation, and coloured with embarrassment, if at all congratulated! Every day at shift-change, scores of miners would silently wait, with eager anticipation and with mounting excitement, on the arrival of the last race to surface! They merely wished, just to be able to say, that they had seen the Tailsman in action! (In fact, the miners knew that it was these precise skills of their Tailsman, that kept the hutches moving and the coal flowing, and their produce carried away!) Time, was always MONEY!)}*

The Tail-puller trembles, then, with a storming rush of warm mine air, the galloping race appears over the lip, the speeding, loaded, race thunders towards him, spewing coal and dust. Fifteen yards! Ten yards! Suddenly, a low rumbling noise starts to vibrate in the depths of Stuartie's chest, it is rapidly growing in volume and intensity! Five Yards! A great sound explodes from Stuartie's throat! He yanks the bell-pull! No change in speed! Stuartie bellows like a rampant bull! The vibrating down-chain slaps into his gloved hand like a bullet and at the same instant, he senses the lessening tension in the tail. His right hand quickly and desperately closes around the dancing chain. He powerfully yanks upwards with all of his great strength, at the same time using his left hand to ward of the increasing pressure of the racing tubs. He then throws himself backwards to safety with the 'tail' clutched firmly and triumphantly in his clenched right fist! *(Stuartie doesn't know why he roars, perhaps it is some "Freudian" pre-action to 'psyche' himself up to deliver the explosion of energy required for only a few milliseconds in time!)*

Stuartie's additional tasks on the dipping head, are to help with the feeding of the loaded hutches through the tumbler, where the filled hutches are individually rotated through 360 degrees and emptied into the waiting coal wagons below. The empties coming through the tumbler are then lined-up and coupled together on a parallel rail-track to form the next down-going rake. *(The last down-going rake was sent on its way as soon as the previous 'tail' had been pulled, with the tail-assembly being quickly re-coupled onto the last hutch in the rake!)* His next tasks, for a short time are much less strenuous! After the empty hutches have been regrouped into a working rake, he can do exactly as he pleases! *(He sometimes sunbathes!)* These are his sole unique tasks. He has developed this unique skill that defies all logic. His steadfastness, his timing, his co-ordination is legendary! Something very, very rare. He is not bound to tackle any other task around him, his position is that of 'tailsman!' His position is quite unassailable, for the simple reason that no one else would want to 'draw the tail' on a regular basis. His worth is reckoned in 'seconds of time' and irrefutable cost! Everyone knows that a 'missed' tail will cost hundreds of points in damage and in lost time! The on-going responsibility that is his, is without peer! There can be no training for this dangerous and suicidal task! It is unique in the extreme! This is what he gets paid for! His total working time during the course of any working day, is probably a deal less that that of other oncost workers, in that he must, perforce, stand-to – while the whilst the 'tail' is running. This juggernaut must not be allowed delays! Stuartie Honeyman is paid the princely sum of Five Pounds and Five Shillings per week. He is an 'oncost' worker in a Nationalised Coal Industry. He is also, virtually Irreplaceable . . .!

Stuartie Honeyman was a very proud old man, but not so proud as to refuse to discuss the few times when, for one reason or another he missed the tail. His particular job came to an end in 1954, when the last of the four foot coal was extracted in the Surface Dipping and the men and machinery were transferred to the pit. Stuartie was well remembered for his undoubted prowess with that tail and was talked about for many years afterwards. After all, he was a 'living legend!'

*(I should know! I was there to see him perform, as were many others!)*

*

** Footnote:- *In the Summer of 1951, a Time & Motion Study Team from the N.C.B. made a monumental decision! They, in their infinite wisdom, decided that the great Steam Donkey Engine was an antiquated relic and had finally run out of 'Puff!' It was to be consigned to the scrapheap! It was to be replaced by a clean,* *modern, more efficient, powerful and flexible, three-phase, Electric Motor and an up-dated Becander Gear! They had decided that this old, steam, donkey engine could no longer cope with the increasing demands of the anticipated larger coal output that was planned. Accordingly, all of the miners in the Surface Dipping were given notice, that all coal-haulage would be suspended for the next week, so that the New Machinery could be installed. Their coal-output would be stockpiled on the mine-bottom until the new haulage was in place. Stuartie Honeyman was now redundant! The new plan called for smaller races to be 'gently landed' to save wear and tear on the fabric of the new Engine House!*

*On the Friday evening, after the 'back-shift' had surfaced, work began on the transformation! Flood-lighting had been erected and the side wall was ripped out of the engine-house. Engineers in their 'tens' moved-in and dismantled the massive old Donkey Engine! The Engineers and the planners worked tirelessly and continuously for a full week, in controlled, twelve-hour, rotating shifts! Everyone then seemingly breathed huge sighs of relief at the completely successful installation of the brand new Machinery. It had been meticulously planned and expertly executed! There only remained the trials and testing!' The 'machinery' lived-up to early expectations. It had been perfectly installed and aligned! No more woofing, shaking, booming vibration with great, white, clouds of expelled steam! This machinery was quiet, smooth and an engineman's dream! This was British engineering at its very best! The engineman was already visualising the amount of time that he would spend, oiling the 'ports' and polishing the brass-work. There was going to be white lines painted around 'his' engine-room floor, with designated 'no-go' areas for visitors!*

*And so started the trials! Anybody who was somebody was there! Sundry officials from the Fife Coalfields were present, all dressed in pressed blue overalls, white collar and tie, and the now obligatory 'white' miner's helmet! All were duly armed with clipped mill-board and pens and the inevitable stop-watch!*

*Let the Trials commence! They did, and with a vengeance! There was a plethora of stock-piled coal on the mine-bottom and backed along the haulage-way, and so, the planners awaited excitedly for the on-rush of 'black diamonds!' They experimented with races of six tubs. Then eight tubs. Then ten tubs, and finally, twelve tubs! Yes!, they could safely land twelve tubs with ease and efficiency. (And without Stuartie Honeyman!) This was a planner's dream! This was the culmination of months of painstaking, minutely defined, criteria! This was the ultimate satisfaction! This was 'O-r-g-a-s-m!'*

*There was only one small fly in this agitated pot-pourri! The coal was piling up on the Mine-bottom! No matter how quickly the passive-tail turn-around was made, the new machinery could NOT keep up with the ever-increasing amounts of loaded hutches below! They even tried to operate a second shift in a futile attempt to reduce the backlog of mounting tonnage! It simply was not feasible! Costs were escalating! This was not the answer! This failed! They failed! The whole system Failed! It was an unmitigated disaster! This new-fanged idea of the 'men in bowler hats' was spreading fear and despondency amongst the miners! There were far too many stoppages! The coal was not being carried away! There was deep frustration! The miner's were losing money . . !*

*The 'Bowler hats' finally admitted defeat! In their terminology, the new system "Did not lived up to expectations!" It was too slow! It did not achieve a fast enough turnaround and the Electric Motor just could not accelerate fast enough with it's inherent load!*

*In the uncensored parlance of the underground miners, - "It coud'ne pu' wurth a fuck!!"*

*

*The Manager of Lochhead Colliery (including the Surface Dipping), a certain William Hampson, was a man in his late forties. He was an unhandsome man of middle build, with sloping shoulders and a slight forwards stoop. He was acerbic by nature and miserable by choice, or so said the many miners who had contractual\* dealings with him. (Of all the miners who had come out of his office after 'contract' negotiations, very few could ever claim to have got a credible deal!) This then, was the Manager who was credited with never having made a popular decision in his professional life! (That was not to say that he made bad ones!) He now made a very critical decision, which if it failed, would surely end his otherwise successful career! He ordered all winding at the Dipping to stop! He then made a monumental decision! He took it entirely upon himself to have the 'New System' ripped out! The rusting Steam Donkey Engine now lying in the wood-yard was to be re-installed! And quickly! Before anyone at the Coal-Board could intervene!*

*This feat was accomplished in five working days and two week-ends! The steam pipes were reconnected and the system was tested! Steam winding was about to re-commence! Stuartie Honeyman was recalled to the 'Front-Line! He was needed! The only trouble was, that Stuartie didn't want to come! He had meanwhile, being just another oncost worker, been given the job of cleaning-up the Dipping Yard, which was just Stuartie's cup o' tea!' (He didn't have the pithead gaffer breathing down his neck!) Stuartie was no fool! He knew that they needed 'him!' Who else would dare pull that Tail!?*

*William Hampson capitulated immediately. He needed Stuartie Honeyman! He was given a grateful bonus and a small pay rise, and he, once again, 'strutted the Dipping's 'boardwalk' as proud as ever! He had thought that his world had come to an end! Now once again, he was 'King of the Hill'. He was still the 'only Tailsman' in the Fife Coalfields. At the end of his shift and after he had been thro' the 'Baths', Stuartie meandered up that pit-road as though he owned it!*

*There were some very 'red' and embarrassed faces amongst the 'men in suits' from the engineering dept. of the N.C.B., almost as red in fact, as the bright red rust on the hurriedly re-installed Steam-Donkey!, which, was still in plain view and would remain so, until such times as they got round to patching the great reopened hole, in the side of the Dipping's Winding House!*

*The Steam, Donkey Engine maintained its performance until the very end, and continued on until the demise of the Surface Dipping in May 1954, when the coal reserves on the 350 level below, became exhausted.*

*The huge unsightly 'scars' on the side-wall of the brick-built, Dipping Engine-house, bore true testimony to an antiquated piece of obsolete British engineering, that just could not be replaced and stubbornly refused to die!*

*The whole of the Coaltown of Wemyss knew what was happening at the Surface Dipping during the transformation. The miner's wives and mother's concerned for their menfolk, kept silent vigil and also their own council! There was only promise of doom and gloom and even less wages! It was however, cause for great celebration after an uncertain month, to see and hear the great, steam, donkey engine's chimney, belching huge clouds of pressurised steam into the still, summer air. It's pure white volumes at first chugging very slowly, then quickly gathering pace and speed! They were 'winding' again at the Dipping! This was cause for celebration! A wee quick nip at the whisky bottle? No! Not at all! It was possibly a time for some silent reflection and perhaps, a collective sigh of relief that the Surface 'Dipping' was back to normal! Most wives and mother's merely continued on with the chores and tasks that were part and parcel of their daily grind. They too! - were on constant 'Day-Shift!'*

*All Stone-miners, (teams of miners who drove tunnels thro' solid rock) 'brushers' and 'back-brushers, (miners who drove the forever, extending, header tunnels, both above and below a working 'Long-wall Face') and developers. (Miners who initially drove thro' and opened-up 'New Coal-Faces!) These 'teams' had a 'Leader', usually a long-term leader, who was perhaps the most experienced and know-ledgeable miner in the group. A man who could be 'trusted' to make the best possible 'deal' for his team! A man that would not be 'faced-down' by the manager, in the 'fight' for a decent remuneration for "Yardage! Friday afternoons was the specified time when 'contracts' were hammered out, and, it was a shrewd and cannie 'team leader' indeed, that left William Hampson's office with a smile on his face and could say to his anxiously waiting team - "I've got a contract!"

* Becander Gear. *This massive piece of 'beautiful' engineering was simply a 'Gearbox!' Except that there was nothing simple about it! It was like nothing that one has ever seem and was completely encased in a very heavy-duty, custom-built casing! It was bolted to the floor of an 'engine-house' and concreted in. It could*

*be described as the 'modem' between the electric driving motor (up to 2,500 r.p.m.) and the 'winding drum'. It had an 'input' and an 'output!' Its inherent 'beauty' was that it would take the 'High-speed' drive from any prime-mover and reduce it to the relatively slow revolutions (10 to 15 r.p.m.) required at the 'Winding-drum' of a haulage motor. It weighed between three to eight tons!*

*Becander Gearings were mostly installed in 'electric motor houses' both above and below ground in much of the Fife Coalfields! They were custom-designed, made to last and were British built!*

\*

# Chapter XVII

## Pit-Head Baths. - *A Clean Environment!*
### *(The Grimy Miner's Salvation!)*

To most miners who worked underground in the days of tallow lamps or even carbide lamps, getting washed clean was always a problem that the coal-owners were fully aware of, but seemingly did very little to alleviate. Pit-head baths were not yet introduced into the scheme of things at the collieries. They were an expensive item to build *(though not to heat, I imagine!),* and would not show up as a profit margin on any private owners books! Most of the private collieries were small concerns, in that they were nothing more than a driven mine into the nearest seam of coal at minimum cost, and the priority was to get the coal out as cheaply as possible. Pit-head baths were a luxury that small private mine owners were not prepared to invest in, or even countenance at that time. As a result, miners had to make their way home, irrespective of the shift they worked, in a state of embedded dirtiness, cold stale sweat and grim, hungry tiredness. Where possible, if a secure hut or room could be found at the mine-head, the miners would keep some sort of old, warm overcoat and a bonnet to hand, so that they could make their way home in bare comfort and with some dignity. Most men had either to walk home across fields, or if they were fortunate enough, they might even own a bicycle. Either way, a good, plain, hot meal and cheery coal fire, were usually quite enough to partially revive a tired and hungry miner. At the pit-head, what they couldn't or didn't get at the end of every shift, was the copious amount of hot soapy water that was needed to give them a thorough clean bath and a re-invigoration! When they arrived home and in order to give their meal some semblance of dignity, the grimy miner would use whatever hot water was available, to thoroughly wash his hands, arms and head completely, to at least feel fresher. However, the main problem remained, the miner was extremely dirty and no amount of cold water, even if it could be faced, would shift this ingrained coal dirt and refresh his body.

His main meal and subsequent sit-down would take place in the scullery or 'kitchen', where the warmth of a coal fire would almost immediately send him to sleep. However, the proper rest or recuperation that his body required, would not be found in the kitchen chair. The miner had to lay down, and lie down he did. Not in a clean warm bed where he ought to have been and where plain dignity required that he should be, but, probably under the high double bed, where he and his wife would normally sleep, she in the bed and him underneath! A man could not, in all conscience, get between clean sheets in this dirty state and expect a wife to lie with him. He had his pride and this was his decision! This state of affairs would happen five nights out of seven, and on the last day of the working week, the wife and family would make every effort to have gallons of hot, soapy water ready for the man of the house, before the Sunday day of rest. The clean miner would then spend two 'comfortable' nights with his sweet-smelling wife in their own double bed, where this rest would have to suffice and sustain him until the following weekend. This state of affairs was rife in many of the poorer, miners cottages, which had only the barest of amenities and was really a thundering disgrace on the parts of the private mine-owners, especially where the Company was a large-sized, profitable concern.

At No. 5 Barn's Row in the Coaltown of Wemyss, where the young Harry was now firmly entrenched as one of this large family, things were just a little different. There was at that time, five working miners living in that house and on differing shifts, there was also six females of different ages including Old Kate. Old Kate would not countenance the thought of any working miners in her house, not being able to get the daily bath that they desperately needed. To this end, being the strong and determined woman that she was, she organised the preparation and heating of copious amounts of hot soapy water on a 24 hour basis. She stoutly maintained that no single person in her house, would ever suffer the indignity of having to sit down to a meal, unless they had been scrubbed clean, and that applied to Auld Tot as well. This water was heated in a great 20 gallons zinc boiler, which was set into the wall and fired by a coal-fire set beneath the boiler, and had to be filled and emptied by hand and bucket.

At bath times, which, during weekdays could be on an eight-hourly basis, the stone-floored scullery was cleared, the tin and zinc hip-baths were filled and the bathing process would commence. The hot soapy water would be splashed about all over the floor without too much concern, and the in-coming dirty miners would take their turn in the baths as they arrived home with hot, clean water being added as necessary. As the miners stripped in the scullery, their dusty and dirty clothes would be taken outside into the back garden by one of the young women, or girls of the family and thoroughly beaten and aired as necessary, or repaired if needs be. The piles of clothes would be kept together and tied into separate bundles, and then placed on clean newspaper before being slid beneath one of the big high beds. Each pile of clothes had their allotted space, so that each miner could gather his own clothes when next needed. As well as this on-going need for constant supplies of hot water, it must be realised that Old Kate had also to organise the feeding of all family members on a rotating shift system, as well as preparing their piece-boxes. It was joking repeated quite often in family circles, that Old Kate had the cleanest kitchen in all of the Coaltown of Wemyss. It was scrubbed-out every eight hours!

\*

When I started work in Lochhead pit in 1950, the pit-head baths were well established, were reasonably well appointed and properly maintained. *(Lochhead Colliery was seemingly on of the first collieries in the coalfield to have customised Pithead baths!)* There was a Baths attendant on duty for 24 hours of every day, including week-ends. The attendant was usually an senior ex-miner who was still fit and active, but who had spent most of his working life at Lochhead Colliery. This was a 'plum' job, and eagerly sought after by many envious, old miners at the end of their active mining days. Their many tasks included the washing and scrubbing of the walls and floor of every white-tiled open cubicle, that had been used by the last up-coming shift, and this of course, was an ongoing process over seven days of every week. A working coal-mine was like the Windmill Theatre in London during W.W.II. 'We never close!' The lay-out of a pit-head baths was quite standard throughout the Fife coalfields, in that they contained an equal number of both 'clean' and 'dirty' twin-numbered lockers, in separate but interconnected 'halls', which may, or may not have been separated by the actual shower cubicles themselves, which in a large coal-mine may have numbered many hundreds. There was, I think, a standard ratio of numbered, locker capacity to the amount of shower cubicles available, but of that ratio, I am not certain! The one thing that most ex-miners remember about the pit-head baths, was the feeling of utter comfortable warmth as they entered the doors at either end. Coming on shift, a miner would enter by the 'clean' end, deposit his street clothes in his 'clean' locker and walk through the interconnecting passageway carrying nothing except his 'haversack!' He would then dress himself in his 'dirty' working clothes and then leave the building through the 'dirty' end. *(And woe betide any miner who dared venture into the 'clean' end dressed in working clothes with pit-boots. Old Abe Leitch\* would turn puce with apoplexy and chase him back through brandishing his four foot tile brush.) \*Yes! Old Abe was father to Bunt Leitch.*

Once, every three weeks or so, both the dirty lockers and then the clean lockers would be flushed out with scalding water and a cleaning agent *(soap)*. *(The floor area of both ends were flushed-out by hose and hard brush on a near nightly basis.)* When locker cleaning was imminent, a chalked notice would appear on a blackboard at the entrance to the baths, to the effect that 'Dirty *(or Clean)* lockers would left open and empty for the week-end of:- *(Days and dates?')* The usual practice, was that the 'dirty' lockers would be cleaned first, so that the miners who had the week-end free, could take their pit-clothes home to be washed, whilst those miners who were working, would be wearing their dirty clothes underground. The following week-end, the reverse would happen and clean lockers would be left open and empty, whilst those miners who were working over the week-end, would ensure that their clean street clothes were placed into their cleaned 'dirty' lockers for the duration of their underground shift.

This system worked very well and ensured that the lockers area was always kept absolutely clean and free from odours. There was of course, a small number of

miners who either forgot to empty their lockers or, were off work through sickness or injury and whose working clothes were firmly locked in their dirty lockers. Not for the first time, did a miner appear for work at the 'dirty' end on a Monday morning, to find the complete contents of his dirty locker to be saturated with soap and water. Needs must when old Abe was on duty, and, though every effort was made to avoid unemptied lockers, hosed water could not help but enter the closed locker through the vent holes. The building that held the baths through necessity, was always very warm, but no amount of heat would dry dirty 'soaked' working clothes hung in a confined locker and as a result, many a miner experienced the discomfort of having to don wet, dirty clothes and travel down a cold mine shaft as a result of their forgetfulness.
*(Yes!, I did suffer the same indignity.)*

*

One of the most dangerous things to have happened in the cleaning of dirty lockers, was when one foolish miner left a one pound *(1 imp. lb/455 grams)* tin of carbide granules in the bottom of an empty locker, which was overlooked both by himself and the baths attendant prior to cleaning. It so happened, that the offending miner had not replaced the tin-lid properly, and the force of the water-jet from the attendants hose had dislodged the full tin and spilled it's contents. The resulting acetylene gas was immediately released in a billowing cloud and issued forth from within the locker, sizzling and spitting with increasing activity. The baths attendant hurriedly removed himself from the scene of the belching locker, with a firm mental commitment to ascertain the name of the occupant of this locker. The emissions of gas did not actually last very long, but the billowing cloud was quite voluminous whilst it lasted. When the emission of gas had subsided and the cloud dispersed, and the attendant was satisfied that all was safe, he then made his way to his office to determine which miner had been allocated this particular numbered locker. To his absolute chagrin, he discovered that the numbered locker was unallocated, and was supposed to be empty. He was not in the least perturbed or even slightly mystified, as he knew that some miners, in order to discourage pilfering, actually hid their personal supply of carbide in an supposedly empty locker, in an effort to protect their own supply. The attendant did have his suspicions, but proving ownership was an entirely different matter. Besides, what miner in his right senses would come clean and admit his guilt?

*

Every three months or so, or when supplies allowed, a chalked reminder would appear on the baths attendants notice board to the effect, that pairs of thick bath towels would be available for purchase at a given time. The price of these good quality bath towels was around six and sixpence per pair and they were eagerly sought after. They arrived at the colliery in large bales or bundles and were stacked in the attendant's office. The distribution was carefully 'rationed' in that every miner and pit-head worker was allowed to purchase only one pair at a time, with his name being entered into the 'baths' book along with the date of purchase. Not all miners took advantage of this concession, and those that didn't were usually pressed into purchasing and extra pair for someone else. The miner's wives, being the thrifty and shrewd women that they were, would sometime remark to their miner husbands, that a concessionary towel allocation at the baths should be about due and to 'make sure' that they did not miss out on the allocation. It must be mentioned here, that the quality of the bath towels 'issued' was much superior to that which could be purchased with clothing coupons after W.W.II., when good textiles were at a premium.

To the casual observer, watching miners walking down the pit road on a Monday, whether it be day-shift, back-shift or night-shift, then that person would invariably see most miners with a large tightly-rolled up bath towel under their arm and containing their clean working shirt *(sark)*, vest *(semmit or peeweet)* and clean working socks. The rolled-up towelled bundles were usually carried in plain view, with no thought at all of trying to disguise their contents. Dirty bundles going home on a Friday, and clean bundles being brought to work on a Monday, were as common as cheese and porridge.

*

One of the most irksome and anger-provoking practices, that sometimes proliferated in the pit-head baths of nearly every colliery in the East Fife coalfields, was the instances of petty pilfering and outright theft! The pithead baths, being the most widely used building on the pit-head, were mostly unoccupied during the working shifts, except for the presence of the attendants somewhere in the building, who had their own work to do. Miners were not keen to be seen visiting the baths during working hours whilst pilfering was rife, as suspicion could fall upon anyone! The miners themselves, were expected to fit their own padlocks onto the locker doors, at least on their clean lockers, provision being made on the locker frame and door handle to accommodate a padlock, through two matched drillings. Some miners did have their own padlocks that were religiously used, but the keys were small, easily lost, and as a result, the padlock had sometimes to be forced. Some miners, experiencing the futility of the 'lost key syndrome', went to the trouble of purchasing three or four digit combination locks, but, either they forgot the combination, or wrote it down somewhere. Needless to say, within days, anyone using the same row who was so inclined, would soon know the combination number. Several such trusting *(or thoughtless)* miners even scratched the four digit number on the upper face of their locker door to ensure that they didn't forget it!

Most miners, ever mindful of the fact that every locker was a potential target, did not carry any great amounts of money to the coal-mine. Most thrifty individuals would purchase a weekly bus ticket on the Monday morning, having brought only enough money to cover its purchase. Very few miners carried more than a shilling or two in the event of any small emergency and many carried none at all! A thief had to be a very desperate or persistent felon, to have to search through endless clean lockers in an attempt to make his efforts produce even a miserable reward, and certainly not worth the damning consequences if he were caught! If a miner had any respectable sum of money in his possession, he would in all probability, transfer it to his working clothes and keep it on his person whilst underground. Some miners actually used the lockable cupboard in the attendants' office, knowing that this was probably the safest cache in the whole of the building.

Payday was always on a Friday and as welcome as this was, it could some-times be a bit of a nuisance to miners not on the day-shift, or employed on the pit-head. The day-shift miner would come up the pit at lunch-time, go through the baths and then collect his wages at the pay office before going home. The night-shift miner would have to disturb his sleep pattern and make his way to the colliery, some time between eleven in the morning and three in the afternoon, to collect his remuneration. All he lost, was probably some sleep! The back-shift miner was the one with the problem. If he lived locally, he would have plenty of time to get to the colliery and return home with his wages, before coming back for his back-shift, but if he lived away from the mine and couldn't or didn't make the prior journey, he was then left with one of two options. After he had drawn his wages, he could wait until almost the last 'tow' before descending underground, with the hope that one of his friends on the day-shift would accept the responsibility of taking his pay-packet home to his wife, or he could take the chance of carrying his wages on his person during his working shift. If he got a friend to accept his pay-packet, well and good! If he didn't, he would be ever mindful of the money he carried with him, which sometimes detracted his mind from his work. This was not at all conducive to safe working, as the miner's thoughts could lie with the safety of his wage packet!

Some back-shift miners, a very few in number, and probably of a trusting nature, actually took the chance and left their wages 'hidden' in their clean lockers, with the fervent hope that it would still be there when their shift was finished. Some of these trusting miners were, on occasion, doomed to crushing disappointment at the incredulity of someone actually stealing their wages and were left to rue their earlier decision. Fortunately, it did not happen too often, but happen it did! There was, unfortunately, the occasional dastardly felon who would observe those miners who collected their wages immediately before the back-shift, and those miners would be the possible targets. They simply had to cache their wages somewhere and, as everyone knew, the choices were limited! There is a case on record at Lochhead pit, where a back-shift brusher, working in the Lower Dysart West, secreted his week's wages in his piece-box, whilst he toiled only twenty-five metres from where the three of them worked. He returned to his belongings, only to find that someone had stolen his pay-packet during the half-shift before piece-time. Neither of the three man had been alone or, out of sight of the other two at any time, so, the mystery remains to this day, as to who could have committed the crime? Thieves, of course, would go to any lengths to achieve their objective and hide their identity, especially when they knew that rough justice would surely prevail, and a thorough beating at the hands of the many would be inevitable, if they were caught.

\*

# Chapter XVIII

## Dauvit's Progress III
## Dauvit goes Further Forth!
### *– and, Into the Dark Depths!*

Monday morning arrived with no small sense of excitement on my part and in the meanwhile, Old Harry had seen to it that I had been supplied with a new carbide lamp from the pit stores. I do remember inspecting it very closely, noticed that it was manufactured by a firm called 'Autolite' and wondered, as to how I was going to keep that large reflector polished. Old Harry had made sure that he was on the day-shift that fateful Monday morning and that at least, gave me some sense of belonging. I was well catered for with a tin flask of tea, a large flask of cold water and a fully extended piece-box, crammed full of buttered toast and the obligatory chocolate biscuit. *(Any fool could see that I was as fresh as a daisy, my pit-boots had been polished black with the leather laces tied in a neat bow!)* The moment arrived, and we, old Harry and me, were in the next group of eight men to enter the steel cage, ready for the rapid downwards descent. The cage arrived with a muted swooshing noise and a great out-rush of cold air, as the vertical trapdoor lifted. Eight, very dirty and weary looking miners, streamed out in a staggered single file and were quickly lost to sight, as the quick movement of the on-going group carried me forwards and into the narrow entrance. This was quickly followed by a pre-emptory warning from old Harry, which did not penetrate my excited mind! My head banged hard on the bottom of the trap-door from which I bounced, not to the ground, but into the cage where I had been carried forwards by the advancing 'rearguard.' Not even started and hurt already! I was not hurt, at least not physically, the only hurt was to my pride with old Harry just shaking his head. I thought at that time, that he just wanted to disown me.

At the start of the 'drop' down, which was just a little faster than I had anticipated, I was aware of the of a distinctly bumpy and noisy ride. This was not as smooth as that of the Wellesley colliery, but certainly much shorter! As we walked of the cage at the pit bottom, I was instructed to follow old Harry and, from the on-coming group of eight men, one of them actually stopped. He took a good look at me, looked at his fellow traveller's and loudly exclaimed, "Just a handful of spunk!" I knew that I was totally embarrassed by that remark. I did know what spunk was! The remark, meant as a singular compliment went totally over my head, and had been delivered in wonderment at my apparent youth and freshness. The miner's name was Ally Rae, a brusher, and he was coming of the night shift.

On the pit bottom, I was taken in tow and delivered in person to a fireman *(deputy)* by the name of 'Chairlie', whose immediate instruction was 'Wait there!' I did! In fact, I waited for about 20 minutes until all of the down-coming shift had arrived and were 'passed through' into the section. Chairlie Sinclair, in company with Tom Coventry *(another fireman),* then picked up their belongings and motioning me to follow on behind, proceeded to walk in the direction of what I now know to be, the Lower Dysart haulage-way. At the start of this walk, it must have been quite obvious to Charlie Sinclair that all was not well with his 'passenger', as I seemed to be stumbling over anything and everything and constantly losing my balance. Charlie finally turned around, shone his spotlight directly into my eyes, and rather exasperatedly asked, "Have you got any carbide in that lamp?" He then removed the lamp from my helmet, deftly unscrewed the bottom, spat into the carbide, screwed the two halves together, and then spark-ignited the lamp. "Didn't they teach you anything at that bloody mining school?" He blurted. I didn't say anything, I felt stupid enough! The walk did seem to take quite a while, with backward cries of 'Mind your head', a very necessary precaution to the unwary. Now I know where this mine got the name, 'The Lower Dysart!' I didn't know it at that time, but old Harry was back-brushing this very haulage road at week-ends!

One little incident that I do remember very clearly *(one of many)* during that first journey into the section, was when we passed the diesel pug sitting waiting at the end of the haulage, with a full rake of empties coupled on. As we passed clear of the manned pug, Charlie Sinclair turned round to me and said, "Nobody is allowed to ride on the pug or the train, see that you stay out of it!" I nodded my head *(and lamp)* in understanding. As we proceeded through the mine, the noise of the on-coming diesel pug grew in volume and Chairlie motioned me into a manhole as an example of how it should be done, and we stayed still as the race of empties rattled by, or at least Tom Coventry and I did. Charlie Sinclair dashed out of the manhole like a shot, chasing and gesticulating at the fast disappearing train and shouting obscenities at the few miners crouched in the moving hutches, all with lamps extinguished, and supposedly hidden below the top rims. Obviously a well practised habit!

As we passed what I now know to be the Lower Dysart Sub-station *(electrical),* Chairlie suddenly stopped, turned around, pushed past me and directed his spot-lamp onto the top of a large, transformer box within the depths of the station, with the shouted words "Wattie! your bloody late again!" and proceeded inwards to the section, without so much as a back-wards glance. Now, what was all that about? I did find out within a few days.

We had arrived! I was given into the tender care of one Jimmy Brown, an oldish, small, slim, frail looking man with white hair whom Chairlie informed, 'That I was to go on the 'couplings' and he was to see that I was properly instructed!', which he then proceeded to do by the simple expedient of instructing a solidly-built young man by the name of Bobby White, to show me how to couple two tubs together 'ad finitum' until a complete race/rake was joined. He also warned, "Stay out of the way of fast-moving tubs!" A warning that I had cause to remember! Jimmy (Paw) Broon, the button pusher, and hence the No. 1 oncost man at the Lower Dysart belt-end, was obviously the 'pin' at this belt-end and the most unlikely worker ever to be manhandling full tubs of coal. He didn't look strong enough or fit enough, to blow up a paper bag. *(An analogy that I had later cause to regret voicing!)*

This belt-end was the terminal point where the incoming empty tubs were lined-up and filled, and this 30-inch wide belt carried the stripped/mined coal from another intermediate endless belt that took the coal from the coal-face via a third smaller, endless belt, than ran the length of the working face. This half-inch thick, rubberised belt was 30 inches wide and was capable of transporting vast quantities of coal in an endless stream. This conveyed coal would pour over the last roller and crash into the waiting empty hutch directly below, which when full, would be pushed clear and replaced by the next empty hutch in line. This process was continuous and depended on the sheer, physical, pushing power of the drawers concerned. The empty tubs/hutches were brought up to, and man-handled past the belt-end, one at a time on the 'empties' side, where their direction was reversed by manually, switched points as the tubs were queued, three or four at a time, to maintain the stream of available empties available to Jimmy Broon. I don't think that Paw Broon actually did much pushing, he was much too old, light and frail! Once this endless belt was started, it ran for something like 15 minutes before anything started to happen.

The first indication that work had commenced on the coal face, was the arrival of small lots of dusty coal gum,* at a frequency that would indicate the shovelling speed of one man. This was quickly followed by an increase in the flowing amount of small coal, until one of the drawers casually mentioned, "That's full flow!" I discovered a little later that what was meant, was, that all of the strippers were clearing their gum simultaneously, in preparation for the subsequent blowing down *(disruption)* of the coal and then, the larger lumps of coal would follow.

*A description of 'gum' is described in the chapter 'Methods of Extraction'. This flow of gum lasted for approximately 20 to 30 minutes or so, and was characterised by the copious amounts of dry coal dust it produced. This cloud of dust was somewhat localised at the point of loading, before being dispersed by the continuous air flow. It was, however, a contentious issue, that the standing load-master, in this case Jimmy Broon, had no option but to stand fast, breathing-in this silent 'killing dust!'\* Breathing masks with renewable filters and safety goggles were available, but even those quickly became clogged up. The drawers themselves were most reluctant to go near to this obvious hazard and often practised holding their breathing, whilst* *delivering the next empty hutch to the waiting queue. It was a brave, yet foolish man, who would endure working immediately behind this spewing monster!*

\* One of the direct causes of pneumoconiosis or, *Miner's Lung!*

As each hutch was filled with coal, the drawer putter would push the loaded hutch up to an ever-shortening distance of perhaps 100 yards/metres away from the loading point. This was to establish the start of a new race of loaded hutches. He would then uncouple and collect an empty hutch from the in-coming race and push it up to the loading point, after dropping the steel coupling onto the middle of the roadway. This would be a continuous process with all of five drawers deployed in turn, to keep the hutches moving and therefore the coal flowing from the belt-end. My task, as it was pointed out to me, was to ensure that each successive loaded hutch was coupled onto the preceding tub, by a multi-linked steel coupling, which in turn, was obtained from the uncoupling of the incoming empty race. All too simple, if there were spare couplings available or, if the drawers were helpful. All too often though, at the start of a new rake, the drawers would not push the loaded hutch the full length of the draw, but merely sent it down to me, at speed and without a coupling! This was their idea of fun! It meant that they did not have too far to walk to pick-up their next empty hutch. As young and inexperienced as I was at time, it soon became apparent to me just who the idle buggers were and, who I could depend upon to send or bring a coupling, along with a full hutch. This Bobby White was a reasonably fit and well-built young man of 'great dreams and schemes', and his recurring theme at that time, was his desire to become a 'Pan-Doctor'. No!, not a doctor in the medical sense *(nor in any sense whatsoever!),* but a face-worker armed with an inch and a quarter spanner, who had to develop the ability and expertise to constantly tighten the forever loosening nuts on a long train of steadily 'jigging pans!' Such is the stuff that dreams are made off!

Bobby often had difficulty in standing his turn at the drawing, he conjured up all sorts of excuses as to why he missed his turn and when questioned and made to stand in line, he vented his surly frustration in sending his mobile torpedoes down to me at 'X' m.p.h.

This first month on the couplings, was the time that I experienced my first lesson on the vagaries of man management. One fine morning on the day-shift, just as the last of the gum was coming of the conveyor belt, young Bobby commenced to perform his usual tricks. Previously he had had a dispute with one of the new drawers, a mature man who had just finished an ex-regular career in the Black Watch and could therefore be described, as a man of the World. Bobby thought that he could play the 'old soldier' with this new drawer and deliberately missed a few turns. He simply 'picked on' the wrong man and was rapidly brought to his senses by one of the most devastating dressings-down, that could be delivered by any self-respecting

Sgt-Major. As a result of the humiliation that he had suffered and due to the subsequent suppressed anger that he no doubt felt, he vented his spleen, first on the loaded hutch and then on me, by sending down his 'wheeled torpedoes!' I then took the only course open to me at that time *(or thought I did)*, I went into the nearest manhole and sulked! I refused to come out and as a result, the growing race of loaded hutches remained uncoupled. None of the other drawers wished to get involved in this turn of events and none of them wished to be seen to be doing my job. Inevitably, Paw Broon soon became aware of the growing tension amongst the drawers, his only concern being of course, was the probability of having to stop the conveyor belt, thereby incurring the predictable wrath of Charlie Sinclair the fireman. Hence his agitation!

The diesel pug turned up with the next race of empty hutches and the hooker-on, one Wullie Jackson from the Coaltown was, after coupling on the first hutch, rather mystified to discover that the pug was chugging away with only one full hutch in tow, with the remainder severely uncoupled. Undeterred, I remained in the manhole, slumped down on my hunkers, with carbide lamp extinguished. Very soon, true to form, Charlie Sinclair came racing up past the belt end, raving and ranting, demanding to know why the belt had stopped, 'reading everyone their undoubted character' and threatening those whose heads were bound to roll! It took all of ten seconds for Chairlie to determine the cause of the stoppage. He came looking for me with a vengeance, spot-lamp probing in every empty manhole. He found me within seconds and attempted to physically pull me out of the man-hole except, that I wouldn't come. At my thirteen stones weight and with a strong grip to match, he just wasn't strong enough. Besides, he was not allowed to rough-handle men! He knew then that he had a problem and very quickly lost his apparent bluster. He quietly asked as to why I had stopped coupling the hutches, to which I fiercely replied, that I had no intention of being 'killed' by flying objects. To his great credit, he seized upon the situation in an instant and stomped away to belt-end, calling all of the drawers together. He let go with a quick, angry tirade, an expletives peppered tongue-lashing that included the direst of threats with men being sent 'up the pit!'. He was greeted with a sullen silence from the wary drawers and at that point, brash Bobby White made the mistake of opening his mouth. Chairlie rounded on the unfortunate young man and delivered a personal tongue-lashing to the effect, that gave warning of instant dismissal on the next transgression. It was a very chastened young Bobby that delivered up the next full hutch to my growing race, but, *without the necessary steel coupling! He was absolutely irrepressible! (I did hear many years later, that he had fulfilled his cherished ambition and did become a 'pan-doctor' on the more modern 'chain-pans'.)*

Around this period of time, the move to convert Lochhead pit a Part II mine was under way, even though the underground workers had not been issued with battery-powered lamps as yet. This did not stop the deputies from issuing orders, that apart from the flame from our carbide lamps, no other naked lights were to be permitted. This meant of course, that there was to be a clamp-down on the heating of tea flask, by any means, anywhere underground. I was at that time, one of the many offenders who usually disappeared for two minutes at approximately twenty minutes past nine, quite 'unnoticed' of course, to set my little candle-wax and waste fire-box alight, usually away from the main roads and within an obscure man-hole. I feel sure that Chairlie Sinclair knew very well as to my small daily subterfuge, but turned a blind eye as long as Jimmy Frew the section gaffer was not in the vicinity of the belt-end. *(Loading point.)* On one such day just as I had lit my fire-box and had returned to my couplings, Chairlie came stomping down to end of the loop-road *(where my fire-box was flickering away in the second manhole)* and said, "That's the last! You can drink cold tea from now on and I don't care if you sulk or not!" That was loudly said and I got the impression it was meant to reach the ears of both Jimmy Frew and Tom Coventry. *(Section gaffer and Old Dook Deputy).* I had been personally warned and took due notice, at least, until the next time. I did however, continue to heat my tea flask, but not every day and always in a different man-hole.

*

Author's Note: - *The heating of cold tea within a tin flask was a daily ritual perform by many underground miners in all parts of the coalmine. (Which, in itself must convince even the most doubting of Thomas's, that the Dysart Main seam was not an inherently gassy seam, i.e. Methane or Marsh gas!) The heated tea was a partial Godsend to the many miners who experienced the blast of cold air that was delivered to some parts of Lochhead pit. The only problem with the heating of tea flasks by the candle-wax method, or even the lengthened flame of a carbide lamp suitable placed, was the amount of black soot that was deposited on the outside of the flask. It was the very devil to remove if not properly cleaned at the time! (Lots of spittle and odd cotton waste!)*

✻

# Chapter XIX

## The New Dook
### *(The Old Collier's Paradise!)*

The New Dook, its name a partial misnomer in terms of origin, was actually commenced in 1904, a mere five years after the commencement of the Old Dook. The Dook itself may have been new in the early part of the first decade of last century, but that was a full ninety-seven years past and several dooks have been driven in Lochhead pit since that time. The head of the New Dook was situated a mere 50 metres from the **Shaft Bottom** and to the north of the pit bottom. The dook was commenced in the middle coals of the Dysart Main seam and driven to the dip, following the Dysart Main seam at an approximate map bearing of 78 degrees. The dook was eventually 'driven' down to a horizontal map length of 950 metres, which would equate to its final, named depth at the **Thousand Yards** level. *(The gradient on this dook varied from 1 in 4.5 in some places, to a rather steep 1 in 2.5 in the worst places.)* The dook was initially driven down to a length of 650 metres by mid-1908, at a datum height of 9250 feet, and it remained at that depth for 28 years until it was lengthened and deepened by its own extension, that was actually **driven upwards from workings underneath**.

The New Dook's extended bottom length of **325 metres,** was actually commenced from the twin **540 headings** coming from the **Old Dook's No. 5 level,** and was driven sometime in 1934. At datum height 9017 feet, at exactly 250 metres above the No. 5 level on the Old Dook, and on the **540 metres headings**, the 1000 yards level was commenced, to both the left and right side from the 540 metres headings. To the left side of the 540 metres heading, two short levels 30 metres apart, were commenced and driven for approximately 75 metres and 110 metres respectively. These were the two, short, parallel levels from which the twin headings were commenced in January 1935, to connect with the 650 metres driven length of the New Dook, at the 9250 feet datum height. The connecting headings, approximately 325 metres long, were in a perfect symmetry with the line of the New Dook, now perfectly straight over its new length of 950 metres. *(1040 yards.)* The levels from which this extension was initiated, was called the **1000** yards level, and would extend for approximately **1000** yards *(north and east)* to the Buckhaven fault. *(The Dyke.)* This connection was completed in January 1936. *(This 1000 yards level was obviously double-named!)*

*

The first levels driven to the left *(north)* from the short 237 metres length of the New Dook, commenced at datum point 9635 feet and in fact, was an extension from the No. 1 level that linked the Old Dook to the New Dook. These levels on the New Dook were named the **200 levels**. The development date for these pair of parallel levels was 1906, and the levels followed the seam of coal for a distance of 875 metres, where they terminated near the Buckhaven fault in 1912. There were exactly 18 short interconnecting roads between the parallel levels, at roughly 30 to 40 metres apart, with both levels for their last 275 metres in length, being incorporated into the last extracted section, as it buffered the Buckhaven fault.

The first section along this **200 level**, was approximately 250 metres along its top length, 140 metres along its bottom length and approximately 115 metres along its two sloping sides. The section was intruded into by two short headings from the bottom 200 level, the first at the 30 metres point, the second at the 85 metres point. The gradient on this section was all of 1 in 4 throughout, and, an unusual feature about this section, was that there did not seam to be any return air ways cut into the pit bottom level, just above in the strata. *(I rather suspect that a return airway was taken back into the man-haulage Dook.)* The coals taken at first, were the Grounds & Nethers and the Myslen splint. What may have been of significance with this first extraction to the left of the New Dook, is that the stoop left between the coal extractions and the left man-haulage Dook was practically non-existent. A mere five metres if not less! *(This was very unusual, in that the on-going practice was to leave a solid stoop of at least 30 to 40 metres in width, to protect the integrity of the main Dooks!)* This section was first stripped using long-wall methods, or something similar, and a re-visitation was made to take the Toughs & Clears along with the Sparcoal. The Head coals and the Coronation coal below were left intact. The section was abandoned by 1911, with no further re-visitations. An estimated 65,000 cubic metres of coal would have been extracted from this small No. 1 section, which would have been transported down to the 200 level, and thence to the main coal Dook and up to the pit bottom.

The second extracted section along this 200 level, was oblong in shape and measured 375 metres by 100 metres. It was intruded into by 5 very, short headings from the 200 level and 3 very, short dippings from the pit bottom level above. The extracted coal would again have been transported down these headings to the bottom 200 level, to await further transport to the main coal Dook. The coals taken were exactly the same as in the previous section on this level and again, this section was abandoned in 1911, without being revisited. The estimated volume of coal that would have been extracted from this No. 2 section would have been in the region of 92,000 cubic metres of reasonably clean coal.

The third section along this 200 level differed from the other two, both in the amount taken and the method of extractions. The first coals that were seemingly taken, were the Grounds & Nethers and the Myslen, but then there seemed to be a series of parallel levels *(as in Stoop and room working)* driven across the width of the section, through the Toughs & Clears and the Sparcoals. After the levels had all been run through, it seems that the side of the stoops were stripped until the topmost Head coals started to crush down, thus negating any further safe extractions. This section lay against the Buckhaven fault, with only a stoop of 30 metres remaining. *(This coal would have been heavily calcinated due to its proximity to the Dyke!)*

The interesting point about this section, is that just on its north-west corner, there was commenced a pair of parallel levels, driven from the pit bottom level, straight through the Buckhaven Dyke, to end up in a new, large section of the Dysart main coal, which was on the north side of the Fault, and which bordered the workings of Wellsgreen pit. These parallel, cross-cut mines, completed by 1905, were cut through the strata above the level of the Dysart main seam. This large section on the other side of the Buckhaven fault, was also intruded into by another heading from the pit bottom level in 1900, at left-angles to the level and approximately 125 metres prior to it. This last section would have produced an estimated 90,000 cubic metres of coal. It was abandoned by 1913 as worked-out.

Author's Note:- *These three sections along this 200 level, were on a gradient of approximately 1 in 3.5 to 1 in 4, which suggests, that all three, were worked as long-wall pick-places to the rise and using aeroplane, or chain braes to raise and lower the hutches. Also, considering the decade in which these extractions took place, and the fact that only the two main dooks and the Lower Dysart haulage were mechanised. I would not hesitate to suggest, that teams of pit-ponies were used to traverse the 875 maximum length of this 200 level. The very length of the 200 level, coupled with the amount of coal extracted, would rule out the possibility of human drawers!?*

*I must also point out, **that very few of the levels** commenced from the left side of the **Main Coal Dook** were actually connected to the left side, man-haulage dook. The left, man-haulage dook was cut through the strata above the level of the main coal dook, and was used as the return airway for some of the coal sections along different sets of levels on the Old Dook! This was out of necessity, because there was no way that coal in hutches could be transported across an open dook, laid with permanent rails for man-haulage bogies!*

\*

The large section to the north side of the Buckhaven fault was probably worked by Lochhead miners around 1905 to 1923. This was a very, large area of extraction and the most interesting point to be realised with its subsequent extraction, was that the top, north-west end of the rising coal seam actually outcropped to the surface, exactly 350 metres to the north side of the Standing-stone road. There were in fact, several short bore-holes to the near surface, as the miners did not actually mine *(extract)* the seam to full outcrop. *(This section shall be described in greater detail elsewhere in this book.)*

\*

During 1907, and into the summer of 1908, the New Dook mines were further extended down to a length of 610 metres *(datum height 9200 feet),* where there were three separate levels commenced to the left side of the coal dook. One at 585 metres, one at 610 metres and the bottom one at 635 metres. The bottom two levels were merely extensions of, and connected with the No. 2 levels between the Old and the New Dooks. These new, triple levels to the left of the New Dook were actually named the **600 levels.** *{A point to note at this juncture, is that the middle No. 2 level coming in from the Old Dook, did not actually connect up with the main coal (right-hand) dook at this junction, but, split into two, separate, curving levels to circumvent and encompass the main coal dook, then re-unite as one level, before running under the man-haulage dook, and thence on, to become the middle of three 600 levels, running on in a haphazard fashion to near the Buckhaven fault}.*

As the New Dook was being extended down to this new depth, five short levels were being commenced to the left of the coal Dook at the 375, the 400, the 460, the 535 and the 565 metres points. *(The second, fourth and fifth levels were commenced from the main coal dook, while the first and third levels were used as return airways to the man-haulage dook! I am also reliably informed that these short levels were collectively known as the **Four Hundred levels** and, with a terrible reputation as one of the 'dirtiest' and most **foul** of working sections.)* This section, as with most others as they were developed along the length of a new 600 level, were not broken up into individual smaller sections as happened elsewhere, but remained as one long unbroken section along its length, from the New Dooks to the Buckhaven fault, a distance of approximately 1035 metres. This was probably its undoing, in that the troubles encountered by the miners trying to salvage valuable equipment as the coal became exhausted, were manifold. Again, I feel that the blame for this sections troubles must lie with the planners/management, in that the extensive taking of the coal and the methods of extraction used, left much to be desired.

The area size of this section was approximately 950 metres long, 200 metres wide at the side of the New Dook, and 475 metres wide as it was worked tight to the side of the Buckhaven fault. There were very few solid stoops left anywhere within this large area, as separations between working parts of the section. Adjoining parts of the section where different coals were extracted, were hard up against each other, with little regard to the use of intervening stoops to delay, or partly negate the crushing process. It is of note to mention that the gradient within this large section area varied greatly, in that even though the **200 levels** and the **600 levels** were actually horizontal in plane, they were **not parallel** to each other. The gradient at the New Dook was 1 in 3.25, while the tangential gradient at the Buckhaven fault was 1 in 5.4. *(This difference is felt not only with the initial transport of coal hutches, but in improved working conditions for colliers!)*

The first coals taken from this section commencing in 1910, were the Coronation coals only, starting from the side of the left dook and slowly progressing towards the north and east in the general direction of the **Dyke.** These extractions carried on moving slowly north-eastwards for a unbroken period of 16 years, until the beginning of 1926, where the colliers struck the brown and burnt, calcinated coals, that signified their proximity to the **Dyke.** Meanwhile, back at the start point of the extractions (*i.e. The area to the left of the dook*), the miners re-visited the area in 1920, and commenced to take the Grounds & Nethers and the Myslen coals, now that the space vacated by the earlier extraction of the Coronation coal had fully crushed down. The Grounds & Nethers and the Myslen were extracted from approximately 90 per cent of the full area, up to and touching the Dyke. This was completed by the summer of 1930.

*(I am rather surprised that coal was extracted tight to the Dyke, in this section, in that the full height of the coals that lay along the length of the Dyke must have been severely calcinated by their contact with this huge mass of igneous rock. Every seam of coal that lay within the upper coal measures in this area, was probably burnt brown to a distance of approximately 40 to 50 metres, to either side of this once molten volcanic rock!)*

Shortly afterwards, the whole area of the section was again re-visited, so that most of the remaining Toughs & Clears and the Sparcoal could be extracted from nearly all parts of the large area, only this time, leaving regular stoops of Toughs & Clears and the Sparcoal to sustain the roof long enough, to enable the safe extractions of the last of the gettable coals. The area covered by the second revisitation amounted to approximately 60 percent of the area covered by the Grounds & Nethers, and was to prove to be one of the most difficult, dangerous and unpredictable sections ever to be worked in Lochhead pit. 'Too much and too soon' was the miner's opinion, but then, how often did the management listen to the miners? The second re-visitations were spread out over a long period of time from 1927 to 1941, which might just indicate, that the management were prepared to tolerate the impossible conditions prevalent in this section, as long as the coal continued to flow.

There were actually ten, separate, short, intrusive headings up into this section, commencing from the middle 600 levels, with the last two intrusions actually being the inner end of the top two, triple 600 levels, that did not quite run into the Dyke. *(The bottom 600 level terminated in a 'blind' end in July 1928.)* There was also, only two short headings driven upwards from this large section in to the 200 levels above. *(These must have been utilised to help the return airflow, from what was an extremely large and difficult section to ventilate!)* If however, there had been case scenario that this large section had developed fouled air problems *(black damp or white damp, not methane!),* then, it could have quite easily been isolated *(stopped-of!)* from the rest of the workings, as it did have solid stoops both above and below, that separated the large, extraction areas from the 200 levels above and the 600 levels below. There was also a solid, 70 metres wide stoop left between the line of the main Dooks and the start of the extractions between No. 1 and No. 2 levels. This whole section therefore, could easily have been totally isolated! (Well, Nearly!)

Author's Note:- *A point of note concerning the driving of the triple 600 levels, was the fact that they were never actually parallel over their driven length. They seem to have been driven in a haphazard fashion. At their closest, the bottom two were only 15 metres apart in places, while, in other places, the distance between was as much as 45 metres. It may have been that many different colliers were actually involved with extension of these levels. After all, their development lasted for 19 years!*

*(This was one of the sections where miners like Old Harry Moodie and Old Hugh Moodie were constantly called upon, to go in and extract electrical and coal cutting equipment and pan engines, that were in danger of being lost forever, due to the instability of the roof and the rapidly deteriorating conditions.)*

\*

The extension of the **New Dook,** from the 9200 feet datum height down to the 9015 datum height, is described elsewhere in this book, but sufficient to remind readers that the dook was actually extended upwards from the **1000 yards level,** as developed from the older and longer **No. 5 levels** coming in from the **Old Dook**.! The **1000 yards** levels did run from the deepened bottom of the New Dook to the Dyke, in a north-easterly direction, but did not actually touch the Dyke, even though the coal extracted in the fourth section along this level, did! There were five level inter-connections between the two parallel dooks along its last extension, and there were twenty inter-connections *(short headings)* between the two levels, that comprised the **1000 yards levels.**

The first two connections were, in fact, made by the **540 metres headings** from the Old Dook's No. 5 level, first to the New Dook's later named 1000 yards levels, then on to the New Dook's No. 2 levels (600 section). *(The distance between the Old Dook's long No. 5 levels and the New Dook's 1000 yards levels, was approximately 250 metres across the strata at their commencement!)* There were four separate coal sections along the 1000 level, and except for the first section, the method of extractions were identical. The reason for the difference in extraction methods, was, that immediately and vertically above this first section, lay the farm buildings belonging to Newtons farm. The extraction process in this first section was limited to conventional Stoop and room with a regular matrix of seven headings and five levels, leaving stoops of approximately 35 to 40 metres square. The coals taken were the Grounds & Nethers and the Myslen, followed by the Toughs & Clears and the Sparcoal. The section measured approximately 200 metres by 160 metres. This section was worked between 1935 to 1937, which means that the coals were taken before the completion of the extended New Dook, which in turn, signifies that the coal was transported down the **540 metres heading** and thence to the Old Dook via the No. 5 Levels.

The second section along this level was more extensively worked, and was separated from the first section by a 40 metres wide stoop of solid coal, lying vertically between the 1000 yards levels and the 600 levels. There were four, separate, short headings into the section from the 1000 yards levels and only one short connection *(airway)* from the top of the section, to the 600 levels above. The section was rectangular in shape and measured 375 metres long by 200 metres wide. The first coals were taken were in 1936, where all of the Coronation coal at 38 inches thick was extracted, except from two rectangular stoops measuring approximately 35 by 45 metres, at the bottom right corner of the area. *(I can find no tangible reason for this omission, there were no surface buildings above this section.)* A return visit was made in 1939, to extract approximately 90 percent of the Grounds & Nethers and the Myslen coals, at 63 inches thick. This was done by dividing the section up into five separate areas, then leaving narrow stoops of the said coals between each extraction. The five areas were again re-visited in 1939, 1940 and 1941, where the 51 inches of the Toughs & Clears and the Sparcoal was subsequently removed from four of the five areas. The bottom two solid stoops at the bottom right of this section, remained forever untouched!

The third section along the **1000 yards level,** was separated from the second section by a 30 metres wide stoop of solid coal, where there was only one single level connection between them, at the mid-point. The section was almost square in shape and measured approximately 225 metres along the top and bottom, 200 metres wide at the south end and 230 metres wide at the north end. There were three, short intrusions into this section from the **1000 yards level,** leaving a 30 metres wide stoop between the 1000 levels and the extractions. At the top end of the area, there was a short level that ran the full 225 metres length of the section, that was paralleled by a second, short level approximately 30 metres above it, but which was driven back to link up with the 'blind', terminated bottom of the **600 levels,** which was just above. *(This obviously had everything to do with increased ventilation in this section, as this short level was driven in 1939, before any coal was taken from this No. 3 section.)* The full Coronation coal was probably extracted in 1940 to 1941, and considering the time period, it was surely under-cut with an Anderson-Boyes 17-inch chain machine, which were making in-roads to Lochhead pit and the Surface Dipping. After the Coronation coal had been extracted, a period of six years elapsed before this section was re-visited for the second extractions. Some time before 1947, the planners decided to extract some more of the remaining coals, which of course, was quite considerable. They divided the section into four, more or less equal squares, leaving long narrow stoops in the form of a St. Georges Cross. From the top, right square, measuring approximately 90 by 110 metres, they took the Grounds & Nethers and the Myslen in 1947. From the top, left square, measuring 75 by 90 metres, they took the Grounds & Nethers and the Myslen, followed by the Toughs & Clears and the Sparcoal in 1948 and 1949. In the bottom, left square, measuring 60 by 80 metres, they extracted exactly the same coals in 1951. Meanwhile, in the bottom right square, measuring 100 by 90 metres, the miners took the Grounds & Nethers and the Myslen coals in 1949 and they split this small square into two parts and proceeded to take a small amount of the remaining Toughs & Clears coal along with the Sparcoal. *(There was one or two, small single faults in this section, which may just have inhibited the planners caution in the subsequent diversity of the extractions.)*

The last section along this **1000 yards level** was in fact, a large, odd, double triangular shaped section, which had they're long hypotenuse's worked hard against the Buckhaven Dyke. The section itself was barely divided into two equal parts, with only a very narrow stoop of approximately ten metres dividing them. That which really separates the triangles, is the method of intrusion and the actual coals taken. Access to the this upper triangle was by three short headings from both of the **1000 yards levels,** through the 30 metres stoop and into the workings. This upper part of the larger triangle measured approximately 275 metres high, 175 metres along its bottom and approximately 280 metres alongside the dyke. The coal initially taken from the complete area, was the Grounds & Nethers and the Myslen, immediately prior to 1948. The area was then allowed a year or so to settle, and then, three further small sections were developed over the same area, but covering only about 60 per cent of the original size, to take the Toughs & Clears and the Sparcoal. This took place in July 1949, October 1950 and in October 1951. No further coals were taken from this upper triangle that lay against the Dyke.

The bottom large triangle was approximately the same size in area covered but this was an almost perfect right-angled triangle, measuring 275 metres and 200 metres on its two sides, with a hypotenuse length of approximately 330 metres, which lay tight to the Dyke. The initial coals taken from this area was the Grounds & Nethers and the Myslen, with the complete area successfully extracted. This took place in 1944-45. After the ground had settled, the miners again went back in to take the Toughs & Clears and the Sparcoal, but this time, the second extractions were absolutely total, in that the whole of the area was successfully re-visited, without the need for additional supporting stoops. This second extraction was completed by 1949, with no further intrusions ever recorded. It is therefore safe to assume, that this last section along the **1000 yards level** was abandoned at that time.

These two triangular sections on the 1000 yards levels that were worked hard against the Buckhaven fault, were memorable for two very good reasons. 1. Along the full length of both triangles, there was an additional inherent single down-throw in the seam as the miners approached the Dyke. This downthrow was approximately 5 to 10 metres from the Dyke and the throw was between 3 to 5 feet. 2. The coals between the down-throw and the Dyke, were seemingly nothing more than compacted brown powder! *(Fully calcinated!)*

Author's Note:- *This side of the Lochhead pit was seemingly not as well ventilated as the other sections in the pit, some of which could be quite cool and airy, especially those parts where there were booster fans installed to suck air in the direction of the up-going air-shafts. Those sections to the left side of the New Dook, especially in the* **1000 yards level** *and more so along the* **No. 5 level** *(from Old Dook), were very warm indeed, so much so in fact, that most miners worked in nothing more than shorts, pit-boots and helmets and sweated profusely into the bargain! The fouled air in the East side of the Pit, had a long way to travel before it was finally exhausted, out and up, the old Duncan shaft!*

*

In July 1933, the three parallel levels comprising the No. 5 levels, *(from the Old Dook)* were further extended north-east in the direction of the Buckhaven fault. The three levels did follow the exact level of the strata, but were driven with a very slight, but consistent incline. At this point, the levels were much nearer to the bottom of the **New Dook,** than to their origination from the **Old Dook**, only 240 metres, as opposed to 1150 metres. The transport problem was, that there was no haulage-way to the bottom of the New Dook, for the coal coming from this part of No. 5 level. Between this point on **No. 5 levels** and all along the remaining distance to the Buckhaven fault *(the Dyke),* the remaining sections, all four of them, were bounded on the top side by the **1000 yards level,** also running to strike the Buckhaven fault. The distance between these two sets of non-parallel levels, was 210 metres at the 540 metres heading, and 500 metres at their maximum length. *(At the Dyke!)*

The first section to the right (north and east) of the **540 metres heading,** measured 200 long metres by 150 metres wide and was a near perfect oblong shape. *(If the New Dook at 950 metres length, had been extended another 300 metres deeper, it would have cut through the middle of this section.)* The section area was intruded into, by 3 short headings up from the No. 5 level and 4 short levels from the right side of the 540 metres headings. The taken coals were the Coronation coal in 1936 to 1937. The Grounds & Nethers and Myslen in 1938, and the Toughs & Clears along with the Sparcoal in 1939. An almost clean sweep and probably inevitable, now that the workings were no longer too close to the sea shore above!

The second section in line to be taken along this **extended No. 5 level,** was approximately the same size, but what seems perplexing, is that the extraction method differed yet again. That was until I realised, that the workings along this part of No. 5 level were directly underneath East Newton, farm cottages, and that the extractions would have to satisfy planning regulations for partial stoops. The section actually measured 185 metres by 165 metres and the bottom two-thirds was extracted using the Stoop and room method, while the top one-third was extracted completely, where the Grounds & Nethers and the Toughs & Clears were taken, leaving the Coronation coal, the Sclits, the Head coals and the Sparcoal intact. This part of the section where these coals were taken, did not lie directly underneath any farm buildings lying on the surface.

The third section along this **No. 5 level** was larger than the first two and yet again, the extraction methods changed. The overall shape of the area was that of an uneven square. It measured 300 metres long, was 200 metres on its left side and 275 metres on its right side. It was opened up on it's left side, by two parallel headings up to the **1000 yards level** in 1939, then by another heading, 150 metres to the right in 1941, and then the third heading, 300 metres to the right, in 1944. These three equidistant headings, had the effect of splitting this largish section into two, nearly equal parts which were 'worked' quite differently. The area on the right side (*nearest the Dyke*) is the easiest to describe, in that the first coals taken were the Grounds & Nethers and the Myslen in their entirety, by 1952. The section was re-visited in 1954-55, where the Toughs & Clears and the Sparcoal was extracted over the complete area.

The left half of this uneven, square, third section, was treated in a wholly different manner. From between the initial two parallel headings, one on the left and the other, up the middle of the section, there was from the left side, seven parallel levels, approximately 30 metres apart, driven through the Coronation coal to link up with the middle heading at the 150 metres point. Several short headings were then driven up from the No. 5 levels, to link up some of those parallel levels. (*This could have only been done to extract most of the lower Coronation coal!*) It would seem that this was the only coal to be taken from this part of this section at this time, and that extraction took place in 1938-39. This part of the third section was re-visited some time later, when the Grounds & Nethers and the Myslen coals were taken in their entirety. (*I have found no extraction date for these coal!*) This area was again re-visited in part, where the Toughs & Clears along with the Sparcoal was partially extracted before October 1962. This third section was surrounded by a solid stoop at approximately 40 metres wide all around, so therefore, it could easily have been sealed-of at abandonment if required!

The very last section along the **No. 5 level**, which I shall number the fourth section, was by far, one of the largest sections in the New Dook, and probably the whole of the pit. This fourth section was separated from the third section by a 25 metres wide, solid stoop of coal, which ran vertically up from the **No. 5 level** to the **1000 yards level**, a distance of approximately 425 metres. The section area was in the shape of a large rectangle, with the right side **slewed down** alongside the **Dyke.** The size of the section was 425 metres long on the left side. 400 metres long along its top side to the Dyke. 475 metres down its sloping right side (*tight to the Dyke)* and 750 metres along its bottom side. (*Terminating at the Dyke.)* The whole of the area was initially worked using regular Stoop and room methods of extraction, where the stoops remaindered measured approximately 35 metres square. At a point on the **No. 5 levels,** just 25 metres to the right of No. 3 section, another long heading was driven up to connect with the **1000 yards level.** This heading was 435 metres long. From this heading and to the right side, there was commenced a series of parallel levels approximately 35 metres apart. The first of these levels was commenced in January 1946 and struck the Dyke at a length of 363 metres. The second level was commenced 35 metres below that and roughly parallel to the first, and struck the Dyke at 388 metres length. In all, there was 12 separate levels commenced from this **435 metres long heading**, all roughly parallel and following the grain of the coal. This final and bottom level, was exactly 763 metres long from it's starting point at the bottom of the **435 metres heading,** and terminated hard against the Dyke. (*A point not to be missed, is that No.'s 10, 11 and 12 levels were actually an extension of the triple levels that formed No. 5 main levels, which were actually 2765 metres long from their start point on the Old Dook.)* These parallel levels were not of course, all commenced at the same time. They were commenced one after the other, to maintain a steady stream of coal from the section and were regularly intersected by a numbered series of headings, commenced from the **No. 5 level,** starting from the right side of the **435 metres heading** and proceeding inwards to form a steadily growing matrix, which, when completed, would terminate the section. The first **ten** headings in this large section, commenced from the **No. 5 levels** to strike the **1000 yards levels,** would be 35 metres apart and roughly the same length. No. 11 heading would strike the Dyke at a shorter length, as would headings No.'s 12 to 19, with No. 19 heading being only 25 metres in length. This **Matrix** of a section took approximately 26 months to extract and the coals taken, were the Grounds & Nethers, the Myslen, the Toughs & Clears and the Sparcoal. At this depth the thickness of coal taken would have amounted to ten and a half feet including approximately 12 inches of stone (*approximately 3.25 metres).* The Stoop and room workings in this section lasted from January 1946 to March 1948.

This section was probably the warmest workings in Lochhead pit, which is not surprising, in that it was also the most distant workings from the pit bottom. This was the deepest and remotest part of the New Dook sections, and its distance from the pit bottom made for a rather poor, slow moving and almost inadequate air supply, which was further tainted with the after-gases of spent explosives. The extremities of these workings actually lay directly underneath the miner's, old, tenements buildings at the Rosie 'T' junction, which lay at the north-eastern part of East Wemyss.

*

The coal extractions from this section did not end there, even though there was a ten years gap before the next re-visitation. In January 1956, the miners were sent back into this section, to the **435 metres heading,** at the start point of this last, fourth, worked section, where they proceeded to completely side-strip the 35 metres square stoops, between headings 1 to 4 and levels 2 to 9. They actually stripped twenty-one stoops leaving only the Coronation, the Sclits and the Head coals. This had been completed by April 1956 and no further attempts were made to strip any other stoops after this date. If the management had intended to completely strip all of the 35 metres square stoops in this large section, it would have made better sense to have commenced at the Dyke and stripped the stoops to the retreat, thereby leaving any roofing problems behind, and the colliers out of harm's way. As it was, I am informed, that shortly after 1956, the triple parallel levels in the large, fourth section began to collapse, so much so, that the innermost 800 metres length of the No. 5 levels were closed permanently and abandoned, thus eliminating the likelihood of any further extractions from this large area.

This then, appears to be the last full section every to be worked in the Dysart Main Seam of coal. From 1954 to 1959, several two-man teams of Stoop and room miners working day-shift and back-shift, were sent to various places down the Old Dook and the New Dook, for the express purpose of stripping and robbing to the maximum, any of the remaining stoops that were accessible and could stand being 'thinned!' This happened mainly down in No. 2 levels between the Old Dook and the New Dook, and in the area of both sides of the No. 3 and No. 4 double levels on the Old Dook. There was also several short headings driven into the solid stoops either side of both No. 5 and No. 6 levels. On No. 6 levels, *(north-east)* at the fifth section, some coal was taken from two short levels, which actually lay parallel to the No. 5 levels, but at 40 metres to the rise. The termination date for these two short levels, is given as 13th April 1959.

*

On the 18th January 1961, William Forbes the colliery manager, signed the 'Abandoned Certificate' stating, that no coal from the Dysart Main Seam had been mined in Lochhead pit since 13th April 1959. The Dysart Main seam of coal, that had sustained Lochhead colliery for a period of seventy years, was deemed to be officially dead!

*

In Lochhead pit and the Victoria pit, the Dysart Main seam had produced coal continuously since being struck into in the 1890's. *(Except for the periods of the three Strikes! – one in 1912, 1921 and the other in 1926.)* The amount of coal actually taken during those 70 years, could only have been but a fraction, of what had been available! Estimates vary as to how much of this rich seam is left, and to all intents and purposes, now unworkable and unobtainable. Extraction figures as low as 10 per cent and as high as 15 per cent have been suggested. My own analysis, based wholly on extraction figures and totals, would seem to bear this out! Sufficient to say, that if this had been a more level inland seam, away from the Firth of Forth and not susceptible to self-ignition and 'discovered' at another time. Then, might not better up-to-date machinery have played a greater part in its possibly enhanced extraction? Might not, much more coal could have been taken, from what was almost, an gross inestimable source?

\*

# PERNICIOUS 'PUNK'!
## *(In a One-Act Play!)*

Back in the middle Fifties, an incident took place on the coal-haulage side of the New Dook that, although it started out as a practical joke, actually misfired so badly, that it nearly ended up as a sacking offence perpetrated on an entirely innocent victim. The participants in this stirring saga were:- Wullie 'Punk' Shields, Bob Anderson, James 'Topscoat' Alderson and Walter 'Wattie' Johnston.

One fine morning on the Day-shift, Walter *Johnston (the Section gaffer/ oversman)* made his way down the Man-haulage companion Dook, stopping of at every 'clipping' level on the Dook, just to satisfy himself that all of the 'clippers' and haulage-men were doing their jobs in keeping the coal flowing! And of course, to show his authoritative face! He was actually making his way down to the **1000 yards Levels,** where he was overdue a visit in the normal course of his once-weekly rounds. Walter never missed an opportunity to observe and comment on how the other *(much larger)* half performed. It was therefore obvious to all and sundry, that Walter was again on his rounds, and word of Walter's coming, greatly preceded his actual appearance. Bob Anderson was the clipper at the **600 Levels** on the New Dook and 'Punk' Shields was the 'bogie' driver. *(Handling the empty in-going and loaded out-coming, 30 plus hutch races on this 600 Level.)* The 'clipper' at the lower **1000 Yards Levels,** was 'Topscoat' Alderson. *(No relation whatsoever to Bob, at the 600 level).* As Walter Johnston passed down the Dook on his way to his destination, a devilish idea took root in the head of Punk Shields, to the effect, that it might cause humiliation and possible embarrassment to the unsuspecting Section Gaffer.

Everyone knew that Walter never took his haversack with him into the working section. That would entail him having to sit beside those horribly uncouth miners of the New Dook.  God Forbid!  Walter always left his haversack in the telephone manhole at each haulage-level, where he could partake of his sustenance in quietish isolation, whilst watching the 'coal go by'!  Everyone on the Dook knew of Walter's solitary habits!  Punk waited until he himself had returned to the 'clipping', point on the Dook with the next loaded race, and whilst Bob Anderson was busily engaged with his 'clipping', Punk made his way into the 'telephone manhole'.  The under-ground telephone was of a type that had a fixed, horn mouthpiece, with two, lift-able, extending earpieces *(donkey ears),* that could be 'clamped tight' to a user's ears to eliminate extraneous sounds.  *(This was only partly successful!)*  Punk had been busily practising and rehearsing Walter's 'dulcet tones' over the past hour, and so, containing his best passable imitation of Walter's voice, he returned four rings on the wind-up handle and waited whilst the clipper at the **1000 yards levels** made time to answer the phone.  He might just not, if he were too busy.  Walter Johnstone himself, might even answer it if he were still there, but in that event, Punk would merely 'drop the phone'!  Punk waited, and within a few seconds the call was answered by Topscoat's raised voice, "Thousant yairds level here, whit dae ye want?"  Punk gathered himself, and in his best Walter Johnston impersonation, replied, "This is Walter Johnston, your section gaffer!  I have been held up in here, so, I will not be coming out to have my piece.  You may therefore eat my sandwiches, drink some of my hot tea, but leave enough so that I may have some when I return!  And be careful with my thermos flask!"  Punk quickly replaced the phone and cut the connection, just in the event of any awkward questions.  He did not tell Bob Anderson of his subterfuge, he would await the outcome, if there was to be one!  'Punk' Shields went back to his labours thinking that Topscoat was no fool and would soon see through his 'silly prank'!

As for *Topscoat, far from thinking that the call was disingenuous, he began to fantasise as to what 'goodies' there might be in Walter's piece-box?  Piece-time could not come quick enough for Topscoat.  He reasoned that with Walter being a section gaffer and therefore, on an altogether higher plane than mere miners, he reckoned that he was looking at a 'picnic hamper' with goodies!, and relishing those mouth-watering thoughts, especially with hot tea or coffee, wished it was nearer piece-time!

Piece-time was signalled by the stopping of the Dook's endless haulage.  Topscoat shucked-off his leather working gloves, dusted his grimy hands on his 'working kerchief' and quickly covered the distance to the manhole.  He lifted down Walter's haversack, sat upon the wooden seat and carefully extracted the piece-box.  He did this rather gingerly, not wishing to damage the imaginary culinary delights that now awaited his pleasure.  His anticipated pleasure did not last long!  Only as long in fact, as to examine the first sandwich that came into view.  Pan bread with crusts and a corned beef filling! - and margarined, not buttered!" - YUK!  The second sandwich was marginally worse!  Soggy, peppered, egg mayonnaise, - YUK!  No more sandwiches, but something wrapped in greaseproof paper?  This would be better?  God's truth!  Dry angel cake with no creme filling!  Hell's teeth!  Was this all the man lived on? There was one other small item wrapped in silver paper! Surely this must be worthwhile? It surely was, but to whom? Two cream-crackers with a lemon-curd filling? Yuch! To Topscoat's thinking, this was really the pits!, and him, a Section gaffer! *(No wonder that he was seen to been swallowing great, big, antacid, stomach tablets on a daily basis!)* Topscoat threw the sandwiches, the cake and the cream-cracker 'melange' into the nearest empty hutch and made to collect his own piece-box. At least, there would be sustenance there fit for a working miner. Topscoat commenced to tackle his own thick-cut, 'plain bread', beef sandwiches with the occasional swig of cold water from his own flask. He then had the thought, that Walter's thermos might provide him with a tasty hot drink. To this end, he obtained Walter's flask, unscrewed the top and removed the cork. A tentative sniff told him nothing, *(as any corked flask wouldn't, having used for both tea and coffee)*, so, he then poured out a cupful of the steaming liquid and re-corked the flask. He lifted the cup to his mouth, gently blew over the surface of the liquid to cool the contents and then, sensing that it was not really very hot in any event, took a fairly good gulp! Just as he was about to savour the anticipated taste of hot sweet tea, realisation hit him like a punch on the mouth. It was sugarless! Not only that, the tea had been stewed! It was awful! Topscoat's jaws inflated and the mixture shot from Topscoat's mouth in a forceful spray. That was terrible! *(Spit)* That was vile! *(Spit)* How could any man drink that . . .? Topscoat then went back to drinking his own clean cold water with the thought, 'If that is how the other half lives? Then let them keep it! No bloody wonder the man eats alone!'

Piece-time came to an end with the re-start of the dook haulage, and Topscoat set to work with the knowledge that he would be kept very busy now that the out-coming coal was at full flow. He also carried the knowledge that the 'Section Gaffer' would not be bothering him until near the end of the shift, he, was obviously otherwise engaged!

About 15 minutes after the start of the haulage-motor, Topscoat was at once surprised to see the searching beam of a helmeted spot-lamp coming out of the haulage-level. This gave him cause to ponder, with thoughts of Walter having changed his mind about sacrificing his 'piece'. As the beam grew closer, Topscoat saw that it was indeed Walter Johnston and nagging thoughts of a doubtful nature began to run through his head. These thoughts became quietly alarming, when Walter stopped at the manhole and made movements suggestive of sitting down and making himself comfortable. Surely the man knows that he doesn't have a piece to eat? He might just be able to avail himself of some of his own disgusting tea, but sandwiches were now out of the question! Topscoat was beginning to get just a little worried and felt a distinct warmth creeping up from his neck! Walter was carrying on as normal and preparing to eat! He was already seated and his piece-box had been extracted. Walter then did what any normal man would do when confronted by a 'light-weight' piece-box. He shook it!, and shook it again! He then laid the box on his closed knees and gently opened the top half of the box. It was empty! Walter looked up, then looked down. It was still empty! No, he had not made a mistake, this was his haversack and piece-box and there was nothing in it! Walter then took a tentative hold of his thermos flask, unscrewed the cup, removed the cork and gave it too, a gentle shake! Walter sat there for a few minutes, his face quickly diffusing! Someone was playing silly buggers with his piece, and Walter wished to know! Right now!

Topscoat watched all of this with a deep-seated sense of foreboding! Walter would surely realise that he himself had telephoned Topscoat with a very specific instruction. *(Which Topscoat had carried out to the letter!)* or, had he? No! Perish the thought. No-one else could possibly sound like Walter Johnston, especially over the telephone! Topscoat began to sweat a little, in fact, he was beginning to inwardly panic. Walter was looking at him in the most peculiar fashion, and with a look that spelled trouble and, not so quiet trouble! Walter's blood was up!, as was his demeanour! In fact, Walter was honking! His mouth was working up to some meaningful questions and his eyes were like gob-stops! - and, Topscoat was the target! Walter's little bristle of a moustache was twitching and, he was spitting! "Who has been at my sandwiches? Who has been drinking my tea? Where has my piece gone?" Walter was quite O.T.T! Topscoat didn't know what to say, he had the answers, but what to do? Obviously, he had got it all wrong! He had been hoodwinked! Walter hadn't given those specific instructions over the phone. Also, Walter was in no mood to listen, and Topscoat was at a loss to explain. Topscoat knew that if he came clean, and said, "It was me!" *(With, or without explanation!)* He would be 'up the pit' on the next tow!"

Topscoat was torn between the 'deil and his henchmen', he just didn't know what to say! Walter Johnston was now demanding answers, and Topscoat was on the receiving end! Topscoat tried to splutter out the content of the earlier phone message, but Walter reciprocated with the words, "What phone message? I never sent any message! I would never tell an oncost miner to eat my piece! What are you raving about?" Topscoat was at a complete loss. Walter Johnston was having none of it! He demanded the truth from Topscoat. "Did you eat my piece? Did **you** eat my piece? Did you drink **my** tea? I **demand** that you tell me **now**!" Topscoat couldn't bring himself to voice the words, it would be too much like an admission of guilt, which he absolutely didn't feel! He looked Walter straight in the eye and did the next best thing. He nodded his head violently! That was enough for Walter Johnston. He made to grab Topscoat with both hands, to perhaps give him a good shaking! - but Topscoat was having none of it and made to twist free! He couldn't! Walter's angry grip was too strong, his spleen needed venting and Topscoat was in the firing line! Fortunately, this did not develop into a fist fight, no actual blows were struck, merely an ungentlemanly grapple, down on, and around the 'plates', with neither man attempting to trade punches. They just rolled around like a pair of adolescent school-girls gathering lots of dirt on their hair, face and clothing. *(The one thing that did quickly spring into the mind of Topscoat at this juncture, was, that his 'job was safe', at least, for the time being. How could Walter Johnston explain to the Wullie Hampson, (the mine manager) that he had initiated an 'attack' on an oncost worker, out of hand and, no matter what the provocation!)*

Punk Shields never did let-on until long afterwards that 'he' was the perpetrator of this piece of devilment and especially not to Topscoat! He kept very quiet about his part in the incident for a long time! If Walter Johnston had ever gotten wind of William 'Punk' Shield's escapade, Willie Shields career as a miner would have surely been short-lived! Walter Johnston could be a very unforgiving man where wayward miners were concerned!

*

* There were two conflicting versions as to how 'Topscoat' got his nick-name. One of which was rather plausible and highly probable and is the version that I shall relate here. James Alderson was not really renowned for his sartorial elegance, but he did have the good taste and extremely good fortune, to be able to purchase an expensive and elegant, light coloured, 'camel hair' overcoat. The coat obviously hid a multiplicity of mediocre clothing, but the 'Topcoat' was everything! James wore this overcoat on every possible occasion, even during those balmy evenings in summer, when normal people were wearing shirt-sleeves. He even wore it when visiting the pub of an evening time, when his mates wore nothing more elegant than a bonnet and neckerchief! James simply loved to parade in his 'pride and joy' and to blazes with dress etiquette. He had something to show, he had something to be proud of, and he flaunted it!

*

# WHITHER SHALL I WANDER?
## *A Coalminer's 'TEMPTATION!'*

Several decades past in the mid-fifties, a certain, unmarried, coal miner from Lochhead colliery, decided to spend a short week-end in Auld Reekie. (Edinburgh) He would catch the Saturday afternoon train from Kirkcaldy, view the Forth Bridge in the passing and stay on the train until it reached Waverley station, by-passing the Haymarket on the journey. He would then make his way up over the Mound, past the looming Castle and descend down into the Edinburgh Grassmarket. Once there, he would take a leisurely walk along one side of the long square, casting lingering, appraising glances at most of the young ladies abroad at this time of day. When he reached the west end of the Grassmarket, he returned via the South side, still casting appreciative glances at the many passing female forms. He was in no hurry to 'chat-up' any of these delightful creatures, the night was still young, besides, many of them would bear closer inspection! He was long enough in the tooth to realise that mutton sometimes comes dressed-up as 'spring chicken'! He then decided that he needed some liquid refreshment and chose a likely looking Ale-house, which he then slowly entered. He strolled over to the bar and ordered a bottle of 'India pale ale'. He had no intention of having anything stronger, at least, not yet!

　　He sat down at one of the corner tables, relaxed and undid his overcoat buttons. He slowly sipped from his schooner of pale ale and, from under his bushed eyebrows, carefully scanned the room for a potential 'click'! The first possibility was quickly rejected, 'that definitely looked like endless port and lemon, and definitely mutton!' His eyes travelled some more and alighted on another possibility. Slightly better, but, No! Too garish, too much vivid lipstick, too old, and too fat! She would look hellish in the morning light! His eyes continued on, his gaze passing over the many improbables. His eyes then fell on a seemingly comely-looking youngish female, who might have been with other female company, except that she did not seem to be part of their conversation. The miners gaze then traversed freely around the room, but no one else caught his wandering eye. He then make show of disengaging himself from the activity in the room and recommenced his sipping. After a few minutes his glance furtively returned to the young female, who, to his quick delight was seen to be glancing back at him. His heart jumped, as did her quick glance away from him. Was this a possible 'click'? Our miner's heart gave another timid loup (jump) and moved up closer to his throat. Was she interested in him? Never venture, never gain! He looked at the glass sitting before her, it was quite empty! He quickly drained his own glass and getting up, slowly walked over to the bar. Sure enough, she had also risen and was also making her way to the part of the bar, where our miner now stood. Our Lochhead miner ordered another bottle of pale ale and covertly glanced in the direction of the up-coming female. She was reasonably well built, had good legs, held herself upright and had a rather pleasant smile. On reaching the bar, the girl sighed, placed her empty glass on the raised bar and spoke a soft "thanks Bill, I'm off now!" and turned to go! Our miner saw his chance quickly disappearing and quickly, with a rather tentative question asked, "Would you like a drink m'dear?" The girl paused, her face broke into a quick smile and a whispered 'Yes' ensued! To the miners' raised inquisitive expression? (Our miner, was now quite unsurprisingly, stuck partially dumb!) The girl softly replied, "The same as yourself if you please!" The miners timid heart jumped anew. She **was** interested and flashed him a warm 'Thank you!'

　　The miner gently guided his companion back to where he had been sitting at an isolated corner table. He invited her to sit down first and placed her drink in front of her, and then sat down himself, not too near to her, but not too far apart either.

He needed to able to converse with her without having to raise his voice. Our miner was not really very experienced in this sort of small talk and, apart from the weather and how nice a day it was, the conversation soon began to flag. The girl sensed this and allowed a single action to replace the silence. She placed her cool hand over that of our hero and demurely looked down into her drink! He knew then that he had 'clicked!'

Our miner returned the simple gesture, looked her straight in the eyes and felt a distinct weakening in his knees. She had returned his gaze with equal warmth and covered both of his hands with hers and, with a slow but positive nodding of her head. Our miners' heart was thumping! All it needed now was for them to agree 'terms' and that meant that our miner was 'fixed-up' for the night!, at least, he hoped so! He was rather excited and keen to proceed, but knew that nothing had been 'finalised' as yet! He then set himself to ask the inevitable question and commenced to broach the subject. The girl, sensing that 'business' was about to rear it's ugly head, silenced him with a quick " Drink up and we'll go! My place is just off the grass-market, where we can have a quiet drink, and a 'chat!" Our miner arose, patted his pocket to make sure that the half-bottle of Haig & Haig was still there, and buttoned up his coat. He then followed his new found lady friend out of the pub, along the market street and on to the lady's flat, with himself being towed by his companion's extended hand.

Our miner was quite proud of himself, he had managed to get this far without any mention of price being made. Perhaps, she entertained from the 'goodness of her heart' and a couple of quick drinks? On arriving at the 'home' of his lady-friend, the miner felt the instant clamminess of the dirty 'close' and the worn stone stairs. He sensed rather than saw the squalor of the building, but followed the outstretched hand. They stopped, a key grated in a lock, then the door was opened, and the miner was firmly pulled into a sparsely furnished bed-sit, which was illuminated by a single down-turned gas mantle. The lady immediately increased the gas flow into hissing yellow light, which showed nothing more than a wooden dresser, a stone sink with cold tap and a bolstered double-bed. There was a single door wall cupboard with a clothes hanger on the outside, on which the lady now hung her street coat. What to do next? Our intrepid miner was rapidly losing his confidence at the sheer reality of the whole thing, and simply did not know how to proceed. The lady did! She motioned him to remove his overcoat, which she then hung inside the wall cupboard, then chided him into removing his jacket, which he did! She then invited him to sit on the bed. Our hero was still at something of a loss, no mention had been made as yet of money, and he began to wonder if he could possibly afford what looked like was going to be an 'all-nighter!' He quickly broached the subject in the only way he knew how! "How much for all night?" he blurted out. "Don't worry about that at this time m'dear, I'll just show you a good time! We'll have a little drink and then a little cuddle, and see what happens after that!" Our hero then succumbed to his lady friend's ministrations and completely lost himself in the most beautiful night of passion, that it had ever been his good fortune to experience! - And it lasted, and lasted, and lasted, it surely did!

\*

Lying in a warm, comfortable bed and surfacing into semi-consciousness was one of the finest feelings ever experienced by man, especially when it was accompanied with the pleasant, boozy afterglow of a dram or two of good Scotch whisky. Coming to, after one of the most beautiful nights of his life, our miner lay with closed eyes, savouring every last detail that floated through his still fuzzy and somnolent brain. The pleasure was still coming in swelling waves, which regretfully, were diminishing as his mind rose to full consciousness and the possibility of a slight headache. He lay with closed eyes in the warmth of the bolstered bed, aware of the feeling of two gentle aches, one in his head and the other, along with a warm afterglow, from around his nether regions! He smiled a wistful smile of great satisfaction and slowly stretched his relaxed limbs. He then slowly placed his arm over to where his recent, loving companion slept, then opened his eyes!

It took our amorous miner just a few seconds to realise that his 'light o' love' was not in bed. It took only a few more seconds to discover that her part of the bed was cold, and a few more seconds to discover that she was gone! She was nowhere in sight! She was not in the room! She was well gone! She had run! She had Bolted! WHY ?

It also took our deflated hero a few precious minutes more to realise that not only had she gone, but so had his clothes! All of them! He had only retained his socks and singlet before diving under the bed-sheets and into ecstasy with his paramour. Now! Everything had gone, had disappeared! Clothes, shoes, wallet, money, whisky!, even the loose change he had placed on the rickety wooden dressing table. He was robbed! He was quite destitute! He hadn't a penny left, she had taken everything that he had, even his overcoat! . . . . His Overcoat!? He jumped out of bed, made a dive for the wall cupboard and wrenched it open! His inhaled breath whistled out between his teeth! There it was! It was safe! She had forgotten the wall cupboard in her damned-able haste! He still had his overcoat! He pulled down the overcoat and plunged his hand into each pocket in turn. No! There was nothing there! He was damned! He had no money, no clothes and no shoes! He was in one hell of a pickle! She had taken nearly everything! Rotten little 'tea-leaf' that she was!

Our Lochhead miner was now in one hell of a quandary, he could hardly go outside without shoes, yet, leave he must! Who did the room belong to in any case? He was quite prepared to believe that it certainly didn't belong to her, or was rented by her. Cuckoo's didn't foul in their own nests. At least, not this close to home! Decision time! But, what to do? He hadn't a clue! He had to get home! He sat down on the rumpled bed and tried to think! After a few minutes, he then came up with what he though was an inspired idea! Firstly, he had to get to the train station! Then, formulating a plan, he would fully button-up his overcoat, turn-up the lapels to hide his lack of clothing, and begin to walk stockinged-feet, all the way to Waverley Station, where he would throw himself on the mercy of the stationmaster, whom, he felt, would be immediately sympathetic to his unfortunate predicament.

Our intrepid hero, now that his decision was made, left the scene of his overnight debauchery and stepped outside, to bravely face Auld Reekie's Sunday morning travellers. No-one paid him the least attention! They were all quite used to the inner-City's eccentrics and by-passed him without so much as a backwards glance. He continued on with growing confidence and head held high. If the locals weren't concerned with him, why should he bother about them? Even then, though, he retained the niggling thought that even the Edinburgh tramps didn't go about without footwear! Thank goodness!, he thought, that he didn't have to mingle with any great crowds as he entered Waverley Station this Sunday morning. The thought of getting trampled feet hadn't escaped his imagination. He quickly found the stationmasters' office, knocked at the door and entered at the peremptory invitation. Gathering himself, he launched into a rapid explanatory diatribe of what had taken place, the situation that he found himself in and, what he felt the stationmaster should do! He felt that he should be given a train-ticket to Kirkcaldy, the price of which could be repaid at a later date direct to the stationmaster! Having finished his outburst, our miner then tremulously awaited an answer! The stationmaster had obviously heard many hard luck stories in his time, especially from hard-up, (penniless) Fife miners. He did not instantly respond. He looked at the miner, his eyes searching from head to toe then back to meet the miners eyes. "So! You want me to give you a free ticket to Kirkcaldy. You are also going to leave me your name and address. You're now going to promise to send me the fare plus a wee bit o' a sweetener 'next week' - and then, you'll go on your merry way home?" "Is that the way of it?" "That's it!" cried the miner, "That's exactly what I want, I knew you'd see me richt!"

> Then silence followed for a while, as the station master pondered,
> He wondered as to where or from, our Lochhead man had wandered.
> But then he thought, for goodness sake' a jaunt's a jaunt for aw' that,
> And every man must pay the price, of a Saturday night being squandered!

**"I don't think so!"** said the station-master, in a voice that brooked no argument, but, then slyly said, "But, that's a fine overcoat that your wearing!" "I'll tell you what I think you should do - and that is to go back down the Grass-market, find an open 'pawn-shop' and see if ye can 'hock' that fine coat!" "You might just get enough money tae buy yerself some claes, (clothes) some auld shoen (shoes) and a train ticket tae Kirkcaldy!" "Now, off ye go, and no waist ony mair o' ma time!"

\*

At this juncture, I have no further knowledge of how our intrepid hero managed to extricate himself from his predicament, or how he managed to get himself home. It was suggested to me that he may have walked home, but that would seem highly improbable, in light of the fact that he returned to work on the following Monday morning and without sore feet! *(He had somehow to cross the River Forth, no mean feat considering that the only vehicular traffic at this point was by ferry-boat!)* He did not describe his outing to Auld Reekie to anyone in general, nor did he intimate as to how he managed to get home. He did however, finally confide in one of his close work-mates giving him the full story, but not before extracting from him, a promise, never to breathe a word to anyone!

His workmate didn't tell a soul, that was, not until the next day! The story then spread like wildfire and, with each re-telling, gained much in lurid detail! Each re-telling of course, came with the dire warning that it must never be mentioned to our returned hero. It never was! There was no need! Each time any workmate or acquaintance of our local hero met up with him, no greetings were exchanged! Instead, all he got was a wide grin, a pronounced wink and the whispering of the emotionally charged words – 'Auld Reekie!'

\*

# Chapter XX

## Diesel Davie *and the 'Turret Syndrome!'*

Diesel Davie was so-called, simply because he was christened thus by the miners, in that this was his full-time underground job! He drove the diesel 'Pug' that hauled the long races of empty tubs from the end of the Lower Dysart endless haulage, into the Lower Dysart west section. He then waited until his sidekick, Willie Jackson by name, coupled-on the waiting loaded rake of hutches, that could number anything up to fifty in a train. I do have certain knowledge of this, in that my first under-ground job in Lochhead Pit, *{The Lower Dysart Section (west)}* was to ensure that all of these hutches were actually coupled together! Yes, I was on the couplings and woe betide me if I had 'missed' any! I was made to travel the long length of the full-race, carrying two couplings in each hand just to be taught the hard lesson! *(Sadists! - All of them.)* I was a tall, well-build lad at 17 years of age, but four multi-linked, steel couplings are bloody heavy!

For a long time, I had the very distinct impression that Diesel Davie was disabled and also gave the impression that he had an enormous 'chip on his shoulder'. *(Discerning readers will instantly realise that this opinion is quoted in retrospect. I was just too green at that time to know any better!)* Davie just did not get out of that Pug! The Pug was diesel-engined of course, had two axles that were both driven, was double-ended and attained the same speed in both forward and reverse gears. *(Come to think of it, I never did find out which was the fore-end?)* This Pug was designed, so that the driver sat side-on in the middle of the cockpit and could see through the perspex windows at either end, his vision being enhanced by two great 10-inch* spotlights. *(One at either end).* The cockpit had no foot controls, *(which enhanced my earlier belief),* but did have three hand controls. The first was the throttle lever, the second was a long hand-brake operated by the right hand, whilst the third was a combined clutch-gearbox lever, operating on the double-U crash principle. This was a very smooth gear-change! The Pug weighed approx-imately three tons and could pull a race of up to eighty, loaded hutches on level ground. The Pug however, did have an unfailing propensity to sudden and unexpected derailment! *(This usually happened when an inexperienced driver on another 'shift' became 'cab-happy' and tried to 'gun-it!')* This is why there just happened to be two 12 ft lengths of steel rail and several lumps of assorted hardwood lying by the wayside!

Diesel Davie was 'custom-built' to drive underground Pugs! He was 5ft - 4ins tall in his stocking soles, *(would you believe he wore suspenders down the mine!)*

and weighed approximately 8 stones. He was slightly built and looked frail! *{In miners parlance:- 'twa-ply reek!' (two-ply smoke!}* He had a sallow complexion, *(what miner didn't!)* and somewhat distended rheumy eyes. All of which seemed to support my earlier conviction! Davie did not seem to wish to speak to anyone, and had the infuriating habit of turning away if anyone even looked in his direction! I have seen 'strippers' coming of the face near 'lousing-time', walk out along the coupling level on their way to the pit-bottom and give a curt nod to Davie. They were usually ignored by Davie, pointedly facing in the opposite direction! If Davie and Willie Jackson were having a conversation as I approached, the talking stopped! If Davy was forced to engage in necessary conversation with any individual, he wickedly turned to face them fully, so that his spot-lamp would momentarily blind them! He was I feel, one of the most morose and moody characters that I came across in Lochhead Pit!

Another reason that I formed the first impression that I did, was that Davy was always amongst the first to get-down the Pit *(sometimes, he was below ground before the Fireman could check him thro' into the section),* and he was the last to leave! He seemingly insisted that he had to service his Pug and put it to 'bed!' To tell the truth - and everyone that I spoke to seemingly concurred, that he simply could not bear the thought of some other 'being' driving his Pug!

If, during the normal course of a working shift, the Pug just happened to 'jump the rails', *(Davy 'never' de-railed his Pug!)* all hell was let loose! 'Chairlie' Sinclair the Fireman, would grab a platoon of 'oncost volunteers' armed with 12ft rails and batons, and lead the charge to the rescue! *(Lost time was lost production!)* With Chairlie in charge, it could guaranteed that he would try to lift-on both ends at the same time! The more level-headed and experienced miners would quietly 'push' Chairlie out of the way, and get it back on the rails in minutes!

During routine maintenance, if the clogged filter-baskets were not changed on the Pug every three days, *(unfiltered diesel fumes are quite toxic in the confined airways of a coal-mine!)* the miners, good-naturedly, always blamed Davy! He usually took this very much to heart! This was a personal insult to him! *(Actually, I feel that this was probably the main cause of Davy's later ill-health.)* Davy usually suffered all of this banter with a sickly grin, a wry comment to Willie and a well-directed, diesel impregnated Gob!

A measure of the not-so-well disguised contempt that some men felt for Davy and his Pug, was exhibited very clearly one day near the end of a day-shift, at about twenty-five minutes past one. The oncost workers were hurriedly streaming through the Lower Dysart Mine on their way to the Pit-bottom. The Pug had just been de-railed! Willie Jackson was frantically, but ineffectually, trying to lift the front end of the Pug. *(Under the expert directions of diesel Davie of course!)* But, to no avail! And would you believe? No-one! Not a single miner stopped their headlong rush to the Pit-bottom! No way! This was lousing time! Leave it be! Let the 'back-shift' lift it!

Within a very short time, Davy did have to abdicate and hand over his beloved Pug to another!  His deteriorating health was obvious!  He knew this and asked to be relieved.  His wish was granted - and he was duly rewarded with one of the plum jobs on the Pit-head!  He became "Sparkie!"  That translates into the 'operative' who ran the lamp-shed.  That little house where all of the miners', electric, battery lamps were daily charged and the miner's safety lamps maintained and refilled!  Miners who initially qualified for the use of an electric lamp *(as opposed to a carbide lamp)*, were charged at the rate of a half-penny per day for its daily hire!  *(That amounted to a whole tuppence-ha'penny per week and the N.C.B. actually had the 'effrontery' to deduct it from the miners' wages!)*  Attempting to get little Davie to hand over a lamp to a unauthorised recipient *(just for one day)*, was akin to asking him for a 'Pound note!  *(He was often described as a true 'company' man!)*

It was rumoured, that Davy often drove the Pug 'at speed' along the Lower Dysart tunnels *(which were in near total darkness)*.  It was said that he did this 'without lights' in a deliberate fashion to put the 'Fear of God' into any wayward travellers, who, invariably dived into 'manholes' simply because they didn't know how near the Pug was!  *(I don't know if there was any truth in this or not!)*

\*

## CROSS SECTION OF LOWER DYSART SEAM
### Showing Three Distinctly Different Leaves

| | Stone | **STONE** |
|---|---|---|
| | Coal – 18 to 20" | |
| | Stone – 2" | **UPPER LEAF** |
| | Coal – 19 to 23" | |
| | Soft Blae – 27 to 31" | **STONE and BLAE** |
| | Coal – 22 to 33" | **MIDDLE LEAF** |
| | Stone – 4" | **STONE** |
| | Parrot Coal – 5" | **PARROT COAL** |
| | Coal – 22 to 24" | |
| | Stone – 2" | **BOTTOM LEAF** |
| | Coal – 18 to 20" | |
| | Stone | **STONE** |

# Chapter XXI

## The Lower Dysart Coal Taking!

### How the West was Won!
### *– or 'Pick-it' and see!*

C-o-a-l for Dysart, Dysart, - coal, - coal, - coal,
C-o-a-l for Dysart, Dysart, - coal, - coal, - coal,
C-o-a-l for D-y-s-a-r-t, D-y-s-a-r-t C- -O- -A- -L,
Oh! - C-o-a-l for Dysart, Dysart C-o-a-l ! !

To get to the Lower Dysart *(west)* section, the route of course, starts at the Pit bottom and travels the length of the Lower Dysart haulage-way, *(originally driven in the Dysart Main seam on an approximate grid bearing of 225 degrees and previously known as Nicholson's level, or the Pit Bottom level.)* At a point approximately 950 metres from the pit bottom, and 50 metres short from the head of Nicholson's Dook, the route takes a 50 degree level turn to the right for a distance of approximately 70 metres, *(this is now the start of a level cross-cut mine with Diesel Davy's pug-house on the left,)* which is followed by another 50 degree level turn to the right, for a distance of 175 metres. This 175 metres length is a straight and level cross-cut mine, cutting through the 25 metres of 'metals', that separates the Dysart Main seam from the Lower Dysart seam. Also, due to the rising slope of the strata at an approximate inclination of 1 in 6, this level cross-cut mine struck into the Lower Dysart seam of coal at the 175 metres point of it's length. At this point, the route takes two directions, this second *(but, original)* route follows a direct line into the old workings of the Duncan Pit, which is barred by double air* doors and is still used as the return airway for the Lower Dysart west workings. The first route now takes a half-left turn for approximately 20 metres, wherein is housed the electrical sub-station that supplied power for the machinery used in the Lower Dysart section. The route takes another half-left turn and continues on following a minute rise, but now driven through the bottom two-thirds of the Lower Dysart seam of coal and 'brushed' at 12 feet wide by 9 feet high. This final level was fully 100 metres long, before it split at the head of the **Lower Dysart Dook** to the left, and the commencement of the straight through **30-inch belt level** leading to the '**Lower Dysart slopes**' coal faces.

The Lower Dysart coal seam as worked on the upper slopes of the named Lower Dysart *(West)* section in Lochhead pit, was a fairly substantial seam of good quality coal that contained three separate, but homogeneously related sandwiches, or leaves of coal. At best, the top leaf showed a thickness of approximately 42 inches of coal, separated by a 1 to 2-inch narrow band of stone at its centre and at it's thinnest, at least 37 inches in thickness. A height fully exploitable by long-wall machine undercut extraction and using a face conveyor system to transport the coal from the coalfaces. The middle leaf, which was separated from the top leaf by a 33 to 36-inch band of soft grey blae, was comprised of approximately 28 to 30 inches of the same type and quality of coal, but was absolutely clean with never a trace of stone. The bottom leaf of the seam was approximately 38 to 43 inches thick, with a single thin streak of stone *(approximately one inch thick)* running through its middle. *(Almost an clone of the top leaf!).* This bottom leaf was separated from the middle leaf, by a 4 to 6-inch band of stone adjoined to a 5 to 7-inch band of parrot coal. *(A visible characteristic of the Lower Dysart coals, was that it showed streaks of white in its make-up. The colliers called this 'water-marked!'.)* As a seam, one of its advantageous features, was, that if it was to be worked as limited height, long-wall face, it could therefore be taken *(extracted)* as two separate entities, with the middle leaf of the seam left in situ as a solid roof, when the bottom leaf was extracted! If the seam was to be extracted as a 'Stoop and room extraction, *(whether it be by old-fashioned 'blast and fill' or by cut by short-wall machine)* then, the decision had to be made as to what two 'leafs' to take simultaneously? Either way, this made for rather uneconomical working, in that large quantities of stone or blae would contaminate the otherwise 'clean' coal. To the best of my knowledge, the initial working of this coal in this section in the early part of 1949, was the start of a short period of experimentation in the taking of the first of the Lower Dysart coals, in the section known as the Lower Dysart West.

The very first area to be opened up in the Lower Dysart west, actually commenced at a small fault-line, at a point approximately 50 metres in from the electrical sub-station and to the right side of the level roadway. The fault showed an upthrow of four feet and ran in a north-westerly direction. *(This could have been the reason that the 100 metres long roadway to the 30-inch belt-end was mined slightly to the rise!)* The section was opened up by the driving of a heading up along the small fault-line, until it struck into the old workings of the abandoned Duncan pit. A distance of approximately 35 metres. This through road now opened up and established the return airway for all of the future Lower Dysart coal sections. The old roads needed re-brushing, but that was always an on-going process in any working coal section.

The coalface was now opened and the face-line would be cut at right-angles to the main roadway. The coal taken, would be the top leaf above the blae, at approximately 39 inches thick and undercut by machine. The first cut was to a depth of 4 feet six inches, along a 35 metres face-line, with the coal being transported along, and down the face, by a 20-inch wide, endless belt conveyor and loaded directly into hutches on the main roadway. With every cut, the face length was systematically extended by a full three feet at the top end, as the old Duncan pit boundary fell away! The system of extraction used at this juncture, was the three-shift system, whereby the wastes were fully packed every day to ensure an even, but regular crush. *(Closure of the wastes!)*

As the main roadway *(Main gate at this juncture!)* advanced, so did the extractions. With the face-line increasing in length with every cut, so did the amount of coal that was taken, also increase! At the 90 metres point of the advance of this face, when the face-length was approximately 85 metres long, the face was stopped dead and terminated! There was to be a change of approach with a marked difference in the extractions, and a change of direction now that the miners were clear of the Old Duncan pit workings! Also, the main-gate would now be stabilised as a main coal thoroughfare and be laid with double sets of semi-permanent rails, to facilitate the running of a diesel pug!

The amount of coal extracted from this section would have amounted to approximately 5,850 cubic metres of reasonably, clean coal. This of course, was the first of the Lower Dysart coal to be taken in Lochhead pit and it had been successfully extracted, without roofing problems!

This 12' by 9', steel-girdered roadway, now 100 metres long, stretched from the electrical sub station, into the start of the section known as the Lower Dysart slopes, was continued on at 12' by 9' from the head of the Lower Dysart Dook. *(This was sometimes known as the **30-inch belt level!**)* On a map bearing of approximately 225 degrees. It roughly followed the level grain of the seam, but was now at a slight dip of perhaps 3 to 5 degrees depression. It was eventually driven to a full length of 375 metres. *(I shall also described this level as the **375 metres level** or, **the 30-inch belt level**.)* The first attempt at a likely coal face in this seam /section, was commenced along this 375 metres level, at a point approximately 150 metres from the belt-end *(loading point)* and to the dip on the laich *(low)* side. The brushers /developers drove a small dipping *(following the coal!)* down to the left for about 50 metres, then turned hard right and continued on for another 130 metres where they stopped. There was no return road developed. Back at the start of this short, 50 metres dipping, the same developers went forwards approximately 10 metres along the level and commenced to dig to the dip and into the 40 inches height of this bottom leaf of the seam. *(This then meant, that the immediate roof comprised of approximately 6 inches of parrot coal, plus 5 inches of stone, sitting below approximately 20 inches of unbroken solid coal, as opposed to the soft blae that lay directly above this.)* This new development would be approximately 8 to 10 feet wide and 40 inches high and, was to be driven to a dipping length of 50 metres. When the development team had cut through to the previously driven bottom road *(at right angles),* this would then become a short sloping long-wall face, 50 metres long and cut by an Anderson-Boyes 17-inch coal cutting machine. *(This machine was 'chain-driven', in that it used mechanically powered sprockets to drag itself along the length of a pre-positioned 'stelled' heavy link chain, as opposed to wire rope!)* The extracted coal from the face would be carried up the slight incline by something akin to a 20-inch belt conveyor, then fed directly on to the newly installed **30-inch belt** conveyor, that would eventually be expanded to the full length of the **375 metres, belt level**. This short run was reasonably successful, in that it took a swath of coal along a length of 150 metres, having a face-length of approximately 50 metres and at a height of approximately 40 inches. This short *(experimental)* face in the bottom leaf of the lower Dysart seam, lasted until mid-April 1949, at which time it was abandoned. It had been a slightly wet run, where the gathered water needed pumping out on an intermittent, but regular basis. The coal produced from the Lower Dysart seam was good, usable and reasonably clean. *(The coal was exceptionally clean, but the inclusion of a thin layer of stone in the filling thereof, made it 'dirty!')* This face would have yielded approximately 7,500 tons of coal at termination. *(This extracted face would have been 'packed' fairly evenly with pillar-wood, blae and redd, to help maintain the integrity of the 12' by 9' steel-girdered main 30-inch belt level, running parallel to its extracted length.)*

Opposite to these workings and at nearly the same time, another group of developers had opened up the first of three headings to the *(high)* right side of the 30-inch belt level. The first heading was driven at a point 70 metres from the belt-end and at an angle of approximately 80 degrees to the right side of the level, and then extended to 90 metres long, to initiate a 'T' junction, that was to become part of the return air-way for the Lower Dysart Slopes section and, leading into the old Duncan pit workings. This 90 metres long heading was numbered 0, with solid stoops of coal being left untouched on either side of this heading, to maintain the integrity of the roadway. *(This heading was approximately 15 metres from, and parallel to, the termination point of the first coalface in the section!)* The next heading (No. 1) to be driven to the right, was at a point exactly 115 metres from the belt-end and at a clockwise angle of 70 degrees. This heading was initially driven to a length of 90 metres. *(This was later extended to 500 metres, to form the centre road (main-gate) of a double, long-wall face, from which coal was extracted from the separate top leaf.)* At the 90 metres length of this No. 1 heading, the direction changed and a 70 degree turn to the left was executed, where the new level continued. *(Readers will surely note that this new level now runs parallel to the 30-inch belt level, but 90 metres apart and to the right of it.)* This new level, I will name the **'upper parallel level'**.

At the bottom of No. 1 heading, development work commenced on both sides of the heading, to develop two, separate long-wall faces in the top 39 inches of coal, which of course, gave the colliers a stone pavement and a stone roof. The coal to the right of this No. 1 heading, was only taken to a depth of about 35 metres along a 75 metre face, whilst the face to left of the 85 metres length, was undercut on the slope and was worked successfully until July 1949, with little or no reports of roofing problems. This 85 metres, long face was undercut to a daily depth of approximately four feet six inches and extracted to a length of 125 metres, where the coalface was suddenly foreshortened by 40 metres from the bottom end, because of a series of four short faults, that suddenly appeared on the coal-face. These 100 metres long faults could not have been predicted, as they developed and disappeared within a distance of approximately 30 metres. This small area of ground was therefore troubled, but the planners decided to cut-through and disregard these faults, which eventually succeeded without loss of face length. This coal-face continued on for another 100 metres in depth at its planned length, before abrupt termination. *(This 85 metres long coal-face to the left, could have been developed for at least another 180 metres in the same direction before termination. The ground was clear and untroubled, with the Randolph Colliery boundary a full 200 metres away.)*

Within this section area, the amount of coal extracted from the top leaf of the seam at this time, would have amounted to approximately 21,000 cubic metres.

*

On, or near July 1949, an assessment was seeming carried out by the planners to the effect, that this was simply not the best way to take the Lower Dysart coal in this area. Out of a possible 10 feet height *(120 inches)* of coal in the seam, they had only managed to extract an average of 40 inches height of coal over a given area, from two small and one slightly larger, section areas. Time for a positive re-think! A change of tack was called for, duly planned and quickly initiated!

The original No. 1 heading was extended in it's same direction towards an untouched part of the seam, and, at the same time, a new heading was started at a point along the belt-level to the right and at a distance of 225 metres from the belt-end. *(110 metres from, and parallel to No. 1 heading)*. This new heading, named No. 2 Heading, was driven up and bisected the upper parallel level, where it crossed it at a distance of 90 metres from the belt-level. This No. 2 heading would later be extended to a total distance of 350 metres *(as the miners took the coal)* in the top leaf of the lower Dysart seam, and absolutely parallel to No. 1 heading. At the innermost dead end of the upper parallel level, *(finished in July 49)* another heading was started to the right side, at a distance of 110 metres from where No. 2 heading crossed the upper parallel level. This heading was parallel to both No.1 and No. 2 headings, but was commenced from the upper parallel level. This new 'attack' on the Lower Dysart seam, commenced at a point approximately 10 metres above and parallel to the 'Upper Parallel level'. A long, 10 metres wide stoop of solid coal was left between the first 'slopes' faces *(below the upper parallel level)* and this new assault.

Meanwhile, back on No. 1 heading and at a point ten metres above the upper parallel level, two new long-wall faces were being developed on either side of this heading, in the top leaf of the seam. The coal-face to the right, would be about 40 metres long, with an 8' by 7' brushed airway *(steel-girdered roadway some eight feet wide and seven feet high)* to it's right, and the coal-face to the left was developed at approximately 110 metres long and connected to No. 2 heading at it's left end. This was now a continuous, long wall face of approximately 160 metres length, including a left side stable-end, an off-centre coal heading, *(main gate)* and a right-side return airway, back down to where No. 0 heading met the return airway to the old Duncan pit. This long-wall face had been developed in the top leaf of the Lower Dysart seam, which showed 41 inches of coal, still with a 2-inch sandwich of stone in the middle. The coal was undercut in the bottom 6 inches of the leaf *(coal)*, and the pavement was the top of the 30-inch band of fireclay *(blae)* that separated the top and middle leaves of coal. The average height of the coal-face was 41 inches, which was a good and 'comfortable' kneeling height. The depth of the undercut was 4 feet - 6 inches on average, and the stripper's stint *(allotted yardage)*, was ten linear yards of coal at this taken height. The downed coal from both coal-faces was fed *(shovelled)* onto two moving face conveyors *(one for each face)*, that emptied onto the wide 'pan-engine' train, in the daily extending No. 1 heading. *(Main gate.)* It was then jigged downhill to the 30-inch conveyor belt in the 375 metre level, and thence to 'Paw Broons' belt-end. To all extents and purposes, this was a very successful double-run, was true to the slight upwards slope and fault-free. This long-wall face continued on for a distance of 400 metres from it's start line, and advanced at a rate of one and a half yards on a daily basis, until September 1950, where it petered-down to a short 40 metres long double run, due to it's having reached it's western boundary, bordering on the workings of the Randolph Colliery

Whilst this ever shortening double face-line was approaching termination, yet another long-wall face was being developed in close proximity and also in the top leaf of the Lower Dysart coal. This new face line, was started to the immediate left of this almost worked-out, double run, and had commenced down at the 10 metre stoop just above the upper parallel level.

This new face was a single run of perhaps 110 metres length, to be undercut in the same general direction as it's predecessor, with the extracted coal being taken down the now extended No. 2 heading *(new main gate)*, again, by a train of jigging pans, to fall onto the 30-inch conveyor in the 375 metres belt-level and thence to Paw Broon's loading point. The coal was again extracted with the minimum of trouble, the only small excitement encountered by the colliers, was that the left part of the run developed a small upwards roll in the strata, that commenced at the left end and rolled gently into the middle of the run, where it petered-out before the run was terminated.

This run was worked to an approximate distance of only 250 metres, where again, its continuance was curtailed by the surveyed boundary with the Randolph colliery. *(This boundary line comprised of a 50 metres wide exclusion zone, that took the form of a long, continuous, solid stoop, that would remain inviolate, when two or more collieries were working within the same seam of coal. The reasons for this are manifold, not the least of which, is that individual collieries do not wish to share each other's problems! i.e. Airflow, gasses, water, etc. etc.)* Within this section, the extraction of the top leaf of the Lower Dysart seam was terminated in mid February 1951.

At this point, the fiendishly clever planning for the remaining untouched bottom leaf of this coal seam, now came to fruition. The developers were now sent back to the beginning of the original double-run *(above the upper parallel level, with No. 1 heading as it's main gate),* that saw the extraction of the top leaf of the seam. They were given instructions to open up a new double run, but to start the development in the bottom 40 inches of the lower leaf of the seam. This would leave a fairly thick band of approximately 25 to 30 inches of unbroken coal as an immediate solid roof, with approximately 30 inches of unbroken fireclay immediately atop the coal. This would serve as a good, solid, unbroken roof, as they extracted the bottom 40 inches of coal, worked as another double, long-wall, undercut face. This time around, the total length of the double face was only to be about 125 metres long, approximately 35 metres to the right side and about 90 metres to the left. This was deliberate, in that they seemingly did not wish to experience re-brushing problems with the earlier, steel-girdered, outside roadways. The new roads therefore, were driven to the 'inside' limits of the previous roads. The Pan-engine train that was used in the original main-gate, was now brought into re-use in the now re-brushed No. 1 heading *(main gate),* and the working and stripping of approximately 40 inches *(one metre)* of the bottom leaf commenced.

This turned out to be quite a conventional face, in that, in spite of the fact that the top leaf had been extracted above their heads, the roof remained solid, and after the daily process of 'steel-drawing', the wastes did fall evenly and systematically as expected, between the man-made 8 feet - 6 inch wide packs and happily, not too prematurely! This long-wall face continued to be undercut daily, with the coal being extracted without too many problems, and this continued until the full face-line had advanced a full 50 metres. The decision was then made, to take the remaining 20 to 25 inches of coal that formed part of the immediate roof, thus leaving the 30 inches of untouched fireclay *(roof)* hopefully intact, whilst they were doing so! The method used to take this coal was simple in context and almost as easy in it's execution. As the face line had advanced this full fifty metres, the roof had performed completely as anticipated in staying up, without appearing to break, even with the complete top leaf of the seam already extracted from above and crushed down.

Over the complete length of the face-line, after a complete day-shift strip, the next long 'line' of roof supports would be fully in place. That meant that there would three complete, but interlocking rows of 'steel straps' set to the roof, over the full length of the coal-face. One set over the face-line, another set over the jigging pans/conveyor belt and yet another set in the wastes (condies). In addition to this, there was a series of 8 feet - 6 inches wide 'packs' placed at regular intervals in the condies, *(wastes)* with a 12 to 15 feet spacing between them, and these packs would each be advanced *(extended)* approximately four and a half feet forwards every nightshift. These packs were set and built from pavement to roof, at the height of the taken coal, and therefore, the space between the packs still showed an exposed coal roof. This 20 to 25 inches of coal was now to be taken out on the 'wastes side', after the long line of the conveyor had been moved sideways and nearer towards the newly-stripped face-line. This meant, that the coal could be stripped from the roof along the length of the 12 to 15 feet strip *(between the packs and, in the wastes),* at whatever height they wished to take, and at approximately four and a half feet broad. The miners who were nominated to take this coal, naturally enough, named it as 'stripping the condies!' This operation did involve strippers /colliers working on both sides of the jigging pans and, with the condies strippers having to set temporary wooden straps and trees, to the now heightened roof. These methods did not pose any great problems to the miners, even with the condie's strippers having to bore and blow down their own given height of coal, where one miner was allocated two full condie-lengths to strip. This system worked well enough, without any fatalities or hiccups and continued on, until the full thickness of the middle leaf was extracted between the packs on this double face-line. This double run was terminated and later abandoned as 'worked-out' in January 1952.

Author's Note:- *To the best of my knowledge, this was the only section in Lochhead pit where this method of 'stripping the condies' was carried out as a planned operation, in conjunction with the simultaneous stripping of long-wall faces! That it was ultimately successful is beyond doubt. The fact that there were no recorded fatalities or serious accidents, would seem to vindicate the planners in this bold, but venturesome experiment!*

Before the completion of the last, double, face-line, the developers had already gone back to the start of the original, top leaf run on No. 2 heading and commenced to develop the untouched bottom coal as a long-wall face, with the explicit purpose of stripping the condies *(middle leaf),* in exactly the same manner as practised on the now exhausted and abandoned, previous double-face. Unfortunately, this long-wall face with it's double, stripping faces only lasted a further three months, until it's abandonment in April 1952, when most of the coal-strippers were transferred to the Lower Dysart Dook. This then, was almost the last of the coal taken from the Lower Dysart Slopes, even though there was some limited, additional output from the rectangular, coal section immediately above the 375 metres belt-level. From this section, the top leaf had already been extracted, but the section had crushed down well as had been expected, with very little breaking-up of the remainder of the seam. The planners in their wisdom, now decided to leave the middle section untouched and concentrated on the total extraction of the bottom leaf. The section was then re-opened in the 41 inches of the bottom coal, where, despite some small coal-cutting* problems, the bottom leaf was successfully taken over an area equivalent to 75 percent of the original section area. *(This time, the planners decided to by-pass the small area of faulted ground and merely shortened the length of the coalface!)* This coal face was terminated and abandoned in August 1953.

# Anecdote or Myth? - *A Chain Reaction!*

Author's Note:- *There was one more attempt to take coal from a previously extracted section, where, on an experimental basis, the bottom leaf had been extracted first! This was the small, rectangular second section area (150 by 50 metres) to the low, left side of the 30-inch belt level. The coal to be undercut was to be the 39 inches of the top leaf, leaving the soft blae as the pavement, underneath which, there still remained the 20-odd inches of the middle leaf, that hopefully, would remain unbroken to help stop any 'pavement heaving!' Shortly after the start of coal-cutting operations in this re-visited section, the machine-men began to experience a rather frightening occurrence, which, at best, stopped the cutting chain dead, and at worse, actually broke the cutting chain, which then sent it spinning and almost locked it under the 'cut!' At the first instance, the machine-man put the stalled machine down to chance, with the unlikelihood that he was over-driving the machine. When the 'stall' happened at the second instance, the machine-man then began to doubt his own handling of the machine, but immediately rejected this as imagination. He was too experienced an operator not to believe in himself. Several days later, just as all thoughts of his 'mishandling' of the machine were pushed to the back of his mind. It happened again! This time, it was serious! In the space of a few seconds, the cutting-chain had stalled, the body of the machine gave an unaccustomed loup, with the electric motor racing and whining with an uncanny piercing sound! The machine-man (No. 1) recoiled with fright, as did his No. 2, him having dived into the condies and disappeared from view. The chain had obviously snapped! But where, and why, and where was it? The machine-man, although badly frightened and sickeningly alarmed, did have the presence of mind to hurriedly cut the power to the electric motor. The No. 2 man couldn't get near to the machine, he had been sprayed by loose gum from the teeth of the horizontal driving sprocket!*

*When the motor was finally silenced, the machine-man's first thoughts were as to what had caused this phenomenon? He hadn't a clue! He had never experienced such a train of unlikely events such as this! He had experienced broken chains in the past, what machine-man hadn't! But this! This was something new and beyond his comprehension!*

*The duty engineer was sent for, along with the duty electrician, just in case! Meanwhile, the machine-man and his No. 2 began to dig out the gum from around the rear-end of the coal-cutting machine. Their prime objective was to expose the driving sprocket and jib, and to locate the broken chain. This would not be too difficult. It had nowhere to go! It would be deeply imbedded in the cut, where the problem would be, to get it howked out! Sure enough, the compressed chain lay deep in the undercut, with the adjoined pick-boxes jammed tight under the coal and held firm with the addition of the packed gum.*

*The first task of the machine-men was to free (unlock) the right-angled jib, and having done so, inch the machine clear of the immediate area. They needed space to work, in order to retrieve the broken chain. This they did and commenced to clear the gum from under the cut. They soon discovered that the cutting chain was indeed compacted and locked under the four and a half feet cut. Undaunted, they set to work with impatience, knowing that the engineer could do nothing until the chain had been recovered, laid-out and the cause of the disruption ascertained! The face of coal had to be hewed and 'holed' for a distance of approximately five or six feet, then painstakingly felled without benefit of explosives. This was time-consuming work, which was partially relieved by the night-shift deputy arriving on the scene and immediately co-opting the six, night-shift packers into the fray.*

*Almost coincidental with the arrival of the engineer onto the face-line, the miners had managed to extract the broken chain from under the cut, and were in the process of laying it out for close inspection. When this was done and fully illuminated by the lamps of all present, a close inspection by the engineer revealed one or two serrated, gouge marks on the underside of the chain, which, in his experience, could only have be produced by 'metal to metal' contact. The gouge marks showed clearly as brighter metal gouges along an already polished surface. "No doubt about that," said the engineer "See for yourselves!" "Bloody unlikely!" said the machine-man. "Did something fall from the machine?" asked the electrician. "Don't look at me!" said the No. 2 man, "I just bloody well work here!" "Get the bloody chain back on!" said the deputy, "Its nearly midnight, and I want this bloody run cut afore the day-shift comes in!"*

*Within 30 minutes, after a few muted expletives from both the engineer and the machine-men, and a few more skinned knuckles, the coal-cutter was once more united with its chain. The cutting-jib was re-engaged in the cut, and, with not a little trepidation, the cutting re-commenced with engineer, electrician, deputy and night-shift gaffer, all in dutiful and nervous attendance. (But, well behind the machine!) The coal-cutting continued on with nary a hitch, until the cut was completed a few hours later. Audible sighs of relief were to be heard from all interested parties as the machine reached the top stable-end, and was switched-of without further incident. The day-shift stripping cycle had not been delayed or interrupted! Unexplained phenomena did not go down well with anyone, especially when the colliery manager demanded explanations!*

*As is the way with unexplained phenomena, it was bound to happen again, just when it was least expected. The following week, on the same coal-face, and during the night-shift, exactly the same thing happened again! This same machine-man, though not actually awaiting a re-occurrence, was again taken completely by surprise. This time, the break was preceded by an extremely loud screeching sound that resembled tortured metals. The machine-man, having not quite forgotten the last episode, acted with spontaneous alacrity and 'killed' the motor. 'Just what the f\*\*\*\*\*g hell was breaking his chain?' He was at once angry, totally dumbfounded, and increasingly frustrated and, at a total loss to understand this turn of events! 'What the hell was happening?' Again, the duty engineer, the duty electrician, the section deputy and the gaffer were sent for! The machine-man was clearly on the point of 'losing it,' but was never-the-less adamant! The reason for these breakages must be explained, and now! This was getting to be a bloody dangerous environment and the two machine-men were being totally unnerved! "Find the cause, else, I'm for up the bloody pit, and so is my No. 2!" loudly exclaimed the intractable machine-man! "Patience, patience" said the deputy, in a most reasonable tone of voice. (He had visions of trying to explain to Willie Hampson as to why the Lower Dysart machine-men had deserted their coal-cutter and come up the pit!)*

*Again, the machine-jib was pulled clear of the undercut. Again, the night-shift packers were co-opted to help down the coal. Again, the broken chain was recovered, only this time it was laid out on a cleared stretch of newly-cut pavement, where everyone could examine the newly gouged welts at, and along the point of break! "That's definitely been caused by metal" professed the engineer. A thoughtful, but suspicious deputy looked directly at the machine-man and slyly asked, "You've no bin using iron paldies - have ye?" "Dinnie be bloody sully!" said the machine-man, feeling heavily insulted! The deputy then turned to the No. 2 machine-man, "You're no stickin steel trees onto the cutting chain - are ye?". . . Longish pause. . . . "Somebiddy wance telt me that ye were a stupit gowk, . . . noo, ah believe it!", said the No. 2 man pointedly. The deputy just didn't know how to deal with such insolence, so, he just closed his mouth!*

*Suddenly, a muffled expletive which broke the silence, was heard from one of the packers who, using his shovel, was attempting to clear the last of the downed coal from the undercut. "What's wrang?", asked the deputy, if only to divert attention from the last voiced remark. "It's ma bloody shovel!", said the packer. "It's catching something, and a cannie lift it!" "It's probably a wee bit aff the chain!" said the machine-man, to which the engineer hesitantly replied, "No! It's no that, it's aw' here!" Slightly puzzled, the machine-man dived into the space created by the shovelled coal and exclaimed, "Whaur is it, what is it that's* *damaging ma machine?" "Look!", said the packer, "It's the tap o' a bloody steel tree!" "What!" said the deputy, looking at the No. 2. "Aye, its a steel tree!" shouted the machine-man, at which point the deputy made a 'breenge' (a rapid grab) at the No. 2 machine-man. "It's the top o' a steel tree richt enough" said a slightly bewildered section gaffer, but how the hell did it get in there?" "Have you been planting steel trees, you bugger?" screamed the deputy, to which the No. 2 vehemently replied, "Whoa the hell ever gave you a 'lamp?" "Just wait a bit!" said the electrician, with his nose against the ground, "It seems to be deeply embedded within the pavement." (Obviously, the only intelligent being on the scene at that time!) "Let **me** have a look!" said the section gaffer, desperately trying to elbow the deputy (who had now released the No. 2 man) out of the way. "God's truth!" said the section gaffer, "It seems to be growing out of the pavement!" The last of the gum was removed from around the top of the steel tree, with the remnants of coal dust being blown directly into the eyes of the nosing deputy, by the blast of air from the machine-man's circled lips. This revealed for the first time, the root cause of the broken chains and the machine-men's mounting anxieties. There, for all present to see, was the roughened and sheared end of a bright steel, two and a half-inch diameter, silver ring of exposed metal, that had once been set by an unknown collier to help support his immediate roof!*

*Those present were absolutely mystified and could do no other, but exchange puzzled glances, without benefit of words. The deputy, a mines official who had been on constant night-shift for years, slowly lifted his head and looked at the section gaffer, with something akin to unbelievable realisation seeping into his mind. "I think ah ken whaur it's come from!" said he, emulating Julius Caesar with clenched fist and extended thumb pointing downwards!*

\*

During the early months of 1949, when the 39 inches of the bottom leaf of the coal had been extracted in this section, seven-foot long corrugated steel straps and locally produced steel 'trees' had been introduced onto the coal face in small quantities. The coal, in both the top and bottom leaves was of such a uniform thickness, that thick-walled, two and a half inch diameter steel tubing, had been cut to a standard length on the pithead, to serve as pit props. The steel trees were obviously much stronger than wooden trees and were infinitely recoverable for constant re-use. They only had one failing, and that was when the steel-drawers went in to recover them! If there had been any down-pressure whatever on these unbending steel trees, they had the inevitable propensity to dig into both the roof and the pavement. *(The direction in which they were forced depended of course, on the consistency of the metals, both above and below the extracted ground!)* If there were lots of steel trees in the wastes, that were deeply embedded, the steel-drawers tasks were then made all the more difficult, in that they had a specific task to do and a limited time in which to finish it!  The upshot of this, was that several of the slightly longer steel trees were left in situ, disguised from the prying eyes of the deputy and hopefully forgotten.  What the deputy's eye didn't see, wouldn't cost the steel-drawer any docked wages!

Naturally, the wastes were packed, but only partially!  Mother nature shall always reassert herself, especially so, with the weight of several, hundred thousand tons of metals above an extracted semi-void.  Within several weeks, the wastes had closed completely and forcefully, but what happened to the upright steel trees still standing in the wastes?  The answer was simple!  They remained standing like sentinels and maintained their stance throughout the subsequent crush to infinity.  Where did they go?  They didn't go anywhere!  They were slowly, but inexorably driven up through the softer coal and then up through the soft blae, without bending!  There they stayed and would have done so, into eternity, had someone not decided to have a go at taking the top leaf of the coal in this section!  The steel trees originally used, had been all of 40 inches long, but on occasions, some of the middle leaf of the coal had been taken, or had fallen, and the colliers had merely stood the 'short' trees on hardwood paldies *(up to six inches thick)* to increase their height.  The distance between the old roof of the bottom leaf and the yet to be exposed pavement of the top leaf, was approximately 40 to 45 inches.  But, after the extracted section had crushed down, the roof and the pavement of the bottom leaf almost became one and the same.  The steel trees still standing, could not be driven down through the very hard rock pavement, but took the only course open to a solid upstanding steel tree.  The hard steel, slowly but surely, poked its way up through the softer coals, eased through the soft blae and barely intruded into the bottom of the top leaf of coal within the seam.  Ninety-nine percent of the undrawn steel trees would not have burst through into the bottom of the top leaf, but those that stood on paldies, or were slightly longer than normal surely did! - and to the detriment of expensive coal-cutting machinery, and to the certain consternation of the night-shift, machine-men . . . *and others!*

Author's Note: - *On the realisation that they had discovered the damnedable reason for the surfeit of broken cutting chains and the fact that it could happen again at any time, and without prior warning, the section gaffer approached the manager with the strong recommendation that coal-cutting in this section should be instantly terminated!  The recommendation was acted on without delay!  Coal-cutting was immediately stopped after only 25 metres of forward travel.  All equipment was withdrawn and the section was completely abandoned!  **Lessons were learned!***

*This was the one and only occasion in Lochhead pit, where the bottom leaf of the Lower Dysart seam had been extracted before the top leaf.  This expensive mistake (blunder) would never happen again, in any other section area of the colliery, under the managership of Willie Hampson! - or Willie Forbes?*

*{Elegant 'purple prose' as used by coalminer's, was sometimes so much more immediately satisfying than articulate meaningful expression, even though its doubtful eloquence did little to convey the guidance needed, nor the explicit directions required! The spontaneous spouting of purple poetry was a necessary steam-release valve for many harassed and trauchled coalminers!}*

\*

The redundant colliers from the 'slopes' were mostly incorporated onto the developing faces down the Lower Dysart Dook, though some of them were transferred to other parts and sections of the New Dook. There were no further coal developments or extractions in this, the Lower Dysart slopes. The equipment was salvaged, the section was abandoned and there was no more development space in this direction, as the workings had come up against the Randolph colliery workings in the West and the Old Duncan pit workings in the East. The remaining colliers in this section would now be used to pursue the coal seam to the dip, as the planners and developers were now actively engaged in doing, in the still relatively shallow depths of the developing Lower Dysart Dook.

The Lower Dysart slopes within the Lower Dysart Section, *(west)* was only ever a relatively moderate, working section, that lasted for approximately three and a half years at most. The extractions were entirely successful, considering the limited potentiality of the area and probably worth the development costs. It was never of course, going to last! It may have been that the section was developed as a stop-gap, until the potentiality of the larger Lower Dysart Dook was fully exploited, and if this was the case, then it worked! The 'Slopes' produced coal in a steady stream, which, when exhausted, was more than compensated for by the increasing output from the larger faces in the ever deepening Dysart Dook and taking the same seam of coal. Much experience had been gained on the 'slopes' testing ground, that would be put to good use down the Lower Dysart Dook!

\*

# NATIONALISATION !?

*When the 'slopes' coalfaces were in full production and the machinery was running well, and when the coal was flowing freely during nearly all of the shift, the drawers at the belt end often indulged in the practise of speculating, as to who would be the first 'stripper' to finish his stint and come through the 'screens' at the belt-end, on his way to the pit bottom. The coal-stripping on the LowerDysart sloping faces was fairly straight-forward at that time, with little or no problems with broken or difficult roof.*

*As on all coal faces, the characteristics, personalities and qualities of the stripper's themselves, varied greatly. Some were young, fast and strong, some were a little older, tougher and just as fast, and some were older and bolder with great endurance, but were slower both in word and in deed! The one thing that they all had in common, was the ability to strip ten yards of standing coal within the allotted time. (If they couldn't, they didn't last overlong on the long-wall, coal-face! Not every coal-getter in the coalmines had to endure working on his knees of course!)*

*The early finishers came from a group of approximately four or five miners, who were men in their physical prime. No individual miner however, had the distinction of always finishing first on a regular basis, but it was usually a miner from this elite group who was first 'past the post!'*

*This particular anecdote has absolutely nothing to do with this group of 'fast strippers', in fact, it also has nothing to do with the 'regular' intermediate group of strippers either! Who it does have to do with, is that small, but regular group of strippers who invariably husbanded their strength and energy, to pace themselves over the length of the shift. This last group usually appeared at the belt-end at around 1-20 p.m. to 1-30 p.m., just when Paw Broon would stop the main conveyor to signify the end of the day-shift. This conveyor could be made to start-up again within minutes, with one or two 'drawers retained', if there was still coal to be stripped on the coal-face and if overtime was to be countenanced!*

*With the appearance of this last, slower group at the belt-end around 1-30 p.m, this meant that the oncost men mingled with the face-miners for the long walk, out through the Lower Dysart mine and the Lower Dysart haulage-way. This walk, more took the form of a leisurely meander than a brisk pace, for the reason that they would be at the end of the 'Tows\*' in any case, being about the last of the shift to reach the pit bottom.*

*\*(Last in the queue to be wound to the surface by cage, before the 'tows' would revert to coal-winding!)*

*This walk was usually accompanied by good, if sometimes mundane conversation, and not at all interrupted by the deadened sound of the endless-rope haulage. This endless rope-haulage would be stopped for approximately three quarters of an hour, to allow the egress and later ingress of miners, to and from the Lower Dysart west! This late group included old Paw Broon, some of the older drawers, the oncost miners from the Lower Dysart Dook and strippers of the ilk of old Jock Sharp, Shuggy Paterson, Wullie Coventry and an ex-German-held, Polish P.O.W. by the name of 'Ted!' This group included youngsters such as 'Tonk' Anderson, Wullie Mathieson and myself.*

*The subject of conversation that day, centred around the fact that government legislation concerning naturalisation in general, had recently been promulgated and was prominently displayed on the Colliery notice-board. This notice informed certain categories of ex- P.O.W's, 'displaced persons' and those that were tolerated 'aliens', that British citizenship could be had after a 'short' examination of determining factors. 'Ted' was married to a Scots girl, had two children and seemed eminently qualified! (Apart from the odd descriptive expletives from his workmates, to the detriment of his character, his Country, his eating habits and his fractured English!) At a point about halfway along the Lower Dysart haulage, after about two or three minutes silence, Shuggy Paterson came straight out with a loud direct question to Ted. "Ted", - he said, "Why don't you get yourself some Naturalisation papers, get them filled in - and we'll organise all of the strippers on the coalface to sign the papers as character witnesses! Then you can become a 'bona-fide' Scotsman?" Silence followed for all of five seconds, then Ted delivered his unique, but unequivocal answer, "Wot is de pont! I mak'e to feel-in de papeers, pays fiftee Pounts, Gets N-A-T-I-O-N-A-L-I-S-E-D !, becomes trews Scotch-man - and yous peoples steel calls me, 'Teddy - de - POLE!".* . . . . .          *End of Conversation!*

## And On - into the Depths!

The development of the Lower Dysart Dook was started almost at the same point as the 375 metres long belt-level, but was driven to the left at an angle of 80 degrees to the level, and down the decline on a map bearing of approximately 150 degrees from its start point. It was eventually driven to a length of 725 metres, dipping down another 240 feet and following the undulating dip of the Lower Dysart seam. This meant that the down-slope was between 1 in 8 and 1 in 10 in this area. The companion dook *(return air way)* was but a short 35 metres to its right side and was not a man-haulage in any sense of the word. The right side of this dook was first developed as a double-sided 'stable-end' *(cut to both right and left of the Main Dook)* in the 41-inch thick, top leaf of the seam, taking only about 5 metres on the left side and an ever deepening intrusion into the right side. The developers in this Dook included miners such as Archie Cooke, Jock Moyes and Bob Ross. They took the downwards developing coal face, now being undercut with an Anderson-Boyes 17- inch chain machine, a full two cuts ahead of the brushing team led by Harry Moodie on one shift and Jim Andrews on the preceding shift, who had the mammoth task of creating a 12' by 9' steel-girdered roadway in following the developers down the newly burgeoning Dook.

The brushing was done in two consecutive stages, where the intervening layer of 30 to 36 inches of redd *(soft blae)*, was first bored and then fired, with the resultant blae *(soft stone)* being used to build and fill, wood and stone pillars on both sides of the 'main-gate', with the heavy remainder being stowed tightly in the left side waste. *(Readers will appreciate that the coal was extracted from the left-side 'waste', in order to accommodate the rendered soft blae!)* The next shift, would allow the following brushers to bore and 'fire' the middle binch of solid coal *(along with some stone and parrot coal),* which would be hand-filled into hutches as production coal. This team would also 'set' the 12' by 9' steel girder, fix the distance pieces and timber the open spaces between the last two girders.

This Dook was probably commenced in the autumn of 1949, in preparation for a new coal section, before the inevitable termination and abandonment of the coalfaces on No's. 1 and 2 headings on the Lower Dysart slopes. As knowledgeable readers will no doubt realise, new developments needed to be up, ready and running, before older faces become exhausted. Forwards planning invariably dictated, that new face development always was an integral part of continued coal production in any thriving coal mine. The top of the Dysart Dook was in fact level, for the first 10 metres, where it then took a dip into the top two leaves of the Lower Dysart seam. This was deliberate, in that there had to be a 'landing-platform' for the five to six hutch races and for the large electric tugger motor, that needed to be installed to one side and clear of the twenty-five and a half inch, semi-permanent rail track, that followed the development downwards.

As the Dook progressed downwards, the amount of coal that was being extracted to the right side, was quickly increased to the extent, that a full face of 140 metres was soon developed and extended to a point, where it almost touched the abandoned 50 metres coalface on the left of the main 375 metres belt level. This coal face was now undercut to a length of 150 metres, to include the main stable end, *(main gate)* but unfortunately, its full length was cut horizontally, but to the dip. It was also slightly wet. I use the word 'unfortunately', for the simple reason that colliers *(strippers)* do not relish having to pick and shovel to the dip, especially when the ground is wet. This makes for heavy shovelling and the face-line acts as a reservoir for standing water.

At a point exactly 95 metres down the Dook and into the coal on the right side, a fully brushed and girdered level, at 90 degrees to the line of the Dook, was taken in to the depth of the coalface, a full 135 metres long. A short time later, when the Dook had been extended deeper, another brushed and girdered level was driven into the coal, at a point 180 metres down on the right side. This level also being driven to a length of 135 metres. Both of these levels, approximately 85 metres apart, ran into a rolling fault that seemed to be about 100 metres long and approximately 25 metres wide, tapering out at each end, back to untroubled ground. At this point, the length of the coalface to the right was shortened to approximately 60 metres in length, ostensibly, to miss and bypass this fault and perhaps force a re-think on how this coal should be taken. *(Extracted!)* The Main Dook then advanced downwards for another 15 metres, where the developers again extended the right side, face-line to 110 metres length and at the same time, drove a brushed and girdered level from the right side of the parallel companion Dook (air way), to the end of the face-line and then slightly beyond. This roadway then took a left turn for about 45 metres, where the road then drove through the bottom side of the fault and into untroubled ground. *(Even though this was not realised at that time!)*

At this juncture, the Dook continued it's downwards passage, with the coalface to the right again being extended to a length that probably exceeded 200 metres. To the best of my knowledge, the face-line was still being undercut at the dip and was still rather wet.

During the summer of 1951, the Dook was transformed from a single track railway with 'tailed' races, into a double-railed, endless haulage-way, with the new haulage, motor-house built in to the excavated solid metals, opposite the head of the Dook. This upgrading had obviously been planned to cater for the expected increase in production, with the new developments that were now in being and the fact that the Dook was now getting deeper and longer, with the inevitable result, that the five-hutch coal-races were experiencing a slower turn-around time.

This method of coal-taking in the upper leaf continued on without change for at least another two years, where there happened an abrupt change in the extraction methods used. The main Dook which was now at a length of 425 metres, was now extended to a length of 575 metres in the same straight line, down through the solid metals, as was the new companion Dook (airway) to it's right, but following behind the face-line as a brushed air-way at probably an 8 feet by 7 feet size. A slightly, sloping level was driven to the right of the companion Dook at the 425 metre point, for a distance of 30 metres and a similar level also to the right, from the 575 metres point for also 30 metres. The two levels were then joined up by a newly, developed coalface, approximately 130 metres long and in the upper leaf of the coal seam. The difference in technique now being, that the newly developed face-line was now on a gradient and lay parallel with the dipping line of the Dook, but with solid coal stoops in between, to protect the integrity of both main Dook and companionway *(return airway)*. The short 30 metres level from the 575 metres point on the main *(coal)* Dook, would serve as the main gate *(loading level),* whilst the short 30 metres level at the 425 metres point on the companion Dook, would served as the return air-way. This face-line was approximately 130 metres long, was machine under-cut on a near level plane, with the machine itself cutting to the rise. The cut face, sloping gently to the rise with each successive, machine under-cut. This method of attack, took care of the small water problem and ensured that it all flowed to the bottom of both the companion Dook and the main Dook. The output from this coalface seemingly commenced in the autumn of 1953, and produced coal for an unbroken period of approximately two years, up to October 1955, when it was halted. The face-line had advanced 650 metres with a broad, face-line length of 130 metres, and, at a height of approximately 43 inches (1.1 metres) of coal. For the want of a proper name, I shall name this coalface, No. 1.A. 1953. *('A' is for above (top leaf), whilst 'B' is for below (bottom leaf).*

Author's Note 1:- *Even by the most conservative of estimates, equating one linear yard of solid coal, standing one yard high and undercut to a depth of four and a half feet, to one imperial ton in weight, this would amount to the staggering total of eighty thousand tons (80,000) of coal from one single production face, over a period of approximately two years.*

Author's Note 2: - *From scientific tests conducted for 'The Second Report on the Coals suited to the Steam Navy', completed in 1849, the average 'weight' of certain Fife coals is given as follows: -*
*Weight of One cubic Foot of coal as used for fuel . . . . . . . . . .    52.6 lbs.*
*Space occupied by One Ton (economic weight) . . . . . . . . . . . 42.58 cu.ft.*
*Therefore, from an average undercut coalface in the Lower Dysart seam, one linear yard in length (three feet), with a four and a half feet undercut, at three and a quarter feet high, would amount to the product of 3' by 4.5' by 3.25' = 43.875 cubic feet. This would therefore weigh, (43.875 multiplied by 52.6 lbs) = 2307.825 lbs, which equates to 1.030 imperial tons, or 1.049 metric tonnes.*
*(One Imperial Ton contains 2240 lbs. One Metric Tonne contains 1,000 Kg's or 2200 lbs.*

*N.B. On the above-named coalface, after about 260 metres travel, the leading edge of a small double-fault did appear and commenced on the low side. It slowly 'travelled' up the face-line in a uniform manner, until it disappeared out of the top end three months later. The second part (trailing edge) of the fault, commenced at the same low side of the face-line perhaps four weeks later and paralleled the first part, following it along the run until it too disappeared out of the top side of the run. This small double fault did not cause any undue delays in the production of coal.*

Approximately four to six weeks before the above face was abandoned, the next full coal-face was already in development. This new face was started directly below that *(further down the Dook!)* of No. 1.A. It too, was approximately 130 metres long. The new coalface also developed in the top 41-inch leaf of coal and was also horizontally undercut to the rising slope. This new coal-face, No. 2.A, commenced output around October 1955, and instantly went into full production along its full face-line. The previous Main-Gate level for No. 1.A. face, now became the top road *(return air-way)* for this new coalface No. 2.A. The Main Gate (coal level) for this new coalface had commenced from the main Dook at the 705 metres length point. This No. 2.A coalface produced coal for a full 23 months uninterrupted period, except for the same fault that had appeared on No. 1.A coalface. This time, the fault appeared only as a single, small jump with no subsequent correction, that started around April '56 and disappeared around August '56. This face advanced a full 665 metres over a 23 month period and produced approximately 78,5000 imp. tons of good coal from the top leaf of the Lower Dysart seam. This coalface was terminated in the last week of September 1957.

*{It was on this No. 2.A coalface, that a certain, newly appointed, mines-official, attempted to cajole the strippers into separating the small amounts of stone from the blown-down coal. (He needed to enhance his somewhat tarnished reputation.) The strippers retaliated by swearing blind, that they could not distinguish the wet stone from the wet coal, irrespective of the different weights! It all looked the same by the light of carbide lamps!}*

Author's Note: - *This Lower Dysart Dook was only driven down to a dipping length of 720 metres. The additional length below the No. 2.A coalface Main-gate, was to house the return wheel and back-balance (on rails), for the one-inch diameter endless-rope haulage system, that ran the full length of the Main Dook.*

Again, before the finish of coalface No. 2.A, another new coalface was also in development. This next face was rather unique, in that the developers went back to near the start of face No. 1.A, but this time, the coalface was developed in the slightly thicker, bottom leaf of the seam, to take only the bottom 43 inches of coal, that included a 1 to 2-inch thick band of stone. This meant that there was a 20 to 25-inch band of solid coal with a thin band of stone and parrot coal above their heads, to act as an immediate unbroken roof, with approximately 30 inches of unbroken fireclay above that, to cushion the broken sandwich of the extracted top leaf. The whole of the top leaf above their heads, having extracted between September '53 and October '55.

The length of this new coalface was somewhat shorter than its predecessor, in that the previous main-gate and top road could not be safely re-brushed to an additional depth of 7 feet below the original roads. This meant that the new main-gate and top road, were to be driven to the inside of the previous two roads that had served coal-face No. 1.A.

This latest coalface in the bottom leaf of the seam, directly underneath coalface No. 1.A, but now named No. 1.B, was approximately 120 metres long with 43 inches height of coal. It was under-cut in exactly the same manner as the previous two faces. This coalface produced coal continuously from it's probable start in February '56, to it's shortened face-length commencing in December '57, when it was gradually shortened down to approximately 80 metres in length before it closed in September '58. The estimated output from coalface No. 1.B would have been in the region of ninety-one thousand imperial tons. *(91,000 imp. Tons).* This was the longest, extracted coalface at 825 metres length on the Lower Dysart Dook, but only by a short head!

Authors Note : *The same double fault again appeared on this No. 1.B face, coming on from the lower end of the face-line around November '56, and disappearing out at the top of the face approximately four months later. Production made have hiccuped, but no change in output was noted!*

Around the months of October and November 1956, the developers had indeed commenced the start of another new face, whilst No. 1.B was still in full production. *(About the same period as the double fault had appeared!)* This new face named No. 3.A, was actually situated immediately above the No. 1.A and No. 1.B coalfaces on the right side of the Dook, and left very little coal indeed between them, to act as a stoop along their respective lengths. Again, it was the top leaf of the coal seam that was taken at a height of 42 inches, and at a face length of approximately 110 metres. The coalface was systematically worked to a depth of 740 metres, over a 24 month period, where an estimated seventy-five thousand, five hundred imperial tons *(75,500 tons)* tons of coal was produced over a twenty-four months, continuous period of time!

Meanwhile, back at coalface No. 2.A, *(the deepest of the three double extractions)* work had been started in October '57, to develop the bottom leaf of coal, not as yet taken. This system of taking the top leaf first and then the bottom leaf approximately two years later, seemed to work very well and safely, considering the fact that approximately 20 to 25 inches of coal, plus stone and parrot coal in the middle of the seam, was left untouched to act as a solid roof, under the 30 inches of blaes fireclay. This No. 2.B face was again by necessity, slightly shorter in working length than face No. 2.A, at about 110 metres long and 37 inches in height. The coalface was steadily advanced to a distance of 675 metres and worked until the end of November '59. This coalface produced in the region of fifty-eight thousand imperial tons *(58,000 tons)* of good domestic coal. *(The inherent single fault did appear again, but as in the past, did not hinder coal production).*

The very last, long-wall, coal-face to be developed down the lower Dysart Dook, was a return to the topmost development to take the remaining bottom coal from underneath the previously extracted coalface No. 3.A. This development, named No. 3.B, was probably started around September '58, with a new main gate and top road cut to the inside of the previous roads, and with a face-line approximately 85 to 90 metres long. It remained in continuous production for approximately 25 months, whence it travelled a distance of 625 metres in length. The height of the coal taken, varied between 40 and 45 inches in natural thickness. This last long-wall face to the right side of the Dook, produced an estimated fifty thousand tons *(50,000 tonnes)* of coal. This was the last coalface to be extracted and exhausted in the West side of the Lower Dysart Dook, and the 'final survey' terminated this last coalface at the end of September 1960. The West side of the lower Dysart Dook now lay abandoned, even though the main coal Dook remained open for a short time and transported other extracted coals.

*(The same two faults that had appeared of coalface No. 3.A, obviously appeared on No. 3.B. Again, the jumps were both only 24 inches in height and did not hinder production.)*

\*

No more coals were taken from the Lower Dysart Dook on the West (right) side at this time, even though an experimental 'shearer' face had been developed from two levels that had been driven into the coal on the East side of the main Dook, at the 615 metres length point. These two horizontal levels were 30 metres apart and driven to a length of 325 metres. At a point on the left level, approximately 125 metres from the main Dook, a small 'shearer' face was developed and experimented with, in the 42-inch thick top leaf of coal. This experimental face covered only a very small area, measuring about 125 by 75 metres and was terminated in January '56. I believe that this small section was used as a 'test-bed' for the modified A.B. 15-inch 'Bluebird' machine, fitted with a vertical rotating drum to its left rear, in place of a low horizontal cutting jib, and that it enjoyed a small measure of success. (*This new innovative 'shearer-machine', was one of the earliest models of its type, having a powered, 30-inch diameter, vertical, cutting drum with a 18-inch wide shearing surface.*)

N.B. *Taking into account the earliest developments on the right (west) side of the Dook, from early in 1950 to the start of face No. 1.A, and the subsequent five additional coal-faces, the grand total of the amount of coals extracted from the Lower Dysart Seam in this Dook, was approximately two hundred and seventy-five thousand, five hundred tons (275,500) of good quality coal, not withstanding the additional forty thousand tons (40,000) from the earlier developments, all in the short ten years of it's life. All accomplished, using initially, a few A.B. 17-inch coal-cutting, chain machines and then later, several of the superior, A.B. 15-inch 'Bluebird' machines, the Mark II, 15-inch high, streamlined, coal-cutters using an half-inch, steel wire rope and flat, rotating, internal drum as it's means of propulsion!*

*These three, double-extracted coalfaces, were the last of their type to be operated in the west side of the colliery. In this type of coal-getting, the undercut standing coal was blown down onto the pavement, where it was manually handled by miners, armed only with picks and shovels. Every pound of coal was physically shovelled, or hand loaded onto the face conveyor belt, by coalminers known as 'strippers'. These stripper's were each allocated a 'place' on the coalface, where their responsibility was to excavate ten linear yards of standing coal to a depth of the undercut, (usually about four and a half feet) and at a height of approximately forty inches. They were also required to set the appropriate roof supports over the length of this place and in a regular consistent fashion. During the course of a six and a half hours shift, they would personally handle around ten tons of coal, and six or seven sets of roof supports. This involved the gathering, lifting and throwing of approximately 2,500 shovelfuls of coal from the pavement, on to the pans. This would be interspersed, with having to lift and hold, a series of seven-foot long corrugated steel straps weighing 90 lbs. (imp) to the roof with one arm, whilst in a kneeling position. The other arm would be used to place the prop in position, prior to its being hammered upright and tight to the steel strap. (If both the straps and the trees were made from steel, then wooden 'paldies' would be placed between them to effect a 'cushion!') In the east side of the pit (within the New Dook), hand stripping would continue to be the main method of coal-getting until July 1961, when the first of the 'power-loading' coalfaces came into being! The demise of the 'coal-stripper' in Lochhead pit, seemingly coincided with the official abandonment of the ubiquitous Dysart Main seam in 1961, even though the manual stripping methods used, had mainly applied to the Lower Dysart seam of coal!*

*The transition from hand-stripping to power-loading did not happen overnight, even though the change did manifoldly escalate through the short period from approximately 1959 to 1964. Slightly before this time and also concurrent to this change, an intermediate method of coal-getting was introduced into one large section within Lochhead pit, around 1956-57. This was the introduction of several of the new A.B. Short-wall, coal-cutting machines, which were used to great effect in the East-Side Mine. This method of coal extraction is fully described in the chapter, - 'Out of the East!'*

*

*Author's Note:- Lochhead pit had a forced air, down-draft shaft, where fresh clean air at great volume and pressure was generated and constantly supplied, by a great 12 bladed, 18 feet diameter, electrically driven, vertical fan. Double sets of air-doors on the pit head, and the 'Low doors' underneath the pithead, ensured that the greater volume of clean fresh air was forced down the 63 fathoms deep, winding shaft. To help channelize and direct the through passage of underground, air circulation, the construction of 'air-doors' in critical roadways was a bounden necessity and doors were invariably constructed and installed in tandem. They were used extensively in underground road-ways, to divert the flow of air into a pre-determined channelled distribution. They were not often visible and many miners were completely unaware of their existence, even though they did realise their significance and purpose. The doors were usually constructed to completely straddle a roadway in such a manner, as to negate (as far as possible) the passage of air through that particular road, mine or tunnel. They were also built in tandem as pairs, or even triples, depending on the degree of insulation required, and their purpose was probably fourfold. 1. To divert the air-flow so that it did not short-circuit the planned route, thereby causing a confused or inappropriate air circulation in deeper and more distant sections. 2. To allow direct access to a section or working, that would otherwise involve miners travelling a longer more circuitous travelling route. (The use of double, or triple air-doors allowed the road to be regularly traversed without interrupting the main air-flow.) 3. To completely bypass a worked-out or abandoned section, or an underground area, thereby maintaining the continued air-flow to other parts of the mine. 4. To give miners quick access to better quality air, if so desired. (Or even use the passage as an escape route in an emergency!)*

When the exact location of air-doors was decided, the usual method was for the resident, underground 'carpenter cum brickie' to come on the scene. He would first construct the heavy timbered door frame, and then build the surroundings brick walls, tight to the roof and the side walls, using wet cement to ensure a close, plastered seal all around the construction. The door frames would be constructed to a uniform size to allow the passage of hutches on rails, whether their use was intended or not! The horizontally, swung doors would be fitted when the brickwork was thoroughly dry and would always be fitted to the windward side of the air-flow, to ensure that there was no likelihood of an accidental opening. Each individual door, whether they were built in doubles or triples, usually had an in-built, sliding trap-door, built into the middle part of the door, which, when slid open, would reduce the air pressure on the windward side, thereby facilitating an easier entry through each door in turn! However, no matter how good the tradesmen were at their profession, such was the volume of moving under-ground air, that there was always some small leakage through the doors. The very strict rules that applied to the passage of miners through air-doors, was, that only one door at a time should be opened and then properly closed, before the single opening of any subsequent doors in the series! Exactly the same rules applied, even when a miner was transporting materials or wheeled bogies through the double or triple doors. It therefore, can be positively assumed, that there was always a respectable 'parking or resting' distance between any two or three individual air-doors.

*One unwritten rule that was never 'posted-up' in the area of air-doors, but which was passed down by hard empirical knowledge. "Never turn your back on a closing air-door! You might just be pushed onto your face!"*

\*

# Chapter XXII

## The Road to Dusty Death!

## Stone Mines & Developments
### – *The Silicosis Maker!*

A Mine, is also the name given to an underground tunnel driven through either solid rock, a mixture of coal and rock, or sometimes solid coal. The strata that the miners would encounter could be hard, soft, brittle, firm, tough or even loose. It could turn out to be a little of everything! It could also be either wet or dry. They may have to drive with the grain of the metals *(strata), i.e.* following the line of a seam of coal *(or solid stone)*, or across the metals in a horizontal line, either upwards or down-wards in a given direction, previously determined by the colliery's resident, underground, mining surveyor. A horizontal tunnel can be termed a Mine or a roadway until it is completed and then may take it's name from the use that it is put too! It may even become a haulage way and be re-named.

A tunnel or roadway that is driven in an upwards direction, can be more readily described as a Heading, whilst a tunnel that is driven in a downwards direction, immaterial of what strata it is driven through, is invariably called a dook, or sometimes a dipping. If a tunnel runs directly down from the surface of the ground and usually in a straight line, it may be called either a mine, a drift or a dipping, or a combination of any two! If a roadway is driven downwards from an underground mine or level, it is invariably called a dook and may be named for the seam of coal that is to be extracted at its destination. Sometimes, the roadway is named after the miner/contractor who initiated the contract. The driven width and height of the mine, heading or dook, will depend on the planned use to be made of the development. If it was an experimental probe, of which there were many, the development would be of the most economical, perhaps an 8 feet wide by 7 feet high mine, with single track light rails and perhaps only being worked on one shift in three. If the development was in earnest with a positive aim in mind and the plan was to cut into new known reserves of coal, then the development work would proceed at a pace which involved 24 hours continuous working on three, consecutive shifts, for seven days of every week, until completion. The chances are that it would also be driven at 12 feet by 9 feet or 11' by 8' or possibly 10' by 7', and steel-girdered along its complete length.

At the period of time of which I write, there was in Lochhead pit, four large dooks of varying depths. I use the term large, not because these were cavernous, but simply because they had been driven at a width and height that was either 12' by 9' or 11' by 8' and steel-girdered accordingly. They were by name: - Nicholson's Dook, the Old Dook, the New Dook and during my time at Lochhead pit, the Lower Dysart Dook. In the case of both the Old Dook and the New Dook, the companion Dooks were laid with a single set of semi-permanent rails *(25.5 inches in gauge)* down into nearly their full depths, *(lengths)* to accommodate the long, passenger-carrying bogies, which were lowered and raised by customised, electric, haulage motors. It would have been too much of a debilitating effort on the part of the miners, to have traversed the length and steep gradients of these dooks on a daily return basis, hence the pressing need for suitable conveyance. These companion dooks were then described as manhaulages, and the passenger bogies were a permanent feature on the steel rails.

*{Nicholson's Dook, previously driven down through the Dysart Main coal in 1917, and located at the innermost end of the Lower Dysart 1000 yards Haulage-way, (old Pit-bottom level/west), did not have a companion dook! The reason for this, was that the driving of Nicholson's Dook was actually an after-though! The section area that the dook was driven through, had previously been extracted from 1910 to 1914 and allowed to close! The dook was then driven as a 'heading' coming up from the Old Dook's No. 1 level, through this crushed-down section, to cut into the Pit-bottom level (west). A distance of approximately 275 metres. This was done to facilitate the transport of coal from the head of the Victoria Sea Dook, along the Victoria level, up Nicholson's Dook and thence to the pit-bottom via the pit bottom level! (Later known as The Lower Dysart haulage-way!)}*

Of the four dooks and their companion-ways *(usually return air-ways)* that had been driven in Lochhead pit at that time, three of them were driven down through the Dysart Main seam of coal and only one of them down into the Lower Dysart seam. The dooks by name and development, are fully described in detail elsewhere in this book, but sufficient to say at this juncture, that mining through solid coal is a whole lot different in most respects to driving stone mines through the solid metals!

When a dook is driven for the express purpose of being the main route for the transportation of the out-going coal, usually within and following the grain of the seam, *(as was the case with nearly every dook ever driven in Lochhead pit)* there is invariably a companion dook driven parallel to the main dook and separated from it by a fixed distance. *(Anything from 100 yards to 250 yards apart.)* The Main Dook and its companion dook is 'brushed' immediately behind the extraction of the usable coal.

The term 'brushing' means, that the 'metals' that lie immediately above and below the extracted coal, are expertly removed, to facilitate the erection of a steel arched girder of a pre-determined size, usually measuring twelve feet wide and nine feet high in the Main Dook, and probably eleven feet wide and eight feet high, in the companion dook. The brushing waste from both the pavement 'lip' and the roof 'overhang', is then stowed in 'packs' built into the spaces left by the extracted coals!

The Lower Dysart Main Dook was driven or 'brushed' ahead of the line of the actual long-wall coal face, using 12' by 9' steel girders. This main 'stable-end' was usually 'cut and stripped', a full three cuts ahead of the main face-line, *(but usually at the same time)* using a separate, coal cutting machine! The brushing work in the main-gate *(stable-end)* on the next shift, which always followed on from the stripping, was always exactly two full cuts behind at the daily completion of the stripping. *(This dook was initially stable-end, intended to be an 'advancing' long-wall face cut to, and at the dip!)* The angle of depression was approximately sixteen degrees down through coal and rock, and the 42-inch thick seam of coal *(which comprised the top leaf)* was cut and extracted in the main stable-end, a full two cuts in front of the brushing at daily completion. This was also a wet coalface, which meant that water accumulating on the sloping run, eventually graduated to the bottom of this main dook. The back-shift half of the six man brushing team, had to drill and 'blast up' the middle two and a half feet of pavement *{stone or blae (redd)}* below the 42 inches of previously extracted coal *{and above the middle coal leaf of seam}* at a width of 13 feet, to effect an advance of four and a half feet. This stone/redd waste was then manually shovelled and 'packed' into the side-space vacated by the extracted coal, to form wood and stone pillars built-up on both sides of the erected girders. When this was done with all of the *redd* removed, the next brushing shift *(the following night-shift),* would commence with the boring, firing and lifting of the bottom binch of the roadway *(actually the middle coal leaf of the seam along with the parrot coal and the 4 to 5 inches of stone underneath)* to the full width, depth and overall 'height' required, to take the next steel girder. This coal would be manually filled into hutches and become part of production output of this dook.

The experienced, senior brusher would then prepare the walls, the roof and the pavement *(with a 'holing' pick!),* for the bedding-in and erection of the next 12' by 9' girder. This girder came in two identical halves and was affixed at the top centre by two 'fish-plates', each of 12 inches long and having four spaced bolt holes. It would take the combined efforts of all of the brushing team to position and hold these two half girders together, whilst the team leader deftly held both fish-plates in position and matched the elusive holes! *(If there were any colourful expletives in the vocabularies of the miners concerned, this would be the most propitious time for their vociferous release!)* One nutted bolt was usually sufficient to initially hold the joined arcs together, thus releasing one or two shovellers back to their own efforts. When all four bolts were in and tightened, work would continue with the addition of distance pieces placed between each succeeding set of girders, with the spaces between steel and roof, being closely and tightly packed with light pillar-wood and heavier wood packing. The idea being, to pre-empt any possible loosening and subsequent falling of any *heavy metals*, onto unsuspecting heads!

Stone miners differ somewhat from 'brushing miners', in that the former would actually take the driveage through untouched and virgin metals, even though both worked with stone or blae's! These 'metals' were not usually comprised of 100 percent coal. *(That sort of development did not employ stone-miners, merely developing collier's!)* The advancing face of a stone-mine could have been 100 percent stone, or a mixed 'sandwich' of coal and stone! If the road to be driven contained a middle sandwich of coal, then the task of the stone miners was made easier by the extraction of the softer, coal sandwich first, followed by the blasting up of the open pavement 'lip' and the blowing down of the stone overhang! *(Explosives always performed better if the 'blasted material' had an open space to occupy!)*

If the roadway was to be cut through solid metals, *(i.e. hard stone)* the miners then had unenviable, painstaking and time consuming task of having to bore holes in solid stone, to facilitate the use of explosives!

If the miners had access to either a compressed air or an electric drilling machine, then the only impedance to the hole-boring, was the need to change 'drilling-bits' rather frequently. If there had been no power source available, then the miner's were faced with the need for one of their number to spend nearly the whole of one shift, operating their diabolical hand-boring machine! *(See Below!)*

Personally, if I were to rate miners according to their experience, knowledge, diligence and work load and capacity for sheer endurance, I would rate the stone miners, the brushers and developers, as the hardiest of a sturdy bunch of men. By hard, I do not mean outward bravado, I mean the quiet mental hardness needed, to endure the daily grind of working with sodden redd on rough pavements, *(a shovelful of wet redd weighs fully twice as much as the equivalent shovelful of coal),* working with heavy steels that needs carrying and manipulating and quite often, not in the freshest of air. New developments sometimes needed to have a forced air circulation introduced into their extending lengths, by additional air-bags and an portable electric fan positioned in an suitable main air-way. At times, this re-circulated air was part-fouled, for the simple reason that the original air that comes down the intake shaft of any coal-mine, gets progressively more foul and polluted as it travels through various workings. The greater the coal-getting activity and the increased amount of 'shots' being fired, consumed great quantities of oxygen from a steadily deteriorating air supply.

*

# A Sometime Incongruity!

*In some of the driest of coalmines where airborne dust was at a premium, the powers that be, would often send large, triple-layered, paper bags of heavy, white, stone dust into the sections. This stone dust was milled into a fine powder and had to be kept very dry. It was economically scattered by handfuls over the walls and roofs of the tunnels, where men worked and travelled. This may seem a little strange to the uninitiated, to actually send more dust into a potentially dusty atmosphere, but the underlying theory behind this so-called 'madness', was that the stone dust being heavier than coal dust, would 'lay' the coal dust thereby rendering it immovable? This actually did work and men would be enthusiastically encouraged to engage in this 'knock-about' chicanery without being told. It did have it's lighter side, in that some of the participants got just a little carried away and eventually came up the pit, a good deal 'whiter' than they went down.*

*It did also have a slightly darker side at times with it's misuse, which when perpetrated by stupid people, could result in severe discomfort to the victims. This was when some unthinking idiot with a warped or malicious sense of humour, would obtain a shovel full of stone dust and trickle it into the air-intake of one of those portable electric fans, which were used to supply a moving air current into a one-way, developing roadway. This stupid action would send choking clouds of white dust through the air bags and into the face of the single road development, forcing quick action on the parts of the miners. It is not the first time that I have been caught in this situation and was forced to flee the heading, gasping for air and coughing-up the effects of the dust. I do remember that it had a particularly vile taste when swallowed and needed copious amounts of water, to dislodge it from parched and burning throats!*

*

# An Infernal Machine!

Before the introduction of compressed air, or electrically-powered boring/drilling machines and tungsten steel, drilling bits, the stone miners and brushers were forced into using one of the most diabolical, hand-operated machines ever devised. The miner's called it 'The Rickety!' This machine differed from the lightweight 'poker-machine', that was used exclusively for the boring of holes in coal! The 'rickety' *(rachet)* machine was designed to be used for the drilling of holes in stone!

The machine itself could be broken down into several parts, all of which were inter-changeable! The drilling part of the machine, comprised of different lengths of one and a half-inch diameter steel augers, having clock-wise, helical, double-twists of approximately 3 or 4 full twists over 12 inches in length! These augers were absolutely straight and were supplied in one, two, three, four, five and six feet lengths. The front end of each auger (drill!) was vertically drilled at five-sixteenths of an inch in diameter and to a depth of 3 inches, and was also cross-slotted, to accept an inter-changeable, tungsten-carbide tipped, drilling bit. (Which needed to be ground and sharpened every day after use!)

To actually bore a hole in solid stone, the machine needed to be fitted together with the drill and bit placed against the face of stone, while the rear end of the device *(which did not rotate!)* was placed against a very solid object, which must not move under the drilling pressure. This was sometimes achieved by the use of a long wooden tree, placed between the roof and the pavement, in such a way that sloped forwards towards the pavement. The rear end of the 'rickety' was held tight against the tree, whilst the fitted ratchet handle was manipulated sideways, to pressurise the drill a quarter turn with each stroke and into the solid face! This was a long laborious process, that entailed the 'driller' having to ratchet the handle through four separate quarter turns, to effect one turn of the drill. The driving threads on the machine were so designed, that it needed one complete rotation of the auger handle to drive the drill-bit - one eighth of an inch into the solid stone!

As each section of the drill reached its maximum depth, the device would then have to be dismantled, with the shorter drill being removed from the 12-inch depth hole! The next longest drill would then be inserted into the hole, the device would then be reassembled and hand-boring would be re-commenced. This would happen repeatedly until the maximum length of hole had been drilled! Also, considering the fact that it needed four pulls on the ratchet handle to effect one complete turn, it is not surprising that it usually took a very fit and strong miner, up to 30 minutes, just to bore one, five-foot deep hole! A solid 'brushing face' usually needed from 6 to 10 such-like holes! *(Just also imagine what may have transpired, if the holes were found to have been drilled in the wrong place!)*

The boring of these holes was usually a shared task, in that after the drilling of just one hole by any one miner, *he would then probably have to revert to the 'comparatively lighter' task of shovelling redd, in order to take a breather, before his drilling turn came round again!*

\*

Author's Tailpiece: - *It was a fact, that each team of stone miner's and brushers had their own personal 'rickety' set! This would be contained, in either a heavy canvas or leather carpenter's bag and when not in use, would be 'secreted' in one, or other parts of the coal-mine, not usually accessible or visited by other miners. However, at times, hidden sets of drills would 'disappear' without trace and a 'contract' would be delayed until another set could be found! Owner's sometimes went to great pains to file secret and personal identification marks on every item in the set, and also tried to ensure that the distinctive marks were not at all obvious to the casual onlooker! 'Twas not the first time that Old Harry had to spend the better part of his borrowed or brushing shift, looking for his stolen set of drills!*

\*

# Chapter XXIII

## Dauvit's Progress IV
## Dauvit is Still Going Forth!
## *– and on to 'PROMOTION!'*
## *(without extra pay)*

At the top of the Lower Dysart Dook which was situated at right angles to, and only a few yards from Paw Broon's belt end, the full hutches of coal were delivered onto a short level standing, by means of a large electric tugger motor, winding-in a steel-wire rope that pulled a race of six hutches up the dook on a 'static tail', affixed to the hooked draw-bar of the leading hutch. The wire rope to which the tail was attached, disappeared down between the rails and around a horizontal return wheel, before being directed and 'booked' onto the winding drum of the haulage motor. This meant that Jock Scott the motorman, was situated in a position immediately adjacent to the first hutch in the speeding race, as it breasted the lip of the short landing-level. He was therefore in an ideal position to quickly de-clutch the motor, throw on the drum brake and de-tension the 'tail', whilst the forwards momentum of the speeding race was realistically maintained. In the space of a few feet *(and in milli-seconds),* the tail-puller had to move rapidly sideways, and synchronise his pace to that of the speeding hutches then smartly 'pull the tail' before the rope disappeared under the pavement. Sometimes, the tails-man 'missed the tail' and as a result, the coupled rake of six hutches (later seven) performed a spectacular cartwheel and scattered their contents both far and wide, with the belt-end drawers 'legging-it' rather rapidly. It must be said though, that I have never seen this clean-up operation take more than nine or ten minutes, with all of the drawers quickly mucking-in with shovels and even the odd bass broom, just to keep the 'slew' plates clean. I mention and describe this operation at this juncture, for the simple reason that I was 'promoted' to be the replacement 'tail-puller' in place of the character whom I shall next describe.

Tayo, and I never knew his surname *(I don't think anyone did, except the manager and pay-clerks!),* was an German ex-P.O.W. who, according to some, overstayed his confinement in this country, and decided that life in a Scottish coal mine, was infinitely preferable than being classified as a Displaced Person. I do not believe that any of his work mates ever discovered just how and why, Tayo managed to stay in this country after the end of W.W.II. He spoke the German language with a thick, guttural accent that I now know to originate in the northern part of Schleswig Holstein. He was a strong, well-built and loud, farmer type of mature man, who swaggered with all of the confidence of the Aryan race. He was a very likeable person as long as he did not have to answer any pointed questions about his particular part in W.W.II. He then became rather morose, bolshie and bad-tempered, and tended to revert to his native German language, whilst mumbling loudly to himself.

There were three men employed at the top of this Lower Dysart Dook. Jock Scott, the motorman from the Coaltown, Tayo, who to the best of my knowledge lived in the miner's hostel at Muiredge and an oncost miner from West Wemyss who, if he had been allowed to, could quite easily have quickly become my underground nemesis. *(He was of less than average height, suffered, I am sure from bad feet, and walked as though he had 'Duck's disease. At least from the knees down. He wore dentures, or at least I thought he did, he just didn't wear them underground! He gave the impression that he was well educated, at least in the rough college of life, was well read, or as much as reading the daily Mirror would illuminate anyone, and he was very rarely lost for words, even if he had to make them up! There was nothing nor any topic that he would not opine too, then simply walk away after having done so, as to effectively have had the last word on the subject. I was not 'allowed' to have any opinions in his presence, nothing that I said or commented on, would carry any weight with him, and he was forever reminding anyone who would listen, that the education system in Fife was not what it used to be! He would then suck his cheeks tight together, flail his extensive and protruding lips and make loud repeated 'clucking' sounds. He was, without doubt, an 'unusual specimen' of West Wemyss manhood! His nick-name was 'Dod!')*

It did not pay to get on the wrong side of Teutonic Tayo, he had the most disconcerting and vicious habit, that if someone asked a question that he did not like, or if he felt that his work-mates were discussing him, he would 'pinch' the biceps of his 'tormentor', using his half-folded index finger and thumb with such a forceful pressure, that it would raise an immediate and very painful welt on the contact point. I have seen his victims actually cringe and cry out in suppressed pain and anger, at his cavalier treatment. No matter the amount of times that Tayo was warned by Chairlie Sinclair against these friendly 'assaults', Tayo would at first show genuine contriteness and then swagger away head on chest and quietly chuckle to himself. The person was completely incorrigible, *but he did pull good tail!*

One Monday morning, I was asked by Tom Coventry *(the dook fireman)* if I though that I was capable of 'pulling the tail'? I must have said 'Yes!', as his next comment was, "Well, lets see you try!". Tom Coventry then had a quiet word with Jock Scott and the impression I got, was that he wanted Jock to bring the next race over the top at the slowest possible speed, so as to give me as much time as possible to make the attempt. I had seen Tayo perform this task many times, and had often wondered as to why he did not actually lift the broad last link over the top of the draw-bar hook, as opposed to just twisting it off.

I reasoned, that my best action was to stand sideways, bending forwards with my long, right arm extended downwards, but with my left arm held horizontally to catch the top rim of the on-coming first hutch, thus helping me to retain my balance. As the six-hutch coal-race came over the lip of the dook and raced towards me, my left hand stretched-out to await the steel edge of the leading hutch of the race, I started to side-step rapidly with long strides, at a speed to match that of the moving hutches. I then firmly grasped the broad last link of the chain with my right hand and paused momentarily, for the feel of an distensioning tail. It happened suddenly, and I quickly lifted the last link clear of the draw-hook and hurriedly dropped it! Success! My first tail, and I was sweating cobs! I distinctly remember starting to breathe again. Jock Scott didn't say anything, he merely grinned in my direction. There was many more 'tails to pull' before the shift was over.

I did discover later, that the drawers *(putters)* had started a small pool as to how many 'misses' I would have, in my first ten attempts.

Within the depths of this dook, and at right-angles to the dook, the coal that was being taken at this time was the top leaf of the Lower Dysart seam which was approximately 42 inches thick with a 1 to 2-inch, thin band of stone in the middle. This coal extraction was by means of a long-wall, undercut face approximately 120 yards long and being undercut to the dip! *(This was also a wet coal-face with water seeping in through the roof.)* This meant that the distance between the coal-face and the head of the dook, did slightly increase on a daily basis, thereby marginally increasing the delivery time between full races at the head of the dook. In order to overcome the increasing need for empty hutches at the bottom 'loading' end, the original five-hutch races had been first increased to six hutches, and then seven hutches, which had the effect of negating the slowing down process of the races as they came speeding up to the lip of the dook. Previously, the five and six hutch races could be slowed down as they came over the lip, in order to give the tail-puller a fair chance at the pull, but now, the risks of a missed tail intensified as the speed over the lip needed to be maintained, in order to minimise the risk of the race/rake stopping and running backwards, for the reason that the last two or three hutches were still on the upwards slope at the moment the tail was drawn. I must admit that the percentage of tails missed did increase slightly with seven hutch races, but this soon dropped with my increasing dexterity and confidence in the abilities of the sometimes apprehensive motor-man.

During the period when these dook races had increased to seven hutches, brushing work had been carried out in the solid rock opposite to the dook landing, where the mine management planned to install a large, electric, haulage motor and mechanical gearing, to convert this dook from a single track 'tugger' brae, to a twin-railed, endless rope haulage. This brushing work had been carried out on both back-shift and night-shift and had progressed rapidly, with the complete excavation being completed in a few short weeks. The large, electric motor and becander gear being housed in an 14 feet by 10 feet girdered and concrete cage, with the haulage motor foundation fastenings embedded in the fabricated, concrete pavement.

This haulage motor was installed in quick time, coupled up and readied and we oncost men *(and youths!)* were informed one Friday morning, that when we returned to work on the Monday morning, I would be a 'clipper' and not a 'tail-drawer', and that Jock Scott the motorman was out of a job and would be transferred elsewhere.

*During that weekend, the road makers (rail-laying miners) and rope splicers were out in force and working on a two shift rotation. (i.e. One twelve hours shift on and twelve hours off.) There was also, many oncost workers needed for the lifting, carrying and positioning of the heavy, semi-permanent steel rails. The electric haulage motor was already in place and bolted down and all it needed was the final coupling up to the High voltage A.C. panel boxes. The work started on the Friday back-shift and proceeded non-stop until the following Monday morning, where the haulage had already been run and tested and pronounced ready for operation. I do recall that two of the brothers Gibb were involved in this operation, namely Laurie and Bob. Rope splicers and roads-men both.*

*

*At the time when the Lower Dysart Dook race increased from five, coupled hutches to six, coupled hutches, there was some small grumbling from the oncost 'coal-drawers' on the near adjacent belt-end level, in that they felt that the addition of the extra hutch, meant an added burden on them. (When I took over as the dook tailsman, the dook's exclusive drawer (putter) had been withdrawn for some reason, and the provision of empty hutches fell to the belt-end drawers, along with myself when I wasn't 'pulling tail!') When the dook race was further extended to seven hutches long, with the 'landing' slightly extended, this grumbling was renewed with a quiet request to the deputy for an extra drawer. The request was first mooted as a suggestion from the men, but was greeted with sarcastic disdain by Chairlie Sinclair. The rumblings soon became more progressive with some dark mutterings about a possible 'go-slow', to make the section's officials sit-up and take notice. I began to feel just a little guilty about the whole thing, simply because they began to exclude me from their mutterings (small conspiracies?), which gave me the impression that they thought, that I was not 'drawing' my share of the required empty hutches. I had previously been* warned by Tom Coventry, (Dook Deputy) that I must be always standing ready to 'Take the tail' as it arrived, and not be the cause of the motorman having to 'kill' the motor if I was elsewhere! As a result, I had to stand fast and be ready to deliver, when the motorman rang the warning ball!

A few days later when it became obvious to all that no new drawer(s) were forth-coming, it also transpired that the stripped (extracted) coal coming down from the 'slopes', was being delivered fast and furiously to the 'belt-end', with the result that the drawers were having a hard time keeping up with the increased frequency of the 'loading!' Old Paw Brown could not really control the great surge of coal that was being delivered from the belt-end to the waiting hutches, and much was being spilled onto the pavement underneath the steel rails. As a result, the 30-inch wide conveyor belt was being deliberately stopped every 10 minutes or so, to allow the scattered and heaped coal to be shovelled back into the hutches and practically, to give the overworked drawers the respite that they sorely needed. They were being massively overcome!

Needless to say! Chairlie Sinclair came storming out through the 30-inch belt level and up to the loading point, shouting at everyone in general and particularly to Paw Broon, demanding to know why the conveyor belt was being stopped so often? (Never before, were so many rhetorical questions, answered by so few dumb replies!) Chairlie Sinclair very rarely listened to answers! He was an action-man! Everyone, except the sweating 'coal-fillers' merely stood and looked away, with Paw Broon ineffectually trying to explain the impossible. Chairlie just stood there, arms pumping up and down, "Get that bloody belt started! - and 'KEEP IT GOING!'". At that precise moment, I was 'standing-to' at my allocated 'Tail' position, dutifully awaiting the imminent arrival of the next rake, but observing events with studied interest. I was standing upright with hands on hips, still watching this 'heated' exchange, when Chairlie came onto the head of the Dook, took one look at me, stopped and shouted; **"What the Hell are y-o-u staring at? Get ON with your WORK!"** At this juncture, and without moving my feet, I bent forwards from the waist, stretched my left arm out to my left, lowered my open right hand to near ground level, and still looking at Chairlie, grinned broadly!! Chairlie stood for about five seconds, grimaced, turned on his heel with the loudly spoken throw-away words; "Another f -f - fuckink comedian!", and then, to the oncost drawers "No! You're not getting another bloody drawer!"* - and then stomped of, rapidly disappearing in the direction of the 'Slopes coalfaces'.

They did eventually get an extra oncost drawer! Jimmy Frew, the section gaffer quietly saw to that!
* I do believe that 'drawers' are known as 'putters' in other coalfields, especially in parts of England.

As I left the section that Friday afternoon, I determined that I would make a point of getting to see David Black, the Old Dook 'clipper' in action. I had previously obtained permission from Tom Coventry, to arrive early at the head of the Old Dook on the pit bottom, to perhaps have a go 'at the clipping', with David Blacks' permission and guidance. I explained to David as to what I wanted, and was invited to 'take' the next full hutch coming over the brow. I waited in keen anticipation at my intended quick and efficient 'taking' of this first hutch, and the thrill of sending the waiting empty over the top and down into the depths. *(But, under control, of course!)* I ended up being rather embarrassed, I obviously did not have the knack nor the expertise at the time, and suffered the indignity of preceding the still unclipped hutch to it's 'stops', where Davie Black was forced to quickly signal the motor-man *(John Burns)* to stop the motor, in order to release the now very hot clip. *(The powerful haulage motor does not 'feel' the braking effect of one locked and captured clip, even though the white metal lining of the wheel-clip quickly becomes red hot.)* I sincerely hoped that I would show a damn sight more expertise at the top of the Lower Dysart Dook, the following Monday morning.

Monday morning came, and I arrived at the top of the Lower Dysart Dook in good spirits and with generous measure of intrepidation, what was a few 'stiff' clips between friends? I need not have worried! The wheel-clips that had been brought into the section, were the oldest and most worn, from both the Old Dook and the New Dook, and some of them had come direct from the blacksmiths shop on the pit-head, where they had been re-lined. The one common factor that pleased me, was that all of the operating threads and screws had been well used and were therefore, not liable to be stiff. This was an advantage, in that the first thing I did, was to clean the threads and screws and give them a good oiling, to ensure a smooth running action.

To start the system running, I was instructed by Tom Coventry to send at least ten empty hutches, each on a single clip, down the dook, this to maintain a constant reserve to the loader working at the bottom belt-end. No more empties needed to be sent down, until the first loaded hutch had been 'clipped-on' at the bottom of the dook. The haulage would continue running non-stop except for emergencies, as there was no resident motor-man allocated to this motor-house. In the absence of an attendant, there was a long, floating wire running above my head, affixed to the stop/trip-switch of the haulage starter-motor, so that missed or jammed clips could be disengaged from the wire rope with safety.

This working operation was designed, so that the clipper at the bottom of the dook, would not be overwhelmed with empty hutches, with nowhere to put them. This was carefully regulated by a dedicated signalling system. The volume of full hutches coming up the dook soon intensified, but, with a spacing of approximately 30 metres between arrivals, I had plenty of time to unclip the full hutch, transfer the clip, and then sent the waiting empty hutch on it's downwards journey.

I soon grew very adept at the practice of 'clipping' and began thinking of ways to develop and hone my increasing skills. To this end, I found that I no longer needed to tighten the down-going clips with the provided tool, but merely gave the wheel a quick last jerk with a strong right hand, which was more than sufficient to clamp the vice-jaws. Tom Coventry didn't like this very much at first, but soon acquiesced after practically testing the tightness of a half-dozen clips or so! This continued for a short time whilst I thought up another idea to boost my expertise. *(Most of the miners around me wore good, stout, well-made, heavy duty, leather boots with a steel toe-cap and studded soles and heels. Some boots even had moulded steel heel plates.)*

With regard to the landing area at the top of the dook, I had previously obtained permission from Jimmy Frew, to use the incoming 2" by 2" by 3 feet long pillar-wood 'sticks', to build a flat wooden walk-way between the sets of rails and the side-walks, to stop the likelihood of my being tripped-up by the protruding rail sleepers. This then, was a near perfect platform on which to work, but meant that I was liable to slip on the solid, dry, level, wooden floor that I had created. I solved this small problem by purchasing a new pair of lacing rubber boots, *(with covered steel toe-cap),* which gave me a new lightness of foot and a non-slip environment. *(They also encouraged sweaty feet!)* I then got an idea that simply suggested itself to me, that, after hooking the loop of the clip on the back end of the empty hutch and dropping the jaws onto the rope, I could use the ribbed sole of my right boot to run-up and tighten the clips on the down-going empty side. I experimented with this, and found to my intense delight, that these new boots gave me the 'feel' of the clip and the dexterity of touch, in that I could run-up the wheel much quicker than by hand, and with more clout! I simply held on the empty hutch with my left hand for balance and 'toed' the wheel to the maximum and with a quick hand grasp, ensured the tight grip of the jaws of the clip. This unorthodox action quickly developed into a drill, whereby I no longer checked the tightness of the clip by hand, but merely 'stood' on the top of the wheel with one booted foot and pushed forwards, using my 13 stones of body weight to ensure tightness. This did not go down at all well with Jimmy Frew, who demanded I use the issued clip-key to ensure final tightness. He was quite visibly surprised to discover that in spite of his heavier weight, he could not gain any advantage with the key, after my cheeky invitation for him to show me how to do it properly!

Around this period of time, where my dexterity with my boots no longer raised any anyone's eyebrows, there came into the section for the first time, a new official whose name was Andrew Wilkie. He was not much of an imposing figure, merely standing about five feet six inches in height, with a somewhat average build and dressed in a clean, but faded boiler suit and with a deceptively soft voice. I did not know who he was at the time, but Jimmy Frew did seem to defer to him a little, whilst both Messers. Sinclair and Coventry disappeared from view at his initial arrival.

This was the new under-manager then, whom Jimmy Frew, polite as always, addressed him as 'Andra!' I had very little experience of managerial officials at this time, but instinctively experienced a slight animosity towards this new under-manager. I also wondered at the time if he was 'well-known?' This Andra Wilkie didn't raise his voice very much, but did seem to spent an inordinate amount of time just casually hanging on to the nearest girder, eyes probing everywhere and watching ever so intently! He observed much, but said very little, at least, not to the oncost men at the belt-end. Very soon, rumours began to spread around that relations between A.W. and the miners, *(brushers and face-men)* were not as amiable as they could be. Brushing miners of long standing were annoyed that this new under-manager was seemingly questioning their skills and work practices and, undermining their relations with the mine manager, Willie Hampson. He was openly critical of their time-honoured methods, and sometimes openly insulting with his comments to very experienced miners. He had an acerbic tongue and used his position to abuse the less forceful of men, who knew damned well that he was acting just like an ignorant parvenu.

On one particular day, he simply chose the wrong brushing team to verbally abuse. This team consisted of Thomas Maethyr, who was just about average height, but built like a small concrete block-house and Albert Jeckson from the Coaltown, who was even bigger! Tam and Bert were back-brushing the return road *(airway)* for the first developed run on the Lower Dysart Dook, (1951) when A.W. happened on the scene. A.W. seemingly attempted some snide remarks at the brushing pair and when that failed, then tried to tell young Tam as to how back-brushing *(in his opinion)* should be done. Young Tam seemingly lost his usually 'calm demeanour' and an quick argument was reputed to have followed. Sufficient to say, that a quick bruising altercation might have taken place. I simply don't know, I was not there! However, nothing took place that in any way altered the progression, or the determination of the back-brushing and I am happy to relate, that no formal disciplinary action was later pursued, thereby quashing the circulating rumour that 'blood had been spilt!'

I mention the foregoing paragraph for the simple reason, that I was not the only miner to fall foul of the waspish tongue of A.W. He did have a large vocabulary of caustic and pointedly insulting remarks.

It soon became quite obvious to both Jimmy Frew and Chairlie Sinclair that A.W. was not particularly enamoured with my method of clipping, in spite of the fact that he had been reassured as to the tightness of the down-going clips. He was *(as he kept reminding every-one)*, a 'hands-on' under-manager, who occasionally would get his hands dirty. When the next loaded hutch approached the lip of the rise, he pushed me aside with the words "I shall decide if the clips are tight enough or not" and attempted to undo the upcoming clip with his bare hands. He quickly made several attempts to loosen and spin the wheel, first with one hand and then with two, but failing, he then made a rapid dash for the 'G' key, but failed again! The still-locked clip ran to 'stops', where the rope disappeared under the pavement, with the still-tight clip screeching noisily and solidly vibrating against the heavy wooden sleepers, with A.W. frantically reaching for the motor-haulage manual trip-wire. A.W did not say anything at all at this juncture, he departed rather suddenly in the direction of the 'slopes', but I was left with the instinctive feeling that I hadn't heard the last of that little incident. *(How right I was!)*

Just prior to the change-over from tail-drawing to endless rope haulage on the dook, a team of brushers had cut a wide, circular, looping mine/tunnel through the now solid metals on the right *(north)* side of the Lower Dysart level, where the rakes of empty hutches waited in a long serried rank, before being systematically and individually fed through to the belt-end, where they were loaded and drawn onto the 'flats', where the full races were built up.

This loop road as it came to be known, was to facilitate a better and quicker flow of empty hutches to Jimmy Broon's belt-end and save the drawers a longish walk as the number of empties diminished on the flats. This loop road method of operation took a long time to develop properly, with much experimentation as to how best to utilize the pulling power of the diesel pug bringing the empty race around most of the loop, and then peeling-off down the slip-road with a sudden change to the switches, thereby allowing the empty race to 'float' down to the loading point. The only problem was that the empty rakes didn't 'float' down to the loading point at all! They either crashed down through, pushing dozens of empties through the loading point, or quite often stopped dead and started to run backwards. Either way, it did involve lots of sweating and swearing manpower to initially keep the coal-loading continuous. I mention all of this because of what transpired next.

As my confidence in my abilities to handle the now increasing output from the dook increased, so did my craving for hot tea at piece-time. My need had now grown to the extent, that I felt that I deserved a flask of hot tea with my buttered toast, but where to heat it? I could not very well trust any of those drawers to take the chance, they too, had been well warned in the past and were not really prepared to take the risk. I decided to take a chance and prepared my 'fire-making box' on a short stub of girder, behind a length of corrugated iron in the nearest man-hole. I could not very well leave the dook or stop the haulage motor so, this near manhole had to suffice. I quickly and surreptitiously lit the fire-box, so as not to miss the next up-coming hutch and experienced a small pleasure, in anticipation of a pint of hot sweet tea with my buttered toast. At exactly half past nine, Tom Coventry came puffing up the coal dook and resting just at the lip, raised his beaky nose to the hot candle-waxed air. He said nothing, but followed his nose to the offending fire-box. On discovering my flask, now almost too hot to drink, he lifted up the flask of tea using it's handle, and raising one large boot, he kicked the fire-makings into oblivion. He came out of the manhole, looked grimly at me, then swung his arm in a wide arc, so that the offending flask sailed over my head and on into the depths of the dook, before bursting against an up-coming hutch in a spray of hot tea. Tom Coventry then turned to me and said, "You may have gotten away with it with Chairlie Sinclair, - but am no sae saft!" *(I am not so soft!)*

It was after just a few short months as a 'demon' clipper, so sure of myself and cocky with it, that I decided to approach Jimmy Frew and ask for 'man's wages'. I didn't really need to psyche myself up to do this, Jimmy Frew was very approachable. It so happened that when I spoke to Jimmy Frew on a piece break, A.W. was also in the Manhole. I hadn't seen him arrive and his presence was unexpected. Undeterred, I voiced my request perhaps a little louder than planned and that directly to Jimmy Frew. The Oversman did not get a chance to put me down to where I should never have left, but A.W. jumped straight in with the words "Your sacked! – Get you round into the loop road! No more bloody clipping for you!" Jimmy Frew placid as ever, murmuring "Now Andra!" Big Jock Napier *(the section's 'spare' gaffer),* lazily hanging on to an overhead girder mouthed, "Serves him bloody right! Uppity young bugger!" *(Old Harry couldn't stand the sight of Jock Napier. He reckoned that he was just idle!)* Chairlie Sinclair going red in the face and wringing his hands, Tom Coventry said nothing! He kept his thoughts to himself! . . . And old Harry? . . . He would have his say just a little later on! *(He knew before I actually told him!)*

And so to the loop road I did immediately go! No clear cut job, just an absolute panic on everyone's part, in the effort to keep the belt-end supplied with empty hutches. This job was the absolute 'pits!' No two, in-coming, empty races ever handled the same. From crashing-on unchecked, into and through the belt-end, to running backwards down the loop road. The supply of empties to the hungry belt-end became increasingly erratic, with the result, that the Oversman and the deputies were forever chasing the drawers up around the loop road, in the quest for empty hutches. During one such morning, when the weight of the in-coming race did not come over the top of the rise in the loop road, but started to run back, A.W. came charging up the loop road to find out what I was doing. I was pushing damned hard in an effort to get about fifteen empty hutches up the slight incline and A.W. demanded to know why I was taking so long? My reply, was of course, not meant for his ears, but it slipped out anyway, to which he began to say something *(for my ears only),* then changed his mind when Chairlie Sinclair came upon the scene, to add his considerable weight to the task. As the small rake got going, with Chairlie racing on with the first few hutches, A.W. turned back to me and 'mouthed' that, which he could not, or would not say in front of a witness. He words were spat out with visible venom and were to the effect, that cast 'very doubtful aspersions on my lineage!' He obviously still remembered my 'attitude' from the clipping job. I made to 'go' at my tormentor just as Chairlie Sinclair reappeared on the scene, who was just in time to get between me and A.W. A.W. just paused, smirked, and left the loop road with a further sarcastic remark, before Chairlie Sinclair could interject with any questions.

On going up the pit that day, I managed to catch Old Harry just as he was going onto the cage *(he was on back-shift that week!),* and with a strong tug, pulled on his arm to the effect, that he left the group and with a heavy quizzical look on his face, followed me over to behind the check-weight-man's office on the pit head. I didn't mince words with old Harry. *(One didn't!)* I went straight into a quick description of the turn of events and concluded with the comment, that I felt that my days in the Lower Dysart section were numbered! Old Harry didn't bat an eyelid. He wouldn't! He merely nodded his head and tried to stammer out an answer. He couldn't! Harry had a quiet anger! I could tell! He merely stabbed a raised straight finger at me. I got the message! This was his pigeon now! I had no need to say any more. Something would now happen . . . !?

Things did happen and events quickly materialised! *(Old Harry's mafia!)* I went to work the next morning as usual, without seeing old Harry that night and on arrival at the Pit bottom, Chairlie Sinclair motioned me to one side and said, "You do not have to go into the loop road today. You're going back up the pit onto the pit head and you are to go and seek out young David Paterson at piece-time". I was quite surprised, I knew that old Harry could throw some clout when the need arose, but this was really fast! I dutifully then did as I was told! I went back up the pit and reported to the pithead gaffer *(George Peggie),* who put me to work in the wood yard for the day, with the instruction to report to Davie Paterson, the under-manager of the Surface Dipping at piece-time - in his pithead office. I could hardly wait and wondered what was in store for me. When piece-time came, I duly sought out Davie Paterson, made myself known to him and the next words he uttered were music to my ears. "You will start work down the Surface Dipping on Sunday night-shift - and in a few weeks time when I can organise a training supervisor, you will start your 'face' training! O. K.?" I surely was, and it must have showed, as his next words were, "I'll see you all right lad!" Enough said! . . . Old Harry never seemed to say very much, but when he did? . . . Things happened, people moved and events took place!

\*

# Chapter XXIV

## The Coaltown of Wemyss
## – *The Middle Village!*
### *(The Prettiest Mining Village in the Whole of Scotland!)*

The 'Coaltown', the middle village of the three villages' Wemyss, stands on the A955 between East Wemyss and the Bowhouse farm, which itself, lies near the top of the B ?, the only road route into, and out of the fishing and mining village of West Wemyss. The Coaltown of Wemyss, commonly referred to as 'The Coaltown', *(pronounced: - Coalt'n.)* is not the largest of the three Wemyss's, but is larger than West Wemyss though much smaller than East Wemyss. The earliest indication of inhabitation is depicted by a carved stone from what was originally 'Barn's Row', and was dated as early as 1645.*

Early in the nineteenth century, the village comprised of two separate parts, named East Coaltown and West Coaltown. That these two separate parts of the Coalt'n eventually came together, was no doubt due to the increased development of Lochhead colliery, the subsequent increase in the numbers of coalminers required and the fact that the two villages were wholly owned by the W.C.C. The two parts of the village inextricably edged closer and eventually merged around the turn of the century. *(By 1900.)*

*(This makes perfect sense, in that in the 1890's, the second Lochhead pit shaft was sunk into the Dysart Main seam of coal, where, at sixty-three fathoms, (approximately 115 metres) it's 22 feet (6.7 metres) thickness was fully exposed, with the promise of greater bounty, as this great seam deepened towards the near coastline and thickened as it did so! The radial expansion methods of working the coal at that time, ensured that additional miners would be needed, as the seam was initially extended and developed on a broad, semi-circular front.)*

*Historical Note:- On the outside of the house at No. 5 Barn's Row, there is a plaque built into the front wall of the house that reads 1645. Alongside this plaque there is also another stone carving that describes 1912 as the date when the houses on Barn's Row were modified, enlarged and refurbished. The original 1645 plaque was obviously retained from the former, stone built houses in the 17ᵗʰ century, and incorporated into each subsequent rebuilding and renovation of the properties. Exactly how much of the very original stone buildings remain is difficult to ascertain, because of the later, brick-built constructions that stand on this site today.*

*(Probably built with house-bricks manufactured by the W.C.C.) This house at No 5 Barn's Row is also of further significance, in that several of the characters that appear in this book, include Old Kate Smart, young Harry Moodie, Belle Smart with her son Brian and David Smart - amongst many others. They all lived in that house for a great number of years.*

From past statistical records, the village population in 1791 amounted to 393 persons. By 1851, this had risen by approximately 50 percent to 600 persons. Over the next 40 odd years, up to the early 1890's, the population again slumped to approximately 380 persons, which quickly rose again to approximately 730 people at the turn of the century. *(1900.)* The in-comers, were of course, mostly miners, their wives and their families, all sorely needed, due to the additional numbers of miners required for the developing coalmine and ubiquitous Dysart Main seam, in the new Lochhead colliery. *(Then named the Lady Lilian!)*

Again, the influence of Randolph Wemyss and his architect Alexander Tod was seen to proliferate, in that they did for the Coaltown that which they later did for the incoming miners families to Denbeath. Randolph Wemyss had his architect design and build a series of commonly fashioned, adjoining, but, semi-detached cottages, that ran in long lines and eventually delineated the outermost limits of this mining village. The names of these two boundary rows were 'Plantation Row', *(probably euphemistically named after the wide expanse of the Wemyss estate immediately behind it)* and Lochhead Crescent *(named for the Lochhead Colliery)*. The single row of houses in Plantation Row, when completed, showed the same identifying characteristics that may possibly have been a trademark of Alexander Tod *(perhaps, on the implicit instruction of the then Laird)*, to have both the front door and the back door and the cottage windows, all facing into the same street, with the back wall of the row a long, solid wall, unbroken by doors or windows. The doors and windows of these houses of course, all faced inwards towards the centre of the village.

The interconnected dwellings of Plantation row consisted of pairs of cottages built side to side, in that the single bedroom of each dwelling shared a common interior wall. Each cottage had a 'kitchen', *(now commonly described as the living room!)* that initially contained a coal-fired range with two bed recesses, each capable of taking one double bed. This kitchen/living-room also housed the family dining table and chairs and the obligatory sideboard. This room was where the family lived and ate! The one bedroom was capable of housing another double bed, in addition to the necessary wardrobes for the family clothing. The scullery, with it's stone floor, *(nowadays, known as the kitchen!)* looked as though it were a later, smaller extension onto the side of the cottage, contained a deep stone sink with running cold water, a water closet without a bath and a coal-fired, cast-iron boiler, brick-built onto the rear wall. There were three chimneys pots to each cottage. One for the scullery boiler, one for the kitchen/living-room range and one for the bedroom. Which meant, that there was three double chimney stacks for each coupled pair of dwellings.

Plantation Row *(surely, a misnomer!)* was therefore, one, long, continuous line of cottages, attached end to end, in the form of bedroom, kitchen, scullery, - scullery, kitchen and bedroom, - ad finitum, along it's full length, that terminated in a short dog-leg at right-angles, that led directly back onto the Main Street, from where it first commenced. *(This common, adjoining, scullery arrangement also ensured common plumbing and waste outlets, at regular intervals along the length of the row.)*

Author's Note:- *At this time, many of the cottages at the bottom end of Plantation Row lie empty and partially neglected and have remained so for many years. (So much so, in fact, that the 'Estate workers' had time to paint black and white, artificial, window boards, to cover the glass, window frames of each cottage!) They are in dire need of refurbishment, modernisation and occupancy! In so saying, I did notice on one of my recent visits to the Coaltown, that this is exactly what is happening! One or more of these solidly built 'one-ended' cottages, are in the process of refurbishment and redecoration and I was delighted to see, that new, full size, rear facing windows have been cut into the back walls of both the 'living-rooms' and the bedrooms. Some of the cottages are having a single, extra bedroom extended onto the rear part of the cottage, whilst several are being interconnected through their common bedrooms to double their original size. (It would seem that the present Laird has no such inhibitions to privacy, as did his fore-bearers!) I also noticed that the single intrusive 'bed-recess' wall in the 'living-rooms' have been removed, thereby ostensibly, enlarging the living rooms! (I also observed that most of the 'original' floor boards seemed to be completely intact!)*

\*

# Young 'Hend' Moodie

*Young Harry was fourteen years of age when his mother of fifty-four years, died, leaving husband William (Bill) Moodie to cope with the youngest, of a family of fourteen brothers and sisters. This was a large family even by the standard of most mining families - and the only saving grace was, that young Harry was the youngest of the brood. He was literally 'taken-in' by his elder sister, known as 'Old Kate', simply because there was nowhere else to go. In those days, blood relations came first and family loyalty was supreme, and no matter that Kate was fully 20 years older than her younger sibling and already had a good family of her own; young Harry's needs at fourteen years old were quite desperate. Young Harry was at once accepted into Kate's family, simply because Kate said so! And that was enough! Young Harry now had the life-line he needed to survive and flourish, which he did, all down to the very unselfish act of his older sister. Old Kate was so named, not because she was old, indeed, she was a fine tall figure of a strong well-built woman. She acquired the name 'Old Kate' simply because she had a daughter also named for her and known as young Kate, - as was the way of things.*

*Old Kate (as I remember her) had, in her younger days, married one William (Wull) Smart, probably around the time that young Harry was born and in the long years that they were married, she bore him four sons and five daughters, William (young Wullie) being the oldest, and David being the youngest. Old Wull was also known as 'Auld Tot.'*

*At the time of the death of young Harry's mother, old Kate and her husband, auld Wull, were living in a two bed-roomed house at No.5 Barns Row in the Coal-town. This type of house also had a living room, in which there was twin wall recesses to house two double beds as was the custom in those days, which was just as well, considering that when young Harry was taken-in at Barns Row, the family was already eleven strong, including young Harry's still single, older brother Andrew. Awn'd had lived with auld Wull and Kate for most of his adult life. Andrew was referred to as 'uncle Awnd,' whilst young Harry being much younger and of an age on par with Kate's eldest boy, was treated as a family sibling.*

*In those days, secondary education usually terminated when a pupil reached fourteen years of age, as young Harry now was and without more ado, young Harry was immediately 'signed-up' at the Pit - and taken directly under the wing of auld Wull as 'trainee' collier. So commenced young Harry's life-long mining career, that started under the experienced tutelage of auld Wull Smart, and continued as boy and miner (collier) in Smart's Dook at Lochhead pit. (A coal section in Lochhead pit probably initiated, driven and named by auld Tot.)*

*Young Harry's progressive training as a coalminer, would be a long, slow, upwards grind, involving dirty conditions, poor illumination, long hours and low wages. (Young Harry's wages would have been paid direct to him by his brother-in-law auld Tot, (mentor /supervisor) from his own coal-getting, paid yardage). At the end of his 'supervised' training, young Harry would have to take his place amongst other 'trained' miners, before he would be able to gain a 'place' on the coal-face. Such places were eagerly sought after, where tried and tested miners always claimed the first options on new developing coal faces. (Pick-places)*

*

On the North side of the main road, the semi-detached, but still continuously joined cottages of Lochhead Crescent *(another misnomer?)* were slightly more spacious, but with still only one living room/kitchen and one bedroom to each cottage. The kitchen/scullery and toilet/water-closet for each cottage was an integral part of the building, but was recessed onto the side of the living room in every case and having it's own 'back' door. Again, both front and back door and most windows faced into the centre of the village with the exception that the 'back' door was actually a side door from the scullery, with the two, opposing, scullery doors facing each other. In addition, the houses on Lochhead Crescent had both a front garden and a spacious back garden, that Plantation Row seemed to lack! *(Lochhead Crescent is nearly twice as long as Plantation Row!)*

In the early part of the 1930's, it became obvious to the W.C.C, that the cottages on Plantation Row and those on Lochhead Crescent were actually too small to house many of the increasingly large families being produced by some of the miners, *(and their wives)* and the families were becoming increasingly cramped. This was partially alleviated by the Estate 'factor' being in a position to allocate two, adjoining cottages to one miner's family, thereby relieving the situation. *(To the best of my knowledge, this occurred with three families in the Coaltown of Wemyss! The Baxter's, the Stewart's and the Blyth's!)* This however, was only a temporary solution to a growing problem, that could really only be satisfied by having a greater number of larger cottages available. There was not much that could be done with the dwellings on Plantation row, as the only possible direction of expansion was onto the Laird's fields, and God forbid, the extensions would have to have windows! That was out of the question! The answer then, lay with the forty-four inter-connected cottages on Lochhead Crescent. *(Lochhead Crescent was split into two separate halves by the intrusion of the pit road, which commenced on Main street at the Co-op and bisected Lochhead Crescent at right angles, then carried on for another half-mile to Lochhead pit.)* In 1936, work began on extending some of the cottages, to build an additional bedroom to the rear of each dwelling. There was a sufficiency of space available within the confines of the back gardens of each of the cottages. The estate architect, Stewart Tod, son of Alexander Tod, carried on the work of his late father, in that he drew up the plans for these extensions and organised their building, all under the critical direction of the Laird of course!

\*

*This 'Coalt'n' o' Wemyss (in the local idiom!) village was a most unusual village, in that there were several features that emphasised it's plural uniqueness. To the visitor who takes the time to wander around it's few streets, that visitor shall surely realise that there are very few building that stand individually by their own right. The few buildings that are taller than one storey are in the minority and consist of Memorial square in the centre of the village and part of Main street North, that runs into Anderson Crescent. The remainder of the dwellings are seemingly all single-storeyed and interconnected in long rows. Indeed, this was the exact premise on which the Coaltown of Wemyss was expanded in the first decade of the 1900's! This village, once described as the prettiest mining village in the whole of Scotland is indeed so! It can be said that it was purpose-built and mainly designed by the*

*Wemyss estates, resident architect and his successors, namely Alexander Tod, his son Stewart, and his son Charles. The village, almost in it's entirety was privately owned by the Wemyss family and was developed to accommodate the increasing numbers of miners being employed at Lochhead colliery, which the village was deemed to serve. The interconnected cottages were built to a simple common design, to include sleeping accommodation for up to six to eight persons (mostly children), running mains water and an inside water closet. The scullery housed a coal-fired boiler to ensure hot water, boiling facilities to each household. If visitors are especially observant, they will also see that each single row of interconnecting cottages, all seem to have their own street name. This was quite deliberate, in that there are several examples of two, parallel rows of semi-detached cottages in the same street and served by the same road, - but, having different street names! Lochhead Crescent opposite Lochhead Row, Plantation Row opposite South Row, Bowling Green Terrace opposite Lochhead Crescent and Main Street north (previously Lancer Terrace), backing onto Lochhead Crescent. There is also a street that does not seem to have a name, and that is the short street that divides Lochhead Row from Bowling Green Terrace, from Lochhead row to the Main Street. (This is the street where Allen the Boot-maker had his premises, next to Peggy's, later Cowan's chip-shop. The corrugated roofed wooden building that housed the boot-repair business, is still standing to this day!) To confuse things even further, part of Main Street incorporates the five houses of Barn's Row, that lie in the centre of the village, and are still referred to as Barn's Row.*

*The village limits that are delineated by Lochhead Crescent to the north-west and Plantation Row to the south-east, were seemingly decided by the completion of these two long rows at the beginning of the 20<sup>th</sup> century. Only a few houses have been built in the village since that time - and those were built at the south-west end of the Coalt'n, to include the six houses of Anderson Crescent, (to house and separate the Lochhead Manager and mining officials from the common five-eights!) and the houses in Hugo Avenue. (Named after the now-defunct Hugo tunnel/ screening plant.) One or two new cottages have appeared within the restricted limits of the village, but these were built on the site of very old dilapidated houses.*

*

One common Family name that did proliferate in the Coaltown of Wemyss, was the surname, Dryburgh! There was so many separate families named Dryburgh in the village, mostly unrelated to the other, that the local people eventually 'christened' each separate family with an additional, identifying appellation, to distinguish and describe the different family branches. One needs only to read the surnames of those from the village who fell during the 'Great War', to appreciate the contribution this family surname made to that war effort. Of the thirty names that appear on the W.W.1. monument to the village War dead, the name of Dryburgh appears six times! *No other surname appears even twice!* That part of the war memorial that is dedicated to those from the Coaltown who fell during W.W.II, shows only seven names. The likely reason for this is very simple! The great majority of the menfolk belonging to the village were probably gainfully employed in digging out the Dysart Main seam of coal in Lochhead colliery and the Victoria pit as well, in addition to the coals being taken from Lochhead's Surface Dipping!

*

# A Coaltown Man! *- or a Character?*

Harry 'Smith' lived in 'The Wemyss' with his wife and their large family of ten children. They occupied two adjoining houses in Lochhead Crescent, so that the full family could be accommodated together - but, with separate entrances. This was fine up to a point, but it very inconvenient to have to use two different front and back doors, especially at meal-times. Harry came up with the perfect solution, he would knock a small hole in the wall between the adjoining rooms of the two houses, and probably install a doorway *(that's if he could lay hands on the right materials),* which he then proceeded to do. The operation was an unqualified success, which was only right. After all, Harry was a coal miner! This solution worked very well for a time, especially in the good, summer weather, where nearly all of the young children though it was great fun to climb through their own 'hole in the wall', indeed, the very original 'Hole-in-the-Wall Gang!, but, unfortunately for Harry, warm summers, no matter how long they are, don't last forever and the cold winds of winter soon began to blow. Miners cottages at that time, were infamous for being notoriously damp and draughty and both of Harry's houses were indeed prime examples. The wind simply whistled through Harry's hole in the wall, much to the howls of the kids, who complained wearily, that they were forever cold and could not sleep, even packed together as they were! Harry solved the problem almost immediately, if rather unconventionally. He obtained two, heavy, army blankets, weighted the bottoms of both and nailed the top end of both blankets to either side of the hole. If it worked down the coalmine, it would work here! The draughts would be stopped, no matter how they blew, and stopped they were! Harry then loudly boasted to one and all, that he had his own set of air-doors between the bedrooms, within his two houses!

Harry, like most miners, did like a drink or two, especially on a Friday afternoon after being paid. Pay day was always on an Friday between the hours of 11-00 a.m. and 3-00 p.m. - and this applied to miners working on all three shifts. Those miners on both day-shift and back-shift were in a position to collect their week's wages, either after they came of shift, or immediately prior to commencing shift and keeping their wages on them whilst working. Those miners on night-shift, had unfortunately, to break their sleep pattern on a Friday, in order to travel to the Pit-head around mid-day to collect their wages. Harry at this time, worked on the day-shift in the Victoria Pit at West Wemyss, and made the daily trek to work at 5-00 a.m. in the morning, through the intervening fields, regardless of the weather. This trek had to be made at that time, for the simple reason that organised transport was practically non-existent! When Harry got paid on a Friday afternoon, along with his mates of course, the temptation to have a cold glass of beer after a week's hard work was very tempting and Harry was no more strong-willed than any other thirsty miner. Harry didn't get the chance to succumb and be led into temptation. Harry's wife was not for nothing, the mother of his ten bairns, she too, made the long trek across the fields to the Victoria Pit-head, just in time to catch Harry coming out of the Pay-shed, wages intact, and be led home by a good and worthy wife, who now carried Harry's pay packet in her own, tightly-clasped hand, deep in pocket.

Harry was a good husband to his wife and often helped with some of the daily chores. He would often, shift permitting, help his wife with the nightly task of getting the bairns bathed and scrubbed and made ready for bed, along with a helping of hot saps for supper. *(Bread, steeped-in, but not boiled, in hot milk.)* The task was completed along production line rules, *(where the bath itself, often took the form of an empty 80 or 100 gallons whisky cask, expertly cut in halves),* where the protests and cry's usually went unheeded and the children were then speedily dispatched to their respective beds through the hole!

*Local legend has it, that on one particular occasion, Harry went into the bedrooms to tuck-in the children and returned rather quickly to his wife and said, "I have just seen them all under the covers - but strangely enough, I counted 'twelve pairs' of feet!" Additional candlelight was hurriedly brought into the bedrooms and, after a cursory but positive inspection, it was sure enough! There were twelve bodies in the beds and surprisingly, two of them belonged to the family from next door! They had been dragged though the washing, drying and nightgown process emitting surprise exclamations, but to no avail! Just two more dirty faces amongst ten others!*

At the time of the first, gas, street lamps, it was rumoured that Harry was seemingly so hard-up, that he once purloined the three gas mantles from one of the street gas lamps outside his house, to use them, one at a time in his own house, and later, actually had the audacity to complain to the local, gas-board fitter, if only divert the blame from himself. In explanation, Harry maintained that he actually saw a strong gust of wind cut into the glass covered lamp and blow the mantles apart. He almost got away with it, until the gas fitter meaningfully asked Harry, 'If the wind had also blown away the missing, three-cornered, mantle-lugs as well!?'

\*

On the First of January 1947, there occurred another milestone that could be added to the village uniqueness, in that it suffered itself to be cut almost completely in halves! Geographically speaking as well as metaphorically! *(The Main road was the 'Mairch-line!')* This was the first, official day of the then Labour government's Nationalisation plan, when nearly the whole of the Country's coalmines were taken into 'public ownership' and gathered-up under the umbrella of the National Coal Board. Up to this time, most of the fabric of the Coaltown was owned by the Wemyss Family Estate, who administered the properties through their own housing Factor situated at Muiredge, Buckhaven. In the months leading up to this January date, the new area officials of the N.C.B. seemingly pressed the Laird to include all of the dwelling houses/cottages of the Coaltown of Wemyss in the 'take-up' agreement/package, that was being hammered out between the coal/land/property owners and the then Labour government. This was necessary to strengthen the hand of the recently created N.C.B, giving them the capability/facility to offer incoming miners to the still expanding Lochhead colliery, the opportunity of living in close proximity to their work-place. The Laird seemingly baulked at this proposal, not wishing to lose possession of the whole of the village. The upshot of the protracted negotiations, was that the whole of the South side of the Coaltown would remain as part of the Wemyss Estate, whilst the North side of the Coaltown would be handed over to the N.C.B. as part of the overall settlement, with the Main road through the village being the undisputed *'Mairch Line'. (Dividing boundary.)* The one exception to this division, was that the five, adjoining houses that constituted Barn's Row, remained under the ownership of the Wemyss estates. *(Cannie lad, the Laird!)*

*

*Of the elements that give the Coaltown its uniqueness, a visitor would have to walk completely around the village, to finally realise, that there is very little that he or she can actually obtain by way of comestibles, consumables, combustibles or clothing. There are no such like shops! There was in the past, a very successful S.C.W.S. that stocked provisions, groceries, hardware, with milk, bread (plain and pan) and tea-breads, but that has been long closed down, even though the original building still remains! Butcher meat at that time, was hawked around the village by a butcher from East Wemyss, named David Dunsire. The horse-drawn wagon was an ventilated, oblong-shaped, enclosed box-cart having a fixed rear pair of iron shod spoked wheels of approximately 4.5 feet in diameter. The front pair of wheels were of the same design, but approximately 3 feet in diameter and, were articulated to the single pair of shaped, wooden shafts, designed to accommodate one middle-weight draft horse. (The front wheels therefore, following in the direction in which the horse was led or driven!) There is at this time, two, small shops that cater for daily papers, milk supplies and limited groceries, but as to supermarkets, mini-*

*markets or small emporiums, there are none! Nearly every consumable has to be transported into the village by visiting traders on a near daily basis, or purchased elsewhere by itinerant villagers. Thankfully, many of the present villagers (many of whom are 'new' to the village) have independent means of transport (as evidenced by the now crowded parking places), which enables them to commute to and from their places of employment. (To the best of my knowledge, there are no commercial or business employer's premises in the Coaltown of Wemyss!)*

*

# Another Coaltown Man ?

## *Soled & Heeled and filled fu' o' Tackits!*

Jim Allen was the cobbler and boot repairer in the Coaltown of Wemyss, and he and his cobbler's performed wonders with torn and damaged, miners pit boots. He re-soled them, heeled them, patched them, and he filled them full of tackits. *(Hobnails.)* He turned them out like new, at a price much below that of a new pair of boots! He occupied a large wooden shed in a short street that lies between Bowling Green Crescent and Lochhead Row, which was sometime extended to accommodate his growing business. At one time, he fully employed twelve cobblers in this shed. His reputation was such, that he was known from the small town of Leven on the coast, all the way to Dunfermline in the West and did regular trade with every town, village and hamlet along the way. He was reputed never to turn away even the most damaged of boots, he would strive to re-build a serviceable pair of boots somehow. He also knew that the repairs had to be accomplished in short hours rather than days. Tomorrow's day-shift did not wait on pit-boots being repaired!

Jim Allen also had working for him at this time, in his shed, a Dane by the name of Alfred Lagergren, who was reputedly described as a "Big Braw Man" of whom it was said, people came from miles around just to see him. He was reckoned to be the very flower of a 'Nordic Adonis', with a natural build that was both powerful and graceful. His passion was in developing and maintaining his strength and to this end, he indulged in wrestling, any form of wrestling, including rough and tumble and with anyone who would stand within the 'ropes'.

One fine Saturday, shortly after midday, Alf Lagergren and some of his friends set out from the Coalt'n and trudged their way over the countryside to attend the Thornton Games, which were held in an open field adjacent to the River Orr, where games of prowess and strength could earn a few shillings for a successful *(or lucky)* participant. Alf and his friends naturally made their way to the roped enclosure, where they found that the impossible virtues of an infamous, London wrestler were being exclaimed to one and all, with the solid promise of a five-pound note to the person who could 'pin' this flabby grappler to the ground, within three, two minute rounds. This was exactly what Alf was there for and the challenge was immediately accepted. Alf was told to strip down to his waist and remove his finger ring, and instructed to stay in his corner while side-bets were made.

At the conclusion of the preliminaries, the 'busker/promoter' then approached Alf, ostensibly to check his dress, but immediately reached out and severely 'pinched' both of Alf' biceps between his bent first fingers and thumbs, in the process. This 'fouling action' incensed not only Alf, but those few who had witnessed the incident. Alf was given no time to protest at the blatantly obvious 'foul' *(whom could he appeal too?),* but was forced to defend himself at the first clang of the bell, signalling the start of the bout. He was immediately caught in an arm-lock by his grinning opponent, forced to his knees with his left arm doubled over the middle rope, pulled-up and twisted, the London fat-man using his broad back and body weight to render Alf immovable. This was stalemate from the first bell, with the contest seemingly over, but for the verdict. Neither man however, had reckoned on the sheer strength and courage of Alf Lagergren. He managed to scramble upright onto his knees, which gave him all of the 'purchase' he needed. He forced his right hand and arm up between the Londoner's thighs and, deliberately seized his opponent's 'wedding tackle' in his cobbler's fingers and squeezed, at the same time heaving upwards with all of his considerable strength. The Londoner quickly abandoned his debilitating hold on Alf's arm in an effort to remain on his feet, but this was not to be. Alf went after him like a shot, grabbed at him, securing a quick, strong hold and bodily lifted him clear of the ground to an overhead horizontal position. At that juncture, the miners, never slow to respond, shouted long and hard together, giving instructive encouragement to Alf, "Into the Orr with him Alf! . . . . . Throw the bugger into the river Orr !". . .

And. . . I am to afraid to say, that I do not know the end of this possibly exciting anecdote. One fine day perhaps, I will find and talk to someone who was actually there, and could relate to me the ending that I personally would wish to hear!

*

Author's Note:- *Alf Lagergren did somehow become part of the Allied Army during W.W.II. He was obviously involved in the fighting at the second battle of the Ardennes in Belgium in 1944, (Battle of the Bulge) where he died as a commissioned officer in the service of his adopted Country! His name appears amongst those of the fallen on the W.W.II. war monument at Memorial Park in Methil, Fife.*

'Lieut. Alfred Lagergren. Army Air Corps'. Died 1944.

*

# A Rare Experience!

On one Saturday morning in the summer of 1948, I struggled onto the rear platform of a W. Alexander & Sons, double-decker bus, carrying two large bakers baskets containing large meat pies, bridies, sausage rolls, cream buns and an assortment of fancy cakes. These 'goodies' were the produce of one William Gillespie, 'Home' baker, whose premises *(bake-house)* adjoined the lower end of Buckhaven Higher Grade School. I was all of fourteen years of age at this time and this was a 'Saturday job', that I had been coerced *(bullied?)* into by one Mary Balfour, of No. 20 Barncraig Street, an acquaintance of my mother. This 'job' entailed my attempting to 'flog' the produce of one of the local bakers, around the not so local, out-lying country areas. This was to be accomplished by my lugging *(by public transport)* these two great baskets in such a manner, as not to damage the pastry and confectionery contained within, before I arrived at my destination! Hence the need to await the arrival of one of W. Alexander & Sons, double-deckers. This was my second Saturday in the job and I had decided to try my 'sales technique' in the Coaltown of Wemyss. *(My earlier attempts along the length of the great, red tenements of lower Methil's 'North British' had been an abject failure. Nobody living there, wished to buy Gillespie's pies and cakes!)* I eventually gained space on rear platform of a bus at the Buckhaven library at 9-30 a.m. and, having paid my fare, sat down immediately adjacent to my two baker's baskets. The journey to the Coaltown took all of twenty minutes, with my eyes never straying too far from the contents and the safety of the baskets. *(Readers will realise that I did feel wholly responsible for the contents of these baskets, even though I had earlier, volubly forsworn the commitment, that I just 'didn't' want to sell Mr 'bloody' Gillespie's pies and cakes!)*

On arrival at the Coaltown, I stood outside of the Co-op for a few minutes to consider my course of action. I reasoned that I did not especially wish to lug these great baskets in and out of front gardens with gates, but should approach a row of houses with back *(or front)* doors leading onto the street. I did know something of the Coaltown, in that both my mother and old Harry had brought me here on several occasions in the past and I concluded, that Lochhead Row would be the best place to start, with my having the option of returning down Lochhead Crescent to complete the 'circuit!' I gathered up the two, cloth-covered baskets, after ensuring that none of the cakes had piled up on each other *(they hadn't)* and proceeded to the first house on Lochhead Row.

I stopped outside the door of the nearest house, placed the heavy baskets down by either foot and paused momentarily. I then timidly knocked at the door, still wondering as to what I would say when the door opened. The knock was answered by a middle-aged, miner's wife, still in the process of drying her hands on her pinny. She looked enquiringly at me, then, getting no response, *(I was at a loss for suitable words!)* looked down at the covered baskets. I may have for the moment, been rendered speechless, but inactivity was not my forte. I whipped of the two dust covers and rapidly managed to blurt out, "Gillespie's home bakery Missus, - from Buckhaven!" The woman took one look at the baskets, her eyes lit up, and she quickly exclaimed, "I'll just git me purse!" As soon as she returned *(I had gathered my wits by then)*, I replied to her anticipated question! "Big pies, sixpence; shell pies and bridies, fourpence; sausage rolls, threepence; cream buns, tuppence, and everything else, a penny ha'penny." That woman, my first customer in the Coaltown had spent all of three and ninepence (3/9d) that day, and that set the tone for the sale of the complete contents of my two baskets. I did not manage to get near to the end of Lochhead Crescent, before the entire contents of my baskets had gone! *(Not quite, in that there remained only two cream buns, a sausage roll and a large meat pie that I had kept hidden. I simply did not have enough left to be going on with!)* I was quite astonished! I had never thought for one moment that it would be so easy! These people must be starved of pies, cakes and cream buns, but I also knew that coalminers were hungry men with huge appetites and, were capable of shifting huge platefuls of good food at one sitting! I could hardly believe my good fortune! I hurriedly gathered-up the two baskets and walked back in the direction of the Co-op buildings, where I would board another double-decker to return to Buckhaven and of course, the bakery. Then, I had second thoughts. If I returned too soon, they would merely re-load my baskets and ask me to make a return journey, back to wherever I had managed to quickly unload the first 'delivery.' I wasn't too keen on that, I just didn't wish to sell Gillespie's pies and cakes every Saturday morning and on into the future! I hesitated at the bus stop and sat down to eat the large pie that I had kept hidden. *(I had superior knowledge of these large meat pies, I had eaten one nearly every school-day at the mid-morning break, whilst at High school!)* I decided that I would stay there for at least another hour before making my way back, but impatience got the better of me and I let only one decker go past, before I lost patience and got aboard the next one. I returned to the bakery to looks of surprise from the manageress, who had estimated my return around lunch-time. To say that she was delighted would be an under-statement. She envisaged my departing with another two loaded baskets after a cup of tea and a pie, to which I registered my reluctance by exclaiming that my dad *(old Harry)* was taking me to the East Fife home game that afternoon! *(As far as I was concerned this was to be the termination of my short career as a pie and bun salesman! But, I had yet to tell my mother!)* After the manageress had tallied up the produce sold against the monies collected. She said, "I think that you are sixpence short!" *(That bloody pie!)* I, very innocently replied, "Yes, you are quite right, the conductress demanded an extra sixpence in fare for the two baskets!" As I was being paid my commission on the amount sold, the manageress was still shaking her head and repeating, "They have never charged us before, I must get on to the bus company at Aberhill!" *(W. Alexander & Sons)* I bid her a quick farewell and beat a hurried retreat, with the certain knowledge that they had seen the last of me, no matter what! I was also somewhat relieved, that she didn't have the presence of mind to ask, if I had been charged another sixpence for the return journey?

*To this day, I can still remember the seemingly, pie-hungry, miner's families of Lochhead Row, who managed to purchase the 'entire' contents of my two large baker's baskets in under half-an-hour! I wonder if anyone can still remember me, Gillespie's pies and bridie's and those two filled baskets? Probably not! - I only ever appeared once!*

\*

Author's Note: - *During the course of the late Forties and into the Fifties, there took place at the mid-morning break, a daily, mass exodus from the grounds of Buckhaven High School. At approximately eleven o'clock when the bell was wrung throughout the school, most of the pupils, spewed out of the classrooms, out of the three exits and made a concerted rush out into the roadways. There was then, a mad, sustained stampede by pupils, ranging from juniors at 12 years old, to seniors at 18 years old, in the general direction of Gillespie's home bakery. They did not attempt to enter the front door of the baker's shop, but converged in great numbers to the rear of the bakery, where an orderly queue, perhaps two to three pupils wide, had already formed. This growing queue snaked from the 'barred' back door of the bake-house, out into the large backyard and was firmly controlled by the hairy, outstretched forearm of the bakery owner, old man Gillespie himself!*

*The large, back, storage room of this bakery, was daily, for 15 minutes sale-time only, transformed into a wide, market, baker's stall, approximately eighteen feet wide, and formed by three, six-foot tables, spread longitudinally across its width. The tables were draped with large white sheets and critically loaded with most of the produce of this home bakery! Behind the tables stood six assistants, each provided with a pocketful of small change. They knew the exact price of each highly edible goodie and were prepared for the most frantic and adrenalised 15 minutes of their working day. The hungry horde of 'starving' pupils appeared every day at the same time, and for the next fifteen minutes there would be bedlam, with no respite, no 'please and thank you', no brown paper bags and no quarter given! Everything that was bought was carried away in upturned hands, with or without hankies! Old Man Gillespie ruled this back-shop with military precision, only allowing the next six pupils to go forwards, after a fulfilled six had disappeared out of the other exit. His staff worked to the absolute limit, but thanks to his 'sergeant-at-arms' drills, they were never overwhelmed!*

*By approximately 11-15a.m, the pupil horde had all but disappeared, and with them, perhaps up to seventy percent of the baker's produce. All that remained, was for the staff to remove that which was left to the front shop, where the much-reduced lunch-time trade would all but clear out the remainder.*

*Old Man Gillespie operated what was described as a 'Home Bakery!' He maintained that his business was to feed the local population, not to cater to their discriminating palates. As a result, the bakery produced plain fare, well filled and cooked, and roundly baked! There was, to my uncertain knowledge, enough fancy cakes produced to satisfy his own regular customers, as evidenced by the fact, that he reputedly employed the finest 'confectioner' in the business, but, the mainstay of his successful business to his daily swarming customers, was his well-filled and baked puff-pastry and the fact that his bake-house lay within two minutes reach of the pastry-hungry pupils, who attended Buckhaven Higher Grade School!*

*(It must be realised, that during the immediate aftermath of W.W.II, many items of food were still rationed and that parts of the population were still 'hungry' for better fare. It would therefore seem appropriate, that a competent baker should command such an lucrative enterprise, especially with having such a semi-captive, daily clientele.)*

\*

With regard to the sports facilities in the village, it seems that if the young men don't play football or cricket, the only other outdoor activity left open to them is open air bowls on the splendidly appointed bowling green, *(with it's own licensed club-room)* or cross-country running, *(of which there is ample opportunity and endless space!).* There is a full-sized, badminton court laid-out in the Miner Welfare Institute, which does get used and this also provides the venue for indoor carpet bowls during the winter months. *(Coal-miners have always had a propensity to play bowls. It's a cannie game played slowly, that suits the sometimes dour and taciturn dispositions of many coalminers!)* Within the boundaries of the village, a visitor would be hard put, not to notice that most of the amenities, services and utilities that make up the social fabric of most fair-sized villages, seem to be missing! There is no resident doctor within the village, and there never has been, to my fairly certain knowledge! Therefore, there is no surgery, either medical or dental, hence the original Ca'hoose! There is also, neither small first aid post or cottage hospital! However, in so saying, there was at one time, two subsequent district nurses who were resident in the Coaltown of Wemyss. They were located at No. 31 South Row during their consecutive tenures in the village. One of their main and prolific duties was to act as midwife\* during local confinements. There is also, neither minister nor priest resident within the village limits and never *(to my knowledge)* has there been a church of any denomination, built to cater for the spiritual needs of the villagers. *(I am informed however, that at one time there was an evangelist of a sort resident within the village, who held meetings of a kind in his home.)* That is not to say that the peoples of the village are heathen, agnostic or even ungodly in outlook. The Wemyss's villages are well served by St Adrian's church in West Wemyss and St George's all denominational church in East Wemyss, (s*adly, St Mary's church by the sea has been decommissioned*) - both situated within one mile, but in either opposite directions. I also sincerely hope that all of the television viewers in the Coalt'n have television licences, for the reason that that is the only 'viewing' that is available to them! There is not, and never has been a local cinema either!

\*

Of the many confinements that Nurse Robertson attended, there is one in particular that stuck in her mind for many long years, that seemingly characterised the courage, resilience and hardiness of most of the people that lived in mining areas at that time. This anecdote concerned a farm workers wife living at Lochhead farm, the entrance of which was situated at the mid-point of the Check-bar road, a quarter distance between the Check-bar road and Lochhead colliery. When the nurse was called out to attend this birthing at one of the farm cottages, she was forced to travel to and from this farm that lay outside of the Coaltown, by means of her bicycle. The mother concerned had already borne several children, now of school age, which, in the estimation of the midwife, should present little problems with her next delivery! The midwife arrived in time, to safely and quickly deliver the new baby, with the result that the nurse was back in the Coaltown before lunch-time, where yet another confinement was imminent. The midwife again safely delivered this second baby of the morning and afterwards continued on home to a light lunch and a short rest. She knew that she had another few miles to cycle in the afternoon, to re-visit the farmhouse location of the earlier delivery of that morning. After lunch and a few routine visits around some of the village old folk, Nurse Robertson again mounted her bicycle to commence the long pedal to Lochhead farm. On arriving at the farmhouse, Nurse Robertson gave a quick cursory knock at her patient's door and entered! The woman was not in bed! She was not in the room! She was not in the house at all! Nurse Robertson made her way out of the house via the scullery, where she saw some cooking pots on the stove, gently simmering away and her eyes also caught the evidence of a clothes washing, that had just taken place. Nurse Robertson quickly exited the back door, where, to her utter amazement, she saw her confined mother of only that morning, busily engaged in the task of pegging out her family's washing on the out-stretched drying-line! The woman had seemingly spent just two hours in bed after the birth of her child, had risen at lunch-time with the full intention of doing the family wash, in addition to preparing the family meal for those children coming home from school and the men coming in from the fields! The new baby was slung in a woollen shawl around the woman's shoulder and waist, sleeping, but in position to be fed natural milk on demand, as the need arose!

# Another Coaltown Man!

Jackie Penman at this time is a spry, cheerful and wonderful old man of 89 years of age. He is keenly looking towards his ninetieth birthday in this coming June and promises me, that he will take me for another brandy and lemonade, (lots of lemonade) and give me another trouncing at dominoes to boot, if I dare to accept the challenge. He, at this great age is 'with it' in every sense of the word and is more than able to hold up his end in the United Services social club in the Coaltown of Wemyss on a Sunday afternoon, where he meets up with his many friends and old workmates, for the inevitable re-hash about every lump of coal that was ever howked from the solid, in the Victoria pit. Not that this is the main subject of conversation as I discovered. Jackie does enjoy a long, slow, drink of a brandy and lemon along with a good small cigar whilst being able to hold a conversation - watch the sport on television and listen to every other happening in the Club at the same time.

Jackie Penman started in the Lochhead pit in 1928 *(two years after the 1926 general strike, when the miners went back to work for 'one shilling' a day less than what they struck for!)*, when he was a mature 17 years old youth. This was perhaps three years later than most others of his ilk, but Jackie insists that attempts were made to keep him out of the coal mines, at least in his less mature years. Jackie's first job down Lochead pit was on the pit bottom, where he, along with a known friend, Peter Ballingall, worked together as drawers, supplying empty hutches straight from the cages to both the Old and New Dooks and alternately to the Lower Dysart Haulage. They were also responsible for the 'drawing' and queuing of full hutches, from all three working sections onto the 'onsetters' starting blocks. This work entailed both dayshift and backshift working on alternate weeks, and paid the drawers the sum of two and sixpence per shift, a total of fifteen shillings for a six day week. A point worthy of mention at this time, was that the name of the onsetter on Jackie's shift was one Charlie Fleming, who alternated day-shift and back-shift, with another onsetter called Bob Taylor. An interesting part of the onsetters jobs at that time, was that they also controlled the running and operation of both the Lower Dysart and Old Dook haulage systems, which will be described in detail in the chapter 'Pit Bottom'.

*(Big Charlie Fleming appears again in another part of this narrative)*

Jackie Penmans' next job took him away from the pit bottom and down the Old Dook to the No. 5 level, where he was introduced to the 'bogies'. This was an endless, rope haulage, approximately 800 - 900 yards long and powered by an electric motor and a becander gear, with the inevitable double-rail track, where the small races of hutches were coupled onto a 'scissors' bogie, which clamped on to rope at any given point, thereby maintaining the flow of both full and empty hutches. This haulage-way took the coal away from the half-dozen or so headings that worked the 'wheeled brae' system, where each heading was driven to a sloping height of 100 to 120 yards and employed roughly about 15 pairs of strippers, each having about 7 or 8 yards length of 5 to 6 feet height of coal, to take to a daily depth of 3 feet. These headings would each be stripped on a daily basis, with the working faces advancing in one direction only, and being 'stowed' daily by a 'flushing' machine, that filled the wastes fully with wet crushed redd, that was brought down from the surface. The coal worked was the Dysart Main seam and this first 'slice' of coal would be mined from the lower half of the seam. Each heading would be fully 'side-stripped', until there was only about a 30 feet stoop of solid coal, between each worked-out heading.

After a spell on the bogies, Jackie was then promoted to be the 'wheeler' on one of the braes, a job of no mean responsibility, that involved some very careful manoeuvring and braking skills, that sometimes needed to be learned the 'hard' way. Time was money to the strippers, and a runaway hutch was remembered 'fondly' with a sharp crack on the ear, by one or more of the disgruntled miners. Jackie reckons that is why his ears seem larger that they ought to be!

After about two years of this type of work, Jackie felt that he was hard enough and fit enough to take on the better paid coal-getting work, and so, as 'boy' to another miner, he began his coal face experience. This paid the slightly better wage of four and sixpence a day as a drawer, that further toughened up the young miner, who by this time, was casting his eyes at the coal face, in that he felt that he too, could make a go at the stripping. A place at the face soon followed and before many months had passed, Jackie was holding his own against the best of the miners, to the extent that his weekly earnings now matched the rest, at approximately four pounds ten shillings to five pounds per week.

By the time Jackie had finished in Lochhead pit, he had performed every face job and preparatory job that there was to do, except 'steel-drawing'. That was a task that involved a great deal courage and intrepidation on the part of the miners concerned. Not for nothing, was this task euphemistically described as *'This is Your Life!'* Thankfully, Jackie was rather on the light-weight side for steel-drawing, which at times, was very demanding on the strength and disposition of the participant. It seems ironical that Jackie Penman did spend a lifetime down the coalmines, in view of the great tragedy that was visited on his family, when his father died in harness in this self-same coal mine, after being overcome by one of the underground deadliest gases known to miners. White damp is an odourless, colourless, tasteless gas, which in it's unpredictability, creeps up on it's unsuspecting victims causing nearly instant death.

It is a silent killing gas, that is however detectable by a modern, miner's safety lamp.

See chapter: - 'It's an Absolute Gas!'

＊

# Chapter XXV

## It's An Absolute Gas!  Part I
## The Air that I Breathe! –
## *or Protocol Harmful?*

It has been said in the past, that out of all the modern day instruments that a Mines' manager had at his disposal, the one instrument that should have been his No. 1 priority, was his office barometer and it's daily readings should have been made compulsory and available to all Mine officials!  *(See 'Prognostics of Weather' at end of chapter!)*

Any self-respecting deputy who had been properly trained and was worthy of his calling, and who performed his duties in Lochhead pit or the Surface dipping, should have in the winter months, cast a jaundiced eye towards the north and ensured that he had taken an up-to-date barometer reading, before he ventured underground. He would have also ensured, that the safety lamp *(glennie)* that he was legally obliged to carry with him at all times, was in full working condition, with an adequate spark mechanism and a full reservoir of naphtha.  He was also duty-bound to test the lamp in the lamp cabin before accepting it, and having done so, would head for the down-going cage, or man-haulage bogies for the start of his shift.

There was a Standing Orders Procedure (S.O.P's), that dictated he contact his opposite number on the preceding shift before entering his section, and take note of any reports that needed to be passed on and/or be acted upon.  This was the normal hand-over/take-over procedure as demanded by regulations.  As a deputy, his was the overall responsibility for the general and individual safety all of the miners and men working in his section, with regard to the state of any new or existing workings, the quality of breathable air and it's content and testing for known, prevalent and unwanted gases.  He was required to know the whereabouts of every manjack on his 'kalamazoo', *(manning register)* and receive prior warning of every visitor/s to his section.

*(During any shift in any section, there would be a procession of Electricians, Engineers and maintenance men, entering and leaving the section at odd times, who were duty bound to report to the deputy, or leave word as to their intentions and probable destinations within the section.)*

In every coal mine of reasonable age and standing, there were a growing amount of worked-out sections where the coal has been extracted to the maximum, and were therefore increasingly by-passed by newer sections being developed, beyond the point of the exhausted extractions.  These worked-out, though sometimes not necessarily abandoned sections, were either screened-off, had double air-tight doors fitted, or more likely if the section has been a troublesome section, they were 'stopped-off!' *(Sealed with solid packings of tightly compressed wet redd and possibly 'bricked-up' as well, or even force-filled with liquid cement if it were a section that was susceptible to self-ignition, or was known to contain poisonous, toxic or noxious gases!)* The main purpose of these precautions, was to 'seal-off' or minimise the ingress of oxygen to such old workings, to negate or reduce the risks of internal or self-combustion, especially within an knowingly suspect or gaseous coal.

It was inevitable, that most working coal-mines did have many old and disused workings near to the pit-bottoms and, were therefore by-passed 'early on' in the long passage of the forced, circulating, air system through-out the coal-mine. It was therefore of paramount importance, that constant checks were made on the circulating air quality, before and after such 'stoppings', especially in conditions of depressed atmosphere when the barometric readings were low.

*{In the case of the 'Baund' section in the Old Dook (where there was reputed to be upwards of 400 stoppings), slow, mouldering unseen fires were known to exist and indeed, were regularly inspected over a long period of time. (Measured in years!) The No. 1 level to the right of the Old Dook, which ran though the middle of Nicholson's Dook and on to the top of the Victoria Dook, was a by-passed section which, in fact, remained a main, underground airway. The large 'Baund' section, which lay between the Old Dook and Nicholson's Dook, which was permanently sealed-off, had a good, steady, air supply passing through the section, which, by virtue of it's regularity, carried away the small percentages of unwanted gasses that did manage to escape into the flowing mine air, thus rendering them ineffective. (The air currents that passed along the Old Dook's No. 1 level through the 'Baund' section, was in fact, one of many parallel air-flows within the Old Dook, that was quickly expelled through the Victoria Mines and up the Victoria shafts, without contaminating the remainder of the Old Dook.) An additional advantage with the planned system of coal extraction in use in Lochhead at that time, was that with the differing levels that were developed in the Old and New Dooks, each level had it's own share of the air supply, that could be exhausted directly to the return air-ways, without contaminating the rest of the workings. In other words, each and every parallel level was being vented individually and systematically.}*

Lochhead pit at one time had four separate return air outlets. Two at Earlseat, one of which was a ventilation up-shaft only, while the other was a stair-pit* escape shaft. The old 78 yards *(approximately 71.25 metres)* deep, Duncan pit shaft *(sunk into the Dysart Main)* near the Check-bar road, could be made to wind men, and the Victoria pit shaft at West Wemyss was also an escape shaft, that was also capable of winding miners. *(Up to 1952 at least!)*

*The over-riding school of thought, was of course, not to allow these mine gasses to proliferate to any great degree. The best and easiest solution was to ensure, that copious amounts of largely, uncontaminated air was channelled or forced into all mine workings, irrespective of their depth or inaccessibility. As already discussed, the chances are, that the oldest and most vulnerable workings were usually in the vicinity of the pit bottom, where, if gasses were present and allowed to proliferate, they would be more likely to contaminate the remainder of the workings, if the circulating air were so conducted or channelled in their directions.*

Breathable atmosphere, or preferably, the breathable mixture of gases that we take for granted on the Earth's surface, is the best quality that there is, for the simple reason that it is forever revitalising itself naturally. The oxygen content, which amounts to near critical 21 percent, is the life-supporting gas that is essential to our continued existence and, this percentage of oxygen hardly varies by as much as point-one (0.1) percent on the Earths surface. When this breathable mixture of air is forced down a mine *(or pit, or shaft, or dipping),* this balanced mixture can no longer be guaranteed or even sustained. Mine air, from the moment it leaves the surface of the ground, can no longer be revitalised! Mine air begins to become contaminated the moment it enters the mine. The amount of contamination that is allowed to develop, or proliferate in the air supply, greatly depends on the volume of air that passes *(is forced)* through the mine, and how fast it circulates. Much also depends on the amount of men *(and animals)* employed within the mine *(using oxygen),* the methods of extraction used within the mine, if there are much and many explosives used *(and the types)* and if there any smouldering, mouldering, abandoned sections, or out-right naked flames used for illumination purposes. *(Naked flames and 'flash-type' explosives were banned for underground use, as the Pit or Mine entered the Part II phase.)*

If comparison can be made with the relatively, still air that a person breathes on the surface with impunity, and the always moving air that a miner feels in the depths of a coal-mine, then one might be forgiven for thinking that the sheer body of moving air in a coal-mine, will surely disperse and render innocuous, all and any extraneous, natural mine gases or man-made gases. Not So !!

Of the most dangerous and poisonous gases that were found in Lochhead pit, there were several that every deputy was constantly on the lookout for, and one in particular, especially at times of low atmospheric pressure. Even though the following names are not technically correct, the inherent mine gases were commonly referred to by their euphemistic nick-names:- i.e. Blackdamp, Whitedamp and Firedamp, and the lesser known gases : Stinkdamp, Afterdamp and Flashdamp.

Why they were all collectively known as 'damps'? I can but make an inspired guess! In the German and Dutch languages, *'dampf'* is the word for 'vapours' or 'steam' or 'smoke!', which, to my mind, adequately describes some, or most of the small amounts of tainted airs that I experienced and breathed in some parts of Lochhead pit, especially when some of the isolated worked-out sections within the Dysart Main began to 'smoke!' *(All too often!)*

**Whitedamp,** is the name given to mine air contaminated by carbon monoxide. (CO) It is colourless and odourless gas. It is also inflammable when mixed with air and other gasses to a given proportion, and highly poisonous to man. It is produced by the incomplete combustion of carbon or carbon compounds, of which coal is the initiating ingredient, especially such loose coals that has suffered crushing and grinding and has been left and 'sealed' behind packed stoppings! *{Within, and near the middle of the Dysart Main seam of coal, there was a 10 to 12-inch band of Myslen coal and stone with a known oily content, that was known for it's propensity to help propagate low-temperature, self-ignition for which two of the named coals within the Dysart main seam were prone to! (The 'Clears' coal, just above the Myslen, and the 'Nethers' coal just below it!) It was scientifically established, that both the Clears and the Nethers coals had a low self-ignition temperature of approximately 172 degrees Centigrade.}*

Whitedamp, in its toxicity, gives little or no warning of its presence. It appears, it is breathed, it is quickly absorbed into the bloodstream and within a short time, the unwitting and unsuspecting victim is dead!

*(Carbon monoxide is an extremely poisonous gas to miners. When breathed in, a miner will act as though drugged, showing signs of quick tiredness or drowsiness, or even showing symptoms of unco-ordinated movement. This is quickly followed by increasing pains in the back and limbs of the victim, who will also experience the rapid seizure of a severe headache. The victim will then suffer acute delirium, quickly followed by the onset of death. If a miner breathes air which contains even the smallest quantity of whitedamp, his lungs tend to absorb the carbon monoxide, rather than the oxygen content of the air. Oxygen is life-supporting! Carbon monoxide is not! When carbon monoxide is breathed, the gas is very quickly absorbed into the bloodstream, to the exclusion of oxygen, where it quickly builds up with a cumulative effect to produce the effects described above.*

*The most damning characteristic of the inhalation of this toxic gas, is that the victim is rendered quite immobile, even before he has realised that he has been poisoned!*

For a deputy to catch on to the presence of whitedamp, he must be very alert and be in a position to inspect a lighted lamp, before the air is even slightly polluted with whitedamp. Whitedamp does have a tendency to slowly rise to the highest point of any given workings, but this really only happens when the circulating air supply is less that adequate. Even a mediocre air circulation, is usually quite sufficient to dispel any pockets from forming.

For a deputy, close inspection of the normal burning flame in his glennie, will show a brighter, burning flame, which seems to taper upwards within the lamp, with a narrowing flare. In greater quantities, the lamp will show a brief, but flaring burn before extinguishment of the flame, with the inability to re-ignite in the presence of this poisonous gas. He therefore, would have to return the lamp to an area of known untainted *(clean)* air, to facilitate re-ignition. Whitedamp is therefore detectable and identifiable by safety lamp!

At Lochhead pithead, there lived and thrived within the Joiner's workshops, a very healthy aviary of yellow canaries, their very presence and existence being for the underground detection of this poisonous gas, amongst many others. Their breathing rate was such, that the merest whiff of whitedamp *(or, almost any damp!)* would topple them from their perch and if the deputy or the miners using them for this very purpose, were as alert as they should be, and instantly back-tracked, or returned them to cleaner air, then, the canary did have a small chance of recovering. A canary or a mouse when carried underground to the possible source of this gas, will succumb up to twenty times quicker to it's effects than a human being, thereby giving a much earlier and timely warning than a deputy with a glennie. *(Safety lamp.)*

**Blackdamp,** *(sometimes referred to as Chokedamp.)* is the name given to mine air *(atmosphere)* tainted by carbon dioxide ($CO_2$), a naturally recurring gas found in small quantities within the atmosphere. It is a colourless, odourless gas, a compound of carbon and oxygen. Carbon dioxide as found in the coalmines, came from two sources. It was formed by the complete combustion of carbons *(burning)* and the result of constant blastings of coal and stone. It is also an expellant produced by the compound breathing of many miners. *(And pit ponies.)* In those old and disused workings which had been taken to the 'dip', and which had been sealed-off, the oxygen content therein was either very low, or had been entirely consumed by the slow process of combustion, which had not been allowed to proliferate, due to the cutting-of, or sealing-of, from the circulating air supply. These old dippings and workings were carbon dioxide rich, partly for the reason that carbon dioxide is heavier than air, and therefore tends to sink to the bottom of the available space, and partly for the reason that these workings were starved of air, so long as the barometric pressure remained between constant narrow levels. Within the workings of Lochhead pit, if the barometric pressure went unusually high, there would be a slight increase in air pressure, with the resulting egress of extra air into these old workings and stoppings, but to no great increase in the production of whitedamp. The incidence of blackdamp was at it's most acute, when the weather was coming out of the North and the barometer was down and falling. The weakening of air pressure as felt underground, would precipitate the seepage of blackdamp out of old abandoned and otherwise sealed workings and to the extent, that the amount of blackdamp released, would be directly related to the depth of the barometric depression. This was the time, when knowledge-able deputies looked to their duties and hurriedly recalled to mind, the standard lamp tests for varying degrees of blackdamp.

*Carbon dioxide is not considered to be a poisonous gas, even though it does not support life. It is also non-combustible! It is heavier than air and tends to gather in the bottom of roadways, in by-passed workings and in closed-off dippings, where there is an insufficient flow of air. It's effect on miners breathing this gas, is to exclude the oxygen content of the air from the lungs, thereby inducing a choking sensation with the inclination and need to breath-in even greater quantities of the polluted air. Air that has been polluted with carbon dioxide is readily detectable by a glennie. The flame within the lamp will grow progressively dimmer until complete extinguishment. Blackdamp is probably the least dangerous of mine gasses, in that it is an extinctive gas that will douse the flame of a glennie lamp, before it becomes lethal to the miners. (Air that will not support a flame, will sometimes support life!) If a miner is overcome with the effects of blackdamp and provided that the miner is rescued quickly enough, then the chances of a complete recovery are good!*

**Firedamp**, the name given to mine air contaminated or tainted with Methane, *(light carbureted hydrogen)* and sometimes referred to as *Marsh Gas,* is a colourless, odourless and tasteless gas. ($CH_4$). It occurs as an occluded gas *(retained, inherent)* to a varying degree, in different types of coal strata. It is not a poisonous gas, but will not support life. It is a suffocating gas if it is breathed in it's pure form, in that it will exclude oxygen from a miners lungs, but it can be inhaled in small quantities, producing no other effect rather than slight giddiness, that will soon disappear when untainted air is breathed.

Methane gas, which diffuses rapidly in the air, can be a very explosive or inflammable mixture, when mixed to certain, wide proportions with mine air. When sufficient amounts of methane gas and air is formed into to a combustible mixture, this then is called Firedamp. Small quantities of methane in the air supply, less than five per cent, will usually not be enough to become inflammable, unless the moving air is so slow as to not disperse the small pockets of methane, that will surely gather in the heights and roof spaces of the workings. At more than 5 per cent, but less than 7 percent, the methane mixture is inflammable and burns quite quietly, *(not forcibly!).* At more than 7.2 percent, the mixture becomes explosive, with the violence of the explosion being at it's most volatile at approximately 9.5 percent mixture. As the mixture increases, the volatility decreases until the mixture approaches 16.5 percent content. At this point, its propensity to explode is nearly nil, even though it still retains its flammable qualities. As the mixture of content approaches 29.5 percent, the methane content tends to extinguish it's own flame, thereby rendering it into an enveloping, damping, extinctive gas.

*(A firedamp mixture is at its maximum volatility, when there is just sufficient oxygen in the atmosphere to ensure complete combustion of the methane content.)*

In addition to the contamination of mine air by methane *(The firedamp mixture.)* There are several other mine gases that can either help or hinder the volatility of the fire-damp mixture. 1. Olefiant gas acts upon the mixture so as to make it 'sharper', as though it had just exuded from the strata, making it more active, with a tendency to agitate the burning properties of the firedamp mixture and render it more dangerous. 2. Carbon monoxide, when introduced to the firedamp mixture, tends to increase the 'explosive' range of the mixture, thereby rendering the mixture more volatile with an increased 'flame length' and linear spread. 3. Carbon dioxide, when introduced to the firedamp mixture, will dampen the volatility of the firedamp mixture to a degree, which depends on the amounts of carbon dioxide introduced. If the firedamp mixture is at it's most volatile point, *(i.e. 9.5 %),* the addition of approximately 15 percent carbon dioxide will render the new mixture non-explosive. 4. Nitrogen, which comprises 79 percent of our atmosphere is a wholly inert gas and does not help ignition of a firedamp mixture. Its effect is exactly the opposite and when mixed with firedamp at its maximum volatility to a degree of approximately 17 percent, has the effect of rendering it non-explosive.

**Flashdamp,** is not a common mixture found in the coalmine and is difficult to detect. It only occurs in certain circumstances, with the immediate presence of both methane gas and carbon dioxide. There are two favourable, but separate unlikely conditions where the presence of Flashdamp may occur. 1. Where carbon dioxide emits from the roof of a seam producing methane or, 2. Methane emitting from the pavement of a section containing blackdamp. In both cases, the mixture is formed when there is nothing to prohibit their intermingling. The theoretical composition of its most explosive condition, is approximately 3 parts carbon dioxide to 5 parts methane.

The Author, J.T.Beard writes - *After a careful investigation of the behaviour of this mixture in the mine, this author has suggested for it the name of 'Flashdamp', because the flamecap (in a Glennie) afforded by the marsh gas, appears only as a momentary flash when the lamp is first raised into the mixture and then promptly disappears. The carbon dioxide present in the mixture, destroys the flame cap that the marsh gas would give, but this does not occur until after the fresh air in the lamp has been exhausted.*

**Stinkdamp,** *(Hydrogen Sulphide, ( $H^2S$) or sulphureted hydrogen is a colourless gas, but having the unmistakable odour of 'rotten eggs!')* It occurs in such small quantities, that it is readily carried away by even the gentlest of circulating mine airs. It is heavier than air and tends to sink to the pavement and into the lowest of workings. The gas in it's natural state does not support life, nor is it combustible. However, this gas when mixed with air, to a ratio of one part in seven, then becomes a highly volatile, explosive mixture. This gas is a particularly noxious gas and when breathed in small quantities, will appear to derange its victims. When breathed in greater quantities, the victims will suffer instant prostration and unconsciousness, followed by death, if not instantly removed to fresh air and treated. The best means of detection of this gas is by its unforgettable 'stink!'

**Afterdamp.** This is not a gas in it's own right, but a heady mixture of gas products that follow an explosion down a coal-mine. The gasses can be indeterminate in nature and volume, much depends on the type and quantity of the explosive used. It is a variable mixture of carbon dioxide, carbon monoxide, nitrogen and small amounts of oxides of nitrogen. It includes water vapour, sometimes methane and occasionally hydrogen.

*

Author's Note:- *The following paragraph is an extract from the diary of David, Lord Wemyss, the second Earl of that name (1649-1679), and often referred to as the 'Great Earl David' of Wemyss. This extract is taken and quoted exactly, from the book 'R. G. E. Wemyss' by Andrew S. Cunningham, published by Purvis & Cunningham of Leven in 1909.*
*This under-mentioned paragraph was seemingly written in 1677.*

"Ther is one thing i most acquente my posteritie with, which is, that when my father left working of the Maine colle, 1616, he left it standing in watter drouned at Methill-hill dores : and when i began to work it in anno 1662 i dryed his Lo. weastes, i commanded boeth wt watter and my oune of yt colle wt a horsse work till 1670 yeirs yt my stone mynd dryed all as it is now going wt 20 collers in itt. But when i dryed my father's wasts, wher watter head stoud so many yeirs, in Sumber in June of 1666, it did tak fire at the crop, so putt itt out befor 1662, and smothered it out att that time. But in anno 1674 the wasts yt i head wrought myself did tak fire yt i lighted a candelle att the hoole colle ualle, upon which i, finding it burning, was sore putt tu itt what to do. In end. I caused all the colle-bearers tak their crilles one their backs and i stoude bessids till i did sea them carie out the small colle burning, and in 24 hours' time i gott itt putt out totally, and then i stoped up from all the aire i could, and so itt stands to this day (dammed up wt claye) and reade. This being done, i consulted with the colliers the resone of ths fire, first in my father's wasts and then in my owne, and we fand it for sertane truthe yt when watter shall stand for a long time, if ye drye yt place and it wear 20 yeirs efter this time, itt will tak neid fire : Therefor noe natter most stand in any weasts of yt colle (or in Methil colles) ever heirefter."

*This author is fully aware that the above paragraph shall probably cause confusion, frustration and perhaps incredulity in the minds of most readers, who are not familiar with the rudiments of the Auld Scots tongue, especially if they are not Scots born or not of mining stock. I have therefore rendered below, a bold, non-grammatical, but free interpretation of the Earl's words. (At least, I hope I have?)*

There is one thing that I must write down for posterity - which is, that when my father (the first Earl) stopped working the Main coal (Chemiss) at Methilhill in Fife in 1616 A.D. - he left the pit/mine lying deep in water up to it's entrance. When I, (*the second Earl*) returned to this same working in 1662 A.D. - I caused the old stone mine and the low level wastes inside (disused or abandoned workings) to be emptied of standing water. By the year 1670 - and with the use of a horse, the emptying and drying out of this mine had been achieved, and I now have 20 colliers (miners) working in this mine. However, after I had completed the drying-out of my father's old wastes, which had stood in deep water for so many years - the dried-out wastes (being a mixture of old coal and slag) commenced to smoke and burn. *(Spontaneous combustion?)* To extinguish the 'fires'- I smothered them by 'stopping-off *(completely blocking)* all of the roadways prior to 1662 - into my father's old workings.

Later on - in the year 1674 A.D., the new wastes that had been produced by the new coal extractions had also 'taken fire'. I then , myself, took a lighted candle up the 'whole coal face' and saw that it was all slowly burning - which, of course, left me with the problem of what to do next? After some deliberation, I caused the collier's to dig-out the smouldering coals and carry the same coals in creels/baskets upon their backs, and out of the mine.

Within 24 hours, all of the 'burning' coals had been extracted, with the 'fires' totally extinguished, with myself standing to one side in overall supervision. On completion of the removal of the burning coals - I ordered all roadways into the new workings to be 'stopped-up' with wet clay and redd, to the exclusion of all air. The mine and roadways therein to remain untouched to this day! When this was completed - I consulted with the experienced colliers as to what may have been the cause of the spontaneous combustions. The answer I received was the 'certain truth', that if old wastes have lain for up to twenty years submerged in water - when that water is next removed, (or drained) the subsequent dry wastes are then prone to catch fire!

The upshot of this revelation was that I then decreed - that in future, standing water would never *(where humanly possible)* be allowed to gather in the waste of any coals, *(extractions)* wrought from the coal seams in the Methil area. (Wemyss Estates). *(An old word of interest in the above original paragraph mentions 'neid fire'. Chambers Scots Dictionary describes; 'Neidfire' as:- 'Fire produced from the friction of two pieces of wood'.)*

It would appear from the above, that the apparent dangers from the effects of spontaneous combustion were realised by the second Earl of Wemyss, who indeed, took various steps to minimise the risks to his mines and his miners. I feel that with such a 'hands-on' coal owner, the Earl's instructions would have been carried out to the letter. Non compliance would probably result in the 'docking of wages!'

Author's Note: - *The following paragraph, is an extract from the Second Statistical Account 1845, (Fife) - and refers to a area on the Fife coast between West Wemyss and Dysart.*

***Prognostics of the Weather:*** - *"*The following prognostics of the weather have been collected from individuals employed in the Collieries and Ironstone works, viz. That before a storm of wind, a sound not unlike that of a bagpipe or loud buzz of a bee comes from the 'metals' in the coal-pits, but that previous to a fall of rain, the sound is much more gentle ; than that twenty-four hours before a storm of wind and rain, there is a blackdamp at the bottom of the ironstone pits, and through the waste; a damp so great, that a lamp will not burn ; but that, before frost, the air below is clear, and that a candle or lamp will burn easily."

\*

## Anecdote or Lochhead Myth?

### *It's an Absolute Gas ! - Part II.*

In or around the mid-sixties, when Lochhead pit was under the management of one Mr William Forbes, there took place a specific incident concerned only a few men and on a single shift, that almost defied logic. It came to light completely by accident and only after prolonged investigations by a determined, learned official, who shall remain completely anonymous throughout this narrative and into the future. This official was one of a very few men who was instantly trusted and respected by the said Mr Forbes.

As a manager, Wullie Forbes as he was invariably called, was the sort of manager who 'led' from the front. He had great confidence in his own abilities, was very vocal in his instructions to lesser beings and suffered no fools, gladly or otherwise! He was a confident self-made man and he was also the absolute manager of Lochhead pit! 'Red Mick' as he was also endearingly known, would listen to advice and then carry on in his own inimitable fashion. He ran that coal-mine (*as far as the Union would allow)* with a rod of iron and 'woe-betide' any miner who managed *(very easily)* to bear-up on his wrong side. He was thought to believe, that he was a man's man, but mostly only succeeded in alienating most of the miners with whom he had dealings. He did had a choice 'turn of phrase', in that his every conversation was liberally sprinkled with meaningful expletives and descriptive, improper nouns of an offensive nature. He could be classed as a one man, debating society with a semi-captive audience!

Mr. Forbes was a man who probably resisted change because of it's negative effect on his overall authority. He liked to feel *(which was perfectly natural),* that he should be the one that should instigate changes in his underground domain, even to the exclusion of sound ideas from his subordinates and especially from outside sources, that were funded and in being, to help and serve the needs of a productive colliery. Wullie Forbes wasn't having any of it! If any 'equipment' was going to be introduced into 'his' pit, he first needed to be persuaded, then convinced that it's operation would come fully under his control and, if it conflicted with any of his 'ideas' or 'views', it would first be questioned, meaningfully criticised and then probably disparaged and rejected. At that point, it's life would be short-lived, and up the pit it would come. Willie Forbes didn't need any new-fangled ideas thought up by some suedo-intellectual 'nerdling', contaminating his pit!

At a time when every thing down the mine seemed to be going well and the production quota was well up to scratch and, at a time when the power-loading miners seemed equally content *(no go-slows, mini-strikes or exorbitant demands for extra money),* a certain outside, supporting, agency official pointed out to Willie Forbes, that his was the only mine in the area not to have this particular item of Mine's Gas Detection Equipment, which he was duty bound to accept and install somewhere within his coal-mines' workings, preferably in the return airway of a mainstream working section. Mick was 'conversationally' convinced that he ought install this automatic instrument, with the advice that it would 'look good' if he acquiesced to it's installation. Mr Forbes, not wishing to be seen to appear openly obstructive, agreed that it should be installed forthwith, but also decreed that he should be the one to chose it's ultimate location and periods of operation within the pit.

Willie Forbes was nobody's fool, much less than to an outside official, whose sole task in life was seemingly to prove that Mick's professional demeanour integrity seemed less than co-operative. Mr Forbes decided that the best place to install this automatic detection equipment, was directly in the return airway beyond a railway level, where the diesel pug operated on both the day-shift and the back-shift. If the air was being contaminated by diesel fumes or any other mine gasses, then this equipment would surely give adequate warning. After some close questioning of the official, Red Mick determined that the device could be adjusted to within quite a wide range of tolerances and readings, from a few parts per million, to approximately many parts per million. Red Mick made his decision, he would allow the installation of the apparatus in the nominated section, to operate within the hours of 10 p.m. at the start of the night-shift and until 6 a.m. the following morning, at the commencement of the day-shift. The device alarm was to be set to operate at a level of air contamination of twenty parts per million, *(20 p.p.m.),* and a repeater alarm was to be extended to the section deputy's work station by means of a long length of cable.

*(Mick knew very well that there was absolutely no coal production, brushing work (including shot-firing) or any transport of coal whatsoever, taking place in the section during the night-shift. All of the coal-producing miners were up the pit. He should know, it was his pit!)*

Mick's choice of this section and shift was a very safe bet, in that the only men working in that section on the night-shift, were the materials supply oncost men and the maintenance crews, neither of which were involved with production work that could pollute the atmosphere.

On a given Friday evening after the back-shift had gone home and the machinery silenced before the night-shift, the technicians appeared *(God forbid!)* and the apparatus was duly installed in it's appointed position. It was calibrated and adjusted to a previously agreed setting, *(Mr. Forbes made sure of this, by ensuring that the night-shift, section deputy was instructed as to it's calibration)* and tested, by the simple expedient of breaking a small phial of carbon monoxide within range of it's sensors. The extension alarm was duly run out, with the cable being hung from appropriate girders, with the repeater alarm installed at the deputy's work station. The alarm itself was actually an electric bell coupled up to a high capacity battery pack, set out of harm's way. After the device had been installed, the technicians duly came up the pit, went through the pit-head baths and supposedly reported to their superior on the Monday morning.

On the following Thursday morning, just after Mr Griller's office opened at nine a.m, one of the technicians was called to the telephone. On lifting the phone and enquiring as to the caller, the technician was subjected to an angry tirade consisting of 'effings' and 'blindings', and a very pointed instruction to "Get this 'bloody-effing' apparatus out of my coal-mine! The bloody-effing thing will not stop bloody ringing!" The technician handed the telephone to his superior *(Mr. Griller),* who patiently listened until the tirade was over and tactfully, but then firmly explained to Mr. Forbes, that since the alarm was constantly being triggered, there must be a reason for it's activation! He also explained, that since this was a registered installation in an N.C.B. colliery, he was duty bound to investigate and ascertain the cause of the automatic triggering. It would not suffice for him to just remove the device without investigation and explanation. Mr. Forbes, realising that he could not bluster his way or intimidate such a senior and experienced official such as Mr. Griller, reluctantly agreed that a small team of technicians would be allowed down the pit to carry out various tests in this particular section, and that they would appear next morning during the most active part of the working day.

On the following Friday morning, the team arrived around eight-thirty, where they were immediately conducted underground and accompanied into the suspect section by a delegated deputy. They ensured that the apparatus was correctly set-up and calibrated, and coupled to its battery pack. All was well, but in spite of the fact that the section was in full production mode with brushing shots being fired, the machine refused to trigger for the simple reason that the circulation air, though slightly tainted, was not badly contaminated. There was no reason why the device should be triggered, the appropriate reading had not been achieved! The current reading was well within the set limits, *(it had been set almost to the allowable maximum!)* and showed no likelihood of rising! The technicians were quite baffled and, in the light of no other explanation, deliberately re-set the device with a further field test and settled down to wait. Absolutely nothing happened! At least, to trigger the device. The technicians then agreed that there was no more that they could do at this juncture and decided to go back to the pit bottom, where they were 'wound up' at the onsetters' convenience. They would not report to Mr. Forbes as to their findings, at least not at this juncture, before they had a chance to report back to their own boss, Mr. Griller. The report duly came back to Mr. Forbes, who seemingly snorted in disgust at their findings, claimed that the equipment was faulty, was absolutely useless and that any one of his deputies armed with a glennie, could do a damned sight better job anyway!

Several days later, a second telephone call was received at the organisation with a demand, that the caller wished to speak with Mr. Griller. On receiving this request, Mr. Griller lifted the telephone and asked politely as to who was on the other end of the phone? The answer was given by a short sharp outburst, "I told you that this 'effing' thing was bloody useless, now get it the hell out of my bloody pit before I throw it onto the bing!". . . Silence . . . . . ! Mr Forbes voice returned "Is there anyone there?" To which Mr. Griller replied, "Yes!, and when you've calmed down, I will tell you what I intend to do". Mr. Griller was not to be bullied, he had not reached this senior appointment in his career to be the butt of Willie Forbes profanity. Besides, he knew quite well, that he could overrule any Mine manager on matters pertaining to the presence of mine gasses in that quantity. He quietly informed Mr. Forbes that he intended making a surprise visit to the colliery without any prior warning, and that he intended asking for complete co-operation from Mr. Forbes and his mine officials.

Mr. Forbes was not particularly looking forward to the impending visit of Mr. Griller and his technicians, and left word with his secretary and at the 'check-box', that he was to be immediately informed of the imminent arrival of this small team. On the following Monday morning and through into the afternoon, Willie Forbes waited in vain, no one showed up! At least, no person that even looked remotely like Mr. Griller. Mr. Forbes then reluctantly decided to go home for the evening and have perhaps, a couple of well-earned pints, there was always to-morrow and that was another day! There was also every chance that the over-sensitive apparatus might just somehow disappear during the night, one could never tell with some of those evanescent, hairy-assed miners in Lochhead pit.

After Mr. Forbes had gone home for the evening, work at the mine progressed as usual and coal-winding continued until 9-30 p.m., when the back-shift would be wound-up and the night-shift sent underground. At approximately 9-45 p.m., the clerk on duty was initially surprised, to see a white, boiler-suited individual come forward to the check-box and present his official N.C.B. credentials, proclaiming the individuals' authority and right to be taken underground without let or hindrance. The clerk issued him with a 'visitors' disc and waived him through, he had no authority, nor the gumption to question an N.C.B. official pass. He hadn't been warned either, as to this impending visit. Mr. Forbes had not thought to have his orders passed on to the night-shift clerk, for the simple reason that he never thought that any such visits would be made outside of 'normal' working hours. How wrong could he be?

Mr. Griller, on arriving at the pit bottom, was immediately approached by the night-shift deputy, who quickly confirmed the officials' identity and purpose and bade him wait until he had checked all of his men through! He would then personally conduct Mr. Griller to the appropriate section and apparatus to be inspected. Mr. Griller was duly delivered to the experimental location, where he was shown the familiar instrument, the power pack and the route to the extended repeater alarm at the deputy's work station. The deputy then left Mr. Griller with instructions that he must not stray too far from this location, as his was the responsibility for the safety of his person. Mr. Griller set to work immediately. He ensured that the power supply was adequate, he checked the calibration of the machine and he re-checked the quite high setting on the internal scale. It was functioning normally as far as he could see, so he then sat down and made himself comfortable in the nearest manhole. He then extracted some paperwork from his hold-all and commenced to read by the light of his cap-lamp. Every fifteen minutes or so, he roused himself to check the reading on the apparatus, but the readings seemed to be hovering around the 3 p.p.m. mark.

At approximately 1 a.m., the deputy appeared and enquired as to developments, only to be given a slow shrug of the shoulders and a non-committal answer. The deputy then suggested that Mr. Griller accompany him to his work station, where they could both enjoy their 'piece' in comfort in a lighted environment. Besides, he said, "There is the repeater alarm at my station, which has always sounded when the primary alarm has gone off! Not that I have ever discovered anything!" He pointedly remarked.

The deputy and Mr. Griller spent an enjoyable 20 minutes eating and sharing the contents of the deputy's piece-box *(Mr. Griller hadn't thought to bring anything!)* and discussing life on the night-shift. The deputy commented, that it really made no difference to him in that his work was already laid-out for him, with specific tasks to be performed on his nightly rounds, but did comment that he, Mr. Griller might find it difficult to keep his eyes open during the wee sma' hours and, that sitting down only realised the need to rest the eyelids for a few minutes. Mr. Griller took the hint and busied himself with the start of an negative report.

After about 30 minutes of relative silence, he was suddenly startled by the strident ringing of the repeater alarm, which sounded extremely loud in the quiet darkness. He roused himself quickly, stood up, and pondered. The deputy was nowhere to be seen and he came the quick conclusion, that he was well out of earshot. Mr. Griller was at a bit of a loss as to what to do, should he follow the route taken by the deputy, or should he follow the repeater alarm cable back to the primary apparatus? He pointedly sniffed the air, taking deep breaths and waited with a sense of quivering anticipation at the effect on his senses and mind. He looked at his out-stretched hands to gauge the possible effects to his vision and, also recited the nine times, multiplication table, to judge his mental competence and his reasoning. He reasoned that he had full possession of his faculties and his breathing seemed to be quite normal. He then made the decision, that whatever was triggering the apparatus was not life-threatening and that he had better investigate whilst the system was triggering! Taking a deep breath *(metaphorically speaking that is!),* he retraced his steps back to where the apparatus was set-up, and sure enough, it was still ringing! He waited, not really knowing what to do at this stage. The system was detecting something and at good strength, but what? Mr. Griller knew that he should do something, and common sense and training dictated that he remove himself from this environment and in to supposedly fresher air, but he was in a positively perplexing situation. The amount of pollution detected, suggested something unhealthy, if not toxic in the moving air, and he thought for a fleeting moment that he recognised the smell of something familiar. He dismissed it from his mind almost at once, he had momentarily forgotten that he was not on the main street in Kirkcaldy.

Against all his training and common sense, he decided to follow the course of the incoming air, knowing that it could only take him back into the working section and thence to the pit bottom. After about five minutes walking, he thought again that he detected the whiff of a familiar smell, only this time, it lingered much longer. He quickened his step and as he did so, he imagined that he heard a deep, rumbling sound that seemed to grow in volume for a minute or two - and then subside completely. The familiar, but slightly obnoxious smell, had now strengthened to the point that it was quite unmistakable, he was quite sure of that! As he progressed on, he came to the start of a roadway that he recognised as being the haulage-way for incoming empties and out-going full hutches, and then, his ears caught the sound of that same deep rumbling that he fancied he had heard earlier. He quickly directed his cap-lamp to the surrounding walls and locating a vertical manhole, moved swiftly within and switched off his cap-lamp.

It was rather an eerie sensation standing there in the darkness, with the deep rumbling sound getting progressively louder and quickly growing in volume. The sound rose to a deafening vibration and then diminished in volume, as the diesel pug hammered past in the complete darkness, amidst shouts of undisguised pleasure. Mr. Griller just simply didn't know what to make of it, but he was absolutely certain as to the cause of the apparatus being triggered. The air was heavy with the stench of diesel fumes, made worse by the high engine revs. Mr. Griller stood still in the manhole, not entirely sure as to what his next move should be! It was just as well! The diesel pug was returning, this time with it's headlight blazing and it's speed more subdued. It rattled past Mr. Griller's manhole with the two occupants of the pug fully engrossed in its progress. The sound of the pug died away and silence descended on Mr. Griller. He rationalised, that the great burst of diesel fumes would trigger the mechanism of the detection device, but he also knew that all under-ground diesel engines were fitted with particle extraction, cotton-waste filters in their exhaust systems, and that they were religiously changed on the expiry of their accountable running hours. That was a mystery in itself, surely the regular driver would log the diesel's running hours?

Mr. Griller listened! The sound of quiet rumbling had started up again and was already growing in volume. Mr. Griller did not have long to wait! Again, the pug was approaching in total darkness, which seemed to magnify it's deafening sound, and again, it was accompanied by whoops of undisguised pleasure, or was it anticipation? The noise of the fast disappearing diesel subsided and quietness again descended on the manhole. This was a bit of a mystery, the diesel was not in train and was not pulling any hutches, either way? What was going on? Mr. Griller had his suspicions - but instantly dismissed his thoughts! Men just didn't do that sort of thing, - or did they? Sure enough, the sound of the pug grew in volume and again, the headlight was on and preceded the pug on its slower, return journey. The diesel was still spewing fumes, but not as much as when previously trundled past, in the dark and at speed!

Mr. Griller was now quite excited, he knew that he was witnessing something rather absurd, but was fascinated nevertheless! He waited with eager expectancy and growing anticipation and sure enough, it was returning! The rumbling and the vibration grew louder and deeper and the pug hammered past in the dark, with a great out-pouring of stinking fumes. Mr. Griller was now certain he had the answer as to what was going on and why the detection apparatus was being stridently triggered. The miners were actually and practically living-out their boyhood dreams and fantasies. They were playing at 'trains' and 'engine-drivers' and they were using the diesel pug to fulfil their long, cherished wish to drive a real engine. No matter that this engine did not blow steam or have a shrill whistle! This was a real locomotive that ran on rails and gave the real impression of speed, especially in the dark and it didn't cost anything, the rides were for free! The oncost miners, supply men all, made sure that the bulk of their work was completed in the first half of their shift, so that when piece-time came and the deputy has passed on his rounds, the miners could then indulge their previously unrequited passion for 'trains' and play 'engine-drivers' till their 'hearts content!'

Author's Note: - *During my time in the Lower Dysart section, the rakes of full hutches leaving the section and the empty hutches delivered up around the 'loop-road', were both hauled in and out of the section by a four-wheel drive diesel pug, capable of pulling up to 50 or more, loaded hutches. The pug could pull in either direction with equal ability and sported a nine-inch headlight mounted at either end. The driver sat sideways within the loco, with observation windows on either side. It had a combined clutch/gearbox with three slots (gears) and could attain 30 m.p.h. The running hours of this locomotive were carefully logged and it was the duty of the regular driver, to ensure that the twin filters (baskets) were regularly changed every three days. (Or, at the maximum running hours!) This was imperative, to ensure that the diesel's exhaust fumes were kept to an acceptable minimum and any delay in the changeover to clean baskets, would result in dangerous fumes being pumped into the mine's circulating air.*

Mr. Griller was faced with a contentious dilemma, if he approached the deputy with his revelations, the miners would surely face the prospects of instant dismissal when the matter was brought to the attention of Mr. Forbes. If he didn't do something about it, then he was failing in his duty. He was for the moment, stymied on the horns of a difficult dilemma. What should he do? The miners could not possibly be allowed to carry on as before, their continued actions might just be the cause of something more serious going undetected and camouflaged *with the diesel fumes,* but he did not wish to be instrumental in the inevitable process of the miners being sacked. He did as any level-headed, sensible person would eventually do, he compromised! He stepped from his manhole when the pug was making its illuminated return journey and focussed his spotlight on the drivers' cab. The diesel pug stopped well short of Mr. Griller, which was just as well, for he had not as yet quite made up his mind as to what to say. Determinedly, he approached the diesel pug, the driver of which had cut the motor. The two miners had removed themselves from the loco as if to disassociate themselves from it, and its actions. Mr. Griller then identified himself to the two miners and politely asked, if either of them was the regular driver? He was greeted with an ominous silence! Mr. Griller then asked them who they were and what they were doing with the pug? The reply he received was totally as expected, the men admitted to being oncost workers who had finished all of their allotted supply tasks and were happily engaging themselves in playing 'trains' with the diesel pug, blissfully unaware that their games were running up and logging illegal hours on the running time of the purification filters of the exhaust system. Mr Griller paused for effect, then delivered of them, a very stern and meaningful rebuke, to the effect, that this would be their only chance and the next time that the apparatus was triggered by excess diesel fumes, the men concerned would be reported out of hand, with the inevitable sackings to follow. Just as Mr. Griller was on the point of leaving the miners, having delivered his uncompromising ultimatum, he turned to the nearest miner and asked, "Why was it that you drove in one direction at a slow pace with the headlight on, and then returned at speed with the headlamp off?" The older miner turned to his companion and nudged his arm, "You tell him!" His companion looked directly at Mr. Griller and replied, "We were playing at 'Chicken'! The idea is to see how far you can drive the pug in the dark, without crashing into the empty hutch stationed near the end of the straight. The driver who stops closest to it, wins the nightly, sweep money!"

*Mr. Griller did not report the miners to the section deputy, nor did he mention them in his report to the manager. He did re-set the apparatus to a more sensible figure to make it even more sensitive and left it ' in situ' for several more weeks.  He never did receive any more phone calls from Mr. Forbes, so he could only assume that either the alarm was not being triggered, or it had been quietly disconnected from the power pack.  He did later send one of his technicians down Lochhead pit to recover the apparatus, which appeared to have suffered no ill-effects, but gave no clue as to it's latter performance.  Mr. Griller could only assume that his timely intervention to the wayward miners had cured their predilection for irresponsible 'engine driving!'*

*Jock Moyes is seen here in a steel-girdered level, sitting on the 'controls end' of an A.B. short-wall coal-cutter. Note the electrical panel-box supplying power to a 'tugger' motor situated nearby in the 'main-gate'!*

*Jock Moyes is attempting to dislodge a damaged steel girder that has buckled under pressure. The adjoining 'fish-plate' can be clearly seen! Also to be seen is one of the steel rollers over which the endless, rubberised belt runs!*

*A coalminer is attempting to set a steel tree on to a wooden strap set against a broken roof. Note the crushed pillarwood pack on the right!*

*Bob Stevens and workmate in the top stable-end. Note the lower long-wall coalface behind them. Geo. Halley, shotfirer, is holding on to a large coal shovel.*

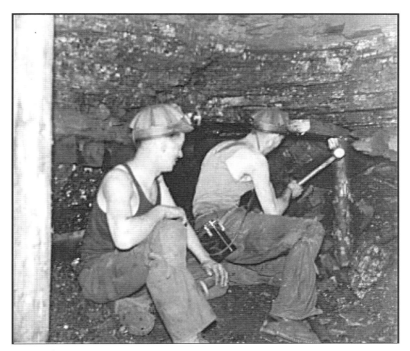

*Bob Stevens is hammering-in a wooden tree to support a wooden strap under part
of the lip of the middle leaf of the Lower Dysart Seam. The six inches of parrot coal
sitting underneath 7 inches of stone can be clearly seen to the left-middle of the picture!*

*Geo. Halley, shotfirer, seen here holding a miners pick, with workmate on the left and Jock Moyes holding
the 'T'-piece of his short coal shovel in his left hand. Their apparent cleanliness would suggest that this
photograph was taken near the commencement of their working shift!*

*A Deputy sitting in a stone manhole having his piece. Note the dirtiness of his hands compared with the whiteness of the bread sandwich! Note also that the Deputy is wearing a dark-coloured boiler-suit!*

*Geo. Halley and miner sitting amongst a plethora of downed coal. Note also the open end of the long-wall coalface immediately behind them. Geo. Halley always favoured the wearing of dungarees and shirt instead of a boiler-suit!*

*This photograph shows part of the middle leaf of the Lower Dysart Seam with the interleaved band of stone immediately above. The Deputy's 'lit' glennie sits before him! Is the miner to the left actually smoking?*

*This looks like a partial roof-fall in a very restricted location.*
*Something is being 'picked out'. Also, note the battery pack belted-on to the miner's rear-end!*
*The lump of coal in the foreground looks rather 'dirty'.*

*A machineman and his No. 2 can be seen, manoeuvring a 'bluebird' coal cutter. Note the horizontal drum of steel wire rope at the base of the machine!*

*Jimmy Hutton and workmates seem to be engaged in the re-setting of roof timber in what looks to be some very unsettled roof. Note the knee-pads and the heavily-studded pit-boots!*

# Chapter XXVI

## The Surface Dipping – *Aye Fond Memory!*

The Surface Mine at Lochhead colliery, commonly referred to as the 'Dipping', was commenced at a point on the East side of the Colliery buildings and driven downwards on a map bearing of 95 degrees. It was driven directly down into the rising metals from a datum height of 10,146 feet. (*146 feet above mean sea level.*) The angle of depression was approximately 1 in 3.8 and, it was driven downwards to take one or more of the shallower seams of coal, that lay well above the Dysart Main seam. The history of the Surface Dipping is of a small, but successful 'one seam at a time' operation, that produced a steady stream of coal for a considerable period of time.

Old Harry spent many years working in the Dipping, first as a coal-getter, but later as a brusher. Initially, the mine was only taken down to the 240 level, (*i.e. 240 feet below mean sea level*) where the first seam of coal was intruded into, developed and partially extracted. When this 240 level was worked out and abandoned, the Dipping was extended deeper down to the 540 level, where another different seam of coal was intruded into and partially extracted.

I have no intention of attempting to describe the coal extractions from either the 240 or the 540 levels, for the good reason that I have no information whatsoever, as to how and when the coals were taken. That the coals were taken is a matter of historical record, and Old Harry would have been the ideal miner from whom I could have gleaned more information than I would have needed. Unfortunately, the initial research for this book did not commence until after the death of Old Harry Moodie and other knowledgeable miners of his ilk. My only regret is that I should have listened more carefully when I was young man, to that which was related to my mother, when old Harry 'worked' the Surface Dipping!

*

## The Water Level - *but Dry as a Bone!*

The 350 level as it was known, was actually 350 feet below mean sea level, but approximately 495 feet below ground level. At underground datum height 9650 feet, at a Dipping length of 565 metres, two level mines were commenced. The first at an angle of approximately 95 degrees to the left of the dipping mine, on a map bearing of 10 degrees (*slightly east of north*) and driven in a straight line for a distance of 80 metres. The second level was at right-angles to this point, but in the opposite direction. This right-angled roadway was at a map bearing of 190 degrees (*slightly west of south*) and driven for approximately 40 metres. These were brushed mines, both driven slightly to the rise, with the left *(north)* road secured with 12' by 9' girders, whilst the right *(south)* road was secured by 10' by 8' girders. They were both taken directly through the seam of coal known as the Four foot seam, with both levels being 'brushed' above and below the coal to achieve the appropriate height.

Author's Note: - *These new level roadways to the left and to the right of the '350' bottom, were not actually driven from the side of the Dipping mine at this level. In order to establish a new, level, mine bottom to serve this 350 level, and provide a 50 metres long horizontal landing, two large tasks were accomplished: - 1. At a point on the Dipping mine and below the rising slope of the four foot seam, at 50 metres or so from the projected start of the 350 level, a horizontal level was commenced in direct line with the Dipping mine but, driven to strike the Four foot seam of coal at the 50 metres point. 2. The waste materials from this 50 metres, new, mine bottom would be used to in-fill and seal-off the lower part of the longer Surface Dipping mine which was still open to the now old 540, level that was now to be abandoned. This was successfully accomplished. The new mine bottom was laid with semi-permanent rails and the old part at 540 level was effectively sealed-off! It can therefore be said, that part of the landing ground at the 350 level had a solid, but false bottom.*

The first area to be developed was the right (*south*) side of the 350 level, but only by a short head, the left (*north*) side being commenced shortly afterwards. I can but estimate, the starting date for these two developments and would hazard a S.W.A.G., (*Scientific wild-assed guess*) that this took place immediately prior to the abandonment of the 540 level in 1949, with the exhaustion of the Branxton* coal.

The level roadway to the south was taken only to a length of 40 metres, where the new developing face was to start. The developers then made a smart right turn into the rising Four foot seam, and drove upwards for a distance of 100 metres, working exactly parallel to the line of the main Dipping mine. This then, was a new, long-wall, machine-cut face (*using a A.B.17-inch chain machine*) cut to the rise on an approximate slope of 1 in 3. This face would support approximately 12 strippers (*colliers*), including the leading main-gate stable-end and the trailing-gate stable-end. The conveyor was probably a train of jigging-pans, with the pan-engine situated at the top end of the run. This particular coalface was never planned to be a long term project and actually finished production on 18th January 1951. The extracted tonnage from this small face would probably have produced something in the region of 17,000 cubic metres of good quality coal, over a nine to ten month period. Not a massive amount of coal, but coal in a steady stream.

As mentioned previously, with the management's realisation that this south-west, long-wall face would probably be short-lived, the north running, level roadway at 12' by 9' was being developed almost simultaneously, and in the opposite direction. This 12 feet wide by 9 feet high level road was taken in to only 80 metres in length, before a left-angled turn was made, directly into the rising Four feet coal, where a new developing face was commenced. This development, (*8 to 10 feet wide at nearly the full height of the coal, and following a straight line*), followed the rising coal for a distance of 125 metres, and again, lay exactly parallel to the Main Dipping mine. This face was also cut as a long-wall, machine face and the slope on the face-line was approximately 1 in 2.5. This coalface did see the introduction of the A.B. 15-inch, long-wall, coal cutting machine, which was superior to it's predecessor in that it was slightly faster, with the travelling drive effected by 50 metres of steel cable housed on a horizontal drum, situated at the bottom front end of the machine. This machine was commonly, but erroneously named the 'Bluebird!' I don't really know why, because the simple fact was, that it was only ever painted brilliant red!

The main gate was 'brushed' a full three 'cuts' in front of the face line and 12 feet by 9 feet steel girders used throughout its length. The trailing gate at the top end, was brushed one cut behind and the girders used were 8 by 7's. (*8 feet wide by 7 feet high.*) This trailing gate was also the return air-way, which also served as the main supply route for the steel and wooden face supports required by the coal strippers on this coalface. *{This also was one of the employment venue of a certain young (but, naval trained), twenty-one year-old, coal-miner, who later went on to become the Chief Constable of the Kingdom of Fife!}*

This face line, (*soon to be named 'The Water Level'*) employed 14 strippers, including those working the two stable-ends. The taken coal on this side of the main Dipping mine (*north*) was just a little thicker, perhaps up to 32 to 35 inches, so the stripper's kneeling position was probably just a touch more comfortable, than on the south-west face and this was a right-handed coalface to boot! This production face probably got started just a few weeks after it's counterpart on the other side, but was already producing more coal, due to it's longer length and slightly, thicker seam. In any event, by the time the south-west face became exhausted, the Water level coalface had progressed a full 150 metres of travel, which equated to approximately 130 working days of production. This Water level, coal-face continued on for 200 metres in length, before both the main-gate and the trailing gate (*top return road*) were turned left through an angle of 35 degrees, and continued on for another 100 metres before resuming their original direction. The reason for this slight deviation in route, was to instigate the solid stoops that needed to be left untouched, as the extractions approached the underground area of the buildings of Newton Farm, near East Wemyss. The face-line itself, remained at its 125 metres length throughout the small, double-change of direction.

Meanwhile, back at the point where the first deviation was made, the 12' by 9' road way *(original main-gate)* was continued on in the same original direction and, following the level of the coal seam underneath Newton's Farm buildings. This roadway was now named 'the Barrier Level' and was to continue on for another 310 metres in the same direction, following the level of the seam, roughly parallel with, but approximately 50 - 75 metres from the new deviated main-gate, that now lay to the left.

*(As a point of interest, this first deviation of the main-gate was where I had my first introduction to working on my own, as a seventeen and a half year-old 'wood-laddie', supplying the bottom four or five strippers on the Water level, coal-face.)*

At the point of the second deviation in July 1951, the coal-face had been extended to 130 metres long and approximately half of a new 'place' *(stripper's Stint)* had been created, which in fact, was not filled at that time. During the course of the next ten months, the face line was further extended to approximately 150 metres, which meant that it was producing more coal than before, and with an additional thickness of a few inches in height showing in the seam.

*{During the month of March in 1952, I was upgraded to coal-stripper and given a 'place' (10 yard stint) on this coal-face by A.W. who, far from blighting my mining career in the Surface Dipping, actually enhanced it! He had actually 'saved' this place for me! I cannot suppose but wonder, if this was a complete 'Volte-face' on his part, or the probability that Old Harry had again exercised the 'Lochhead Mafia?'}*

When this face started being 'cut' on a regular basis, the coal produced from the coal-face, was conveyed down the run by a 20 inches wide, rubberised, endless belt. This worked well during the first 200 metres travel of the daily cut face-line. The coal came of the face and was loaded directly into a constant stream of empty hutches, one at a time, each hutch taking approximately 20 to 30 seconds to be filled. As this main gate was being extended daily, the distance from the mine bottom also increased. As soon as it was feasible, a twin, rail-track, endless-rope, haulage system was installed in the main-gate, which also ran round the 90 degree corner and onto the level mine bottom, thereby delivering races of full hutches right up to the start of the inclined mine and bringing races of empty hutches back to the loading point. This system worked very well, right up to the time when the main-gate made a 35 degree left turn for 100 metres and the another 35 degree turn to the right, to bring it back on course. *(Readers will remember, that the line of this coal-face was on a slope of 1 in 3, so, that in turning left and then right with the Main-gate, it was now virtually impossible for an endless haulage system to follow suit.)*

When the ever-extending main-gate made it's first deviation to the left and was therefore slightly inclined, the point where the coal came off the coal face was getting progressively higher and further away from the original loading point, on the level part of the main-gate. To overcome this, the final loading point on the level main-gate was retained and made permanent, and several lengths of 10 feet long, static, 'jigging-pans' were utilised to facilitate the passage of the extracted coals. These sections of pans were laid along the pavement of the deviated, inclined, main-gate and bolted together, so that the coal that came of the face, fell directly into the trough of the static pans and was whisked downhill by the force of gravity and into the hutches at the static loading point. The system worked so well, that the oncost* miners were forced into hanging lengths of heavy chain from the roof above the loading point, to impede and greatly reduce the speed of the gravitating coal! These 'pans' were left 'in situ' over the full 100 metres length of the deviation and when the extending main-gate made it's second deviation to the right, a thirty inches wide endless belt was installed in the new main-gate, to take the coal from the extending coal face, back to the top of this static pan-train and thence to the permanent loading-point!

*(This 30-inch wide, endless belt with it's daily extending 'barrel-end', did finally extend to a maximum length of 575 metres, until it too, came hard against the yet to be discovered, in-coming, geological fault, that terminated this successful coal-face!)*

Author's Note: - *During the life of this particular coal-face, approximately two and a half years, there were three, new, young, individually trained, strippers introduced to this coalface, who had all been nurtured in the Surface Dipping. These young men by name were Willie Briggs (23), David Moodie (18) and Willie Moodie (22). The names of the miners who stripped coal by hand on this run, were, in numbered order from the Main-gate: Danny Gough from the Rosie, Charlie Slaven from Denbeath. On the face-line:- Jimmy Hutton from Woodside, (Later Denbeath), Jimmy Higgins from Denbeath, Dave Brown from Kirkcaldy, Jimmy Rutherford from Kirkcaldy, David Moodie from Buckhaven, Jock Ritchie from Kennoway, Andy Cunningham from West Wemyss, Wullie Aitken from Buckhaven, Willie Briggs from Smeaton, Watty Gilbert from the Coaltown, Jock Sharp from Leven, Bill Moodie from Buckhaven and Davie (young Tot) Smart from the Coaltown. (First cousin to Bill and myself!) The shot-firers were George Halley from Methihill and Durham Dryburgh from the Coaltown. The alternating day-shift and back-shift firemen (deputies), were 'Chappie' Cunningham and newly appointed Jock Biggins, both from the Coaltown.*

As mentioned earlier, the 'Barrier' level *(mine)* was developed to the 310 metres length point, having been driven clean underneath Newton's Farm buildings, and the way was now clear to redevelop the ground that lay between the Water level and the Barrier level, a width of approximately 80 metres. *(The solid stoop that protected the Newton farm was now in place!)* At the 310 metres point, a development was commenced upwards to the left, and a 10 feet wide swath was taken up through the coal seam, to connect the Barrier level to the Water level. This then, was an additional 80 metres length of face-line to be added to the existing face line (*which was now in advance of this face*), ostensibly creating an 50 per cent addition to the coal output. This plan, if plan it really was, never actually materialised, in fact, it was barely commenced. This short developing face hit problems almost from it's first cut! Almost as soon as the coal-face opened, there intruded onto the face, a 40 metres long fault in the form of a hard stone dyke. The under-cutting machines could not cut this stone, it was far too hard and then, various efforts were made to circumvent the problem, even to the extend of drilling and blasting, but to no avail, and the decision was made to by-pass the fault. After about 50 metres travel, the fault disappeared as quickly as it had appeared and the face was re-extended again to 80 metres long. This face was cut and operated as an independent coal-face, with it's production being transported out along the Barrier level to join the main haulage-way at the first deviation point. (*The permanent loading point!*)

On the 5$^{th}$ of January 1954, this short coal-face was abandoned as a viable proposition, even though some, small, profitable amounts of good coal were realised, up to the abandonment date of the Surface Dipping in April 1954.

(*The original, longer, 'water-level', coal-face immediately to the left and above this short face, had been abandoned in October 1953, because of the permanent intrusion of a serious fault (stone dyke), which eventually intruded into the full length of that working face.*)

The amount extracted from these two runs would have been approximately 95,000 cubic metres of coal, over a two and a half year period of near continuous production.

On the 15$^{th}$ of October 1953, the final* coal face in the Surface Dipping officially opened for full production. (**The management however, did not seem to release this at this time!*) This latest coal face, was again a long-wall face and it had been fully developed before the abandonment of the 'water level' run. This coal-face was again on the south side of the 350 level, but this time, the four-foot seam to the east of south was to be taken. {*This new face lay immediately adjacent to the now abandoned south-west extractions. In fact, the original Main-gate for the abandoned south-west run, now became the top road (return air-way) for this newest coal-face. This however, did change a little later on, in that a new top road (trailing gate) was developed as the face-line progressed, but did have the knock-on effect of shortening the face line by about 15 metres.*} This coal-face was about 110 metres long, cut on the slope of approximately 1 in 3.5 and had a 'Huwood' under-load face conveyor. (*This was the left-handed run of previous notoriety!*) The slope on this run was consistent over the whole of its short life and I can not remember ever having had any trouble with an unpredictable roof. The average height of the taken coal was approximately 28 to 32 inches. Not very much as a vertical height, but considering the fact that the slope on the coal-face was 1 in 3.5 and that strippers wrocht leaning forwards, that in itself, gave an apparent 3 to 4 additional inches of head room. This coal-face lasted only until 23$^{rd}$ April 1954, which is given as the abandonment date. The total output from this face would have amounted to only 11,500 cubic metres of clean coal.

Was it worth the development costs? The answer must be yes, in that considering the relatively shallow depth of the seam, the very short transport distance to the surface of the mine, and the fact that this was the third series of developments that had materialised within the depths of this Dipping!

Immediately prior to the above date, all development work in the Surface Dipping was terminated. This was the last of the production faces to be worked at this 350 level. All of the miners (*save for a few on salvage work recovering used steel girders*) were to be transferred to the pit, where work of a *similar* nature would be found for them! Similar indeed! From working on my knees, (*which took weeks of unspeakable agonies to acclimatise to!*) to getting used to 'filling and drawing' in a 'stoop and room' environment, where Jackie Dryburgh (*my neighbour/ workmate*) had sometimes to sit on my shoulders just to reach the roof to set supports.

*

# 'Wullie' McD. Moodie

## *(An Exercise in Work Experience!)*

*\*Amongst the half-dozen or so oncost workers who worked at the Loading point on the 'Water level', there were three that I remember very well. The first was the loader (therefore the 'Pin!'). He was named Dod Salmond and he lived in West Wemyss. (He was also a keen, sea fisherman!) The second was a slightly-built smallish man who also came from West Wemyss. His name was Jackie Anderson, the very same old Jackie, who nowadays, is well retired, absolutely 'compos mentis', practises local and limited anthropology and frequents the West Wemyss Arms in the village of that name! The third oncost miner, one of the drawers, was an auburn-haired, well-built, twenty-one years old, young man by the name of 'Wullie' Moodie. Willie Moodie had originally joined the Police Force in Kirkcaldy, after doing his eighteen months National Service in the Royal Navy. Willie's initial short career in the Police service lasted all of fifteen months, when he suddenly announced that he was unsettled, and that he intended joining the coal industry and would seek employment at Lochhead colliery.*

*At this juncture, it seemed that Willie Moodie just didn't know exactly what he wished to do in life! He had been an apprentice butcher in Methil, a national service sailor in the Royal Navy, a probationary police constable in Fife, and now, a potential, trainee, coal-miner at Lochhead Colliery? It was not that young Willie was unsure of himself, quite the reverse. He was absolutely full of enthusiasm and brimful of self-confidence. It was just that at that time, he didn't know what he wanted from life! Now, he was to attempt a career in the coal-mines! To this end, he then approached Wullie Hampson, the manager at Lochhead and unsurprisingly, was immediately 'signed-on'. (Wullie Hampson would probably have done this with eager alacrity, if only to 'please' old Harry!)*

*To give Bill Moodie all due credit, he did not hang around. As soon as was feasible, he applied for and was selected for underground training, which was accomplished with almost indecent haste and before long, he was allocated an underground 'oncost' job in the Surface Dipping, as a 'drawer' on the water level.*

*Within a few short weeks, it became apparent to some, that far from being content with his lot, Willie's thoughts were already elsewhere. One of the diversions that Willie soon introduced to the other drawers at the loading point, was one that got right up A.W.'s nose. Amongst the supplies that often came into the section, was large lumps of pure, white chalk that was used extensively by deputies, shot-firers and brushers. Willie commandeered some of this chalk and used it to initiate a mental diversion amongst the on cost men, that involved them exercising their minds for a while, in trying to remember the names of all of the forty-eight states of America. Every remembered state was carefully and gleefully chalked on selected steel girders, the number of which, soon grew to magnificent proportions. The 'game' was played whole-heartedly by all of the miners involved and was much enjoyed, with all of the drawers plus many more miners, happily indulging in the cerebral, time-passing teasers!*

*A.W. did not like this turn of events one little bit. He was of the opinion that fixed wage miners were there to work all of the hours that they were underground and, that 'mind-games' could only distract them from their work. Besides, they were making one hell of a mess of 'his' girders! A few quick words and a few implied threats, were enough to put stop to this 'foolish' chicanery and return the drawers to more mundane labours.*

*Willie Moodie however, was not to be thwarted! He simply devised other mental games and puzzles that again involved the use of great quantities of chalk, but this time, spread over the sides of metal coal hutches. (I felt at the time, that this was a deliberate ploy on Willie's part to further frustrate A.W.) This of course, got so badly out of hand, that very soon, other colliers were beginning to believe that the meaningless hieroglyphics on the sides of the hutches, were a part of a diabolical conspiracy to under-cut their wages! (At least that was what A.W. implied!) Again, came the veiled warnings with the threat, that any miner caught 'defacing' the sides of hutches with chalked symbols, would be made to wash it of before being sent up the pit! (This man was entirely predictable as well as being an utter killjoy!)*

*It was then that A.W. seemingly got to thinking that the best way to deal with an 'young upstart' (who had only been down the coalmines for 'five' minutes!), was to attempt to isolate him and place him in a work environment where he could not possibly 'contaminate' the workers.*

*The top road (the 8' by 7' return air-way) onto the 125 metres long coal-face where this author worked, was long, rough, weighted, laid with temporary six-foot long rails and undulated madly! The returning air was hot, tainted and heavy with the rank stink of spent explosives. The working conditions for the wood and steel supply men involved much lost sweat, uncertain footing on the steep ascent with a loaded bogie and a hazardous slow journey, along the length of an ever-extending torturous route. This was very physical work in a hot, fetid, unpleasant atmosphere, where water needed to be drunk in great quantities! This then, was the environment where A.W decided to place the independently minded, Willie Moodie.*

*The small team into which Willie Moodie was introduced, were already a hardened bunch of tough cynical miners, who knew every inch of this impossible route and played the 'oncost game' to the full. Neither Wullie Hampson, nor the resident parvenu, could tell them exactly, just how much work needed to be done to achieve the end result, but the very fact that they now had an additional miner co-opted onto the team, was the result of weeks of lobbying for extra help. Now they had received it and A.W. 'secretly' wished them well! As for Willie, within a few days of joining this happy band, he seemingly found identification and almost instant rapport with all of them. He was perhaps the second tallest and heaviest, was more than capable of pulling and pushing his full weight, and much to their surprise, he was just as 'bolshie' as the rest of them when 'push' came to 'shove! He fitted in an absolute treat!*

*After several short months on the 'wood supply', Willie felt that the time had come to move on to better paid work. (This author, a coal-stripper at eighteen, his younger sibling by three years, was earning approximately double that earned by Willie at this time!) He felt that he was ready for face-training and quite relentlessly pursued the under-manager until he was accepted and, much to my initial surprise, was allocated an experienced supervisor in the top stable-end of the coal-face, under the tutelage of our first cousin Davie (young tot) Smart. (At the top stable-end of this coal-face, there was, over the three shift cycle, brushing, packing, steel-drawing, coal-cutting, coal-stripping, support-setting and filling! There was also five unallocated yards of coal on this face-line! Was this the use that it was to be put too? It would seem that there was more to A.W.\* than I had previously given him credit for!*

*Willie Moodie took to this training like a duck to water. He was strong and fit, keen to learn and positively burning with the desire to succeed. That what his younger brother had already achieved, also served as an added spur! - Incentive even!) He was also very aware of the fact, that there was more money to be made in stripping coal, than heaving wood and steel straps. Willie's training was progressive, thorough and successful, so much so in fact, that he was immediately given a new place on the coal-face, that had actually been slowly extended by another five yards to create another ten yards stint! (This meant of course, that A.W.'s production total must surely rise!)*

*Willie Moodie, like every other new stripper faced with the daunting task of a long, ten yards of standing coal, joined that group of semi-frantic miners who faced wide-pans, broken belts, lack of supports and tardy shot-firers (sorry George!) on a daily basis. He was now up against the clock and the pressing need to shovel coal while the pans jigged! He never faltered! He was often hard-pressed, was initially often late in finishing and sometimes had difficulty with steel straps, but he persevered and he succeeded!*

*Within a few months, it also seemed that there was a developing situation of sibling rivalry in the air. Geordie Halley, the shot-firer, on his way off the coal-face one day, casually mentioned to me, that I ought to crawl up the run and have a look at my brother's 'place'. I retorted "Why?" George merely smiled, saying nothing more and continued on his way. I was intrigued to say the least, but hesitated, for the simple reason that 'what could Willie possibly have done, that I hadn't done in the past?' I was soon to find out, for the reason that curiosity simply got the better of me and after a few minutes, I found myself making (crawling) tracks towards the top end of the coalface.*

At this juncture, I make no excuse for the wonderment that I initially beheld, even though I was quite dumbfounded! After I had taken a good look around his fully stripped place, I could barely countenance what I saw. His place had been 'brushed' clean like a billiard table, with all of the loose coal picked from the vertical face and cleanly dressed, and every scrap of coal and dust removed from the pavement. Also, every particle of coal had been removed from around the base of the 'trees'! The area had been hoovered! - or so it seemed! *(At this point, there was much 'trooping' of colourful, purple prose from my lips!)* The pit props had been lined-up like a troop of soldiers and formed two perfect ranks as though they were on parade. I shook my head in wonderment at the pristine image of this ten yards length, before the implication of the possible repercussions began to dawn on me. I certainly didn't wish to have either Jock Biggins or 'Chappie' Cunningham *(deputies both)* coming onto my place, and regaling me with insulting taunts of 'how clean was my brother's place in comparison to mine!'

I said very little to Willie Moodie at this time, but unusually, made to leave the coal-face by the top road. I didn't! I doused my carbide lamp and quietly waited in the return air way, until such times as I saw him crawl down the face-line, with his helmet illumination subsequently disappearing from view. I re-sparked my carbide lamp and crawled back onto the face. There was nothing I could do *(or would do!)* about the serried ranks of pit-props, but I could surely change my brother's painstakingly created image of 'bull-shit!' I crawled over the pans and into the condies, then scooped up several handfuls of spilled coal, which I then proceeded to throw and scatter all over Willie's ten-yard stint. I spent a few minutes gathering and carefully throwing a liberal sprinkling of coal and waste over its full length. Not enough for him to have his wages docked for an un-stripped place, but just enough *(I reasoned)* to earn him a reprimand from the back-shift fireman.

The next morning, I later learned, the back-shift fireman of the previous day had annotated his 'kalamazoo' with a remark, that the place second from top was a mite 'dirty!' This had of course, been passed-on to the day-shift fireman, who had attempted to take Willie to task for leaving a 'dirty' place. The comments fell on deaf ears, with Willie laughing his head of at the seemingly 'negative praise', with him thinking that this was only Chappie Cunningham's warped sense of humour! *(I never did tell him the truth! I am positive that he would never have believed me! - or would he ?)*

Unfortunately, and for reasons that I will not elaborate on, Willie's time in the coal-mines came to an abrupt end and was just as short as his time at any other of his previous occupations. All in all, this amounted to something short of two years. Willie had finally decided that the wearing of a dark uniform with a black and white checkered cap, was infinitely more rewarding in the terms of career prospects and job satisfaction, than ten yards of standing coal every day into the foreseeable future. And so, William MacDougal Moodie somehow managed to rejoin the Fife Police Force.

The rest of his long, on-going career, his progress, his success, is now recorded history, but he will still remind anyone who cares to listen, that once upon a time, he was a working Coalminer!

\*

*\* I distinctly remember being told by 'Big Sam'- leader of the upper 'wagon-train,' of an ugly incident that took place on the supply route to the top end of the coalface. This verbal altercation took place on a day when the down-coming supplies were late in arriving and the team were doing their very best to transport the much-needed supplies to the top stable-end. At a point halfway along the return airway, where the team had stopped for a well deserved short break and some refreshing water, A.W. suddenly arrived on the scene in what could be described as, 'a very breathlessly state and in a seemingly vile mood.' Such was the mindset of the man, that he instantly began to berate, first the team leader, and then the sweating men for their supposed tardiness and lack of commitment. He was spitting venom in an endless tirade, with never a chance for Big Sam to retaliate. He finally stopped, ordered the men to 'bloody-well hurry-up' - and then paused for breath! He then had the bloody gall to ask no one in particular for a drink of their precious*

*water! Needless to reiterate, no one volunteered their water flask, which only made A.W. all the more angry at their supposed dumb insolence. None of the miners determined to say anything at this time, probably for fear of instant reprisals (like being sent up the pit!). One young miner however, much to the astonishment of his workmates, did open his mouth and spoke very calmly and pointedly to a rather taken aback A.W., "Mr Wilkie, you may be able to teach me a tremendous amount about coal-mining, coal-taking and mining technology, but I feel that I can teach you a darned sight more about manners, civility, courtesy and man management!" Andra Wilkie seemingly did not reply!*

(That Willie Moodie never could learn to suffer bullies, hypocrites or charlatans ever! - and after that little episode, I am quite surprised that A.W. ever consented to Willie Moodie's face-training!)

*

# Chapter XXVII

## The Gentle Giant! – or Local Bogie Man?
### *(Anecdote or Woodland Myth?)*

Sandie Honeyman was once described as a man who merely wished to permanently commune with nature. He was a man who was large and well-built, was strong and healthy and by all accounts, a veritable 'gentle' giant, or at least, until such times as to when he was under the influence, which again, by all accounts, was fairly often. *(At least once per week.)* Sandie was a well known, local character who, in spite of his much-travelled appearance, remained something of an mysterious character to most of the youth population of the surrounding areas of Thornton, the Gallatown, the Boreland, West Wemyss, Dysart and the Coaltown of Wemyss. He was a frequent visitor to all of these places, as he did his weekly rounds in pursuit of only he himself, knew what for. He was seemingly a man of unfailing courtesy, at least to those few people that he deigned to speak to, and then, not for very long! His clothes were old and roughly patched and occasionally he would be seen sporting a new or clean item of apparel. His clothing, though never filthy or ragged, always seemed to be care-worn or slept in and always smelled of wood-smoke. His skin and countenance was ruddy and weather-beaten and in the summer weather, his outwards appearance gave the picture of glowing health. However, a close look at Sandie Honeyman bespoke an altogether different story. His eyes were not particularly clear, they were tired and almost listless. His large frame was just a little stooped and his good shoulders were just a small bit hunched. His gait, though strong and mile-eating, was somewhat forced with little spring in each succeeding step, and his expression when properly read, told a story that was somewhat different from the usual six days a week, down one of the local coal-mines.

Sandie Honeyman was different to other men, he was also quite unlike to his brother Stuartie, who was gainfully employed as the Dipping tailsman at Lochhead colliery. Sandie made no bones about the fact that he wished to commune with nature, and true to his beliefs, that was exactly what Sandie did! Sandie's abode was not merely a *pied-de-terre*. His permanent home was actually in the woods! He lived there! He slept in the woods, his 'shack' was in the woods. His everything was in the woods, and not just in the hot months of summer, but in the rain and winds of spring, the balmy months of autumn and in the deep frosts and wet snows of winter. Not for him, a warm caravan, or wooden hut insulated against the weather. Not for him the convenience of a flushed toilet, or even a kitchen sink. His home, deep in the 'firs', was as far removed from the small comforts of even a miners cottage, as it was possible to be. Sandie's home was entirely portable, in that it was constructed *(if indeed, this word is appropriate),* from old wooden fence-posts, broken timber, old railway sleepers, old sheets of corrugated iron, old and worn coal-sacks and scraps of hessian and rope, to hold it all together. It was reasonably waterproof in that it was not entirely out in the open, but situated between convenient fir trees. In fact, its location gave questionable rise to Sandie's address as regards the local 'dole' office, where was issued, Sandie's only means of support . . . His address was: - Sandie Honeyman, 'The Firs,' Earlseat. Also, to give Sandie his due credit, he was neither tinker, tailor, vagabond or thief. He was in fact, a proud man. He was as straight as a die and did not beg. He did not covet any of the material riches that he might have obtained had he worked, but remained a total enigma to those whose paths he crossed. Everyone in the surrounding area knew of Sandie Honeyman, knew that he lived in the Firs, and indeed, many of the older local children made it their business to visit the area of the lair of the local 'bogie-man', who was often seen on his many travels, collecting and 'purchasing' his meagre needs, usually around the closing times of the various co-operatives that abounded in nearly every village. No one really knew as to what Sandie lived on, but staleish bread and buns and tinned food, did seem to be figure enormously in his diet. He was to be regularly seen coming out of the local Co-op at closing time, with a small hessian sackful of 'cutting' bread *(day-old unsliced bread)* slung over his shoulder, and not only bread, but broken biscuits, scragends of bacon and bashed tins of meat and corned beef. This was how Sandy lived and he was well known to those managers who ran such Co-ops, from the Coaltown of Wemyss to the Gallatown.

*

*There is a fragmented story that was told many years ago in the Coaltown of Wemyss, and in the depths of the woods where Sandie had made his home. That this story is true, there is no doubt! It concerned only a few people, but some of those persons still living in the Coalt'n, can bear witness to its authenticity! Two of those persons are Jack Birrel and his lady-wife - Mary (Tod) Birrel! The other was the Coaltown's district nurse, Miss Robertson!*

*Not too far from where Sandie had made his den, there were one or more static caravans, where a few hardy, or dispossessed families lived! (Probably tinkers.) They lived there on a permanent basis, all through the heat of summer and through the very depths of Fife winters. At least one of the female residents was still of child-bearing age!*

*The winter of 1946-47 was an especially bad winter, with both day and night temperatures well below zero. It had snowed very heavily during the month of December, where the Coaltown area had suffered one of the severest snowstorms in recent history, with snowdrifts up to fifteen feet high against solid obstacles. Fortunately, this had the unusual effect of making some parts of existing roadways quite passable, with great stretches of the open fields relatively clear of snow, but frozen hard! The weather outlook was not predicted to change within the immediate future!*

*On one very cold night, just as Nurse Robertson had banked-up her small coal fire for the night at No. 31 South Row, she was alerted by a loud consistent banging at her front door! She knew that she had to answer it! That was her job in life, but she wondered just who it might be and what for! Nurse Robertson was well aware of her responsibilities and instinctively knew that there was no patient treatment pending, or that she had inadvertently missed-out on any patient or expectant mother visitations. To her intuitive knowledge, there was no female in the village who was remotely near her time. It must therefore, be a call of an emergency nature! She then answered the door!*

*On opening the front door, Nurse Robertson was faced with an ghostly apparition. It was tall, well muffled-up against the weather and looked all the world like a huge snowman! It wasn't! It spoke! - and it was Sandie Honeyman! - The very last person that Nurse Robertson ever expected to see! The man was near frozen stiff and appeared to be dog-tired. Nurse Robertson immediately beckoned him inside, but he was reluctant to enter. He slowly undid the wrappings from around his head and with difficulty, spoke! "You're needed at the Firs! - croaked Sandie, "One of the women from the caravans can't get her baby to deliver! You're sairly needed!" Nurse Robertson clutched her hands to her chest, "I can't come!, she soberly admitted, "I don't have any transport except my bike!" "You're badly needed!, repeated a shivering Sandie, "The womans' in great pain! - She needs you!" Then, to her eternal credit, Nurse Robertson realised instantly that she must attend! But how?*

*Sandie, on quick reflection, now saw the dilemma posed by his urgent request. He had managed to trudge the two miles over the snow-covered fields, passing through Lochhead Pit without losing direction, but could the district nurse manage to achieve the near impossible return journey? Nurse Robertson quickly made up her mind. She would attend! She was morally bound to, irrespective of the blinding weather conditions! She would get there somehow! She then checked her medical bag, ensuring a ready supply of necessities, and obtained and thoughtfully attached the long leather carrying strap to her bag. She then made preparations to leave. This was not going to be an easy journey and well she knew it! She did however, have faith in Sandie Honeyman. She quite correctly reasoned that any man who had the courage to turn-out in these conditions for a perfect stranger, would not fail her in this emergency! She therefore dressed accordingly and having left a note pinned to her inside front door, left her house, thereby placing her life in the doubtful, but capable hands of Sandie Honeyman!*

*At the commencement of this epic journey, Nurse Robertson could but only, hold onto the arm of Sandie Honeyman for the first two or three hundred yards, after that, she felt herself lagging behind and then being dragged along with Sandie's arm around her. Her strength, she knew, was not really up to this sustained exertion and she began to have serious doubts as to how far she could go!*

What followed on, has been carefully pieced together and is therefore, not pure conjecture! I have no certain knowledge of what actually happened during that terrible journey, but subsequent events stand for actuality. That she did arrive at the 'Firs' is beyond doubt! The tinker woman was safely delivered of her baby in the middle of a very cold night, and mostly due to the skilled efforts of Nurse Robertson.

*Nurse Robertson would never talk of her experiences of that terrible night, nor would she ever describe the part played by Sandie Honeyman during that awful journey. Sufficient to say, that after the first half-mile of travel, Nurse Robertson had no further recollection of events. As to how she actually got there, she claims that she has no knowledge thereof and no memory whatsoever! She has, several times since that night, imagined a vague recollection of Sandie Honeyman having stopped for a minute or two, of having her long scarf being wrapped around her head and face, and then being hoisted 'piggy-back' onto the shoulders of Sandie Honeyman. She was vaguely aware of two strong arms pulling her legs tight around his waist, and of her own arms being clutched around his neck. She can also recall hearing the sounds of hard, tortured breathing emanating from his great chest. Nurse Robertson maintains, that she must have been carried for a distance of one and a half miles, in the dead of a frozen night, to a destination of which she knew not where!*

The next day, during the hours of daylight on a cold, but clear day, Nurse Robertson was walked and escorted back to the Coaltown of Wemyss via the main roads by Sandie Honey-man, who seemingly turned on his heels as they reached the outskirts of the village. Sandie Honeyman, duty done, then departed. He wished for no thanks, received none, and returned to from whence he had come, in the dead of night - Back to the Firs!

\*

Of those who felt that they might know just a little of Sandie's hermetic life and habits, the village bobby in West Wemyss might have laid some small claim to having had close regular contact with Sandy. Once, every two weeks or so, Sandie would put in an appearance at one of the village pubs, usually on a Friday or Saturday evening. He, initially, would not make any trouble and indeed would sit and steadily sup his ale, until the effects of a warm coal fire and a heated atmosphere would make him quite garrulous. He was amiable at first, then, as the ale began to have its effect, he would become obstreperous when asked the sort of questions that everyone was dying to know. This would end up with Sandie being understandable upset and getting completely out of hand until the village 'bobby' arrived on the scene. Invariably, when Sandie got to this inebriated state, he had a propensity to remove his jacket and invite his tormentors to a fist-fight and sometimes, depending on his opponents equal state of inebriation, the challenge would be accepted. These loud, but not damaging altercations, would mostly end with both participants sitting on the ground unable to move, with one or more of them becoming sick with their unaccustomed physical exertions. At times, the village bobby did get involved with Sandie Honeyman and thought nothing of removing his helmet and tunic, and 'squared up' to Sandie, in an attempt to pacify him. He didn't want Sandie cluttering up his cells for the night with the inevitable repercussions the following morning. However, this 'camaraderie' that Sandie had developed with the village 'booby', did not extend to certain small misdemeanours that were sometimes committed by Sandie, in that one morning, the village residents were much amused to read of a small report in the local broadsheet, that a certain 'Sandie' Honeyman, unemployed, of 'no fixed abode', had been fined the princely sum of 'half-a-crown' *(12.5 pence)* at the borough Sheriff Court in Kirkcaldy, for urinating *(whilst under the influence)* into the doorway of the village branch of the S.C.W.S. *(Scottish Co-operative Wholesale Society.)*

*(The report did not state as to whether the offence was committed into the doorway of the Butcher's, the Baker's or the General provisions entrance!)*

Sandie liked to swim in the sea and thought nothing of running into the water at West Wemyss and swimming a full mile out into the Firth, where he would make like a basking grampus and 'blowing' like one too! He was as bald as a coot and sported a flowing black moustache, and with his long underpants, large physique and adorned upper lip, he looked the very epitome of an old-fashioned, beach guard.

'At home', he had an old, iron stove situated just at the opening of his dwelling, that served as cooker, heater and glowing 'camp-fire'. He spent many hours out and about in the surrounding woods cutting, breaking and gathering a sufficiency of both solid and dead wood for his personal needs. He had his own 'wood-pile' which he replenished on a regular basis and was not likely to run short of fuel. He did, of course, have access to an unlimited supply within his unlikely surroundings.

\*

Author's Tailpiece: - *I know not know of what eventuality became of Sandie Honeyman, perhaps his old age finally forced him into some sort of civilised accommodation. I simply have not been able to discover! Perhaps he faded away somewhere out there in the woods. I rather suspect that if he was not finally overcome with some malady like pleurisy, pneumonia or something similar, his natural instincts would cause him to seek a less demanding way of life and give him some degree of warmth and comfort in his later years. Who knows? Perhaps he just expired!*

\*

# Chapter XXVIII

## Dauvit's Progress V

## Dauvit continues to Go Forth!
*– and into 'An Early Posting!'*

I had never been on the night-shift before, in fact I don't think that I had ever been very late in getting to bed of a night-time. This was a new innovation to me and I was so excited at having to go to work on a 'mans' shift for the first time, that I disregarded some well-meant advice to try and get some sleep before hand, or failing that, some quiet time before I left home at 9 p.m. It felt odd having to get ready for work just as the darkness was coming down, but having said 'goodnight' to my girlfriend at the bus-stop, I was soon on my way to the Coaltown of Wemyss. Travelling down the pit road in the darkness *(not illuminated),* I could hear the jumble of voices, but could recognise any of the miners at this time. It was not until I had reached the warmth of the pit-head baths, that I recognised the clean faces of men whom, I would usually meet coming off the night-shift as I descended the pit on the day-shift.

I went through the baths fairly quickly *(out of street clothes and into pit clothes)* and made my way to the Surface Dipping. *(But not before helping myself to some of Old Harry's carbide, I had forgotten mine again!)* There were no man-haulage bogies as such at the Surface Dipping. The standard rake of twelve empty hutches were standing idle on the slope, just a little way down from the level landing, with each hutch having a wooden duck-board laid in the bottom. The only concession to the comfort of the travelling miners! Luckily, there was only a limited number of miners *(this was a preparatory shift of the working week)* descending underground on this night-shift, for the reason that the Surface Dipping had only one, newish, production long-wall face at this time at the 350 Level *(in the east side,* although the place that I was to work was in the new, developing, west side. I was taken in hand by an oldish miner and his partner who, as it turned out, were brushers cum developers. *(Stone-miners as it were, but driving down through coal and stone.)* They were the night-shift pair of a team of six men, who were developing the dipping main-gate for the second west side long-wall face at the 350 level. *(Opposite side to the Barrier level, that itself lay below Newton's farm.)* This level roadway to the west side, had only been recently developed and was but a few metres long (35 to 40 metres). The developing dook to the left, had only gone down about 55 to 60 metres and as yet, there was no return airway, so therefore, clean air was forced down the dook by means of a large, electric, portable fan sitting in the main air-way. There was a continuous column of ten feet long, 15-inch diameter, heavy, but flexible, canvas air-bags connected together, which stretched from the electrically-operated fan down to the working face, to ensure a circulating air current to the brushers and remove the fumes from spent explosives. Also running down the length of the dook, was a single set of light six feet rails, which the brushing teams would extend on a near daily basis as the dook was extended.

At the top of this dook on the high side of the level, there was set into the metals a large, electric, 'tugger' motor. This free-wheeling, tugger motor was used to lower the empty hutches down to the brushers on the end of a running tail, and under control! This being determined by a very substantial braking system. This brake was foot controlled, where the weight of a mans body on the pedal, was more than enough to slow and lock the winding drum and, before an empty hutch was lowered over the edge, two items were paramount! 1. Ensure that the running tail was actually coupled onto the empty hutch! 2. Ensure that the 'Jock' was placed within the down-going, empty hutch. *(The 'Jock' was an unique piece of equipment, that looked all the world like a giant, steel, darning needle with the middle of one side of it's eye removed. The 'Jock' was perhaps three feet long, made of heavy steel and weighed perhaps all of 25 lbs. (approximately 11.5 kilo's). The gap on one side of it's eye, was just enough to allow it to be inserted above and dropped over the rear axle of the last hutch on an up-going rake. It was also used on single, loaded hutches drawn by a large tugger motor. It's use ensured, that if the loaded hutch were to start to run backwards downhill, the point of the heavy trailing 'Jock' would immediately dig into the pavement or wooden sleepers, and cause the loaded hutch to instantly de-rail and probably turn turtle, thereby retarding it's downwards flight.)*

There was a primitive bell system for signals, which was nothing more than a long piece of flexible, seven-strand, bright wire, with a small four or five-inch brass or steel bell attached at the motor end. *(My end!)* They, the brushers at the bottom of the dook, would signal when the hutch was full and this required me to power-up the tugger motor, then commence to draw the loaded hutch up the dook in such a manner, as to ensure that it remained on the rails throughout its ascent. This I managed to do quite successfully, but the only problem was how to land the loaded hutch? If I retained the 'drive' too long, the hutch would smash into the steel crash-girder situated cross-wise in front of the motor, which meant that I would have problems uncoupling and getting 'slack' on the rope or, if I cut the drive too quickly, the loaded hutch was in danger of rolling backwards down the dook! *(But, for the fitting of the 'Jock')*. Now I realised why they needed someone else for this job!

It did not take me too long to get into the swing of this job and I quickly developed the skill to 'land' the heavy hutch with a skill and dexterity, that I became rather proud off. The only small fly in this 'ointment', was that I had nothing to do between the lowering of one empty hutch to the raising of the next full one. (*The brushers used the 'dot and carry one' system, where the down-coming, empty hutch was 'couped' to one side, to allow the sitting, loaded hutch to be immediately removed. This method ensured that they did not have to wait for an empty hutch!*) As a result and with my being on my own in the quiet darkness, I inevitably felt the inclination to 'rest my eyelids' and on more than one occasion, nearly succumbed to the 'black art'. (*I tried hard not too!*)   There was approximately 20 to 25 minutes between the raising of each loaded hutch, which meant that I had ample opportunity to 'hunker-down' and rest my eyelids.  There was little danger of being caught, as I soon sussed-out that the twice nightly visits of the 'fireman' were quite predictable and he could be quite readily heard approaching, in the stillness of the section.  The brushers below only needed to jerk that bell-rope once and I was instantly to the fore and raring to respond!  . . and that, was nearly my undoing, or should I say, my likely prosecution!

During the last few weeks, I had more than often, forgotten to remove the 'Jock' from the rear axle of the up-coming hutches and, as a result, had omitted to send it back down in the next empty hutch.  This was deemed as sacrilege to dipping brushers, who, quite correctly, insisted that all tail-end hutches in an up-going race must have the 'Jock' attached to the axle of the last hutch (*if only for their own peace of mind!*), and here was me constantly forgetting it's existence.

The two contracted brushers had been none too pleased with me in the recent past, in that I had been previously upbraided for it's omission!  I knew that I was in the wrong, but when you are seventeen and dog tired, things sometimes just do not penetrate!

This night-shift was beginning to be somewhat of a bind, with no immediate respite and no mention as yet of my promised 'face-training'.  My daily sleeping routine at home had changed somewhat, in that I found that I needed less sleep during the day, knowing quite well that a few cat-naps at night could be relied on! Again, my undoing!  On this particular night-shift after a rather hectic Sunday afternoon, I was less alert than I should have been.  On arriving at my Tugger-motor, I discovered that the 'tail' was down at the bottom of the dook with an empty hutch attached.  I immediately succumbed to temptation, partly hidden behind the tugger motor and rested my heavy eyelids. (*I also took the precaution of purposely vibrating my right leg and ankle, in an effort to keep myself awake!*) Nothing untoward disturbed my reverie and I was quickly brought to my senses with the insistent jangling of the bell.  I immediately started the tugger motor and commenced to 'raise' the loaded hutch. On its arrival at the top of the dook, it did bang into the steel cross-member, but what-the-hell, it had arrived! I removed the tail, wrestled the 'jock' from the back axle and slewed the heavy hutch onto the rails, before wheeling it the short distance to it's rest. I returned to the next empty hutch, slewed it across the plates into it's grooves, threw the 'jock' into it's interior and carelessly pushed the hutch over the edge as always . . . . . . . . . . . . Yes! That's what I did! I had pushed the empty hutch over the edge, with **no running tail attached! Realisation hit me like a lightning bolt! What had I done . . . . . . . . . . !?** Fortunately, I did not panic! I very quickly raised both hands up to the sides of my mouth, moved to the lip of the dook and '**roared and roared**!' I was now fully awake and certainly alarmed, and could only hope that the sound of my stentorian voice would reach the two brushers, before the careering, heavy, steel hutch struck the solid face at the bottom of the Dook! *(It must have been absolutely flying by the time it struck bottom!)* I listened intently, heart beating heavily, but heard no response from the brushers below. Seconds later, I heard the loud, but muffled crash of the empty hutch smashing into the solid face below. **There then followed absolute silence!** My thumping heart threatening to jump out of my chest. Still, I heard nothing and I dare not shout down again. The awful deed was done and I was convinced that I had killed them both. I waited in the darkness and wished that the ground would swallow me up! Then, I imagined that I heard the sound of heavy breathing and the sounds of someone scrambling up the steep dook. I imagined that one of them was killed and the other one was coming up to wreak a terrible vengeance. I waited and wondered as to how I would respond. Would I defend myself or just take the onslaught?

The old man finally arrived at the head of the dook, his eyes were staring out of his head. He was puffing and blowing, but spluttering with rage, his lungs severely constricted by the need to draw deep breaths after his maddened, headlong rush up the dook! I just stood and waited! This miner needed to get something big of his chest and I was to be the recipient! He gradually regained his breath and his wits, and proceeded to berate me in a fashion that even Old Harry could not match! He was livid, he was spluttering with unsuppressed anger and he was incoherent! He delivered his forceful tirade and I merely stood still and took it! I could do no other. I had nearly killed him and still I waited apprehensively on being told the dreadful fate of his neighbour. Gradually it dawned on me, that he was speaking in the plural, using lots of we's and us's, to which the realisation slowly emerged, that they were both all right, but frightened shitless! (*Indeed! - if the truth be later told!*) I had been delivered! The old man sat down rather quickly after his outburst and I said absolutely nothing! I intuitively guessed, that silence was the better part of valour and any words on my part might just induce heart failure. He sat there for a full five minutes with nary a word, with his bowed head in his hands, thinking, I don't know what? He then got up, looked directly at me and said, "Don't ever do that again son!, but if you ever do, just make sure that you shout even louder!"

On reflection, I will say this to his eternal credit, that I never ever, even to this day and to the best of my knowledge, ever heard the story of this incident being repeated anywhere, or at any time. It was certainly never reported! The memory of this incident often comes to the forefront of my memory, for the reason that I do not know the name of the brushers of whom I then nearly killed that night and to this day, fully expect to be confronted by the family of one of the many miners that I am presently interviewing in the course of my research.

One fine day, - I may even discover their names. . . .?

*

*A little while later, after the old man had manually pulled the empty tail all the way down the length of the Dook, there followed a longish pause and then the belled signal to raise the hutch. I tentatively powered the winding drum, but did not accelerate the motor, I gingerly balanced the clutch on the slip. This was just as well, as the hutch now coming up was the damaged unit, which had somehow developed a twisted chassis and lost one of it's axles. The second man of the team was having to hold the front wheels onto the rails (with his body weight) and support the rear of the hutch, to prevent it from tearing up the light rails. As he and 'it' came into view, I could only but marvel, at the ingenuity of the developers in the 'raising' of this severely damaged hutch. The running tail was affixed to the hook on the front draw-bar (as it should have been) and a length of bell-wire had been attached to the rear, draw-bar hook, slung over the top of the hutch, fastened and tensioned onto the last link on the tail-chain, where it was joined to the wire rope. The hutch was therefore held tight and balanced horizontally. The hutch was thus being pulled up the steep slope like a two wheeled chariot, with the 'driver' standing crouched and balanced on the leading chain of the tail, using his body-weight to ensure stability. Thank God, I managed to land him safely. After he uncoupled the hutch and helped me push it into the dark depths of the level, he delivered me a very large wink that suggested to me, that they had experienced this before. He then showed further confidence in my abilities as a motorman, by jumping into the newly-coupled empty hutch going down, but not before mouthing the silent words . . . "S-L-O-W-L-Y  N-O-W !!*

*

# Chapter XXIX

## Into (my) Kingdom Come!
## *- or How did Old Harry get his Stutter?*

## (Anecdote or Lochhead Myth?)

A story once did the rounds of Lochhead pit describing exactly how Old Harry got his stutter and how it developed and, with each subsequent repetition, it came back to him slightly more embellished and even more distorted. The story gained credence with each re-telling and by the time it got to me many years later, I could only shake my head in wonderment at the vivid imaginations of the originators of this marvellous tale.

The story opens with Old Harry as usual, in the middle of a brushing contract, with a team of nine men working the three shift system, in the development of a 12' wide by 9' high mine. Harry's team was on the back-shift at the time and, at a point during the working shift, where the last girder had been set and secured and the holes in the brushing face had been bored and charged. The shot-firer was in the process of ensuring that the brusher's were well clear of the brushing face, with the miner's having been ushered into the safety of man-holes, prior to the detonation of the shots. The shot-firer, whose name was Davie S . . . . s, had completed the connections to the wires of the electric detonators hanging from the brushing face, and had now retired to the safety of a manhole to connect the 'running wire' to the exploder box. After ensuring that good, electrical connections had been made to the box, he did the customary check for electrical continuity. All was good! Continuity was apparent! Now to proceed! "Firing in the HOLE!"

One final check to ensure that all was clear, and then he closeted himself into the safety of the manhole before the initiation! Half -a-dozen turns on the winding key of the exploder to charge the capacitor and . . . . 'FIRE! . . . Nothing happened! Men relaxed their anticipation and waited for the next warning order. . . . . .? Sure enough it came! . . . . FIRE! Again, nothing happened! The shot-firer, realising that he had a probable fault somewhere in the line, then disconnected the running wires from the exploder and removed the winding key. He then went along the level onto the brushing face, closely followed by Old Harry. A thorough check was made on all of the connections, but nothing untoward was found! Every connection was clean and good and Old Harry and the shot-firer slowly retired to the man-hole, double-checking the condition of the running wires as they did so. They arrived back at the man-hole, rather puzzled as to what the reason could have been for the double 'misfire!. The shot-firer again, coupled the running wire to the exploder and again, checked for continuity. Sure enough, the red light was illuminated, showing continuity. (*A clean unbroken path for the conduction of electricity.*)

Once again, the exploder was wound up to contain a charge, and again the brushers pressed deep into the man-holes in anticipation of the detonations. Nothing happened! Again, the system had misfired! This was becoming a little too much! The occasional misfire was acceptable, but three in succession was quite abnormal. The exploder had on three attempts been fully discharged, so, no residual electric energy remained in the exploder. Old Harry then gave a quick glance in the direction of the exploder to satisfy himself that the running wire had been disconnected and with the shot-firer trailing him, they both made their way back to the unblown, brushing face.

Another thorough check ensured that all of the connections were still coupled and neither Old Harry or the shot-firer could fathom out just what the cause of the three misfires might be? Whilst Old Harry was staring at the face and touching each of the wire connections in turn, Davie S . . . . s mumbled something to himself and crouched down with his back to Old Harry, with his voice now getting louder and clearer and repeating the words "Continuity! . . . I've got continuity!" Old Harry quickly glanced down and to his astonished horror, saw that the shot-firer had re-coupled the end of the trailing wire to the exploder, both of which he had also brought with him. He was now in the process of winding-up the exploder to re-charge the capacitor!

(*Some knowing miners claim, that it was at this juncture that Old Harry's thinnish grey hair actually stood on end and remained forever so, until he left the coalmines several years later. Other 'knowing' miners claim that Old Harry was rendered absolutely speechless, at the thought of being blown to kingdom-come by a thoughtless shot-firer, whose stupid actions, if they had been allowed to continue, would have created two widows that very day. Either way, the final fatal push to the exploder button did not materialise and both men lived to tell the tale, not that Old Harry or even the shot-firer would later ever publicize the event, but miners being what they are, can never resist reiterating a good strong anecdote, and word of the near fatal incident soon got around, no doubt highly embellished by competent story-tellers.*)

I was later told that Old Harry completely lost faith in that particular shot-firer and when he next turned up to 'fire' Old Harry's brushing face, Harry kept well out of the way until the 'shots' had been successfully fired! It was also rumoured, that by the time Old Harry managed to get the appropriate words of condemnation into his head and voiced, the aforesaid shot-firer was half-way to the pit bottom. Old Harry rarely showed anger either by word or deed. His anger took the form of cold rage, which invariably rendered him not entirely speechless, but struggling to convert his thoughts to the spoken word - and woe betide the fool who attempted to put words in his mouth!

*From then on, and up to a few weeks afterwards, every time that Old Harry came across this shot-firer in the pit-head baths, he would chase him out of sight whether he was dressed or not! It soon got to the stage that Davie S . . . . s used to sneak into the Baths without speaking to anyone, in case old Harry heard the sound of his voice. Old Harry could not bear to be in the same location as his near fatal nemesis. Harry would simply not let old Davie S . . . . s forget his incompetent actions and crass stupidity. To make matters worse, old Harry was still quite speechless at the sheer on-going effrontery of this mine's official, who later carried on as though nothing had happened, which in no way served to pacify Old Harry's unrelenting spleen. Needless to say, this particular shot-firer did not last too long in the Lochhead pit environment!*

\*

# Chapter XXX

## Full of Eastern Promise
## *– or, Lower Dysart Delights*

## A Savage Undertaking!

In my original research into the exact underground location of the **East-side mine** *(and the subsequent areas of the taken coal)*, I was initially quite flummoxed as to where it actually was? Far from being able to identify the location of the east side mine, I found myself looking in the wrong direction and area. *(It is still very strange how one can be intimidated by a place-name.)* In light of my findings, I must first of all, disillusion all of those miners who worked there and indeed even extracted it's multipartite bounty, in that even though the 'mine' did prodigiously produce 'lots of coal', it was not at all in the east! It did of course, exist, as many miners will confirm, but it's location as a 'coalfield' was actually to the north of the shaft, with most of the extraction taking place to the west of north. I must admit to partial confusion in my research, in that I simply could not locate the east-side mine on the first preliminary examination of the coalfield area. There were subsequently, much and many extractions of Lower Dysart coal to the geographical eastern area of the shaft, but those extractions did not conform to the criteria of the methods and the machinery used in the east-side mine and, are covered elsewhere within this narrative.

To get to the start point of the area covered by these extractions, a level cross-cut mine was commenced from near the head of the New Dook, on a map bearing of approximately 20 degrees and for a distance of 175 metres on a near level plane, to cut clean through the metals. At this point, the mine changed direction to the left and, was continued on at a map bearing of 325 degrees for another 140 metres on the same plane, until the full rising face of the Lower Dysart seam was exposed. The vertical downwards spacing of the Lower Dysart seam from the Dysart main seam at this point, was approximately 23 metres, with the intervening 'metals' showing no coals. The size of the planned area *(section)* of coal to be taken, measured 350 metres wide and approximately 410 long. The Lower Dysart seam in this area, showed a top leaf of 36 inches of coal with a 2-inch band of stone in the middle, and with a 30-inch band of soft fireclay cum blae immediately underneath it. The middle leaf showed approximately 14 to 18 inches of coal, sitting atop 4 to 5 inches of grey stone, with 4 to 5 inches of parrot coal underneath that. The bottom leaf showed approximately 38 to 44 inches of coal, again with a 2-inch band of harder grey stone at it's middle. *(This Lower Dysart seam did contain three separate, but closely sandwiched leaves of good coal!)*

This then meant, that Lower Dysart seam in this area consisted of approximately 100 to 105 inches height of coal *(approximately 2.6 metres),* 30 inches of soft blae between the top and middle leaves, and approximately eight inches of interleaved grey stone. This was not an entirely 'clean' layered sandwich, even though the coal itself was of the best quality!

The decision as to how this coal would be taken was probably made before this time, but contrary to the coal extraction methods used in other parts of the colliery, there was a new, influencing factor being introduced to the coal extraction equation with the proposed development of this section. There was a new 'giant' in the East Fife coalfields, namely, the Anderson-Boyes, 'short-wall', coal-cutting machine. This new machine was surely a 'first among others' in that it would do so much more than its smaller brother and much, much quicker! This machine was box-like in appearance, measuring approximately five feet long and four and a half feet wide, at a high of eighteen inches. It had a long, forwards, extending jib which, at a basic 7 feet long, could actually be extended to 7 feet - 8 inches if needs be. It had two powerful winding drums, one at each side, and a centre drum at its rear for added mobility. It could of course, be made to cut in both directions *(with a reversal of the ripping picks),* and did so with equal dexterity. The new machine was a revolution in progressive coal-cutting and this East-side mine was to bear witness its absolute savagery.

At some time, prior to August 1953, the decision was made to use the awesome power of this machine to drive a straight heading up through the length of this new field. The slope of this heading (one in seven) was eminently favourable for the installation of a pan-motor *(engine)* and an extending set of jigging pans*. *(To transport the extracted coal.)*

At this juncture, I must, for the benefit of non-miners and interested readers, explain the underlying premise behind the experimental exploitation of this method of coal-getting within this seam, and in this section.
*\* Pan-engines with their attendant 'trains' of jigging pans, were first introduced into this country by the firm of Mavor & Coulson, immediately prior to W.W.I., (The Great War) and initial used by the Lothian Coal Company!*

The heading or level to be driven, is usually at a width of 12 to 14 feet and probably 8 to 10 feet high. The central, ripping jib of an A.B., short-wall machine can be likened to a tree-cutting chain saw, in that the principle is exactly the same, except, in that it's cutting and ripping power will actually decimate and rend solid stone! To effect it's cutting momentum, two, one inch, diameter holes are bored vertically down into the pavement to a depth of 24 inches, one at either corner of the short face-line. Two intermediate holes are then bored near the midpoint of the face to the same depth. The end of the steel cable from the right drum is fixed at ground level to a heavy, steel pin, sunk into the right extreme borehole, and the end of the cable from the left drum is fixed to a steel pin, sunk into one of the intermediate boreholes. The machine-man then applies power to the machine, then slowly engages the individual dog-clutches, driving the separate, revolving drums to tension the steel cables, thereby drawing the whole machine bodily to it's starting position at the face. The cutting chain is then engaged *(clutched-in)* and almost instantly reaches its cutting speed. The machine-man, now with positive control of both clutches, drives the ripping chain and jib into the bottom face of coal, at the same time maintaining his correct line of approach onto the right-side boundary line. The machine is slowly ripped-in to its full depth and forwards progress halted. The ripping chain is stopped!

The steel cable from the left drum is now extended and attached to the pin on the extreme left, and looped round the guiding pulley on the lower, left, front side of the machine, to effect a sideways, propulsion to the steel cable and thence the machine and cutting jib. The appropriate clutches are engaged and the cutting jib commences it's sideways ripping cross-cut along the bottom of the face of coal. On reaching it's left-side boundary, the machine and thence the jib, are withdrawn from under the cut, and the machine is parked and stopped clear of the train of pans, with the now stilled cutting-chain and jib held tight to the left side wall. The pan engine that is situated at the bottom of the pan-train is then started and the 'coal-gum' quickly shovelled away. No time is lost on this 5 to 10 minute task. *(In point of fact, the pan-train (without the duck-billed pan attached) is kept going during the cutting operation, so that the two other members of the team can be gainfully employed in removing the coal-gum.)* The undercut is gummed absolutely clear and the next stage of the cycle ensues. One man will operate the 'ram's head' and using the short penetrating drill, shall commence boring the necessary sequence of holes in the 'face', whilst the other two miners will attach one of the two, fore-shortened pans onto the now stopped pan-train, in order that this 'duck-billed', flat pan is at least positioned halfway underneath the cut, if not more! When this has been completed and the 'shot-firer' is sent for, the long seven foot drill is then attached to the ram's head and each hole bored to the maximum depth. By this time, the shot-firer should already be 'charging' and stemming the holes in preparation for a sequential firing programme. The whole principle of this type of operation now depends on several factors. 1. The pan-engine is stable and the pan train bolts all tight. 2. The charged holes have been strategically placed and that there are no dud detonators in the starting sequence. 3. The firing sequence is adhered to, with due consideration given to possible damage to the pan train. *(The initial starting load and subsequent coal loads on the pan-engine and pan-train can be colossal!)*

The pan engine is started, the empty pan-train happily jigging the duck-billed pan, to and fro, underneath the under-cut, but still solid coal face, and at this point, the shot-firer now plays his part. This sequence, which is electrically detonated, is coupled up individually by the shot-firer, who now couples up the trailing wires of the charges placed directly above the duck-billed pan. He retreats to safety, checks for continuity, and detonates the first shots. There is no need to see the effects of this first shot *(or shots)*, the changed sound of the pan engine tells all. The blown coal falls directly onto the duck-billed pan, breaks up and is slowly jigged away downhill. The shot-firer advances quickly to the face, couples up the next opposing pair, then retreats and 'fires' again! The principle of a balanced sequence of 'firing', is also to throw a regulated load on the pan train, so that it 'sees' and feels an even load. *(Too much too soon!, will cause an overload on the 'face' end of the pan column and could result in the pan train 'heaving' in the centre of the 'train', causing untold damage.)* The coal starts to flow in volume, which increases as the subsequent pairs of shots are fired, to the extent, that for something like a twenty minute period, the coal capacity of the pan-train is at 'full flow'. On completion of the firing sequence, the pans will have jigged away something to the order of three-quarters to four-fifths of the blown coal. There is now an almighty scramble by the three man team to deliver the remaining coal *(at both sides and still heaped)* onto the emptying pans. The face of coal is 'picked' clean of any loose or overhanging lumps and big shovels are wielded with an hectic urgency. The coal-cutting machine can not be brought forward until the remaining coal has been cleared, the pan engine stopped and the duck-billed pan disconnected. When this has been completed, so ends one complete cycle and the next 'undercut' commences as before!

Author's Note:- *As mentioned elsewhere in this narrative, if a miner from one of these short-wall, coal-getting teams did not turn up for work on any given shift, the other two miners were not at all keen to indulge the temporary presence of any other coal-miner from any other type of coal-getting into their team, even for one day. They would of course, countenance the temporary addition of a coal-miner from another team working in the same environment. The reason for this, was that this type of work was very demanding, ceaselessly unremitting, and continually dependent on the endurance of the miner, who must be capable of long sustained outbursts of physical effort. Besides, these short-wall face coalminers were on to a 'good thing' as regards wages, and they knew it! The attitude of some of them seemed to be, with it's xenophobic connotations:- "I spy Strangers!"*

\*

The first heading to strike deep into the new section area, probably commenced at a point on the south-west side of the area, approximately 60 metres from the left boundary. The heading progressed at a speed that would equate to perhaps, 14 feet per shift. *(i.e. two by two, two cuts and two strips per three man team in the course of a single shift!)* This therefore, would advance the heading at least 26 to 28 feet in 24 hours, working day-shift and back-shift. That would add up to a total weekly advance of approximately 45 to 48 yards, *(41 to 43.8 metres),* barring mishaps, break-downs and maintenance. This then meant that the northern extremity of the *section* area would be reached in approximately 8 to 9 weeks. This first extremity was actually reached in early September 1953. Two other headings were driven by exactly the same methods, as two, new, similar machines were brought into the section. The middle probe struck the extremity in mid October 1953, and the right side heading struck in early January 1954. This then, was the planned layout of three separate headings, almost the same distance apart and roughly parallel to each other. The top and middle leaves of the seam had been taken and the roof was reasonably secure, but not as perfect as was the Dysart Main with its solid Head coal. Care would have to be taken, and tall trees and short straps would have to be set to the roof, to maintain it's integrity, at least, for a short time after extraction. The stage was now set for the grey, speckled,* black mineral extravaganza!

*Unfortunately, the 30 inches of soft blae that lay between the top and middle leaves had to be taken at the same time, with no attempt being made by the miners to separate it from the coal. The 'redd' was jigged away down the pan-train along with the coal!*

There were now three A.B. short-wall machines available and two of them were now put to maximum use in the taking of coal. *(The third machine would be used to develop the second, parallel heading to the right of the initial intrusion.)* At the extremity of the original *(left)* heading, two of the machines were brought back to a point 50 metres from the boundary end, *(extremity)* where two horizontal levels were started, directly opposite each other, one to the right, and one to the left. These two, opposing levels were each taken to a length of about 50 metres, where the pan-engines had been re-set up at the commencement of these levels, to 'jig' the Pan-train *(coal)* directly onto the newly installed, 30-inch wide, conveyor belt, now situated along the length of the main heading. At the 50 metre point, these two levels were stopped, where a change of direction was executed. At the end of each level, the miners turned their machines to the slope and began an incursion to the rise. *(A turn to the right in the left side level, and a turn to the left in the right side level. This meant, that these new, developing headings were approximately 100 metres apart, parallel in direction and both intruding to the rise.)* As these headings progress, it shall become apparent to the inquiring and slightly confused reader, that getting the coal out of these angled headings, in volume and at speed, might just pose a problem. It did! And it was not so quickly solved! Along the 410 metre length of the original heading where they had installed a 30-inch endless belt, this would now convey all of the coal that could be produced from either side of that heading. As the side levels were driven in, it was a simple matter *(in principle)* to install a pan-engine and a train of jigging pans, the ever-extending end of which, was always kept up to the face of coal, with the pan-engine motor in a static position near the bottom of the pan-train, almost touching to the main 30-inch belt. As the coal was jigged away down the train of pans, it was emptied onto the moving 30-inch belt whence it was carried away. *(By divers means to the pit bottom!)*

The problem now faced by the planners, was how to get the coal out of these two new headings, now that they had each taken another 90 degree turn to the rise, without installing any more expensive machinery. Also bearing in mind, that as soon as each of these headings reached a length of 50 metres, they would be abandoned and a new heading started, by the simple expedient of withdrawing back a distance of approximately four or five metres along each level and developing new headings to the inside of, and parallel to the newly abandoned ones. This method then, was in principle, the same, old-fashioned method of coal extraction as used extensively in Lochhead pit, when they extracted the Dysart Main coal 50 years beforehand. It was still called 'Stoop and room!' (or Pillar & Stall). But now, working to the 'retreat', the coal was being nearly fully extracted from pavement to roof *(an approximate height of 9 feet/2.75 metres)* and up to 14 feet wide *(4.25 metres),* with only just enough of a long, solid stoop of coal left between successive parallel headings, to initially absorb the intermediate roof weight, whilst the other 50 per cent of the coal was being quickly ripped-out, and, at frightening speed! This indeed, was the Stoop and room method of extraction, but with 100 percent mechanisation and at bewildering speed, compared with hand-loaded hutches!

However, the basic problem still remained! How to get a pan-train to jig coal round a 90 degree bend, with just one pan engine doing the work? Not possible, said the miners! "Don't ask us!" said the deputies. "Can't be done!"- said the Oversman! "We've got a problem!" said Willie Hampson! "How say the engineers . . ? ?"

This problem was overcome, and then quite quickly. There was, newly on the mining market, and possibly manufactured by either Goodman's or Anderson-Boyes, a fixed, metallic, trough installation, that could be fitted either left or right-handed within almost any part of a 'pan-train', and would conduct the full forwards and reverse thrust of the pan-engine train around a 90 degree turn, without loss of momentum! *(And, without undue spillage of the load!)* This innovative installation negated the use of a second pan-engine, where the coal needed to be transported by conveyor around a left or right angled roadway! This piece of equipment saw regular and sustained use in the east side mine *(section)* and was quite dependable, provided that it was properly installed and solidly stelled to the roof and pavement!

*

# A Dastardly Equation ! - *or 'Simplified Arithmetic?'*

This method of coal extraction did depend on several inter-linking factors, in that the wear and tear on all of the machinery was quite prodigious, in that both electrical and mechanical engineers were always on-call - and maintenance was only possible on the non-coal getting shifts. (i.e. nightshift.) The pan engines and pan trains needed maintenance men on the spot, as under the heavy loads, the pan bolts were forever slackening, and where a pair of loose bolts at the same point, would nullify the propulsion motion at the face end of the train. The great outlay in expensive equipment seemed however to be practical, as long as coal flowed and production was continuous. The one imponderable factor with all of this increasing coal production, was, as to how the miners were to be paid and how this was to be calculated. There were now six A.B. short-wall machines in use *(four Anderson-Boyes and two American Goodman's)* and six pan-engines, two of which were Goodman's. All six sets of machinery were now in full use and the coal was freely flowing from six, separate sources in the east-side mine. The volume of coal from each individual 'cut and strip' was approximately 25 cubic yards, therefore, each shift would produce approximately 50 cubic yards at two cuts and two strips. The total amount of coal from six such faces, would produce roughly 3000 cubic yards of coal in any given week. Considering also, that the amount of coal produced was measured in cubic yards, it seemed entirely feasible that this type of measurement *(which could not be gain-sayed at that time),* should be the yardstick by which the contract price was to be negotiated. This was agreed between men and management and also, considering that the coal was to be cut at not less than 12 feet wide and 9 feet high, the only measurement needed, was the total length advanced during one week and measured in yards. With the width and height at a standard 4 yards by 3 yards, the agreed, weekly, advance measurement in yards, would be multiplied up by a factor of 4 *(for width)* and then 3 *(for height)* to give the weeks total volume. This volume, averaged out over a few weeks, would be used to calculate the rate per cubic yard to be paid to the contract. *(Six miners on two shifts.)* A monetary rate was soon agreed and fixed, and then the coal simply flowed.

However, in negotiations for any new contract, the miners usually held back from giving their very best, knowing that more money could be made from a lucrative contract after the price was 'fixed'. The manager, holding the position that he did, was also not fooled by this obvious ploy and did his best not to be fleeced. So, after a statuary, negotiatory period, both parties were usually satisfied. At times however, one side or the other would make a mischievous miscalculation, either to the detriment of the miners in receiving lesser wages, or to a supposed reduction *(miscalculation by the deputy)* in coal production. As the coal began to flow from the east side mine, it soon became apparent to all that the earlier, estimated, volume of coal was now being greatly exceeded. The miners wages had increased beyond expectations and were becoming slightly obscene. 'Twas time to act! . . . ? Or, at least, someone obviously thought so!

Author's Note: - *In relating the following narrative, I have been asked to name no miners, nor mention dates or places. I was given the information on the strict understanding that no clues, suggestions or insinuations would be made, or any effort made on my behalf to identify any of the participants.*

NOMENCLATURE.
(T) Tapes, Measuring. - metallic, woven linen, stitched leather cased. 30 metres /100 feet.
**Upside:-** *Graduated in metres, decimetres and centimetres.*
**Downside:-** *Graduated in yards, feet and inches.*

*One Friday night (end of the back-shift) after the last cut and strip had been cleared and the last of the coal had gone down the pans, the section fireman and the team leader measured and agreed the week's progress (in linear yards and feet), which was written down in the fireman's work-book and signed. As the fireman (deputy) closed his kalamazoo, he casually mentioned to the team leader, that next week's measurement would be done using the 'other side of the tape'. On being questioned, the fireman replied, that a message had been left on the 'slate' in the pit-bottom howf, that in future, all linear measurement in the east side mine were to be taken using metres and decimetres, as opposed to yards and feet. (Metric measurement in place of Imperial.) This would make the calculations easier, in determining decimal points of a metre and if the men had any questions, would they please go to see the manager's male clerk. The team leader decided to call in on David Anderson on the following Monday afternoon, when this team would be on the day-shift. This he did and was immediately received. The secretary explained to him, that in keeping all other records, all future measurements would be in the metric scale and to compensate for the change, the present rate of yardage remuneration would be increased by a whole ten per cent (10%). This was because there were 39.4 inches in a metre and only 36 inches in a yard. The 10 per cent increase should only have been 9.5 per cent, but the miners would seemingly benefit from the extra one half per cent? There was no need to negotiate a new contract, as only the criteria of measurement would change. This was put to all of the team and heartily agreed, since they felt that they had nothing to forfend, especially when the management seemed to be giving away something for nothing, - or, were they?*

*

*Work went on as normal, except that just a little more coal was extracted as the miners conditioned themselves to the intense work cycle and no real problems were experienced in measuring their advance in metres either.  One week later, when the pay-slips were available to the miners on the Thursday preceding payday - the team leader eagerly scanned the payslip to find the 'yardage' (Now metre-age!) - and sure enough, the correct measurement was there in metres - it even had the yards equivalent in parenthesis to simplify and confirm the earlier measurement.  The team leader even took the time to do a quick arithmetical sum, to verify the Coal board clerk's calculations - and found that they were correct! The slightly puzzling thing was, that even though they all opined that extra volume had been extracted during the past week - it did not seem to be reflected in the total amount to be paid.  The last weeks' progress was re-measured and re-confirmed and the men had to be satisfied! . . . . At least for the time being ?*

*This team leader, though saying very little to his men before the money was due to be paid out, was still rather sceptical of the motive behind the change and reasoned that there was something wrong with the calculations.  He pondered for a while - then quietly and unobtrusively made his way into the mine surveyor's office, where, to his delight he found his 'quarry'.  He quietly asked for some help and quickly explained his dilemma.  He explained the change to metric measurement from yardage and with it - the very perceptible difference in their expected total wages.  The surveyor instantly digested the proffered information, made a few quick calculation on his slide rule and then to the consternation of the team leader, unexpectedly left the room.  He returned just as the team leader was on the point of leaving, thinking that he was being 'passed-on' to the male secretary.  The surveyor merely grinned and by way of explanation, showed the miner his calculations on paper.  The secretary had been quite right in his calculations that One Metre was almost 10 per cent (9.5) longer than a Yard in distance, but what had not been explained to the men, was that payment for 'all three' measurements - width, height and length, should have been increased by a factor of 10 per cent, for the simple reason that - **One cubic metre of coal contained exactly 31 percent more in volume than one cubic yard of coal.** They had been given an extra 10 per cent for length only and were about to be 'screwed' by a short volume of approximately 2 times 10 percent on the miscalculated width and height!  In other words - a overall short-fall of approximately 16 percent!\**

*\*(They should have received 131 percent of the cubic yardage rate,  - but, they*

*were only being given 110 percent!  110 divided by 131 - multiplied by 100 equals 84 percent.)*

*Needless to say, as soon as this was discovered and re-calculated, the team dropped all pretence of assumed civility and angrily trooped into the manager's office - where, no doubt, accusations of swindle, chicanery and cheating were bandied about, until common-sense prevailed.  The miners were determined that they would not draw any monies from the pay office, until a extra 'white-line'\* payment was authorised there and then.  Most miners didn't believe in 'getting it all sorted out by next week!'*

*\*White paper 'demand' signed by the manager and directed to the 'pay-office,' requiring instant cash to be paid to the bearer!  (I am 'unreliably' informed that this practice may have been peculiar only, to the management and miners of Lochhead colliery.)*

The short-wall, face colliers concerned, never did find out if this was a deliberate ploy to reduce their hard-earned wage packet, or whether it was a genuine mistake on the part of some clerk or other!  It mattered not, simply because it was also agreed between miner's and officials, that the linear measurement system would revert back to yards and feet, with no more talk of metres and decimetres and the matter was quickly and quietly dropped, never to rear it's ugly head again! .   .   !?

\*

# Chapter XXXI

## The New School of Pitch and Toss
### *(Your Wages or your Wife!)*

To the old and bold, there is absolutely no need for me to explain the simple, yet sometimes highly controversial rules that applied to the often argumentative, compulsive addiction of Pitch and Toss. To the young and uninitiated, it was a very serious game of chance, played with two, often old or favourite pennies of the now defunct imperial currency. The outcome of a series of high, twisting tosses, thrown into the air, of two pennies balanced carefully on the two outstretched fingers of one hand, would determine the loss or winnings of many a miners' weekly wage packet. These falling pennies, if the toss was good - would determine or not, whether a miner went home flushed with success, or home to face an angry wife, a cold supper and a cold bed. It all depended on a pair of heads at a probability of 4 : 1, a pair of tails at 4 : 1, or at odds of 2 : 1, a mixed pair.

When I first became aware of these gatherings, at an age of about thirteen, it was by listening to adult social conversations at home and in the company of my mother and father (old Harry), when he was not actually engaged in shift-work down the Pit, or in the noisy construction of garden sheds in the back garden. I was unwittingly made aware that the miner's gambling 'school' took place in the region of the 'Dubbie' bing down near the sea shore, well away from prying eyes and certainly out of sight, but not out of reach of the local constabulary. This form of gambling though certainly illegal, was well attended by 'regulars' and rather notorious for the unusually, substantial sums of money that constantly changed hands. Sums approaching two to three hundred pounds would be a likely bonanza to a miner on a winning streak, which happened on quite a regular basis. This amounted to something akin to six months wages and often more. A single miner winning this amount of money, would result in approximately thirty to fifty or more miners going home absolutely penniless, having lost their whole weeks wages with nothing to show. This gambling 'fever' affected lots of miners on a weekly basis, 'feeling the need' to recoup last week's losses at least!

This 'school' finally got so completely out of hand, that short, but noisy altercations proliferated on a near weekly basis, with fisticuffs being the instant response to cries of subterfuge and cheating, though it must be said that the miners concerned usually 'policed' their own fracas and 'condemned' the cheats. Things eventually got so much out of hand, partly due to the many local complaints, that the local police force became more and more involved in what was of course, highly illegal gatherings, with the participant miners being constantly chased by the local constabulary nearly every Friday afternoon. This weekly illegal gathering was now such common knowledge, that it attracted miners from other collieries in the Wemyss area, which in turn, became an increasing source of embarrassment to the local community. The gatherings were raided time and time again on the instructions of the local authority, until the miners concerned finally had enough and by tacit agreement, decided that the efforts of being chased onto the beach and beyond on a weekly basis, was not really very dignified. As a permanent result of the so-called harassment, the school of 'Pitch and Toss' at the Dubbie bing was terminated!

Or so everyone thought! But, the men concerned, irritated by the thought that the authorities had spoiled their fun, merely bided their time. A new venue had to be found, somewhere in the reasonably immediate area, well away from prying eyes and snoopers and with good all-round access. The miners did not want to be caught again in the situation, where the escape route was along a wet beach or over the top of a smelly bing! Again, needs must, and a new, good location was subsequently located. It was found to have good access and was out of sight. It was surrounded by woods and a lookout or two could be posted, to give early warning of impending intrusion. This new site was ideal and was situated in the depth of the Boreland wood. *(On the Wemyss-field estates.)* It had paths coming in from all directions *(and, by implication, leading out!)* and had a small clearing in a suitable location, that seemed to be acceptable to miners from Dysart, West Wemyss, the Boreland and the Coaltown of Wemyss. It was also accessible to miners from the 'Randy' and 'Earl-seat', which could only serve to swell the numbers attending. This new 'school' developed in a very short time and the numbers soon swelled to that previously experienced at the fateful 'Dubbie' bing. I have been told, that small schools did gather on both Friday and Saturday afternoons, but the main gathering usually took place from around 10 a.m. on a Sunday morning and was in full swing by midday. This would progress non-stop through mid-afternoon, with only the die-hands and potential winners in the final 'pots'.

Those who had lost and those who sometimes knew better, would stand back and await the final outcome of the days betting and often, a miner's friend's and mates would hang-on, just to see fair play and suppression of possible threats and 'due consideration' to an ultimate winner. Skulduggery was not tolerated and transgressors would be dealt with, but the temptation was there and often quite manifest, especially as sums approaching a year's wages could quite easily be the worthwhile outcome. There were usually some children playing in the woods on a Sunday, but it must be said, that these children were always kept well back from the large group of miners participating in the play. Indeed, some of these children did belong to some of the men involved, but they themselves were under the watchful eyes of older youngsters, detailed to keep the younger siblings away and dutifully 'uninformed'. I mention this for the simple reason, that the miners did have their own moral code, and it was totally unacceptable to allow any non-mature person or callow youths under the age of about twenty years or so into, or view the nefarious doings of grown men.

\*

On one such Sunday morning directly after breakfast, Old Harry asked me if I would like to go with him up to the Coalt'n? I can't remember as to my response, but I soon found myself being treated to a long bus trip all the way to the Boreland. That was way beyond the Coaltown, the farthest I had ever been in my 11 short years of life! This was certainly a mystery trip, as Old Harry just would not tell me where we were going, in spite of my repeated requests. I do remember asking myself as to why we needed to take a bus trip, just to go for a walk in the woods. My questions were soon answered, even though I still did not know what all of this was about. Old Harry took me up to the edge of a large crowd of men and told me to stay put, and slowly nosed into the active centre of the large group. He did not disappear from sight and indeed, did move himself to keep me within his vision. He was obviously there to have a small flutter and I had been brought along to add some legitimacy to what might ostensibly, have been a Sunday visit to Harry's older sister in the Coaltown of Wemyss.

At Harry's continued glances in my direction, one or more of the men at the centre of events made comment to a few others and more and more suspicious glances were cast in my direction. Harry by this time was a little more involved in the activities and had for the time being possibly forgotten all about me. He was brought up short by one miner, who pointedly asked him if that youngster in the crowd belonged to him? He could not very well deny me and reluctantly admitted that I was indeed his and, as a result, he was very pointedly asked to leave! It was a very embarrassed and chastened Old Harry that walked me out through the Boreland woods, down into the Coaltown of Wemyss and thence to No. 5 Barnes Row, so that he could then justifiably claim to have visited Old Kate, his sister. I never did let-on to my mother, even though she had her suspicions and I have kept old Harry's little secret until this very day!

Authors note: - *At the time of the Summer holidays in the coal mining industry, the annual miner's break amounted to two, full, weeks holiday with pay. On the Thursday before pay-day, the miners knew exactly how much money they were bound to receive, as the pay-slips were handed out by the section Gaffer or Fireman. The largish sums of money were the equivalent of three weeks wages and were so keenly anticipated by the miners, (and their wives) to the extent, that as much overtime 'as and where possible' was worked during the preceding week. This so-called holiday pay often* amounted to a small bonanza, which was jealously guarded until safely delivered home. Most miners could be depended upon to go home and deliver an unopened wage packet to their wives, but a smaller number of miners were met by their wives and 'escorted' home, wage packet having been previously inspected and found to be intact. However, a very small, but determined few would evade all attempts at 'ambush' and find their way to the Boreland woods. This 'School' at this time of the year was infamous for the large amounts of money that were wagered, occasionally won, but mostly and irretrievably lost!

Needless to say, outright winners were in the small minority and there was some very disappointed, disturbed and chastened losers that eventually found their way home. With regard to Lochhead miners, there was one single consolation that gave them the hope, that not all was lost. Willie Hampson, the Manager at Lochhead pit, would not turn away any holidaying miner who wished to work during the Summer break, as long as they turned up at Lochhead on the Saturday morning and saw him personally. He would countenance no other approach. The coal mine was all but deserted, but for inspecting deputies, duty engineers and electricians, and maintenance work was always a necessity in any coalmine, provided the miner concerned was a capable and efficient worker. Notwithstanding of course, that Willie Hampson did something that, to my knowledge, no other Wemyss colliery manager did! - and that, was to give any deserving miner a 'white line' advance of up to one third on the next weeks wages, provided a full weeks work was in hand by the following Friday. Willie Hampson saved many a miners 'bacon' with this initiative, that was received with thanks by not a few miner's wives over many years!

\*

# Chapter XXXII
## The Black Art!

## The Sleeping Beauties!,
### *– or A very Gentle Practice!*

It is there, clearly defined in black and white for all to read in the Coal Mines Act, and in not so many words; *'Thou Shall not sleep down the Coal Mine'* and if the offender is 'caught', there is no reprimand, at least not down the coalmine. The penalty is usually instant dismissal 'on the spot,' with the offender immediately sent to the Pit bottom and thence 'up the pit!' Part and parcel of the transgressors shame is having to walk in ignominy, all the way to the pit bottom, trying desperately to field awkward questions as to why a perfectly 'healthy' miner should want to go topside in the middle of a working shift?

Not many miners have experienced the shame of being sacked on the spot, even though there has been in the past, lots of border line cases. Much depended on who the miner was, his length of service, his experience and character. *(And his worth to the manager.)* Much also depended on who did the 'catching!', whether it be deputy, oversman, under-manager or God forbid, the Manager himself. In the case of the latter or his deputy, it was certainly 'curtains!'

It's was not the first time, that a miner who had been officially caught sleeping under-ground, had been returned to the same work within the next week. The manager of the coal mine was always duty-bound to sack the transgressing miner and have him clear his things from the pithead baths. Managers were sometimes very reluctant to have to do this, especially where they knew the miner to be a sober dependable miner, who had merely been unlucky enough to be caught having a cat nap, but he was morally bound to back up his officials. He also knew that some underground officials would use this charge to remove a miner with whom they personally, did not have agreeable working relations. This didn't happen very often, even though not every deputy was well-liked! Every mine official knew that the practice was rife in one form or another, if only for a few minutes during the course of a shift. The men concerned did their level best not to succumb completely, where, but for the awareness of others, some men would even start snoring! 'Twas not the first time that, at about a quarter to ten on the day-shift, Chairlie Sinclair would roar-up from the Gaffer's manhole on the Lower Dysart Dook, "I don't hear any sounds coming from up there, are you all asleep or what?" They probably were! Or nearly all were dozing, and this was

Chairlie's way of warning everyone that Jimmy Frew *(section oversman)* was about to make a move!

If any miner who had been caught and sacked at Lochhead, had the temerity to go back to see the manager and apologised meaningfully, or even grudgingly for his misdemeanour, the chances are that he would be reinstated. The manager in question, whose name I shall not mention, did, on one occasion to my knowledge, reinstate a transgressor outwith one week, who actually returned to his old job a much wiser and chastened miner, at least until the next time, but vowing never to get caught again!

Taking into account the working shifts cycle in any coal mine, the piece-time shift *(which is an oddity itself)* followed by the back-shift, would appear to be the most civilised of workings hours. Getting up at a 7-30 or eight o' clock in the morning and having breakfast before going to work, is the most natural thing in the world and should not interrupt normal sleep patterns. Neither should going to bed at 11-00 p.m. or 11-30 p.m. of a night time after coming off a back-shift, and even getting up late the next morning and partaking of a late breakfast. None of these shift patterns should cause undue tiredness to a fit miner, if this pattern were regular. Unfortunately, there was nothing of any regularity that lasted overlong in any coal-mine. At sometime or another, every miner had to take his turn at shift work, be it by necessity or Hobson's choice. The dreaded night-shift, was in my opinion, the shift most likely to sent a weary miner nodding off just after his piece, followed closely by those miners working on the day-shift.

The day-shift in most small coal mines is the coal getting shift and the one most likely to be fully manned, where miners often worked together in differing sizes groups all over the colliery. When break-time came and the machinery was silenced, the men would sit down, usually each to his favoured spot in the company of mates and the men's piece would be quickly eaten. The piece-box and flask would then be put to one side and the miners jacket then reached for! The jacket would be pulled smugly across the chest and shrugged around their shoulders and the miners would either smoke, chat, stare introspectively into 'space', or follow the 'gentle art' of 'resting their eyelids!'

Some miners, especially those of long standing, including those who said that they didn't need very much sleep, would never dream of closing their eyes down a coal mine and these were the men who would try to keep some conversation going, if only to make it appear that the men were all 'with it!' It was however, inevitable, that after a mouthful of food and possibly a warm drink, the very silence itself was an inducement to close one's eyes and think of payday, or a week's holiday in the daylight and sunshine, only to be rudely elbowed awake just as your eyes had closed. Some miners had this 'gentle art' so finely tuned that they could 'hover' between awareness and unconsciousness, without actually closing their eyes. They were 'gone' to the world as regards workplace normality, but would become instantly awake and alert at the first sign of anything untoward, like the sound of a deputy's footfall and they had the ability to sound absolutely normal if questioned, albeit belatedly! Other miners who actually felt weary or tired, often felt like dozing-off and would often rest their heads on their folded arms and momentarily close their eyes, but felt conscience bound not to give in to temptation. They would devise a small personal distraction, that involved movement of a limb or part of their bodies, to deflect the onset of 'heavy eyelids'.

One such movement on my part that developed quite alarmingly and initially caused much amusement amongst my fellow workmates, became so manifest as to lead to disbelief, then later on, exoneration, amusement and embarrassment. It's automation in later years, did provide me with a classic defence, in that when I was a Warrant Officer in H.M. Forces, my Battery Commander pointedly asked me if I had fallen asleep during my 'night watch' in the command post? My G.P.O. *(Gun Position Officer)* came to my rescue when questioned and confirmed that I could not have possibly been asleep, as my right leg was trembling 'ten to the dozen'. Later on in life, this 'affliction' turned into a source of mild opprobrium, when my wife, having been wakened from a sound sleep by a mildly shaking bed, seemingly reproached me with the accusation of never seeming to have had enough, only to discover that I was in a 'dead sleep'. I had in fact, whilst down the coal-mines, developed the ability to rest my left foot flat on the ground and, with my right foot slightly to the rear with the heel raised, I could 'tremble' the calf muscle to the extent that my heel would begin to 'drum' on the ground. This did eventually in later life extend to my left leg as well, much to my embarrassment and to the great amusement of other people, who thought it was a relic of my near fatal R.T.A. in 1975.

During one piece-time whilst I was on the night-shift *(1-30 a.m.)*, working hard at setting two pillar-wood and stone packs on a long wall face, I discovered that I was almost out of carbide. My lamp had dimmed down to a feeble glow, which necessitated my searching out an oncost worker, whom I knew to be in the vicinity of the trailing tail-gate *(return airway)*. His task was to supposedly supply me and one or two others with pillar-wood to be used in the erection of my two packs, so my journey was doubly purposed. I crawled and fumbled my way up the coal-face and out into the trailing tail-gate, to discover the 'wood-man' lying alongside the long flat supply bogie, groaning loudly and in apparent agony. I quickly scrambled down beside him and demanded to know what was wrong? He moaned and groaned and thumped and massaged his legs in turn, and complained that his legs had seized solid and could not be bent. He had tried to stand up and had fallen down alongside the bogie. He was certainly in great pain, but after about ten minutes of massage and with my help, he managed to stand up and take a few tentative steps, where his legs seemed to respond to this treatment and within 15 minutes, all feeling returned to his legs. He then sheepishly explained to me, that he had fallen asleep lying face down along the wood bogie, where his long legs had protruded way over the end with his knees clear of the bogie's lip. He had been jerked awake by the sound of my scrambling up the face and immediately thought that it was the fireman. His knee joints had actually 'frozen' in the straight position, due to the unaccustomed weight of his legs and heavy pit boots and simply would not flex when asked to do so, as he tried to stand up. He certainly suffered for his misdemeanour, which made me think as to why he was on the night-shift at all! Maybe for the same reason as myself!

In Lochhead pit, one particular miner did have a mild affliction that would send him to sleep at a moments notice and without warning. This would not happen of course, during the hours of manual activity, but given a slack period of enforced idleness, or when piece-time came round, this miner would sit down to start eating and before the sandwich had reached his mouth, he would be sound asleep! Everyone who worked beside R. J. knew as to his impediment/affliction and would keep a watchful eye on R . . . . . for any sign of laxity. A slight nudge was all that was needed and R . . . . . would then open his eyes and respond. He did have the ability in the early stages of unconsciousness, to stay perfectly upright, for the simple reason that he was quite a heftily, built lad with a low centre of gravity.

On one such day-shift as the second rake of miners were being lowered down the man-haulage in the New Dook, the bogies were slowed down as the rake passed over what the night before, had been a derailment and rail displacement on the dook. The new rails and sleepers had been re-laid in haste and were not as yet bedded-in, so the motor-man had slowed the speed to walking pace to negotiate the previously damaged section of rails. At that point, the whole bogie gave a solid lurch as it mounted a proud joint, and men held their breaths. . . . ! Just then, R . . . . . suddenly shook himself and shouted, "I'm awake! I'm A-W-A-K-E! I wasn't sleeping . . . .!?"

There is a story that has done the 'rounds' and is well remembered in Lochhead pit, where the manager accompanied by one of his more disagreeable deputies, were doing the weekly rounds in a warm section down the New Dook. The manager had timed his visit rather particularly, in that they arrived in the section just five minutes before the machinery was due to start, just after day-shift piece-time. They had walked in quite noiselessly and with dimmed battery lamps *(mine officials were usually issued with helmet-held spot-lamps),* when they came upon a scene of utter tranquillity, where almost every oncost worker was 'resting their eyelids' and those that weren't, were gazing into space. The manager stopped and the deputy made to arouse the nearest miner who was sitting in the 'hunkered-down' position, in that his feet were on the pavement, his back was supported against a side-wall and his arms were folded across his knees, with his head resting on his forearms. The manager restrained the deputy with the quietly spoken words "Let him be, he's still go a job so long as he's sleeping!" Just then, one or two of the somniferous miners quickly opened their eyes and sensing the impending doom, stared defiantly at both the officials and the luckless victim. A few moments passed and just as the manager made a small move towards the apparently dozing drawer, the miner slowly lifted his hands, meaningfully pressed them together, raised his head ever so slightly heavens-ward and quietly, but forcefully intoned: - "Thank you Lord. A - M - E - N !" . . . ? Wullie Hampson slowly tipped back his head, turned to the deputy and said, "Well! That takes the bloody biscuit! There's just no come-back to that . . . is there !?"

*

# Chapter XXXIII

## Dauvit's Progress VI

### Dauvit Goes even Further Forth
### – and 'A Change of Venue'

Shortly after this incident on the night-shift *(or more likely, because of it!)*, I was transferred to the day-shift and given a new job. This was a different kind of work, but more familiar territory, in that I was to collect a fixed amount of seven long feet, wooden straps and four feet long, wooden trees, then bogie them along the daily extending belt-level of the main-gate, in order to directly supply the bottom three strippers on the coal-face with roof support materials. I was pressed to ensure, that I had a flat, wood bogie loaded with straps and trees sitting at the bottom of the coalface *(at the inside end of the main-gate!)*, awaiting the first call from any one of the bottom five coal strippers. On the first call, I then proceeded to throw ten straps and twenty trees *(pit-props)* on to the first place on the run, which was then stripped by Jimmy Hutton. Jimmy was always very accommodating and helped me throw the remaining wood upwards to the next place, after keeping two sets for himself. The next place was stripped by Jimmy Higgins, who was probably the dourest man imaginable, a man who never smiled and never wore 'knee-pads'. Jimmy would help with the wood, but only to get it out of his way! Jimmy Higgins was a man of very few words! The No. 3 stripper on this run was Dave Brown from Kirkcaldy, who was not a fast worker, but could be depended on to keep going and get finished. *(I reckon that I stripped more of Dave Brown's coal than anyone else's!)* The No. 4 stripper was Jimmy Rutherford, also from Kirkcaldy, a tallish, well-built man, who never cleaned up the loose coal lying behind him until the very last and, always wanted extra 'straps' to act as coal barriers as he stripped upwards. The No. 5 stripper, who was the last of my 'men', was a smallish mature man from West Wemyss, who was as good as any stripper, and who months later, always 'saved' me a jam sandwich at the end of the shift, in the hope that I had 'saved' him a cigarette. His name was Andy Cunningham.

This job on the 'wood' did give me ample time and opportunity, to actually get on the coalface and help the tardier miners strip some of their coal. I so arranged my work, that I always had a surfeit of trees and straps available at the belt-end, so that on arrival in the water level, I could get straight onto the coalface, grab a shovel and get dug-into someone's gum. *(Small pulverised coal produced by the coal cutter.)* This became my daily task and I divided my attentions fairly between the first, five strippers on the face-line. This gave me the practical experience of constantly picking and shovelling in a confined space and on my knees, with time to recover before being called to the 'wood-front!'

As the weeks past, I found myself being hardened to working on my knees for longer periods of time, without having to stop, lie down and stretch my long legs. *(This acclimatisation takes much longer than readers might suppose.)* I must admit that the initial hours spend working on my knees, gave me the urge to try and stand up on this 3 ft - 3 inches high face, which of course, was quite impossible, but the urge still remained.

One morning at 6-30 a.m. just as the strippers were climbing onto the coalface prior to cleaning up the gum, I noticed that I had built up a sufficiency of trees and straps, that negated my having to go and collect some more. I had supplies enough on hand to satisfy everyone's needs until after piece-time. I patently waited until Dave Brown had cleaned up his gum and called for the shot-firer, and when we were sheltering from the blasts, I casually said to Dave Brown that I would strip his first two fired shots. He readily agreed and pointed to his shovel and pick and said, "On you go!" I knew exactly what to do and commenced shovelling the loose coal onto the moving conveyor and rapidly cleaned up the front of the fired coal. I then began to dig into the standing, but broken coal and using the shovel as a capacious lever, soon began to make inroads into the fired heap which was approximately three yards long. *(Approximately three tons of coal.)* I soon began to sweat profusely, but did not slacken pace as I was working well within the bounds of my strength. I picked at the downed coal to free the largest lumps and happily shovelled away the freed coals. My arms and legs did begin to ache, and comfortably so, but I did also have this awful urge to stand up to continue. I couldn't of course, but the progressive aching in my legs was beginning to overwhelm my determination to continue. My legs were soon screaming to be stretched, but doggedly I persevered until I had cleaned up the last of the downed coal. At that point, I threw down the shovel and dived down the run with the intention of getting to the main-gate, where I could stand up to my full height to relieve my aching legs, only to hear the mirthful mocking voice of Dave Brown shouting, "Aren't you going to set the strap and trees as well - you idle bugger, you'd be as well to finish what you started ?"- He, laughing heartily! All of this of course, stood me in good stead and helped the long process of developing muscular endurance and the necessary inurity to sustained hard work.

*

*A few months later, I did hear a first hand account of a story that was being told by Dave Brown, in a miner's social club in Kirkcaldy. The story, was of course, being related to other miner's and their wives, as a coal-mines anecdote from the Surface Dipping. The story opens with Dave Brown on the point of clearing away the last coals of his opening shots and calling down the run for 'wood!' (Wooden straps and trees). Ninety-five percent of the time, the 'wood-laddie' would be an average sized 17-year old boy, who instantly obeyed every instruction given to him by the strippers and always said 'p-p-please' when he wanted to pass the 'wood' up through a strippers place, as this involved the stripper having to stop his own work and help the wood-laddie to manhandle the straps and trees.*

*A certain few of the bottom-end strippers could get very greedy and take more than their share of the initially restricted supply, with the result that the miners on places No. 4 and No. 5 were caused unnecessary waiting. (The shot-firer would not 'blow' a strippers next shot until the appropriate roof supports had been set!) I had previously been told by the Fireman, that I must ensure that each and every stripper to whom I supplied wood, only received their proper share at the onset and that I was not to be intimidated by selfish strippers. Easier said than done! - but I did follow this instruction to the letter, so much so in fact, that I had miners like Jimmy Higgins complaining, that he was never getting enough wood to keep him going. (A gross distortion of the facts! Dave Brown's version of events was also a great distortion of the facts, but what is the point of letting a few simple facts get in the way of a good anecdote. Dave's version of events reads somewhat different to my recollections.*

*Dave relates, that as he called down the run for 'wood,' he did set the single strap with the only two trees that he had available and then called for the shot-firer. George Halley duly arrived, charged the hole and fired the shot! Just then! Dave relates, this great, hulking brute of a 'wood-laddie' arrives on the scene, throwing seven foot straps and four foot trees before him, like they were giant match-sticks. This 6 feet 2 inch, 14-stone brute, faced with the barrier of two yards of downed coal, elbows Dave Brown to one side, grabs his shovel and tears into the coal with the words, "When are you bloody stripper's going to stop getting in my way when I'm here to deliver the wood?" then quickly stripped two yards of Dave's downed coal to make a through path for the wood! David Brown has obviously embellished the story, but it makes for good telling in any case! And, it did give me a small insight as to how some of the strippers viewed my bold efforts as a 'wood-laddie!'*

\*

One small incident at that time, one out of many such-like, did stay in my mind even unto this day, and that was one day-shift when there was lots of problems, with the main belt having suffered three, full breakages in one shift. The last of the three breakdowns lasted a full 45 minutes (*almost unheard of !*), where the deputy had to send to the mine-head for another 'monkey' machine. *(A portable hand operated, mechanical machine, for attaching and clamping the steel wire, inter-locking grips onto the newly cut, squared end of the rubberised, conveyor belt, to enable a join to be manufactured.)* The stripper's were a long way from being stripped, even to the extent that not one of them was anywhere near completion. Everyone and anyone who could handle a shovel, was co-opted onto the run to shovel downed coal onto the conveyors. Almost every stripper had yards of downed coal piled up against the pans, waiting to be cleared. Most of them were pretty well advanced with the setting of their wood, but the coal production for the day was well down. I myself borrowed a shovel from Danny Gough and crawled onto the run to help whatever stripper I could, and immediately attacked Jimmy Hutton's downed coal and thence, some the coal of the next miner on the run. The shift eventually finished, but ran to the maximum when the conveyors stopped at 1-30 pm, as the oncost men were bound to stop work. I didn't think any more about this incident and my stripping of some of Jimmy Hutton's coal, but the following Monday afternoon as we were walking out the water level to the mine bottom, Jimmy stopped me and pulled out his pocket watch box. He produced a florin and a single shilling, which he pressed into my hand with the words, "You stripped the coal, you earned the money!". He refused to take back this money and said, that he would be much offended if I insisted on returning it! I was as pleased as punch, this was my very first reward for stripping coal and it was totally unexpected. Jimmy Hutton, coal stripper, a gentleman indeed! *(At that time, a long-wall stripper was paid exactly two shillings and eleven-pence half-penny for every yard of coal stripped.)*

✳

# Chapter XXXIV

## Mary's Transgressions! – *and One to You!*

### (Anecdote or Lochhead Myth?)

Amongst the many improbable and sometimes unbelievable stories that were often told about some of the men and characters that worked in Lochhead pit, there is one amongst many that has survived the vagaries of the near forgotten and has been brought slowly back into my own memory. A particular and unusual event that happened all of fifty years ago. This story was told to me by Bob Stevens, now living in Glenrothes, who stripped coal in Lochhead pit at that time and directly concerned his close friend Jimmy Hutton, who at that time, was part of a nine-man 'contract' group of brushers, led by one John Nichol of the Coaltown. John Nichol of course, would have been the miner who initially agreed the 'contract' with the mine manager Wullie Hampson. During every Thursday lunchtime, whether he be day-shift or back-shift, John Nichol would receive the weeks pay-slip, describing in full detail the amount of yardage advanced and therefore, the total amount of money to be paid to the leader of the contract for the previous week's work.

The subsequent division of this money amongst the nine miners was not simply the total amount divided by nine, things were not quite so straight forward as this simple arithmetic division. In a contract of this type where there would be nine men involved, they usually worked in three shifts of three men, each with a 'team-leader', who would invariably, be the most experienced and senior miner. There would be a No. 2 man and a No. 3 man on each of the three shifts. This would be the normal method of team composition and made for good training and hard experience for the future team leaders. The miner *(contractor)* heading the 'contract' would, as his due, probably expect and 'demand' a small, additional remuneration over and above that of the other two team leaders who, in turn, would expect a smaller, extra remuneration as was their respective due. The 'payout' therefore, would reflect the small differences in remuneration due to each miner according to his position in the set-up, with each of the No's. 2 and then each of the No's. 3 being paid exactly the same amount, provided that they had all worked the same amount of hours and shifts. When each team leader was informed of the total amount of the week's remuneration, it was then a slightly, involved matter for each individual to calculate his own share for that week. Men were usually scrupulously fair in their dealings with each other and good working relations were especially important within each contract team.

On one such Thursday, John Nichol collected the pay-slip for the previous week's work and went home as usual, to a hot meal and some much needed sleep. The rest of the miners in the contract had not seen or, spoken to John on that Thursday after he had collected the pay-slip, the on-coming shift had somehow missed him in coming through the 'baths'. His own two workmates had also 'missed' him and, as a result, no-one else in the group had any idea as to the amount of the next days 'payout!'. *(Expectations always ran high!)*

Friday morning arrived and John Nichol did collect the groups wages as was usual. He then went back home to have lunch before coming back to the pithead and thence to the bike-sheds, which was the appointed place where the contract would gather for the subsequent share-out, with all of the men in attendance and waiting with eager anticipation. *(And the later possibility of a pint or two of cold beer).* What happened next on that fateful afternoon, was colourfully painted into the minds of the men concerned, was remembered in all of its detailed glory, and was told and re-told even to this day, especially in light of it's delicious aftermath!

John Nichol did not turn-up at the bike-sheds that afternoon. The men waited until the back-shift was due underground and then, John's wife Mary, turned up at the bike-shed. The men thought that Mary had come to explain John's absence and confirmation that they would have to wait for their wages, but not so! Mary demanded that they all re-gather at the appointed place and then bluntly stated, that she was going to handle the 'payout!'

The miners were rather surprised and taken aback, but grudgingly accepted the female presence and gathered around in a tight circle, 'hunkered' down onto their knees, in anticipation of a simple, quick explanation of the yardage recorded and satisfactory division of the moneys. This was not to be! Mary Nichol had somehow managed to change all of the paper money into single pound notes, which she now held in her left hand, and naming Jimmy Hutton as the first in the circle - proceeded to intone and deal out the paper money: - **"That's one pound to you, and one pound to you, and one pound to you and . . . . . !"** "Whoa there! Just whoa there!", - voiced Jimmy Hutton, the first to recover his wits, "Just what do you think your doing ? - Who told you to do this?" To which Mary nonchalantly replied "Well, no-one exactly, but John did say to me, that as he had turned quite ill after lunch, to take the money down to the bike-sheds and 'see' that the men get their wages," he himself, having had taken his due share. Mary, dutiful wife that she was, had taken this instruction quite literally and taken upon herself to pay the men according to her definition of equality. At this point, Jimmy Hutton stopped Mary Nichol dead in her tracks, gathered up the distributed pound notes and gave them all back to Mary, with the pointed instruction to take the lot back to her husband, get him out of bed and back down to the bike-sheds 'tut-suit', where the angered men would still be impatiently waiting! This time, with given offence, and more than just a little suppressed anger at Mary's cavalier attitude and bland assumption of power!

So, Mary was sent packing back up to the Coaltown of Wemyss with a flea in her ear, a very red face and a pocketful of the 'contracts' money. John was turfed out of bed by a very embarrassed and irate Mary and sent on his way down the pit road to the bike-sheds, where a satisfactory division of the wages did eventually take place, embarrassingly supervised by a rather chastened John Nichol.

This story does not quite end there! These men were not the only 'contracts' to be paid-out in the bike-sheds on that fateful day, and word of Mary's bland assumption of the post of 'paymaster' spread quickly though the pit, so that by the following Monday morning, there was nary a man who did not know the story of Mary's 'faux pax' and this included most of the pit-head workers. That very day as John Nichol and his men were working underground and had come to the point of requiring a tub-full of pillar-wood to build the next two pillars, his No. 3 man discovered that the pillar-wood had indeed arrived into the section and into the road, but emblazoned across the front and back of the hutch, printed in bold white chalk lettering were the words: - 'To the 'MARY NICHOL' Section!'. Folklore does not record the comments made by John Nichol at that time, if indeed there were any, nor does it record just how long the charade actually lasted, or of the hurt to his feelings. It was certainly a long time before John Nichol lived down the embarrassment and dent to his pride, caused by a 'dutiful' wife, who thought she would, or could, take the place of her husband in such a solemn ritual as weekly paymaster to the 'contract!'

∗

# Chapter XXXV

## A Miner's Piece!
## *– or, "What's the Recipe Today, Jeem?"*

In days gone by, a miner's wife or mother, would invariably ask her man or son/s as to what he/they would like for his 'piece' for the on-coming shift, knowing full well that there was rarely anything different to be had in the way of 'sandwich' fillings. There were only two types of real bread available at this time and the lesser, finer one, was always a poor second choice. The mainstay of the miners piece-box was the type of heavy bread, shaped in its baking, to fit easily inside the custom designed metal box *(or was it the other way round?)*, that was readily available in some hardware stores, but always available from the itinerant hawker Pete Dickson, who carried plentiful stocks and visited the Coaltown of Wemyss once every week. This metal box was perhaps10 inches long by 5 inches wide and approximately 3 inches deep, it was square-shaped at one end and half-round at the other end. It was designed in two halves, that separated to open into near identical halves, but with one half slightly smaller than the other, to facilitate a dustproof container when filled and closed. This metal box was a very necessary part of a miners underground accoutrement, along with some sort of screw-topped, metal container of large capacity, to ensure a clean and sufficient supply of fresh water. Most miners usually survived on nothing more than bread and water whilst underground, the main reason being, that with a time limit of 20 minutes only to devour any sustenance, steak and kidney pudding with pie-crust and double veg was not an option!

The type of bread that most miners favoured was called 'plain' bread or batch bread, as it was baked in large batches at most bakeries and was sold by every bakers van and shop that sold bread. The Co-op in the Coaltown of Wemyss *(S.C.W.S.)* sold vast amounts of plain bread on a daily basis and was usually sold-out by mid-afternoon on any given day. This shape of this bread was the same shape as the metal piece-box, but slightly smaller. It was brown crusted at top and bottom, the top being rounded and the bottom being flat. Depending on the baker and the bakery, these crusts could either be slightly burned and crispy or less well-fired and more chewy. Either way, this type of bread was the favoured 'plain' choice of most miners. Of the mine officials and most of the older oncost workers, the second type of bread which was known as 'pan' bread, was much favoured, simply because of it's softer, lighter texture and it's soft, light, all-round crust, which was to men with dentures, a much more viable proposition. This bread was smaller in size, could be cut into finer slices and would take up much less space than its 'plain' counterpart. *(This was probably the type of sandwich that the older and less active miner might prefer!)*

The 'fillings' favoured for the most part were, butter *(margarine)* and jam, *(the butter was needed to ensure that the jam did not saturate the open-pored plain bread, so rendering the 'buttie' uneatable,)* cheese in its many forms *(no matter what age, it could always be toasted),* or spam in any of its forms. Eggs were a favourite if they could be had, and many miners did keep one or two hens, whose eggs were a very welcome addition to an otherwise limited menu. Sometimes, if money was short, or if there was no bread available, it was not uncommon for a miner to open his piece-box, to reveal nothing more than some porridge-cake,* or one or two cold, boiled potatoes. One such miner who was just a little ashamed at having to eat cold potatoes in the presence of his workmates, was instantly mollified by the comment of his friend, "That's no sae bad Tam, yer wife's given ye a wee twist o' salt to gang wi' yer tatties, and here's a wee bit o' cheese tae gang wi' it an aw!" Such exchanges were readily practised amongst miners and the recurring thought was never very far away, 'There but for the grace of God, go I!'

Amongst other delights that sometimes filled a miners piece box, were often last night's fish and chips. Fish and chip shops were prolific in the thirties, forties and fifties and North Sea haddock was plentiful. *(Except for the war years 1939-45.)* A family could be fed a good and satisfying meal of haddock and chips relatively cheaply. Usually, there was just enough to satisfy all of the family, but if there happened to be anything left over, then that would not be wasted and disposed of, but would appear as a cold 'supper' in the miners piece-box. This would sometimes actually be a welcome change to a diet of bread sandwiches.

There was one, particular, single miner called Benny, who lived in West Wemyss and worked in Lochhead, who perhaps prepared the most bizarre concoction of all on his sandwiches. He obtained the 'meat' quite easily, as it was not rationed at that time, was in plentiful supply and was quite inexpensive. He spread the contents of these tins quite liberally on his sandwiches and at piece-times, often offered them around to unsuspecting recipients, who quite unwittingly ate them. Most of his acquaintances knew the contents of these tasty morsels and shudderingly declined the proffered sandwiches. After all, it's not everyone, no matter of how hungry they were, that would enjoy a slice of fresh bread liberally spread with a generous helping of 'Kit-e-Cat!'

Water would be used as the main thirst-quencher to most miners, but cold tea, mostly without milk, was carried and drunk by most miners. Those fortunate beings that were able to afford a 'thermos' flask, did enjoy a welcome hot drink at piece-time, but these men were in the minority. Thermos flasks were very fragile and needed lots of cosseting and miners who used them, carried their haversacks as though they were transporting eggs! The average life of a 'vacuum' flask was approximately two to three days and it was a fortunate miner indeed who could make his flask last longer. Not many miners could afford the luxury of a new flask every month or so, and often reverted back to the one pint tin flask. Cold tea can be very refreshing if sipped in small quantities, but invariably, was usually swigged with great relish along with bread and jam at piece-times. A few colliers, mostly those who worked in the sections and, in out of the way workings, and many oncost workers carried with them, or secreted in the local area, their own personal fire-box. This consisted of a small tin box, usually with a closing lid about the size of an elastoplast box. The miner would obtain ordinary candles which were in plentiful supply ( yes! from Pete Dickson or the Co-op!), and a small amount of clean cotton waste the size of a present day, two-penny piece. This small fire-box would be opened-up, clean waste placed in it's centre and melted candle grease dripped over the cotton waste. At this juncture, the fire-box would be placed on the flat inside of a short piece of 'H' shaped girder, strategically sighted within the confines of a manhole. (Out of the regular air-flow.) The cotton waste would be fired-up with a naked flame and a two-inch length of broken candle used to fuel the burning waste. The tin flask with it's cork fully loosened, would be placed directly over the flame and left to heat. Within ten minutes, the contents of the flask would be more than hot enough, to burn both lips and mouths. The down-side to this highly illegal practice was, that no matter where this fire-box was sighted, the distinct odour of the burning candle wax could be detected without even trying. It was the one obvious give away! The trick therefore, was to organise the brewing either before, or after the Fireman made his rounds. Where one individual was involved in this practice, it was quite usual for him to be approached by another miner who would also 'like' to share the fire-box. That was all very well if the individual was willing to go half-shares with the purchase of the candle wax, but that was usually not the case! He would only want hot tea for that single day and of course, every 'single' day that followed. The only way to deter such scroungers, was to accept the flask for heating, add an extra length of candle wax to the fire-box and 'forget' to loosen the others mans cork! This usually resulted in the tin flask literally exploding apart because of the steam pressure from the boiling tea, as 'his' flask was left on the flame just a little while longer!

Some miner's, especially those that did very heavy and strenuous work and those that worked in dusty conditions, consumed copious amounts of water. This was needed to first cleanse the coal and dust from a man's mouth and throat, before actually drinking and swallowing, as most men often needed to work with an slightly open mouth, in order to process the large amounts of oxygen needed to sustain the prodigious work rate. It was therefore quite usual for a miner to carry a full gallon of cold water underground, and still have to ask an oncost worker if he could spare some much needed water at the end of a shift. Many underground workers doing repetitive, but less energetic work, purposely carried full, half-gallon, water flasks just for this very reason. It was often a small and keen pleasure to a thirsty miner to be offered and accept, a welcome drink of someone else's water.

When piece-time came, it was always a thought to have to dig-in to a piece-box with hands that were absolutely black. The sandwiches would invariably be of a thick cut, sometimes wrapped in tissue cloth or greaseproof paper, but mostly without. The bread would by comparison appear to be clean, white and in pristine condition - and needed to be eaten! Needs must, so they just got stuck-in, sandwiches getting blacker by the second, with the dirtiest being those with thick crusts, that needed manipulating and tearing, before being reduced to bite-size. Most everyone ate their share of coal dust, whether it was with cheese, butter and jam, corned beef or even spam!

At a normal piece-time when the 20 minute break commenced, all machinery was stopped, which meant that for the first few minutes the silence was deafening! It was of course, that men's hearing required a few minutes to adjust to the veritable silence. Men would gather in small groups and usually sat close together, side by side where possible, not to be especially chummy, but merely to stop the cold air-flow from each others bodies. For the most part, conversation would be rather desultory, with each man concentrating on his sandwiches and masticating forcefully or not, depending on what type of crusts were the order of the day and whether a man wore dentures or otherwise. Quite often a shouted reminder would issue forth from the No. 1 manhole near the top of the Lower Dysart Dook, "Put those damned crusts back in your piece-boxes! Either that, or be sure to throw them in an empty tub. You know as well as me that there are rats about this place!" This shouted instruction was heard at least once every week and emanated from the 'private' manhole, where the section gaffer, one James Frew and the two section firemen sat in isolation, to have their respective snacks. This shout invariably came from Chairlie Sinclair, the Lower Dysart 'slopes' fireman as opposed to Tom Coventry, the dook fireman, whom I personally never heard raise his voice. It is still very strange to relate, that the incidence of the rat population in the Lower Dysart section, seemed to be at its most prolific in the region of this same manhole, where all three members of mines officialdom did partake of their daily bread, and even more significant, that all three men were the proud possessors of a full set of false knashers!

One of the most absent-minded of men that was known to some miners, at least as regards his piece, was Bob MacDougal from the Coaltown. Bob was a very conscientious man, was rather mature and was very diligent in his responsibilities as a Fireman. This particular anecdote which involved Bob, was an on-going act of devilment perpetrated by Tam Mathers, and was one that lasted for many weeks.  This was related to me by Tam Mathers himself, amidst bouts of unsuppressed laughter, a full fifty years on after the events.  Tam was originally native to West Wemyss, but is now lives in the Coaltown of Wemyss.  The story opens when Bob MacDougal was one of the Fireman in the Lower Dysart Section at the time, and he was of course, much concerned with seeing that all of the coal-carrying machinery remained in motion and was properly attended.

 If any one part of the 'chain' broke down or even stopped momentarily, Bob dropped everything and rushed to the scene of the de-faulting machinery, no matter how far away it was.  Bob just had to be in attendance, if only to be able to report that the breakdown was being investigated.  For some obscure reason Bob did not take his piece at the recognised time, but usually consumed it shortly afterwards.  His duties obviously took him to all parts of the section which he would visit more than once during the course of the shift, and with reasonable regularity.  He usually appeared at the 30-inch belt-end just as the belt-motor was due to start-up again after the mid-morning break, where he could have his piece and monitor the amount of coal coming of the belt at the restart.

Bob, more often than not, did not actually sit down to have his piece, he was more concerned with being in a position to hurriedly drop everything, including his sandwiches and fly of to the source of a potential breakdown.  When Bob could be persuaded to sit down and relax and enjoy his food, he would do so in the most peculiar fashion, where he bent one leg and foot underneath him in such a way, as to give him instant leverage to an upright position, in the event of a lack of coal over the belt-end, or a shout of mechanical breakdown.  Either way, Bob would leave both sandwich and mug of tea exactly where they were, and in an exposed state.  When things were back to their normal running state, Bob would return to his piece quite oblivious of the fact that his sandwiches were full of coal-dust and his tea was stone cold.

On one such day-shift, just when Bob had just opened his piece-box and extracted his first sandwich, he had barely taken a single bite, when the call came up through the mine that the pug was derailed.  Bob dropped everything and dived past the couplings and on out to the stone mine, as usual, his tea and sandwiches were left in limbo.  This time, without a word to anyone, but in full view, Tam Mathers emptied Bob's piece-box of its sandwiches and substituted the one, half-bitten, sandwich back in their place, he also half-emptied the warm tea by the simple expedient of throwing some of it away.  Bob returned sometime later, the de-railed pug now back on the tracks and made to resume his interrupted piece-break, only to find the remnants of one sandwich and an near empty mug.  Bob merely mumbled "Aye! Ah must have eaten it!" - and proceeded to place both box and flask back in his haversack, without so much as a raised eyebrow.  Tam Mathers was beside himself with barely suppressed mirth, as were most of the other miners.  They simply could not understand how a man could accept having consumed his piece when he obviously had not. This did happen several more times with more than one culprit involved, and Bob seemingly never did catch on! If he did, I am positive that he would never say! Bob's wife must have thought it rather odd at times, when she realised that Bob was eating far more than usual at his next meal time at home.

Bob MacDougal was one of the most dedicated, dependable and equitable of men that it has ever been my privilege to know. An opinion, I know, that was justly shared by many miners within the colliery.

*\*Porridge Cake?:- Oats or oatmeal made with water (sometimes with a helping of milk) and a good pinch of salt, cooked to a thick consistency and when ready, poured into a side-board or dresser drawer and allowed to cool into a firm 'cake'. When cold, the cake could be cut into firm 'squares' in situ - and easily handled. It could be eaten with milk (hot or cold) or sprinkled with scarce sugar to make it more palatable. It could even be fried in butchers dripping! It was eaten by many miners and their families at that time, simply because it was very nourishing, relatively cheap and easy to prepare and it did fill empty bellies.*

*Also, and quite surprisingly, there were two fish and chip shops in the Coaltown of Wemyss at one time, both of them doing good business. It therefore must speak volumes for the amount of fish-suppers that were consumed by the local residents. (The Coaltown was too far from both East Wemyss and West Wemyss (in terms of walking) to have supplied outsiders!)*

\*

# Chapter XXXVI

## The Lochhead Syphon – *Old Harry's Intuition!*

### Anecdote or Lochhead Myth?

In most of the Coalmines within the area of East Fife, there were sections that were powder dry with never a trace of dampness, some sections which were quite damp, but with no actual running water, and one or two sections and parts of some new, developing sections that actually had standing water, simply because it could not gravitate to lower workings. This was not because the water was seeping through in great quantities, but simply that the 'standing' water could not drain away because of the down slope of a developing heading or section.

At the start of development of new 'working', the contracting Brusher/ Developer often had to argue forcefully with the management in order to get mechanised tools into the development. It was sometimes an achievement just to be able to get an electrically operated or compressed air, drilling head into the new workings. If the heading was deemed to be just a narrow probing exploration, then power tools were a last priority.

Once a section was fully opened-up and ready for coal-getting, any incoming water would usually be controlled and channelled to even lower ground, where slowly, rising, water levels were kept to a minimum, by the use of powerful electric pumps, where the water was either diverted to older disused workings and thereby neutralised, or if needs be, could eventually be pumped to the Pit-head. *{Or, in the case of Lochhead pit, it was internally pumped to lower workings and allowed to gravitate to the Michael Level via the three main dooks, and flowed (under control) to the Micheal colliery!}* Pumping water to the surface was not an option that was needed at Lochhead pit, as the flow of seeping water never really reached critical levels. If the mine workings were under sea level or indeed, were under the sea waters of the Firth of Forth as was the case with the Francis, the Michael and the Wellesley collieries, this then necessitated long columns of three, four or six-inch water pipes to carry the ingressing water back to the area of the shafts, where it could conveniently pumped to the surface. *(There were no lower workings or elsewhere to which the water could gravitate too!)*

In Lochhead Pit at this time and even before the start of the 50's, there was a standing convention that was later converted to a standing written rule, whereby all miners who worked in what was designated a 'wet' section or wet conditions, were granted immediate and undisputed access to the next up-wardly, mobile cage. This in fact, meant, that a 'wet' miner would come hurriedly walking out of his employment section at the end of his shift and walk straight to the head of the standing queue for the next cage!  This was a very necessary concession, in that these miners worked in conditions where they were sometimes virtually up their knees in cold water and, that water was actually falling onto their persons as they worked.  They did complain to each other at times, but they just got on with it, knowing full well that they must keep working and therefore remain 'warmed-up' until the end of the shift, where the prior knowledge of a forced walk and a quick 'tow' up the pit, would not give them time to 'cool-down' and therefore, be highly susceptible to colds and chills, or even a bout of pneumonia.  This concession was at times abused by men, who worked beside other men who actually got wet, and because these men were their workmates, tried to pull them 'though' with them.  This gave rise to barely suppressed animosity by those miners who were partly wet, but not wet enough to qualify for an early bath!  To overcome any possibility of 'cheating', wet miners were then issued with a chit by the manager, which had to be produced on demand to the pit-bottom onsetter, in this case Charlie Fleming, who was a veritable giant of a man at 6feet - 6inches tall and build like Garth!  Charlie was a man of few words *(he was also partly deaf! - or so it seemed to Wattie Johnston),* and he ruled the bottom of that shaft like a fiefdom.  If Charlie said you weren't going on that cage.  You just didn't get on!  Even the pit-bottom gaffer, Wattie *(call me Walter)* Johnstone didn't bandy words with Big Chay.

In the year of 1948, the management and planners at Lochhead decided to open up and breach the Lower Dysart seam in the west side of the pit, in order to gain access to this thick and rich seam of coal.  This was the start of a brand new section within this slightly deeper, seam that was to be named the Lower Dysart west.  This section was to be a fully mechanised with long, wide conveyors all the way to the coalfaces, where the coal was to be under-cut with the new 'bluebird' machines on several long-wall faces of varying lengths.  The main gate's would be brushed either at 12' by 9' or 11' by 8', with the return roads cum airways to be brushed at 8' by 7'.  Harry Moodie and his team of brushers had completed their last brushing contract and were therefore available for a new contract.  Negotiations for yardage were then commenced with Wullie Hampson, volubly discussed and a price agreed!  Work on this new contract then proceeded!

This new contract, this heading, was to be driven through coal and stone for a given distance to be later determined and, through slightly rising strata.  'Ideal conditions' thought old Harry, no water* problems here, or if there were, it would surely run away downhill.  The contract had been decided on an 12' x 9' dry level, with nine men working in three shifts, six or seven days per week, depending on the physical state of the men and the work involved.  The contract included 'power' drilling and free 'powder' (explosives), and oncost drawers after 15 yards of advance.  Work commenced and progressed smoothly with all three shifts 'pulling their weight', in that a four yards advance was made though the 'solid' every 24 hours. The only small fly in the ointment, was that the heading seemed to be getting a little more wetter with each successive round of 'shots!' This did not make too much difference to Old Harry and his men, except of course, that the coal and redd, especially the redd because it absorbed more water, was understandably heavier and therefore required more lifting. This was all part of the day's work and coped with, as were all things and treated as yet one more, adverse, working condition.

*(Up until then, all of the underground, encountered water found down to and within the Dysart Main seam, had been slightly alkaline in nature. Now the miners were faced with encountered water, that was actually acidic in nature and decidedly corrosive on their metallic tools! It was a known phenomenon in this coalfield, that every coalmine within the East Fife area that took the Lower Dysart coal, were faced with this problem! It was seemingly inherent to the make-up of this seam of coal! Water of this nature will eat into the seals and packings of powered water-pumps and quickly corrode their innards!)*

As this level progressed inwards, it soon became apparent to Harry that the pavement seemed to be levelling out. *(Levels such as this, follows the strata to maintain contact with the coal.)* This did not present a problem, as the water level could not rise to any degree even on a longish road, but what did begin to worry Old Harry, was the fact, that the road had now taken on a slight downwards slope, as was evidenced by the fact that standing water was now beginning to accumulate at the lower working face, thereby making it difficult to site the lower bore-holes. This would not effect the detonation properties of the explosives, as they could quickly revert to waterproof strum detonators. The growing problem was the increasing flow of seepage from the exposed strata, which now resulted in 'standing water' that was rising to unacceptable levels at the working face. The men were now working almost knee-deep in cold water, which meant that every shovelful of redd was 50% heavier and water was being sprayed all over the place at every shovel stroke. Serious as the problem was becoming, Willie Hampson would still not acquiesce to Harry's request for a water pump. *(For some reason, not even a hand-operated pump was available!)* For whatever his motives were, he maintained that the work must go on at an even pace, regardless of a 'little' drop of water. *(Even though he did concede a 'wet-chit' to Harry's brushing team).*

The miners, were at that time, frustrated in their efforts in trying to come up with a working solution to the problem, and it was then that Harry came up with a possible solution to the problem, but even then expressed reservations. He explained to his fellow miners that he had once seen a device that sucked water out of a hole through a pipe and lifted it over a rise, where it ran down a longer pipe and away into an old working, but he also recalled, it had to be recharged every time it sucked air. It was called a syphon!

The section gaffer was approached, was convinced that it was worth a try, and quickly demanded that a wooden, bogie load of twelve feet, lengths of 3-inch flanged pipe was hurriedly despatched into the section from the pit-head. This load to include a sufficient number of appropriately sized flange clamps and the necessary tools to tighten them. This loaded bogie arrived in the section within the space of one shift, and the on-coming shift set-to under Harry's somewhat tentative direction, to construct this device and set it in motion. The pipes being only twelve feet in length, could of course, be 'bent' over the rise, as the clamps allowed for small changes of direction and so, this was not to prove a problem.

The men went to work with a will and soon laid the necessary lengths as directed by Harry, so that the 'drain' side was twice the length of the draw side, and the open end of the short draw side was totally immersed in the rising water. That much Harry knew! On completion, the next problem was to 'charge' the system to get it to 'draw', but how? How could they get water into a long length of pipe that had no openings in it? The answer of course, was to break into the pipe at its highest level and fill it up with water! Again, work stopped on the syphon whilst the necessary valve was sent for, whilst the necessary brushing work resumed. Yardage and tonnage still had to be completed on a daily basis. Next morning, the 'inspection-port' valve duly arrived in the section. The miners immediately set-to, the pipes were split, a gap forced and the valve inserted. Now to 'charge' the system.

The men concerned now knew enough about the doings of syphons, as by this time 'the word had spread!' They knew to plug both ends of the contraption whilst the pipes were fully charged from the highest point. This was achieved by the simple expedient of filling several buckets with the standing water at the face and then 'charging' the system at the highest point. When this was done, the inspection port was swung closed, nutted tight and all was ready. Harry went to the low end of the syphon and instructed the man on the valve to throw it open, which he did! The water simply poured out of the down-pipe! This lasted all of a few seconds, then it suddenly stopped! No more water! The pipe was empty! No change in the water level at the face! What had gone wrong? Harry Moodie and his 'bloody' syphon. *(Muttered the disgruntled miner's.)* Waste o' bloody time! Disappointed miners continued working with ever-increasing water levels and the certain knowledge that an electric pump was their only salvation. As piece-time approached, Old Harry was unusually quiet, he just could not figure out why the syphon would not draw, so the 'piece' was taken in silence. Talk was desultory and spiritless and the men were wet. What to do?

Just then, one of the miners made a rather contrary, but ultimately profound remark, "It's bloody easy tae blaw air up through a water pipe, but it's no sae easy tae blaw it doon!" Harry looked at his workmate for a few long seconds, grunted something in his direction, then rose quickly and hurriedly departed, leaving them sitting there still eating their piece, with old Harry quite obliviously to their astonishment! Harry returned sometime later, clutching a short length of right-angled pipe, but saying nothing at all. He then proceeded to clamp the right-angled bend to end of the down pipe so that the open end was turned vertically upwards. He then organised another pail-gang to refill and 'charge' the syphon once more. When all was ready and the top valve re-nutted and tightened, Harry made sure that the miner operating the bottom valve, clearly understood that the 'right-angled bend' was absolutely filled to overflowing with water *(not air)* before closing the valve at his end.

It was with a great deal of anxious trepidation that Harry's workmate now opened the bottom valve. At first, nothing very much, then a quick gush of frothy water, then, an absolutely full-bore column of water hit the roof in a solid jet that was instantly converted to a torrential downpour as the water, robbed of its power quickly dropped to the pavement and nearly drenching the operator. The syphon was working, and water it drew! So well in fact, that within just a few minutes, a very loud and unusual sound was heard at the stone-face. It was the active end of the syphon drawing air as the water rushed through the down pipe! The syphon had done its work and drawn the standing water!

The miners were rather jubilant, with none more so than Old Harry. He 'knew' it could be made to work, but the one item of information that Harry never did pass on to his men, was that he had clean forgotten, that the curved piece of pipe was always needed at the bottom of the syphon, to stop air egressing into the bottom pipe as the water started to flow. The small volume of water at the bottom bend had stopped the air from getting back up the column of pipes, thereby negated the pulling power of the down column of water! Harry didn't tell his workmates that he had actually forgotten that this bent piece was needed, he just let them think that he had discovered the solution! Such is the stuff that reputations are built from, and that they would remain none the wiser for Harry's lapse of memory!

Note: *Unfortunately, for the miners involved. If the syphon was allowed to 'run-dry' and allow the ingress of air at the 'draw' end, which was usually allowed to happen, then the miners on the on-coming shift were forced to re-charge the system before hole-boring could begin. If the bottom valve on the syphon could be closed before the intake end of the syphon 'drew air', then the system would remain 'charged', with only a small amount of standing water remaining at the brushing face.*

\*

# Chapter XXXVII

## The 'Baund' Section!
## – *A Dreadful Orchestration!*

(Author's Note:- *It was within this section, that one old collier belonging to the Coaltown of Wemyss was reputed to have once said, that during the night-shift - 'He could hear the coal breathing!'*)

The rather infamous Band section, that probably took its name from the fact that most members of the Lochhead contingent of the three collieries, brass band worked in that section, was situated in the roughly, oblong-shaped area between the Old Dook and Nicholson Dook and above No. 1 level, that in fact, ran through from the south-western edge of Lochhead's boundary with the Francis Colliery, through and past the top of the Victoria Sea Dook, through Nicholson's Dook, the Old Dook, the New Dook and on to the Buckhaven fault to the north. A total length of approximately. 3,450 metres. *(3.45 Kilometres.)*

This largish 'Baund' section, *(as it was known)* was sandwiched between part of the length of No. 1 level and the near parallel length of the Lower Dysart haulage-way. *(Cut through the Dysart Main coal and originally named Nicholson's level)* The gradient at the north end *(the Old Dook)* was approximately 1 : 4.5, while the gradient at the south end *(Nicholson's Dook)* was approximately 1 : 7. The approximate size of this area, measured 650 metres long, 375 metres deep *(wide)* at Nicholson's Dook and 175 metres deep *(wide)* at the Old Dook. The section covered an area of approximately 36 acres or 18 hectares. The estimated total reserves of the Dysart Main coal in this section alone, amounted to approximately 1.2 million cubic metres.

Allowing for the premise that the top 1.18 metres *(approximately 4 feet)* of Head coal would be left untouched to form a solid roof, this still left the staggering total of nearly one million cubic metres of workable, coal reserves in this one section. The general premise or aim, seemed to have been, that the planners in their wisdom had decided to try to take most of it! The coals that were finally taken out of this section to a very large degree, were the Spar Coal, the Toughs & Clears, the Myslen in parts, the Grounds & Nethers and the Coronation coal. These coals were removed over an extended 21-year period with different layers of coal being progressively extracted with each successive visitation. The average total thickness of the separately taken coals was probably in the region of 17 to 19 feet, and the methods used, were possibly long-wall pick-places and Stoop and room levels and headings, taking coal, both to the advance and to the retreat. In any event, the amount and thickness of coal extracted was to prove too much for the minimum small stoops left standing, and this section became the focal point for great and sustained, area-wide crushing of the remaining coals.

There were main three characteristics that behove the Dysart Main coal. They were:-

1. The roof did take kindly to weighting pressure, indeed, it became quite obvious to all miners who worked the seam, that the roof did 'bend' quite dramatically in places.

2. This was deemed **not** to be an inherently 'gassy' coal, in spite of the slow mouldering burnings, (*not open fires in the sense of flaming beacons!*) that were prevalent in only **some parts** of the colliery.

3. Some named* coals within the seam, after being exposed, or left as unclosed or exposed 'waste', or where an extracted section had not 'closed' properly, did have a propensity to self-ignition on being subject to crushing pressures that raised the inherent coal temperature to above 172 degrees centigrade. *(*Clears & Nethers coals within the Dysart Main seam!*)

(*In J. T. Beard's 1921 erudite volume 'Mine Gases and Explosions', under 'spontaneous combustion' - he writes and I quote "It has been suggested, and with much reason, that the movement of the strata incident to the extraction of the coal from a seam, accompanied as it naturally is with the evolution of heat, contributes its share towards spontaneous combustion occurring in abandoned mine workings").*

On being exposed to the atmosphere and being subjected to a crushing effect after the extraction process, the tendency was for the coal on either side of a heading, dipping or level (*if it were not steel-girdered*) to crush down, break-off and then lie piled up, close to the solid mass of the stoops. This downed coal needed to be left severely alone, in that the more that it was removed, the more that would slowly crush down by that, creating more downed coal. Its constant removal would result in the 'thinning' of the stoops and, unwanted increase in down pressure and therefore, greater 'weighting' and crushing on the whole of the section. One inherent advantage in the working of the Dysart main seam of coal was, that in using, and not abusing, the stoop and room method of extraction, the miners had very little need to set any sort of roof, as long as a sufficient thickness of hard coal was left in situ and untaken, to act as an unbroken roof. This meant that a band (*layer*) of coal known as the 'Head coal' was left untouched, while the

Sparcoal, the Toughs & Clears, the Myslen and the Grounds & Nethers would be taken, either in one leaf or two separate leafs, depending on the prescribed method used in that section, leaving the thick band of the Coronation coal lying below the approximate four feet thickness of the Sclits and the Fireclay.

The Dysart main seam was an easily worked seam, in that all of the coal was highly usable except the Sclits. *(An even mixture of coal and stone that was fully integrated and utterly inseparable!).*

The seam as named from the top down, consisted of the following described coals along with their respective average thicknesses: -

*(The individually named coals within the Dysart Main seam thickened appreciably, with the addition of thinner unnamed seams as it descended and deepened, to spread out and under the 'Frith' of Forth!)*

1. Head Coal (upper) ................................................................... 32 - 40 inches
   Head Coal (lower).................................................................. 15 - 20 inches
   Stone ..................................................................................... 2 inches
2. Sparcoal ............................................................................... 18 - 22 inches
3. Toughs (coal) ...................................................................... 22 - 26 inches
4. Clears (coal) ....................................................................... 18 - 22 inches
   Stone ..................................................................................... 5 inches
5. Myslen (coal) ...................................................................... 12 inches
   Stone ..................................................................................... 6 inches
6. Nethers (coal) ..................................................................... 32 - 38 inches
7. Grounds (coal) .................................................................... 24 - 32 inches
8. Sclits (50/50 coal and stone - inseparable!) ........................ 28 - 32 inches
   Stone ..................................................................................... 13 - 17 inches
9. Coronation (coal) ............................................................... 52 - 68 inches

The strata immediately beneath the bottom-most Coronation coal was grey fireclay/stone.

Very few attempts were made by the underground colliers to separate the small amount of stone from each separately taken layer of coal. *Unless, of course, the miners were 'filling and tagging' hutches under the watchful eyes of some or other Under-manager or Gaffer, wishing to vent their spleen on unsuspecting miners, and verbally threatened them with the ignominy of 'couped' hutches on the pithead and 'docked' wages, in an effort to get them to separate the waste (redd/stone) from the coal.*

This separation was done on the screening tables at the pithead and by the youths and women pickers, *(all above sixteen years of age since 1947)* whose working conditions in winter were almost indescribable!*
*See Chapter: - Pithead Tables.*

Authors' Note: - *In the course of my research for this book and during my long interviews with the many miners who gave me some of their very precious time, I repeatedly heard tell of an ongoing nefarious practice, perpetrated during the time of the managership of late George Welsh, who reigned supreme at Lochhead for many years and was by all accounts, a very successful manager. It was almost certain, that George Welsh actually detailed a collier\* to visit a particular section in the pit where the Dysart Main coal was being worked, and his instructions to this miner were, to fill as many sandbags as would fit into an empty hutch with the cleanest of the 'Clears' coal as it was being mined! This hutch to be 'addressed' to the pit head to await collection. This miner's task on arriving at the pit head at the end of his shift, was to arrange for these sacks of 'Clears' coal to be delivered to the home of the colliery manager for his family's domestic use. George Welsh knew his Coronation from his Clears, and his Clears from the Toughs and 'woe betides' the said miner, if he had botched the 'fillings' in the darkness. Not for nothing did George Welsh gain the reputation that he thoroughly deserved. (I knew his name before I discovered coal. At home, at dinner time, his name would probably come up just after the soup course. If it came up before this, Old Harry would be rendered nearly speechless with indignation and would probably lose most of his soup!)*

*In an interview with ex-Lochhead, coalminer, Bobby Grubb from Buckhaven, Fife, in February 2000. I was assured that Bobby himself was once 'detailed' by his section deputy, to handpick and fill a 10 cwts. hutch with fist-sized nuggets of 'Clears' coal. This 'perfect' coal was to be taken from the Stoop and room heading in which he was working the Dysart Main seam with his 'neighbour.' Bobby's neighbour, (workmate) on hearing to whom the coal delivery was intended, literally smashed the bottom 90% contents of the hutch into smithereens with his 'mash' and covered the top four inches with sparkling, fist-sized nuggets. The up-going hutch was heavily chalked, 'To the Manager' and sent on its way. History does not reveal the outcome of this miner's subterfuge. Sufficient to say, that never again, were that swingeing pair of colliers ever 'requested' to deliver-up the 'clear' sparkling black diamonds onto the pithead! {I can also guess, that this was the reason that later consignments of 'Clears' coal ear-marked for the same recipient, was delivered-up to the Pithead in 'tied' sandbags! Feeling (lumps) is believing!}*

In this 'Baund' section after the extraction of the maximum amount of coal, the exhausted sections would be partly abandoned and left to 'weight'. These abandoned parts had to be carefully watched and regularly inspected by competent deputies, especially at times of low atmospheric pressure, where the tendency was for some, or lots of the gasses present in these old workings, to filter out through the man-made 'stoppings', into the current, regular workings and thence into the main airways. As already described within another chapter in this book, these gasses were very harmful to the miners, even in small doses. The older these original workings became, the greater became the coal extractions, and the subsequent expellation or leakage of their accumulated gasses. So much so, in fact, that it became a necessary and common practice to seal and 'stop-off' some of these old workings, to limit both the ingress of air and the egress of toxic and oxygen replacing gasses. Many miners, over the course of many years, were engaged in the 'sealing-off' process, but no matter how well the work had been done, the seals, the packs, or even the cemented brick walls would eventually break down, with the pressing need to be quickly resealed. The men engaged in this work were efficient, dependable, sensible and usually very careful. Mine gasses were wholly indiscriminate and very unforgiving!

At some of these 'stoppings', any miner could probably stand close to a pack, a wall, or a steel door and feel the heat coming from behind through sheer heat radiation, without ever touching a stopping. Indeed, with some stoppings, there was built into the wall of the stopping, a steel trap-door that could be gently opened, where the escaping air could be tested and a visual inspection carried out if anything untoward was suspected. Not for the first time, was there a dark red glow discovered, where ignition had already taken place and slowly expanded, but needed no immediate action, as it was 'under control!'

This Baund section became very well known amongst some of the older colliers and achieved lasting notoriety for the amount of 'suppressed' fires that were slowly moldering within its boundaries. The section was however, reasonably well ventilated, being on a main airway, but 'stoppings' were many, with the 'gob' area behind the 'stoppings' being quietly active. Of dire necessity, a 24-hour watch was kept over a seven-day period, even through official holidays. Over the course of many years, there was reputedly, upwards of 400 official stoppings build within, and around, the inlet peripheries of this large abandoned section. This may have seemed to an interested spectator, to be a great burden on the mine manager and his deputies, who were duty-bound to maintain a twenty-four presence in this section and, the question may have been asked, as to why this section could not have been permanently sealed and by-passed?

To my mind, the answer is not an easy one. So therefore, I shall not attempt to do so! Instead, I shall put forward several reasons as to why this might not have been feasible.

1. The section was far too large, with too much coal having been extracted in a rather haphazard manner - and with too many visitations. But only in light of present day extraction methods!

2. This section formed part of the 'early coals' extracted from Lochhead pit and

very near to the pit bottom.  Its upper boundary (the Lower Dysart haulage-way) formed the main air-way for that section.  Its lower boundary (No. 1 level on the Old Dook) formed the main airway to the Victoria Sea Dook.  It would have difficult to by-pass this section.

3.  The Dysart Main seam of coal as broached in 1890, was the only seam of coal that Lochhead pit worked for approximately sixty years, up to the opening of the Lower Dysart west around the late forties!  Therefore, every working (extraction) was in the same seam of coal, with literally hundreds of interlinking roads and air-ways that needed to 'breathe!'  (A maintained air-flow!)

*{J. T. Beard also writes in the same chapter of his informative volume, "Gob (deep wastes) fires may be the direct result of the spontaneous combustion of fine coal and 'slack' in the Mine wastes.  The subject is of importance here only with respect to the gasses produced and the resulting increases of danger in the workings.  Carbon monoxide is produced in considerable quantity where the fire has been deep-seated and the combustion has eaten its way well under the gob, especially where the circulation of air is slow.  In a seam generating marsh gas (methane), a gob fire is a serious menace to the safety of the mine, owing chiefly to the carbon monoxide (white damp) produced increasing the explosive condition of the mine air.  Moisture in the strata is favourable to the rapid extension of a gob fire.}*

In retrospect, and in speaking thus, I have formed the opinion that as far as management and planning was concerned, they too, must have come to the conclusion that they had been a little too greedy, in that far too much of the thickness of the Dysart main coal had been removed from this section area, by haphazard and irregular slow methods.  The height *(thickness)* of the coals eventually taken, along with the very limited 'stoops' that were left, resulted in a long, sustained, but irregular 'crush' on the section, over a long period of time, which inevitably, resulted in many further 'hot-spots' being initiated and igniting.

Many and prolonged attempts were made to suppress, both the ingress of air and the egress of poisonous gasses, to and from, the extracted areas of this Baund section.  Most attempts were successful, but only after protracted and numbing labours on the parts of the miners, who braved the dangers and rigours of this volatile section.

In no other large section in Lochhead Colliery, was the same thickness and volume of coal extracted from any part of the Dysart Main seam by the then, same, slow, manual methods.  The sharp and painful lessons exacted on management and miners in the taking of this coal, must have surely affected the future planned extraction of other sections, although it was not until twenty-five years later in 1948, that a grim tragedy* was visited upon three miners, who lost their lives in trying to control the irrepressible gob fires, with numerous attempts to suppress the forever leaking gasses in this, the Baund section, that continually needed bigger and better 'stoppings' to contain its lethal poisons.

* See 'In Memorandum'.

*

In some sections and at one time, there was a method of 'packing' that was used to partially replace some of the extracted coal within days of its removal, and before the roof had time to collapse and break. This method used recycled, inert, waste material in a form that was easily worked, was quite portable and could be readily manipulated. It was called the 'Stowing', or the 'Flushing!' The basic premise was to take the 'waste redd' that was extracted from amongst the coal at the pit head, and then powerfully crush it to chip size of approximately 'one-half' to 'three-quarters' of an inch, then mix it with dead ash from the pithead boilers (only sometimes). It was then sent back down the pit in hutches to the section where it was to be used.

At a strategic point in the section, close to the area to be stowed and preferably in a main airway, there was a large and powerful, electric air-compressor with an attached overhead hopper for the input of the crushed redd mixture. Attached to this machine and leading into the recipient section, there was a continuous, shortish run of 8-inch diameter, airtight piping leading to the point of ejection. (As from the muzzle of a gun!) When the machine was started and running and a sufficiently high pressure built-up in the reservoir, the crushed, waste mixture was emptied into the hopper and the controller at the 'firing' end would open the 'gun'. The crushed redd would be forced through the pipes at a great rate of knots and be forcibly directed into the recipient empty 'wastes', thereby packing the space vacated by the extracted coals. The open end of the pipe-gun had limited movement from side to side, and could be partially raised or lowered. (The velocity and weight of the compressed air and crushed redd, tended to force the end of the piping out in a straight line!)

A manually-operated, double-handled, moveable shield on the end of the 'gun', could be made to divert the powerful stream of wet redd to its intended destinations.

The 'stowing', usually commenced at the bottom (farthest away) end of the coalface, with the piping being systematically shortened as the wastes were filled to a given depth! In the case of long-wall operations, it would take a complete shift to fill the full face-length to a depth of four feet, although the roof space had somewhat crushed down, perhaps leaving a remaining height of only 50 percent of the original. The limited, directional capabilities of this machine was also used to build near airtight, compressed packing's in levels, headings or dooks, which needed to be positively sealed-off, to negate the ingress of air, or the egress of unwanted gases.

The 'flushing/stowing' machine as used in Lochhead Pit, was of German origin, design and build. It was actually operated and maintained in Lochhead pit by three German engineers, brought over specially, to initially operate and instruct the Scottish miners, who were destined to take over! This happened much sooner than anyone had anticipated or expected, in that in the year of 1938, the three German engineers were suddenly recalled to the Fatherland, without reason or explanation being given. (W.W.II., was ominously looming!) Fortunately, enough technical 'know-how' and empirical knowledge had been garnered by the attendant miners, so that continued 'flushing' or 'stowing' was able to be carried on with, without too much of a hiccup!

The one difference that the Lochhead system had to the original German concept, was that the Lochhead piping was made from steel, whereas the German pipes were made from specially toughened glass, so that any 'blockages' could be quickly detected.

The Lochhead miners had no such luxury and breakdowns became synonymous with lost time, as the steel lengths of piping were systematically dismantled in efforts to locate blockages! Within a short time however, the operators discovered after some experimentation, that it was just possible to locate the start point of a blockage, through listening to the resulting resonance, at points where the pipes were struck with a steel object. (Could this have been because of the bandsmens 'ear' for music?) This system was used successfully in various parts of Lochhead Pit, especially on several long-wall faces where roof-weighting was visibly manifest and, a plentiful supply of inert crushed waste material could be made readily available from the pithead crusher. (This crusher was situated adjacent to the picking tables on the pithead, into which the lumps of waste 'redd' was hand-loaded by shovel - see Chapter XIII, Dauvit's Progress II.) This system was used only in one section of Lochhead at a time, due in part to the limited availability of crushed redd and, simply because there was only one such stowing system imported into Lochhead Colliery.

*

Author's Note: - *In the above narrative, I have said that the Dysart Main seam was not a gassy coal! (Meaning, the expulsion of methane, or marsh gas from the coal seam, in dangerous quantities!). I did, at one time, believe the opposite to be true, for the simple reason that many old and not so bold miners, had sworn (and meaningfully spat!) that this was the case and that I must not light a match underground. (Naked flames from carbide lamps were exempt from this belief!*

*I carried this notional belief for many years (in spite of never having experienced its presence) and casually mentioned as much to a learned gentleman, whom I interviewed during the course of my research.*

*This individual had a quite superior knowledge and practical experience of these matters and asked of me, as to what areas of Lochhead Pit did I work? When told, 'The Old Dook, working 'Stoop and room' - he further asked of me, "How many miles of levels, headings and dippings had I travelled through during my time, that were still standing open, which had not fallen and were not stopped up?" After some quick reasoned thinking, I replied with the answer that he obviously expected. "Miles and miles, all still standing, and NOT on Fire!" There was the answer! These many miles of still open, coal-encompassed roadways and headings, had been left abandoned, were completely open and exposed to many decades of fresh oxygenation - and over all of that long period of time, the negligible amounts of methane gas expelled was probably immeasurable!*

*

Many valuable lessons were learned after the debacles of the Baund section, with it's many points of self-ignition, it's slow moldering fires and the subsequent multiplicity of its many and varied stoppings. The main lesson however, was that the Dysart Main seam needed to be treated with the greatest of respect with regard to the method of extraction and the amounts of coal taken! In those times, it would appear that this seam was just too thick to be safely and comprehensively extracted, considering the era in which it was taken and the semi-primitive methods at the disposal of the miners over that forty-year period. *(From 1900 to 1940)*

The Stoop and room method of extraction, as practised down both the New Dook and the Old Dook, was where a 'Section' consisted of a large block of standing coal, measuring anything from 200 to 600 metres long and approximately 100 to 200 metres wide. This method of extraction, after which the section was 'abandoned', left something like 85 percent of the coal seam still intact, but with the whole 'block' evenly matrixed, both laterally and vertically, like a draughts or chess board. *(With parallel levels and headings!)* It must also be remembered, that frequently the Sclits and the Coronation coal were left intact, along with the Head coals and sometimes the Sparcoal. Therefore, in calculating the overall percentage extracted, especially by the method known as Stoop and room, one needs to realise that perhaps only the middle 40 to 45 percent of the coal height was taken, with approximately 25 percent above and probably 30 percent below, left untouched! (Including the Sclits!)

Author's Note:- *It has been latterly brought to my attention, that of the nine named coals within the Dysart Main Seam, the two coal layers that were most susceptible and therefore, the most likely to low temperature ignition when pressurised after partial extraction, were the remaining Clears and the Nethers coals!*

Before the advent of 'pneumatic' stowing or 'The Flushing' as it was commonly named, there was a previous method of 'packing the wastes' *(mainly roadways)* that involved using the same sort of materials, namely broken-up stone and redd and possibly soft blae if it could be had! In the first three decades of the taking of the Dysart Main seam, at a time when most of the early mistakes were made in the extraction of the Dysart Main coals, the miners/colliers were made to separate every last piece of stone from the mined coal, on the threat of the ignominy of a couped hutch on the 'pithead landing', along with the pain of docked wages. Some small percentage of stone did manage to escape the eagle eye of the check-weight-man on the pithead, but this enforced, underground practice did ensure that there was a near sufficiency of waste material left down the pit, to help cope with essential packing's/blockings near the points of need!

If the colliers were taking the middle coals of the seam, i.e. *Grounds & Nethers, Myslen and Toughs & Clears.* Then, the two bands of Myslen stone, the five to six inches sitting above the Myslen coal and the six to seven inches immediately below, would be painstakingly removed from the coal and deposited immediately nearby. If the roadway was a straight level, that meant that there would be both a high side and a low side to the road, with the face of coal having something like a one in four slope from right to left, or vice-versa. The coal to be taken, was invariably 'blown' and extracted to its full height, both left and right, with the Myslen stone being used to fill the pavement on the low side, to 'level' the laid light rails as the miners progressed inwards. If the colliers were driving a heading or a dipping, the pavement would then be 'level' both left to right, so then the residual stone would be stacked to other side of the advancing roadway, but to the rear.

It was very unusual that a driven **Level** in the Dysart Main seam needed to be 'stopped-off'. *(The longest levels were invariably used as airways!)*

As was the principle of Stoop and room, the long levels were driven as access roads and served as main airways where the coal sections were to be worked. Headings could then be driven upwards, and dippings could be driven downwards from the parallel levels.

*{The overall strategy in taking the coal, was to open-up a section by working to the rise, (headings) where the use of simple 'gravity operated, aeroplane braes' would work to the advantage of the developing miner's. This principle simply does not work in dippings!)*

Where a 'sealing-of' or 'stopping' any kind was needed or required, quantities of redd, stone and waste materials of almost any kind were found and used. Often, it would have to be sent down from the pithead as crushed waste, because this was the type and size of waste material needed and desired, to construct a near airtight stopping. If a stopping were to be commenced on a level, a heading or a dipping, the larger pieces of stone would be used to commence the building of a 'dry-stone' dyke, directly across the roadway to be 'stopped'. This dyke would be started and built up from pavement level and, packed on the 'retreat' side with the crushed stone and small redd. The dry-stone dyking would be raised up to roof level, with the smaller, packing material systematically built up behind to support it. Every so often, the 'packing' would be thoroughly wetted down with unpressurised, hosed* water, to help pack the smaller waste into a better concentration, or airtight mass. *(Never really perfect though!)* This packing material would be hand-shovelled onto the growing mass until it was tight to the roof and sides, and would be extended to the proposed, solid depth of the Stopping, whether it was three, five, or even ten metres in depth. The outside wall of the stopping would also be built up with an integral dry-stone dyke to terminate the pack/stopping. At each stage of its construction, *(if water were readily available?)* when new dry material was shovelled into the great pack, the new material would be heavily wetted, to ensure good bedding down within the interior of the stopping.

Later, when the pack had settled, it was not uncommon for the outside part of the pack to be partially dismantled and inspected, with regard to further settling of its contents and then tightly re-packed if needs be! This would usually be done if a permanent, double cement and brick wall was to be built to the outside of the substantially constructed, stone pack.

These stoppings, wherever they had been constructed underground, had to be inspected and examined on a near, daily basis by a competent fireman/deputy, whose sole duty it was, to ensure that 'readings' *(air samples)* were taken and recorded at each and every stopping within the section.

There were several parts of Lochhead pit that were heavily sealed and packed against the probability of leaking gases and smouldering fires, but this was not really prevalent in the greater numbers of the other different sections in the coalmine. As the declivitous Old Dook and New Dook were gradually extended deeper in the taking of the Dysart Main coals, the favoured singular method of taking the coal, was limited Stoop and room extraction, where only the middle coals of the thickening seam were taken, leaving larger and greater, near square, stoops between the cross-matrixed extractions. Even at this greater depth, there was very little crushing effect on the extracted levels and headings, so little in fact, that many of these roadways survived and remained open for many decades, without even a single support ever being set to the roof! The principle being, the greater the depth, the larger the stoops, with relatively less of the coal seam being taken! *(The same approximate amounts of coals were being extracted on a daily/weekly basis, but the taken percentage within each section area was becoming smaller!)*

*\* To the best of my knowledge, no water of any kind was ever piped down into Lochhead Pit. Lochhead colliery was basically a 'dry' pit except parts of the Lower Dysart Dook (west). The water from this dook was pumped up onto the '30-inch belt level' via a four-inch steel column of pipes, where the water flowed out of the Lower Dysart section through a narrow cut Gauton, back out through the Lower Dysart mine and thence onto the head of Nicholson's Dook. All of the dooks in Lochhead Pit had narrow gauton's cut into the low side, of either the coal dook or the return/man-haulage dook. There was therefore, a small but consistent supply of clear, but mineralised, acidic or alkaline based water, available to the miners (not for drinking!) at nearly every level on the dooks, that could be used for dust suppression if needed!*

*To produce a 'water-pressure' at any part of any long level, the miners simply obtained and laid, a jointed column of narrow-gauge water pipes up, along, and into the water carrying gauton, thereby filling the pipes completely and creating a self-pressurising head of water, capable of being carried an indeterminate length along any level, even to the extent of forcing it up into a short heading! At an average slope of approximately 1 in 3 on any Dook, a considerable head of water could be obtained with a comparatively short column of pipes. (Jackie Dryburgh and myself used this very method to obtain dust-laying water, when we were contracted to 'fill' from the downed Sparcoal at No. 4 level (west) on the Old Dook. Jock Suttie was the fireman who obtained the necessary pipes and joints and had them delivered to us, strapped onto the man-haulage bogies! We did not get paid for laying of the pipes, or the laying of the coal dust!)*

This then, was an early form of 'hydraulic stowage' or packing, as used in Lochhead Pit. It was not a general means of packing the wastes after extraction of coal. It only ever had limited use! It was time consuming, greedy on trans-ported waste material and definitely not cost effective. In both the New Dook and the Old Dook, the **Stoop and room** extraction methods left nearly all levels and headings as air-filled voids, usually passable and mainly bereft of roof supports. Fully seventy percent *(my estimation!)* of the **taken** Dysart Main coals in the Old Dook, the New Dook and the Victoria Dook, were mined by this sustained, but economic\* method. Having to 'flush' or 'stow' every coal-extracted section would have been virtually impossible and wholly cost prohibitive. There was simply no need, within limited Stoop and room workings!

*\*Economic, in that few (if any) wooden supports, or steel girders (if they had been available) were needed during the extraction process. General mechan-*

*isation and electrification of the workings was practically non-existent. The only form of basic mechanisation available, were aeroplane braes and hand-operated 'poker-drills!' This method of extraction did however, leave far too much of the available coal still standing! Perhaps forever!*

Author's Note:- *The village of the Coaltown of Wemyss stands (not sits!) over a worked-out underground section, that suffered limited Stoop and room, matrixed extraction at a depth of 700 feet/212 metres beneath the village. Currently, I am not aware of any known 'on-going' substantial subsidence within the village limits. (The bowling green situated within the centre of the village and donated by Randolph Wemyss, remains reasonably flat to this day!)*

# Chapter XXXVIII
## Dauvit's Progress VII

## Dauvit's nearly There! – *and 'Into the Coal!'*

The time had come, the news was good and Davie Paterson had timed it with precision. My eighteenth birthday was exactly nine weeks hence and I was to start my face training almost immediately. I was to appear on the following Monday day-shift, replete with a new pair of knee-pads, where I was to report to the under-manager, *(Young David Paterson)* who would formally 'introduce' me to a stripper by the name of Danny Gough *(who had been cursing me quite roundly these past weeks, because I was forced to interrupt his work with the passage of 'straps and trees' onto the coalface)* who, with Bobby Slaven as his No. 2 man, cut and stripped the bottom 'stable-end' in the Water level. *(The name given to the single working face taking the Four foot coal at this depth!)* This main gate was a 12 feet by 9 feet, brushed, steel-girdered roadway, with the coal-sandwich cut and taken, exactly 3 cuts in front of the main face and brushed daily, to one cut behind that. *(This meant that the back-shift brushers did not take any coal, but took the stone above the extracted coal and 'lifted' the 2 to 3 feet 'binch' of pavement redd, before setting the next 12' by 9' steel girder.)* This also meant, that the pan-engine train jigging it's load of coal down the length of the face, could empty it's load directly onto the 30-inch conveyor belt that carried the coal down to it's loading point. *(Fully described in the chapter - 'The Surface Dipping!')*

The 'barrel-end' *(daily extending, belt return roller, double-sylvestered in the working end of the Main gate,)* of this 30-inch conveyor belt was where Bobby Slaven shovelled and loaded the coal stripped by Danny Gough.

Every morning on the day-shift, Danny, with the help of Bobby, would start up the A.B. 17-inch chain, machine (A.B. *long-wall coal-cutter. Mark I)* which lay parked on the high-side of the stable-end, parallel to the roadway and behind the set girders. The machine would be 'flitted' down across the face of coal and set into the 'laigh' *(low)* side of the face, approximately 6 to 7 feet below the line of the roadway. The coal-cutting would begin and Danny with the help of Bobby, would cut up and across the stable-end, and finish the coal-cutting on the high side, with the machine stopped and re-parked parallel to the road. The 'cut' would be furiously gummed by Danny, with Bobby shovelling the 'gum' from the portable, steel ground plates onto the belt-end. This completed, they would then bore the necessary holes in the face as a precursor to the blowing of the coal and then start the 'strip'!

When the first shots were fired, Danny Gough would then furiously tear into the coal, (*he was a veritable, shovelling machine*) throwing it wildly in the general direction of the pre-positioned, steel plates. Danny did not care just how much coal piled-up on the plates, his main task was to strip the seven yards width of the stable-end and set some supports to the roof before the top leaf of stone collapsed. Danny usually managed to 'fire' and strip, about half the bulk of the coal by piece-time, which meant that they both could slow down after their 'nose-bags,' with only the laigh-side to strip and the plates coal to clean up, before 'lousing-time'! This then was where I was to spend the next seven weeks or so, learning to strip coal, set supports and clean-up the plates, all in the seven hours available. This was going to be a dawdle, - right-handed shovelling from smooth steel plates (*either standing or kneeling!*) and helping Bobby Slaven.

Somehow I knew, that this was not quite the 'picnic' that I had first imagined. Day two, saw Davie Paterson asking Danny Gough if I was 'up to the mark', to which I saw Danny Gough gainfully nod his head. Davie grinned in my direction and motioned to Bobby Slaven, "Top stable-end for you Bobby, off you go!" Then I got it! I was the new No. 2 to Danny Gough. Nothing changed for Danny, he worked as furiously as ever and the flying coal was being scattered all over the plates. Danny merely grinned, "You've got longer legs and arms than little Bobby - chase it!" (*The coal that was!*) I worked solidly and hard, but soon realised that I needed to pace myself over the time allowed - and of course, when Danny was completely stripped with the 'wood all set', he would then help me to clear any of the remaining coal from the plates and over the 'barrel-end' until we were finally finished.

Within about three weeks, I felt that I was ready for some more involved work and mentioned this to Danny. He said "All right, after piece-time!" Sure enough, that was when he pointed to the 'laigh' side and motioned me under the brushing lip. He reckoned that with my new found strength, increasing dexterity and my long reach, that I could stretch down into the seven feet depth of this low side and strip the coal in one sweep, without having to use an intermediate shoveller. I found that I could do this quite easily and from then on, this was my task, the bottom three yards after piece-time, with Danny on the plates. I would like to say that I scattered the coal far and wide, but I couldn't, the working space was restricted with each gained shovelful, a measured throw. Danny Gough knew exactly what he was about! (*He was the 'strip-meister!'*)

My face-training proceeded with nary a hitch, with Danny Gough teaching me some of the finer points of coal-cutting, coal-stripping, stone packing and the setting of pit-props, (*steel trees and straps actually!*). He was more than pleased with my progress and even encouraged me to further my training (*only when we had finished*), by allowing me to venture onto the main coalface in the event of any stripper being held-up! *(I also got the impression that I was, on occasions, being loaned-out!)* However, on the Friday before my training finished, I came in to work with a deep apprehension in my gut and a measure of grim fore-boding. I had learned along every other miner in the Surface Dipping, that Davie Paterson would not be returning to the Dipping on the Monday morning. He had been promoted and 'appointed', and was being transferred to another larger coalmine as colliery manager within the East Fife coalfields. His successor was none other, than my erstwhile nemesis from the pit, Mr. A. - bloody - W. himself! I experienced a terrible feeling of grim *deja-vu,* with the onset of a temporary mild depression coming on and resigned myself to my imagined, inevitable fate! Bang!, goes my face-place. Bang!, goes my prospects, and 'bang' goes my future in the Surface Dipping!

I was unprepared for a total surprise! I could not have been more wrong! Whilst shovelling manfully on the plates, I was greeted very pleasantly, and paused for a few minutes by A.W. himself, who assured me that my face-training would be recognised and that he personally would sign my Face-training, proficiency certificate.

In fact, a little later on, I was a little amazed and quite gratified, to overhear a conversation between A.W. and Dave Brown, who had asked A.W. for a move to the newly vacant No. 6 place on the coalface. A.W. replied, "No! - That place is reserved for David Moodie who has just completed his face training!" I could hardly believe what I had just heard and as if to compound the statement, A.W turned to me and said, "That No. 6 place is yours! Get yourself some 'graith' before Monday morning. *(Pick, shovel, saw and short-shafted 5lb hammer.)* Me! - imagine me! - a coal-stripper! And only eighteen tomorrow Saturday! I rubbed my hands together. Face wages in prospect and me only eighteen! I was to say the least, pleased! Danny Gough was standing hands on hips and grinning from ear to ear, and I must mention here, that the first person to come up to me and shake my hand, was none other than my other 'face-trainer' Jimmy Hutton, who stripped the first place on the run. A man whom I, along with many other miners, regarded as the finest stripper in Lochhead colliery. A man whom I also regarded as a friend! Danny Gough then said by way of advice, "Don't worry too much about keeping up with the rest of the stripper's when you get started. Pace yourself to last the shift. You're a big, strong lad and certainly worth all of the training, but you need to develop lasting endurance!"

*

Author's Tailpiece: - *During the previous eight weeks face training, I had been especially pleased to be now included in the general conversation that took place during piece-times where, in addition to Danny and myself, there was also the oncost man who operated the face conveyor (pan-engine) and the three strippers who had the bottom places on the run. They obviously thought it worthwhile to crawl down to the main-gate to have their piece. They were Jimmy Hutton, Jimmy Higgins and Dave Brown. Danny, Jimmy Hutton and Dave Brown were the men who kept the conversation going, with any contribution from Jimmy Higgins about as rare as getting a pay-rise from Willie Hampson. One day, I felt bold enough to inject a general question into the conversation, but when I did, it was greeted with a long silence and quite unexpectedly snorted at by Jimmy Higgins "Huh! you'll soon find out!" Which, of course, left me none the wiser. I was soon to find out just what Jimmy meant, and when the question was quite practically answered, it did quite surprise me, leaving me quite bewildered and just a little speechless! The question I had asked had been, "Why does everyone call the day-shift fireman 'Chappie?'" "You will find out!" - I was informed, and strangely enough, did get the answer much sooner than expected and quite out of the blue.*

*Of course, I did not realise it at the time! It happened at the start of the last week of my face training, just after we six had sat down to start our piece. The fireman had arrived upon the scene, but had not as yet, made to move past us on his journey up the face line. He was standing to the down side of the small group, with one arm extended to a girder and the other hand resting on his thick, numbered, deputy's stick. I had just opened my piece-box, placing the top half on the pavement by my feet and incidentally, had not noticed that I was the only one who had done so. Just as I opened the tissue paper that covered my pristine sandwiches, Chappie dashed forwards as though to pass, then suddenly jabbed at the solid redd above my head with his official stick. This dislodged some loose chips and dust that fell directly onto my clean white bread sandwiches. "You have got to watch that you know, you can never be too careful!" - and with those parting words of wisdom, dived onto the bottom end of the coalface and nonchalantly crawled away upwards. I was rendered quite speechless at the seemingly provocative act and was somewhat taken aback! I looked at my near ruined sandwiches, looked in the direction of the departing fireman and then at my workmates, who in turn, looked back at me with pitying glances. Jimmy Higgins quietly broke the silence. "Now you know why he is called Chappie!"*
*{Note:- In the Auld Scots Tongue - 'To chap! means 'To knock!}*

On that last Friday, Jimmy Hutton gave me his spare hammer and Danny Gough gave me a spare saw. Old Harry contributed the head of a beautiful 'holing' pick on the Sunday morning *(he was not aware of it at the time!)*, and I purchased a new size 5 shovel and a pick-shaft from the stores. I already had the knee-pads and the helmet, therefore I was ready! *(I didn't know it at the time, but I had no need to purchase a new pick-shaft from the stores. There was a standing convention/ concession in the Surface Dipping, that all strippers were given a 'free chit' for new pick-shafts.)* When Andra Wilkie became the under-manager of the Surface Dipping in place of young Davie Paterson, he would not issue this free chit unless the pick-shaft was actually broken! *(Readers may well imagine, that nothing other than a pick-head with a jagged stub of a shaft, was ever presented for his inspection!)*

\*

# Chapter XXXIX

## Happy Elder, - *Too'loose Latreuc, with boot-laces!*

## (Anecdote or Lochhead Myth?)

In the early 1950's, the Lower Dysart 'slopes' *(west)* were going full blast! This was the time when there was two, adjoining, coal-faces in production in the top leaf of the Lower Dysart seam, which was then undercut with two A.B. 'Bluebird' 15-inch long-wall, coal-cutting machines. The continuous face-line was split, *(but not broken!)* in that there was a long face of approximately 120 metres at left-angles to the main-gate and a short training-face of approximately 40 metres length at right-angles to the main-gate! At this time, many of the future shearer-cut, long-wall coalfaces in Lochhead colliery hadn't been thought of, or were still at the planning stage, but development work was proceeding apace, with the new coalfaces in the neighbouring Lower Dysart Dook, becoming a 'breaking ground' for newly-qualified strippers. To this end, a few, mature, experienced strippers on the Lower Dysart slopes were selected to be taskmasters, supervisors and mentors to the mostly young trainees. These trainees were all trained underground workers, but 'face-training' was a separate entity, with generous working time needed to build a trainee's skill, endurance and confidence. Their on-going progress was closely monitored by the management, with weekly reports from both supervisor and deputy, with recommendations from the section Oversman. The training period could last anything from six weeks to three months, with much depending on the health, strength and commitment of the trainee.

Most of the recruits for 'face-training' came from the ranks of the fit, young, oncost workers in their early twenties, eager to earn more money and who were quite happy to accept the grinding work-load, in return for the added remuneration it brought. Most of these young men did not have very much to say for themselves, they were too young to be men of the world and probably not well educated or travelled. Besides, they were all much too tired with the unaccustomed workload to cause too much trouble. The mentor or supervising stripper as he was known, was given very clear instructions as to how he was to proceed with his 'charge'. The trainee was not allowed to venture onto the run without his supervisor, nor was he allowed to leave the run unaccompanied. He must remain within direct communication distance of his supervisor and must not carry out any independent work action without permission. His every movement on that coalface was the responsibility of his supervisor. To most young miners who aspired to greater deeds and higher wages, this was no great problem, they gladly succumbed to the temporary restrictions and the on-going personal instruction - and eagerly followed the guidance and advice of their elders and betters.

Some recruits to the coal-face were just a little older than others and a few of course, were men who had gone into the war-time Army and returned to the coal-mines, but were out of touch with the latest methods and machinery. A very few of these trainees were actually mature miners *(oncost)*, who were married men with grown-up families and were probably around forty years of age. These men, were of sound health and had developed the necessary strength through years of hard oncost work, and they had the necessary endurance. Such men, once trained, were more than capable of sustaining the increased workload and were eagerly included in the training programs. Some small problems did arise with this type of trainee, in that they, being mature men, did find that their supervisors were often somewhat younger miners than they themselves were, which sometimes was a little awkward, or even embarrassing to the supervisor. Where possible, mature trainees were allocated to the more senior of strippers, who were themselves willing to partake in the training scheme.

Work-wise, a stripper who agreed to take a trainee, was allocated his own 'stint' of ten linear yards of undercut coal, plus two extra yards only. This small extra yardage was added to cater for the fact that the trainee was actually an extra 'pair of hands', but as yet, untrained hands! The management in their wisdom, reckoned that if there was no added work-load on the supervising stripper, then, every stripper on the face-line would inevitably clamour for an 'assistant!' After the 'gum' was lifted and the undercut gummed, the trainee's instruction on the coalface would commence, when one, or more 'shots' were fired on the supervisor's ground and they both then could muck-in *(one at either side)* to clear the downed coal. This system worked well for the initial weeks of the training, but soon got to be abused in that when the trainee's confidence grew, the stripper would *allow* his 'mature' charge to 'break-in' to a different part of the 12 yards stint 'on his own', thereby dividing the work load. This, in fact, meant that a full 50 percent *(i.e. twice by three yards.)* of the coal was taken at the first bite and cleared well before piece-time. With any luck, a second round of two separate 'shots' *(only with the acquiescence of the shot-firer)* could be fired before piece-time and cleared away, so that approximately three yards of standing coal would be all that was left to take, after the men had eaten.

One such stripper on the Lower Dysart slopes who was selected and had agreed to take a trainee, was one Jock Ritchie. *(The same Jock Ritchie who had stripped coal on the same coal-face as this author, on the 'water level' in the Surface Dipping.)* The trainee to whom Jock Richie was to supervise, was a mature family man by the name of 'Happy Elder' and, that name itself should have warned the Deputy, if not the Oversman, of the temperament of the man. Happy was a man who was at peace with the world, and everyone else as well! Not for nothing was he named 'Happy!' He was completely irrepressible and always looked upon the brighter side of life. He also had a grown son who stripped ten yards of coal on the 120 yards long, coalface, adjacent to the training face.

After a few weeks 'training' with Jock Richie, Happy was more than ready to take his place on any long-wall face. His 'training' was almost a farce in that he already had 20 years experience in, or near coal faces and had the empirical knowledge to match. *(The reason that Happy Elder was in 'need' of face-training, was that he had served in H.M. Forces during W.W.II. He had also been awarded the Military Medal!)* But, needs must, and he had his time to serve. The upshot of this arrangement was, that they both had time to spare and true to form, Happy would get up to all sorts of tricks and small scams on the coalface, just to brighten up the miners days, or so he thought! Some miners though, could not be bothered with his continuous antics and, a few may have be a little envious of the 'easy strip' that was the lot of Jock and Happy.

During one week day, the miners were warned that the Mine's Inspectorate were sending a senior official into the Section, and probably onto the double coalface as part of the yearly inspection. The strippers were reminded that they must stick rigidly to the standing safety instructions and ensure that precautions were strictly adhered too - with regard to trainees. All trainees! Jock Richie also ensured that there was only one 'break-in' to his coal that morning and that Happy Elder was suitably warned! Everything was in hand as far as the section deputy was concerned and the mines inspector finally arrived at the main-gate, albeit rather late. *(Near the end of the shift as it transpired!)* His main concern was to seemingly investigate the working conditions on the 'main' long face, onto which he then crawled. *(This face was approximately 42 inches high.)* He therefore disappeared from the sight of the main-gate strippers, and mostly out of the minds of the 'training-face' strippers. The supervising strippers on the short face had received the 'word', that the inspector had gone in the other direction, and they merely carried on with their coal-getting, or cleaning-up as needed.

Suddenly, a quick, subdued warning was delivered onto the short face. The Inspector had changed his mind and had decided to revert to the short, training-face instead. *{A very 'sleekit' (clever, plausible) fellow no doubt!}* Jock Richie hurriedly looked around 'his' place and finding nothing untoward, motioned Happy Elder to 'do' something! Unfortunately, there was very little left to do! Nothing that is, that would fool a government inspector. Jock Richie was at something of a loss. All of their coal had gone and all the necessary steel straps and trees had been set. Needless to say, the irrepressible Happy Elder rose to the occasion. Wordlessly, he sat down on the pavement and began fumbling with his knee-pads, which he then removed. He fully undid the leather laces on his pit-boots and removed them from his feet. He opened wide the eye-holed, jaws of the boots and forced the back upright seam forwards onto the inner sole, almost converting them into a pair of Dutch clogs. He quickly shook out his moleskin trousers, hitched up the knees and, kneeling fully upright, proceeded to 'walk' into his opened boots. He then tied the long laces around the back of his knees and tentatively proceeded to 'shuffle' the short distance down the face, to where the Mines Inspector had now arrived. *(Non-miners may realise that men with dirty, sweat-stained faces, show very little of their features and thus their true age*

The mine's inspector took one longish look at Happy and with eyes nearly popping, burst forth!, "What the hell are you doing on this coal-face son? . . . . . and who the hell sent you here?" Happy, for once, was quite speechless! This was not the reaction he had expected! Hell's teeth! This inspector had taken Happy's size and appearance at face value and was looking for answers. No one spoke, the silence was almost palpable! Happy quickly realised that he had better come clean, but unfortunately, chose the wrong method of approach! He thought that he could bluster his way out of this dilemma - and took the initiative!

He raised his voice slightly and in a tone of outraged dignity, voiced his retort, "Son!, is it? I'll have you know that I am a short, mature man with a grown up son, who is actually stripping coal on this very run! Son indeed! You need your eyes tested!" A short pregnant pause followed, with none willing to break the silence. The inspector looked at Happy Elder, his gaze anything but friendly and then looked at Jock Ritchie. He turned quietly back to Happy and told him to get off his knees, replace his boots back on his feet, and replace his knee-pads to where they should be. He removed a small covered note-book from his inside his boiler-suit pocket and made a short entry. On returning the book to his pocket, he looked directly at Happy and quietly said, "Now that your boots and knee-pads are in place and properly fastened, you can remove yourself from this coal-face and from this coalmine. You have disregarded the Coal Mines' Act at several instances and broken mining regulations by removing your safety boots whilst underground, and increased the risks of knee damage. *(The affliction known as 'beat-knee'.)* I consider your behaviour to be frivolous and infantile and your actions to be unsafe. You will remove yourself from this coalmine as of now and report to the mine manager on your arrival at the pit head!" Jock Ritchie suddenly found his voice and rounded on the inspector, "You can't do that to him, he's my trainee! I tell him what to do!" The inspector now looked directly at Jock and reiterated; "He's your trainee is he?" To which Jock replied, "He certainly is, and he's my responsibility!" The mine's inspector showed no hesitation whatsoever and in a more meaningful tone retorted, "In that case, it's more your fault than his, so you can go up the pit as well, where the manager no doubt, will instruct you as to the real meaning of your responsibilities!"

Jock Richie made no sort of reply, had no option, he was utterly and irrevocably in the wrong and everyone knew it! He motioned Happy Elder to follow him off the coalface, still belatedly mindful of his responsibilities and locked away their 'graith', still smarting at the supposedly needless indignity of his dismissal from the face-line. It was therefore, a sorry and dejected pair of miners who then made there way out to the pit bottom and thence up the pit, to await the wrath of Wullie Hampson.

Jock Ritchie was not too worried! They had actually been 'stripped', so no wages would be lost. They would surely be allowed down the pit again tomorrow, but not before the same Wullie Hampson had given them a shameful bullocking and a severe reprimand, with the possible termination of Happy Elders face-training. They took their medicine as was only just, and fortunately, both of them were allowed to descend the pit and return to the coalface the following day. Understandably, it was a more contrite and chastened Happy Elder that followed Jock Ritchie the next morning, onto the Lower Dysart 'slopes'. Jock Richie could sometimes enjoy a laugh and a joke as well as any man, but when daily wages were in jeopardy and a miner's livelihood was threatened, raw humour and irresponsible clowns could be quite easily transported down the 'pans' just as well as large lumps of coal!

Happy Elder did manage to complete his face training without any further 'recorded' incidents, and gained a 'place' of his own on the adjacent long-wall coalface alongside his oldest son. He soon gained a reputation as a 'character' on the 'slopes. His existentialist behaviour could not be repressed! Not everyone, however, appreciated his exuberance and quirky zest for life! Which is not surprising really, considering that many old coal miners are considered to be dour and taciturn individuals by nature! Probably an awful lot to do with the experience of the grim realities of a lifetime's work down the coalmines!

*

# Chapter XL

## Pithead Tables!

### The Lochhead Lassies!
### *– or Patiently Waiting on Tables!*

The screening plant at Lochhead pit as I knew it, was an later addition to the pithead superstructure and was designed to 'clean' the upcoming coals before their transportation by coal-wagon along the mineral railway, to the 'washer' and grader at the Wellesley pit at Denbeath, Methil. As the pit cage was drawn *(wound)* hurriedly to the surface of the pit-head landing, the upper air-doors were automatically and noisily raised, and two filled hutches of coal were air-rammed out of the open cage, by the sheer force of the two, empty hutches being rammed in. The Banksman quickly gave the necessary signal and the now, down-going cage, quickly changed direction and hurtled back down the shaft in the direction from where it had just come. Fifty seconds later, the opposing cage would surface with the same unrelenting urgency, and another two filled hutches would be air-rammed out of the darkness of the cage. Every fifty seconds, the two fully-loaded running hutches that were discharged from the cage would be expertly 'snibbled', to bring them to a hurried stop before the check-weighbridge.

*('Snibbling' or 'spragging' was a rather dangerous, but required art form, practised by only those with a calibrated eyeball and a co-ordinated arm. It involved the controlled, but contained throwing of a two-foot long, steel spike, into the spokes of the rapidly, rotating, bogie wheels of a full coal hutch, thereby instantly 'braking' the captured wheels.*

*The locked wheels are held immovable until such times as the snibble/spragg is removed at stops. This hazardous practise can quite easily result in damage to the snibblers/spraggers wrist, arm or shoulder. There is on record at Lochhead pit, the case of an oncost worker who was thrown into a double-cartwheel, simply because he didn't, or couldn't, let go of the snibble!)*

These full hutches would be briefly queued at the weigh-bridge before being dutifully weighed, with any individual miners' numbered, tab being removed from the lifting rings and slipped throw the 'post-hole', for the immediate attention of the check-weight man.

The weighed hutches were then taken around a 180 degree turn and again, briefly queued before the 'tumbler'! At the tumbler, the full hutches were wheeled singly into this contraption, where the loaded hutch, after pushing out the preceding empty hutch, would come to an abrupt stop. The operator would then pull an impossibly, long, activating lever, where the tumbling machinery and the hutch would then describe a lateral 360 degree rotation. The contents of the hutch would be emptied through a four-foot, diameter hole in the pavement, directly onto a wide, moving shaker pan, *(jigging motion)* which automatically separated the dross from the coals. Meanwhile, the emptied hutches ran on around another 180 degree turn, where they were again briefly queued up before the cages, to recommence their under-ground cycle. All of the coal that was brought up Lochhead pit was subjected to this procedure, and this momentum was maintained throughout the coal-winding shifts.

The screening area was fairly large and consisted of three, five-feet wide, long, moving tables. The 'loading' ends of these three, separate but parallel, moving tables, lay directly under the large shaker pan and at right-angles to it. The shaker pan was designed with three different sets of apertures, each one directly over the loading ends of each moving table. This had the effect of separating out the finer small coals, before delivering the larger, remainder coals above a certain size, to one or more of the parallel tables. The tables, which were three metres apart, were built from metal, with a solid raised base standing approximately 30 inches high and level, over their full twenty-five yards length. The construction of the moving tables consisted of a series of five feet long, three-quarters inch diameter steel rods, interspersed with thinner rods and interconnected with a two-inch diameter solid, steel roller wheel at either end of the thicker rods. These roller wheels ran along a track at either side of the table, with the thicker five feet rods approximately 12 inches apart. These thicker rods were connected to the thinner intermediate rods by five separate rows of inter-linking chains, which separated and spaced the five feet rods. The spacing between the rods was approximately one inch, which allowed the smallest of the coal and any remaining 'gum', to fall though into an large intermediate 'small coal' wagon, sitting on permanent rails directly underneath the appropriate opening. All coal above the size of 'chirls' *(trebles)* would be carried along the tables. The space along both sides of the tables was occupied by pickers, whose task it was to spot, identify and locate all of the stone, redd, sclits and unbroken blae that was mixed in with the coal, and then hurriedly remove all of it from the moving tables, so that only 'clean' coal was passed over the open end of the tables, and into the large, open, waiting coal wagons below.

*(This redd/waste/stone etc., was thrown into an ever-growing heap, that needed to be shovelled away by hand by the very fittest of workers, who were sometimes seconded from the last cages to go underground. The manager of the colliery had the right to second, late-arriving, underground miners onto the pit head, without warning, in an effort to keep the tables churning!)*

The amount of pickers as they were known, would be directly related to the amount and percentage of rubbish that the up-coming coal contained. This did not vary overmuch. *(These women and girls would you believe, could, after a while, actually identify the section that the coals came from!)* The pickers themselves were a hardy bunch of workers. They included older miners who were well past their prime, disabled miners to whom the company 'owed', and women both young and mature. Many young girls from fourteen upwards found employment on the tables, and spent many, hard years doing manual work before they too, found an eligible miner. These girls rarely married outside of the mining industry. They found very little opportunity! Such girls, when 'caught', usually proved to be a very worthwhile wife, partner and companion to young miners, in that they were understanding in the ways of the coalmines, the needs of their men, their eating and sometimes drinking habits, and the vagaries of the 'shift' system. It always helped, when the daughters of the house were brought up in the 'coalmines atmosphere.'

That area of the pit-head that housed the 'tables', was large, airy, overpowering hot, or bitterly cold in turn, and usually quite dusty. It was also a very noisy place constructed from a steel girdered framework, built-up with brick walls and having a bare, corrugated, tin roof. It was constructed one high storey above three, parallel sets of permanent railway lines and was not compartmentalised. Therefore, every sound emanating from within, whether it was loud or louder, was magnified numerous times. Conversation could be conducted, but only in American Indian fashion, literally touching and face to face.

In summer weather, the steel roof retained the sun's heat and turned the shed into a veritable hothouse, where the dust rose in near vertical columns. There were very few doors in the place and even though there were plenty of glassed windows to admit daylight, most of them were thick with grime, ingrained with years of coal dust and much too high to be cleaned. They were never meant to be opened! The men employed there, could strip down to shirt sleeves or semmits (vests), safe in the knowledge that a hot shower awaited them in the baths, but the women and girls had no such luxurious facility to look forward to! There were no shower facilities for females at Lochhead pit! They simply could not afford to even bare their arms or open their neck buttons, for the reason they had to head for home in nearly the same state, in that they finished work.

In winter time, the very air was almost too cold to breathe, in that everything around was constructed from steel and brick. The cold air however, was no barrier to the raising of the coal dust, which often made the air painful to breathe, without some sort of mask. Quite often, it was a small blessing to have a winter wind blow through the structure, just to clear the dust-laden air. Sometimes, it was almost impossible to tell the sex of the pickers, in that every-one was so wrapped up against the winter weather, with scarves and mufflers around their faces. The winter weather was also played merry hell on the hands of these girls, who went through gloves at the rate of one pair every week, grudgingly purchased by themselves when they could afford the expense. These women and girls for the most part, suffered the torn hands, ripped nails and chilblains and endured their lot without any word of complaint or bitterness. They worked, they earned their wages, and they got on with their own lives!

*(In my younger days, I did at times wonder, as to why some of the more healthy and robust of young girls, were not employed in the wood-yard during the hours of daylight, being part of one of the small 'wood' teams engaged in lighter, more amenable tasks, out in the fresh air. But now, on reflection, in my more mature years, I do realise why young, robust, fit and healthy girls were **not** employed in the wood-yard, in daylight, out there in the fresh air!)*

Some of the women and young girls who worked on the tables, did not live in the Coaltown of Wemyss and were forced to travel to and from work by public transport, much to their embarrassment and discomfort. Another problem was, that they had to keep their hair completely covered at all times for obvious reasons, not withstanding, that long hair might just get caught up in the moving tables. The hours that permanent pithead workers were contracted for, was exactly 50 minutes more than the average, underground miner and in the case of the 'tables' workers, they were in fact, given an extra 30 minutes break on the day-shift when the man-winding took place from 1-30 p.m. until 2-00 p.m. They returned to the tables exactly at 2-00 p.m. and worked until 2-20 p.m., which was the official finishing time. During this break-time, the women and girls often locked themselves in their pit-head bothy and made valiant attempts to get themselves clean and presentable, before travelling on public transport. Indeed, most of the girls who did travel on the buses, did have a presentable change of top clothing which they wore exclusively for travelling.

The conditions in the bothy were very spartan, with nothing more than bare wooden six foot tables and six foot forms to sit on and a single wooden cupboard for over-clothing. There was also a deep sink and a cold water tap, which inevitably, froze solid in winter. The women employees on the tables were also responsible for the weekly scrubbing out of this bothy, which had to be done with carbolic soap and hot water. There was nothing in this bothy that could be damaged by this treatment - as even the floor was made from concrete. Two of the women collected the 'hot' water every Friday lunch-time, by the simple expedient of carrying two galvanised buckets of cold water to the blacksmith's shop, where the water was heated almost instantly by the blacksmith. He would, using long tongs, withdraw several blocks of red-hot iron from his furnace and systematically douse them in the waiting pails, thereby instantly raising the inherent temperature very quickly. The water was not particularly clean, but it was piping hot! This was a standing weekly ritual and the women took turns to collect the hot water.

During the late thirties and early forties, both of my cousins worked on the tables at Lochhead pit. They worked on both day-shift and back-shift all the year round. They travelled that pit road in all weathers, leaving home in the Coaltown of Wemyss at 5-30 a.m. on the dayshift, and not arriving home from the back-shift until approximately 11-00 p.m. These two girls had also to get scrubbed and bathed on a daily and nightly basis, but not before the men of the house coming of shift had been bathed first. That was the rule and the way of things at No. 5 Barns Row. The girls wages at that time at Lochhead pit tables, was 13/2d *(thirteen and twopence/66 new pence}* for a six-day week on the day-shift, and 11/- *(eleven shillings/55 new pence}* for a five-day week on the back-shift. *(Lochhead pit did not wind coal on Saturday afternoons.)* It usually transpired, that the two young girls of the house found themselves on different shifts, which of course, was just as well. Old Kate, as fit and strong as she was, did need lots of help with four or five grown miners to feed and care for, so, both of these daughters of the house, worked a 'double-shift' every day except Sundays and alternate Saturdays. *(Old Kate Smart was also the unofficial mid-wife to Dr Khambatta in the Coaltown of Wemyss and was liable to be called to a confinement at any time, with, or without the supervision of the East Wemyss doctor. The latter was usually the case!)*

The work on the tables was grinding, back-breaking and unforgiving. It was also dirty, wet at times and mind-numbing. It was basically relentless! The workers had to stand on the cold stone floor, or sit on the edge of the narrow steel sides of the tables. Either way, it was a very uncomfortable position in which to work. If a person stood and faced the tables, this meant constant back-bending with the stones, redd and rubbish having to be lifted cleanly, and be thrown directly away over the tables to the picker's front. If a person ever managed to sit precariously balanced on the edge of the tables, this then meant that only one hand and arm could be used, with the restriction, that if anything heavy was to be lifted, the lifter ran the risk of falling over onto the tables. *(Readers will no doubt realise that working stances were changed very regularly during the course of a shift.)*

If these girls survived long enough at these tables and had the mental grit to carry on, they developed a strength, fitness and work orientation, that was the equal of any working man. The imposed discipline on the tables could also rather harsh. On every table there was an older worker, charged with the unsung duty of ensuring that no one person skived-off. He was usually positioned at the control point of the table, where he had immediate access to the operating lever. He was armed with a steel-headed tool on a four-foot shaft, which enabled him to spread the down-coming coal evenly over the width of the table, thereby allowing easy access for the pickers.

His task was three-fold, he spread the coal, he operated the control lever to start and stop the tables and he rode 'shotgun' over the pickers. If one of the girls 'needed to go!', she usually sought to first to catch his eye and gain acknowledgement. If she didn't, his pocket watch would then be produced and the absence timed! On occasions, these girls could sometimes appear to be somewhat vulnerable and subject to intimidation, or even mild bullying at worst, but fortunately, this did not happen very often. If it so happened, that there were mature, older women employed at the tables, especially women who had brought up large families, then, the status-quo changed somewhat. Mature ladies were not about to be told by a mere male whether they could take a pee or not, and these ladies would 'face-off' any male from the manager downwards. If one of these women made a stand on a point of contention, the mere sight of an large, irate female, standing fists on hips, would be enough to make our hitherto unsung hero, take a most peculiar interest in the nearest, most unlikely, lump of coal!

I knew of two girls from the same family, who worked on the tables at Lochhead pit, whilst I was doing my best to develop a hernia from shovelling redd onto the conveyor, that fed the stone-crusher on the pithead tables. These two girls were rather young, with the younger one exactly one year younger than me. The younger girl joined her older sister on the tables, around the time that I spent on the pit-head - and I was most surprised to see just how shy, quiet and utterly reserved she was. These girls were actually well shielded from the attentions of any predatory males, and some older men made it there business to see that they were 'looked after'. They were, after all, the daughters of a working miner, who lived at home with their parents in Denbeath. They both usually travelled on the same bus that I used in the mornings whilst on the day-shift and the interesting thing was, that after I had gotten home from work around 2-30 p.m., I would see them walking down the full length of Barncraig street to go into Denbeath. They always left the bus a full two stops earlier at College street, in order not to be seen leaving the bus at the Wellesley Colliery at Denbeath. Their odd behaviour puzzled me for many years, until my cousin Belle unwittingly gave me the answer. What grimy and tired young female would wish to be seen, coming of the same bus, with the same clean miners on a near daily basis?

Belle Smart and young Kate Smart both endured the rigours of Lochhead pit road and the filth, dirt and grime of the tables for a good few years. It certainly did not do them any lasting harm. How could it? They were both very well known, liked and well respected, and the daughters of old Tot Smart and as ever, always under the watchful care of Old Kate, their mother. Young Kate did marry, and to Joe Blyth who was an engineer at Lochhead pit. *(Who did heat up those pails of water?)* They lived in the Coaltown of Wemyss and raised a large family, some of whom went to the mines. Unfortunately, young Kate has passed on, but her older sister Belle, now in her eighties, is still reasonably fit and well able to get onto a bus and travel. Belle did have a family, namely a daughter Evelyn, of whom she is very proud and a son Brian, who ran the Coaltown United Services & Social Club to the good and benefit of it's members. Belle is to be found on most Sunday afternoons, either in the Bowling club in the Coaltown of Wemyss, or sometimes in the United Services & Social Club, playing and holding her own in the afternoon, dominoes tournament. 'Tis not the first steak-pie that she has taken home as a prize in the late afternoon and maybe feeling just a little flushed from a very welcome brandy and lemon. *(Or two!)*

\*

# Chapter XLI

## Old Harry – *an Institution!*

### Harry, 'Hend' or 'Hairy!' – *What's in a Name?*

This part of old Harry's life picks up when I was approximately 12 years of age. All before that, concerns my own formative years and schooling, without knowing too much about what went in the upper reaches of family life. When active boys are 'running wild' in childhood, all that mattered in life was the 'Adventure and Wizard' on Tuesday's and the 'Rover and Hotspur' on Thursday's, and two clean and empty 'jam-jars' on a Wednesday evening, being the entrance 'money' for the matinee on that late afternoon after school.

If we were especially lucky, we might even be given one of those new-fangled chocolate biscuits 'on a piece' *(chocolate biscuit sandwich)* before we went! All discussions regarding household, domestic and work related matters were discussed by a committee of two, and behind closed doors! This was the 'distaff mafia' at work and all discussions seemed to end with the words from my mother, "I'm sure your right! - Harry!, but we will soon see!" With old Harry being left quite speechless, but not convinced as usual!

I must make mention here, that my father did have what could be described as a small speech impediment at that time, in fact, I can not recall him ever not having it! He was a quiet man *(in his middle and later years)* who smoked a pipe, and did not have too much to say by way of social conversation at any time. He would greet people with a wave of his hand, a nod and a grin, but without much spoken words. If he did engage in conversation, it would be with people of long acquaintance, who understood Harry's 'problem' and knew not to interrupt Harry's sometimes laborious discourse. Just how sensitive old Harry was to this 'small problem', was made very clear to me one summer's evening, when I plucked up enough courage *(prompted by my mother)* to ask him for sixpence for the 'pictures'. *(That allowed entry to the front wooden seats of the Globe cinema in Buck'hind for an evening performance.)* He replied, "J - J - Ji - Ji - Jis - Jis - Jist - t - t - ta - tak the ni - ni - nails oot o' tha - tha - that b - b - bit o' wi - wi - wo - wu . . . . ? "Wood Dad!" I interjected, in a hurry to get my sixpence. My head banged and rung! I just didn't see it coming, but I sure felt it! I landed on the ground about six feet away, having been propelled quite violently by one of the biggest and callused hands I had ever seen! It wasn't so much of a blow, but more of a powerful push. I had just learned to my painful cost, just how much old Harry realised his impediment and I also quickly realised that nobody, but nobody, ever interrupted Harry in the middle of a delivery, not even Wullie Hansen, the manager of Lochhead Colliery. *(As I later discovered!)*

Old Harry had the sort of quiet anger that moved mountains, and almost rendered him speechless at times! This was why I am sure, he was so very successful at getting good well-paid contracts for he and his men. He was bloody stubborn with it! Wullie Hampson was once reputed to have said to Harry, "You are a stubborn man Harry Moodie! You've got a singular-tracked mind!" To which old Harry was reputed to have replied: - *(Eventually!)* "Aye, . . . . . j-j-jist as lang as that sing'ol rail-track I'm hae'in tae lay, t-t-tae git the redd oot o' that b-b-bluudy stane mine, that yer trying aw-aw-awfu' hard no tae p-p-pey me fur!"

That translates: - *Yes! . . . Just as long as that single-track railway that I am having to lay, in order to transport the stone-waste out of the b\*\*\*\*y stone-mine that your are most reluctant to pay me for driving!*

During the week when Harry was on the night-shift, his sleep pattern would vary, according to whether he had outside joinery work to do, and whether he had enough 'coppers' to visit the W.C.C. sawmill at Muiredge. If he had wood to hand, he would then commence outside work around eight in the morning after breakfast and continue on until lunch-time, where, after a good dinner, he would seek seven or eight hours sound sleep, before leaving home in the late evening, to start his night-shift at Lochhead at 10 p.m. We, my brother an me, on coming home from day school, always knew when our father was in bed snoring his head off, we would be greeted at the back door of the house by our mother with one finger held up across her mouth, with the whispered words "Quiet, your dad's in bed!" That was warning enough! Old Harry did not take kindly to noisy youngsters, hell bent on making the sort of whoopee that disturbed his much-needed sleep! If old Harry had cause to raise his voice from the depths of his bed, to either of his noisy offspring, then repercussions would surely follow!

Old Harry never seemed to say very much at the best of times, though he obviously kept his plentiful thoughts to himself. This was evidenced by the fact, that, during the very few occasions was called upon to arbitrate *(dispense rare punishments to wayward progeny)* on domestic disciplinary matters, he always seemed to be in full possession of the relevant facts, without ever being told! Most of the time, Harry kept his distance from his children, leaving the actual day to day 'behavioral education' in the hands of my mother. This was inevitable, because of the fact that old Harry was either at work, *(seven days of every week!)* in bed snoring his head of, or, in the back garden, busily banging nails into yet another wooden construction. His meal-times therefore, rarely coincided with that of his offspring while we were at day school. When old Harry sat to eat, he usually had the table to himself and ate with a single-mindedness that almost defied description. Harry did not waste words! Our mother, having had her meal, would be busy in the scullery, listening for the sound of Harry's soup-spoon rattling in an empty soup plate. That was the signal for mother to appear and ask old Harry if he wanted 'a wee drap mair soup?' Old Harry would seem to ponder the question, but the answer was invariably the same, . . . *p - a - u - s - e* . . . **Aye!** . . . and lifting the empty place with both hands. Old Harry would rarely waste words between courses. *(Usually two! - but always three on Sundays.)* This was eating time! The vegetable soup would always be followed by a great plateful of well cooked meat and boiled tatties. The red meat would either be a great lump of boiled beef *(done in the soup pot)* or stewed steak with minced beef, braised steak with mince, or casseroled steak with minced beef and 'doughboys!'.

Old Harry loved his beef steak so much, that he would actually raise his eyebrows if something other than cooked beef were placed before him. He would not actually say anything, but would devour everything that was placed before him by my mother. He just did not complain! At the end of such a meal my mother would always ask of Harry, - "Was that all right then Harry? To which, old Harry would always reply in the same noncommittal fashion, . . . p - a - u - s - e . . . **Aye!** . . . **No bad!** . . . And that was as much compliment as old Harry ever allowed, about each and every meal that he completely demolished every time that he sat down to eat!

*

*Every Saturday night around 11 p.m. at No. 34 Barncraig Street, a seven-pint capacity, cooking-pot - half-filled with clean water would be placed on the gas stove. Into this pot would be placed, one and a half pounds of lean, stewing steak, one and a half pounds of lean, minced beef and a full pound of separated, beef sausages. The pot would be topped-up with water to within one inch of the rim, and the lid would be closed down. The gas-ring underneath would be lit and turned down, to what my mother described as 'a wee peep!' She would then retire for the night.*

*On the Sunday morning around 7 a.m., my mother would head for the scullery, where her first action was to lift the lid of the stew-pot, where, to her obvious satisfaction, the contents would be done to a turn. There then followed the sounds of something being stirred and mixed in a jug, followed by the sound of this 'bisto mixture' being poured and added to the readied contents of the pot. This was the last part of the Sunday morning ritual to prepare 'the Sunday morning breakfast!' A large plateful of thoroughly cooked, stewed streak, minced beef and sausages, eaten with nothing more than fistfuls of 'plain' bread and a fork. (The distaff side of the family, were far more genteel!) This was the standard Sunday morning breakfast in many a coalminers' household in the Kingdom of Fife and for many years, I knew of no other! My mother, on several occasions spread over many years, attempted to satisfy old Harry (and us!) with grilled bacon and sausages, scrambled eggs and toasted bread, only to be pawkily asked after its rapid consumption, as to when the 'beef-stew' was ready to be served . . ??*

If cheese was described as 'Miner's Wedding Cake', then surely, a large plateful of stewed-steak, served first thing on a Sunday morning and eaten with plain bread, could be correctly described as:- 'The Miner's Wedding Breakfast!'

\*

If it were the case, that Harry did not have any outside work to do whilst on night-shift, then his sleep pattern would subtly change. He would disappear to bed directly after breakfast around 7.30 a.m., where he would probably take advantage of the fact, that while we boys were at school all day, he could look forward to having a long, undisturbed sleep for up to nine hours, before we returned home. Harry did this, probably twice a week, to ensure a sufficiency of good rest. After Harry had risen and consumed his main meal, he then had perhaps four or five hours leisure time, before having to collect his piece and head to College street, in time for the 9.15 p.m. bus to the Coaltown o' Wemyss. Harry invariably did one of two things, or he said that he intended doing one of two things.

He always informed mother on leaving the house by the back door, that he was going for a 'wee daunder!', or he was 'awa tae the corner!' In either case, the end result was always the same. The 'slow meander' commenced with Harry standing outside the back door for a few minutes, whilst one hand fished out the short, smelly 'stonehaven' pipe from an outside pocket, which then was stuck between his teeth. The plug of familiar Condor tobacco would then be produced along with Harry's little pocket knife. One of the ragged blades of the small knife would be carefully rasped against the smooth brickwork of the door jamb for a few strokes, which would be followed by a few moments silence, while a good few parings of carefully, sliced tobacco was cut from the plug. With the small, tobacco knife temporarily returned to his bottom right, waistcoat pocket, the parings would be held in the middle in the middle of the palm of the left hand, while a carefully-aimed small, clean spit would be directed onto the waiting tobacco - that was now gently and lightly kneaded between Harry's two, great, hard hands. On being satisfied that the mixture was kneaded to the correct texture, the pipe would be filled to capacity, tamped down with the tip of a middle finger and a trial suck would determine the density of the packed tobacco. *(It was very rare occurrence indeed if the pipe needed re-packing!)* Fire could now be applied! A 'swan vesta' would be scraped along the same brickwork, followed by gentle sucking sounds as the firing process commenced. A few more gentle puffs and the fire-box was drawing satisfactorily. Another gentle tamp, this time with the squared end of the tobacco knife and when the pipe was going to Harry's satisfaction, the smoking bowl of the pipe would be partially stopped with the nickel-silver 'crown', thus enclosing the burning embers and terminating the ignition process. The slow, measured enjoyment would then begin, with the sweet taint of Gallacher's condor plug, fleetingly lingering around in the still air at the back door of the house!

Harry would slowly walk out through the 'close', where, on reaching the side gate to the street, he would stop, look first 'up' the street, then 'down' the street, as if to decide which direction to walk. It mattered not, for the simple reason that Harry, like many other coal miners of his ilk, usually headed for and ended up at the same place and did exactly the same thing, as many of the Denbeath miners did on a near daily basis. They congregated 'at the corner' beside the Goth!

The 'Goth', or, to give it its full title, was actually one of the chain of *Gothenburg Public Houses that proliferated in and around, nearly all of the sizable mining villages and towns in the Kingdom of Fife. This Gothenburg at Denbeath lay opposite the Wellesley colliery, where Ward Street joined the fifty-foot *(15 metres)* wide Wellesley Road. The paved walkway outside of the 'Goth' *(as it was universally known)* was equally wide and also served as a bus-stop for W. Alexander & Son's omnibuses to Leven and all points around the east Neuk of Fife. This wide, pavement area terminated at an open corner and this was where many miners gathered for a 'crack' *(chat)* or a blether, *(and often, many over-heated discussions on the belaboured state of the coal-industry!)* - or a pipeful of tobacco and a 'spit', away from the confines of their living-rooms. Coal miners would gather at this location almost at any time of the day, or evening - and split up into many little groups of two or three, or even more, if the subject under discussion was of general interest. *(The subjects that would always attract large discussion groups was the mention of union activity, local politics and politicians, government attitudes towards the coal miners and the seemingly unfeeling section gaffers and under-managers at the different coal mines. Old Harry occasionally came home, all fired-up about the latest injustice perpetrated on some luckless miner, who was too chicken-hearted to take his so-called grievance to the proper source. When this happened, my mother usually let Harry digest this on his own, his slow discourse was such, that the laborious re-telling sometimes took rather a long time.)*

Sometimes, the miners just stood or hunkered** down against the wall in rows, with no conversation in progress. 'Tis not the first time that I have joined the 'queue' at the bus-stop at the 'Goth', only to be left standing as the bus pulled away, through 'no-one' having made an attempt to board.

* *See chapter at 'Addendum'.*

*(At the end of WW.II, at the time of the general 'blackout' and the first few months afterwards, before any form of street-lighting was re-introduced, the miners that gathered at this corner were often the unwitting cause of a few 'heart-flutters'*

*amongst some of the unsuspecting inhabitants, who just happened to be abroad after dark and not consciously aware of this habitual gathering. During the evenings of the non-summer months when the streets were dark and the cloud-base was low - and the twilight was simply non-existent. Total darkness pervaded the streets. It was literally impossible to see even the most indistinct shapes. Hand-torches of any kind were few and far between, even if dry batteries could be had. The darkness could be totally pitch, blinding and quite unnerving. The inhabitants of the village had not as yet, got out of the war-time habit of shuttering everything tight and keeping 'black-out' curtains fully drawn, besides, most houses were still lit by either paraffin oil, candle or firelight, with a very few people using coal gas and white mantles. Mains electricity for general domestic use was still well into the future.*

*When out and about in this environment, the one human sense that each and every one of us depended most upon was neutralised! No matter how hard one strained, there was very little light reflection whatsoever. Everyone was blind! No matter how long a person stayed out, the blackness did not lift! This, in itself, to some people who were forced to be out in this environment and of a nervous or timid disposition, was rather frightening. Some people hated the darkness and would not venture forth, whilst others simply had no choice, but to brave the real or imagined horrors of the 'nocturnal demons' that awaited outside! If one's sense of sight was all but neutralised, the one sense that seemed to be exaggerated and amplified was everyone's hearing. Unfamiliar sounds, both real and imagined, caused many a strong heart to flutter and beat much faster, along with the conscious desire to move more quickly and even quieter, if that were at all possible. Quite often, the only sound that could be hesitantly recognised, was that of one's own increased heartbeat, going ten to the dozen! Many a youngster could be heard passing outside, either singing at the top of his voice or madly whistling of-key, usually to boost his courage or make others aware of his forced bravado! Either way, the dark of the war-time streets was no time to be abroad, especially for the faint of heart, with a great horde of hairy-assed coalminers waiting quietly at this street corner, just to 'frighten' old dears and unsuspecting youngsters. (Neither women nor youngsters, were welcome 'at the corner'!) This was a time, when people would quite literally and unwittingly, bump into each other in the blacked-out streets, for no other reason that they simply could not see each other!*

\*

\*\*Hunkering Down:- This method of 'squatting' was common amongst coal miners who, over the years, had developed the technique of sitting down and balancing their body weight on nothing more than their own two feet! This was achieved, by completely folding their thighs against their calves, and their chests between their opened knees. The flat of their feet would remain in contact with the pavement, while their arms would be placed around the front of the knees to counter-balance their body weight. This position, to the uninitiated, is very painful to imitate and and was only possible because of the very lean, supple and extremely flexible nature of some coal miners bodies. The fact also, that their backs, legs, ankles and feet had been slowly conditioned to their performing a complete shifts' worth of heavy work on their knees, greatly contributed to they're being able to perform and maintain this rather painful contortion!

\*

# Chapter XLII

## Escape Routes!
## In Extremis ; exeunt omnes!

### (Hoo' dae a' git oot o' this?
### *– or, Aye! You lead the way Chairlie!)*

The three, later two, escape routes/roads from the Lochhead pit were used when, for one reason or another, the main winding gear, the cages, or the pit-shaft itself was out of commission. There was always considerable wear and tear on the vertical guide-rails *(sometimes called 'slides')* that ran the length of the shaft. These were usually set in single long columns *(or more)*, on the opposite side of both cages. Very often, because of the constant, fast, winding speeds involved, the slides were constantly being knocked and, as a result, sometimes became so loose that they got caught-up somewhere in the shaft, or more often, were knocked loose and fell down the shaft to the Pit-Bottom.

If such an event took place, everything would come to a halt and the first action to follow, would be for an inspection team to be called upon, who would either be gently lowered, or raised to the site of the problem whilst in verbal contact with the winding engine-man. Their task was to thoroughly investigate the degree of damage, assess the amount and type of materials required and make a qualified prediction as to how long the repairs might take. This was not because the mine manager wished to 'chivvy' the shaft repair team, but to help the manager make a decision as to how the up-coming shift were to be brought to the surface - and whether to divert the on-coming shift or not!

If the winding shaft was out of commission, those miners who had completed their shift's work, were invariably brought to the surface by both of the two 'escape' routes. Both routes involved a considerable walk from the pit-bottom, but if the stoppage occurred whilst the miners were still within the Lower Dysart section, those miners could be intercepted in time, to divert them into travelling down Nicholson's Dook, *(at the inside end of the Lower Dysart haulage-way)* and thence by divers means to the Victoria pit bottom.

The miners from both the Old Dook and the New Dook would be directed to use the escape route that actually commenced behind the 'howf' on the pit bottom and, cut though the old Dysart Main workings, to eventually ascend the stair-pit at

Earlseat, on the north side of the Standing Stone Road. If the stoppage occurred near man-winding times, when all, or most of the up-coming shift would already be on the Pit-bottom, their numbers would be evenly split between the two routes, as would the down-coming miners on the next successive shift.

If the stoppage happened during normal coal-winding times and the length of time needed for repairs could be estimated, the miners in the 'sections' nearest the escape roads would be warned in time and held in their sections for two good reasons. 1. So that an unnecessary long walk could be avoided and that the section Fireman (deputy), would be in a position to lead (guide) his miners along the strange and unfamiliar routes. 2. The Fireman could list and check the names of everyone following him to the surface!

In Lochhead Pit, both escape routes were used at the same time and for good reason. Men had to be brought-up to the surface at locations sometimes a long way from the pithead baths, with transport having to be organised and sent to the correct destinations. *(Discerning readers will realise that the type of transport used had invariably to be in keeping with the state of the miners clothing and dirtiness! Luxury holiday coaches were not an option!)*

* Authors Note:- *The N.C.B. in the Fife Region, did at that time, have one or two, wooden seated, single-decker, Utility Coaches available.*

In the case of those miner's from the Lower Dysart section, the route was back out through the Lower Dysart mine to inside the end of the Lower Dysart haulage, down the deep slopes of Nicholsons Dook, along part of the length of the Old Dook's No. 1/300 level, out through the length of the Victoria's cross-cut mine and then queued, to be hoisted up to the surface in a small, cramped, four-man cage, to find themselves standing on the beach at West Wemyss! *(In days gone by this route was favoured by some of the 'locals', as the resident publican could sometimes be persuaded to dispense 'welcome sustenance' at all sorts of odd hours.)* The end of this journey terminated at the bottom of the shaft of the Victoria Pit at West Wemyss *(worked out),* where the winding gear was kept in good, operating condition and consisted of the customary 'one up and one down' cages. *(But only up to 1954!)* What made this small shaft utterly unique was the fact, that each cage was only large enough for one hutch, which meant that it could only hold four miners at each winding!

It must be realised however, that in walking through the escape route, men could only travel in spaced-out single-file. Which was just as well, otherwise, there would be a queue of cold and hungry miners queuing at the shaft bottom. The journey by coach or lorry back to the pithead baths at Coaltown of Wemyss, only lasted about 10 to 15 minutes, but the whole 'escape' process by this route could quite easily add a full two hours to a mans' working day.

For those miners who had gathered at the Pit bottom in anticipation of using the Earlseat route, they too, had to wait until a fireman or other mining official was ready to lead the waiting columns through the not so long, but also tortuous underground route. This route was not altogether favoured by the miners, for the reason that it terminated in a vertical climb up the disused shaft at Earlseat. *(For tired and weary miners this was quite an effort, considering the difficult walk and fetid state of the very warm air.)* The climb to the top of the shaft was strictly controlled, in that the means of escape was by fourteen, separate flights of near vertical ladders, each having ten rungs placed one foot apart. Each ladder was supported by a horizontal wooden platform at intervals of ten feet. The standing rule was, that every person ascending the stair pit, must ensure the miner before him had reached the safety of the next platform, before he himself could commence his next ten feet climb. No two miners were allowed on any one ladder at the same time! Not for nothing was this rule strictly enforced. All it needed, was for one item of equipment to fall on the head of a following miner, to send him to his doom - and secondly, due to the constant humidity of the spent air regressing from this air-shaft, the wooden framework was susceptible to premature rotting and needed constant attention. This disused mineshaft was also in the middle of a field, enclosed by a small, brick-built compound and topped off by barbed wire and a lockable steel door. It was permanently covered over by a heavy, wooden platform with a protected air-vent, to allow the egress of humid, tainted, ventilating air from the sizeable New Dook section.

Transport from this escape shaft back to the pit was not usually available, as both the shaft and Lochhead Pit baths were in the middle of different adjoining fields, separated by the Standing Stone Road. It was therefore, a rather tired and weary bunch of miners that would trudge their way across the fields, to finally arrive at the very welcome 'hot showers' of the Lochhead baths.

Author's Note:- *There was one other, early, escape shaft that bears mention and indeed, most miners in Lochhead pit will remember it by name, as the Duncan shaft. This shaft actually was the Duncan pit at one time and did produce and wind coals taken from the two Dysart seams. At the time of which I write (late forties/early fifties), it did have a small, steam driven, engine at its pithead that was maintained in good working order, at least up to 1955. This was the Duncan pit and also worked the Lower Dysart seam in the years from approximately 1921 to 1928. Amongst the coals that was taken earlier, was much from the Dysart main seam.*

*It is highly probable, that the Duncan\* shaft was sunk as a result of the Mosswood test drilling, which was carried out only 400 metres from the site of the Duncan shaft, that was itself situated near the north end of the Check-bar Road, where it connects with the Standing Stone Road. The Mosswood test bore had gone down to a depth of approximately 3380 feet below the ground surface and so established the presence, depth and thickness of the strata and coals in that* *particular area, including those seams of coal within the Limestone Coal Group.*

*At this point, both the Dysart Main seam and the Lower Dysart seam were at no great depth, with the Lower Dysart seam only about 50 feet under the Dysart Main. The Duncan pit was originally sunk to a shaft depth of approximately 48 fathoms, whence, two, separate, shallow, underground dippings, roughly 12 metres apart, were driven in a south-easterly direction, on a map bearing of roughly 150 degrees and to a length of 450 metres. These two dippings eventually linked up with the **Lower Dysart cross-cut Mine** in Lochhead pit, that itself connected the main 1000 metre Lower Dysart haulage level (in the Dysart Main seam) to the Lower Dysart section. (If readers can imagine the Lower Dysart, level, crosscut mine being extended another 500 metres in the same direction and following the grain of the coal to the north-west, it would then finish at the base of the Duncan shaft.)*

*The coal taken from the Duncan pit was from either side of these shallow dippings and mainly to the east and west. The extractions covered a ground area of 600 metres wide and 400 metres north to south. The coal was taken mainly by 'Stoop and room' methods, with a few wheeled braes, (resembling non-mechanised long-wall faces). These Duncan pit workings bordered Earlseat pit coal to the West, Earlseat Mine workings to the north (within 30 horizontal metres) and the Lower Dysart (west) section in Lochhead pit. All within 25 to 50 metres along all of their respective borders. I estimate that this Duncan shaft stopped winding coal about 1930, but the shaft and it's main dippings to the south-east, were maintained on a regular basis, up to the closure of Lochhead pit. (In March 1970.) The open route through the original dippings and the Duncan shaft, were used as an return air-way to ventilate the Lower Dysart (west) section workings in Lochhead pit. David Davidson, an engineer employed at Lochhead at that time, remembers having to travel the short, overland distance to the head of the Duncan shaft, where his routine maintenance task was to fire-up the small boiler, to raise enough steam to exercise the winding engine, and effectively raise and lower the opposing, twin cages several times. On one such occasion whilst firing-up the boiler and waiting for steam, he overheard what he took to be human voices filtering up the shaft. He cupped his hands and shouted down several times - and, much to his delight, an answering voice responded. He said that he was about to test run the winding gear and cages and requested the miners to stand clear. This he successfully did - and then asked the miners if they would like to be wound up to the surface to eat their piece? They willingly acquiesced and duly arrived at the surface of the shaft, more than willing to breathe the much cooler and cleaner air. {The air coming up through the Duncan shaft was fetid, hot, slightly gaseous (afterdamp) and reeked of spent explosives.} The two miners were engaged in 'back-brushing' the old, dipping, passage ways, in an effort to keep them open and passable and were working in the most appalling conditions. Their names were Harry Moodie and Hugh Gilbert, both stone miners and brushers of long standing and great experience, with many years of hard empirical knowledge between them.*

Author's Note No. 1:- *There was an old hoary legend that was forever circulating amongst miners in just about every coal-mine in the East Fife area - that it was entirely possible to walk the full distance between the Wellesley colliery in Denbeath and the Francis colliery in Dysart, and never have to surface at any coal-mine en route. It was therefore said, that every colliery was linked by underground passage/escape routes to it's immediate neighbour on either side. I have personally spoken to Firemen/Deputy's who had made such journeys between three collieries in the course of weekly inspections of these routes (Victoria pit to the Micheal pit via the Old Dook in Lochhead), - but I have never met any one person who has claimed to have made the whole underground journey.*

*To anyone who has even the most rudimentary knowledge of the underground topography of either the Lower Dysart or the Dysart Main coal seams and, considering the fact that all of the colliery's mentioned in this narrative probably worked these same seams, and, irrespective of how deep they had to go to get to them. It is entirely feasible, that with underground workings spreading out in all directions from any Pit bottom, that workings from two, adjoining coal mines are eventually going to end up in the same 'field!'*

*Underground survey is just as meticulous as ordnance survey, they have more to lose! It is therefore entirely plausible, that mines from adjacent pits could be driven to converge at a given point underground - after all, opposing mines could be driven at the 'same' level within a given seam. Direction would then be the only imponderable!*

Author's Note No. 2:- *Immediately behind the cottage gardens at the north-west end of the Check-bar Road, between the Coaltown of Wemyss and the Standing Stone Road, the old brick and concrete remnants of the Duncan pit-head can still be seen. The raised brick and concrete platform is overgrown with grass and bushes and it is not so long past, since engineers were called to the site of the old shaft to organise some more refilling of the old sinking. Weather and time had caused partial subsidence of the previous in-fill. To the best of my knowledge, no efforts have been made to date, to level, enhance, or clear the site of the old Duncan pit. The ruins stand quite mute, but intrusive, to that which was once a small, thriving coal mine belonging to the W.C.C.*

\*

# Chapter XLIII

## The IRON MEN!
## *– or, The Sylvester Logarithm!*

Those miner's whose work takes them in daily contact with falling strata, needs to think long and carefully as to how they go about it! There are such miners and I don't mean those steel-drawers who practised this science on working coal-faces, but those salvage men who go into old, abandoned, girdered roads to recover the still-standing, but bent and twisted steel girders that no longer serves any useful purpose, except to slow the unrelenting, crushing pressures that will inevitably close these roadways forever.

Most of these steel, arched girders were crushed down from the middle and were greatly distorted from vertical pressures, that tended to collapse them from the centre down-wards, where they had been joined by 'fish-plates' and bolts. All standard-sized, steel girders *(in the Fife coalfields)* were manufactured and formed from 'H' profile-shaped channel iron *(steel)* having a width and depth measurement of approximately six inches by five inches. Irrespective of their overall size, *(except for 14 ft by 10 ft extended three-part girders)* the outline shape of steel arches was uniform, in that when joined at the top centre, the top part of the 'set' was in the shape of a perfect semicircle, whilst the two sides were usually straight and vertical, when erected! *(Except where they were erected in a dook or heading!)*

Steel arches were always described by their erected size in feet, both in width and height, so that a 12' by 9' steel arch, actually measured 12 feet wide and 9 feet high when erected. Girders set in any roadway, were always set apart at the given distance as proscribed by their attendant 'distance-pieces' *(approximately three to four feet long!)*, - or, in some cases, by the 'wording' on the miners contract.

When a coal section was worked-out and abandoned and if the Main-gate and the trailing-gate had been steel-girdered, *(unlike most of the levels and headings in the Dysart Main Seam),* it may just be a viable proposition to the management, to have these steel arches recovered if they are at all accessible and not too badly distorted. *(If the girders are only partly bent out of shape, without damage to the 'H' profile, this girder can be re-shaped on the pithead. If the girder has been subjected to distortion with 'kinks' in it's 'H' profile, it is then graded as quality steel scrap!).* Along the length of a Main-gate where the girder size is superior, it may be that parts of its length may have been back-brushed. *(Distorted girders removed, down-weighted roof strata selectively removed and new steel arches erected. It may also have been, that the roadway had been brushed with either 10 feet, or 12 feet long flat girders, sitting on constructed packs. Either type of steel girder is replaceable!)*

It may also have been that the extracted coal seam was not of a great thickness and therefore, the subsequent crush of no great depth. In the event, even after the section has been abandoned, the chances are that the roadway was still standing and eminently passable. This roadway is therefore ripe for salvage, especially if the recovered steel arches are not too badly distorted and can be reconstituted back to their original shape, by the mechanical steel press on the pit-head. If the girders were too badly distorted, they still had a very worth-while scrap value. *(Miners were always very wary of this type of abandoned section, there was always the ever present likelihood and danger of pockets of 'Blackdamp'.)*

During, and for a long time after W.W.II, steel of every, and any kind was obviously at a premium, and it was a case of salvage and recovery almost at any price! This did not mean that the miners were any more reckless in their efforts, it did however mean, that management would pay a better rate to salvage miners who were not actually engaged in the production of coal, which, at that time, seemed to take precedence over any other mining activity. These miner usually worked together in the smallest of teams, in that standing, damaged girders could only be 'removed' one at a time, and only from the innermost depth of the targeted roadway and working to the retreat. The teams were kept purposely small, in that space was sometimes limited and 'too many' miners would tend to get in each other's way and, most important, the least amount of miners in a 'contract', meant a better share-out of the hard-earned remunerations.

Fortunately, for the miners engaged in this type of work, most of the abandoned sections where they plied their craft had long since 'weighted,' with the extracted coal-faces that the road-ways had previously served, now tightly closed. There were no 'steels' to speak of, on the collapsed coal faces, so the closures would have been uniform, even and complete. I mention this at this juncture, for the reason that no extra weight would have been liable to be thrown onto these roadways during salvage operations, which might have been exacerbated by the uncompleted closure of the old faces. Therefore, the salvage miners had only the immediate roof to look out for, as they removed the old girders, one at a time!

The tools that the miners used, were mainly heavy, steel pinch-bars, miners double-ended picks, heavy hammers and several lengths of strong, one inch, iron chain. *(The miners would not countenance the use of wire-rope, as it had the grievous propensity to stretch and break and 'whiplash' at speed, thereby causing probable serious injury. Unyielding link chain has no such characteristics!)* The one item of equipment that was essential to the operations, was the standard 'Sylvester!' A piece of equipment, that when used properly, could multiply the pulling power of a miner's strong back and arms, by a factor of between 25 and 30.

The Sylvester was designed, so as to convert a leverage stroke of approximately 36 inches in length, to a power stroke of approximately 1 inch in distance, thereby increasing a single 150 lb pull, into something like **one and a half** to **two tons** pressure, which could be consistently delivered in easy single stages of **one inch at a time** and safely **locked,** with the load pressure maintained!  The Sylvester itself was manufactured from duralium, a strong, tough, light-weight material from the aluminium family of alloys and was used extensively in all of the East Fife coal-mines.  It consisted of three parts.  1. The 40-inch long serrated *(with dog-teeth situated 1 inch apart)* sword with a closed eye at it's rear.  2. and 3. The thirty-six inches long operating lever, which was actually attached to the travelling, enclosed 'box', which housed a spring-loaded dog that engaged in the serrated teeth of the sword.

    The action of the Sylvester was very basic, in that the 'closed eye' tail-end of the sword was attached to a chain, which had been previously anchored to an immovable strong point, somewhere to the rear.  The operating box was released to the forward end of the sword, with its dog-tooth re-engaged in one of the deep serration's.  The 'load' chain would be firmly attached to the 'load' requiring extraction, with the loose chain tensioned by hand and manually anchored onto the operating box.  Several oscillations of the operating lever would tension the load-chain and 'machine', in readiness for a more concentrated pull!

    The operating miner would then sit down on the pavement *(or on a small wooden sleeper)* and position himself like a oarsman having a single oar, and then, with his feet braced against an immovable object, would begin exerting the pulling pressures on the operating lever needed to 'take the load!'  *{As any discerning reader will appreciate, the Sylvester is strategically positioned, so as to give the user safe overhead cover and a good, operating position to boot!}*

    Given suitable experience on the job, each salvage team developed their own inimitable method of extracting and retrieving steel girders with the best degree of safely.  I can however say, that no two teams performed this dangerous task in exactly the same manner.  Much depended on the state of the overhead strata, the degree to which the girders were distorted and twisted - and the capabilities and confidence of the miners concerned.  There was no conventional way in which steel girders were salvaged, as the word itself might suggest, but the one over-riding factor did come into every equation with no matter who calculated the odds, was that it had to be done steadfastly and safely!  It was often the case, that when the connecting plate between the two girder halves had been safely undone, the roof would slowly commence to trickle down and when the attempts were made to extract one of the halves, the roof would decide to dump it's not inconsiderable weight on top of the partly freed half girder and the still-tensioned sylvester chain.  This sort of thing was apt to give the salvage team a bit of a headache in more ways than one. The management only paid for recovered, not buried steel girders!

    A likely scenario for the salvage and recovery of a 'set' of used and damaged girders could be as follows. . . Given an abandoned Main-gate where the coal seam has been extracted and roadway has been terminated at a solid end. *(All underground straight roads have to end somewhere!)* The chances are, that if it had been a substantial working section, the extracted coal face would have weighted and subsided long since past and the strata would be well settled, thereby easing, if not actually removing, most of the overhead weight from above the girdered roadway. The arched girders along the length of the road would have absorbed the weighting pressure and gradually buckled from the centre downwards, without actually breaking. That means therefore, that there is a great pressure being exerted upon the distorted steel, but it is not an unremitting pressure! As soon as this pressure *(not weight!)* is relieved in what, is gradual steps, the chances are that the roof, broken as it might be, could be self-supporting, at least in the short term! The miners are fully aware of this! But even then, speed is essential.

The only miner whom I can recall who permanently worked at this job, was a miner by the name of Harry Walker. I do remember him being employed 'on the steel' in the old 240 level in the Surface Dipping, whilst I stripped coal in the 350 water level further down the Dipping. Not very many miners were actually aware of what Harry actually did down the Dipping, he, along with his neighbour worked in near isolation, having only the visiting deputy for company perhaps once only during the course of a shift. Certainly not the most salubrious of working conditions!

Harry Walker was actually quite well known in Lochhead pit in spite of the fact, that he worked in near isolation. He was known for two quite separate, but regular occurrences in that, in the first, he was rarely seen and in the second, he was highly visible! In the first instance, when he worked on the pit bottom, he somehow qualified to be on the 'first tow' up the pit, at the commencement of man-winding! His particular 'forte' was to be up the pit, through the baths and up the half-mile long pit-road to be in time for the 2 p.m. bus to Leven. His chances were always 50/50 in that if the bus was one minute early, he would miss it! If it was 'on-time or a little late, then he would 'catch it!' He was the only coal-miner in Lochhead pit to attempt such a daily challenge, which, if he missed the bus, would result in him having to wait all of 15 minutes until the next scheduled arrival. Enough time in fact, for the first lot of the day-shift miners to catch-up with him at their more leisurely pace!

In his second more visible endeavour which Harry did voluntarily, he stood patiently for all of four hours nearly every Friday pay-day outside of the pay-office, with a small money bag in his hand. He was not begging, but was collecting donated money for the family of some miner who had not worked for a considerable period of time, through being broken, bashed, or otherwise damaged down the coal-mine! At Lochhead pit, as at other mines in the Fife coal fields where a miner had been incapacitated for a continuous period of 10 to 12 weeks, an approach would be made to Harry Walker, with the information that a disabled miner would qualify for a 'drawing' on a given Friday! This being a genuine case, a chalked notice would appear outside the pay-office on the Wednesday or Thursday prior to pay-day, to the effect that a collection/drawing for a named miner would take place on pay-day! Harry Walker would stand, and hopefully, the paid miners would deliver! Most miners made a point of being 'seen' to contribute to this collection in that, 'There but for the grace of God - that could be me!' Small accidents were many, but thankfully, bad accidents were not too common. Even though, it was of some comfort to an injured miner to know, that when his turn came, his workmates and friends would 'come through' for him, if the need should ever arise!

A fair 'drawing' a Lochhead pit would raise anything from ten to twenty pounds sterling, much depended on how long the miner had worked at Lochhead, how well he was known, how many of a family he had and to some extent, his own record at contributing! Many a Lochhead miner has had cause to be grateful to Harry Walker and his unstinting help in all weathers! To the miner's wife, these collected monies would be in her hands within an hour of being donated and, to the full amount collected!

*

Author's Note: - *It was sometimes inevitable, that two or more, disabled coal miners would qualify for a 'drawing' on the same week. This was unfortunate, in that if two separate 'drawings were to be held on the same Friday, then, inevitably, either one or both would suffer a shortfall, in that contributing miner's could not be expected to support two different causes at one go! The policy therefore, was to allow only one drawing on any one Friday, to allow the maximum contributions to each deserving case. This then hopefully meant, that the unlucky miner in the case of week's delay, would not suffer a diminished offering! The criteria in deciding which miner's family would suffer a small delay, usually rested on a combination of factors, not the least of which was, how many of a young family did the miner have and how long had he served at the colliery!*

*

# Chapter XLIV

## Why Goest thou?
## *– or Three wheels on my Wagon!*

## Anecdote or Lochhead Myth?

On one, rather bleak, weekday morning at about twenty past five, just as the first threads of dirty daylight appeared in the Eastern sky, Tam Gibb and Bob Stevens were proceeding north-east along a very wet and windswept Standing Stone Road, to the junction of the Check-bar Road, just to the north-west of the Coaltown of Wemyss. They were sitting side by side in Tam Gibbs motor car, a Ford Eight with just barely enough headlight power to see along the black tarmac road. They were both on the day-shift and were fellow strippers on the Lower Dysart slopes in Lochhead. They were driving along at a steady 40 miles per hour *(just about all-out for a Ford eight)* and had just passed the Pheasantry at the eastern end of the 'Randy' Pit, when Tam suddenly peered forwards and immediately began to slow down. He thought for a second or two, that he had glimpsed a long stretch of water being reflected within the limited range of his headlights. Tam stopped the car at once, which was just as well, the road at that point seemed to take a slight dip. *(The Standing Stone Road was infamous for it's 'switch-back' properties, being totally subject to the many and various subsidence sequences caused by the underground workings of the Randy, the Earlseat Pit and Mine, the Lochhead Pit and Wellsgreen colliery.)* Tam estimated that the road seemed to be completely under water for a stretch of approximately fifty yards, even though there was no quick way of confirming this in the uncertain dawn light. Tam and Bob did know this stretch of road, and decided that this standing water was far too deep, even to wade through, let alone attempt it with a temperamental old boneshaker. The decision made! They both got back in the motor-car where Tam then executed a multi-point turn, and then made to drive-of, back in the direction from where they had just come.

As Tam revved the engine to effect a take-of, there appeared in view a single steadily wavering headlight, accompanied by the distinctive sound of a single cylinder, motorcycle engine. The headlight grew stronger as did the engine noise, and both Tam and Bob got out of the car and started forwards, to warn the rider of the impending water hazard. As the motorbike drew closer, Bob recognised the rider as Dave Harrison, and with him was his friend and 'mucker'- Willie Fields, who occupied the side-car. These two close friends also worked at Lochhead Pit and on the day-shift as well.

Tam and Bob explained to them the reason for their stopping and drew their attention to the length of standing water. Tam also expressed their intention to go back to the Boreland, travel down to the Dysart road end and thence to the Coaltown of Wemyss. Dave Harrison turned to his side-kick Willie Fields and said, "What do you think Willie?" To which Willie replied, "A dae think it's aw' that deep!" David Harrison seemingly concurred with that somewhat optimistic view, in that he turned back to Tom and Bob and said, "I think we'll give it a go!" To which Tam Gibb answered, "I don't think that you'll get through it Dave, but it's your motorbike and I think that you are being a little bit fool-hardy!" *(In other words, - 'bloody stupid!')* "What do you mean, we can't go through it?" voiced a slightly belligerent Dave Harrison, - "Just you turn your car around again and shine your headlights on that puddle and I'll show you how to do it!" Whereupon Tam did as he was requested, turned the car around and played the car feeble headlamps along the water, and keeping the engine revolutions just high enough to give the maximum power to the lamps.

Dave Harrison had meanwhile turned the bike around, and with Willie Fields firmly ensconced in the side-car, had driven back a full 100 yards before turning around again. David gave a final wave of his hand and gunned the engine hard, up through the three gears, until maximum thrust was developed. He crouched low over the handlebars revving to the utmost, and with a last twist of the throttle, made to sail over the surface of the water and be carried onwards by the sheer speed of his careering combination as it hit the waters edge. It did hit the water, as did the side-car, as did David Harrison and Willie Fields. And, . . . the water hit him! As it did the motor-bike! As it did the side-car and it's unfortunate passenger! The side-car was the first to go under, as did Willie Fields, as did the madly revving motor-bike, as did finally, David Harrison. David's efforts had propelled the speeding machine exactly five yards into the dirty rainwater, which was already several feet deep at that point and rising. The rain had never let up! Both Tam Gibb and Bob Stevens moved quickly to the edge of the water. Dave Harrison, having held a extremely tight grip on the handlebars at the point of impact, had described a seemingly perfect somer-sault through the air and plunged into the cold water feet first, where he simply disappeared for a second or two, before emerging waist high in the freezing water, spitting forcefully. Willie Fields had also taken a severe ducking and had been slower to respond to the initial impact, but the sheer coldness of the water had quickly brought him to his senses. Both David and Willie had received a sudden shock to their systems and a soaking to boot, but, to the credit of both of them, in no small part due to their respective health and fitness, their initial recovery was very swift and their reactions positive. On surfacing, David's first thoughts were for his passenger whom he made efforts to reach, and on seeing him surface, his thoughts turned immediately to his conked-out motor-bike, on which depended their prompt arrival at Lochhead, before the man-winding on the day-shift had stopped.  Tam Gibb and Bob Stevens gave Dave and Willie a willing hand to recover the motor-bike *(after the sorry pair had pulled it out of the water!)*, which was successful, after which, Dave Harrison made a valiant attempt to re-start the bike, but to no avail!

Both David and Willie, realising the awful state they were in, decided to quickly walk the one mile across the fields and down the mineral railway to Lochhead pit *(there was no way they were going to be allowed in Tam's Motor-car - even if it did have the 'guts' to carry them!)*, knowing that they would be late, but better that, than catching a dose of flu.  They both knew full well, that their only salvation at that time, was to get into the warmth of the pit-head baths and the likelihood of a quick, warming shower, before they went down the pit.  As the intrepid pair set off down the road, Tam and Bob got into the car, turned it around again and resumed the drive to Lochhead pit by the longer, but safer route.  They had promised that they would report the incident to the 'check-box' and assure the duty clerk, that both Dave Harrison and Willie Fields would turn up for the day-shift that morning, albeit late and half-drowned, but fit for work.  This they eventually did!

This anecdote does not quite end at this point.  As to when Dave Harrison and Willie Fields arrive, I do not know!  But arrive they did and they were not too late for their day-shift, as they were allowed to go down the pit at half-past six.  On arrival at the pithead baths, they had been met by old Abe Leitch who was the baths attendant at that time, and who, surprisingly, offered to have their sopping wet clothes dried in the boiler-house before the end of their shift.  This was a very generous offer on the part of old Abe who, it must be said, was not the most communicative of men.  The offer was immediately accepted and the wet clothes handed over to Old Abe, who disappeared in the direction of the boiler-house, dragging the untidy bundle along with him.  David and Willie, being the hardy miners that they were, didn't have a warm shower, but changed into their 'pit-clothes' and proceeded underground, with the happy knowledge that their sodden garments were being looked after.  Good old Abe!  There was hope for him yet, him being a 'Hallelujah' man and all that!

At the end of their respective shifts *(they did not work together)*, having both managed came up the pit at the same time, they then made their respective ways through the 'dirty end' of the pit-head baths and thence to the hot showers.  After the very welcome soaking and a thorough drying, and having suffered no ill-effects whatsoever despite their near drowning, they, both of them, made their way to the boiler house that lay midway between the clean and dirty ends of the baths.  Sure enough, their clothes were hung-up and strung-out all over the boiler room and their clothes were pepper dry.  Unfortunately, not only were the clothes absolutely dry, they were also 'board-stiff and rock hard!'  They had been hung and spread to dry in their original filthy state, without due thought having been given to their drying properties.  They had dried like starched shirts and were totally rigid.  David's thick canvass coat was by far the worst, it simply stood by itself and needed to be beaten with a thick stick, before it collapsed on the concrete floor. Undeterred, they both set to, and managed to 'break' the rest of their clothing before attempting to dress themselves. Even then, it was a sorry pair that finally made their way up the pit road to the Coaltown of Wemyss, just about chafed raw over most parts of their bodies, where the unrelenting fibres of the 'blow-dried' clothing touched their scrubbed flesh.

*

Authors note: - *A much wiser Dave Harrison, along with a much chastened Willie Fields, did manage to get the Velocipede up and running again with one week. After all, they did have their part-time, window cleaning business to consider and needed the side-car to transport the buckets and ladder. But, that is another story that will be told at another time!*

*

# Chapter XLV

## Dauvit's Progress VIII
## The Arrival! March 1952!

### – *and into* 'Mine Own Ground!'

It may be difficult for non-miners to understand, especially in the light of my proposed, voluntary elevation to the ranks of the 'harder worked', but getting a position on a working face where the very amount of coal that is stripped determines your wages, this to me, was what I had been aiming for since my inevitable 'enlistment' in the coal mines. A chance to pit myself against ten yards of under-cut, standing coal, with the realised option of getting a 'mans' wage at the age of eighteen! I was green, partly inexperienced, but most willing - and I had the health, strength and gumption to tackle anything. (*Besides, Old Harry was minutely following my progress.*)

I arrived at the pit-head baths as usual that Monday morning with a tremulous feeling of elation and the knowledge that at last, I was to be my own man with a numbered place on the run, but would have to prove beyond any doubt, my ability to hold such a place and vindicate my 'supervisors'. I need not have worried! Within the short space of one month, I became very adept at clearing away a 'fired shot' within 15 to 20 minutes (*as long as the pans continued to jig!*), and soon mastered the art of raising a 90 lbs (*41 kilo's*) steel strap to the roof with one hand, and holding it there, while positioning a previously cut wooden tree to it's middle as the forward support.

One encouraging thing that did happen almost on a daily basis during my first two weeks, was that either Danny Gough or Jimmy Hutton, would call up the coal face after they had 'got stripped', just to see if I needed any help. I can safely say that I managed to keep pace with the 'clock', if not with the best of the strippers and invariably managed to get cleaned-up in time, before the pans stopped at the end of the shift. Within three months, I was stripping coal with the best of them, even though I don't think that I was ever the first stripper to come of the face-line. (*That was usually Jimmy Hutton, for the simple reason that even if the face conveyor stopped, he could always get rid of his coal by throwing it down onto the main belt!*)

During the time that I first started stripping coal on the Water level (*Four foot seam*), - my older brother Willie started as an 'oncost' worker on the 'drawing,' (*manhandling full hutches of coal away from a loading point and replacing it with an empty hutch, a seemingly endless task as long as the coal has flowing!*) a few hundred yards from where I was stripping. *(Willie's short career and progress in the coal mines is described in a different chapter within this narrative.)* At that time, I must admit to having had a sense of sibling superiority, knowing that my older brother was helping to *draw* the coal that I was stripping!

I had been on this long-wall face for about one year, when I was casually informed by the under manager that my older brother was to join me on the run. *(He had successfully completed his face-training.)* I remember being asked by A.W. if I wished to have him given a 'place' next to mine, to which I can remember having replied, "Not if I can help it!" Willie deserved a place at the other end of the run *(well away from me!)* which he actually got! There was no way that he was going to be intimidated by me, his younger brother! Willie did progress well, though it took him a little while to master the art of 'raising' steel to the roof in one combined operation. *(Willie had not gone through quite the same hardening-up process as I had done on the pit head, and therefore, had not as yet developed the 'explosive' strength needed to sustain a single-handed, 90 lbs lift to the roof.)*

One morning, when I had finished my strip quite early, 12-15 p.m. as I remember, I meandered *(crawled)* up the run, ostensibly to have a pee in the trailing gate, but obviously, to spy on my big brother. *(He was an inch and a half smaller than me!)* I waited until he had cleaned up his last-fired shot and made preparation to set the next steel strap to the roof. As he positioned himself for the 'lift', having sawed the supporting tree to the correct length, I slowly crawled down onto his 'ground' with my lamp doused. He knew I was there, crouched, watching him, and re-sparking my lamp, casually said, "I must see this!" Willie didn't move! He absolutely refused to attempt to set the steel to the roof whilst I sat there watching! I waited, and watched, and grinned, and he just sat there and looked at me! He would not budge, I could see that he was beginning to lose patience with me, but out of sheer devilment, I remained immobile and waited for at least five minutes before deciding to move on. To this day, I still wonder just how long he would have delayed his strip, just because he thought that he might lose face, if he bungled the tricky operation in the presence of his younger brother. *(There's nowt as queer as folk!)*

Another incident that took place on this long-wall face before the intrusion of the dyke *(the exposing of the fault),* was very nearly the undoing of the complete working face and to the near detriment of every stripper on the run. This near disaster *(through curtailed wages)* that was visited on the miners on this coal face, was wholly man-made and could have been easily averted, but for the ambitions of one man. The seam of coal that was being worked was known as the Four-foot seam, which actually consisted of nearly forty-four inches of coal at several points. *(It varied in thickness from top to bottom of the run and was at it's thickest, during the working month of January 1953.)* Nothing unusual in that, except that the whole of the coal was not taken. The top six to nine inches of the seam was quite unlike the bottom 27 to 35 inches, in that it was a very hard coal that did not split easily, nor did it bore easily, but made for a near perfect, immediate roof because of those very properties.

A.W was not altogether happy with the taking of this coal, he wanted the strippers *(every man-jack of us!)*, to go back to our start point after stripping and using picks only, to attempt to split this hard roof coal thereby dropping the bottom half, to add to the general production total from this face. This task was almost impossible in that the hard coal resisted even the most sustained efforts on our parts. Besides, any self-respecting coal-miner on a working face needs must have faith in the roof, so most of the attempts were rather half-hearted.

At this point, the hole-boring pair of miners who were on the night-shift, were instructed to drill their holes higher-up and into the middle of hard top coal, so that it could be forcibly split with explosives. The problems that arose were two-fold. The drilling of the harder coal took two to three times longer to bore, thereby causing the hole-borers much unpaid overtime and the amount of sharp, drilling-bits required, all but doubled! When the raised holes were next 'fired' by the strippers *(shot-firers)*, the expected results did not materialise, in that the whole of the hard top coal actual shattered with the slightly increased explosive charges needed - and the whole of the top coal parted company with the soft redd roof. Needless to say, the soft redd did not stay up for very long and many light falls were experienced all along the face-line, much to the disgust of the disheartened strippers. *(The fallen redd had to be heaved over the conveyor into the wastes, before downed coal could be got at! An unpaid, time-consuming inevitability.)* That particular part of A.W's abominable experiment was abandoned within three weeks, with the return to holes being drilled in the softer coal below the harder 'dugger' coal. The good, hard, safer coal roof was soon reinstated, much to the relief of the coal-strippers. *(Including myself!)*

Not to be outdone and suffer his authority to be undermined by a bunch of 'bolshie' strippers, A.W. approached the blacksmiths workshop on the surface and had them manufacture specially shaped, hard steel wedges, that were to be used in an effort to split this nine -inch thick, 'dugger' roof coal through it's middle. This again, to satisfy the demands of this stupid man, who insisted in putting the supposed increase in coal output, before the integrity of the immediate roof! He obviously visualised that these steel wedges would successfully split the roof coal, hopefully giving him the credit for increasing the coal production from an already viable run. The strippers were persuaded, coerced and finally threatened with the direst of consequences, if they did not comply with this foolish, stupid diktat.

The subsequent splitting of the roof coal again commenced after each miner's full strip, amid much swearing and knuckle skinning. The strippers who could, did manage to make inroads into the roof coal, but it was hard work and not very successful. The coal did not want to split and showed little propensity to do so evenly. Some miners like myself, delayed our stripping so that no time remained to make the attempt, a chancy gambol at the best of times. To those that did make the attempt, it ended in disaster! The thin uneven band of coal that had held the immediate roof in check, started to give way, and it spread! It spread to nearly every place on the run except for both stable ends. Production was immediately retarded. Great amounts of redd was passed down the conveyor in place of coal, *(the condie's could not take any more)* and the coal-face was in a shambles. The immediate roof was simply out of control and coal production halved. Regular conventional supports were out of the question and much time was lost by the strippers in setting clumsy straps with longer trees and having to shovel needless amounts of heavy redd. What the hell was to do?

The miners were by this time feeling quite dismayed and rather disgruntled, and wages for some men were down. They just could not comprehend the fact, that they were doing the work of back-brushers on a coal face, and only being paid for the little coal that they could get at! Something had to happen and very soon! It did, and not before time! A.W. made the decision *(or, some say, had it made for him by Wullie Hampson),* that they would abandon the attempt to take, or split the top coal and return the face-line to its original state. The hole borers were, once again, instructed to drill immediately below the hard top-coal, in an attempt to leave it intact and firmly attached to the roof! This they did, and within days, the face-line soon returned to it's original viable state, where it was noticed, that the thickness of the coal seam was getting slightly thicker, if only by the odd inch or two. This was the sort of thing that is quickly noticed by miners who are forced to work on their knees, the difference in fact, of being able to raise one's head or having to keep it bowed. For a few months more, the 'four foot coal' on this face-run did get back to full production and the strippers 'improved' weekly wages soon regained some regularity.

In the month of September 1953, a fault line *(a stone dyke)* appeared at the top of the run and with each successive daily cut, it progressed slowly downwards. *(No one, neither the under-manager or the surveyor, would tell us where the coal seam had disappeared to, either upwards or downwards!)* The face length at this time was 130 metres long and was shortening at a rate of approximately 3 to 4 feet daily. This went on for about 30 days, reducing the face length to approximately 105 metres. At this point, the fault held firm for a few days, then receded backwards during the course of the next month, until the coal face again measured approximately 120 metres in length. At that point, the very next cut brought the fault line back with a vengeance, where it advanced down the face-line at a consistent 45 degrees *(four feet, six inches of cut - four feet, six inches of advancing fault!).* Each stripper knew, that when the fault appeared at the top of his place, he had exactly eight working days left on this run, with an ever-decreasing yardage to strip. This working face was officially abandoned on the 14th October 1953, when the fault line struck the bottom stable end and the last of the machinery and men were withdrawn.

\*

How this particular coal face came to it's conclusion did always intrigue me, in that I was fortunate to be on a long-wall face that terminated rather visibly and slowly, from top to bottom, in that a single, stone, fault line (*grey stone dyke*) appeared at the top of the run and advanced, to encroach deeper and deeper into each mans ten yards stint. The fault took about eight working days to intrude fully though each place with the face-length getting progressively shorter every week. The miner who stripped the place above mine was called Jock Richie, (*yes! - the same Jock Richie from the Lower Dysart slopes*) and when the fault gravitated to the bottom end of his place and on to mine, I was the coal-stripper who was moved out, with Jock Richie working the coal against the fault in my place. I was simply not experienced enough (*or so they thought!*) to handle the extraordinary conditions. I knew where the other strippers had been sent and now, I was to join them on the newly developed, coal-face on the west side of the 350 level. (*Readers will remember the incident and place, where I had previously (and accidentally) pushed an uncoupled hutch over the lip of the new developing dook.*) This new face was ready for working, and indeed, coal was being taken along an ever-increasing length, as more strippers were being released from the previous 'faulted' face. This new face was slightly steeper than the previous one, in that the slope was approximately 1 in 3.5. and much to my chagrin, (*and others*) the coal face was a 'left-handed run!' This meant, that as a stripper crouched down and faced up-hill, the coal conveyor was on his right side and moving down-hill, with the under-cut face to his left. It would have been possible to retain a right-handed shovelling momentum, but that would have meant stripping (*picking and shovelling*) down-hill! Only a fool would have considered that!

The options were to attempt to work left-handed, or to shovel over-handed. I chose the latter and to my surprise, found that it did not make one iota of difference! In fact, because of the installation of the new 'Huwood' under-load belt, which travelled only inches above the pavement, it was much easier to use the shovel as a veritable scoop in getting rid of the coal, as only partial lifting of the shovel was involved. I became so adept and quick at stripping coal in this environment, that it eventually led to a small altercation with a shot-firer, that resulted in a coldness that was never to be resolved between us.

At a time, soon after this west side face had been fully manned and into full production, I encountered a shot-firer by the name of 'Durham' Dryburgh. (*Another 'Dryburgh' from the Coaltown of Wemyss!*) He was an average sized, youngish man who was dark of hair (*that which he had,*) and (*to me*) even darker in temperament. He was a dour character even to the point of 'dumb insolence' (*in my opinion, due to his non-committal answers and lack of manners*) and quite disliked by some. My place on this run was exactly in the middle of a 115 yards face-line, which of course, included two stable ends, one leading and one trailing. The two shot-firers on this run, were George Halley and 'Durham' Dryburgh and I was positioned exactly between them. I invariably called for George Halley to 'fire' my ground, but occasionally took the 'services' of 'Durham' Dryburgh when he condescended to answer my call.

On this particular morning, I was really up 'to the mark', feeling very energetic and raring to go! I had cleaned up my gum in record time and was strategically replacing the wooden gibbs (chokes) under the cut-face, in preparation to firing my first two shots. George Halley then appeared on my place just at that time and commenced to 'charge' and stem my opening shots, with the comment that he felt like 'starting at his top end that day'. This was to my advantage, and sure enough, my shots were the first of the morning, which gave me a flying start. The conveyor belt kept rolling and my coal was soon removed, with a double set of roof supports going up in record time. *(These two first shots had actually removed four yards of standing coal.)* Just as I had finished setting the steel, 'Durham' moved on to my place with the words, "Take cover, I'm firing in the next place!" - to which I replied, "That's good! You can fire my next shot whilst you are here!" He grudgingly did so, with a few muttered undertones, but he could not in all honesty refuse. My place had been 'made safe!' He quickly fired my third shot and left me to get on with the subsequent clean-up and steel-setting. This coal I soon got rid of and just as I had finished setting my third set of steel, George Halley appeared from below and announced his intention to 'fire' the place below mine. I quickly collared George and motioned him to fire my next shot. George looked at my stripped ground and replied "I will fire your top shot for you, but I feel that you ought to set another set of close steel at your bottom end before I fire that!" This was fair enough comment and George duly fired my top shot. I again, quickly cleaned up the blown coal within 15 minutes and set another steel strap and two trees to the roof at the top end, which would be my last set of supports at the high end on my place.

At this juncture, I fished out the tin box from my trouser pocket and parted the cotton wool that safe-guarded my wrist-watch. I was pleasantly surprised to see that it was only twenty past nine. I thought at the time, that I had done extremely well that morning, with still 10 minutes to go before piece-time and then, I had a rush of blood which suggested to me, that if I could get my last, shared yard of standing coal fired, I could then boast of having been completely stripped before piece-time. I knew that George Halley was down the bottom of the run, but Durham Dryburgh was within earshot, and I asked the stripper above me to call for the shot-firer. This he did in the usual manner, by shouting the call upwards. Within two minutes Durham Dryburgh crawled onto my place, stopped, looked pointedly at me and casually inquired, "Where's the shot?" and then, more surprisingly, but in a louder voice, asked, "Where's your coal?" gazing around the wide empty space. He repeated

"Where's your coal?"- seemingly very puzzled! "I think that there's something funny going on here!" I couldn't reply. I was speechless! "Where was my coal indeed?" "Probably on the pit-head by this time!" I hesitantly replied. To which, he turned on his hands and knees and crawled away upwards muttering "There's something strange going on here and I'm going to find out all about it! - and I am not going to 'fire' any more of your coal until I do!"

He had absolutely refused to fire my last shot, *(a yard and a half of coal)* of which the coal was only half mine, the other half of the coal belonged to the stripper next place below me on the face-line. I was completely bewildered! Where was my coal indeed? What a stupid question to ask and what a perverse attitude to take! To my mind, the man's actions were stupid, and entirely incomprehensible!' I could not imagine just what was going through this shot-firer's mind in uttering such a stupid statement. I could have mentioned exactly within my anatomy, precisely where the coal was not!, and used a couple of descriptive adjectives to enhance the statement, - but I didn't!

I did have to wait until after piece-time to get my last shot fired - and by George Halley, to whom I did not mention his colleague's earlier refusal. Needless to say, I never did give that man the 'time of day' ever again. Old grudges die hard!

This coal face though very successful as regards continued output did not last too long. This was a great pity, considering the fact that it was a very stable face-line with little, or no production time lost through bad ground. The face itself was only a short distance from the mine bottom and delays and breakdowns were few and far between. Even a broken face-conveyor belt could be repaired in 10 to 15 minutes. This coal-face opened up in the middle of October 1953 and terminated in April 1954. The termination of this coal face, finally saw the abandonment of the whole of the Surface Dipping as a coal-getting mine. All of the coal miners were to be transferred to, and employed in the main Lochhead pit. Not that that inconvenienced anyone to any degree, the miners merely had to walk approximately 200 metres in a different direction to the Lochhead pit cages when they left the pit head baths!

*This first long-wall face (the water level) to the north side in the 350 Level where the permanent fault appeared, was officially closed on the 15th October 1953, which was the date on which the new face in the west side officially opened. In truth, in the closing of the first face and the opening of the second face, the transition actually took about six weeks to complete. The strippers were seconded onto the new coal face one by one, with the decreasing output from the original run being compensated for, by the increasing production from the new south-west side face. The transition was painless, which gives rise to the theory that the planners knew all about the impending fault and were properly prepared!*
They could have informed the miners!

\*

Thus ended my time as a long-wall face 'coal-stripper', I was never to work consistently on my knees again as a stripper, except for the odd shift in various other sections in Lochhead pit, where temporary absenteeism necessitated *'stoop and room'* miners from the Old Dook being drafted on a daily basis to various parts of the mine, to fill vacant positions on a standing face-line. None of us particularly enjoyed the problems caused by those daily transfers. It meant coming under a different Deputy and Oversman, who sometimes got the miners names mixed up, with the inevitable confusion over who did what, and when, and what rate was to be paid, which invariably and detrimentally, was reflected in our weekly pay packet. That would often result in a wasted Friday morning or afternoon, trying to catch up with Deputies or Oversmen, in an effort to refresh their memories and obtain our hard earned remunerations.

Author's Note:- *Daily wages. Under these conditions, where a face-stripper could not complete his daily allotted task because of adverse or abnormal conditions, this was when the 'minimum wages agreement' was brought into play! A coal face miner, through no fault of his own, was bound to receive a fixed, minimum, daily wage in the event of his being thwarted in the pursuit of his 'piece-time' wage. This agreement was certainly brought into play on this coalface during this uncertain period. I do believe that the agreed 'minimum daily wage' for face workers at this time was twenty-six shillings (£1-6-0d) per shift. The stripper's agreed wage in 1953 was 2/11½d. per yard of coal stripped, plus the 2/9d. 'Porter Award'. A likely daily wage of 32/4d. (£1-12-4d.) per shift! The implementation of the 'minimum wage agreement' therefore amounted to a daily shortfall of approximately 6/4d. to every coal-stripper so affected!*

N.B. *This 'Abnormal Conditions' minimum wages agreement, was one of the primary causes of the miner's 'Minimum Wage Strike' in 1912!*

\*

# Chapter XLVI

## A Sobering Enlightenment !
### – *or Thar She Blows!*

## Anecdote, - *or Lochhead Myth?*

Whilst Old Harry was still a relatively young man, he wrocht *(worked/laboured)* in the 240 level in the Surface Dipping on the coal, *(stripping)* and on the brushing. Whilst he was on manual stripping, the coal-getting was mostly in 'pick-places', where the coal was under-cut by means of the miner 'holing-out' the length of his 'place' using a sharp holing pick, usually on a wheeled brae or heading, or indeed, an advancing level. Harry soon became one of those unfortunate miners who, like many others forced to work on their knees, was very susceptible to an inherently, vocational condition called 'beat-knee!' Most miners who worked on their knees, or even sat on their splayed angles did wear thick knee-pads, but the condition once experienced, did have a propensity to return, which in turn, caused many strippers to abandon getting down on their knees and opted for brushing and back brushing, that involved heavier lifting and probably more physical exertion in working with heavier redd and steel girders.

This type of work was usually done standing on two feet, thereby saving some small wear and tear on damaged knee cartilages and ligaments. The wear and tear was then usually transferred to disc problems with the miner's back and included pulled biceps and shoulder strains. I mention all of this, simply because Harry did have his fair share of industrial injuries, both as a stripper and a brusher, and what was at first an irregular series of damaged knees, later progressed to an irregular series of 'racked-backs' and even slipped discs. A condition which rendered many fit miners, to at least, a forced convalescence of approximately six weeks.

After one such 'racked-back', Harry returned to his brushing work on a Monday afternoon, back-shift in the Surface Dipping. By all accounts, this promised to be another warm, summer's afternoon and the on-going shift gathered at the top of the mine, awaiting the last coal race coming up, so that man-haulage could commence. There was no special man-haulage bogies available to travel down in, the men merely climbed into the standard race of empty hutches, three men to a single hutch. This was no mean feat, especially if big Wullie Aitken attempted to climb into a hutch already containing two miners. *(If I was in a hutch with one other miner and Big Wullie tried to climb in, I always got out! I, like many others had no wish to be crippled by the sheer body-weight of Will Aitken, before the shift had even started.)*

Also at the mine head waiting to go down, was a preparatory worker by the name of John Parks. John had in the past, been a pan-shifter and probably still was. He was one of the none coal-getting, face workers *(preparatory miners),* who job it was to help prepare the coal-face for the next cut and strip! John was so dedicated to his job *(either that, or he had been unaware of the artistry at the time!),* that he had image of five, jigging pans tattooed on his fore-arm, spread in the open fan shape of a poker hand of playing cards. Since that time, and by virtue of this tattoo, he then quite naturally, acquired the nick-name, 'Five-pan' Parky.

Now, Parky had a skill, a very well developed and honed skill that he practised nearly ever day of his life. He made the instrument of his hobby from the finest of materials, and indeed, finally had it made to his own design in slim, but solid steel. Parky became a living legend in the developed skill of the heavy catapult. The 'Y' shaped body was made to his own design and the eighth-inch *(3 mm.)* thick rubber was cut by hand, from the best of punctured motor car inner tubes. The sling part was fashioned from stiff, leather tongues cut out of his own pit boots, and the whole weapon was immaculately held together by bindings of strong black waxed cotton. Parky always selected the best and rounded of small pebbles, so that his ammunition was at least dependable. Most everyone knew of Parky's skill with the catapult and he was never backward or shy when asked to demonstrate his marksmanship. Everyone knew, that up to ten yards at least, no target was safe and Parky was not in the least coy, in duping unsuspecting miners into making rash bets on the outcome of any challenge.

During the period of Old Harry's enforced absence, his old, battered, carbide lamp had finally cracked at it's soldered seams, *(Harry's lamp had lain in his hot dry 'dirty' locker during his enforced absence)* and was leaking beyond repair. He was forced therefore to go to the pit-head stores* and purchase *(sign for)* a new carbide lamp, which he grudgingly did. *(The cost of which is always debited from the following weeks' wages.)* The lamp was not of the British 'Premier' make, but one of the better American 'Auto-Lites'. This lamp was a little more expensive than its British counterpart, but boasted a much larger and shinier reflector plate. Most miners after purchasing a new lamp, would borrow someone else's, already burning lamp and use the black, smoky soot to 'kill' the effects of the reflecting plate, so as not to 'blind' anyone caught in the dazzling beam. Not Harry! His lamp was new and he was not going to disfigure his brand new purchase! Harry had already filled his new lamp with carbide crystals and filled the reservoir with clean water. He did not even spit into the carbide to 'start' the gas. He wished to keep everything pristine and clean!

As Harry came up to where the back-shift were sitting, one of the sitting miners asked Harry if he would like to borrow his lamp to darken his reflector, to which Harry replied or merely shook his head in the negative. Nothing much was said, the relaxing miners' merely enjoying the warm sunshine. At this point, Parky casually remarked to Harry, that his new lamp was simply not as good as the 'premier' lamp and that if it fell off his helmet and hit the ground, it would easily damage or break. Harry, in defence of his purchase, immediately disputed this in his very quiet way and told Parky in no uncertain terms, that he simply didn't know what he was talking about. Parky appeared to think for a moment and then said, "I'll bet you, a brand new Premier lamp, that your lamp can not stand the effects of being knocked of a fence post and falling onto the ground". Harry thought for a short moment, and then said, "All right! - but I'll take you at your word and when I win, I'll want that new lamp today!" "Fine!" said five-pan Parky, "you put your lamp up on that fence post and I'll find a wee pebble!"

Old Harry was no fool, he knew of Parky's prowess with that damned catapult, so he moved away from the line of the near fence and placed his 'Auto-lite' on a post on the back fence, all of 25 yards away. Harry stood back, and invited Parky to do his worst, knowing full well that Parky's shot would have lost nearly all of its velocity even if he was lucky enough to get anywhere near the target. Having selected his 'pebble', Parky pulled back his catapult, took careful aim and 'let-go!'

The small missile sped true to its mark, struck with a muted smack and dislodged the 'Auto-lite' from atop the post! The lamp fell to the ground amid a quick silence, and before anyone could open their mouths, the lamp exploded with a bloody loud bang and completely disintegrated. The missile had struck the lighted lamp so violently, that the water reservoir had quickly flooded the carbide chamber and the subsequent, expanding, acetylene gas pressure finding no release, had completely disrupted the structure of the new lamp. Harry was at first incredulous, then dumbfounded, then angry with himself for being intimidated in the first place - and finally embarrassed in that he was forced to return to the stores to purchase another new lamp, much to the puzzlement of the slightly confused, company storekeeper.

*

Author's Tailpiece: - *Five-pan Parky never did tell Old Harry, that far from using a small pebble to attack his new lamp, he had in fact, used one of a precious small supply of five-sixteenths, steel ball-bearings that he had secretly obtained from the engineer's shop the week before. Parky did not even tell his close mates of the deception, just in case they mentioned it to Old Harry. Quiet man that Harry was, Parky knew that what Harry didn't know, wouldn't cause Parky any grief! Not only that, but Harry had the task of explaining to his wife Maidie, as to why he needed to buy two new carbide lamps in one week!*
(That's if he ever did!)

*

Author's Note:- *In the not so distant past, there existed two separate, but wholly iniquitous systems, whereby the mine owners did their utmost to ensure that the miners in their specific employ were not actually paid in cash - but in kind! The first was known as the 'Truck' or 'Truckage' system, whereby each individually bound miner, and by implication his wife or mother, would obtain the daily necessities of life from the company owned and managed 'general store', whether it be in the form of comestibles, cleaning materials or hardware, in order to maintain a semblance of existence. The system was operated and conducted by the Company agent or officials, whereby each miner's 'account' was weekly credited with his earnings for the previous week. Before any monies were paid direct to the miner on pay-day, his weekly (or otherwise!) 'tab' at the company store would be finalised for the week, and all monies due to the store would be deducted in full from his earnings. As happened in most cases, the weekly amount due would usually exceed the miners pitiful earnings, so that, as opposed to the miner receiving any cash remuneration whatsoever for his week's work, the chances were that he would find himself in hock to the Company store by the smallest margin. Company clerks would simply not allow a miner or his family to run into anything other than 'small debt!' - although, if a miner owed monies to the store, he usually just had enough in newly earned wages to cover the small debt.*

*In the year of 1798, 'The Society for the Betterment of the Poor' seemingly wrote - and I quote:- '**Thus, the collier, is not able to squander the 'mass' of his gains to the injury of himself and his family!'***

*There were those disinterested outsiders who looked at this system and saw what they thought was a good and benevolent act on the parts of the mine-owners, who supplied their workers with basic housing at a fixed rent and a general store, where all of the few needs of the miner and his family were catered for! They did not see the low wages, the long hours in underground purgatory and the twelve-years old children that accompanied the father underground. They also, did not see the grimy village cottages, the lack of sanitary conditions and the amount of family members sleeping together in the same small rooms.*

*There was also some very interested outsiders who saw this system exactly for what it was, a thoroughly iniquitous and binding system, whereby the miner became completely dependent on the so-called 'benevolence' of the Company. Those observers saw the 'puppet strings' of near slavery, which would never allow the miner to rise above the bonds of his employment, in an effort to gain the recognition of his worth to the coal-owners. The 'truck' system was roundly condemned from all sides with the belated recognition, that 'company' prices could be fixed upwards as well as being 'fair', - depending on the degree of 'commitment' the Company wished to exert on its workers.*

*(The continued usage of the word 'truck' or 'truckage' has found it's way into the modern vocabulary of everyday day use in that the expression:- "I'm having no*

*truck with that!" is still very commonplace!)*

*The second system described here, was rife in such parts of the Country where the Mine-owners and the 'Iron masters' still held sway. This system, was at times operated alongside the 'truck' system and often in some cases, replaced it! This system was known and described as the 'Tommy-shops' system. With this system, the coal/mine-owners and the Iron-masters setup a series of shops within, or just out-with the confines of the villages, that sold the exact necessities required by the miners and their families. The miners were actually paid in cash at the end of each week's work, but they were almost duty bound to spend this same money in the company shops, for the simple reason that the Company had a monopoly on the location and operation of such 'businesses!' The shop prices would be pitched so that the families had to spend all of their wages, and then some more, on their overall requirements. (The overriding premise from the Company's point of view, was to keep every miner every so slightly in debt!) Again, the 'Company' was in the iniquitous position of deciding how and when a miners hard-earned wages were to be spent!*

*This system was also roundly condemned by many interested outsiders, who saw the naked power of the coal-owners as almost unbreakable. These two iniquitous systems as practised for a long time by both the coal-owners and the Iron-masters, lasted well into the middle of the nineteenth century, when the practise was officially terminated. The relics of this practise however, was still apparent, up to, and including the closure of the coal-mines in East Fife, where 'store purchased' hardware and the 'concessionary coal allowance', was always deducted in full from the miners' next weeks wages!*

*

# Chapter XLVII

## INSIGHT – Part II
## A Fearsome Indoctrination!
## *(or, A Visitor's Guide to the Nether Regions)*

I would boldly venture to suggest, that ninety-nine percent of the population of this country have very little idea of what it is like to feel as a coal-miner does, as he enters the bowels of the Earth for the very first time, whether it happens by walking down an open mine-shaft, or being 'dropped' at speed down a vertical shaft. It can be a mind-numbing, yet neck-prickling experience to those who are aware of it at the time. Descending the shaft of a coal-mine has been likened to descending a lift from the top of a tall building. That is absolutely so! I have experienced both many times. But there I am afraid, the comparison ends! The pitcages are open at both ends, are very dirty and cold, and the internal safety of the miners is guarded by removable steel trellised gates. The interior is usually not lit. *(Except for the miners' lamps.)*

The cage drops alarmingly in the first few seconds, with the feeling common to all that 'something' has been left behind, and the ride is usually quite rough and bumpy with the cage being judderingly banged from side to side with nauseating regularity. The rush of air as the cage travels though the restricted space is very audible, with a distinct 'whoosh' as the two cages pass each other at speed. The descent is as fast as safety permits for persons, a speed that may double or even treble when coal is being wound. If a visitor were to play his cap-lamp on the side walls of the shaft, then he or she may get an impression of sheer speed that may even make them dizzy. *(I have been reliably informed and have personally witnessed, that many visitors, experiencing this first plummet into the black depths of a coal-mine, actually close their eyes very tightly for the duration of the drop!)* Fortunately, not too many people are aware of the coming metamorphoses as they leave the wide open spaces of the pit-head - and descend into completely enclosed, black box of a world having apparently six, separate sides. *(The not unnatural feeling of claustrophobia, becomes manifest with many first-time visitors!)* A left and a right, a forwards and a backwards and an up and a down. Movement and travel are restricted in every direction, except in following the many tunnels, roadways and mines that have been painstakingly cut through the solid metals. Roadways that turn left and right, tunnels that head up then drop down and mines that seemingly go on forever. All for one single objective. To get to the waiting coal!

To the itinerant employee who works outside of the coal-mines, their journey to their place of work is probably quite uneventful, in that they usually travel the same route, day after day, with the same familiar noises and the same familiar faces, without too much of an awareness as to every-day events and small happenings. He or she can probably afford to read the daily paper or even a book, or even engage in various small talk to friends or acquaintances to while away the travelling time! How different to descending down to, and travelling through a coal-mine! From the moment a coal miner walks down the mine, or is dropped into the depths of a pit shaft, a different set of developed senses come into play. It may be that these senses have been there all along, but after a few weeks on a coal face or in a developing mine, these senses are activated, heightened and sharpened to a fine pitch and usually within a very short space of time. Sometimes, the miners' very safety depends on this increased awareness! Miners who travel through underground roads, know full well that any roadway, at any time, can instantly be subject to the vagaries of nature and be closed tight, without prior warning. Fortunately, this does not happen too often, as in ninety-five percent of cases, small warning signs are there to be read, felt, or even intuitively sensed, but only if someone is in the vicinity!

After a few, short, months working in this sort of environment, miners will naturally develop an awareness and a feel for the environment in which he works, and his senses become so finely tuned that he can invariably second guess and anticipate small, almost imperceptible movement in the nearby strata. Sometimes, the only warning, if the movement is small, is nothing more than a slight puff of dust landing on a bared arm or shoulder. This is sometimes the only pre-emptive warning of a localised roof fall, that may just be enough to bury, suffocate, or even kill an unwary miner. *(This will usually only take place on a working face, or within a development where miners are cutting into the solid coal or strata.)* Quite often the warnings are plentiful and expected, as with regular and planned roof falls. Sometimes, the warnings are quiet roof movements that are sensed or felt, rather than heard. To a perfect stranger being taken down a working coal mine for the first time, as soon as he stops hearing the sounds of quietened machinery, and the sound of his own foot-falls, he will be absolutely amazed at the absence of familiar everyday noises. The absolute silence in the interior of coalmines on the non-production shifts, when all machinery is stilled, is nigh 'deafening'. A non-miners' ears are so strained to catch the sound of something/anything, that all he will initially hear is the sound of his own heartbeats and his own breathing. To the underground visitor with untrained ears, there are very few sounds that can be readily identified in the depths of a quiet coalmine!

*(There was a working section in the Francis Colliery at Dysart, that was named 'The Ghost Section'. The strata in and around the daily mined coal face, was forever on the move, with all of the accompanying strange sounds and unexpected noises emanating from all quarters, and being amplified, distorted or even echoed by the surroundings. This happened even more-so in deeper pits, when a greater thickness*

*of coal was being rapidly extracted! To the initiated, the sounds are sometimes heart-thumping! To the uninitiated, the sounds are simply bloody terrifying!)*

There are of course, plenty of recognisable sounds even in a quiet coal mine, but they need to be listened for very carefully and in quietened circumstances. At piece-times, when all machinery is silent and the men are well away from seemingly noisy electric panel boxes, it is often possible to hear the tiny squeaks of the resident grey rats. Some of the older and bolder and most eloquent of miners, would have visitors believe that they can even identify individual rats by their squeaks and squeals. This is probably stretching the truth, in that the miner will probably only recognise the rodent that he feeds on a daily basis. (Some of them did!)

In the area of a working face, especially on a long-wall face where the wastes are 'packed' on a nightly basis, the strata is forever on the move, and indeed, the method of extraction will have already pre-determined, that the roof area in the G.O.A.F. *(wastes/condies)* will break and fall in a controlled manner and on a regular basis, after the planned withdrawal of the last redundant line of steel supports. To the first-time visitor hearing this for the first time, it is not at all surprising that the unfamiliar sounds and their unaccustomed volume, may cause instant fear and mounting apprehension and result in a *'crise de nerfs'* on the part of the uninitiated.

A very large number of mines/tunnels/headings/dooks and roadways in a coal mine remain abandoned and unused. This was uniquely planned and happened continually and repetitively in the workings of both the Old Dook and the New Dook, where the Dysart Main seam was extracted by the method known as Stoop and room. *(Pillar and Stall!)* Many of the roadways that were cut though solid strata *(especially through solid coal in the case of the Dysart Main at between 12 to 32 feet in thickness),* were found to be passable even after seventy years had passed. The reason being, that the strata on either side of the roadways was solid and had not, and would not move. It had nowhere to go! In the case of workings initially commenced in the first decade of 1900's, the original roof line in the first 300 metres of the Old Dook remains as was! The original 'Head coal' as left on most parts of the 1500 metres long dook, was still there to be seen in 1970. The question therefore begs itself; *'Is it still in place at this time, (30 years later) albeit submersed in either salt, fresh, alkaline, or acidic water?'*

In the relative shallow depths of Lochhead pit in the Twenties, Thirties and Forties, it was possible for a inquisitive visitor to get an indication of the use that a roadway was to be put too, as this could usually be determined by the 'size' of the roadway. Roads were usually driven for a specific purpose and their ultimate use would determine its dimensions. *(And therefore, it's development costs.)* It was very unusual to travel a permanent road that did not have some form of coal-haulage or material transport running through it! *(Unless it was a permanent air-way!)*

It may have been an endless-rope haulage that would have supported a twin-tracked, semi-permanent railway of 25.5 inches gauge. In that case, there would have been a steel, haulage rope running between each separate track, with the loaded coal hutches always running in the direction of the pit-bottom, with the empty hutches running in the opposite direction. It may also have been that a diesel pug was used to pull the long trains of hutches in other direction!

Along such roads/levels, a casual visitor will see 'refuge holes' cut into the side-walls at regular intervals on both sides, and to a depth and width of approximately one metre. They will also see that the outline of these 'man-holes' are fully delineated by a thick band of white paint from roof to pavement for easy recognition. These manholes *('person' holes does not seem quite right!)* are of course, for the safety of officials and miners who are authorised to travel these roads whilst the transport means are active.

*(Manholes were cut into and along both sides of a permanent roadway if there was any mechanisation whatsoever installed, or if the road was designed to be used by locomotion, whether it be a dook, a heading or a level. Manholes were not cut or needed in roadways to and from a developing section, which were to be allowed to close after the coal had been extracted.)*

Another thing that may strike a visitor, when he or she has descended to the depths of the pit bottom, is that not only is the bottom usually lit by electric lighting, but the walls are invariably bricked-up between the heavy-weight steel supports and heavily coated with white-wash, to enhance the lighting effect. In all, it may even seem quite bright and give the totally illusionary impression, that coal-mines aren't too bad a place after all! That impression has got to be short-lived! Total darkness pervades ninety-nine percent of any working coal-mine. However, by law, enshrined in the various and constantly up-dated, Coal-Mines Acts, areas such as the pit-bottom, where there is mass of movement of hutches and a preponderance of heavy machinery, these regular working areas needs must be permanently lit for reasons of absolute safety. Also, at any point underground where there is permanent loading point, a motor-house, an electric sub-station or permanent gearing machinery of any description. That point shall suffer to be lit by permanent and protected electric lighting. These regulations in no way negate the requirement for every underground miner, each to carry, maintain and use his own personal lamp at all times, and in the case of carbide lamps, to keep them fully charged and ready for use!

After the Second World War (1939-45), the coal mines saw some great changes both in terms of machinery and money spent, and the working conditions of coal miners and work practices. January 1$^{st}$ 1947, saw the nationalisation of the coal mines in Scotland. During the years of W.W.II, many miners believed at that time, that it was a case of coal at any price, but not at great cost! The coal mines in general *(in East Fife)* were run down, in need of maintenance, were deteriorating in the absence of long-term planning development, and suffering through the lack of up-to-date machinery and dearth of investment.

The coming of nationalisation and the National Coal Board, did slowly provide better working conditions for the miners and the start of a limited flow of more modern equipment, but that of course, did not happen overnight. One thing that did improve quite quickly, was the amount of supplies needed to improve the support systems in existing road-ways and new developments, even though many miners still claimed that more than just some of the increasingly available materials, were of an inferior quality and of pitifully poor inherent strength.

To the visitor about to descend Lochhead pit in the years immediately after the Great War, they might have been aware of a great, clanking beast of a steam engine on the immediate pithead, that did not seem to serve any function in the vicinity - and indeed it did not! After W.W.II., the underground Old Dook haulage was served by it's own motor house within the 'howf' on the pit bottom, and the Lower Dysart haulage-way was also served by it's own motor-house, just off the pit-bottom. But, just after Great War, this powerful, steam engine on the pithead, did actually run both the Old Dook and the Lower Dysart underground, endless, haulage ropes. At this time, there were two powerful enough, electric-motored, haulage-ways installed underground, namely, at the top of the New Dook and at the top of Nicholson's Dook, but not on the pit bottom. The 'drive' power for both the Old Dook and the Lower Dysart haulage-ways originated on the pithead, where the steam engine rotated a large drum, on which was wound a steel endless rope, having a diameter of one inch and a quarter (32 cm.) inches. This double rope *(drive and return)* was tensioned down one side of the winding shaft, diverted under the rails and pavement of the pit-bottom, and ran into the open 'howf', where it connected to two fixed, but separate driving drums for each of the two haulages.

If either of the haulages was forced to stop, then the pit-bottom onsetter had to signal the pit-head to have the steam engine stopped, or disengaged. The reverse of course applied, when any one of the haulages needed to be re-started. *(The policy at the time was to run the system continuously, as this would effectively cause less wear and tear (and strain) on the running gear, than irregular stopping and re-starting!)* This was of course, not an ideal arrangement, but it did have the advantage of using readily available steam power at a time when the Coal Mines were in private ownership, and where the proposed costs of powerful, electric machinery was usually set against the profits gained from the extracted coal.

In the days after W.W.II, when mines' officials and visitors went underground, the common practice was that they wore the standard miner's safety boots, *(usually stiff, leather, lacing, ankle boots with a covered steel toe-cap)* a white painted helmet fitted with a portable electric cap-lamp, and a suit of overalls commonly called a boiler-suit, that was usually blue in colour. This form of dress usually distinguished the mine officials and visitors from the common 'five-eighths' *(coalminer's),* in that the men in boiler-suits and blue overalls rarely thought to perspire, whilst the common man invariably did so, and very profusely at times! Clothing, and the need for several changes of clothing, was always a bone of contention with those miners who worked in warm or hot sections. On going down the pit on a cold winter's day, the miners invariably needed warm clothing in order to stay comfortable until they reached their workplace. At the commencement of work, each miner would divest himself of sufficient clothing to ensure that his movements were not restricted. Very soon, more clothing would have to be discarded as the miner began to sweat freely and until such times as the sweat soaked his remaining clothing.

There was a limit as to how much clothing a miner could remove and be safety conscious, but in some parts of Lochhead's New Dook, it was common practice for a miner to strip down to nothing more than boots, shorts and a soft canvas 'yankee' cap to hold his lamp. If the miner were to retain any more of his clothing, then the garment(s) would soon become absolutely saturated with running sweat, coal dust and dirt, and rivulets of coal dust and water would begin to run from his body. This process could happen several times during the course of a single shift and very often, the miner would have to wipe his face with a clean piece of washed towelling, to get rid of the salt and dirt that would soon begin to irritate his eyes. Some miners carried an extra semmit or two, *(sleeveless vests)* so that they had a degree of dryness and comfort during and after piece-time, but those too, soon became soaked.

The classical type of trousers worn by many miners was the material called 'mole-skins', a hard-wearing, close woven, thick, brushed cotton trouser produced in different sizes and several shades of grey. This material was strong, hard-wearing, infinitely washable, and most of all, sweat absorbent! The material was very comfortable to wear when dry and soaked-up the excess perspiration as it built up on the miner's body! Moleskins trousers however, had one uncomfortable characteristic, and that was to dry out like tough old leather, after being exposed to constant dirt and the excessive daily sweating of miners. Especially, when dried in steam-heated, drying rooms, or miner's dirty lockers. Miners socks often came into the same category, if the miner had sweaty feet or worked in wet conditions. At the start of a shift, sweat soaked, trousers and socks had to be 'cracked' open into a malleable state, before they could be worn. Discomfort, was not quite the description that I remember!

One rule that most miners did not forget, and that was never to place wet boots into the bottom of the 'dirty' locker. They just did not dry out! The boots were usually laid side-on in the running framework underneath the long aluminium seats, specially designed to hold pit-boots. To help keep the leatherwork of pit boots in good, serviceable condition and to inhibit the ingress of water, there was in a designated area of the pit-head baths, an electrified system of rotating, circular-shaped brushes, where the clogged dirt and coal mud could be scoured from leather boots and in addition, a few specially prepared pots of leather grease that could be brushed on the boots before going down the pit. The only trouble was, that if the grease got on to the moleskin trousers in any quantity, it had the effect of 'rotting' holes in the material over a short space of time!

When it was known that special visitors were to visit a coal mine or a section of a mine, it was not unusual to find hutches full of filled, paper sacks of white, stone dust being sent into the section immediately prior to their visits. This stone dust, being heavier than coal dust, was sprinkled liberally over the dark walls and roofs of roadways and haulage-ways, where coal dust was seen and known to gather, to in effect, 'lay' the dust for a short period of time. This was done by miners armed with shovels and wearing masks, *(sometimes!)* and throwing limited quantities up and around the walls in an advancing wide arc. The total effect of this *white-washing* was quite weird, in that it highlighted the contours of any roadway and gave added illumination to a weak light source, because of it's reflecting powers. It also produced the illusionary effect that mine tunnels went on forever! *(Which some of them seemed to do!)* There was a standing joke that was often circulated and nodded on, amongst the oncost miners, who, contemptuously noting the behaviour of these inevitably tired visitors, noticed that when they came out from a working coal-face and into an illuminated area, that they all without exception, needed a steel arched girder to hang on to, and cheekily requested a long, thirsty drink of another miner's precious water!

*(It was therefore inevitable, that sarcastic remarks of such calibre as:- 'They didn't 'hang around' long enough!' - were often heard when the visitors departed.)*

It also was cause for utter amazement on the part of the casual visitor, who, after travelling what seemed to be an seemingly endless, tiring, dark and dirty journey, to be suddenly led inside an underground motor-house which, to all intents and purposes had taken on the pristine mantle of their mother's front room, where everything is exceptionally neat, tidy and spotlessly clean, where the floor space was delineated with white lines and carefully painted. The brass railings were smooth and gleaming and devoid of fingermarks and the machinery housing (casing) was highly-polished, so much better in fact, than when it came from the factory. The machine-man was dressed in clean overalls and even wearing a collar and tie. *(Albeit brown or navy blue.)* This is the motor-man's personal fiefdom. He is there to start and stop the haulage motor when necessary, which is usually twice or three times during the course of a shift. Most of these haulage motors are designed to run for many hours with the minimum of maintenance and often run for a complete shift without stopping. The motorman does not have a mate, or an assistant, and is therefore trusted and depended upon to be alert and awake at all times! This is not an easy task, in that the warmth, the constant hum of the machinery, and his own comfortable 'armchair', are all conducive to somnolence, especially after piece-time and during an uneventful shift. All of this is known to the motor-man and to the mines officials - and so the motor-man is encouraged to be as active as possible within the confines of his fiefdom. He therefore, usually goes about the business of brightening up his surroundings, first by keeping the place very clean and tidy, by perhaps obtaining a sweeping brush to remove the invading coal dust and dirt from the flat concrete floor, and then perhaps, introducing a spot of paint in the form of white lines, to define or outline the most dangerous parts of the machinery.  He will then paint the insides of the white lines with red paint to delineate a no-go area, and this is usually followed by the independent purchase of a tin or two of 'brasso', to brighten-up the brass-work and bright steel of the surrounding railings.  As any visitor will soon surmise, the amount of time needed to maintain all of this 'bull-shine' *(spit and polish),* is sufficient to keep any conscientious motor-man sufficiently awake/alert during the progress of any underground shift.

Some of these conscientious motormen were sometimes apt to lose sight of their remit, in that they would not allow anyone but a mine's official into 'their' motor-house, and only then, with a curt reminder to wipe their feet on the conveniently situated bass mat *(personal purchase!)* sitting in the underground doorway.  Quite often, a curious miner would find his passage barred by the body of the motorman, if they attempted to enter uninvited.  As a measure of the commitment that some of these men devote to their job, I once knew of a motorman in Lochhead pit, who returned to his motor-house one Monday morning, after the engineers and electricians had spent the week-end doing maintenance work and carrying out modifications to the machinery.  He stood in the door of 'his' motor-house and burst out crying at the apparent damage to his beloved paintwork and state of the filthy brass-work!  He was quite inconsolable at the 'outrage' and wailed like madam at her best whore's funereal.  So much for the image of a strong, tough, hairy-arsed miner!

If a visitor were to travel through more than one section in a coal mine, he would invariably meet more than one fireman/deputy.  Every large separate section in a coal mine has one deputy at least and if it is a particularly large section, it might even have two or more.  The deputy's task is quite separate to that of the section gaffer, even though their respective responsibilities often coincide.  The section deputy is the person who has the safety responsibility for the section and the miners employed therein and in matters of overall safety, he invariably has the last word.  Deputies came in all shapes and sizes and with various degrees of general education, and usually progress from the ranks of the miners themselves, coming up through the position of shot-firer, followed by specific attendance at mining school to take the deputy's qualifications.  The newly qualified deputy, having worked amongst his fellow miners for long enough, is usually assigned to another section on appointment, so as not be beholden to, or even intimidated by old loyalties from his erstwhile workmates.  Deputies, like people from all other walks of life, either take to their new appointments, or find the transition quite beyond their immediate capabilities.  Man-management is something that does not come naturally or easily to some people, even though not a few did have the ability to effect the transition through natural or possibly emulated ability.  Some deputies were liked and respected, whilst others were disliked, but never-the-less respected, because of their position and efficiency, whilst some deputies, a very few, were thoroughly disliked, disrespected and actively ignored where this was possible. Some of the older miners would simply not trouble to hide their dislike and contempt of some deputies, whom they thought were idle, inefficient, or thoroughly ignorant.

*(There was once a very worthwhile anecdote that was bandied around in Lochhead's Surface Dipping, where the concerned wife of one such deputy was prompted to request of her seemingly bored husband, during one long summer's evening, 'To go out and have a couple of pints with his friends', to which he moodily and tersely replied, - "What bloody friends!?")*

Some deputies were very approachable, they would listen and then act if the request was reasonable and legitimate. They would sometimes, also go out of their way to obtain supplies or equipment for needy miners, if it was within their powers. If they had been warned that important visitors, or coal board officials were to be expected to the section, they did not get into a 'mild panic', for the simple reason that they knew that they were doing their job properly and things for the most part were running efficiently and smoothly.

Other deputies, especially those who were not quite so confident, or were not quite so efficient or capable, did tend to 'flap' a little, perhaps needlessly, and it showed! They would run around at a great rate of knots, warning all and sundry that 'so and so' was coming and they had better 'look-out!' For what? - I don't know! On one particular occasion whilst I was still an 18-year old stripper on the water level in the Surface Dipping, a certain fireman who shall remain nameless, came rushing down the coal-face on hands and knees at a great rate of knots, at around twenty minutes past one, and shouting, "Quick! Spread yourselves out! The manager's coming!" I didn't quite grasp the significance of that forceful command, I was the only coal-stripper left on the run!

*(The lack of my wood saw had delayed my wood-setting, as my saw was still in the 'joiners' shop for sharpening! Lazy Archie again! Also, I haven't mentioned the intrepid fireman's name. I don't wish to 'knock' him!)*

\*

# A 'Clog Dance' to the Music of 'Lousing' Time!
## (Finishing time!)

*In the immediate grim years after W.W.II, when coal exports re-commenced to other countries, the merchant ships often came back loaded with all sorts of timber, cut to many and standard sizes for building sites, construction work and industry in general. Much of this timber was of good quality, having been stripped of its bark and seasoned. A large percentage of these timber imports were destined for the coal mines, where the wood was sorely needed and generally put to immediate use. Being seasoned and straight, without too many knots and basically lacking in 'resin', the wood was eagerly sought after by miners who, previously, had to rely on the 'green' wood from the Scottish forests, which was usually wet, rough, knotty and covered with weeping resin. This latter wood being virtually useless because of it's soft weak core. Just into the early Fifties, there came into the coal-mines, the lovely, smooth, strong, dry wooden 'trees' from Scandinavia, which were totally seasoned with nary a knot to be seen! It almost felt heavenly to cut-into, in that a hand-saw would glide through it with comparative ease, leaving only a small tell-tale heap of dry sawdust that the merest puff of air would blow away. Not like the heavy, wet, sticky, Scottish pine and fir that clogged a miner's saw and caused the teeth to rust solid.*

*Within a very short space of time, it quickly become apparent that just about every coalface in the coal mine that used wooden pit-props, seemed to be consuming this superior wood to a much greater extent. The increased usage seemed to settle down to about five to ten percent more, that could actually be counted 'standing to the roof' on any face-line!*

*\*(Individual orders from each underground section were sent to the pithead on a daily basis, for the attention of the men in the wood-yards, giving the sizes, description and amounts of all types of wood that was required! Therefore, a comparative check on totals could easily be made by comparison with previous consumption.)*

*This in itself was rather mysterious, because the 'off-cuts' from the trees which were produced when the face-strippers cut the trees to size to set to the roof - those off-cuts were seen to disappear also! In the past, when Scottish pine or fir was used, these off-cuts were seen to be disposed off, either by being thrown into the condies, or by being dumped on the coal conveyors and thence into the surface-bound, loaded, coal hutches. (These off-cuts could have been dumped in the extending roads leading up the coal-faces, to be used as packing-wood by the relevant brushers.) It therefore, took a long time for the mine's officials, mainly deputies, to realise what was happening to the new off-cuts, and more importantly, the whole trees! (The shot-firers on the long-wall faces knew exactly what was 'happening' to these off-cuts! They were not blind. But who wanted to fall foul of half a dozen hairy-assed strippers?) After a while, one of the alternating, day-shift Firemen (deputy), Bob MacDougal by name, came up with what he thought was the solution to the missing trees. Whilst doing his daily rounds along the 'wood-trail' up to the Lower Dysart slopes, he came across a small pile of dry sawdust in the 'main-gate' leading to the double-faces, (inc. the short training face). He quietly and systematically searched for further evidence of any other little piles of new sawdust, and sure enough, there were considerably more! Bob quickly looked up into the spaces behind the weighted girders, and 'low and behold', there was several cut lengths of this beautiful new wood, jammed between the steel girders and the broken strata! Bob investigated this discovery just a little more, and yes!, there was more off-cuts to be seen! There was the answer, plain to see! The brushers were cutting (sawing) up and using this new wood rather liberally, to tighten up the packings between the girders and the overhead broken strata. That was the answer! Mystery solved! The miners had even 'chalked' the girders that had been so packed, so that they knew where to 'pick-up' (commence) again!)*

*Bob MacDougal felt quite pleased with himself at this discovery and casually mentioned his observations to Jimmy Frew\* the section gaffir. Jimmy looked a little askance at Bob MacDougal. He had been down the coalmines far too long to share the same faith in human nature, (even though he was a 'Hallelujah' man and stalwart o' the kirk!) - and was immediately suspicious of any miner who did something for nothing, or didn't request extra payment. Jimmy Frew came from the west coast, an area that reputedly bred 'cannie' men.*

*Jimmy Frew pondered the revelation and considered a possible course of action. He then instructed Bob MacDougal to 'wait-on' after finishing time and then go back to the same wood road, (main-gate) to carefully look to see if the same short cut lengths of tree were still in place behind the chalked girders. Bob dutifully did as requested, and, after 'Paw' Broon's belt end was silenced and the drawers departed, Bob made his way along the belt level and up the slight heading into the longish main-gate. He meticulously picked his way along the heading, carefully examining every girder to locate those with the chalk marks. On coming to the area where the girders were chalked, Bob directed his spot-light on to areas where he had previously spotted the short cut lengths of tree. Bob looked in vain! There were no sawed-off lengths to be seen! Bob took another look at the places marked by the chalk, but this time, gave the suspected areas more than just a cursory glance! Sure enough! There were some small scuff marks on both the insides of the girders and on the exposed redd, where 'something' had been jammed in-between. However, there was not a single off-cut to be seen anywhere!*

*Bob sat down on the now immobile belt and simply gazed around him, his eyes following the direction of his head-mounted spot-lamp. What could the miners be doing with the sawn off-cuts? Yes! Yes! - of course! What could the miners be doing with perfect off-cuts? The answer was staring Bob right between the eyes. What could they be doing indeed? The answer was very, very simple! Didn't Bob himself have to do the self same thing ever weekend at home, simply because his wife didn't like to handle the axe! What better materiel to make perfectly chopped kindling wood than a short length of clean, dry, knot-free, seasoned pit-prop? There was the answer!, - and fool that he was, he had just not given this likelihood any thought whatsoever!*

*Bob made to quickly rise and report this discovery to Jimmy Frew, but then abruptly sat down. Why had Jimmy Frew sent him up to the main-gate in the first place?  Did he know something that Bob MacDougal didn't?  Bob pondered this and slowly came to the conclusion that the section gaffir had already discovered, that which Bob had just found out!  But why the subterfuge?  Was it because that Jimmy Frew didn't wish to be the first to approach the miners, or was it his way of giving the credit for the 'discovery' to Bob the fireman, who, justifiably, should had tumbled to it in the first instance?  Either way, it mattered not!  The truth of what was happening to the wood was now, or was soon to be common knowledge, and this had to be conveyed to the culprits (miners) in such a manner, as to negate any resultant opprobrium or alienation on the parts of the strippers!*

*Bob MacDougal slowly made his way down to the belt-end which was now deserted!  The section gaffer had not waited on Bob returning - and the back-shift had not as yet arrived into the section.  Bob gathered up his belongings from the No. 1 manhole on the Lower Dysart Dook and made his way out along the Lower Dysart mine.  He had some thinking to do!  Bob knew that the average length of the 'trees' coming into the Lower Dysart on the supply bogies were of a standard four feet length.  The standard three feet long (36 inches) trees were just a little too short to help support the top leaf of the seam, at approximately 42 inches high.  Better trees that had to be cut, than trees that were too short and useless!  Bob did some quick and simple calculations.  Each seven foot steel strap set to the roof by the strippers needed two trees, cut to length to support it!  Forty, to forty-two inches of height, minus the one inch thickness of the steel strap, meant that each tree had to have approximately 7 to 9 inches lopped its end to ensure a perfect fit.  (48"- 7" or 9" = 39" to 41".)  Each stripper had ten yards of coal to strip, which meant that there were seven straps and fourteen trees used on every ten yards stint!  That also meant that each stripper had access to fourteen off-cuts during the course of every working shift!  These off-cuts were not being sent down the conveyor amongst the coal, or were they? - and were they possibly being snapped up and ostensibly hidden by the oncost miners?  Bob's simple sums told him, that on a long-wall face length of 120 metres plus the 40 metres length of the training face, that would amount to probably sixteen times fourteen, a grand total of approximately 224 off-cuts per day-shift.*

*Where, in the name of goodness was all of this wood disappearing too?  Bob's intuition also reminded him that the coal mine operated on a three shifts system - and that the back-shift and night-shift miners also had coal-fires at home, and more than one fireplace!  Bob's mind began to boggle at the sheer veracity and the cunning of the section's miners!*

*

*\*\*At this point, I must admit that part of the above narrative describing the 'exact conversation' between Bob MacDougal and Jimmy Frew is part imagination and part supposition.  However, a meaningful conversation did take place at that time,*

*and between those three officials in that manhole, with regard to the sudden increase in the amount of Scandinavian pit props being used! The frustrating thing was that I was not close enough to overhear what was being said! I was trying hard enough! But, when the mines' officials were discussing matters of import, the conversation always became muted. (I particularly noticed this when I was 'drawing the tail' at the head of the Lower Dysart Dook). This did happen around the time when the beautiful wooden trees from Scandinavia were being imported into Lochhead pit, and most important, a percentage of the said trees were not being delivered onto the face-line, but were being covertly sequestrated by all and sundry and cut into 5 or 6 equal lengths, to be secreted about various parts of the mine for almost immediate or belated collection! This supposed 'off-cut' was later placed in the 'piece-box' or 'tin-flask' compartment of each of the culprits army haversacks, and carried home on a near daily basis. These 'off-cuts' were not termed, named or described as such, but were euphemistically known as 'CLOGS!'*

*One of the methods by which this minor transgression was discovered, was the fact that the up-coming miners as they gathered on the pit-bottom, were seen to be carrying either piece-boxes or tin flasks in an outside jacket pocket in addition to having a 'full' haversack, that should by all accounts be near weightless! The practice finally became so blatant, that some miners were seen to be carrying both piece-box and tin flask in jacket pockets whilst their haversacks held two or three 'clogs!'*

*The petty theft, as it was described by the management, became an prevalent occurrence and very soon became a 'crime' in the eyes of the management, who eventually went to great lengths to stop this 'pilfering', even to the extent of searching the miners haversacks as they queued-up on the pit-bottom to ascend the cage! Most married miners worth their salt, took home at least one 8 to 9 inches (20 - 22.5 cms.) long clog, every day of their working week!, and those that say they didn't, shall never go to Heaven! Hallelujah!!*

*(Even Old Harry was known to carry home the odd clog or two on a daily basis, and he was on back-shift and night-shift. He did this to such an extent that my Mother used to ask him if he had remembered to bring home a clog! Old Harry usually just smiled, and dived his hands into the deep inside pockets of his custom-made, pit-road jacket and placed the proffered clogs in the hearth near the fireplace, to await the wood-axe!)*
*\*I learned to chop 'nice' sticks before I learned to use a knife and fork, or I think that I did!*

*Author's Tailpiece: - Jimmy Frew was the section gaffer in the Lower Dysart section, whilst I was 'drawing the tail' at the top of the Lower Dysart Dook. He was a tall, heavy man, much given to enjoying food. He was also a keen Churchman who never seemed to raise his voice in anger, nor did he swear! Neither would he countenance any official or miner doing so in his presence. He was happily married with four grown daughters at this time, of whom he was particularly protective. During one day-shift, when a certain back-brusher (who was built like a brick outhouse and named, - Tam!) came up to the belt-end, he said 'hello' to all present and then 'good morning' to the section gaffer. Jimmy Frew returned the compliment and asked as to how Tam was faring? Tam replied in the affirmative and asked after Jimmy's family - and especially his daughters! Jimmy Frew took one, long, serious look at Tam - and quietly said, "Young Tam, if I ever see you sniffing around any of my daughters, at any time! . . . I'll break yer legs! . . . Aye, - and her's as weel fur even looking at you!"*

\*

# The Route to an 'Upstanding' Wage!

A very few of the more ambitious young miners quickly saw the 'way the cookie crumbled' as regards possible 'promotion' with a guaranteed wage and voluntarily attended Mining School in their own time and at their own small expense. *(Classes were at nil cost to the students, either during the weekday mornings or in the evenings.)* The classes, which catered for nearly all aspects of mining engineering and general qualifications leading to mine managership, were designed to cater for miners working on any conceivable shift during the working week and even for those miners working the rotating shift system. Day-shift miners could attend in the evenings, back-shift miners could attend in the mornings and night-shift miners had the alternative of attending either in the morning or early evening, depending on their sleep pattern. The same part of the syllabus was taught at both morning and evening sessions, so that no student missed out on instruction.

If a young miner was keen enough and dutifully attended a given number of lectures over a complete season, and if he was good enough to pass the subsequent end of term examinations, he would then qualify for 'day-release'. This was a method, whereby a committed student would be sponsored by the N.C.B. to the tune of one day's wages *(at his average daily wage whether it be 'oncost' or 'piecework')* every week, whilst he attended mining school. This sponsorship applied so long as the student religiously attended his 'one day a week' classes and at least one evening class, and also appeared for work at the coal-mine for the other four days of the week, irrespective of the current shift that he happened to be on. Also, providing that the mining student made satisfactory progress and proved it by reasonably progressive results in examinations, then, the N.C.B. was quite prepared to continue the sponsorship, almost ad finitum! I do remember attending classes for a time on successive Wednesdays, which had the effect of breaking a working week up quite nicely, with the added incentive that I getting paid for doing so!

A large part of the small percentage of miners who did 'sandwich' courses at

Buckhaven Mining School, were actually mature men who had probably served something like two, or three, or even four lustrums in the coal-mines as boy and man, and were either persuaded or cajoled into 'Taking the Lamp'. *(This phrase is probably derived from the required duty of every deputy, to carry and be properly trained in the testing procedure for mine gases, using the standard issued safety lamp.)* A fair number of mature miners classed this as 'selling-out' or 'deserting their mates', or even 'crossing the rubicon' and opting for an 'Up-standing wage packet!' *(Mine deputies and shot-firers on appointment, were guaranteed a fixed salary irrespective of their hours/days worked.)* In point of fact, if the miner gained the necessary qualifications, he would then probably make a better deputy than someone coming up through the mining surveyors' route. He certainly had the empirical qualifications and experience, which hopefully, he would apply in fair and just manner in his dealings with his erstwhile workmates. However, sometimes the metamorphosis did not work very well and the new deputy would feel that he owed too much allegiance to his ex-workmates, then bent over backwards to accommodate their every whim, either that, or they executed a complete 'volte-face' and turned against them, with little, or no common ground recognised!

During the course of my research, I did interview a few miners who had 'taken the lamp' and were appointed to the same section where they had previously worked. In this circumstance, old mates and acquaintances were sometimes at something of a loss as to how to greet the new 'lamp-carrier', and were therefore quite reticent and even suspicious of how they should be treated. Some were acidly rude, while others often showed their downright embarrassment or ignorance - and refused to met the eye of the new deputy. A small number of miners actually resented the elevation of their ex-workmates, they themselves being purposely blind to the fact, that the same qualification was open to them and theirs for the taking, if they were so inclined!

\*

James 'Piper' Lawson was one such miner who become a qualified deputy and graduated as most aspirants to the 'appointment', in that he was employed for a spell as a shot-firer, in the developing, short-wall headings above the **three parallel levels** down the New Dook. In one particular heading, the short-wall machine-man on the three man, day-shift team, was a miner by the name of 'Bunt' Leitch, *(Old Abe's loving son!)* whose sole aim as a miner, was to make as much money as he could by dint of the sheer volume of coal that could be produced in any given shift. Time, or the lack of it, was anathema to Bunt, who never wasted a single, physical movement in his working life. He had a mind like a rat-trap, which was invariably several steps ahead of anyone else where work progression was concerned. Bunt always called for the shot-firer long before he was actually needed, so that he was on hand as the holes were being bored. He would not even waste time in talking to the shot-firer, in the event of the limited conversation distracting him from 'charging' his bored holes! He also, was not above attempting to 'help' the shot-firer, if he thought he detected any tardiness in the preparation of the 'firing' sequence.

During one such firing sequence, where the miner's had retired down the heading, the shot-firer, Piper Lawson, had good cause to believe that one of a group of four shots had failed to detonate, and that the uninitiated detonator along with the eight ounces of unexploded 'gelignol', was therefore still deeply embedded within the depths of otherwise blown coal. He immediately ordered the Pan-engine to be stopped, so that the unfired detonator could not be jigged-away with the partially downed coal. After waiting the statutory time period, Piper and Bunt Leitch advanced up the heading to the coal face, where, after a cursory search amidst the lower downed coal, Piper ordered Bunt to carefully pick at, and clean out the area of the suspected malfunction. Bunt, already spitting impatience, swore loudly with the comment, **"I bloody well won't, you know! . . . .** I heard **four** shots going off!"  By this time, the other two men in the team had arrived at the face, and Bunt, quickly rounding on them, said, quite pointedly, **"You two heard four shots!, . . . didn't you?"** Piper quickly interjected, **"I'm the bloody shot-firer here,** and it's my hide that gets nailed to the wall if someone finds my numbered detonator and two sticks of gelignol *(permitted flashless explosive)* going up the pit! So, . . . dig out that coal, because I'm not firing any-more shots until you do!" At that juncture, Bunt Leitch completely 'lost it' and unexpectedly launched himself at the shot-firer. He did not strike him!, but threw a powerful arm-lock around Piper's neck and proceeded to wrestle him to the pavement shouting defiance, where, to the amazement of the other two miners, Bunt refused to let go! Bunt had utterly and completely 'lost it!' His frustrations at the time delay and lost tonnage had wrecked his nervous tension and he simply 'blew!' The altercation seemingly only lasted a few minutes, with Bunt immediately and shamefacedly, but grudgingly, apologising for his outrageous attack on an old friend, who was now a mine's official. But, to Piper's great surprise and heavy disappointment, this unwarranted action was quite unforgivable behaviour on the part of a experienced miner, who, just a few short months before, had been a fellow stripper on the Lower Dysart slopes.

Piper thought prodigiously about this incident and carefully pondered his two obvious choices. He knew that if he pursued the proscribed course of action, Bunt Leitch would be hauled up the pit, taken immediately before the manager and dismissed instantly, without course to redress. Striking a mine's official whilst underground is regarded as one of the most heinous of crimes! Piper was loath to be the instrument of a good miner's downfall and decided that discretion was the better part of enmity. He just did not inform Bunt Leitch of his decision, but allowed him to stew in his own misery for several days before it became apparent to him, that Piper Lawson by his inaction, had declined to take the incident any further. Piper could have made a 'name' for himself by pursuing the incident, but didn't, knowing full well that 'pit gossip and rumour' would probably stand him in better stead in the long run!

*

On one Saturday afternoon in the summer of 1955, 'Piper' Lawson was walking down Kirkcaldy High Street, when he was accosted by an recent ex-workmate who had crossed the street to surprise him with the somewhat surly greeting, "Hello there! You rotten basket!" To which Piper calmly replied, "Hello yourself, and I probably am, but we are 'well met' and you haven't ignored me, which you might have done! That alone must tell you something about our working relationship!" Piper turned on his heel and walked away, leaving a thoroughly bemused Lochhead miner trying to fathom out the exact meaning of the short exchange!

*

# A Well-Earned Extra!

*One week-day morning whilst I was on constant day-shift, and still employed in the Lower Dysart Section, I descended the cage as usual along with another seven miners, some of whom I knew worked in the Lower Dysart, and some whom I didn't know that worked in other sections of the pit. As the cage 'struck' bottom, several miners 'peeled off' in the direction of the Sections where they were employed. Three of us who were bound for the Lower Dysart mine, made to 'report to, and pass either Chairlie Sinclair or Tom Coventry, the Section's deputies, when we became aware that nearly all of the Lower Dysart miners who had preceded us down the pit were standing and sitting around, not making any move to enter the haulage mine. I made to ask Tom Coventry the reason for the supposed hold-up, when Chairlie piped up, "Jist wait! - ye'll bi telt efter!" (Just wait! You shall be told in good time!) I looked around the gathered miners. There were brushers, strippers and on-cost men all intermingled. I also spotted a few men in blue boilers suits whom I knew to be electricians. Amongst them, the journeyman Walter Lavery, and a young apprentice of my own age, called Jackie Dryburgh. (No! Not my later work-mate, even though they both had the same name!) I then spotted Bobby White, slowly sidled up to him and made to ask the inevitable question! He looked directly at me, raised a forefinger to his mouth and 'hushed' me to silence, with a heavy wink and a knowing smirk on his face. He obviously knew!, and so did just about every miner in that large group. I was soon to be initiated into the very loosely, guarded secret.*

*After waiting another ten minutes until the 'last tow' (last group of miners) of the morning had descended, Chairlie Sinclair nodded to the Chief electrician, who quietly called for a little bit of hush from what seemed to be the whole of the Lower Dysart section workforce, gathered at the entrance (pit bottom) to the 1000 yards long, Lower Dysart, endless haulage. Walter Lavery (senior sparkie) explained in his soft Irish brogue, that we were all gathered here to act as packhorses, to carry a 150 metres length of heavy, armoured, electrical cable all the way along the 1,000 yards haulage-way, into and along the full length of the Lower Dysart mine, onto the coupling level, and into the full length of the 350 metres long Belt level. This was to be accomplished if possible, without rest or stoppage. Chairlie Sinclair then mildly reproached Walter Lavery with the words, "You have about sixty men here, they're all yours!, - and don't take all day!" Walter nodded his head and visibly did a quick mental calculation, "O.K. lads! Just line yourselves up at about a three yards (metres) interval along the length of the cable starting from the back end, but don't pick it up just yet, I don't want some men to be picking it up when other men are putting it down! So!, wait for the words of command." (Pause) "There will be two words of command, the first is 'Lift' closely followed by 'Walk!'" "Are ye all ready now?" The miners moved and quickly spaced themselves out along the length of the cable without the need for pushing and shoving, suggesting that many of these co-opted miners had done this before. When the last of the miners had reached the head of the cable with Walter Lavery, all that could be seen along the length of the cable, was a long, evenly-spaced row of naked lights illuminating the darkness. When all was stilled, Walter cupped hands to mouth and bellowed the order, "Lift!" Miners then simultaneously laid horny, callused hands on four-inch (10 cms), diameter, tarred cable and heaved the heavy cable onto a common shoulder. (This would change intermittently and independently as the miners tried to ease their aching shoulders!) When the complete length of the stiff unbending cable was raised shoulder-high, (not for everyone) the order was given to "Walk!"*

*I would dearly like to relate that the one mile long (approximately 1.56 km) inwards journey to the section was uneventful, but this was not usually the case! There were always several embuggerance factors to impede the smooth transportation of this long heavy cable. It was hard, heavy, sticky in parts and positively unyielding. It could be correctly described as multi-copper-cored, heavily insulated, high tension, electric cable with heavy, double spiralled, steel-wired, armoured windings, twisted in opposite directions. It was also very, very expensive!\**

*As the 'trachelled' (burdened, hindered, struggling, fatigued) miners began the inwards trek, it soon became obvious that some of those miners who had done this previously, had chosen their positions well. It certainly did not occur to me to seek a favourable position on the 'line-up', but it did to some! I found myself lined-up amongst the oncost men and drawers from the 'belt-end', (what is more natural than staying together?) who had quickly positioned themselves at the rear end of the cable. (Crafty Bobby White, he knew that the foremost miners would have to walk all the way into the 350 metres belt-level.) Once this cable was hoisted onto the shoulders of the long 'snake' of men, there would be no respite until the cable had been borne all the way to its destination and fully positioned, before the miners* would be ordered to 'drop' the cable. The dead weight of this armoured cable would preclude any further longitudinal positioning, once it had been deposited at its destination, well clear of the light rail tracks and onto the 'high' side of the level.

The 'walk' began, and with it, the dead weight of the cable pressing down onto the miners' unprotected shoulders. Within the first hundred yards of travel, the miners were soon fumbling for some sort of padding to place between the unyielding cable and their soon-to-be bruised shoulders. (The cable had to be prevented from slipping, by the use of one or other, of the miners hands and arms). As is the way of this type of load-carrying, the tallest men in the line found that they unwittingly bore more of the burden than did those of lesser height. The men were spaced out at approximately every three yards (or metres) along the cable, which could be accurately described as unyielding. It just did not bend sufficiently over such a shortish length, with the result, that if a tall miner were flanked by two shorter miners, he would inevitably be carrying his own portion, plus half that of the man in front, and half that of the man in rear. It did not pay to be positioned between two short miners! In addition to that, the pace of advance had to be steadily governed to accommodate those with the shortest legs. (I had midi-sized Bobby White in front of me and small-sized Jimmy Broon in rear of me!) Also, as I have already mentioned elsewhere in this book, the traversing of the Lower Dysart haulage-way was by no means an idyllic ramble! It was less than five feet high over at least 50 percent of its 1,000 yards length. All too soon, the sounds of cursing, swearing and cries for 'divine intervention' were to be heard from some parts of the long line. (Others were too busy gasping and grunting!) Heads were being banged against solid girders, exposed flesh was being bruised and lifted from aching necks and shoulders, shins were being barked against full and empty hutches, and boots were being kicked against protruding railway sleepers. Men were losing their balance and falling down, unable to regain 'their' place on the moving 'snake!' Sometimes, the downing of one man would 'trip' the falling of another, which, in turn, would initiate a 'ripple' effect. This did not stop the juggernaut, but merely caused the other miners on the 'train' to dig-in and pull stronger. The snake must not be allowed to stop! Sometimes, in parts of the haulage-way, there would be ten or fifteen yards of cable with no miners left attached to the cable. One, or two, or even three miners falling at the same time, would cause the dead weight of the cable to fall onto the shoulders of others who were still walking, but were soon brought to their knees by their reluctance to 'let-go'. There was usually a frantic haste for these downed men to reposition themselves on the still-moving cable.

As the head of the snake approached the right part-angled turn into the Lower Dysart Mine, one of the deputies, usually Chairlie Sinclair, would forcefully warn the men to transfer the cable to their right shoulders and 'drive' the cable to the outside of the turn. (To the left side.) This meant that the men on the forward part of the snake, would slow down ever so slightly to allow the cable to be forcibly 'bent' around the 40 degrees slow turn. If this were not done, the cable would be dragged to the inside curve of the bend where the sweating miners would be forced to relinquish their grasp and 'throw' the cable! If this happened, the 'braking' effect would be almost instantaneous with the foremost line of miners falling like ninepins! A situation to be avoided at all costs! The first turn to be negotiated, was followed within seventy metres by a second 40 degrees turn, also to the right with exactly the same procedure being followed. By now, all of the miners on the cable were sweating profusely, in spite of the cold air sweeping in from the haulage-way. Undaunted, the miners 'walked-on', they still had a full shift's work in front of them and there was no time to dawdle! (The deputies were fully aware of this!)

The long 'snake' was now in the 175 metres straight length of the Lower Dysart mine. The passage through this mine gave the miners a chance to re-group and re-position themselves. It was straight, level and high, and whitewashed for nearly it's full length. (This was where diesel Davie practised 'Zen Buddism' to come to a 'oneness' with his beloved pug!) The mine was clear with no standing rakes of either full or empty hutches. This actually gave the miners a much needed respite, in that they were able to walk fully upright, with an even distribution of weight over the full length of the snake. Not for long though! There was another double bend directly ahead with two 50 degrees turns to the left, within 25 metres of travel. Chairlie Sinclair barked again, "Cables to the left shoulder"- but not so loud that everyone would do it at the same time! That would only end in an unmitigated disaster. (One of the belt-end oncost drawers attached to the snake, was an ex-Black Watch, drill Sergeant 'marching' at attention with his free arm swinging, shouted, "By the left!" An expression that was used extensively in the future, as a cry of 'surprised exclamation!') Such was the weight and inflexibility in the cable, that the miners could not raise the cable above their heads to effect the transfer, but in turn, merely grasped it in both hands to perform a 'duck under' and across the shoulders movement, without any loss of momentum, or a break in step. Also, there was such a sharp turning moment on the double bend, that the snake was purposely slowed down at the front, to minimize the chances of accident in negotiating the sharp double bends, with great resistance on the part of the cable to conform to the near about turn.

Thankfully, those miners who had experienced this hurdle before, instilled a sense of caution in those who foolishly wished to press ahead. The snake then continued onto the coupling level with the additional hazard of a narrow, two feet wide centre walkway, between long two rakes of hutches. One full rake awaiting entrainment, and the other one consisting of waiting empties. This was where the miners had to be extra careful in ensuring that everyone remained true-footed and upright, with no imbalance to one side. (The lack of space limited everyone to carrying the load on the same shoulder, usually the right!) This was where a short break could be taken if needs be. The cable could be gently lowered onto the length *of either long rake without any undue strain, with the re-lifting process therefore made easier. The last stage of the 'snake-train' was nearing completion with perhaps only two hundred yards to travel. All hutches had been cleared away from the belt-end (loading point), to ensure an unobstructed path through into the 350 metres long belt-level, where this new 150 yards extension was needed.*

*The positioning of this cable had to be quite precise, in that once it had been dropped/placed by the roadside, its location had to be such, that its trailing end showed approximately two feet overlap to the panel box, to which it was to be connected. This measured feat was accomplished without any undue hassle, in that every miner on the snake, did, at the shouted command - Stop!, (pronounced -Whoa!) Lowered the cable to arms-length, and awaited the order to shuffle a few paces left or right, to accurately position the cable. On the shout of 'Lower!' (pronounced - Doon!) The cable was grounded, clear of the rails on the high side and temporarily tied to sufficient steel girders to ensure its stability. Thus, the task was completed, at least for the miners on the snake! They each, of course, still had a complete shift's work to produce - and on time!*

For this task alone, each and every participating miner, *(both piece-workers and oncost men)* was additionally paid the extra remuneration of an extra 'half-crown!' (2/6d. or 12.5p.) Boys were paid on a pro-rata basis according to their ages. This boy, earned 'one and nine-pence.' (1/9d. or 8.75p.) *(It probably cost just as much in chemical astringents to remove the stubborn tar stains from the miner's flesh!)*

*

# A Copper-bottomed Felon!

*Whilst I was employed in the Surface Dipping as a 19-year-old coal-stripper, I overheard a true story of how one particular duty electrician spent at least part of his night-shift stint, engaged in a rather nefarious, but ultimately, rather lucrative pursuit. He had seemingly pursued this activity for many months before he was accidentally discovered, and then, only though his own carelessness. His felonious actions had remained undetected over the course of many months, for the simple reason that him being the duty electrician, he was completely free to wander unchallenged into almost any part of the working section. If he was not required for routine maintenance, he was usually to be found at a fixed station somewhere within the section. If he was not at this station, he was most likely on walk-about around the section, ostensibly 'looking' for the source of potential trouble. He was accountable to no one in particular, as long as everything electrical was running smoothly! Needless to say, he usually had 'time' on his hands. That this particular duty electrician certainly put his night-shift, free time to lucrative, if not highly improper use, is beyond question!*

In the depths of the Surface Dipping, now working and taking the Four-foot seam at the 350 levels, (previously, this mine had taken the coal seams at the 240 level and the 550 level) there lay, running down the length of the mine and in the old 240 level, several hundred feet of disused, but not abandoned - 10 centimetres diameter, armoured, electrical cable. This cable had either three, four or five internal, high-grade, copper conductors, (or cores) running through it's full length. These ductile conductors were approximately half-inch to three-quarters inch (12 cms. to 18 cms.) in diameter and made from the purest of copper. This valuable non-ferriferous metal when stripped free of it's insulation, would fetch a premium price with any metal broker, if it were delivered in 'clean' and sufficient quantities.

This electrician working alone on the night-shift and armed with his own toolkit, needed only a hacksaw with spare blades and a pair of eight inch pliers, to help cut through the steel, armoured strands. He seemingly selected a long length of armoured cable, that had lain untouched and half buried in the redd and coal dirt that had accumulated over the years, and commenced his activities by measuring a length, that when cut, would lay snugly along the bottom of his haversack! Probably around ten inches. (25 cms.) The cutting through of the cable had to be done as quietly as possible, in short bursts, with enough pauses for him to check that all was clear, with no unwanted visitors in the vicinity, especially inquisitive deputies. (His on-going problem was how to hide evidence of the newly hack-sawed cuts in the ever-shortening cable! He did this by rubbing wet coal-muck from the pavement onto the exposed copper ends!)

This ten-inch length, though quite heavy, could be easily concealed and contained within his haversack, without arousing anyone's suspicions. All that remained for him to do when he returned home, was to carefully strip the outer insulation from around the thick stands of copper, and then get rid of the incriminating evidence. He could do the stripping in complete secrecy in his lock-able, garden shed, with the strippings being carefully collected and disposed of during his daily travels. His only small problem was back at the site of his nocturnal activities, where the discarded ten-inch lengths of protective steel wire were becoming more difficult to conceal. He knew he had a small problem in deciding exactly how to conceal the cut wires. Should he bury the short lengths altogether in one place, or should he spread small amounts in different locations, thereby increasing the chances of the strands being discovered?

As the months went by, with this duty electrician still employed on rotational night-shift (once every four weeks), he became quite emboldened with the apparent ease of his clandestine operation and the apparent lack of his being suspicioned to date. Previously, he had only taken one cut-length during the course of one week, but now decided to attempt two 'cuts' per week. He had previously decided to dispose of all of the steel wire off-cuts, into the same dusty hiding place along the old 350 disused coal level, that had been abandoned several months prior to the

*Water level (new coal section) opening up. (What he didn't realise at that time, was, that even though an coal old section may seem to be abandoned and disused, as long as it remains open and not collapsed, it is the duty of one of the mine deputies to effect a regular inspection of all old workings, and submit a written report as to his findings.) Now that he had decided to take two 'off-cuts' per week, the risks of someone finding the steel wire strippings loomed even greater. Someone was bound to discover either the strippings, the clean, newly cut cable, or even his very presence in a part of the mine where he no legitimate right to be! He then decided to effect a straight-through off-cut on his next visit, without stripping off the wire armouring. This could be done perfectly well in the privacy of his garden shed, thereby minimising the risks of detection. This would also be far better than visiting the abandoned level where he had previously hidden the newly severed evidence.*

*What the undiscovered felon didn't realise, was that the game was nearly up! The inspecting deputy had discovered that someone other than himself, was visiting a certain part of the original 350 level on a semi-regular basis. He could easily tell from the foot-prints that had tramped a faint, but discernible path to almost the same point on every visit. He had thought nothing of this for a while, for the simple reason that some parts of many open but abandoned levels, were regularly used by miners to relieve themselves. (There were no toilets down the pits in my time!) This was quite apparent by the forever lingering smell of fresh excrement. This fireman was no fool, (him being a reluctant student of human nature) he knew that miners do not defecate in the same spot twice running, nor do they squat near to any other fresh turds. Therefore, if a set of footprints stopped at the same place continuously, there had to be good reason. It did not take the fireman long to discover just what was cached under the loose, dusty, fallen waste. It equally, did not take him too long to discover just where the newly-cut short wires had come from? He knew the whereabouts of every piece of mining equipment that was yet to be salvaged from this level, and that included disused, but not abandoned armoured cables!*

*He did not report his findings to anyone at that stage, but continued on with his investigations over the course of the next few weeks. He had discovered the newly-cut end of the disused cable, which had been re-buried in dirt and waste and re-levelled, and he had taken the cannie precaution of measuring the linear distance from the cut-end, to a ground point that he had fixed. He then began to wonder, first 'who', and then 'when?' and finally, 'How often?'*

*It was at this point that this night-shift electrician made his first mistake. He got too greedy! He daren't go near the abandoned level during his day-shift nor his back-shift. He obviously had no chance to get down the mine during his monthly, one week stint in the surface workshops, which meant that he had to take a double opportunity as it presented itself, on his monthly, week-long, night-shift. This was his primary undoing! The inspecting fireman had kept up his solitary watch on the cut end of the cable and could not help but notice, that the pattern of theft had occurred on a four-week cycle! The question therefore begged itself! Who worked on the night-shift down the Surface Dipping on a four weeks cycle - and who had the necessary tools? The answer seemed to present itself! The chances were, that it must be either an engineer or an electrician, even though that was merely inspired guesswork! It could be almost anyone! However, both of these tradesmen had the time, the tools and the opportunity, and it fitted in with their shift patterns. Therefore, to catch the miscreant red-handed, would mean keeping a near permanent watch at the scene of the theft, an activity that would not remain undiscovered for very long. (Who would wish to sit in the pitch darkness for hours on end, even if such a course of action could be countenanced?)*

*The problem of what to do next was placed squarely in the hands of Wullie Hampson, the colliery manager. He, on realising the gravity of the on-going thefts and the length of time involved, did the only thing possible. He called in the police! The police wasted little time. Their suspicions fell immediately onto the most likely suspects with the result, that a day-time call, 'in small force', was made to a certain house in Kennoway, Fife, where a quick visit to the locked shed in the back garden, provided the police with all the evidence they needed to initiate the instant arrest of the culprit. In the words of the arresting officer, "The lazy bugger hadn't even tried to hide the evidence of his criminal activities!" The one precaution that he had been taking, showed that he was perfectly aware of the ductile and malleable properties of undrawn copper, in that he had hammered the soft copper strands into an unrecognisable shape, before moulding them into compacted cubes which, in no way, would reduce their 'scrap' value! (This disfigurement would also help to disguise the original source!)*

*When news of this electrician's arrest reached the miners and tradesmen working at the Surface Dipping, and the pit in general, the miner's, in view of the fact that he had been caught red-handed, opined that they had seen the last of him! Not so! He turned up for work as usual and most surprisingly, was allowed to carry on as before. Several of his colleagues seemingly tried to question him, as to what was likely to happen to him, both at work and with regard to the forth-coming court case, but he simply brazened it out, giving nothing away, not even the date of his court appearance. A few weeks later whilst he was on the back-shift, he failed to appear for work one Tuesday afternoon. The general opinion amongst the 'informed' miners, was that he had found another 'disused armoured cable!'*

*He had of course, been summonsed to appear in court on this date, to which he dutifully attended. He did not appear for work on the Wednesday back-shift, to which the miners again opined, 'Aye, we've seen the last o' him this time! They've locked the thieving basket up for six months!' Which seemed to gain credence in the absence of any other information. However, the felon did turn up for work on the Thursday back-shift, with not the least sign of contrition - or so it seemed! He had not been sacked! He retained his status and he did not suffer incarceration! He had* seemingly pled guilty, accepted the charge in full - and threw himself on the mercy of the court. He was (I am informed) fined the sum of One hundred and fifty pounds, which he cheekily proceeded to pay on the spot. (Which probably did not endear him to the court.) He then took his leave!

When later asked by a curious miner as to how he viewed the judgement of the court, along with the heavy fine (approximately three months wages!) he disdainfully replied, "I was fully prepared! I had all of three hundred pounds with me!"

Author's Note:- *I was recently told during the course of my research, that this 'Taffy-bach' continued to be employed at Lochhead pit after this incident, but had returned home to his Welsh homeland in later life! I am quite positive that when he returned home, it was to something other than a rapturous welcome in the valleys!*

\*

# Chapter XLVIII

## Coalminers at Play! – *At What?*

If a stranger were to ask a tired miner what he did by way of recreation, he might just receive a very sharp non-committal answer, that is if he managed to illicit any answer at all! Not that the miner would be engaged in any illegal or otherwise nefarious pursuit. It was just that the miner would feel that it was no business of anyone else! For some reason or other, most miners like to keep things to themselves. They are by virtue of their profession, a very self-reliant breed, the type of men who just get on with most things in life, mostly without help from others and with very little complaint. Most of their underground work is carried out amidst dust, noise and concentrated activity and in near darkness!

Men will often work a full shift and if closely monitored, will appear to have uttered no more than a score of words during that time. Near silence was a way of life to some miners who had spent many years underground, and merely getting them to engage in any sort of conversation was often a bone of contention to many suffering wives. It is not the first time that I have heard the expression; "Haud yer wheesht, woman!" *(Literally, - Hold your tongue, woman!).*

Ask any miner's wife of reasonably long standing and she will claim if pressed, that her husband had the ability to sit in an armchair, seemingly void his mind of everything and descend into a near catatonic state, with his wide open eyes staring fixedly into space. Repeated questions on the part of the wife would simply 'bounce' of him, only to be answered much later - and in a monosyllabic fashion, when the husband returned to full 'consciousness!' This could have been an early form of 'Transcendental Meditation', either that, or *sheer, bloody, pig ignorance* on the part of a tired miner, bored by his wife's incessant chatter!

The accepted regular hobbies and interests of working miners, were horse racing, dog racing, pigeon fancier, gambling on all three and then some more, and not forgetting the pleasures of the bowling greens. A few more indulged in some of the more active sports and hobbies such as fishing, photography and the rearing, keeping and showing of caged birds and even beer-drinking. In some cases, the latter was foremost and a few indulged to the absolute limit, even to the extent where they have been refused access to the 'cage' on a Monday morning, simply because they were still 'reeking with alcohol!' Thankfully, these men were in a very small minority. Picks and shovels and a 'skinful' of ale, did not mix well with a Monday morning on the day-shift. This was not a good time to be working in the middle places on a long-wall coalface. The smell of vomited rum, whisky and stale beer was not the most pleasant of odours in the restricted air flow.  Thankfully, it did not last too long before 'clean' air removed the awful smells.

Most miners indulged in outdoor activities that kept them out in the fresh air and with a hobby that didn't involve too much expense, they were a 'cannie' lot that knew exactly how many pennies were in a half-crown.  One of the most unlikely activities, and one of the least known secondary professions, was practised and maintained by some of the miners who lived along the sea shores between Leven and Kirkcaldy.  This was the designing and building of fishing, sailing and rowing boats.  There was a wealth of expertise and tradition in places such as Buckhaven, East Wemyss, West Wemyss and Dysart, to name just a few villages.  In most cases, this stemmed from a long tradition of fishing and seafaring and was simply in the blood.  This boat building took place, not in established workshops equipped with modern power tools, but in small backyards on the open beach and in small wooden sheds.  Some of the tools used were rather old and traditional, but were nevertheless skilfully handled by miners, whose natural skills were unquestioned.  These boats would be manned and sailed by their individual owners and sometimes even powered if the boat were big enough, even though most of them were oar powered, row-boats, but perfectly sea worthy.

Fish, of the type of cod, rock cod, the occasional haddock, mackerel in season and flounders would be caught by line, and shellfish such as crab *(parkin)* and lobster would be taken, by the simple expedient of running a line of lobster pots or creels on the sea bed, and baiting them with severed fish-heads.  This could produce a limited bonanza from certain known fishing grounds in the Firth of Forth and each fisherman/boat had their own selected grounds, which they kept to themselves.  'Twas not the first time that a given boat would 'fish' in barren waters just to fool their workmates/competitors!  Extra fish and shellfish could and would be taken by the local inhabitants or busy hotels in the towns, where hard cash was the order of the day!  This was the freshest of sea-food and was quickly snapped-up by some hotels and catering establishments.

It was not every day that the fish were 'running' and at times, boats would return quite empty-handed with men not making excuses, but just accepting the fact that fish were not biting!  The fish was there, but how to catch them?

At such times, various little ploys were sometimes used to encourage the fish 'into' the boat, but it must be said that only a very few 'fishermen' used the one that I shall now describe.  This method was available to fisherman/miners who had access to the cut-off unused 'strum' *(safety fuse cable used to ignite detonators),* that came in six feet lengths attached to the detonators.  Strum was manufactured with a continuous, single core of black, gunpowder granules, bound within a spiral of tarred, cotton strands and wound in a clockwise direction.  A second, tarred spiral of slightly heavier, tarred cotton strands was bound in the opposite direction and covered in an inert, white powder.  *(Probably to negate the sticky effect of the tar!)*

This constructed safety fuse was meticulously manufactured and had two inherent properties: - 1. It had a uniform burning rate of one foot per 30 seconds in time. 2. If kept dry, (i.e. not damp) it would burn underwater. These properties were quickly recognised and utilised by one or two of the said fisherman, and a effective little device was constructed from the simplest of materials. The only original piece of materiel required was an empty matchbox, which would hold approximately one and a half ounces of light powder. *(Black gun-powder granules.)*

To manufacture the 'squib', a length or two of cut strum would be obtained and a measured length of approximately nine inches cut from the whole and set aside. The remaining strum would be held over a sheet of newspaper and the outside layer of cotton would be unwound in such a way as to be about six strands wide *(approximately six millimetres)*. This would be carefully straightened out and put to one side. The inner core of the strum would then be unwound in the same manner, with the now exposed black gunpowder granules being carefully collected on the newspaper. When sufficient 'powder' had been collected and kept dry, this amount would be poured into the matchbox until filled and the matchbox closed. The nine-inch length of fuse would be inserted into one end of the box via a small tight hole, and the lengths of tarred cotton would be systematically wound around the matchbox, so that it was completely covered in several layers and made waterproof! The longer it lay unused, the more water tight it became, the sticky tar seeping through and into small spaces between the windings. All that remained to do, was to attach or tie a small, stone weight to the device to make it sink, but not too fast, in order not to frighten the fish! *(A simpler Mark II version could be made using a small aspirin bottle in place of the matchbox!)*

This device was nothing more than a large squib or firecracker. This was not a hand grenade nor a bomb, but merely a cheap and effective method of making an 'low', disruptive, underwater device, using the natural properties of water to make it effective! As to how effective it was? Well! A few isolated fishermen could have vouched for its efficiency! If, for one reason or another, the fish were nibbling strongly, but not biting, then, the dubious means would be employed. The perpetrator/s would ensure that no other boats were in the immediate area and the device would then be quietly employed. It would be lit and surreptitiously dropped overboard whilst the men continued fishing. It's effect would be felt within about 25 seconds and would sound and feel like a dull, but recognisable slow 'boom' from underneath the boat, with an patient, but anxious wait on the part of the perpetrator until any stunned or dead fish rose to the surface. When this 'catch' was presented for sale, it was a really clever fishmonger or hotel owner, who could tell the difference between one sort off caught fish or another!

This extra mural activity did not of course, always take place during the hours of day-light. Time and tide wait for no man, as do also day-shift, back-shift and night-shift. Fishing took place as and when, shifts, tides and weather permitted.

Fishing in the Forth was often curtailed for days on end, and sometimes weeks at a time, because conditions would not allow the launching of small boats. At other times, the sea could be like a giant millpond with only a small swell and slightly broken water. At these times, boats would put out in the summer 'darkness' and the miners would take with them, their underground carbide lamps to illuminate the immediate vicinity of their boats and of course, their cigarettes and pipes if they were smokers. Carbide lamps were much more dependable than matches or petrol lighters and lasted much longer. These lamps were also very handy to light strum fuses, as they did down the mine. Boats with three or four of a crew, all fishing in the darkness and having lamps lit, could be easily identified from the shore and of course, this worked in reverse and could be used to bring a boat ashore at the correct place.

*

# An Altogether Fishy Tale!

Fishermen in general are famed for their tall stories and sometimes notorious in their descriptions about the 'one that got away'. Coal miners are no exception, in fact, they are even more imaginative with their having fished in the darkness! Each and every fishy tale is superseded by an even 'taller' story.

One of the best and brightest anecdotes that gets repeatedly and joyously retold is still related in the Wemyss's of today, and concerns a miner by the name of Jimmy Gorrie from MacDuff *(East Wemyss)*. One evening, he and one of his friends both went out fishing in the Firth of Forth, but in separate boats. When his friend returned, he is reported to have said; "Man Jimmy, you should have seen the one that I just missed! It was t-h-a-t long!" - indicated by two outstretched arms. Jimmy, not to be outdone replied; "Well! - I didn't catch any fish, but when I bent over to look down into the water, my cap-lamp came free, fell into the water and sank to the bottom. So, what I did was, I threw my line over the side, managed to hook my lamp on the line in the shallow water and brought it up to the surface, and guess what?" "The lamp was still burning!" Jimmy's friend looked back at him with rather an askance glance, and then with an incredulous expression on his face, said, "That takes a wee bit o' believing Jimmy, are ye shair *(sure)* that it was still lit . . .?" This was then followed by a cannie pause for all of two minutes, then Jimmy pawkily replied:- "Weel! If you tak twa fitt aff the end o' yer fish, . . . . . I'll blaw oot the lamp!" *(Well! - If you 'cut' two feet from the end of your fish, . . . . . I will blow out the lamp!)*

*

# Chapter IL

## The Dysart Main to Outcrop!
### Onwards and Upwards – *or, "I can see daylight!"*

Since the breaching of the target seam in the 1890's, in the 80 years working life of the second Lochhead pit, the greater bulk of the extractions came first from the Dysart Main seam and latterly, the Lower Dysart seam of coal. Most of this coal, especially since the late Forties, was progressively extracted by machinery that was developed to be more powerful, more sophisticated and less dependent on the expertise of the solitary miner with a pick and shovel, a hand boring machine and poor quality explosives that were supplied in powder form.

The great bulk of extractions were taken from within the nominated and partly natural boundaries *(barriers)* of Lochhead Pit, where it's workings did not encroach on the territory of any other pit or mine. *(This was strictly observed by all competing coal companies!)* However, the early planners at Lochhead did encroach beyond one of the natural barriers of the area, and subsequently went on to take coals from one of the most prolific sections ever to yield its multiparous bounty. The great irony was, that this series of extractions was planned and commenced around the year 1900, and continued on until 1923 at the latest, where only the most basic of hand-operated machinery was used by the colliers within the large targeted area, and at no great depth!

At a surveyed point on the high side of the long Pit bottom level to the north, and within the Dysart Main coal, at a place approximately 740 metres from the pit bottom, a heading was commenced be driven on a map bearing of 320 degrees *(almost due north-west)*. This heading was taken up through the middle coals of the Dysart Main seam, taking the Grounds & Nethers, the Myslen and the Toughs & Clears coals. *(Readers are reminded that the Dysart Main seam did get thinner as it approached the surface to the North side of the Standing Stone Road. Therefore, the estimated thickness of the seam nearest to outcrop in this area was probably around 12 to 14 feet.)* The slope on this heading was exactly 1 in 4 and the commencement datum height was exactly 9778 feet. *(422 feet or 128.5 metres below ground surface. Ground surface was exactly 61 metres above mean sea level at this place).* Almost at the same time, a second parallel heading was commenced, approximately 25 metres further along to the north and to the right of the first heading, and from the same level. These two parallel headings would follow the same 320 degrees map bearing and would be interconnected at approximately every 25 metres of advance. These twin headings were driven exactly 200 metres in distance and interconnected for the last time at datum height 9919 feet. The decision was then made to carry forwards development of the right heading only. The miners had now arrived at the start point for single, heading assault on the hard rock Dyke *(The Buckhaven Fault),* that separated Lochhead pit from the potentially rich pickings in the one area of the Dysart Main seam on the other *(north)* side of the Dyke, but within the periphery of the area, that was claimed by Lochhead Colliery. The work would now intensify and be concentrated on breaking through the sort of ground, that had not been attempted in Lochhead pit previously.

The Dyke was actually 35 metres thick (wide) at this point and the stone miners were attempting to attack it at a oblique angle, that would mean an approximate thickness of 45 metres of hard rock mine to cut through, if they were to be successful. They were! The break-though was achieved sometime in December 1901, at an estimated datum height of 9999 feet. *{This meant that this part of the workings was only 201 feet (61.4 metres) below ground level!}* One curious observation that I have made during my research, was that there were two 'blind pits' developed to the surface, during the driving of this 'stone' mine. *(A 'blind pit' is the name usually given to a shaft that is struck vertically down (or up!), strikes nothing in particular, and is then abandoned! Often without being filled-in!)* These blind pits may have been struck down directly over the Dyke, to determine the depth of the fault, either that, or they may have been used as air shafts, to dispel the fumes from the great amounts of explosives that they would have had to use on the hard rock. *(A point to note, is that the upwards slope on this short, stone mine was now at 1 in 4.5. As a further point of coincidence, it is also interesting to note that the break-through in the Dyke, occurred directly underneath the point where the Dyke cuts across and underneath the Standing Stone Road, the A915T.)* Also, as far as I am able to ascertain, the break-though was also entirely successful, in that the stone-miners struck the middle part of the Dysart Main seam which reappeared at the other north side of the Dyke, just as quickly as it had disappeared. The Dyke therefore, appears as a 30 metres wide, long and deep, vertical intrusion at this point, with the lie of the coal seam relatively undisturbed on either side of this long fault line.

*(All of coals within the seam that lay against the length of igneous stone on either side of the dyke, was calcinated (burned brown) for a considerable distance into the seam, on both sides of the fault! This is one of the more unfortunate consequences of direct contact between underground seams of coal and igneous rock!)*

This first single intrusion into and though the Dyke, was followed reasonably quickly with a second pair of intrusions that burst through the Dyke in February 1905. They are described as follows: - At another two, separate points on the right side of the pit bottom level, *(north and nearer to the Dyke)* the first point at 840 metres from the pit bottom, and the second point at 35 metres further along the level, there was commenced another pair of parallel levels. This time, they were actually two, cross-cut, level mines driven on a map bearing 35 degrees. These level, cross cut, mines were exactly 15 metres apart and interconnected on an irregular basis as they were extended. These two parallel, mines were initiated at the same time in October 1903 and were developed on an equal basis. The left, western mine struck the dyke after 175 metres of straight line development, at an oblique angle of 45 degrees. This meant, that given a Dyke thickness of 25 metres at this point, the stone miners had actually to cut through 40 metres of solid stone to make the breakthrough. This was achieved in February 1905. At the point of the break-through, the miners turned the mine though an angle of 30 degrees to the right and drove on for another 60 metres, before they struck into that part of the Dysart Main coal which was to be the eastern approach to this section area of the Dysart Main, as well as being the most easterly part of the future extractions.

The right-hand eastern mine actually travelled 225 metres before it too, struck the Dyke at the same oblique 45 degree angle. This eastern *(right)* stone-mine, then continued on through the Dyke for a distance of 40 metres, but, before they were clear of the Dyke, they changed direction by 40 degrees to the left and carried on for a further 12 metres, before they effected a break-though. Why this was done is open to conjecture, but at this point, the miners dipped the mine and went **under** the 30 degree turn of the **left** mine, so now, the mines had actually crossed without joining up! This right mine carried for approximately 60 metres as a coal level, until it connected with a short heading coming up from the termination of it's sister mine. Both mines were now firmly embedded in the heart of the eastern end of this large section, that was about to be opened up and developed into one of the largest sections in Lochhead pit! It was also the shallowest section ever to be extracted! *(Apart from a surface coal-heugh!)*

This large section area which, in the past, had obviously been given a name, but of which, I know not, was not only one of the largest in the area and probably one of the most successful of sections ever worked in Lochhead Colliery, but was worked over a longish, unbroken period of time. The shape of the section as seen from a plan, was actually a fat, egg shape with an overall length of 700 metres and an extended width of 400 metres.

The depth of the coal at it's deepest point at the south-east end was approximately 65 metres below the surface, whilst the depth at it's north-west boundary, was a mere 7 metres below surface level. The section was quite dry and latterly well ventilated, but most of all, was not subject to the crushing weight of the many thousands of tons, that could be expected and experienced at greater depths. At the north-west limit of the section area, there were three, short, vertical, irregularly spaced bore-holes taken up to the surface for reasons of probable air circulation, and in addition, there was also a short mine approximately 20 metres long driven to the surface, where it was possible for the miners to walk directly into the daylight and fresh air! However, these were in the future and played no part in the coal transport from this **Dyke Section,** as I shall now name this working area.

In 1905, just after the second eastern mine had been intruded into the Dyke section, the collier's would then start extracting the coal, which would have to be transported back to the pit-bottom level (north) and thence to the pit bottom. Once there, it would come up Lochhead pit as part of its overall production. Coal from the south-east part of the section would egress via the eastern, cross-cut, level mines, whilst coal from the west part of the section would exit via the original western heading, come dook. The north-east part of this 'Dyke section' actually bordered with the workings of Wellsgreen pit to the north and west, but there was a 25 metres, boundary stoop left in situ between the two coal-mines, and of course, Lochhead pit was part of the Wemyss Coal Company, whilst Wellsgreen pit belonged to the Fife Coal Company *(and never the twain shall meet!)*.

I have no doubt, that small, smuggled amounts of Dysart Main coal *(to feed the 'home' fires!)* may have found its way up through the small, 20 metres long, egress mine to the surface, but as to the commercial probability of great amounts being transported out by this route, this was simply was not feasible for the plain reason, that there was no road, no railway and no mechanical means of portage to do so!

I cannot say for certain as to how the coals from this Dyke section were wrought or taken. I also, do not know how many men and boys were actually employed in the section at any one time, nor do I know just how basic the machinery was that may have been used to work the coals. And I can, but only guess, as to how much the miners were actually paid for every ton of coal that was laboriously 'howked' from this section. What I do know, is that many of the pick-place miners actually employed the boys themselves *(with the permission of the then Colliery Manager),* and paid them out of their own remuneration from their accumulated weekly tonnage. I also know that the softer, bottom coals of their measured stint, would have been meticulously and painstakingly 'holed-out' by the miner to a given depth, before he commenced the hand-boring of carefully sited holes in the standing coal and drilled deep enough, to 'down' the now precipitous face of coal, using only the minimum of 'powder!'

*(In some of the collier's places, it was not necessary to use any explosives, or bore any time-consuming holes at all! The undercut (holed-out) coal would be 'stelled' or 'ranced' from the pavement, using strong wooden supports, until a suitable length and depth of 'holing' (under-picking) had been accomplished on a miner's stint. When this had been completed, the 'stells' would be safely knocked-out one by one, until the full volume of the 'undercut' coal crashed to the pavement under it's own dead weight - and with the help of the ever-so-slight downwards pressure!).*

I also know, that sometime during the previous evenings, the collier, his wife or members of his family, would have been occupied in the making of a quantity of brown paper cylinders, measuring approximately 6 to 8 inches long, about 1 inch in diameter and closed with candle-wax at one end. *(This was usually stiff, brown, manilla wrapping paper.)* These empty cylinders would be carefully filled by the miners *(or their wives!)* with the previously purchased, explosive powder *(Gradely Powder)* and then, when needed underground, would be sparingly used to charge the bored holes. A measured length of 'doubled-over' safety fuse *(strum),* would have been inserted into the filled cylinder to initiate the powder, prior to the now closed and sealed 'charge' being inserted into the bored holes. *(Detonators were not in common use in the coal mines at this time).* The remaining length of the charged hole being *stemmed* with *fire-clay,* in one form or another. The powder and strum *(safety fuse)* would have had to be purchased from the Company on a daily basis, with the cost thereof being deducted from their weekly wages! The miners in those days had to pay for their explosive powder and also for the hanks of safety fuse used!

That which does seem to be conclusive with regard to this large Dyke section, is that the coals that were extracted over the whole of the area were uniformly the same. The Head coal remained untouched over the complete area, as did the Coronation coal. The coals that were taken in their near entirety, were the Grounds & Nethers, the Myslen, the Toughs & Clears and the Sparcoal. A total thickness of approximately 10 feet, including around 10 to 12 inches of stone.

*{This stone would not have been filled into the wooden hutches and tagged. The boys would have had to separate this stone from the coal and build the stone (redd) into temporary heaps. This broken stone would have been used to help fill the pillar-wood or stone 'packs', built to support the immediate roof. If a miner's tagged hutch was suspected to contain stone, through overweight at the check-weight bridge, then that hutch would be 'couped' (overturned) on the hard-standing at the pit head, (in full view of all of the miners entering and leaving the cages) with the culprit publicly castigated and humiliated, and the full weight of its contents being discounted against his wages!}*

From such information that I have been able to glean from the underground plans, I have every reason to believe, that nearly the whole section would have been divided into a series of sub-sections of approximately the same size *(area)* and worked as long wall pick-places, using the 10 percent stone along with 'waste' coal to maintain regular packs along the length of the face-lines, in order to ensure a 'bending', but not broken roof. I also feel that in each sub-section, that the whole of the Grounds & Nethers were removed and the area given a year or two, to weight and settle, before the miners returned to the same sub-section, to later remove the Toughs & Clears and Sparcoal. Common sense also tells me, that the miners may have just left the Myslen coal under their feet as a pavement when they took the Tough & Clears and the Sparcoal, but of that, I can not be certain! I do know that not all of the sub-sections were worked at the same time, this I can glean from the extraction dates. It therefore makes economic sense, to assume that the whole of this large section area was never fully manned at any given time.

With regard to the transport of this taken coal, the drawing *(putting)* of both full and empty hutches, to and from this section, must have been a logistical nightmare, as there does not seem to have been any permanent roadways left open after extractions had been completed. Also, considering the early period when these extractions took place, and the fact that the section was never very deep, I would surmise that nearly all of the set roof supports were comprised of cheap wood, which of course, is biodegradable, and would not have withstood any pressure whatsoever after a relatively short period of time. After the total extraction of the Grounds & Nethers *(and probably the Myslen)* from each of the sub-sections, and after the weighting process had settled over the extracted ground, the miners would have then gone back in to take the Toughs & Clears and the Sparcoal in one fell swoop.

From what I have been able to further glean from still existing plans of this area, it is obvious that the section's miners were able to take almost all of the Grounds & Nethers and the Myslen from the complete area, using methods akin to long-wall pick places, without resorting to the wasteful 'Stoop and room' methods of extraction. This also meant, that with the general use of wood and/or stone packs in the condies *(extracted wastes)*, that the 'weighting' roof movement would be controlled, in that there would be an even 'crush' or 'bending' of the roof strata without breaking. This in turn, meant that the miners could return to the same workings at a later time, to further extract coals such as the Toughs & Clears and Sparcoal, which lay directly above the previous extractions.

In the subsequent taking of the Toughs & Clears along with the Sparcoal, the miners did resort to the Stoop and room method of operation over parts of the section, after the total extraction of the Grounds & Nethers. It would seem, that contrary to an expected 50 percent maximum extraction that this method would have yielded in this area, the extraction of the Toughs & Clears amounted to something like 90 percent of the total coals available. I can only surmise, that this unexpected yield from a revisited section, was due to the fact that the three-foot thickness of the strong, solid, Head coal was left untouched, but more likely, because the area worked was so close to ground surface, with the resultant vast reduction in the overhead weight of the thin layers of intervening strata.

*

Author's Tailpiece:- During the long period of time before the coal-mines were taken into Public ownership in 1947, at a time when the coal-mines belonged to the land owners, the miners themselves were really on a hiding to nothing with regard to their *wages, in that nearly everything the miner needed to use underground had to be borrowed, leased or purchased outright from the 'Company' by the miner concerned. (Not for nothing did Slim Whitman sing about 'owing his soul to the Company store!') A pick-place miner had to pay for all of his tools, all of his clothing, all of his wet working protection, (if indeed at all!) and the weight of the explosive powder used. He even had to pay his boy-filler out of his own yardage. He even paid into a hospital/doctors fund on a weekly basis in case of under-ground accident, of which there was quite a few and on a regular basis. The miners of course, tried to keep their operating costs to a weekly minimum and they also endeavoured to spread the on-going costs so as not to suffer too many deductions from a single week's wages. Personal tools and belongings were therefore carefully looked after and meticulously maintained, and most miners/collier's were rather adept at sewing great, cloth patches of any size or description, onto their forever wearing-out mole-skins and pit jackets. Standard underground dress was boots, (many times repaired/mended) often without socks, moleskin breeks, semmits or peewits, old discarded jackets and thick, wide, multi-oddments, woollen muffler in Winter. Shirts and jerseys were luxury items not well afforded! (miner's hard safety helmets were rarely worn). Every last item of footwear and clothing and especially picks and shovels, were allowed to be worn down or reduced to uselessness before being discarded.

On the social side, young miners, like any other healthy male, didn't need any encouragement to seek out the fairer sex, even though most of then had probably never been farther from home than the nearest sizeable town. Mining families tended to stick together in their communities, with the simple law of averages dictating that there would probably be a miner's daughter somewhere, for every single eligible miner within the area. One thing that was rather apparent, was that the young people tended to wait a little longer before they actually got married, for reasons that housing was in short supply, wages were very low and young miners were not in a position to support a wife until such times as they had qualified for a 'place' on the coal-face. When they eventually got married, it was very rare for the wife not to produce a child within the first year or two of co-habitation. With more mouths to feed, the miner would seek to work harder, 'fill' more coal and utilize every moment of his working time in the pursuit of greater (coal) weight every week. At the end of every pay week (Friday), the young miner would scrupulously hand over all of his hard won wages to his lady wife, and perhaps receive some small returned pittance intended for the purchase of tobacco, or even a 'pint or two' on the Saturday night.

This then, was probably the enduring 'lot' of the great bulk of miners before nationalisation, television, child benefit and Social security. Most miners happily accepted their lot without question, were reasonable happy without being ecstatic and enjoyed the comforts of married life with the responsibility of supporting a wife and children. All of this, along with the keen pleasure of having two pints of good ale at the weekend, and a pipeful or two of 'thick black or bogie roll' tobacco!

However, a small few amongst the many in the mining community, did not take to their changed married status with any grace whatsoever and soon began to cheat, lie, and generally conspire into hoodwinking their naive young wives into believing the most outrageous of penalties, that were supposedly being heaped upon them by the 'Company'. Some of them even attempted to embroil their workmates and their wives into believing their wretched lies. Some disbelieving wives even went as far as to approach the Mine manager, to demand retribution and full repayment of the company's supposedly pusillanimous deductions.

What was it then that caused so much grief, disbelief and sometimes opprobrium on the parts of a small number of miners wives? The answer is very simple! - Although the outright reasons for them doing so, are way beyond my remit. The miners' simply cheated on their wives, in order to keep as much of their pay packets as 'their decency' would allow. They either drunk to extreme, or gambolled heavily on things such as card games, pitch and toss, the horses or the greyhounds. They wanted, needed, extra money to fritter away on 'dogs' or such like, or piss it away against a convenient public house wall!, - and this they did, on a regular and sustained basis! Most miners knew who they were and probably confided as much to their own wives - with the caution never to repeat it in public, but such goings-on in a small, closed community could have never have remained secret for very long.

How was it done then?, - and again, it's simplicity beggared belief! It was simply a case of a young (or, not so young) wife, implicitly trusting her 'rogue' of a husband to deliver a whole, pay packet into her capable hands on pay day. It just didn't happen! The offending miners did not allow their wives a 'look' at their pay-slips on a Thursday, nor did they allow the wives to see their pay packets on a Friday. They simply handed over an amount of money that they thought sufficient, and pocketed the remainder, the amount of which was unknown to the wives. The excuses that were made to 'mitigate' the company's deductions were absolutely classic and often barefaced, and even though the general education of most of the miners was very basic at this time, their imaginative machinations knew no bounds.

I have listed underneath such nebulous excuses that I can remember, but no doubt there are many, many more!

They have taken another hutchful of pillar-wood off my wages again this week!

I gave my pony an extra nose bag of oats last week and they made me pay for it!

I wrocht two extra yards of coal last week, (for which I have not been paid) and they made me pay for the extra six feet rails!

They had a 'drawing' for a hurt work-mate this week and I gave 'two' half-crowns!

Some thieving dog stole all my tallow supply and I had to buy more!

The roof 'came in' and I had to 'buy' extra larch bars! (Heavy beam-like roof supports.)

We had to pay extra for the new Scandinavian pit-props!

They have started to charge us for 'new' rail sleepers!

Our Pit pony dropped dead and we had to kitty-up to get a new one!

They have increased the charges for rail dogs-spikes! (Metal nails used on rail sleepers.)

I shall have to buy a steel girder next week! *(This excuse was used very sparingly!)*

Somebody's been stealing my numbered tags from my filled hutches! (Again!)

They've taken that same set of six feet rails from my wages again this week!

We're having to pay a lad to waft fresh air along our level!

The blacksmith accidentally burnt the (wooden) shaft of my 'holing' pick! (Again!)

The Manager's cut our filling rate! (Again!)

They have run out of six feet rails - I had to employ two boys yesterday!

* During the first two decades of the last Century, I have estimated that a pick-place miner could earn anything from Four and sixpence *(twenty-three and a half pence)* to Seven shillings *(thirty five pence)* per day, if he was fortunate enough to fill his daily quota. He also 'paid' his boy approximately one third, or one quarter of his daily earnings.

\*

# Chapter L

## ZEITGEIST!
### *In the Spirit of the Times!*
### *(Wark on, wark on, ye Wast Wemyss Loons!)*

In the West Wemyss of today, yesterday and the previous sixty years, developments have not really moved very far forwards. There are very few new modern buildings in this small, almost land-locked, mining village and even the fine looking Belvedere hotel is really no more than a completely refurbished old Miner's Institute. A few of the old* houses are boarded up, but not quite abandoned and some of the private dwellings have been partially modernised. The general layout of West Wemyss is much the same as it has always been, except for some small cosmetic changes in the narrow streets and at the tiny harbour. There are of course, nowadays, more motor vehicles in evidence around the narrow village roads, with even the odd, abandoned, motor car, but that in no way alters the general feeling of having entered a slightly different world that lies within the confines of the village. Even in the height of summer when a few holiday makers and tourists do find their way down into West Wemyss, the place could never claim to be a holiday venue. There is only one 'iron' road down into the village and the same iron road takes travellers back up on the way out!

*(There is however, a very interesting coastal path from West Wemyss to the old Wellesley bing, that passes through old Buckhaven and allows walkers and hiker's the chance to see and visit the remnants of the famous old red sandstone caves abutting the sea-shore. This coastal path did at one time, have the distinction of being the main route between the sea-side towns of Dysart and Leven. Indeed, it was also the main 'letter and parcel' route for the daily postal service!)*

The people of this village were like the good folk of many small villages, in that they were a fairly, close knit community, mostly engaged in what local industry there was to sustain the inhabitants and their families. The village was also additionally unique, in that there was only one main industry, closely supported by one or more secondary industries, in which many of the latter were joined by many of the former, when shift-work allowed.

The main occupation that supported the bulk of the small community, was the getting of coal by one means or another, and secondly, the taking of fish and other sea bounty when time, tide and shift working permitted. The fact that nearly all of the inhabitants were employed in one industry and probably in another *(or more!)*, meant that very few people ever felt the need to leave the village from one year's end to the next. One additional, traditional industry that did exist and flourished well, involved the construction of quite a number of wooden boats. Fishing smacks, sailing yawls, sailing dinghies and rowing skiffs and dinghies. By all accounts, many of these boats were very successful and subsequently became quite famous and, were frequently named as winners in the local regattas.

Much of this boat-building was a full-time occupation to some, whilst much of it was practised by working miners in addition to their daily coal-shift down the Victoria, or Lochhead pits. Public transport in and out of West Wemyss did exist, but was infrequent and probably inconsistent.

*(There is a story that emanates from the folklore of this village, that I have heard repeated many times over, that revolves around the first ever Iron-clad ship that sailed into West Wemyss harbour. The local housewives in experiencing complete wonderment and disbelief at seeing something made of iron actually floating, all took turns at throwing their door-keys into the harbour to see if they would float too! History does not record as to how the keys were recovered!)*

In this sort of environment, church or chapel was part and parcel of everyday life and in this small unchanging community, the resident minister would no doubt know the names of every one of his parishioners and indeed, every family in the village, whether or not they attended church on Sundays. In this sort of community, those that did attend church on a regular basis, usually got dressed up in their Sunday best, and indeed, it was probably a point of honour, that they were absolutely clean and tidy and seen to be dressed in their best and seated in church in good time. It may also be said, that the minister would have been able to cast nothing more than a casual all-seeing glance around the wooden pews and note the non-attenders on the day. Many of the congregation were good, clean living, God-fearing, pious people, whose weekly attendance at church was the culmination of a hard week's work, and whose presence there on a Sunday, was as much part and parcel of their lives as going to work on the other six days of the week. Some of the older men prided themselves in being solid and devout members of the church and even part-time officers during Sunday services. These pious men found no conflict in doing their duty on a Sunday, and working alongside some of the most uncouth and graceless of miners during the rest of the week, whether it be in the Victoria pit or Lochhead colliery! The swearing habits and foul language of some of these 'uncivilised' miners, was as much anathema to these stalwarts as 'picking their nose' in public - and even the mildest of swear words, would sometime result in the gentlest of rebukes to the offender. *(They didn't push their luck too far!)* In such a coarse climate as the coal mines, these men went about their daily business quite unabashed and were often wholly impervious to the devious, dreadful and dastardly deeds that were sometimes concocted, manifested and perpetrated in the darkest recesses of the coal mine. *(Much of these goings-on were nothing more than the product of a frightened and timid imagination!)*

It is also quite incidental, that many of these men were not actually coal-getters themselves, even though they might have been in the past, but were now part of the mines lower-archy.  Foul language and swearing, or taking the Lord's name in vain was therefore taboo, and gentlemanly expletives and worldly expressions on their part, took the form of: - 'Oh Dear!,' - 'Oh Gosh!,' - 'Oh My!' and, 'Michty Me!' *(Such worthy characters were often meaningfully described as 'Hallelujah' men.)*

This enforced isolation was a way of life to most of the inhabitants of West Wemyss and indeed, if the truth be known, many of them would not have it otherwise.  It could therefore be said of many of the people of this village, that certain aspects of life and living had not changed at all, since the days of their grandfathers being employed in exactly the same industries and occupations, and that certain old Victorian values remained in-being and indeed, were practised daily.

*(Another tale that emerges from the folklore of the village, is that when W. Alexander & Co. sent the first omnibus down into the street of West Wemyss, the wives' and mothers of the village were so terrified, that they scooped up their children, took them inside and bolted their doors. (They couldn't lock them, as their iron door keys were all lying at the bottom of the harbour!)  The story goes on to relate, that the oldest and boldest of the women formed a delegation to approach the Omnibus company, whose name appeared on the side of the omnibus, to come and remove this belching monstrosity from the sight of the villagers.)*

It was therefore not uncommon for a wife, never to undress in the presence of her husband, or if in the same room, under the cover of total darkness.  It would also be unseemly for a wife to discuss womanly things with her husband, and a wife's next door neighbour would be the wife's most likely confidante.  In matters concerning things like his wife's confinement, the husband would be pushed out of the way and indeed, would probably find the door locked in his face!  In this climate and environment, it is therefore not surprising that some miners when troubled by small discomforts, would certainly not confide in their doctor, much less their wives, but would invariably turn to their work-mates.  Men with whom they worked with on a daily basis, knew very well, and with miners whom they often shared their piece!  As a result of this inherent preoccupation, one embarrassing incident did however occur in Lochhead pit, as late as the early Fifties, that nearly ended up with the participants having to swear an oath in Court and explain their actions before a Judge and jury!

*

In the second half of the Forties, just after most of the Victoria miners were transferred to Lochhead pit, the managerial policy of the time was to try and keep most of the ex-Victoria miners together in their small teams, where familiar working habits and mutual trust usually produced the best results.  This however, was not always possible and eventually, the Victoria miners were completely integrated into and amongst the slightly more itinerant Lochhead miners, who daily, travelled from much farther afield. In transferring from the Victoria pit to it's larger parent Lochhead, some of the miners chose not to take advantage of the meagre transport that was available, and either walked across the fields, a distance of approximately one mile, or they got on their bikes. Either way it was hard going, as there was a long climb to be negotiated coming up and out of West Wemyss. In winter, the journey was especially rough if they didn't stick to the roads, with the inevitable result that wet feet, chafed skin and chilblains became commonplace, especially amongst those few miners who doggedly refused to use the pit head baths.

Author's Comment's: - *There was at this time, a strong commonly held conviction amongst some of the older miners, especially those from the Victoria pit, that the constant washing of a miner's dirty back with hot soapy water would only serve to induce lumbar enervation! This belief was so inherently strong, that even when such die-hards were finally persuaded to use the pit head baths, they would forcibly deter other miners from attempting to wash their backs in return for having their own done! (Mutual back-washing was one of the more civilised rituals commonly practised amongst 'coal-dirty' miners.) 'Twas not the first time that one of these old miners was 'reminded' when coming out of the shower cubicle area, that they had 'forgotten' to have their backs washed. (This was manifestly apparent in that the water rivulets had left water trails down their still dirty parts of their backs!) This polite reminder was usually accompanied with a disdainful, "I'll wash it when I get home!?" I can actually recall the names of several such old miners from the village of West Wemyss! (Old 'Hair' (ninger) Salmond springs to mind! - along with a few others.)*

*There were also some miners, who, because they didn't like, or didn't wish to be touched by other men, (or maybe, men not from West Wemyss!) would actually rub soap over an area of the white-tiled wall of the shower cubicle - and perform a series of protracted contortions and gymnastic gyrations with their backs against the wall, in an effort to rub away the dirt and grime. They did not seem to be the least self-conscious about their solitary 'Rock and Roll!' . . . . . . It could have been that they were humming - "Onwards Christian Soldiers!" "Hallelujah!!"*

\*

## Anecdote - *or Lochhead Myth?*

### *"In Flagrante Delicto!"*

In the late Autumn of 1955, two experienced miners equipped with a A.B. short-wall, coal-cutting machine, were employed in the driving of a heading up into the Lower Dysart coal seam, from the bottom of the **three parallel levels,** *(at datum level 9558 feet)* situated between the 'New Dook' in the Dysart Main and the Newer Dook in the Lower Dysart seam. The start point of this heading was exactly 21 metres directly **underneath** the New Dook man-haulage. The map bearing of this heading was approximately 330 degrees, a direction that would take the miners in a straight line, to the abandoned, termination point of the recently worked-out area of the **East-side Mine.** This developing heading was to be carried forwards using 10 feet by 8 feet, arched, steel girders, for the good reason that this particular heading would soon become a coal transport outlet for series of parallel workings, that would soon be developed into a small, working, **short-wall** section. The extracted coal from this new heading, did not travel down a column of 'pans', but was transported by one of the new, narrow, 'Cowley-Shaw', scraper conveyors to the bottom of the heading, where the coal emptied directly into waiting hutches at a loading point controlled by an oncost miner.

The loaded hutches were to be transported back along the bottom of the three parallel levels, and through the connecting mine to the **10 metres diameter loop road,** then be taken up the New Dook, endless rope, haulage-way and thence to the pit-bottom.

The development miners/brushers employed in this heading on the day-shift, were Bob Stevens from the Gallatown and Jock Moyes from Kennoway. The oncost loader at the bottom of the Cowley-Shaw was Wullie 'Shah' from West Wemyss, who was in fact, being helped by a young, fresh-faced lad called Jackie Ireland, who was all of seventeen years old and about to experience his third day down the coal mine.

On this fateful morning, Bob Stevens came down the pit as usual, clutching his haversack on one shoulder and a steel box containing explosives *(5lbs./20 by 4 oz. tubes)* in his other hand. He was waiting on the pit-bottom for the arrival of his neighbour, Jock Moyes, who was also required to collect the same weight of explosives from the pit-head magazine. Jock Moyes duly arrived, and Bob and Jock made their way to the New Dook man-haulage, where they were trundled down to their destination at the 9545 feet datum height, loop-road. They alighted from the man-haulage bogie and proceeded to make their way through the short, cross cut mine and onto the bottom level of the **parallel three**. On nearing their destination *(i.e. the bottom of 'their heading' at the Cowley-shaw )*, they beheld the vision *(in the near darkness!)* of Wullie Shah awkwardly perched on a make-shift seat made from pillar-wood, sitting head in hands, with *one bum* parked off to one side. Wullie was slowly shaking his head from side to side and moaning softly, blissfully unaware that he was being watched. Jock Moyes silently nudged Bob Stevens and both men stopped short of Wullie's predicament, so as not to intrude on whatever was the matter with Wullie Shah. Wullie did look up to see both Jock and Bob, and immediately but slowly arose, in a rather stiff-legged stance and proceeded to walk around in slow, short circles.

"Whatever is the matter with you Wull?" cried Bob, to which Jock echoed, "Are you all right Wull?" Wullie drew a sharp breath, but said nothing!, - but stood with legs apart holding on to a steel girder. "There must be something the matter with ye man, tae be staundin there like that!" "Come on man! Out with it!" voiced Jock Moyes. Wullie paused, and then softly said, "It's me bum Jock! - I've got a sair bum!, - and I don't know what it is! I've had it for a few days now and it's getting worse!" "Hell's Teeth man!" voiced Jock Moyes, to which Bob Stevens quietly said, "Easy there Jock, the mans' in pain!", but Jock, undeterred, but in a quieter voice asked, "Could ye no get yer wife to look at yer bum Wull?" To which Wull replied, "A man cannie ask his wife to look at his bum, it's no richt, it's no decent!" Bob and Jock just looked at each other, paused, as if to come to a mutual decision, then Jock said, "O.K. Wull! Bob and I will have a look at yer bum if you don't mind, and we'll look and see if there's anything that's no supposed tae be there!" "But mind you, Wull! We're no goin to touch anything!"

Wullie Shah considered this generous offer and reluctantly agreed to the two men having a cursory look. He looked along the level and satisfied himself that no one else was approaching, then undid his heavy, leather belt. He wincingly eased down his moleskins and gently slid down his navy blue underpants. He slowly bent forwards, holding on to the side of the Cowley-Shaw with the words, "Oh dear - Oh dear! - If it's bad, just tell me the worst - I'm fu'll prepared!" After Wull had gotten down to the examination position, Jock Moyes looked at Bob and silently invited Bob to make the first move. Bob Stevens demurred with a shake of his head and said, - "No Jock, I shall be quite happy to go along with your findings. Please go ahead!" Jock Moyes adjusted his helmet lamp and directed its beam to the near recesses of Wull's bum. There followed a brief silence which was followed by positive diagnosis, "You have got a cluster of ripe, dingle-berries attached to your fundamental orifice Wull, and I think you need a surgical procedure!" "I've got what?" said Wull, in a heavily resigned voice. "You have varicose veins on yer bum", said Jock. "In other words man, you've got piles!" Wull turned his head with the words, "My varicose veins are in my legs, but what's the matter with me bum and what's wi' aw' them big words your using?" Bob Stevens chipped in at this time and said, "Never mind thinking to understand all of them fancy words Wull, we ken what you've got, and noo, so dae you! So, Jock and me, we'll try and help you oot, if you're willing!"

Meanwhile, unbeknownst to both Bob and Jock, the young Jackie Ireland, quite unnoticed, had quietly arrived on the scene and when he spied Bob, Jock and Wull and, what they were doing! - he obviously didn't know what to think, he just stood still with eyes wide open and mouth agape! He has obviously never before in his young life seen such goings on! Bob Stevens sensing the disbelief and incredulity on the part of the youngster, quickly reassured him that everything was all right, but his 'specific help' was needed at this time!

Jock Moyes also assured Wullie Shah, along with the nodding of Bob Stevens head, that there was something that could be done to ease the situation, and since all the necessary equipment was to hand, emergency first-aid could commence! Two things then happened simultaneously, Bob instructed young Jackie to stand in front of Wullie Shah, who was directed to open his mouth wide when the 'operation' started - and young Jackie was instructed to look down Wull's throat and quickly 'shout out', if he saw anything moving. *(That took immediate care of young Jackie Ireland!)* Meanwhile, Jock Moyes had opened up the first-aid box and extracted a single, clean, white, 2-inch bandage, a tin of antiseptic cream and a box of sticky plaster. He prepared the wooden, seven-foot long, stemming pole, by wrapping some of the clean bandage around one end, - and using the sticky tape, he lightly affixed the remaining 'plug' of still-wound, bandage to the 'clean' end of the stemmer. He then opened the tin of antiseptic cream and liberally spread most of the contents over the two-inch long plug!

Jock then advanced over to the rear end of the still prostrated Wullie and said, "Are you ready for the treatment Wull?" Wull's answer was lost amidst his coat and scarf! Bob Stevens moved to the side of Wullie and held his arms tightly around Wull's waist, whilst Jock Moyes positioned the treated plug. "I'm going to start the treatment now Wull! I'm going to push these piles o' yours - back up your bum!", and then commenced the somewhat painful operation. Old Wull's loudly spoken words could be heard quite clearly: - "Oh dear! - O-oh dear! Oh- oh - oh dear - dear - dear!!" O - o - o - o - h - h - H! Just then, young Jackie shouted, "Stop! - Stop! - I can see something moving!" Bob Stevens cried, "What's wrang? - what's wrang?" Young Jackie replied, "Thair's twa things moving at the back o' his mooth!" Old Wull croaked, "Oh dear me, - 'shairly' not, - s-u-r-e-l-y not! - - Oh dear! O - O - O - Oh- dear!" Jock Moyes from around the area of Wull's rear end and expressed a loud sigh of satisfaction. "I've managed to push your piles up Wull, but keep that plug in place where it will do the most good, but dinnie move about ower much! Let the lad do the drawing!"* At that juncture, Bob and Jock turned to go up the heading to start their own day's work, but not before hearing Wullie's mournful sobs. "It's awfu' sair! It's awfu' - awfu' S-a-i-r! O - oh - it's burning! O - O - O - Oh - it's B-u-r-n-i-n-g!"

*Manual pushing (drawing/putting) of hutches from point A. to point B. along a light steel rail-track.*

Bob and Jock hurried up the short heading and prepared to get on with their shift's work. They were just a little late in getting started, and were quite also dismayed to observe that there was no 'standing' coal left in the heading. Their place was clean! *(Bereft of downed coal!)* The 10 feet wide, face-line needed to be undercut and the 'barrel-end' of the Cowley-Shaw needed to be advanced six feet nearer to the coal face. This however, was what they were contracted to do and both men set to work each at their agreed tasks. Bob Stevens commenced to 'break' the scraper chain, fit another 6 feet extension pan into the column and link-on another 12 feet of scraper chain. Jock Moyes bored the necessary holes into the pavement at the bottom of the face, in preparation of moving the coal-cutting machine and its cutting jib, under the coal face. Only one remark was made during this time, and that was to the effect, that old Wull would have fifteen minutes to rest his sair *(aching)* bum before the coal gum would begin to flow.

Within thirty minutes, both tasks had been completed with Jock Moyes just 'parking' the machine at the low side, when, up the heading came John Panton the deputy, who also acted as shot-firer. Bob Stevens had meanwhile, just finished shovelling the last of the gum onto the Cowley-Shaw. "Morning Bob!, Morning Jock!" said John Panton. "I saw the thin stream of the gum coming down the conveyor and I reckoned that you would just be about to start boring!" "Morning!" chorused both Bob and Jock, and "Aye" said Jock, "You have come just at the right time tae start stemming. I'll hae the first hole bored in a couple o' minutes!" - And with Bob's help, the drilling operations commenced. There was not too much conversation at this time, what with the vibrating noise of the 'Ram's head' and the constant spitting out of coal grit, from both the rotating drill and their dry mouths! Besides, neither Bob nor Jock took much interest in John Panton's extollations of the 'good clean life' - and his ripping experiences with the collection plate at last Sunday's church service . . . . . . *(Hallelujah!)*

As the last of the 10 holes was being 'charged' and stemmed, Bob Stevens made his way down the short length of the heading to warn both Wull and young Jackie that shot-firing was about to commence, thereby giving old Wull plenty of time to seek shelter from the blasting. A quick casual remark from Bob to old Wull, ascertained that the 'plug' was still in place and actually getting more 'comfortable' - if you please! . . . *And, let's say no more about it at this time!*

Within ten minutes, all 10 shots had been successfully fired and the portable electric powered fan and air-bags soon cleared the heading of its noxious fumes. John Panton then went on past Old Wull with a quick word of departure, nodded to young Jackie and proceeded on his way to continue his rounds. He would not reappear at this heading again until the end of the shift. This development only got 'fired' once per shift! Work then began in earnest. There was approximately 11 to 12 tons of coal and redd to be howked and shifted during the next five hours - and that would amount to probably 22 to 24 hutches being loaded from the Cowley-Shaw. Jock and Bob set-to with a will and with two No. 5 shovels being wielded non-stop, they reckoned to have it's 'back' broken before piece-time! The one advantage in working in this type of environment, was that they could change their working position from kneeling to standing, with no loss of momentum!

Piece-time came around *(this was signified by the sudden stopping of the Cowley-Shaw by old Wull)*. Both Jock and Bob decided to take their piece down beside old Wull and Jackie, if only to enquire as to Wull's present 'condition'. On arriving at the bottom of the heading, Bob and Jock sat themselves comfortably down, with their backs resting against suitably placed, pillar-wood sticks and opened their haversacks. "Before you ask", voiced old Wull, "Everything's all right - and it's a darned sight more comfortable than it's been for ages! . . . And that's enough said about that for the present, thank you very much!"

Sandwiches were munched with noisy mastication and tea was drunk with many appreciative slurpings. As was usual, there followed about 10 minutes worth of absolute calm and voluntary silence, as each miner savoured the next few minutes alone with his thoughts.

A few minutes later, both Jock and Bob dived back up the heading, they had approximately 4 or 5 tons of coal and redd to shift *{at this time and in this contract, all of the waste material (redd and blae) was carried away by the conveyor}* and then 'set' an 10' by 8' steel girder, before the end of the shift. The work proceeded apace and within an hour and a half, the coal was all gone and the Cowley-Shaw again extended. Only the girder to set, a couple of 'distance-pieces' to bolt between them and some odd bits of pillar-wood to tighten the girder to the roof . . . And that would see the end of another shift!

They had just set the last of the 'chock-wood' between the girder and the roof when they heard a loud, but tremulous voice calling up the heading - "Come quick! - come quick! Its come down, I can feel it! I can feel it in my drawers!" "Jesus wept!" cried Jock, "Is there nae end tae it!" Bob, sensing Jock Moyes's frustration, called down the heading. "Just hold on Wull! We're coming down! We'll be there in a couple o' minutes! Just H-O-L-D  O-N!!"

Bob and Jock locked-away their respective graith *(Pick, shovel, saw and heavy hammer)* and gathering up their haversacks, quickly made their way down to the loading point. Old Wull was just standing there holding on to the last filled hutch with a pained expression on his coal-dirty face. "It's come down", he whispered hoarsely. "I can feel it in my drawers!" Jock Moyes simply sighed! "There's no-one about" interjected Bob Stevens, "Let's have your drawers down so we can see what's up?, - or for that matter, what's not up!" - Laughingly! "It's no sae funny", lamented Old Wull, but at the same time gently easing down his moleskins and then his drawers. "I'll hae another look" said Jock Moyes and hesitantly peered at Old Wull's rear end. Old Wull turned his head, "Tell me the worst" he intoned. "Has it all come away?" "Behave yourself man!" - admonished Bob Stevens, "Jock will soon get you sorted out!" Jock Moyes straightened himself up and quietly said - "It's no that bad Wull, it's no all the way out. It just needs another wee push tae put it back". "O-o-o-h-h dear" said old Wull, I doubt if I will be able to stand it this time!" "Nonsense!" said Bob Stevens, "It's only got a wee bit to go, so haud yer wheesht!" *(Literally: hold your quietness!)* Jock Moyes had meanwhile taken hold of the stemmer again, liberally anointed the bandaged end with some more fresh vaseline and prepared to advance!

Just then, as Jock raised the stemmer to the horizontal position, there came from behind them, a very loud exhalation of released breath! Bob and Jock turned simultaneously, to be confronted by the sight of the deputy John Panton, standing with stick raised, his mouth gaping and his tired, old eyes staring out of his head. He was more than utterly speechless! He was agog! His mouth started to work open and shut, but nothing emerged. He appeared to be struck totally dumb! No one moved, not even old Wull, still leaning over the loaded hutch with his drawers around his ankles. Bob recovered first, and made to offer suitable explanation as to the simple explanation for the suspicious looking tableau. John Panton was having none of it, he was seemingly in no mood to even listen! But, recovering his wits, then broke into the most fearsome diatribe imaginable. He cursed them as instruments of the devil, he roundly accused them of the vilest of practices. He accused them of evil deeds and the grossest of obscenities, and heaped calumnies upon their heads. He also called for the wrath of the good Lord to be heaped upon them! He then made to move of, as if greatly disturbed by the spectacle of what he imagined he saw, and his parting words struck small fear into the hearts of the three bewildered and innocent miners, 'That in all of his years in the coal mines, he had never witnessed the likes of this!' - and that he was going directly up the pit to report every one of them to the manager forthwith and may the devil take the hindmost! *(Not intended as a pun!)* With that quiveringly spoken, damning threat, he quickly departed the section and disappeared from view, leaving all three miners' almost speechless and with heavy sinking hearts!

Bob Stevens, Jock Moyes and old Wull made their weary way up to the pit-bottom, closely followed by the young Jackie Ireland, and with a deal of nervous anxiety awaited their turn to ascend to the pithead. On arriving at the pithead, they made their way off the pithead landing and passed the Sparkie's shop *(electrical engineers building)*, heading in the direction of the check-box, but with somewhat heavy hearts. Sure enough, as soon as they started to cross the open ground, they were accosted by a stern looking, John Panton - who said nothing, but pointed meaningfully in the direction of Wullie Hampsons office. "Now we're for it!", muttered Jock Moyes. "We'll be sacked for sure!" "We'll just tell him the truth!", countered Bob Stevens, "Old Wull will back us up!" "That's enough! - you two!, screeched John Panton, "You can tell your pack o' lies to the manager. Now get over to that office!"

Jock preceded Bob into the office, where John Panton closed the door from the outside. Before either Bob or John could open their mouths, Wullie Hampson pre-empted them with the unimaginable accusation, **"You've got exactly one whole minute to explain and totally convince me as to why I should not send for the constable and have you both charged with attempted Buggery!"** Jock Moyes and Bob Stevens were rendered almost speechless, but were instinctively aware that Wullie Hampson was in no mood to listen to any explanation. They wisely kept their mouths shut until Wullie Hampson had vented his uncalled-for rage. After about five, full minutes of furious, non-stop, tongue-lashing, Wullie Hampson quickly sat down, semi-exhausted by his unnerving tirade. He was still angry, but looked as though he would now listen to explanations... Bob looked at Jock Moyes, but only saw helplessness in his eyes, but never-the-less motioned Bob to do the talking. Bob, given the only chance that he was liable to get, took the initiative and with growing confidence, heatedly explained about old Wull's embarrassing dilemma and his earnest request for immediate help. Jock Moyes repeatedly nodded his head and 'chipped in' to substantiate Bob's description of the episode, with Bob suggestively finishing up by saying, that if the deputy John Panton had been on the spot as a trained, first-aid, operative, the whole episode might never have happened! John Panton would have been in here instead of them, writing up the 'accident' report and the treatment administered!?

Wullie Hampson glared at the two miners, paused for an interminable moment or two, then snorted with what sounded like a sigh of relief, perhaps relieved that he wouldn't have to call the police after all! But, then gathering himself again to act the angry manager, he sternly admonished the two miners, to never again partake in such a practice and leave the administering of first-aid to them that knows better! He then ended the interview by the final gesture of opening his desk drawer, diving his hand into its depths, clutching and then throwing a thin, paper, booklet at Bob Stevens and shouting, - **"If you sorry pair wish to practice proctology down my bloody coal mine, then you'd better be prepared to write out the bloody 'insurance lines' as well!"**

*(Required official Certificate of miners' compensation entitlement in the event of a witnessed accident in the coal mine, and signed by the Colliery Manager.)*

\*

# Pastiche'

In the early Forties, the small village of West Wemyss may have been nearly isolated in terms of only one road in and out of the village, the lack of regular transport, and a very small population, but the worthy citizens of this quiet village knew exactly what was taking place in the Country and the World in general. They knew all about the 'Munich' accord and the meaning of that dubious word 'Appeasement!' They had heard about Neville Chamberlain and *'Peace in our Time!'*, on the wireless. They knew all about the Austrian upstart, Adolf 'Schickelgruber' and his dastardly henchmen, and they had heard about his designs for the greater good of Deutschland and the *Dritter Reich!* They had also heard of - and knew the meaning of *'Lebensraum'* and the demagogue's plans for future expansion. They also knew that they were having none of it!, - and just in case that pseudo Teutonic *'House-painter'* had any dastardly designs on the vulnerable and exposed coastal village of West Wemyss, the good folk of the village rallied together and showed the county, that they had the patriotic fervour and the 'willingness to serve', and to raise their own Local Defence Volunteer Force! They would show Herr 'Pickel-hauber' that it didn't pay to mess with cannie Fifers! It was a fact, that many local coal miners couldn't resist the 'Call to Arms!' and responded with steely resolution - *and a canny enquiry as 'what the wages were*?' Brushers, strippers, developers, roadsmen, motor-men, clippers and oncost workers, all became 'privates' in the embryonic Home Guard! They would all help to thwart the cowardly, devious and nefarious plans of Herr 'Schickel-gruber' und der *'dick-hund'* Herman!

## No! - Not my Dad's Army!

*Two members of the same family (father and son, and known as old Willem and young Wulson), worked together on the developing and brushing (in both coal and stone) in Lochhead pit - and were just as well known for the fact that they couldn't bear to work apart, as they were for the God's truth, they could not stop arguing with each other. Many a shift's work had been accomplished, with nothing ever having been passed between them except a pick or shovel, either that, or the mine air would be heavy with the sound of strained, unremitting, loud voices. Each one of them, loud in their belief, that they knew better than the other!*

*Usually, they were contracted to brush or develop the smaller size, levels or headings, working both stone and coal, where a two man operation was called for! They were very good craftsmen and could be trusted and depended upon to work unsupervised, usually well away from the hustle and bustle of the coal-producing shifts. Their work was very satisfactory and they were more than capable of resolving any unforeseen, work-related, problems that they might encounter. They lived in West Wemyss and were indigenous to the village.*

*Old Willem had not always been a coal-miner and seemingly spent some time in the Royal Navy prior to W.W.II. - where of course, he had acquired some basic grounding in discipline, small-arms handling and limited foot-drill. Old Willem obviously knew his 'straw foot' from his other one, and when the village L.D.V. was formed at the commencement of W.W.II. - old Wullem immediately volunteered! Very few of these volunteers had had any form of military training, even though, they were all to a man, keen as mustard. It therefore fell to the visiting regular Army* sergeant, to instruct them in the basic drills required of a foot soldier and rifleman. (Even though pick-helves were sometimes substituted for rifles!) As the drilling progressed, each and every potential, junior N.C.O. was given the chance to 'drill' and 'exercise' the squad, under the senior N.C.O's watchful eye, which of course, was a ploy designed by the regular senior N.C.O. to show them just how good and well-trained he himself was! He obviously took great delight in watching members of 'his' squad make absolute fools of themselves, in their efforts to assume the role of 'platoon sergeant!'

Old Willem stood nervously in the ranks, impatiently awaiting on his turn to be called out to 'drill' the squad. He felt, that with his 'background' and superior naval knowledge, that this army drill-thing would be as easy as 'Swabbing the Decks!' Old Willem's turn came at last, and he confidently marched to the front of the squad. Willem stood facing the squad whilst he himself stood at attention! (Or what passed for 'attention' in the Navy.) The squad were at the 'stand easy' position, and just as old Willem was about to issue his very first order - a small, but known voice behind him whispered loudly, - "You show them Willem!" Willem didn't turn round, nor did he acknowledge the offered encouragement. 'Why didn't the wives stay at home and let their menfolk get on with it!?'- thought Willem. Willem then raised his voice to what he thought was parade-ground volume, and hollered, - "A-T-T-E-N - - - - - SHUN!" ..... "S-H-O-U-L-D-E-R - - - - ARMS!"

How smart this initial movement was - I have no way of telling! And again, a loudly, whispered voice was heard to say - "Aye Willem! You show them!" Old Willem did not bat an eyelid, nor did he acknowledge the presence of his espoused supporter! Old Willem, still under the watchful eye of the regular N.C.O. then decided that it was time for some simple drill movements. Still standing at 'Attention', he ordered - "Right TURN!" - and 'intoned' to the Regimental 'pause time' - a shouted:- "ONE, two-three, ONE!" At which, some of the quad stopped dead during the execution of the order, not knowing what the 'numbers' meant! Old Willem grimaced! Not because half of them had got it wrong, but because half of the village was looking on! Old Willem then gave out with a very unmilitary order in that he quickly and loudly shouted at them - "All of you men face me - and stand still!" This time, a loudish voice was again heard from behind, that intoned, "You tell them Willem - yer dae'ing fine!" Willem merely shuddered at this third interruption. Old Willem then decided that he better do something quick to redeem himself in the eyes of the professional senior N.C.O., who looked as though he might just intercede at any time.

The squad were now standing loosely 'At Attention', facing the front with 'shouldered arms', patiently awaiting Willem's next command. Willem thought that his next command had better work - or this would be the end of his ambitions as a leader of men. He then ordered the squad to - "STAND at EASE, . . . . . . . STAND EASY!" Some of them did! Some of them didn't!, - but most of them just looked sheepish and did nothing! How could they possibly 'Stand Easy' with their 'rifles' *on their shoulders? Old Willem, was by this time, getting just a little hot under the collar, especially when the voice behind him was heard to remark - "They're just making you look saft (silly) Willem!" Much to the growing laughter of the 'civilian' half of the village.*

*Old Willem gritted his teeth! It was now or never. Everyone made mistakes! - but he'd show them! He ordered the quad to "Stop laughing!" He made them stand to attention with their 'rifles' tight to their right sides, with the butt-heel resting on the ground, and still in three ranks. Willem then decided that his next order would be the one that would make the squad work together as 'one' man without actually moving! He had decided to make them 'Present Arms' as a general salute and gave the necessary order, - or thought that he did! What he actually ordered was, - "Officer on Parade! . . . . . . . To the Front - SALUTE!" Which they all DID! Only this time, they all to a man, after just the tiniest pause, dropped their 'rifles' onto the ground in front of them - and raised their right hands to their foreheads! (Some of them with very beautiful, silly grins on their faces.) This was immediately followed with much heavy sniggering and suppressed laughter from the other half of the village, when that same known, but now plaintive voice was loudly heard to exclaim, **"Fur Christ's sake Willem! Can ye no get it Richt?"** Old Willem had now taken enough! Someone needed to be taught a lesson, - and now! He then did the one thing that he knew would work perfectly well. (He had been on the receiving end far too often in the past!) And to 'hell' with the grinning platoon sergeant. He stabbed two fingers forwards at the nearest two men and shouted, "You TWO, - Forwards March!" HALT! - and then pointing his arm towards his grim-faced wife, ordered the two men - **"Put that Woman under 'Close-arrest' - and mairch her to the Gaird-Hoose!"** Which they dutifully did!*

And thus forever blighted Old Willem's budding career as a potential N.C.O. in the embryo 'Scottish Home Guard!'

Author's Note:- *I have taken some small liberties in the re-telling of this anecdote, for the simple reason that some of the facts seem to have become distorted or forgotten. That the incident described in this story did take place and is basically true, there can be no doubt!*

*I have been told this same story by so many different old miners, that I have no reason to doubt it's veracity or it's pertinent facts. I have therefore, composed this anecdote from the many, slightly different, versions that I have listened to, with each teller swearing that his version was the correct* one. *I have also couched the appropriate dialogue in the local vernacular of the place and time! Also, in order to protect those family members still living, I have changed and 'disguised' the names of the principal characters.*

*

\* Author's Tailpiece: - During my last series of visits to this village in the spring of 2000, I could not help but notice that a few of the boarded-up houses in the narrow 'high' street were in the process of being partly demolished prior to their being re-built. Some of these dwelling houses were of course, very old buildings which were built from stone and mortar with very few house bricks in evidence. *(Some informed readers or villagers may find it strange that I should make comment on this, but I do so, bearing in mind that the W.C.C. did own and operate a successful 'brickworks' at Aberhill at one time, as well as owning most of the dwelling houses in this village!)* This new building and refurbishment project is probably long overdue, but is welcome nevertheless. This village is not quite as quiet as it may seem, as evidenced by the amount of motor vehicles parked along the narrow streets. Indeed, local parking seems to be something of a small problem, which inevitably, can only get worse within the confines of the narrow streets. For tourists to the main Belvedere hotel, *(the old ex-miners institute)* there is ample parking facilities, as there is for visitors, tourists and sightseers to the small, enclosed harbour along Coxtool road *(named after one of the seams of coal that lie directly underneath the village!),* with a beautiful view over the waters of the Firth of Forth. *(But only when the seasonal sea-haar allows!)* To the more adventurous of tourists who can tear themselves away from the sea front, a short walk along the narrow main (high) street will bring them to face with the village clock that seems to have stopped permanently at twenty-three minutes till two, on all faces. *(In reply to what seemed like a very innocent question on the part of a inquisitive tourist, one of the local worthies when asked what was wrong with the clock, was heard to droll what must have been classed as the most profound of answers, - 'Aye Lad! - **It's Stoppit !**')*

A little further walking along this road will bring a visitor to the single, village, nondescript ale house. *(Apart from the modernised Belvedere hostelry.)* If a visitor is looking for a bright, airy, modern, spacious ale house with up-to-date amenities, and an *al-a-carte menu* with customised lager-louts, then this is not the place to go! If however, a visitor is looking for a small, cramped, old-fashioned, dimly-lit ale-house with dark, varnished, wooden panelled, surrounds, then this is the ideal place to find a pleasant welcome, some good quiet company, a roaring wood and open coal fire, and a full measure of quality ale in clean glasses. A visitor to the 'Wemyss Arms' could also enjoy the company of old Jackie, a local who purports to be an anthropologist of the first grade and never tires of telling people that he is all of 85 years of age, - you know! *(Another tourist, hungry for local colour, once asked of a young village woman, "What is the age of the **longest standing edifice** in the village?" The young woman seemingly hesitated, then positively replied, "Aye, jist bide there, I'll gang into the 'Wemyss Airms' and fetch me Grandad !").*

\*

At the north-east end of the main street, the road terminates with a sharp right turn, which brings motor-vehicles onto a small round-about cum car park, that offers an unbroken, but limited view of Auld Reekie, Portabello and the wide expanse of the Forth estuary.

If visitor's were then to do an about-turn, they would see a single, iron gate leading into the small church graveyard, where a inspection of the headstones of the long-dead, will give a good indication of the common surnames of the past and present inhabitants of the village. Surnames such as Christie, Salmond, Coventry, Dryburgh, Anderson and Allen, to name but a few. This casual reading shall also give visitors an inkling as to just how long some of these families have remained within the village. Visitors shall also, without doubt, read from these same headstones, the names and ages of some of the very young, local lads, who lost their lives working underground in the nearby, but now defunct Victoria pit.

*

Part of the following short anecdote was 'passed on' to me by Mr George Gillespie, retired mining engineer (C.M.E.) with the N.C.B., now doing volunteer work at the Scottish Mining Museum at Newton-grange in the Lothian's.

Scenario: - *During the course of a visit to Lochhead colliery by George Gillespie, he casually asked of Willie Forbes the manager, as to the general character and quality of his inherited West Wemyss colliers, to which he was subjected to the last two lines of this under-written prose, which served as the inspiration for all of the fore-going verses!*

Come on, list on, ye Wast Wemyss folks, and harken tae this tale,
Pit doon yer pints and listen weel, while still yer minds are hale!

Hear this, hear me, ye Wast Wemyss joak's, you're entered tae the frae,
where none at all has ought to spear, which none at all gainsay!

Fill on, fill up, ye Wast Wemyss dours, all ye that gaither'd coal.
Nae mair yer neighbor's hutch tae tag, or the deil tae tak' yer soul!

Fecht yon, fecht yet, ye Wast Wemyss loons, its coming tae the day,
When aw' the coals is doomed an' lost, an' you're bereft yer pay!

Live on, live up, ye Wast Wemyss men, for ye shall coal no more.
From Pilkembare to Lethamwell, it's gonne' for evermore!

And so, and noo, ye Wast Wemyss folk's, its past the witchin' oor,
When aw' yer pits are closed an' filled, and coveret ower wi' stoor!

Fish on, fish far, ye Wast Wemyss trolls, till aw' yer lines gang rank,
Fur aw' the fish have upped and left, or swum tae Dogger's Bank!

Sail on, sail up, ye Wast Wemyss loons, for ye shall fish no more,
For aw' the bounty's deed an gone, or wash'd up on the shore!

And so, its true, ye Wast Wemyss folk's, thair's nothing mair tae say!
Of coal, and fish, and crusted salt, it's gone the spindrift way!

Swim on, swim doon, ye Wast Wemyss joak's, for ye shall swim no more!
For what ye thocht was 'Haich-two-Ess' in fact, was 'Haich-two, Ess-Oh-four!

\*

# Chapter LI

## Ex-Military Provender! - or, MINER'S *(canned)* WELFARE!

In or around the year 1946, just after the end of W.W.II, while I was a young lad of around 11 or 12 years old and living at home, I do remember a time when there was an excitement around the house, especially on the part of my mother. At this time, I did know that general food-stuffs were in very short supply and that everything except cold water seemed to be rationed. *(I distinctly remember having to get up early on some cold mornings, to take my place in a queue outside of a baker's shop, so that my mother could arrive just before the shop opened, in order to claim our daily bread ration!)* There was little of the everyday things that we seem to take for granted nowadays, things like bread, butter, eggs, potatoes and especially fruit. I seem to remember that the war had been over for just a short time, *(I can distinctly remember the jubilation felt by adults when my father came home with the news that tomorrow would be V.E. day!)* - but most comestibles were still severely rationed with no sign of the stable foods becoming any easier to get! I did not exactly know as to what this small excitement was about and was not really old enough to realise the significance of what was to come! Indeed! What was to come? I simply did not know and our mother would not tell us! *(I thought at the time, that she did not know!)* The coal-miners at Lochhead pit had seemingly been first asked if they wanted this 'small offering' and if so, then the basic cost to each miner would be approximately twelve and sixpence (12/6d or 62.5p).

This notice had no sooner appeared on the notice board in the pit head baths, when it was immediately oversubscribed with potential recipients. It seemed that the whole of the coal mine wished to benefit from this once-in-a-life-time offer. Because of this over-subscription, the management in their wisdom, then decided that the best method to ensure an even and fair distribution of the one-time offer, was to allocate the primary shares to those miners who were married with large families, with the remainder being divided evenly amongst those newly-married and single miners. The management was informed that no guarantee could be given as to the exact contents of the goods, only that a certain weight of goods would arrive at a time to be specified later!

Like everything else that was promised by the Government *(Ministry of Food)* at that time, waiting always seemed to play the largest part of any promise, with people 'going off the boil' in their anticipation. I know that Old Harry was constantly being asked as to just when the 'surprise' was due in at the pit baths? More often than not, old Harry just shook his head!

Then, one day it happened! On coming home from the night-shift early one Friday morning, we siblings were awakened by muted sounds of delight coming from the kitchen. We listened intently and discovered that the 'goods' had arrived at the pit head baths during the Thursday afternoon, and were to be distributed and collected on the Friday, from one o'clock onwards at the 'attendants room' within the baths. They needed this extra time to inspect the contents of the goods, stack them up into their respective categories and count the numbers in each large stack. That *count* would determine the exact allocation to each miner, who would then collect his share and sign the 'off-take' sheet. *(Payment to be deducted from their individual wages.)*

Old Harry went to bed just after 8.00 a.m, with the knowledge that he would be wakened at midday, just in time to have a large bowl *(or two!)* of thick vegetable soup, before he made his way back to Lochhead pit by bus. *(Not by bicycle. Old Harry didn't want to be involved any unforeseen accidents on the way home!)* Just after his midday meal, old Harry then made his way to Lochhead pit to collect his twelve and sixpence worth! As the afternoon wore on, my mother then began to glance more often in the direction of the mantel clock. *{I was attending Denbeath public school at this time, the gates of which lay exactly ten yards (nine metres) across the road from our front door.}* My mother was still clock-watching when I came home at 4-15 p.m. I remember being given instructions to 'wait!' when I asked what was for tea? By this time, I was beginning to sense that something was about to happen, nothing bad, but there was an air of expectancy that could almost be felt - and I had the feeling that it could involve the begetting of the family tea meal!

We were all sitting quietly in the living-room watching the bright flames of the coal fire, when my mother said 'Wheesht!' *(Hush!)* She sensed, rather than heard heavy footsteps coming through the enclosed space between the houses. Sure enough! It was Old Harry and he was carrying something quite heavy across one shoulder. He came into the house beaming wildly, his face split in to an enormous grin. "At last!" he said, "That was one big scramble! Everybody wanted some of everything and they aw' started fechtin!" *(arguments!)* "So! What happened, was that one of the gaffers took charge, to ensure that everyone got their fair share and no more! Some of the men wanted nothing more than cigarettes and sweets and no one wanted dried egg, spam or corned beef!" "The gaffer then said that each miner would only get one tin of cigarettes, one tin of sweets, one tin of every kind of meat and that they all had to take the dried eggs, corned mutton and spam as well, or they would end up with nothing!"

Old Harry then slowly tipped out the complete contents of a new hessian sandbag onto the living room floor. Every tin-can was exactly the same colour of Army khaki and showed a thick black arrow along with a description of the cans' contents! For the princely sum of twelve and sixpence *(62.5 new pence),* each miner had received approximately 25 one pound *(14-16 ozs # 390-448 gms.)* tins of Army rations, including a sealed tin of 50 Players, navy cut cigarettes. My mother would not allow anyone else to touch these tins, until she had inspected every one of them for they're labelled *(stamped)* contents. There was Irish stew, mutton - Scotch style, English steak and kidney pudding and Corned mutton *(produce of the Argentine).* There was dried eggs, potato powder and green and orange lemonade powder in heavily waxed, green packets. There were cans of meatballs in gravy and bacon and sausage, all duly stamped with the Stars and Stripes of the U.S.A. There was jam, marmalade, something called 'preserves' and a tin of grapefruit. There were boiled sweets covered in white powder, but without covering paper and 'cocoa' chocolate *(full of bits of grit!).* There was a tin of American peaches, a tin of grapefruit 'bits', a tin of evaporated milk - and two small tins of sweetened condensed milk. There were cans of Ginger pudding with fruit, American spotted dick in fat and a tin of cornflour. There was even a can with matches, water sterilisation tablets and 'toilet paper!' I had never seen to much food, all at the same time and, in our house! No wonder my Father was beaming, my Mother was smiling and we, the boys, just wanted our tea-meal! Old Harry took one look at the scattered tin cans, stretched out his hand and gathered up the can with the distinctive blue top. He would smoke one or two of the Player's navy cut now and eat the hot 'Irish' stew when it was served up with boiled cabbage and boiled potatoes, in some half-hours time! Old Harry was well content!

I do believe that this 'once only' inexpensive 'gift', offered to the Country's miners and their families, could have been described as belated recognition of their particular on-going exertions in the getting of the Nation's coal, and its contribution to the War effort, at a time when there was barely enough food to sustain the miners in their continuous struggle, to maintain coal output on such a meagre diet! I am fairly positive that it was a politician by the name of 'Ernie' Bevin, who had everything to do, with regards to whom the extra thousands of tons of surplus military rations would be offered to! I can however say, that in our home at least, the gesture was greatly appreciated and thoroughly enjoyed!

✳

# Chapter LII

## DANSE MACARBE!

## Willie's Pirouette!
## *– or Dancing in the Dark?*

At the time when Laurie Gibb worked on the Pit Bottom as a rope splicer and roads-man, when Walter Johnston was oversman to both the Old Dook and the New Dook as well as the pit Bottom, there took place an incident, that was nearly the undoing of Wattie and probably even threatened his 'appointed' position. At this time *(and during my employment in Lochhead pit),* the large, electric, motor-house that powered the Old Dook haulage was situated in the Howf at the Pit Bottom and, was a enclosure, separated from the dirt and dust and noise by its own tightly fitting door. The inside of this motor-house was painted in bright white, was spotless in appearance, with polished brass-work and white lines painted around the floor. The inside looked and smelled like the engine-room of a small passenger ship. The motor-man's sole official task was to respond to the bell signals received therein, and act accordingly in the starting and stopping of this stepped electric motor. He did have a good sufficiency of time between starts and stops, etc. So, spent and filled his working hours in maintaining the interior of the motor-house and keeping the brass-work polished. As to be expected, the conclusion that any discerning reader will immediately jump too, is that any such motorman must, even in the most generous of descriptive terms, be grossly under-worked. He probably was! Therefore, it can be more than readily guessed, that any miner who is fit and able and willing to labour, would not be employed in such a mundane, but enviable task. This was the sort of job that was given to a older miner with long years of dedicated service who had served the company well, without hassle and was probably disabled to some small extent. Such a miner was fully deserving of such a position and equally deserved the security, steadiness and clean environment, that was part and parcel of the reward for long service and loyalty.

Willie Anderson was such a miner. He had worked in Lochhead Pit for many years, until an unfortunate accident left him a little bereft and quite unable to follow his usual line of work. As befitting his long service *(and the need to earn a living),* he was probably given the motor-mans job in the 'howf' at the pit bottom, simply because the position was becoming vacant and he was readily available. By all accounts, he was a good and attentive motor-man who responded promptly to the belled signals and did not fall asleep. *(Induced somnolence was the main hazzard in this job!).*

Also at this time, there was serving in the New Dook, a deputy who had also given long and faithful service in Lochhead Pit. He was a very stout mature man, over-weight to say the least, who now found the sometimes 1 in 2.5 gradients on the New Dook, getting to be quite beyond his exertions. He was not yet of retirement age, but needed to slow down and still his overworked heart and lungs. He was past the prime of life, but was reasonably sound of wind and limb in normal conditions. He needed to do something less strenuous with much less ambulant activity. Feelers were put out, and a not so subtle approach was made to a certain oversman, who then approached someone just a little bit higher up. The wretched result, was that the disabled miner was turfed-out of the Old Dook Haulage motor-house, with the retiring deputy being installed therein. By all accounts, and not surprisingly, this retired deputy made an efficient motorman and did keep this motor-house well up to pristine standard set by his predecessor. This ex-deputy was the occupying motor-man during my time in Lochhead Pit, and was very adept at keeping out tired, dirty and cold miners at the end of their shifts, as they waited to be wound topside. This erstwhile deputy, now newly installed motor-man, shall remain nameless and plays no further part in this anecdote.

The question now arose as to what do with Willie Anderson and where to place him! This decision came under the jurisdiction of Wattie Johnston, so, it can be assumed that he either allocated an oncost job to Willie, or that he acquiesced in its offering. Either way, Willie found himself down at No. 5 level in the Old Dook. His task was that of assistant to the main dook clipper, in that he helped unclip the down coming empties and binched (slewed) them into the level. He then helped to 'binch and clip' the up-going full hutches of coal from the main level-haulage to the main dook endless rope haulage. This was the sort of task that was very work intensive and could be sweatily arduous during the out-rush of coal, and did involve much protracted leg-work of a straining nature. Willie did seem able to cope with this change of venue and occupation - and any lingering doubts that Walter Johnston may have had initially, were soon dispelled.

However, during the course of one fine morning on the day-shift, just as Walter was finishing his hot cup of coffee in the oversman's cabin in the Howf, one of the telephones rang. Walter reached out and picked up the appropriate telephone from the wall, well aware of the fact that the call was coming from somewhere on the Old Dook. He had barely time to place the instrument to his ear when he heard the agitated voice from the earpiece shout; "Get an ambulance-man down here quick! - There's a man with a broken leg!" Walter's senses immediately went into a shallow spin and he hurriedly asked, "Where?" to which the reply was, "No. 5 Level, on the plates". Walter wasted no time, he quickly grabbed the 'surface' telephone, got a quick reply and began shouting for an ambulance, a doctor, and a stretcher team to get down the pit and, 'Hurry! - Hurry! - Hurry !' Laurie Gibb, who by this time, had been alerted as to the unfolding events *(Laurie was a trained, underground, ambulance man at this time),* quickly went to the Old Dook telephone to ascertain the name of the injured man and the degree of damage to his limbs. Laurie had a level head and a deal of knowledge regarding mining injuries, and had a nagging thought, that the level of hurt and damage might not be as bad as first assumed. Men hurriedly reporting accidents, usually and unwittingly exaggerated the degree of damage at the unnerving sight of running blood. Laurie, quickly but calmly, first asked of the injured man's name, which he repeated, and was immediately overheard by Walter who had finished his call to the surface. "Willie Anderson" -"W-I-L-L-I-E A-N-D-E-R-S-O-N !" he stammered! Walter was rendered speechless! He also went into a partial funk! His mind was probably reeling! The disabled miner whom he had withdraw from the motor-house and sent to labour on the plates at No. 5 level, had now been involved in another accident and now, his good leg was broken and probably permanently damaged as well! Willie had obviously struggled valiantly on the 'plates' with his artificial right leg, but had thrown too much of a strain on his good left leg. But now, through his own sheer determination and physical effort, he had unwittingly succumbed to this devastating accident that was surely going to rob him of his future mobility. Walter's mind was racing and dejectedly, he sank his head in his hands and hid his face. This would be the end of his job, his career, his pension and probably his reputation. He was shattered!

Laurie was still on the phone, he was calmly but purposely, extracting information and details, but seemed to be taking an increasingly relaxed attitude to the whole accident, which did not fit the urgency of the situation and further exasperated Walter Johnston, who had not entirely 'given up the ghost'. Laurie finished up on the phone, turned around to face Walter and pointedly said, "You had better get back on that phone to the surface and quickly cancel all of your previous instructions, and when you have done that, I will give you the real details". Walter soberly did as he was directed, and after a hurried apology with more explanations to follow, he turned to Laurie and demanded to know exactly what was going on! Was there a miner with a broken leg? - And was it Willie Anderson? To which Laurie replied, "Yes! It's a broken leg. Yes! - It is Willie Anderson! And Yes! - He is all right! Laurie paused, whilst Walter recovered some of his composure. And grinningly, Laurie went on, "It's his wooden leg, he twisted it slewing a heavy hutch on the plates, and in doing so, the metal framework at the knee twisted and fractured, causing Willie to fall heavily. Therefore, to the clipper, his leg appeared to be broken". Walter Johnston seemingly breathed a huge sigh of relief with all of the images and thoughts of disaster magically disappearing, and re-asserting himself, forcefully grabbed the written details from Laurie's grasp. He was once again master of all he surveyed, and with returning efficiency and aplomb, rapidly dictated telephone instructions to the pithead, wherein, a motor vehicle was dispatched to Willie's home at the Earlseat cottages on the Standing Stone Road. This vehicle would be used to transport Willie's spare appendage to the pit head, where it would be placed in a down-coming hutch, suitably addressed to the pit bottom, where Walter would then despatch an encumbered oncost messenger down to the No. 5 level on the Old Dook. (tut suite!)

The motorised emissary to the Earlseat cottages was entirely successful and Willie's spare leg did appear post-haste at the pit bottom. Willie received his spare appendage well before 'lousing-time', and appeared at the pit bottom in grinning good spirits, with his severely 'broken leg' under his arm and in good time to be wound-up at the normal, shift change-over time. Walter Johnson did not appear on the Pit Bottom that afternoon to enquire as to Willie's health. 'Twas just as well, Willie's remarks might have just earned him the sack !

Author's tailpiece: - *I would like to think that the National Coal Board paid for Willie's new leg, but I just don't think that they did . . ! As for Willie's job on the plates . . . !?*

\*

# Chapter LIII

## PANZER POWER!

### An Awesome Extraction!
### *– or 'Begone thou Pick and Shovel'*

In the days when the hand-pick was 'King' and every shovelful of coal was counted, and twenty-three hundred-weights constituted a ton, the collier's depended on their own strength and that of their 'neighbour' (workmate) to fulfill a daily quota. If they didn't work hard enough, or couldn't get enough coal to fill, their daily wages suffered. They bored the holes, they 'blew' the coal, they hewed and broke coal, and they shovelled away this same coal. They set wood to the roof, they laid their own rails and they 'drew' their own hutches. They were, in effect, 'Universal miners!' They came up through a hard, but empirical 'school'. They could change from one type of work to another, almost on a weekly basis - without losing out on wages due to incompetence. They could be strippers one week, brushers the next week and preparatory, face workers the following week. They took the work as it came, gladly and without question, their main concern being steady work and a respectable wages packet the following week, whether it be day-shift, back-shift or night-shift!

Then came creeping mechanisation, followed by specialisation, followed by realisation and unification, and then, inevitably, - Demarcation!

*

The Dysart main was finished! No more Grounds & Nethers! No more Toughs & Clears!, and no more named coals within a seam! The 'King' that was the Dysart Main was dead! Long live the Lower Dysart Seam of Coal!

The Lower Dysart seam had been started into, within the section known as the Lower Dysart West in 1947-48, where both the top leaf and the bottom leaf of the seam had been successfully extracted over a wide area. Lochhead's continued future seemingly lay in the further development of the unfortunately split, Lower Dysart seam of coal. {*Split, or separated, in that there was a 30 to 33-inch band of soft blae (shale or redd) lying between the top and middle leaves of the 12 to 13-foot thick seam, with thinner streaks of stone and parrot coal interleaved between the middle and bottom leaves. (The seam showed three distinct leaves.) As the seam deepened, it also thickened, with up to two, extra, thin layers of coal intercalated into the seam, each with an accompanying thinner layer of stone.*}

As various sections were opened up in different parts of the pit to take this coal, the decision on which part of the seam to take, *(upper leaf, middle leaf, lower leaf, or a combination of any two?)* was seeming determined by the new coal-cutting/taking machinery, now in use and the proposed newer, more sophisticated, foreign machinery to be phased into Lochhead colliery. The A.B. 15-inch, long-wall, coal-cutters *(Bluebirds)* were being phased out as the main cutting machinery on long-wall faces, and use of the A.B., Short-wall, coal-cutters was being curtailed. *(This was not because of any inadequacy on the part of the machine, but merely because the 'Stoop and room' type of extraction in which this machine excelled, was being phased-out, as not being appropriate for the continuous extraction of the Lower Dysart seam! - See chapter on East Side Mine!)* These were being replaced initially by the revolving drum, shearer machines of various marks and sizes, *(initially British)* and eventually by the German 'Panzer' System. The A.B., short-wall machines, though retained for 'main-gate' development use with the new Panzer runs, would never again be used primarily as they had been, in the extraction of the Lower Dysart coals in the East-side mine. In fact, hand stripping as it was known, would almost become a thing of the past, as 'picks and shovels' for the most part, were almost becoming obsolescent. The 15-inch, 'bluebird', long-wall machine would still be needed and used in the stable-ends of 'Panzer-runs', where the coal was conventionally cut and stripped, but 95 percent of the flowing coal would eventually and finally, come from the teeth and jaws of powerful machinery such as the German, mechanical behemoth, known as the Panzer-loader, or 'power-loading' as it was soon to become known.

No more the three-phase system of under-cutting, boring, firing and stripping! All was to be combined in one fell swoop on an unsuspecting face-line. Miners were now to be grouped into select teams, with the strippers, packers, pan-shifters and steel-drawers now becoming 'Panzer technicians'. They were all there to serve the coal-eating monster that produced a never-ending stream of 'gum-like' coal and chirles, with nary a lump in sight! *(Later machines did produce larger coals!)*

This would also create a new breed of face workers, who, no longer were described by the face task that they originally did, but were grouped together under the collective title of 'Power loaders!' with each man in the somewhat, select team being paid the same ubiquitous rate! *(This included engineers and electricians who were part of the maintenance team!)* Obviously, this did not happen overnight, but the relatively short, transition period was almost frightening in its brevity - and far underscored that which had volubly transpired in wordier, former times. Miners had also latterly, developed the tendency to vote with their feet in respect of their claims and needs.

The coal developments within the Lower Dysart seam, almost encompassed the complete area of ground that had been covered by the underground extraction of the Dysart main seam, but with one main exception. To the best of my knowledge, there was a minimum of coal taken from the Lower Dysart seam within the rising ground to the west of the original **Pit-bottom level,** (*which lay in the Dysart Main seam)* and covered a large area, stretching from the Francis Barrier to the Buckhaven Dyke. Therefore, the areas in contention were, four, separate, coal faces (*runs/sections)* that were developed in the area of the Old Dook, and eleven different sections (*runs/sections)* in the area of the New Dook. The Lower Dysart West sections had finished producing coal in September 1960 and been left abandoned.

{*There were three, separate, working sections (coal faces) developed underneath part the area covered by the original Victoria pit workings in the Dysart Main seam, but these new, deeper extractions only covered a very small part of the whole of the area taken by the original Dysart Main extractions. In addition, the Lower Dysart coal taken from this area, was actually taken by miners working from the Francis Colliery in Dysart. These three sections were worked out by: - September 1956, February 1958, and April 1959 respectively. There were no through-road connections from these sections into Lochead colliery workings.}*

*

To put things into perspective for non-mining readers, who just might ask why variously named worked-out sections and roadways in different parts of the pit were being revisited, the answer is very simple! That part of the Dysart Main seam of coal to which the Pit Bottom was sunk into, remains 'as is'! The Pit bottom has not moved, nor has it been sunk deeper into the Lower Dysart seam, which sits approximately 25 metres below the Dysart Main seam. The steadily rising metals to the north-west (and therefore, the seams of coal) will ensure, that any new mine driven horizontally from the Dysart Main seam, in a north-westerly direction, shall positively strike into the Lower Dysart seam as it rises at an approximate slope of 1 in 4. This is exactly what the planners did, as there was only a 23 to 25 metres vertical difference between the two seams at that point in the strata. (*And within any of the workings of the coal mine!)* Therefore, once the planners and the miners are into a seam of coal, new tunnels, mines and roadways can be developed in any given direction required, to fully open up a new seam of coal anywhere within the mineral boundaries of the colliery. Obviously, existing furnished roadways already cut into the Dysart Main seam around the area of the pit bottom, were utilized and used to the full.

*

In the summer-time of 1955, at a point 375 metres down the coal haulage side of the **New Dook,** at datum height 9545 feet, there was commenced and driven, a straight and level, cross-cut mine on a map bearing of 335 degrees *(grid).* The cross-cut mine was driven through the metals for a distance of 125 metres, until it struck and intruded into the Lower Dysart seam of coal. Right from the outset, it was plain to see that the planners had envisaged this mine being used for the egress of coal, by rail and hutches 'to and from' the existing New Dook, haulage-way. This cross-cut mine had been commenced from the left side of the coal-haulage dook and turned through an anti-clockwise arc of approximately 100 degrees at a ten metres radius. It then continued on as the straight 125 metres long, cross-cut mine! However, this would only serve as a single-tracked route to supply empty hutches into the new section. It therefore still needed an egress route for the loaded hutches being returned for upwards transportation to the pit-bottom. This was accomplished by the commencement of an anti-clockwise, three-quarters circle, level loop-road, started from the right side of the coal-haulage dook, again on a 10 metres radius arc and being driven over the top of the main coal dook, until it arched left into the start of the straight length of the 125 metres long, cross-cut mine! Thus, a full, circular, loop-road had been developed. One quarter of which handled the empties traffic, while the other three-quarters loop handled the out-going loaded hutches.

At the 125 metres length point, where the mine had struck into the Lower Dysart seam, the miners/developers then changed direction and commenced to develop a level roadway, both to the right and to the left in the Lower Dysart coal, at the 9555 feet datum height. This level did of course, follow the exact, level grain of the strata, and extended 150 metres to the left, *(south and west)* and 300 metres to the right, *(north and east).* Back at the 125 metres strike point *(the innermost end of the level 125 metres mine),* a short heading *(at right angles to the new level at 9555 feet)* of approximately 50 metres length was driven up through the coal, from which two parallel levels were to the right were commenced, one at 25 metres and the other at 50 metres. This now meant that there were three, parallel, developing levels approximately 25 metres apart and running south-west/north-east. These three levels were at slightly different heights, with the first level at datum height 9555 feet, the second level at datum height 9575 feet and the third level at datum height 9595 feet.

These three levels, *(now to be known as the **Three Parallel levels**)* including the 125 metres long, level mine back to the New Dook, *(main endless-rope haulage)* would be the means, by which the start of the coal from the first three shearer sections *(coalfaces)* in the Lower Dysart seam in the east side, would be transported. This would only be a temporary transportation route at the start of these developments, until a more direct and shorter route and means of transportation could be devised. But first, the new development had to prove its worth in terms of large quantities of gettable coal, preferably in fault-free seams.

Before any of the yet to be developed 'shearer' or 'power-loading' faces opened up, there followed a short period of uncertainty of how to develop this Lower Dysart, coal seam between the New Dook and the Buckhaven fault, and stretching down *(and under)* to the north end of East Wemyss. This was a large triangular shaped area, that covered approximately 205 acres of coalfield. *(An estimated 2,500,000 cubic metres of quality coal!)* This development would surely prove to be the largest, single, continuous operation ever tackled in Lochhead colliery. . . . And so it was to be, but not just yet!!

The uncertain start to the opening up of this Lower Dysart coal section in the east side of the pit, took the form of two separate, but later inter-linked developments. The first development started from the extreme left end of the bottom parallel level at 9555 feet datum height, *(which was twenty-three metres directly underneath the line of the New Dook)* where the miners commenced to drive two headings exactly 30 metres apart, on a map bearing of 335 degrees. These two headings were driven to a length of 325 metres and touched the eastern edge of the now abandoned, east-side mine, extractions. *(Completion date January 1956).* At separate points and at left-angles *(west)* to the left-side 325 metres heading, the miners, using several A.B. short-wall machines, commenced to develop six parallel, full head-room headings at 60 metres, at 70 metres, at 90 metres, at 100 metres, at 130 metres and at 180 metres. *(These six headings were all parallel to the line of the new Dook!)* They took the full height of the top and middle leafs, including the 30 inches band of *redd* in the middle. *(This taken height amounted to perhaps eight and a half, or nine feet including the 30-inch band of blaes!)* These headings varied from 100 metres to 180 metres in length, from which everything *(including copious amounts of redd)* was sent down the conveyors. When the headings were terminated, the miners then turned sideways and extracted the coal between the headings in a series of parallel Stoop and room operations, leaving only a minimum stoop between each successive cross-cut. This then, productive as it surely was, could surely have been nothing more than an exercise in quick coal getting, whilst other developments were progressing. The rough date for this extraction was probably late autumn and winter of 1955. *(All of the coal extracted from the three parallel levels and the six short parallel headings, was brought down to and transported back through the 125 metres cross-cut mine, to be taken up the New Dook haulage-way.)*

Meanwhile, the other second development had already taken place. At the right extreme end of the bottom parallel level *(9555 feet datum height)*, a left-angled heading had been commenced and driven to a distance of 175 metres. *(Completed in September 1955.)* At a point 30 metres from the start of this heading and to the right, a second heading was commenced from the right end of the middle parallel level, *(datum height 9575 feet)* to run parallel to the first and driven to the same length. (*Completed in October 1955.*) A little later, a third parallel heading was commenced from the 9595 feet datum level, at a point exactly 25 metres to the left of the first heading and driven to the same approximate length. These three headings did not develop into a coal section, but they did pave the way for the first, small, shearer section to be developed in a northerly direction following the level of the metals.

The first long-wall coalface to be developed in this area, had its main-gate at the 9585 feet datum level and this face was only 75 metres long, but with a slope of 1 in 3.5. Each successive cut would follow the level metals which would barely rise, but the cutting slope itself was quite steep for bi-directional cutting. My information on the characteristics of the coal-cutting, was that this face-line was only ever 'cut' to the rise, with the shearer machine being 'flitted' back to the bottom of the run before the next cut was taken! This face did not have the more modern 'Panzer' machinery, but did have a system where the cutting machine and the endless steel conveyor (*chain-driven pans*) were separate entities! The coal-cutting machine (*shearing machine*) had a revolving drum firmly attached to the inside trailing end of the machine body and was 36 inches in diameter and 18 inches wide. The machine itself was held hard against the face of coal and sat hard to the pavement. As the coal-cutter was slowly powered up the coalface, the revolving cutting drum ripped the coal from roof to pavement and to a depth of 18 inches, leaving a smooth clean vertical face. This meant that the machine could be held tight to the face with each successive cut, thereby not losing valuable inches from the depth of the next cut!

Following along immediately behind and firmly attached to the coal-shearing machine was a metal 'plough'. This plough was held hard against the pavement (*by the sheer weight of the ripped down coal*) and 'lifted' the broken coal in a clean steady stream, raising and tipping it sideways into the ever moving conveyor, whence it was transported down the face to a loading point. As the machine and plough passed any given point, a few miners armed with shovels, cleaned away the extraneous coal that had either been missed or overflowed from the plough, in preparation for the machine being 'flitted' down the run prior to the next cut or rip. When this had been completed and the pavement adjacent to the cut face was clean and bare, the miners working in small teams and using portable hydraulic rams, would then, starting from the top, proceed to 'push' the flexible conveyor over to its new position, approximately one machine-width from the newly cut face-line, where it remained until after the next full machine cut!' Then the cycle of operations would begin again!

I do not know the exact date when this No. 1 face first started to produce coal, but an inspired guess would be during the spring of 1956. This coalface did last until April 1957, where they fully extracted the top leaf of approximately 43 inches of coal, including approximately 1 to 2 inches of stone. The length of the coalface did shrink to a minimum of 60 metres at its half-way point, but did extend to a full 110 metres width at its 325 metres, advance, termination point. When extraction of the top leaf had been completed, the developers went back to the half-way advance point of the previous extraction and commenced to go slightly deeper, *(underneath the previous top-leaf extraction)* to take the slightly thicker bottom leaf for a distance of 150 metres, at a face length/width of 110 metres. *(This was to done to by-pass having to work the face through the three fault-lines that had plagued the early part of the top leaf extraction!)* This bottom leaf extraction finished in October 1957.

This face was plagued by three separate faults, that showed-up as the face developed. This was the reason that the first face-line was reduced to a 60 metres length halfway through its life! It is also ironic, that out of the 11 coalfaces that were to be developed in this general area, this face-line was the smallest and the only one of two to be affected by faults! The amount of coal taken from this small section would have amounted to approximately 40,000 cubic metres. This however, was not really mixed *(lumps and such like!)* coal, except for what was taken from the two stable-ends. The coal would have been in the form of smallish and 'chirled' coals, *(smaller than mixed trebles)* probably best suited for power station consumption. This first coal section in the Lower Dysart East was terminated and abandoned in October 1957.

The coal that was extracted from the above described coal-face was probably transported to the Pit-bottom via a roadway at the 9555 feet datum level, out through the 125 metres level mine and thence to the loop-road on the haulage-way, at the 9545 feet datum height on the New Dook. *(Coal Dook).* All of the coal being carried and transported by 10 cwt. capacity, steel hutches. This method of transportation was to change in the not so distant future, when hutches were to be rendered almost obsolescent, with endless belt conveyance being the primary source of transportation all the way to the pit bottom. *(But, not yet awhile!)*

\*

Whilst the above described shearer-run was being developed, there was another development that was already underway, which was to negate the use of the New Dook, haulage-way, loop-road *(at the 9545 feet datum height),* and be instrumental in opening-up the larger 'power-loading' runs that were yet to be developed. At two surveyed points on the 9555 feet, datum level *(bottom level of the **parallel three**)* exactly 180 and 230 metres to the right of the end of the 125 metres level cross-cut mine, two new separate headings/dippings were commenced. The map bearing of these two headings/dippings was approximately 75/255 degrees. I describe these diametrically opposite bearings for the reason, that these parallel, twin dooks, approximately 38 metres laterally apart, were to be driven Up, as well as Down!  The driving of these newer dooks in the Lower Dysart seam was commenced in November 1955 and was partially completed by October 1959.

The development work to this dook's ultimate depth was not continuous, because two other sections on the upper part of the dooks were to produce coal before the lower coal sections were to be developed.  The line *(bearing)* of these two *(newer)* parallel Dooks in the **Lower Dysart coal,** lay approximately 220 metres to the north of the **New Dook** *(in the Dysart main)* and almost parallel to it.  These two separate pairs of parallel dooks, were of course, in different seams of coal, *(and separated vertically by approximately 23 to 25 metres of intervening metals!)*  The length of these newer dooks were 725 metres and 765 metres respectively.  The left hand dook *(looking down)* actually 'commenced' *(had been driven up too!)* at the end of the **dog-leg, level mine** that had originally been driven into the now abandoned **East-side mine** workings!  This now meant, that there was a direct communicating transport roadway between the newly opened-up sections in the East-side Lower Dysart seam and the Pit-bottom.  The right-hand dook *(approximately 20 feet lower down in depth)* was continued upwards for a further 75 metres, before making an acute left and then a small right turn, to join onto a previous roadway already cut through the **Dysart Main coal**.  This old, used roadway was part of the old return airway to the old Duncan shaft, from previous Dysart Main extractions.

*(The old **Dysart Main** workings to the north-west of the original Pit-bottom levels, was where Old Harry and Old Hugh were gainfully employed in the 'back-brushing' thereof, in order to maintain and re-build old roadways cut through the Dysart Main seam, at the commencement of the Century.  This roadway was one of the return air-ways from the original New Dook sections in the Dysart Main seam, but was now being used to vent the new Lower Dysart sections in the East side!)*

The second coal-face to be developed in this area, with the Main-gate at the 9472 feet datum height, was immediately to the east of No. 1 coal-face and fully adjacent to it!  The intervening long stoop left between these two coal-faces along their worked lengths, was to be between 10 to 20 metres.  The newly developed face-line was approximately 75 metres long and probably commenced production by late October 1957.  *(It was developed out to a full 110 metres at termination.)*  The chances are, that the same machinery that was used on this coalface, came from the previous coalface, along with most of the face workers.  The slope on this face-run was much steeper at 1 in 2.7, which indicates that this face-line was also cut only to the rise.

There are predictable, inherent problems with attempting to 'cut' downhill, not the least of which, is the tendency for the machine to run out of control to the detriment of men, machinery and face-supports.  This coalface, taking the 41 inches height of the top leaf, lasted from early November 1957 until early January 1959. - a total of 14 months.  The distance travelled over this period was 375 metres in length, which equates to 1231 feet. The amount of continuous days from start to finish was 426 days, of which there were 304 working days, minus approximately 21 days for holidays. That leaves approximately 283 working days, to travel 1231 feet of cut length, which equates to 4.35 feet of forward travel per working day. This would suggest, three separate cuts *(traverses)*, each of approximately 1.45 feet depth, with an 18-inch wide drum, all within each 24 hours period. Which of course, makes perfect sense! *(18 inches equates to 1.5 feet.)*

On the 9th of January 1959, the developers went back to near the start point of this extracted coalface, and dug into the bottom leaf of the coal at a point approximately 125 metres from the start of the top leaf extraction. *(This was to by-pass the same three fault-lines that had again appeared in the taking of the top leaf of coal!)* This lower coalface was only 85 metres long and travelled 225 metres from January 9th to October 8th, 1959. It had lasted just nine months. The total amount of coal extracted from both upper and lower leafs would have exceeded 60,000 cubic metres of coal, which contained about 5 per cent stone. An interesting point to note, is that on the working of this bottom leaf, the rate of advance amounted to an average of only 3.9 feet on a daily *(24 hours)* basis. This coalface was terminated and abandoned on the 8th October 1959.

The third coalface to be developed in this area, actually lay to the left *(west)* of the No.1 coalface, was higher in the strata and lay immediately adjacent to the eastern boundary of the now abandoned East-side mine. It's main-gate was commenced at the 9760 feet datum height, and its 120 metres long opening face-line lay exactly 40 metres from the line of the left hand dook, and parallel to it. This No. 3 coalface was quite unique, in that it was to advance across the rising metals at an approximate angle of 45 degrees. This meant, that the coalface would be cut to the rise and the line of advance would also be to the rise. *(This was most unusual, but not unsurmountable!)* The incline on the coalface was 1 to 4.5, whilst the slope of the advance was to be approximately 1 in 5.2. The development date for this third coal-face (No. 3 face), would suggest that it was commenced immediately after production had started on No. 2 coal-face. The start date for production however, *(the actual cutting and taking of the coal!)* was not until after the starting date on the bottom leaf, on the No. 2 coalface. The approximate date of commenced production on No. 3 coal-face, was the 9th of April 1959. The coalface had actually advanced a full 110 metres in length with limited coals being taken, but this was over a 15 month period of time, which suggests that the planners were in no hurry to open up this latest face. I would venture to suggest, that this was because there was an impending change of route for the transportation of coal to the pit bottom being developed, and this was to be at the top of the left dook, where it met the dog-leg mine into the old east-side coal section. At the point where the top of the left dook met the old dog-leg mine *(initially developed to take the coal from the east-side mine/section!)*, a 15 metres, radius, quarter circle roadway had been cut, connecting the top of the dook to the dog-leg mine, it being already connected to the pit bottom. This suggested the smooth turn of new railway track, or the gradual turn of new conveyor. Either way, a new coal transport route was in the making.

Initially, only the top leaf was extracted on this No. 3 coalface, with the face-line limited to a length of 120 metres. The coalface was advanced to a distance of 350, metres until it came up hard against the Buckhaven Fault-line *(the Dyke again!)* at the 9883 feet datum height. *(At the other side of the 30 metres thick Dyke, the workings of Wellsgreen Colliery within the Lower Dysart seam, had been terminated against the Dyke!)* The extraction of the top leaf, after it finally went into full production, covers an approximate total of 210 working days, that suggests an advance rate of approximately 3.9 feet per 24 hours.

The bottom leaf extraction commenced on, or about May 1960, with a face-length of approximately 100 metres. The main-gate remained in situ, but a new trailing-gate *(return air-way)* was brushed parallel to, but below that of the original one in the top leaf. This coalface in the bottom leaf, advanced only 150 metres over a five months period, before it too was abandoned, a full 75 metres short of the Dyke! *(There were reports of the extracted coals from the vicinity of the Dyke being too calcined to be of practical use!)* The rate of advance over this bottom leaf was approximately 5 feet every 24 hours. A decided improvement on the rate of advance over the top leaf? The amount of coal extracted from both coal faces in this No. 3 section, would have amounted to approximately 57,000 cubic metres.

As I have mentioned elsewhere in several chapters within this book, when a permanent level, a heading or dook is driven through a seam of coal, the policy is to leave a solid, continuous, stoop of coal running parallel to the development, which, to all intents and purposes musts remain inviolate. There are very few exceptions to this rule for the simple reason, that if 'weighting' trouble is experienced in such coal routes or haulage-ways, lost production is the first casualty. When two thick seams of coal are perhaps only 25 metres apart in the strata *(and covering a wide area),* the same rules still hold firm. The previous New Dook, cut through the Dysart Main seam of coal, had parallel stoops a full 50 metres wide, running either side of the twin dooks, and the new twin dooks (approximately 35 to 40 metres apart) in the Lower Dysart seam of coal, had parallel stoops running to the outside of each, to a width of approximately 35 metres. Therefore, the 'proximity' rule was adhered too! The resulting 'space' between the two sets of dooks was therefore, only a near parallel 150 metres wide over their respective lengths. The area to the rise *(west)* above the three parallel levels, had already been loosely worked out by an indeterminate 'Stoop and room' section, but there remained a small area undeveloped, below the three parallel levels and above *(west)* the Newton Farm stoop. *(A measured circular area immediately under the farm buildings left in situ within the coal seams so as not to cause subsidence and damage!)* There was the possibility of a small production coalface being developed in this area - and the planners went for it!

At a distance of 550 metres down the right side dook at datum height point 9475 feet, the miners commenced to dig a level in the top two leaves of the seam, at less than right angles to the dook and following the exact level of the coal. This level was started in January 1958, and driven only 40 metres in length. This would eventually be the trailing-gate/return-airway for this new development, but not at this time! The main-gate level, also to the right, was to be was commenced in August 1959, at the 675 metres point down the right dook, at the 9300 feet datum height. This roadway also followed the exact level of the coal and was driven in to 80 metres length exactly. At this point, the miners turned right and drove a ten-foot wide heading up through the 41 inches of the top leaf of the seam, to connect with the first level lying above, at its 40 metres intrusion point. This low heading was not exactly parallel with the right-hand dook, but it was now an 110 metres long, coal-face ready to be cut to the rise. *Discerning readers will immediately realise that this could only be another left-hand run. Not that any left-hand strippers would ever be needed! They were already a dying breed of miner. The coal-cutting machines (shearers) however, as they had been deployed on runs No's. 1, 2 and 3, had all cut to the right side, (looking from behind), but unless this new small run was to be 'sheared' to the dip, (a slope of 1 in 2, or 27 degrees!) it would have needed an modified coal-cutting machine.*

This, the smallest coalface to be developed in the east side, was opened up during October 1959, which suggests that some of the miners who had been employed on the now abandoned No. 2 coal-face, could be used to produce coal in the short term, until the development of the new, larger and longer 'Power-loading' faces, yet to be developed! This coal-face remained at a production length of 110 metres and over a period of nine months, it advanced a distance of 160 metres. The extracted coal was taken out by the lower main-gate, which had been extended to the left coal dook, whence it was transported up to the top of the dook and around the 15 metres radius, quarter circle, roadway, and out through the dog-leg mine to the pit bottom.

The amount of coal taken from this small section was approximately 18,000 cubic metres. The rate of advance of this coalface was approximately 2.9 feet per 24 hours, which suggests single shift working only, with two full cuts per shift from an 18-inch, shearer drum attached to the machine?

*

In April 1959, at the same 9300 feet datum level, just prior to the driving of the main-gate *(lowest level)* into the previous small section, another level had already been driven to the left side of the left dook, which was in fact, a pre-extension to this main-gate level. This new level *(now to the left of the coal dook looking down!)* was driven to a length of 80 metres where it was stopped. At this point the developers turned right and drove a short dipping to a depth of 40 metres and again stopped. At this 40 metres point, the developers turned right again, and commenced to drive back in the direction of the coal dook for a short distance, until it struck an existing roadway.* Simultaneously, back at the 80 metres point of the short level, directly opposite the short dipping, the developers had also turned left and commenced to drive a heading up through the coal, so that the new heading *(and new short 40 metres dipping)* became one long heading exactly 150 metres long, which was to be the new No. 5 face-line. The top *(high)* end of this new face-line, would cut directly into the old main-gate of the abandoned No. 2 coal-face, which would now become the return air-way for new No. 5 coal-face.

*\*This new, main-gate serving No. 5 coal-face, did not actually strike the left (coal) dook. It struck an 18 metres radiu,s loop road which had been commenced from the left side of the coal dook at the 9260 feet datum point, and described a 15 metres radius clockwise circle to terminate at the 9240 feet datum point on the coal dook, but on the right side. (A full 20 feet dropped difference in height, between the start and finishing point of the circle, with the two ends never meeting.) The loop-road was designed with slight clockwise declination, but it still looped over the top of the coal dook at its bottom-most point of arc. It was to be used for the ingress of empty hutches coming down the left side of the coal dook, and on to the first quarter of the loop-road, at a slight downwards slope. The clock-wise continuation of this loop-road was used for the egress of full hutches coming from the main-gate, around the remaining three-quarters of the route and also at the same slight decline. This meant that there was no interruption in the flow of either full, or empty hutches, with adequate space for the accumulation of either! This loop-road was completed by January 1960.*

Production from No. 5 coal-face commenced in September 1960 along an 150 metres, long, face-line in the top leaf of the seam, an increased thickness of approximately 44 inches, of which the middle two inches was grey stone. The upper leaf was successfully extracted from September 1960 until June 1961. The coal-face advanced a distance of 410 metres over approximately 190 working days, that showed a rate of advance of 7.13 feet per 24 hours.

*(This rate of advance was certainly the fastest to date and over the same sort of ground, but the question seems to be: - 'Was this the same coal-cutting machinery as used on the previous four coal-faces and now being used at maximum efficiency, or was there a new type of shearer machine being introduced to the coalfaces?')* Was this 'five cuts per 24 hours' with a 18-inch

diameter drum, or more likely, two and a half cuts, per two, shearing shifts?

On the day the top leaf was abandoned, the miners commenced to extract the bottom leaf, starting at a point approximately 40 metres inside the start point of the top leaf. This coal-face was approximately 10 metres shorter in length than the top leaf face, but with the same main-gate, and with a new trailing-gate/return air way a little below and inside that of the top leaf. This face advanced a total of 300 metres and was abandoned in April 1962. The rate of advance on this bottom leaf was 5.6 feet per 24 hours period. The estimated amount of coal taken from both leaves of this No. 5 section would have topped 110,000 cubic metres, with perhaps 4 to 5 percent of the top coal being stone and probably up to 6 percent of the bottom coal being stone. *(This was a partially 'dirty' seam in that the stone had to be taken with the coal, even though the coal itself was a clean good burning coal.)*

*

# The Advent of the Power-Loader!

## *(The Best Years of our Working Lives!)*

The two parallel dooks in the east side in the Lower Dysart coal, were now at a datum level depth of approximately 9220 feet, and they had served the transportation needs of five, separate, coal-faces. The depth of the left coal dook was terminated at this point and would not be extended deeper at its present direction. There would be a deeper extension, but the deeper continuation would actually be commenced from a point on the main-gate belonging to No. 5 coal-face, at a point exactly 20 metres to the north *(to the left!)* of the Loop-road. This extension road was to be driven much deeper on a map bearing of 112 degrees, and would eventually be driven another 635 metres longer, down to a datum height of 8770 feet. It would become the main transportation route for the greater volume of coal that was yet to come! *(Most of it untouched by human hand!)*

The right hand dook, that was approximately 750 metres long and also reached down to the 9220 feet datum height, was to be extended 'as is', but with a smaller change of direction - in that for the first 175 metres it would follow a map bearing of 105 degrees until it was approximately 40 metres apart from its companion dook, thence it would then run parallel to it, until it reached another dipping 620 metres distance, in addition to its original 750 metres length. This right hand dook now reached down to the 8770 feet datum height.

These two extensions were commenced in January 1961 and August 1962 respectively. The left dook would serve the needs of a further four, very large coal-faces and the right dook would cater for another two, large, coal-faces, but those developments were yet to be opened up!

\*

In January 1961, on the No. 5 run main-gate *(datum height 9255 feet),* at a point approximately 20 metres in from the Loop road, a dook was commenced as an broken extension *(i.e. not continuous.)* to the left side, 725 metres long, coal dook and continued downwards into the dipping metals. The map bearing of this extension was 112 degrees and the full, nine-foot height of the top, two leaves seam were taken, at a width of approximately 12 feet. The deepening of this dook was reasonable progressive in its development, in that it took six months, up to mid-July in fact, to develop a 200 metres dipping depth down to datum height 9083 feet. The declination on this stretch of the extended dook was therefore 15 degrees or approximately 1 in 3.8. The probability was, that this dook was developed by blasting from the 'solid' with hutches being filled by hand, as in 'Stoop and room' operations. *(Short-wall coal-cutting machines can be deployed in this dipping environment, but there is always the problem of the need for increased safety, in the 'parking' of the heavy machine during blasting and filling.)*

At this same 200 metres extension point during February 1962 *(at datum height 9083 feet),* another team of developers turned hard left and commenced a 'level' which they then drove in a straight line for 75 metres. This level road-way was taken to the slight rise and would eventually continue on in a straight line, to a total length of 575 metres, with a difference height of plus 92 feet. Therefore, the slope on this level would be approximately 1 in 21, *(sufficient to maintain the forwards motion of a loaded hutch, or to encourage the flow of water).* At a point on this level, exactly 70 metres from the left extended dipping, which the miners would now be under-cutting with an A.B., short-wall machine, the developers temporarily stopped the driveage. This 'to be extended' level was now to be the Main-gate for the new No. 6 coal-face! *(This was planned to be the longest coal-face to date in this Lower Dysart, east side!)*

Back at the No. 5 level main-gate at around October 1961, a different team of developers had commenced a dipping/dook, at a point approximately 75 metres in from the dook extension. This dipping was driven down in the relatively short time of three months, to meet up with the level that had been developed from the 9083 feet datum height. *(No. 6 main-gate.)* This dipping was fully 200 metres long and ran parallel to the dook extension, with a 75 metres solid stoop between them. This 200 metres long dipping, now that it was connected to the bottom level *(No. 6 Main-gate at 9083 feet datum height),* would now form the basis for the start of the next new coal face to be commenced, using the new bottom level as the main-gate, *(No. 6),* with the old No. 5 main-gate as the new return air-way. The No. 6 main-gate level would be cut, stripped and girdered and taken forwards well in advance of the coalface  using an A.B. short-wall machine, and the two stable-ends would also be cut and taken, slightly in advance of the coalface.  The reason being, that this was to be the first of the 'Power-loading' coal-faces, where 95 percent of the coal would be cut and stripped mechanically, with the mixed-size, ripped coal being conveyed down the wide, chain pans in an never-ending stream.  (*As long as the machine didn't break down!*)

This, the first of the 'Panzer' runs (*No. 6 coalface in the Lower Dysart East side*), commenced operation on 2$^{nd}$ April 1962, where the top end of the coal-face was shortened by ten metres, so that a separate, trailing gate/return airway could be developed away from, but parallel to the '*old*' No. 5 Main gate.  (*The 'trailing-gate' was actually driven ahead of the run, in that the Panzer machine needed an 'open' end of solid coal at both ends of the coalface in order to commence it's shearing cut! I do believe that later marks of this type of machine could be made to cut in both directions!  As to whether this particular coalface was cut using a bi-directional machine, I simply have no knowledge!*)  The coal to be taken was the 42 inches of the top leaf only.  The top 'trailing' road (*return airway*) of the run also took the 42 inches of the top leaf, with the addition of 25 inches of stone above and 32 inches of stone and coal below.  This meant that the top road was 'brushed' to a full height of 9 feet 4 inches, which suggests that 11' by 9' steel girders were used along its length.  (*This was somewhat unusual in that trailing-gates were normally brushed with 8 feet by 7 feet steel girders!  What therefore, was the reason for this?*)  The lower (*in the strata*) No. 6 main-gate was cut through the full height of the seam to include both top and bottom leafs, with the addition of an extra thin seam of coal that had appeared within the seam.

(*At this depth of strata in the east side of the pit, the make-up of the Lower Dysart seam of coal was now as follows, from bottom to top :   Bottom leaf:- 14 inches coal; 3 inches stone; 20 inches coal.  Middle leaf:- 7 inches stone; 14 inches coal; 16 inches stone.     Top leaf:- 26 inches coal; 2 inches stone; 12 inches coal.*)

The main-gate (*roadway*) always had to have a lower pavement than the top road, for the reason that extra depth was needed to accommodate the next (*daily extending*) conveyor system in the transport chain, on which the face coal was poured onto!  (*This would have been an 30-inch wide rubberised endless belt conveyor!*)

From the April 1962 until January 1963, this coal-face advanced a total distance of 500 metres over an 190 metres face-length.  This would have amounted to approximately 185 working days.  The rate of advance would have been nearly 9 feet per 24 hours, which equates to a daily production rate of approximately 600 hundred cubic metres of coal.  Considering the total length of the coal-face and the now increased rate of advance, it can now be seen that the installation of the new 'Power-loading' equipment was beginning to pay dividends.  The total volume that was extracted from the top leaf on this coal face, was approximately 110 thousand cubic metres over an nine months period. This allows for a five day week with 15 days holiday *(non-working week-days)* over this period of time. The miners on this coal-face were all now known as 'power-loaders!' They were all paid at the same rate - and in principle, were supposed to be somewhat ubiquitous, in that all tasks associated with the exigencies of 'power-loading', were deemed to be in the province of all members of the team, with the exception of course, of the machine-man who had his own deputy.

*(Events as later transpired, proved that this objective was somewhat of a farce in that some miners paid greater 'lip-service' to its principles, than to its actualities. Demarcation was again rearing its iniquitous head!)*

\*

The 'panzer-loading' equipment was deemed to be a great success. The volume of coal being produced came in good, mixed size and shape and was fairly easy to control. Lochhead colliery was now winding coal on both day-shift and back-shift, and the pit-head baths were fully populated and full to overflowing. Indeed, some miners were having to share lockers and the colliery had taken on a new lease of life! The coal was flowing, the miners were making money - and the manager was reasonably happy. He, the manager, no longer had to deal with small groups of strippers, pan-shifters, packers and steel-drawers on an individual basis. The contracts called for all miners on a power-loading 'agreement' to be treated in exactly the same manner, with an elected leader who dealt directly with the management! Could this for better or for worse? Coalminers had by this time realized - and were beginning to practise, that two moving feet cast a much stronger vote than a flimsy ballot paper!

\*

During the month of August 1962 when No. 6 coal-face was halfway advanced, the developers had commenced another dipping at a point on No. 6 main-gate level, at exactly 90 metres from the left dook extension, which itself, had now been extended and reached a new depth at datum height 8930 feet in August 1962. This dipping was taken down to a length of 210 metres and was completed in November 1962, where it met a new level that had been driven in from the left dook extension at the 8930 datum point height. This level had been driven in slightly beyond the junction point where the dipping had arrived, and this was to be the new No. 7 main-gate level. This 210 metres long dipping, between the No. 6 main gate and the newly driven No. 7 main gate as it now was, would now be the start of the next 200 metres long 'power-loading' coal-face, with the existing No. 6 main-gate being converted to the No. 7 coal-face return air-way, after the No. 6 coal-face, had been fully extracted* and abandoned.

*(There was a 10 metres wide stoop of solid metals (in this case, coal!) left between the advanced No. 6 coal-face and the advancing No. 7 coal-face, along the full length of the division. The old No. 6 main-gate roadway would be separated from the No. 7 return, extending air way by this 10 metres wide stoop, along their complete lengths. This was and would continue to be the standard practice in future, between each subsequent new, developing and advancing face-line!)*

*Author's Note : Only the top leaf of the Lower Dysart seam was now being taken in the East side with the advent of the new 'Panzer-loading' equipment, with no attempt being made to extract the bottom leaf of coal. This would be strictly adhered to, on all coal-faces from No. 6 to No.11 in this the East side. This meant that only 40 percent of the available coal would be taken! Whether this was a deliberate policy on the part of the management - I have not been able to discover!

No. 7 power-loading, coal-face commenced in the first week of January 1963 along a full 200 metres long face-line. The new 'top' road was also commenced exactly 10 metres below that of No. 6 main-gate and would run parallel to it, but using the start of No. 6 main-gate level as the return airway. The solid stoop of 'metals' that had been left between the line of the left dook extension and the start line of this No. 7 coal-face, was exactly 90 metres. The reason for this, was that the 200 metres long, start line for this face, lay exactly at the northern end of the untouched solid stoop, that safeguarded Newton Farm buildings on the surface.

The main-gate level for this No. 7 coal-face was also taken slightly to the rise with a slope of 1 in 26, which again, was ideal for the short-wall machines and the transport of coal. The incline on the coal-face was 1 in 4.2, which again, would help with the downwards conveyance of the cut coal. This No. 7 face was in production from the second week in January 1963, until the end October 1963, a total of approximately 202 working days. The total advance of the coal-face over this time was 563 metres, which gives a daily advance rate of approximately 9.2 feet per twenty-four hours. This in fact, was a small increase on the productive rate of advance from No. 6 coal-face, but I do not think that this was anything to do with a greater work rate, more likely the vagaries of statistics and how they were interpreted. The amount of 'coal' extracted from this No. 7 section would have been in the region of 124,000 cubic metres, which included approximately 7 percent stone. The main-gate at the bottom end of this face took the full height of the seam, which showed a breakdown as follows: -
*Bottom leaf:- 12 inches coal; 3 inches stone; 22 inches coal. Middle leaf:- 10 inches stone; 16 inches coal; 12 inches stone. Top leaf:- 26 inches coal; 3 inches stone; 12 inches coal.* The topmost leaf of the three yielded the most coal,

but only by a small margin.

*

During the months of April, May, June and July 1963, the left dook extension was lengthened down even deeper, from datum height 8930 feet to datum height 8765 feet, an additional distance of approximately 250 metres and in exactly the same direction. The right dook, parallel extension had been continued down only to datum height 8906 feet, where it was temporarily stopped. The left dook extension had now reached a length of 650 metres from the left coal dook loop-road *(at datum height 9250 feet)* and was stopped at this point, *(datum height 8765 feet.)*

Meanwhile, back on the main-gate level on No. 7 coal-face *(datum height 8930 feet)*, at a point exactly 50 metres from the left Dook extension, the developers turned right and down *(parallel to the dook extension)* and commenced, in the first week of April 1963, to drive another dipping down to a length of approximately 215 metres, where it was stopped and surveyed at datum 8788 feet depth on 30th August 1963. *(This dipping was not abandoned, merely checked!)*

At the 8780 feet datum height on the left dook extension, the developers turned 90 degrees left and commenced a new level, that was to be the new main-gate for the No. 8 coal face. This advancing level again took the full height of the Lower Dysart seam, using a A.B. short-wall coal-cutter to undercut the level. After only 50 metres of travel, this level met the 210 metres long *(checked)* dipping at the 8788 feet datum height. Thus opened up the 210 metres length of the new No. 8 coal-face. The shallow slope on this new No. 8 main-gate level was approximately 1 in 13, whilst the slope on the coal-face would be approximately 1 in 4.4. The first part of the old No. 7 main-gate level was now to be used as the return airway for the new No. 8 coal-face, with the return air now being channelled along a 'backwards' extension to the old No. 7 main-gate, along a new level cut below and underneath *(and at right-angles too!)* the left dook extension - and into the right dook extension and thence to be carried away.

Author's Note: - *The opening of this No. 8 power-loading face in the east side was the deepest the planners had gone to date with any fully mechanised section - and using this method of total extraction of parts of a seam over a wide area. The fact that the coal continued to flow from this long face without hindrance, was testament enough to the methods being used. Sufficient to say also, that no attempt had as yet been made, to re-visit any of the areas where the top-leaf extractions had taken place. (No's. 6 and 7 coal-faces.) It seemed, that no attempts to date were made to take the bottom leaf of the seam! It must also be remembered, that the metals (strata) between the two uppermost leafs of the*

*Lower Dysart coals were composed of a soft blae's and most unlikely to form a solid, strong roof at this depth. There was however, as much good coal below the soft fire clay as there was above it! There was also, an extra band of stone that had intercalated itself into the middle of the seam and, it seemed to be getting thicker! The coal therefore, was getting progressively dirtier! (Readers are reminded that this does not mean that the coal is of any less quality, but merely that the percentage rate of stone to coal has increased slightly!)*

\*

This No. 8 face-line commenced production in or around the first week in November 1963 and produced coal in a near steady flow until mid December 1964. The total distance advanced with this coal-face far exceeded everything else so far - and was therefore regarded as being the most productive 'power-loading' face-line to date. The total advance amounted to 765 metres travelled, with a 210 metres long, production face at a taken height of 45 inches, (1.15 metres). The thin band of stone within the top leaf of the seam had now thickened to a full 5 inches in thickness and had seemingly also moved to nearer the top of the leaf. The 'sandwich' now read: - *Top leaf:- 10 inches coal, 5 inches stone, 30 inches coal.* The volume extracted, therefore amounted to approximately 185 thousand cubic metres of coal, of which approximately 8 percent was waste. *(Redd)*

The rate of advance on this No. 8 coal-face was approximately 9.25 feet per 24 hours, which was slightly more than the previous average and showed a slight increase over runs No's. 6 and 7.

The extracted coal and stone from the main-gate was as follows: - *Bottom leaf:- 12 inches coal, 3 inches stone, 27 inches coal. Middle leaf:- 8 inches stone, 16 inches coal, 9 inches stone. Top leaf:- 28 inches coal, 4 inches stone, 12 inches coal. (These figures for the No. 8 main-gate extractions are in fact meaned figures, in that the thicknesses varied along the complete length of the main-gate and therefore, the entire 210 metres wide line of advance.)*

Author's Note: - From tests carried out on the relative weight to volume ratio, using the Lower Dysart coals, it would seem that it needed a solid volume of approximately 44 cubic feet of coal to weigh approximately One Ton! (2240 lbs imp.) [There are approximately 35.4 cubic feet in One Cubic Metre!]

\*

Back at the 8780 feet datum height on the left dook extension, where the developers had turned left to develop the No. 8 main-gate, another group of developers now turned 90 degrees to the right and commenced to develop the same No. 8 main-gate level in the opposite direction. This meant that the level road they were now driving was taken slightly to the dip, but followed the grain of the coal seam. This level to the right of No. 8 main-gate commenced development in July 1964, whilst No. 8 coal-face still had 6 months to run. At a point 50 metres along this new level (No. 8, but to the right), the developers again turned 90 degrees right and commenced to drive a heading, which looked as though it might connect with the right dook extension at datum height 8906 feet where **it** had been stopped! This was not to be, the heading was driven 205 metres upwards, where it stopped at a point adjacent to the right dook extension and was then joined to it, with an extra short level road approximately 20 metres long. (*There was no need for this new 205 metres heading to be in a straight line with the right dook extension, for the good reason that it was only ever going to be a return airway, and **not** a coal transport dook with a haulage system.*)

The development work on this the new No. 9 coal-face was completed by late October 1964, and it was immediately readied for the incoming miners and machinery from the No. 8 coal-face. No. 8 face was then abandoned on 14 December 1964, after the most successful of production runs.

Note : *The Main-gate for both No. 8 and No. 9 coal-faces (on opposite sides of the dooks) was one long continuous level, which was eventually 1735 metres in length and cut in an almost straight line, with a mid-point at datum height 8780 feet, on the left dook extension.*

Author's Note: - Of the eleven production coalfaces developed in the Lower Dysart seam in the east side of the Pit, eight of them were to the left side of the Lower Dysart twin Dooks. Every one of these coalfaces were terminated at their particular length, for the reason that they had run up against the Buckhaven fault. (The Dyke!) This fault line was of course, the natural boundary between Lochhead colliey and Wellsgreen pit in this area!

No. 9 coal-face *(Der panzer werkstatt!)* commenced slow production in December 1964 and then swung into full daily output by the end of that month, with the coal being taken out along the No. 9 main-gate and up the left dook extension. The downwards slope of this main gate level was approximately 1 in 12, and the slope on the coal-face was approximately 1 in 4.7. The **lie** of the coal-face was not at right-angles to the rise of the strata on this run, but at an angle of approximately 35 degrees to the left. This would mean that the face-line would always be sheared to the dip, whether or not the actual cutting was done to the rise, or to the dip. This would have made no difference to the coal-shearing machine, except perhaps in having the advantageous effect of 'pulling' the machine even closer to the face. *(Time would tell!)*

With regard to the developing main-gate on this No. 9 coal-face and its direction, it may be of interest to note that the line *(bearing)* of the main-gate, ran directly to the west of the fixed, underground stoop, that was left below the entire village of East Wemyss. This of course, meant that there would not be any more developments in the Lower Dysart seam, directly to the east and south of this No. 9 section. The planners had reached their maximum eastern boundary, that ran almost parallel to the advancing length of the No. 9 main-gate roadway!

This No. 9 coalface was by far, the longest advanced run to be extracted in this east side. It was also the first, great, coal-face to be driven to the right side of the dooks *(in the Lower Dysart seam!)* and taken slightly to the dip! The sheared length of the coal-face was exactly 205 metres and would remain so for an advanced distance of 675 metres. The main gate roadway then made a 15 degree turn to the right, to continue on for another 185 metres length, in this new straight line. *(This advancing coal-face had been in danger of encroaching onto the Michael pit boundary line. Both collieries did belong to the W.C.C., but the great stoops protecting the Michael colliery's deep shafts needed to be protected!).* This had the effect of gradually shortening the coal-face to a length of 165 metres over this 185 metres distance of travel, which took approximately four months.

This No. 9 coalface was in production from mid-December 1964 until the end of April 1966, a total of sixteen and a half calendar months. That would have equated to approximately 330 working days, including all holidays. The total ground extracted during this time, was over the longest distance that any coal-face had advanced in the east side to date - and indeed would remain so! The distanced advanced was exactly 860 metres, with a 205 metres face-line over the first 675 metres and an average face-line of 185 metres, over the final 185 metres length of the main-gate. The amount of coal extracted from this coal-face would have exceeded 198 thousand cubic metres, of which 7 to 8 percent was stone. This volume was therefore the largest total to date, extracted from any coal-face in the east side within the Lower Dysart seam.

The rate of advance on this No. 9 coal-face averaged out at 8.6 feet per 24 hours, which of course, was slightly below average, but still a good rate of advance at this depth. Again, the miners in the driving of the main-gate, took the full height of the seam, which, from  bottom to top, broke down as follows. *Bottom leaf:- 8 inches coal; 2 inches stone; 18 inches coal. Middle leaf:- **4 inches parrot coal**; 8 inches stone; 25 inches coal; 27 inches stone. Top leaf:- 28 inches coal; 3 inches stone 14 inches coal. These measurements are of course, an average over the full length of the main-gate and, it is interesting to note - the intrusion of a thin, but returned band of parrot coal at this depth.*

*Part of the reason for the continued good production rates for the last four power-loading, coal-faces was, that the miners did not encounter even one, single, fault line at any stage over their long advances. (The underground plans show 'clean' ground on both sides of the dooks!) As to whether the miners encountered any great roofing problems (localised falls of roofing metals), this is certainly not borne out by the production figures, which usually are the first casualty when extracting coal through bad ground!*

Author's Note:- The advancing main-gate from this No. 9 coalface (south and west), was driven on for another 75 metres length (as a steel-girdered MINE!) in the same straight line, where, in June 1966, it broke through into the 'Michael Colliery' twin dooks, that came from the direction of the Old Dook (in the Dysart Main coal.) and from the bottom of the new twin dooks that had already been developed* in the Lower Dysart seam to the south (left and parallel) of the Old Dook. These twin dooks, 50 metres apart and driven in the Lower Dysart seam and, coming from under the line of the Old Dook, were in addition to the two dooks already driven to the Michael colliery, in the Dysart Main seam. As already stated, all water from the Lochhead pit gravitated to the Michael colliery workings, via these pairs of twin dooks. Any water that might have collected, or was 'running' in any of the developments in this east side, was now free to gravitate down to, and along the length of the new, dipping, No. 9 main-gate, and thence to the Michael Colliery workings. This No. 9 coalface (main gate) being the deepest development to date in the east side of the pit and in Lochhead Colliery!

\* See: The 'Old Dook Re-visited'.

\*

The penultimate coal-face to be developed and opened up in this area, was not adjacent to No. 9 coal-face as might be supposed, but actually adjacent to and below *(length-wise)* the now abandoned No. 8 coal-face, back on the other side of the left dook extension. Back in July 1963, the developers had actually reached down to the 8765 feet datum depth with the left dook extension. At this point, the brushers/developers turned half-left onto a map bearing of 60 degrees and, continued to drive the dook downwards in a straight line for another 325 metres. This continuation was not as steep as previously experienced, in that the dip was only 1 in 20.

This 325 metres long underground roadway, ran almost parallel with the main road in the Village of East Wemyss, but was of course, 231 fathoms underneath it and 175 metres to the north-west.

At a point along the still open *(not crushed down)* No. 8 main-gate, exactly 260 metres from the left Dook extension and at datum height 8847 feet, the developers turned 90 degrees to the right and commenced to drive a dipping, which would be the start point for the new No. 10 coal-face. This dipping was started in October 1965 and was driven down for 205 metres to *(datum height 8710 feet)*, where it was stopped in the first week of December 1965. *(The 260 metres length along the old No. 8 main-gate and the 205 metres length of the new dipping formed a near perfect right angle, with the addition of the 325 metres dook length, which joined up the two ends. (The hypotenuse!) This then formed a right-angled triangle of approximately five and a half acres in area, which would be left completely untouched and unworked. I can find no reason for this triangular stoop being left unworked, but it may have been under the site of a potential new housing development at the north-west end of East Wemyss!)*

Author's Note: - In the driving of the 325 metres long, final extension to the left dook, the developers noticed a marked change in the make-up of the Lower Dysart seam. Several decades earlier, other developers and colliers had seen marked changes in the make-up of the Dysart Main seam of coal as it deepened out and under the Firth of Forth. Now, other changes were noticeable in this seam, in that other and mixed thin strata had been intercalated into its makeup! The following is a cross-section of the strata at the 8730 feet datum height, as discovered in the depths of the new twin dooks: -

*Bottom leaf:- 16 inches coal, 2 inches stone, 24 inches coal.*
*Middle leaf:- 5 inches stone, 2 inches coal, 3 inches stone, 18 inches coal, 7 inches stone.*
*Top leaf:- 27 inches coal, 6 inches stone, 2 inches coal, 2 inches stone 14 inches coal.*

It can therefore be seen, that the middle leaf had lost much of its coal thickness, with the addition of two extra layers of stone. The top leaf had also gained two extra leaves. *(i.e. one each of coal and stone!)* However, the seam itself seems not to have thickened, merely that some of the interleaved coal was being replace by useless stone!

Also in January 1965, there happened a strange little development for which I can find no answer, and a second, developed, heading parallel to No. 8 coal-face opening line, for which I can give good reason. At a point 60 metres along No. 8 main-gate level, the miners had driven a short dipping down to the right, for a distance of 20 metres. They then turned left and drove a level, parallel to the main-gate for 35 metres, before they turned left again with a short heading, until they re-emerged on the main-gate at a point exactly 35 metres from where they had commenced. This was a near square, loop-road for which I can find no tangible or logical reason. It must surely have been developed for a purpose, but for what? I know not! *(This little development **did** intrude into the five and a half acre right-angled triangle, but it was the only little development that ever did so!)*

*In April 1965, a heading was commenced from No. 8 main-gate (level), 12 metres to the left (south) of the start of the original face-line. (And approximately 35 metres from the left dook extension!) The original coal-face-line development (return airway) had obviously collapsed and the miners needed a new, return airway back to the right dook extension. This was it! - and it parallelled the old face-line, until it coupled with a connecting level between the high side of No. 8 and No. 9 coal faces. This connecting level ran underneath the left dook extension, giving a through return airway up the right side dook.*

This No. 10 coal-face commenced production in the first week of May 1966, immediately after the abandonment of No. 9 run, on the other *(right)* side of the right dook extension. Nowadays, there were very little teething troubles with these 'power-loading' runs, with regard to the technical side of operations. Things did go wrong and equipment did break down, but both engineers and electricians along with the miners themselves, were now very experienced in the vagaries of the systems, and very little time was lost that actually retarded production.

This particular coal-face was again slightly different from any of the previous runs, in that the rise of the metals did not really favour, either the machinery or the miners. The face line *(i.e. the cut!)* ran approximately 50 degrees across the grain of the metals *(and coal)*, whilst the advancing face *(inc. main gate at bottom end)* was at an approximate angle of 40 degrees to the level of the strata! The incline on the shearing face was approximately 1 in 4.5, whilst the slope on the main-gate roadway was 1 in 3.75. This was probably the worst possible scenario with regard to the cutting machinery, even though a slight declivity will always help the flow of coal. The face-line was 190 metres long, with the main-gate roadway at the deepest point in the strata. This was where the full height of the seam was again taken, using an A.B. short-wall, coal-cutting machine to undercut the short face. This was not the deepest coal that was taken from the Lower Dysart seam, but it was certainly the farthest distance from the pit-bottom. This coal-face at abandonment, was only 100 metres from the Buckhaven **Dyke** and lay directly underneath the **Rosie Village** at the north-east end of East Wemyss.

This coal-face produced coal in a steady, continuous stream from the first week in May 1966, until the first week in May 1967. That would approximate to 245 workings days, over a total face advancement of 675 metres, giving an average rate of advance of 9.1 feet per 24 hours. This rate of advance equates most favourably with previous figures, and clearly signifies, that both cutting and advancing into a regular slope had very little effect on production totals. If anything, it would have thrown an added burden on the 'power-loading miners' themselves, in having to constantly labour to the rise and manipulate heavy steel supports to boot!

This No. 10 coalface was abandoned on 9th May 1967, after a successful advance of 675 metres, through the 44 inches of the top leaf of the Lower Dysart seam of coal. The coal-face length remained a constant 190 metres throughout its advance, with a 10 metres wide, solid stoop left between No. 8 and No. 10 sections, which of course, ran parallel to each other. The output from this coal-face would have exceeded 145 thousand cubic metres of coal, although the stone content might have approached a full 10 percent, in that the top sandwich of the top leaf did thin down to approximately 9 inches of coal at the conclusion of the run.

This then, was the final coal-face/section to be extracted on this left side (*north*) of the left dook extension. The workings here had reached the maximum boundary.* There was but one area left untouched, where a sizable section could be developed, and that was where the planners next sent the developers. The area in question was to the right of the right dook extension, and lay, in geographic terms, adjacent to the west side of No. 9 section and immediately 'above' it in height. The initial problem with the earlier development of this section was, that its proposed starting point lay within the restricted stoop underneath Newton's farm buildings. However, the planners obviously obtained permission from somewhere, to have the restrictions lifted and a revision of the area took place, in that permission was given for the developers to commence a coal-face underneath the south edge of the Newton Farm buildings.

*\*This boundary line was in place to negate any further coal extractions to the east of No. 10 extraction area. Any further intrusions into the untouched reserves of deepening coal, would have eventually resulted in much structural damage to the dwelling houses at the north side of the A 955 trunk road, opposite the East Wemyss cemetery and the Macduff's, long ruined castle!*

\*

In mid-February 1967, a level roadway was commenced to the right side of the left dook extension, at a point exactly opposite to the old main-gate belonging to No. 7 coal-face, on the left side of the dooks (*8930 feet datum height*). This level road, a reverse extension to the old No. 7 main-gate, was driven at right angles to the dook, past the right dook extension and on to a length of 175 metres, where it was stopped, (*8885 feet datum height*). Prior to this drivage, a parallel road had also been driven to the right side of the dooks in December 1966. This previous road was exactly 210 metres above and driven to a length of 165 metres before being stopped. (*This upper level was at datum height 9045 feet.*) These two levels were parallel to each other and would form the main-gate and return airway for the last coal-face to be developed in this east side. At a point on the top level *(datum height 9045 feet)*, the miners had actually commenced to drive a declivity *(dipping)* to join up these two levels. This 210 metres long dipping, would of course, form the basis for the development of No. 11 coal-face. The very last in this area! *(Why this was developed as a dipping does not make sense? It is always much easier to work to the rise than to the dip, especially if large amounts of material are to be shifted!)*

This No. 11 face commenced production almost as soon as No. 10 coal-face on the other side has ceased functioning. The main-gate on this No. 11 face, did as normal, take the full height of the seam which had thinned somewhat, in that the Middle leaf now showed a reduced thickness of blae's. The breakdown of the seam in this area is as follows: -

*Bottom leaf: 11 inches coal, 2 inches stone, 20 inches coal.*
*Middle leaf:   7 inches stone, 22 inches coal, 17 inches blae.*
*Top leaf:     30 inches coal, 3 inches stone, 12 inches coal.*

These are of course, an averaging out of the coals over the length of advance, with the coal content of the top leaf gradually getting slightly thicker. It is however worthy of note, to record that the coal content of the 'middle leaf' did increase on this the right *(south)* side of the dook extensions, not that it mattered very much, the middle and bottom leafs of coal were never taken, on the last six coalfaces in this east side. The full height of the seam was taken only in the respective main-gate's of each new coal-face. (And sometimes in the trailing-gates/return airways.)

This No. 11 coal-face produced coal from May 1967 until July 1968, where the 'Final Survey' was carried out to end production in this east side. This face advanced a total of 630 metres over this fourteen months period, with a full 210 metres long face line. The incline on the face-line was 1 in 4.3 and the slope of the advance was approximately 1 in 25 to the dip over the first half of the advance, with the remaining half-distance levelling out completely. There were no fault lines whatsoever anywhere within the vicinity of this coal-face, the face slope was perfectly acceptable and indeed, was such that it would speed the egress of the sheared coal. The miners were of the most experienced in the colliery, having successfully extracted five, separate sections of this ilk with the same type of machinery, and all previously in the rich, top leaf of the seam. But, how would the rate of advance and the production totals from this section compare with previous extractions?

No. 11 coal-face produced coal from 10th of May 67 until the 12th of July 68, an approximate total of 285 working days including holidays. Over a total distance of 630 metres advanced, this would divide down to an advance of 7.3 feet per 24 hours working!? Was this lowered average a by-product of enforced demarcation, or did the miners see the writing on the wall, or was the machinery on its last legs? The extraction methods had served the management well and, had given a goodly number of miners some very welcome pay-packets over a fairly long period of time, and with some measure of security. But, had the bubble burst and had it all turned sour? Production was down by 20 percent with the inevitable rise in production costs, for no discernible reason at all, on this, the last coal-face in the area.

What had gone wrong? . . . If anything?

*The total amount of coal extracted from this, the final coal-face in the east side, exceeded 150 thousand cubic metres, of which 7 percent was stone. In production terms, this was a successful coal face in the amount that it produced, but it was far from being the most cost effective as far as the management was concerned. However, such things are beyond the remit of this book and needs must be discussed elsewhere, - if at all!*

The culmination of these extractions in the Lower Dysart seam in the east side, greatly reduced the coal output from Lochhead colliery. This east side had been a veritable bonanza while it lasted, for the six and a half years of the power-loading runs. It had produced something to the order of nearly 1.2 million (1,197 thousand) cubic metres of coal, of which approximately 7 to 8 percent was stone and fireclay. (*Some of the fireclay was of a quality that could be used in the production of light-weight, building blocks.*) Lochhead colliery would never again produce coal to the volume that it did, when the east side was in full production. The only section to have come anywhere near this in terms of overall production, were the three, long, coal-faces in the Lower Dysart Dook which were double-stripped, in that both the top and bottom leafs were taken. However, in terms of sheer output over the shortest time, the Lower Dysart West, Dook cannot be made to compare with the 'Power-loading' coalfaces in the Lower Dysart East - for the reason, that all coal from the Lower Dysart West, Dook was stripped in the old-fashioned way, by an A.B. 15-inch, 'bluebird', undercutting machine and by miners armed with picks and shovels!

*

Author's Tailpiece:- It remains something of an enigma to the miners who took the coal from the top leaf of the Lower Dysart Seam in the east side (New Dook area), as to why 'ways and means' could not be found as how to safely extract the equally rich, bottom leaf within the seam? (Leaving the hardish thickness of the middle coals as a solid roof?) The planners had managed to do this in the Lower Dysart West, Dook with good success - so why not in the equally rich east side, Lower Dysart coals? This could have added five or six good, productive years to the life of Lochhead Colliery, without the need for further new equipment investment.

(Could it have been because of the great variations in the thickness of the bottom coals and, having to use machinery that was designed to take an even thickness of between 43 to 45 inches in height?, - or could it have been that the bottom coals were just a little too dirty? I have no other explanations to offer!)

This was the very last of the Lower Dysart coal taken from the 'New Dook'. What further extractions there were to come, would now be 'taken' from the Lower Dysart seam directly underneath the worked-out Dysart Main seam in the area of the original Old Dook!

*

# Chapter LIV

## Hoch Dutch!

## The Power that Preserves!
## *- or, When in Rome . . ??*

When miners were required to approach the same onerous, mind numbing and sometimes dangerous tasks, every day of their working lives *(even though, no two working days can be described as being of a sameness!),* - it is not surprising that different miners approach nearly the self same tasks with utterly, different attitudes and varying degrees of forbearance. Some miners show unbending stoicism, completely unfazed by whatever fate chooses to throw at them - and simply accept the trials and tribulations of their daily grind. They have experienced mostly everything, that both the underground manager and mother nature has thrown at them, by way of low wages, inadequate materials, hellish conditions and sometimes exceedingly 'bad' ground. They endure this with unremitting patience and stout hearts. They do not waste breath and energy, railing at the inevitable, nor do they persistently oppose the near impossible! They get on with their daily tasks in the only way they know how and that means flexibility, adaptability, innovation and sometimes a modicum of dourness. All of that, along with keeping their own counsel.

However, there were a few miners, who, through knowing perhaps that there were no women employed down the coal mines when they were conversing, gave gesticulative and descriptive vent to their thoughts and feelings, no matter what! Their every utterance was invariably and liberally sprinkled with the most colourful expletives. It seemed that every second word that issued forth from their mouths was either an expletive, a curse or a blasphemy! This colourful language was not delivered in hot anger or unrelieved frustration, nor was it intended as any form of insult, either to the recipient or any listener to their purple emissions. It certainly was not meant to compensate for any lack of volubility or vocabulary on the parts of the protagonists, it was simply an unfortunate method and means of forthright expression. I would venture to suggest that most of the time, the culprits were only vaguely aware of their highly unsocial mannerisms. It was also not the case, that this profanity was reserved for just a select few. This purple tirade was directed at everyone with whom they came into contact with in their underground environment, whether it be fireman, gaffer, friend or foe! Everyone received the same treatment! It became such a habitual way of underground life to them - that they themselves, were hard put to realise that they were doing anything wrong, hence, never the need for an apology!

*(Strangely enough, it is not the first wife or mother who, when casually asked about their husband's or son's temperament (language) at home, replied, that they had never heard a wrong word uttered in anger within the confines of the family home!).*

This purple prose that emanated from these few miners was not in fact, confined to the depths of the coal mines. It may have been that most of these miners were somehow able to 'switch-off' from the coal mines 'habits' when they had left their place of work, in that the colourful language all but disappeared after the occasion of their metamorphosis through the pithead baths, at the end of their shift's work. In some cases, a contrary word would not pass their lips, until they had returned to the underground environment the next day. However, sad to relate, that there were a small number of miners who simply could not help themselves. It may have been that they never uttered a wrong word at home, but being out and about and 'socialising' was a slightly different environment, especially if they with some of their mates in the 'social club' or local pub, where they were known to be 'regulars!' When the beer started to flow, so inevitably, did the linguistic, purple prose!

In the area of the Wemyss's, coalminers were no different to their counterparts in any other coalfield. They worked hard, they played hard and they all, to a man, looked forwards to a weekend's rest, with some at least, looking forward to a few pints of cold, clear ale.

Conversations in the club, or pub, could be about anything and everything. No topic was sacred - and would be discussed with knowledge, forthrightness and vigour *(especially local politics).* Sometimes, conversations/discussions would get so heated, even to the extent, that the participants would stand up and forcefully gesticulate with constant repetitions in the direction of their argumentative adversary, just to make a point! This was usually when the strong expletives were added to loud retorts to emphasise an opinion! This purple prose was not usually only contained within the body of a phrase or sentence as an enhancing adjective, but was used as a noun in exclamation, to initiate each, and every, forceful retort! e.g. You C***!, You W**re!, Ya B*****d! Ya F****r! - would be the commencement of every sentence, with colourful adjectives interspersed for added dimension.

\*

Last year, on a warm summers' evening in the midst of the tourist season, there was several Dutch guests in residence at the Belvedere Hotel *(the refurbished and renamed old miners institute)* in West Wemyss. These guests consisted of several gentlemen and their ladies, whose knowledge of the English language was somewhat limited. *(Their knowledge of the local vernacular was even less!)* They were from that part of Holland where 'High Dutch' was generally spoken and where landed gentry built large country homes. On this particular evening, the lounge bar of the hotel was occupied by a small group of retired ex-miners, who, though no longer employed in the coal mining industry, had not lost their zest for life or living *(or politics)* - and unfortunately, had by no means lost any of their habits from former times. They could still out-drink any tourist and swear like troopers!

This small gathering was fairly innocuous at the outset, with contained and reasoned conversation within the group and little sign of outlandish language. The beer then began to flow, with everyone taking turns to 'dip their hands' in keeping everyone supplied with fresh pints. The conversation then, inevitably, became a mite more animated and a little more boisterous. At this juncture, there came into the lounge several of the Dutch hotel guests. At first, they quietly ordered drinks for themselves, without paying too much attention to any of the other 'guests'. After they had seated themselves comfortably with a 'drink' in front of them, their attention was curiously drawn to the group of ex-miners, who were obviously enjoying themselves and in the early throes of being 'taken' with their beer! The Dutch guests could not really follow the cross-conversations that were taking place, partly due to their lack of understanding, but mostly due to the rapidity of the Fifer's tongue, already heavily peppered with descriptive accusatory nouns! *(Expletives really!)* The guests seemed to be fascinated with the conduct and friendly banter of the miners, so much so in fact, that one of them attempted to join in their conversation and follow the 'discussions!'. He introduced himself as best he could, and was welcomed into the group with open arms. He did not say too much, but listened carefully! *(Trying very hard not to match these miners, pint for pint!)* He seemed enjoyed the company very much and laughed when he was supposed to! *(Even though no doubt, he was the butt of a few sly jokes!)* This Dutchman even paid for a round of drinks which, no doubt, made him 'one of the lads'. *(At least, for the evening!)*

The evening reached another very successful conclusion when most of the ex-miners who had consumed a 'skinful', took leave of their new found friend, who had merely to walk upstairs to *'get home!'* He had not really consumed too much of the Tartan Special, he had been pre-warned with regard to the potency of Scottish ales and had been 'advised' about the drinking habits of the 'locals!' He had however, added several new words of the 'English' language to his limited vocabulary and steadfastly, hoped that he would still remember them in the morning.

Morning came, clear and sunny and with it, a fresh breeze from the south-west that struck the West Wemyss shore with strong salt air. The Dutchman, having enjoyed a thoroughly good rest, felt really good this morning. He had risen early, taken a warm shower, had given himself a close shave and was ready for breakfast. He had deliberately risen just a little earlier than his friends this morning - with a view to tripping downstairs alone, to practise his new found English on the pretty receptionist, whose office was adjacent to the dining-room. When fully dressed, he left his room and proceeded downstairs with the hope that the receptionist was behind her desk. She was! He strode smartly up to the desk, stopped several feet short, clicked his heels together, and with his arms to his sides, produced a stiff little bow from the waist. In his best, newly acquired 'English', he then completely 'floored' the young woman with the fractured, but greatly misunderstood statement:-

"*Ya Hoor!* - Goot Mornink!, - und ja! - a *'Blooty'* vine day. Fur meine *'fookkink'* breekfust I haf komm". . . . . *Y-a-a  H-o-o-r !* ?

. . . . . . *As I Live and breathe!*

\*

# Chapter LV

## Dauvit's Progress IX

### The Return! - *Back to the PIT!* 1954

The transition from the Surface Dipping back to the Pit was utterly painless and merely entailed going onto the pit head at the commencement of the shift and then down the pit cage on the Monday morning, as opposed to being taken down the Dipping mine in a rake of rattling hutches. Our only previous given instruction was to appear in the 'Howf' on the pit bottom and report to Wattie Johnston, the Pit bottom Oversman. (*The Ex-Dipping miners had been split-up into various small groups over day-shift and back-shift and told to report to different deputies/ oversmen, after they had descended the pit.*) I had no idea about what sort of work I would be offered. I imagined that they would allocate work to suit each miners capabilities, thus precluding the need for short, initial readjustment or retraining!

When my name was called, I responded with a 'grunt', which was my first real introduction to the face of Wattie *(call me Walter)* Johnston. He informed Jackie Dryburgh and me, that we were now 'neighbours' and we were now also be classed as 'Stoop and room' collier's. We to be initially employed in the No. 3 *(water)* level down the Old Dook. We would continue to work on constant day shift into the near future, but would work alternate back-shifts as the work-places developed.

I didn't know Jackie Dryburgh at all, and he didn't know me. Our paths had never crossed prior to this meeting and I cannot remember if we had ever nodded to each other in the passing. We were an unlikely twosome, Jackie was all of 5 ft - 7 inches and I was all of 6 ft - 2 inches. I did hear someone pass a remark, (*I was later told that it was Wattie Johnston of all people!*) that, as a mismatched pair, we were the perfect combination for the 'high and low sided' levels, as commenced from both sides of the Old Dook.

I did know where the Old Dook commenced, but was quietly steered by Jackie, past the head of the Old Dook, on through the start of the Lower Dysart mine to the head of the old companion dook, in which lay the man-haulage, that commenced adjacent to the first adjoining No. 1 level between the two dooks. (*The main coal-dook and its companion man-haulage dook.*) I had never been down this dook, but Jackie obviously had and knew his way down to No. 3 water level. After we had descended and, as we were the 'new miners' on the block, the

Old Dook deputy followed us through the interconnecting level at this juncture, to show us the exact point of intrusion into a near virgin, coal stoop to the left side of the dook, where we were to commence work.  A horizontal level had been commenced from the immediate side of the haulage-way, but, by the looks of the downed coal and the amount of gathered dust, it must have been in the year 'Dot!'  The 'level' was exactly five metres into the solid and had lain dormant for 'X' number of years, but was now to be opened up by Jackie and myself, using nothing more than a 'poker-drill', explosives and picks and shovels.  At that time, I simply could not believe that this was a viable proposition, where we were expected to 'make wages!'

The five metres, initial intrusion to this level was just enough to enable us to lay down two, six-foot lengths of light rails and sleepers, with enough space to unclip two, down coming, empty hutches and 'binch' them inside this short level. (*This was very easy to do, I used to be a 'demon' clipper!*)  One hutch would be 'couped' to one side, while the other hutch was placed ready for filling, but only when we downed enough coal to do so!  The 'plum-lines' had already been surveyed-in, so it was a simple matter to drop the twin lines from the roof, look through the line of the vertical plumbs and mark the centre point of the new, coal face.  This done, the plumb-lines were then tied clear to the roof and Jackie then measured off six feet to either side of the centre point, to delineate the left and right (*the higher and lower*) limits of beyond where, the shot-holes must not be bored.  Jackie carefully eyed the face and with my 'holing' pick, selected ten carefully placed sites where the holes were to be bored.  The incline on the main dook at this point was downwards from left to right, at approximately 1 in 3.5.

The coal face to be 'bored and fired' was approximately 12 -13 feet wide and approximately 7 feet high.  (*We were to take the approximately 65-inch thickness of the Grounds & Nethers, a few inches of stone, the 10 inches of Myslen and a further inch or two of stone.  The Toughs & Clears, the Spar-coal and the thick Head coal, was to remain in situ as a solid and unbroken roof.*) Using this antiquated 'poker' machine - (*this was something akin to slowly pedalling a bicycle uphill with your hands and arms!*) and expending a little sweat, Jackie bored two holes, approximately two feet apart, right in the centre of the coal, with a further two at the same height, but near the extreme edges of the measured width.  He bored three holes along the bottom, one in the centre and one at each side, and then finished with three holes along the top, just under the 1 inch of stone, one hole again in the centre and one hole at either extremity.  He explained to me that the different coals were not all hard, some merely brittle, which made them susceptible to crushing.  He also explained that the less hard coals would be blown out first.

Jackie then explained to me, that calling for the shot-firer in this environ-ment was not quite the same as being on a long-wall face!  Here, the deputy was the shot-firer and he visited each pair of collier's as part of his twice-daily rounds. Usually, each working face only needed to be fired once during the course of a shift, with sufficient coal blown out and down to occupy the miners for the whole of the shift and produce enough coal to fulfil each pair of miners' daily quota. All it needed, was for us to fit our firing sequence into his proposed visits, so that he need not alter what was a specific daily routine. If we did not acquire sufficient coals from one round of 'shots', the answer was to drill the holes perhaps six inches deeper!

*This was my first experience of this type of firing sequence and I must admit to more than passing curiosity. The Deputy's name was Bob MacDonald and, as he reamed each hole to test for breaks, he explained exactly what he was doing and how he calculated the correct amount of explosive for each hole. He also demonstrated how to puncture the grease-proofed wrapped 'sticks' of gelignol and how to insert the detonator. I noticed that these were not electric detonators, but came with a six feet length of safety fuse (strum) inserted into the detonator, to initiate delayed disruption. When all ten holes had been carefully charged and stemmed (packed firmly with malleable fireclay), Jackie and I were sent to separate points, both up and down the dook, acting as sentries, to deter and deny entry all intending trespassers and bar movement until he (the Deputy) gave the all clear! What made this event all the more interesting was, that the deputy retained my carbide lamp and I was left sitting in the dark in a suitable man-hole, twenty metres down the Dook. (I did not need to 'see' to carry out these duties for the simple reason, that if there were any miner's abroad at this time - they would be carrying personal lamps!) He had also explained to me that the firing sequence was very important and showed me exactly how it was to be perpetrated. The two centre holes in the 'Nethers' would be fired first\* - followed almost immediately by the two centre side holes in the Nethers. This would 'remove' the centre of the 'sandwich!' The three bottom holes in the Grounds would be fired next and almost simultaneously, to heave up the 'binch'. The last three shots in the Myslen would also be fired together, bringing the whole lot crashing down into the already large and broken heap of mixed coals. This made perfect sense to me and needed no further explanation, but I was rather curious to find out just how long it would take for him to traipse in and out of the level to 'fire' all of these shots? Especially where he had to light (ignite) them by hand! My unspoken question was answered within minutes. Three to be precise!, - and the job was completed!*

\*Author's Note : *This deputy and others of his ilk, especially those involved in Stoop and room working in the Dysart Main coals, had a vast empirical knowledge of how a whole solid coal face should be fired to obtain the maximum volume of broken coal.*
    *The miner's themselves would site and drill the holes, the knowledge of which,*

again, came from long, hard-won experience. The miner's would also know exactly to the ounce, just how much explosive was needed to disrupt the coal without smashing it to smithereens. The six feet lengths of strum (safety fuse) attached to each detonator had a very regular 'burning rate', which could be timed almost to the second. That burning rate was 12 inches every 30 seconds. Therefore, a six feet length of strum with a detonator attached - would take exactly 3 minutes from 'lighting' to detonation. This regular burning rate of the strum, was therefore utilized to initiate the sequential firing order, determined by one visit. The six feet length of strum on the middle four shots (Nethers coal) would be trimmed by 24 inches with the ends slightly, but deliberately frayed. The bottom three lengths in the Grounds coal would be trimmed by 12 inches, with the ends again slightly frayed. The length of the strum on the top three holes in the Myslen coal would be left at their full length, again with deliberately exposed frayed ends.

**The Coal-face was now fully prepared for blasting!** The deputy would now post sentries at separate fixed points, both up and down the main dook, where they would take appropriate shelter and remain clear of the moving endless rope haulage. The shouted alert from the deputy, "Firing in the Hole" would be echoed by the sentry's - "Firing in No. 3 Water-level!" At that point, the deputy would 'light' the 10 dangling lengths of strum, in strict sequence and with naked flame. (*Successful ignition of the strum would be signified by the classical spurt of flame from the frayed ends!*) This was commenced with the centre two shots in the middle layer, followed by each of the middle side shots, followed by the three bottom shots. Then lastly, by the three top shots. The deputy would then quickly confirm that all safety fuses were burning - and then, he would 'walk' out of the level, out of the firing line, and then take shelter to await the numbered explosions. He would wait until the first explosion, where we would all individually, count the number of detonations as they happened and hopefully agree on a correct total as the reverberations ceased. We usually had to wait for about five minutes until the stink of explosives and fouled air had dissipated into the airflow, before we could re-enter the level. The deputy would always accompany the miners into the face of downed coal. He needed to see the effects of the placing of the holes and if the correct amount of explosive had been used. He would also check for misfired detonators and unburnt explosives. (*This was a very rare occurrence!*)

This was the point in time when we really got mucked in! There was probably up to fifteen, cubic metres of downed coal, heaped and scattered across the sloping face, with most of the coal lying on the laigh (*low*) side. Our first task was to clear the rail-track, to allow us to get an empty hutch into the heart of the heap, to ensure the shortest distance between coal and hutch. The heap of coals, if the shots had done their work properly, would consist of large lumps down to small coals, with all sizes in-between. The general idea when filling each hutch, was to try and balance the filling, so that no great empty spaces were left in the heart of the hutch through loading too many large lumps together. This balance ensured that the hutch's capacity was fully utilized to carry the maximum weight, which was approximately ten hundred-weights *(10 cwts.)* for a level hutch, or up to twelve hundred-weights *(12 cwts.)* for a 'built-up' hutch, using middle sized lumps around the inside periphery. This was where Jackie and I chose to differ. I preferred to fill a hutch fully, but only just above its level rim. *(The contents of the loaded hutch would shake down during the journey to the pit-head!)* This ensured that we were paid the absolute maximum for the minimum weight of 10 cwts of coal. We could not be paid less for a topped level hutch. Jackie preferred to line the rim with medium lumps, then fill-up the inside, to ensure a given 12 cwts. at the surface, check weigh-bridge. We compromised by filling two level hutches, then followed that by two built-up hutches in turn.

*(It needed to be done this way, as each individual miner has his own identifying numbered tab, which is alternately attached to each filled hutch before it leaves the coal face. This loaded hutch, along with many others, is transported to the surface, where it is wheeled onto the check-weight bridge on the pit head. The hutch is weighed, the tab is removed and the weight credited to the hopefully mounting total for each, individual collier employed in this type of contract. It can be therefore realised, that the hutches were filled as matched pairs!)*

One of the daily rituals practised by all collier's working in this environment, was to visit the numbered rack on the pit head, standing adjacent to the check weight-man's office. This was where all the colliers' recovered tabs were hung in bundles, to establish *(hopefully)* that each and every tagged hutch had been accounted for. Quite often, tagged hutches 'slipped' past the weigh-bridge without the tab being removed, this of course, to the detriment of the wages of the individual miner. Jackie and I were never too troubled by these oversights, but were sometimes forced to stand our ground with a recalcitrant check weight-man, who occasionally doubted our 'honesty!?' *(He had the most disconcerting habit of looking directly at any doubtful claimant over the top of his half-moon glasses. Just like the 'beak' trying hard not to disbelieve an habitual offender appearing before him!)*

The change from 'kneeling' stripping to standing 'Stoop and room' mining, actually took a little time to get used too! For a start, it was much more work intensive, more exhausting, with many more muscles being brought into play! A miners' waist and back suffers much more hard usage, and his arms and shoulders take the added strain of constantly lifting heavy shovelfuls of coal, sometimes up to neck height, if a miner fills on the laigh *(low)* side! *(As did Jackie Dryburgh.)* I was very fortunate in having Jackie as a neighbour when we drove levels to the left side of the Old Dook. I always ended up on the high side, where I barely had to lift my filled shovel to thigh level!

During the first week in this new level, we both discovered a degree of creeping tiredness that did not occur on a long wall, coal face. This must have been due to the extra energy that needed to be expended with the unaccustomed change of environment. This was not really significant however, for the simple reason that a fit miner will adapt to the inherent work load and changed conditions, in a very short space of time. What was uncertain and that which had to be decided, was exactly how *little,* Willie Hampson was prepared to pay for each ton of coal mined and in turn, what minimum would we, the colliers, grudgingly accept? *(The old two-headed monster again!)*

This was a game that was traditionally played to the to the bitter end, with neither side willing to concede the proverbial inch until either, or both sides had obtained the maximum concessions. There were six pairs of Stoop and room miners involved in these negotiations and we all were all of the same mind. No great production targets and the liberal extraction of coal until a 'sensible' rate was agreed. The miners, once hardened into their new environment, would not give of their best, but keep something like 20 percent in reserve. Willie Hampson, no man's fool, knew exactly how well to play the waiting game and deferred from setting a substantial rate. Thus the game was played! The miners would initially, only produce *(fill)* a maximum of 24 loaded hutches of coal between each working pair, and every pair would conform. This would credit each miner with six or seven tons of coal per shift, which was what Willie Hampson would base his projections on, and hopefully offer a substantial rate per ton. This would 'to and fro' for only a few weeks, before the Manager would fix a compulsory rate, which inevitably meant, 'take it or leave it!' The rate he eventually fixed was six and sixpence per ton, with a 10 cwt. minimum on every level hutch.

Needless to say, that after a few weeks, we colliers had hardened to our new tasks and the coal began to flow. Within two weeks, 26 hutches per day. Within four weeks, 28 hutches per day. Then finally, within eight weeks, Jackie and I had gone up to 30 loaded hutches per day! At six and sixpence per ton and with an average daily weight of approximately 16 tons, we were earning in excess of fifty shillings *(£2-10/-d, or £2.50p)* per shift, which was approximately 8 to 10 shillings *per diem* more than that earned by a long-wall stripper!

*(Unhappily, this 30 loaded hutches per day could not be achieved every working day, but was certainly the target number! This was killing work and besides, the level was getting longer as was the time taken to 'draw' the loaded hutches.)*

As this level progressed beyond the 25 metres point, it transpired that we were entitled to 'oncost' men *(fixed wages)* as drawers. This of course, meant, that we did not have to waste valuable filling time away from our coal face. The drawers did not have an easy task, in that their job was to keep us supplied with empty hutches, and remove and wheel away those that we had filled. At first, they could rely on having a short rest between filled hutches, but as the level advanced, so did the length of their 'draw' *(putt)* between coal face and the coal haulage dook! Very soon, they too were working at their maximum, with little or no actual rest time between hutches. Something had to give, and it did! The drawers simply went on a planned go-slow in every working level and limited their charges to exactly 24 hutches, for each pair of collier's. Their complaint was, that whilst the collier's were making good wages, they the drawers, still received the same fixed wage, no matter that they were now working like cart horses. So!, the carefully considered ultimatum! - and they were determined to stick it out! Which they surely did! We, the colliers, did not retaliate, not against them! They were our means to make good wages! Jackie and I, and the other Stoop and room miners did nothing to jeopardise the drawers stand, *(in fact, we were quietly livid!)* but we reasoned that we could fill every hutch to maximum capacity to obtain extra tonnage. This worked for exactly 24 hours until 'word' got around and then, suddenly, one or two of our overloaded hutches began to mysteriously derail. This meant of course, that we both had to leave the coal face, travel to the point of the derailment *((which was usually at the maximum distance from the face)* and help lift the heavy hutches back onto the rails. We soon got the 'message' and quickly reverted back to filling, levelled hutches of coal.

Willie Hampson held out for approximately ten working days, which was actually a long time in production terms. This was totally unheard of, for oncost drawers to demand additional wages under any circumstances for a fixed job of work, and Wullie Hampson paid little attention to this state of affairs. For a while, it looked as though he would not surrender. Wullie Hampson however, did go to the length of coming down to the Water level to see for himself, as to the conditions and the legitimacy of the drawers complaints. He, along with another official, spent time in observing the cart horse conditions that bedevilled the drawers, and after a meaningful time spent in discussion with his companion, made a sensible offer to the drawers. This offer was eventually accepted! In retrospect, if either Wullie Hampson or his companion, had stopped to consider and observe the greatly increased work-rate from the miners themselves (us!), he might have realised that it equated to a rate of perhaps 40 filled hutches in one shift. A work-rate and impossible total, that neither Jackie nor I could have sustained for very long!

\*

One day, just after piece-time, whilst we were still driving one of the levels at No. 3, Jackie and I were alerted to the sound of one of our drawers running hard the long level to our coal-face - and without an empty hutch! He seemed a little agitated and breathlessly informed us, that the drawer from No. 5 level had rushed up the coal-dook to inform them, that the two miners in No. 5 level had been caught in a 'fall' - and that help was needed! We hurriedly downed tools and rushed out of the level - and then headlong down the coal dook to No. 5 level. We both knew that the two miners in No. 5 level were actually brushers and they were in fact, father and son. Their names were Wullie (*Lots\* o' coal.*) Allen (*Father*) and young Jim (*young Lots o' coal*) Allen. Old Wull Allen was lying well away from the fall, to where he had managed to stagger - and Jackie made a bee-line towards him, with me following close behind. Old Wull quickly indicated inwards to where the fall had happened, and to the effect that young Jim was still inside. I ran quickly into the level with back and head slightly bowed, as the roof was only about 5 feet - 6 inches high along its length. *(I believe that they were 'brushing' the Toughs & Clears coal to the retreat, which had been left in situ with the original intrusion.)*

As I approached the site of the fall, I saw that it had been quite extensive with lots of downed coal and stone, but young Jim had managed to crawl clear of the worst of it and sheltered under the unsupported lip of coal. Just as I was about to ask him where he was hurt, the troubled area gave another menacing rumble and more stone fell away from the unsecured roof and began to tumble down towards us. I simply did not stop to think, I darted forwards, grabbed Jim's right wrist and threw it over my bowed head and neck. I simply heaved upwards, yanking young Jim's body over my shoulders in a fireman's lift - and using my right hand and arm to steady his legs, ran out of the level for all that I was worth. I vaguely remember Jim's voice roaring with what I though was pain, but kept on running! I was having to run quite doubled up, because of the roof restriction, but that proved no obstacle to a fit young twenty-one year old like me. On arriving at the mouth of the level, I gently placed the groaning Jim Allen down beside his father and breathlessly asked him, as to where he was hurting? To which I received the most unexpected of answers, "It's my back, you mad basket! You obviously didn't feel it - but you rattled my back against every set of wood on the way out. I wasn't too badly hurt in there, but I think I'm much worse now!" I could not help but think to myself, 'How's that for gratitude?, - and there was me just trying to help!'

*\* Old Wull Allen got his nick-name from the fact, that every time he bored his holes and had his 'shots' fired, he would rub his hands quickly together and gleefully exclaim ; "Lots o' coal . . . . . . . lots o' coal!" His son was therefore known as:- 'Young Lots o' coal!'*

Nothing of any great importance happened between then and 1957, but then I realised that the amount of workable coals remaining in the Dysart Main seam seemed to be dwindling. We were being shifted from one old level to another, at different points on the Old Dook to take downed coal, which was surely the last ditch attempt by the management to gain tonnage. The writing was surely on the wall with regard to the available remnants of the Dysart Main seam of coal in the Old Dook - and in Lochhead pit as it transpired!

It was whilst I was still employed down the Old Dook, that I decided that I needed a change of general surroundings and a different working environment. This was becoming apparent in that I began to lose time and money with the ever-changing work-place venues, and daily recurring thoughts, that I just did not wish to spend all of my working life in the coal mines. One Saturday morning whilst I was in the 'Lang Toon' of Kirkcaldy, I chanced to meet an acquaintance of whom I had lost track of in Lochhead pit. His name was Marion, and he claimed to be a Ukranian who had come to Lochhead pit as a 'displaced person' around 1947-48.

Marion had done well for himself and was now working as a short-wall machine-man in Earlseat Mine. He was part of a six man, two shift team, working alternate day-shift and back-shift. Earlseat mine, situated just to the north side of the Standing Stone Road, was a small drift mine taking the Lower Dysart seam as it out-cropped towards the surface. It was therefore not very deep, but rather long. To get to their place of work, the miners actually walked down the length of the mine, made one right or left turn, and found themselves at their place of work. *(The coal face!)* Marion very quickly sounded me out as to what I was doing, and receiving a rather negative reply - immediately made me the offer of second placed miner on his shift. His second man had simply quit, because he could not stand the pace and wished to revert to less strenuous employment. This was a great opportunity for me to get near to, and perhaps operate a coal-cutting machine. A job I had long coveted, and working as part of a strong, three-man team on a short-wall face. There was still good money to be made in short-wall places and positions were jealously guarded. *(Some of the machine-men/colliers within the short-wall places in Lochhead's east-side mine, would rather work short-handed than allow a 'stranger' a foothold in their contract!)* I told Marion that I would think it over during the weekend and promised him, that if I was still like-minded on the following Monday morning, I would appear at Earlseat Mine before I went to Lochhead pit on the back-shift.

Monday morning came and decision made, I appeared at Earlseat Mine at mid-day. I went directly to the managers office, knocked, and found Mr Kennedy sitting behind his desk still sporting his favourite bonnet, *(flat cap)*. I mentioned my name, told him where I currently worked and mentioned the 'place' offer from Marion. Mr Kennedy's next remark was that 'he' was the manager of Earlseat mine!, and he did the hiring and sacking, but I had come to the right place. His next remarks were somewhat strange in that he asked to look at my hands and then remove my jacket and turn around. Having duly presented my two opened 'shovels' and flexed my deltoids, he grunted "You'll do! - When can you start?"

It was as easy as that! I did of course, hand in a week's written notice to the check-box at Lochhead pit as a matter of form, but obstacles were never really put in a miners' path when he wished to transfer to another coal-mine. *(Meaningful shades of a new era!)*

This transfer to Earlseat Mine in 1957 ended my association with Lochhead Colliery, manual Stoop and room mining and the Dysart Main seam of coal, even though Old Harry continued to be employed at Lochhead pit until his redundancy at 63 years of age, in 1967.

*(I never had any intention of attempting to emulate his 49 years in the Coal mines!)*

# The Leaving of Lochhead!

Little did I know then, that within one year, the Coal mines would be well behind me and that I would soon be enjoying a life in the fresh air, jumping out of perfectly serviceable, R.A.F., Transport Command aeroplanes, *(driven by the Brylcream Boys)* with a slide-rule in one pocket and a set of log-tables in another. I would be wearing a red/maroon beret adorned with the Gunner's cap badge *(ubique quo fast et gloria ducunt),* and dressed in a camouflaged, Denison, jumping smock sporting a set of parachute wings, with the very famous, Pegasus, winged horse emblem patched onto each upper sleeve. This was certainly a far cry from a pair of pit-boots, knee-pads, a 'holing' pick and a No. 6 size coal shovel!

*(But, - back to having a dirty face, playing at Red Indians and shouting 'GERONIMO, as I jumped out of the aircraft!).*

\*

# Chapter LVI

## The Old Dook - Re-visited!
### (*Hail to the new King! - and Farewell!*)

During the month of April 1956, Jackie Dryburgh and myself were employed on the Old Dook and engaged in the driving a coal heading up through part of the **Dysart Main seam** of coal. We had commenced this heading on **No. 3 companion level** on the left (*north*) side of the dooks, at a point approximately 50 metres in from the Coal Dook. This heading was to be driven parallel to the **Main Dook** for a proposed length of 100 metres, to strike into the next level higher up and also leading in from the Main Coal Dook. At that time, the heading 'belonged' to us in that we were the only pair of collier's taking coal from this working. It was therefore 'our' aeroplane brae, complete with a single oncost 'hanger-on', who uncoupled our down-going loaded hutches and attached our up-coming empty hutches. (*He also, along with the on cost miner working in the level below, attended to the 'clipping' of both loaded and empty hutches for each of the working levels.*) As this heading advanced upwards we gradually became aware of some 'goings-on' in the near vicinity. This took the form of deep rumblings that seemed quietly resonate and echo within our very workings. I didn't pay too much attention to this at first, but merely thought that this signified the presence of another pair of miners who were also 'firing in the solid', in a level or heading quite close to us.

After a short while, the rumblings began to grow in volume and intensity, and to sound very close to us indeed! Within a very short space of time, a few days perhaps, I actually stopped what I was doing and voiced my concern to Jackie. Jackie also stopped for a few moments, looked at me and said, "I did tell you that the Allen's (*Old Wull and young Jim*) were driving a cross-cut mine from the **level below** (*on the same dook*) and into the Lower Dysart seam of coal, but you merely shook your head and carried on!" I then replied to the effect, "That you didn't tell me exactly where they were!?" They were in fact, directly underneath us, but mining within the **Lower Dysart seam** of coal. They were only twenty-three metres down below us in the strata, and engaged in the driving of a horizontal **cross-cut** mine that lay exactly parallel to the one we were driving - and only fifteen metres to our immediate right! So much for my being aware of what was actually going on in the vicinity of our particular workings! I was only interested in getting coal!

\*

**No. 3 water level** to the left of the Old Dook *(datum height 9250 feet)*, lay exactly 25 metres further down the dook than did our **No. 3 companion level**. In that level, at a start point approximately 65 metres from the Main *(coal)* Dook, old Wull and young Jim Allen had commenced to drive a short heading, that had been planned and surveyed to lie parallel to the Main coal Dook, as it progressed. They were working on the day shift, whilst another two-man team worked the back-shift. *(This would alternate on a weekly basis.)* They had commenced this heading up through the grain of the **Dysart Main coal** for a short distance of approximately 15 metres, then changed their vertical angle of progression. The upwards slope of the metals *(and therefore the coal seam)* on this part of the dook was approximately 1 in 4, which they had followed for the first 15 metres. This was about to change! It was common knowledge amongst the experienced colliers, that the Lower Dysart seam of coal, lay only 22 to 25 metres below the Dysart Main seam - and followed the same slope *(incline)* of the 'metals!' Their newly, surveyed instructions was to cut into the rising metals with a **level, cross-cut mine** and continue on, in the same straight horizontal line. This they did, by cutting down through the 30 inches thick Sclits, the 15 inches of Stone, and the 55 inches of the Coronation coal at the bottom-most layer of the **Dysart Main seam**. Then cutting into the mixed strata and stone, that separated the two different seams of coal. This cross-cut mine needed to be regularly steel-girdered and heavily packed with wood along its length, for the reason that it would eventually become a temporary, but busy, coal haulage route in the near future, and that the roof was being cut through variously different and unpredictable 'metals'!

The cross-cut mine progressed steadily during this double-shift cycle and its length rapidly 'over-took' that of our heading, in the above Dysart Main coal. This level mine was wrought a horizontal distance of exactly 85 metres, before it struck in to the full ten and a half feet height *(3.2 metres)* of the rising Lower Dysart seam, which in that area, showed two separate sandwiches of mixed coal, separated by approximately 28 to 32 inches of redd, *(stone)*. The developers continued on in this same straight line, but now that they were into the Lower Dysart coals, the slope of the mine would now change to that of the **rising seam** *(approximately 1 in 4)* and carry on up through the virgin coal for another 180 metres.

This new mine that had been commenced from No. 3 Water level, was now exactly 280 metres long and was now joined-up and established into the targeted Lower Dysart seam. This meant that there now was a viable connecting coal outlet, cum materiels route, from the inside end of the 280 metres long mine - to the established coal haulage outlet on the Old Dook, via No. 3 Water level. This mine had commenced at No. 3 Water level *(datum 9255 feet)* in the Dysart Main coal and was now terminated in the Lower Dysart seam of coal, *(datum 9393 feet)*.

The profile of this mine, was that it was on an incline of 1 in 4 for the first 20 metres, then basically level for the next 85 metres of its length, until it struck the Lower Dysart coal at a datum height of 9260 feet *(a rise of only 5 feet)*, where it then followed the rise of the coals for approximately another 180 metres to its termination point, *(datum height 9393 feet)*. The extracted coal and waste materials from this new mine, was taken back down the mine to No. 3 Water level, where it was loaded into hutches to be transported along the short length of the level and onto the main coal dook. The hardware and materials needed for this new developing section, were being brought down the Old Dook to No. 3 Water level and delivered into the new, expanding section by the same route.

*(This new, 285 metres long heading would have been equipped with a 30-inch, rubberised, endless belt for the downwards transportation of coal. One of the inherent features of this type of conveyor, was that it could be made to operate in reverse, so that in-coming materiel could be transported inbye, providing that there was someone at the receiving end to off-load the supplies!)*

At this juncture, during the month of March 1957, the planners and developers were now almost ready to open up this section into a new coalface. At a point on this 285 metres long mine, approximately 115 metres from its commencement, just above the point where the Allen's had struck into the Lower Dysart seam, *(datum height 9262 feet)*, a second team of developers had been brought in. They, at this point, turned half-right and commenced to drive a level following the grain of the coal, but not at the full height of the seam. This level was to be the new **main-gate,** serving a long-wall undercut coal face, taking at first bite, the top leaf of the seam. *(40 inches in thickness and containing a narrow 2-inch thick band of stone. Approximately 1 metre height of extraction.)* This slightly rising level was driven *(in a straight line)* to a length of 100 metres where it was halted, *(datum height 9263 feet)*. This main-gate level *(as it now was!)* had been commenced shortly after the original developers had driven the straight *(285 metres)* mine up through the coal.

When the first, original team of developers reached the driven 285 metres length of this *(mine)* heading *(datum 9393 feet)*, they too turned half-right and commenced another level roadway, exactly parallel to the bottom one, *(main gate)* and again following the grain of the coal. This top level was only taken in to approximately 30 metres where it too was halted! This top level would become the trailing gate or return airway for the new coal face.

Author's Note: - *The opening up of this coal face was somewhat delayed, because both teams of developers encountered a single, unexpected, jump-fault of approximately 12 inches, at right angles to their line of advance. The planners decided to ignore this and instructed the developers to continue! At a point exactly five metres beyond this jump, the second team of developers turned left and commenced to drive a developing heading up through the coal, at approximately 10 feet wide and in the top leaf of the seam. This developed heading lay exactly five metres to the north of the jump-fault, over its 130 metres driven length!*

Down on the bottom, 100 metres long, main-gate, at the innermost point, the second team had made a left-angled turn and had now struck into the top leaf of the seam where, following the steepest rise of the seam, they commenced to drive a 3 metres wide, 1 metre high developing face, up through the top leaf of the seam. This would be driven in a perfectly straight line and would hopefully emerge out into the top road *(trailing gate),* after a distance of approximately 130 metres. This it did within a few feet! Both teams of developers were heard to produce deep sighs of relief just before the developing face was opened. It transpired that they had both encountered a small, but long, fault line in their respective roadways, immediately prior to their halted levels. It turned out to be just one, isolated, small jump and had not appeared on the developing face-line!

The date of temporary completion of the main-gate, before the face development started, was April 1957. By the 8th of July 1957, the embryonic coal face had been cut through and was ready to be furnished with the necessary hardware needed for production.

There then followed a longish delay before this face-line produced coal. . !?

\*

One of the main reasons that this coal face did not start production at once, was the fact, that the return air way back to the companion side of the Old Dook had not been completed. To effect this, the first team of developers now went back to near the mid-point of the 180 metres sloping mine, *(at datum height 9306 feet and approximately 65 metres above the start point of the lower main-gate)* and commenced to drive a mine at **left-angles** to the main heading. This mine was driven in a straight line for **25 metres** and then made a **half-left** turn, before it was continued on at an incline, to strike back as a **cross-cut** mine into the **Dysart Main seam,** at a point on, or near, to the **companion Old Dook**. This **cross-cut mine** actually crossed underneath the **Main coal Dook.** This second, dog-legged, cross-cut mine, linking the two seams of coal was exactly 90 metres long. This then was the completed **return airway** for this new, developing, Lower Dysart section.

It was not until March 1958 *(a delay of eight months?),* that this section began to produce coal as a conventionally cut, long-wall, coal face. The length of the face-line was approximately 130 metres, taken at 40 inches *(approximately one metre)* high, with very little attempt by the strippers to separate the 2 inches of stone from the coal. *(I was later informed that attempts were made at separation, when either the deputy or the gaffer was on the face-line!)* This coal face advanced a total of 350 metres over approximately 285 working days, which would give a rate of advance of roughly 4.1 feet *(1.25 metres)* per 24 hours cycle. The total coal output from this top leaf would have approached forty-six (46,000) thousand cubic metres, including the top and middle leaves taken in the main-gate and probably the trailing-gate as well!

During the months of June/July 1959, just after the top leaf was abandoned, the developers went back to the start point of this coal face and, commenced to drive a new main-gate and a new trailing-gate in the bottom leaf of this same section. The new main-gate and the top road were both cut to the inside of the previous 'gates', thereby shortening the new coal face by approximately 20 metres. By the end of July 1959, this new face was ready for production on a foreshortened 110 metres, long face-line, where the extracted coal was conveyed out by exactly the same route. This face advanced a full 400 hundred metres where it was halted, so as not to compromise the integrity of the standing stoops of the New Dook, a mere 40 metres away, but in the Dysart Main seam! This coal face would have produced approximately fifty thousand (50,000) cubic metres of coal, with a stone content of about five percent.

The average slope of both of these advancing faces was something to the order of only 40 feet, over an advanced distance of approximately 350 to 400 metres, which works out to a gentle rise to the order of 1 in 32. The incline on the coal face was completely different, in that, in the taking of the top leaf *(and by implication the bottom leaf also)* the difference in height between the main-gate and the trailing gate amounted to 140 feet, which works out to a rather steep slope of 1 in 3.2. This sort of slope *ensures* that blown coal is probably scattered far and wide and needs must take a lot of shovelling!

Author's Note 1.:- *This inevitable scattering of the blown coal could have accounted for the fact, that the rate of advance on both of these coal faces, was slightly less than would be expected on a conventionally, undercut, long-wall face. It may also have been that the depth of each succeeding undercut was deliberately withheld!*

*At the 300 metres advance point on this long-wall coal face, a single, long, two feet high, jump fault appeared over the full length of the coal face. This fault did slightly inconvenience all of the miners working on this face, but the 'machine-men' merely cut through and over the fault without any loss of production. To the best of my knowledge, no additional time was lost through roofing problems.

Author's Note 2:- A point of general interest with regard to the above coal section within the Lower Dysart seam, was that it lay in its entirety, directly underneath one of the most prolifically worked sections in the Dysart Main coals, that was situated exactly between the Old Dook and the New Dook, between No. 1 and No. 2 connecting levels. These two seams were only 23 metres vertically apart in the strata at this point, but the near total extraction of this Lower Dysart section seemingly made very little impact on the integrity of the extracted section above! It would not have mattered in any case, the Dysart Main section above, had been closed and abandoned since 1931. Notwithstanding the fact, that nearly the whole length of No. 2 level between the Old Dook and the New Dook, was re-visited and extensively robbed of standing coal, from 1955 to 1958!

*

Sometime just after the start of the taking of the bottom leaf of the above No. 1 section to the left of the Old Dook, the planners and developers were already turning their attentions elsewhere, with a view to the planning of the next section to be opened in the Lower Dysart coal, in this immediate area. Back on the original 285 metres long mine *(at datum height 9261 feet)*, at the start of the No. 1 section main-gate, the first development team now did an about turn. They commenced to extend the line of the main-gate for the No. 1 section, in the opposite direction and in the same straight line. This new mine commenced around the month of September 1958. The miners carried on in this new direction taking the roadway slightly to the dip for a distance of approximately 140 metres, where the surveyors created a datum height measured at 9257 feet. Fifty metres further on, the same survey team created another datum height measured at 9258 feet. This new mine had taken the developers directly underneath the line of the **twin Old Dooks** and the two, newly created datums, were exactly 50 metres and 100 metres to the right of the Old companion Dook, but approximately 23 metres underneath it in the strata. The positioning of these **two datums** *(points)* is **significant**.

Meanwhile, back at the point on the **285 metres long mine/heading,** where the original developers had joined-up the mine to the Old Dook *(return airway)* with the 90 metres long mine through the metals, *(at the 9306 feet datum point)*. A short dipping had been commenced at the point, where the return mine had turned left for 25 metres. This new dipping roadway was taken down approximately 40 metres in length, where it was turned right, to parallel its neighbour *(the now reversed main-gate for No. 1 coal face)*, which had already passed the 9257 and 9258 feet datums *(points)*. This **Upper level mine** was exactly 40 metres to the west and to the rise of its predecessor, *(now known as the Lower level mine)*. This upper mine would parallel its lower neighbour in the same direction, for an extended distance of approximately 525 metres, where they would both be terminated at a given point in the metals, without striking anything! These levels I shall name the **Twin 525 metres levels** for want of a better name and refer to them, as the **Upper 525 level** and the **Lower 525 level.**

On the bottom level of these twin 525 metres levels, at datum heights 9257 feet and 9258 feet, which were approximately 50 metres apart, there was commenced in October 1958, **two separate headings** to be driven upwards through the Lower Dysart seam *(at 4 metres wide and 3 metres high),* and taken up to an approximate length of **663 metres** for the right side heading *(looking upwards)* and **625 metres** for the left side heading. The **left** heading would be driven upwards and underneath the **old Lower Dysart haulage-way** *(in the Dysart Main coal),* and to a point that was 40 metres beyond its line, where it would reach the pit bottom area, via a **short mine** that was yet to be driven towards the pit bottom. The **right** heading would also taken up and under the old Lower Dysart haulage-way, but was also continued up into the old **Dysart Main workings** to the west of the pit bottom. This was to be used as the new, return airway for the remaining, new developments within the Lower Dysart seam, to the right of the **Old Dook**.

When completed and developed to their full length, these two headings would be re-named as **dooks.** These headings/dooks were completed in February 1960 and joined at right-angles to a **275 metres long level,** that was created in the Lower Dysart seam, to run parallel with the Lower Dysart haulage-way *(within the Dysart Main seam),* but approximately 40 metres to the west of it. This level followed on to a 100 metres long, 1 in 4 slope to the north, that transported the coal from the right dook *(looking down)* and delivered it to the pit bottom.

At a point on the lower of the 525 twin levels, 60 metres to the south *(left)* of the start of the 625 metres left heading, there was commenced another heading at datum point 9265 feet. This heading, No. 1 *(west)*, was commenced on 25 May 1960 and progressed up through the Lower Dysart coals, cutting through the **upper** 525 level at the 35 metres point on its progression. This new No. 1 *(west)* heading ran parallel to the new twin headings (dooks), but was of course, to the left side *(south-west)* by a distance of 60 metres. It was developed upwards for a distance of 175 metres to clear the solid stoop of metals that needed to be left under the Coaltown of Wemyss. This No. 1 heading would become the main-gate for the next, new coal-face to be developed at the 9400 feet datum height. At this point on the heading, the developers then turned hard left and commenced to drive a steel girdered roadway, that was to serve a double purpose. It was to be the start point for a new coal face, but before this face could be developed, a return airway would have to be cut through to an existing, return airflow! This area in the Lower Dysart seam had as yet remained unworked, so the answer to the air problem, seemed to lie in the driving of a new road from this present No. 1

*(west)* development into a previously worked, but still open old working! The nearest old section in the Lower Dysart seam, actually lay to the south and west and was in fact, the twin 325 metres long twin levels that had been driven from the east side of the Lower Dysart Dook, from datum height 9580 feet, during the last months of 1955. The bottom level of the two had been steel-girdered - and was still open!

The developers at the 175 metres point on the No. 1 heading, *(datum height 9400 feet)* now commenced to drive a girdered mine in the general direction of the old 325 metres levels, on the now abandoned Lower Dysart Dook. This mine was initially driven on a map bearing of 230 degrees for 155 metres, then on a bearing of 240 degrees for another 170 metres. At this juncture, it struck the bottom side of the old, second, experimental shearer face, then veered left on a bearing of 210 degrees for approximately 90 metres. At this point, the mine turned right, onto a bearing of 250 degrees for a distance of 160 metres, where it struck into the old bottom 325 metres level. *(This bottom 325 metres level had been driven through virgin ground with no coal extractions whatsoever taken from either side, therefore, the roadway was in a good state of repair!)* This zig-zag mine now served two purposes, one of which, was, as a return airway for the new section to be developed and the other, was as a possible escape route in any eventuality!

Meanwhile, back on the **lower** 525 level, a different team of developers had commenced yet another heading, a full 175 metres further along the level to the south, at datum height 9288 feet. This No. 2 heading *(west)* was driven upwards and parallel to No. 1 heading, from the 9265 feet datum height. This No. 2 heading also crossed the **upper** 525 level at the 35 metres point and continued on for a total of 250 metres until it too, cleared the solid stoop under the Coaltown of Wemyss. This **No. 2 heading** *(west)* needs must be remembered, in that it was first taken upwards through the strata, and altogether past the new section it was meant to serve *(as an initial airway),* and carried on upwards until it had reached under the Lower Dysart haulage and beyond, and later still, it was extended downwards in the opposite direction, into the depths of the strata, to develop a deep section to the West of the **Wemyss Castle Home farm.** This **No. 2 heading, cum dipping** was ultimately **1035** metres in length!

(Author's Note: - *I shall, with the remainder of this narrative relating to the 'New' Old Dook, refer to this development as the **1035 metres dipping** or **heading**!*)

By the end of September 1960, both No. 1 and No. 2 headings from the 525 twin levels, were connected by a newly developed, face-line starting at the 9400 feet datum at the bottom right end. *(This start-line for the new coal face was actually developed parallel too, but fully 10 metres above the first 155 metres length of the 'zig-zag mine!' This was standard practice to preserve the integrity of the girdered mine!)* This face-line was of course, not level and would also be advanced to the rise. The rise along the line of the coal face would be approximately 1 in 7 from right to left and the gradient of the advance would be 1 in 5, all well within the capability of an A.B. 15-inch long-wall coal cutting machine. However, this section did not open up at this time, the developers were to be otherwise employed and no other 'strippers' or preparatory miners were available.

In April 1961, the newest coal face *(New/Old Dook No. 2 section)* at the **9400 feet** datum level, commenced production over a 155 metres length. The coal face had barely commenced cutting, than a sizeable fault appeared at the higher, south end of the face-line. This necessitated the face-line being shortened by a full 40 metres from its left top-end with the inevitable consequence, that the return road *(top stable-end)* first took a sharp right turn at the 9480 feet datum height and ran parallel to the face line for 40 metres, until it made another left turn to parallel the main gate. This resulted in the face-line being now only 115 metres long. At this juncture, the whole coal face stopped its limited production for approximately 6 to 8 weeks, for reasons that I do not know. At the start of June 1961, the face opened up again and then continued to advance onwards and upwards until September 1961, gradually shortening its 115 metres length until it was only 105 metres long.

*(I can find no tangible reason for the ever-shortening of this coal-face at this time, even though I shall advance a theory, later on in this chapter!)*

The length of this face continued to shorten until the last week in October 1961, where it was down to only 75 metres wide. The face continued to advance, but getting slightly narrower in width through the month of November and into December. The face wide had now been reduced to only seventy metres and there was trouble ahead! Suddenly, around the middle of December, the machine-men began to cut into a series of jump faults. These faults, initially about four in number, took the form of either up-jumps or down-jumps, some as much as three feet at a time, with one particular fault looking as though it was there to stay! This was certainly troubled ground! As soon as one isolated fault was worked through, another one appeared at another point on the face-line. These small isolated faults continued to appear at regular intervals along the coal face until the second week in January 1962, when the machine-men once again found themselves cutting into clean ground. *(Except for the one 4-foot fault that remained almost immobile near the centre of the coal face!)* At this juncture, the width of the coal face was once again extended to approximately eighty metres and for the next three weeks, coal production and coal-cutting continued without hindrance!

Suddenly, at the end of January 1962 and without warning, another series of small jumps appeared at three, separate places on the face-line. They were however, of no great consequence, merely an additional embuggerance factor, that the machine-men and the colliers took in their stride! By the last week in February, the main fault that had plagued the miners for many months, had all but disappeared out to the right side of the coal face and the last of the small faults had also disappeared! The last month of coal production from this No. 2 Section was relatively trouble-free and indeed, the face-line was actually extended out to ninety metres width, up to the termination point of the coal face on March 29th 1962! This face had continued to produce limited coals, until it had advanced to a point where it terminated at datum height 9690 feet, which placed the termination point approximately 20 metres due west *(above)* of the Lower Dysart haulage-way, but 21 metres beneath it in the Lower Dysart seam. The total distance covered by this No. 2 section in the New/Old Dook *(Lower Dysart coal),* was approximately 435 metres of advance, with a face-line that changed continually throughout it's advance.

*(Readers are reminded that in working the top leaf of this Lower Dysart seam, both an up-throw and a down-throw of between one to four feet, would result in the miners having to cut into and dispose of all of the redd and blae's that appeared on the face-line! If this had been a conventionally undercut, coal face working the three shift cycle, the redd and blae's would have been used to build supporting packs in the wastes. If this had been a power-loading run, then, all of the taken coal and stone would have been conveyed along the coal face and thence to the pithead!)*

The amount of coal extracted from this No. 2 section is difficult to access, but a conservative estimation would probably determine upwards of 45,000 cubic metres of coal, containing approximately five to ten percent stone, *(redd and blae's).* The rate of advance on this run would have been approximately 7.2 feet per 24 hours period, which poses the question:- 'How many production shifts were worked on this short face-line? Was there a 'round- the-clock' agreement where coal was produced non-stop?' - or have I got it completely wrong and was this a 'shearer-cut' run? I simply don't know, but I do feel that the 'hand' of Willie Forbes might just have just been of influence in the determination of this limited coalface!

Author's Note: - *As to why this particular No. 2 coal face within the Lower Dysart coal was ever developed is something of a mystery! It was in partially faulted ground. It was never a very wide coal face. The slope along the length of the cut face was approximately 1 in 7, whilst the inclined slope to the advancing cut was 1 in 5. This was certainly to the advantage of the coal-strippers in their working conditions, but had the disadvantage of having to work coal-cutting machinery at right-angles to the steepish incline. A rather unsafe practise to say the least!*

Author's Comment: - I do have my own theories as to why this width of this limited production face-line was seemingly limited, and I advance the preposition, that it was because the colliery manager in his infinite wisdom, saw the likelihood of the greater area to the south-west and immediately adjacent to this No. 2 coal face being taken and comprehensively extracted by a superior means, i.e. power-loaders, or combined mechanised shearers. Also, there is no way that anyone could have foreseen the amount of troubled ground that this No. 2 section was worked through! If this had been previously known, the chances are that the section would never have been developed in the first place, what with the troubled ground encountered and the limited amount of coals eventually taken!

*

The planner's approach to the development of No. 3 section in the New/Old Dook *(Lower Dysart coal)* was somewhat lengthy, in that the No. 3 section was an isolated development well away from No. 2 section and much, much deeper. The area to be developed, lay to the south-east of the Coaltown of Wemyss solid stoop and to the immediate west of the Wemyss Castle Home farm solid stoop. The initial opening dipping roadways into this section went deep into the Lower Dysart seam and, actually commenced from the **upper 525 metres level** in July 1960 and was driven upwards, as well as downwards. This **heading/dipping** is described in this narrative as the **No. 2** *(west)* **/1035 metres heading/dipping** and was one of the longest, single roadways ever driven within Lochhead pit, to open up a coal section. It was driven to develop one, single, isolated, section/area of coal. *(Albeit a fairly large one!)* The upwards *(inclined)* driving of this **No. 2 / 1035 metres** heading/roadway had taken place up to the end of September 1960, where it had been temporarily stopped, after being driven approximately 250 metres to the No. 2 section coal face. *(It was to be greatly extended at a later date.)* The downwards projection of this road *(dipping/dook)* from the **525 metres lower level** did not commence until December 1967, a full seven years after the commencement of No. 2 section. *{This actually meant that there was no coal developments of any kind in the Lower Dysart coals of the New/Old Dook, for a full six years period of time!}*

This **dipping** mine was taken down to a distance of 305 metres, where it was terminated, *(datum height 9080 feet)*. At this point, a level roadway was commenced in both directions and roughly following the level of the coal seam. The level to the right *(south)* was taken forwards, following the level of the seam for only 65 metres, where it was temporarily stopped, *(datum height 9081 feet)*. The date was mid-June 1968. This level was now to be the main-gate for a new

No. 3 coal section and this point also, was to be the start point for the development of a new, long, face-line. The return road *(airway)* for this latest section was commenced from the **Lower 525 metres level,** at a point approximately 50 metres from its southern end, *(datum height 9290 feet)* and was completed by the end of February 1968, *(datum height 9230 feet)*. This short dipping was exactly 75 metres long. At this 75 metres point, the developers turned right and commenced to drive a short level, that was taken exactly 60 metres forwards and stopped, *(datum height 9234 feet)*. This upper 9234 feet datum height was exactly 175 metres in distance from the 9081 feet datum height.

This new No. 3 face-line was a little unusual, in that it was developed from both top and bottom, as the development dates would seem to indicate. The developing face was cut through and joined by the end of May 1968, and the coal-face was then 'furnished' with the appropriate machinery. This No. 3 face-line did not begin to produce coal just at this time, as there seemed to be a small delay in equipping this section. It would also seem from my research, that the delay was seemingly caused by the installation of two, separate, but connected 30-inch belt systems, that would carry the coal all of the way up the **1035 metres long heading** and onto the **275 metres level,** running into the pit bottom. I do believe that initially, the coal from this No. 3 section went up part of the **1035 metres heading**, along the **lower** or **upper 525 metres twin levels** and the up the **left coal dook.** Unfortunately, I cannot find enough evidence to confirm which direction the coal did take. What I do know, and can legitimately claim to know, is that the coal most certainly did arrive at the pit bottom by either, or both of these means. *(According to my research, the use of hutches for the transportation of coal to the pit-bottom had been superseded by the extensive use of belt conveyors, nearly all the way to the pit-shaft.)*

This No. 3 section *(coal face)* commenced production, on or around the last week in July 1968, along an 175 metres face-line. The taken coals were again the upper leaf, which showed a height of 42 inches, that also included a 2-inch band of stone. After just one week of advance, the top return road was diverted slightly to increase the face-line length by another 10 metres, where it continued on at 185 metres wide and then being very slowly extended, so that its termination width was measured at 205 metres.

The upwards slope on the face-line from left to right was 1 in 3.8, and the gradient of the advance was virtually nil, with the barely perceptible change of height over the whole 375 metres length of the advance, amounting to only 12 inches or so. This face was not trouble-free, but the initial faults were so minimal, that I would hazard a guess that they were not anticipated. The seam itself however, was becoming quite unpredictable as to the inherent formation of the coals and stone. The top coal was thickening and the sandwich of grey stone was also becoming slightly thinner! There were also thin streams of stone being intercalated between the bottom coals. The main-gate brushers were taking approximately 10 feet of height, which included 26 inches of soft blae and 12 inches of stone in 4 separate layers. However, by September 1968, there was a marked change in the top leaf of the coal. The coal content had thickened out to 50 inches thick, with only 2 inches of sandwiched stone and everything was taken. This continued until the beginning of October 1968, when a series of four, small faults began to intrude onto the top end of the run and encroached downwards for all of 50 metres, the fourth fault in the series extending down for a full 150 metres. *(These four fault-lines showed up as jumps of only up to three feet in height, so, no part of the seam actually disappeared!)* None of these fault-lines however, seemed to slow the rate of advance and extraction totals were maintained. *(As to how much of this extracted material was actually good coal remains unknown. If this was, as I suspect, a power-loading section, then every last piece of material would be transported down the conveyor. A dirty cocktail indeed!)*

This No. 3 section progressed well enough until the third week of April 1969, when a long fault suddenly developed on the top half of the face-line, and within a few weeks, had travelled down to the bottom end of the coal face. Luckily, the fault was an upthrow, which meant that the taken material was either soft blae, or the middle mixed leaf of the seam! Seventy percent of the face-line had now passed through troubled ground and unfortunately, there was more to come! For the several weeks that the fault persisted, the miners thought that the coal face was heading for more trouble and were more than pleased when the fault disappeared! The coal ground immediately behind this fault was still troubled, but work still proceeded and any extra discovered redd or blae was taken with the coal. Within one week of the coal fault disappearing - disaster struck! The bottom two-thirds of the coal face again struck into troubled ground with the addition of the commencement of a 'roll' in the strata. Then, at the end of May 1969, a great spike of a stone fault intruded in from the bottom of the run and within days, had travelled more that half-way up the coalface. At the same time, a few other fault-lines had also developed, running obliquely across the stone fault, which was to prove even more troublesome. The long spike of stone proved to be about 10 metres thick and was very hard to cut as they discovered, but the decision was made to cut, or blast directly into it, with the hope that the coal seam would be located behind it! The coal seam was eventually located, but the ground was still heavily faulted. The colliery manager then made the bold, but inevitable decision, to close and abandon this No. 3 section and withdraw the coal-cutting and conveyance machinery, and all electrical equipment. This coal face could have advanced a further 200 metres if the ground had been clear!
*(The colliery manager, Willie Forbes, must have heaved a great sigh of relief, in that the next coal section to be worked was almost ready for production.)*

The rate of advance over the first ninety percent of this No. 3 section, was approximately 7.6 feet over a 24 hours period. *(It must also be remembered, that the height of coal taken was considerably more than first anticipated and this may have slowed down the rate of advance, but probably not the overall amount of coal extracted.)* The advance rate over the remaining ten percent was slowed down considerably, with the onset of the faulted ground, in that an advance of only 50 metres was achieved in the three months period from May 1969 until August 1969. Also, the stone to coal ratio near the latter end of production, would have risen dramatically to about 6 to 4. Very dirty coal indeed! The amount of coal extracted over the initial 90% advance, would have amounted to approximately 100,000 cubic metres, of which approximately 4% was stone. The last 10% of 'advancing production' is not included in these figures for obvious reasons.

*

*This No. 3 section on the 'New' Old Dook was an oddity, in that the development work needed to actually get there, was considerable to say the least, especially in light of the fact that only one section was ever developed. In this vicinity, there was the possibility of three further sections of equal size being opened up, one to the south of the Home Farm stoop and two to the immediate north. It mattered not, that the bottom leaf of the coal was becoming irregular and unpredictable. The top leaf was actually thickening. I also feel at this juncture, that the writing was on the wall with regard to Lochhead colliery and, any further coals that could be taken! The remaining coals needed to be taken and 'got out' much more cheaply and much closer to home! i.e. nearer to the Pit bottom.*

*

# A Valedictory Statement!

## *(or, Go for It Lads! - We have nothing to Lose!)*

The very last coal face to be developed and wrought in Lochhead pit, was about to be commenced. The developing colliers had cut the 180 metres long, face line through by May 1969 and it was equipped almost immediately. This section would be the final coals to be taken from this, the Lower Dysart seam, and again, it would be the top leaf of the seam that would suffer the extraction. This would be the fourth section in the 'Old Dook' to take the Lower Dysart coals and, as befitting any successful Army that had taken fully of its spoils, the 'commander' intended to take its full bounty on the retreat! The chosen ground for this final assault was immediately adjacent to the ill-fated No. 2 section *(that coal face of indeterminate and ever-changing width)* and lay approximately 50 to

60 metres to its left *(south)*.

The retreat was to commence along a broad, 180 metres wide, face-line and the 'retreating' advance was almost directly to the incline. The right to left rising slope along the face line, was approximately 1 in 10 from the main-gate, which itself, was the **1035 metres heading** which, at its top end, connected directly on to the **275 metres level,** that ran directly to the pit bottom. The gradient of the retreat was approximately 1 in 6, well within the capabilities of the current, coal-shearing equipment *(Yes! This final section had been equipped with the best of the remaining Power-loading equipment in the coal mine. The miners were the last of the most experienced power-loaders in the pit, and they knew within their hearts that the Lochhead bonanza was nearly at an end!)* There would be no 'stone packs' built in the wastes of this coal face as the face-line advanced *(to the retreat!)*, the only roadway to be kept open would be the **1035 metres heading** up to the pit bottom, and that, because it was still the main airway for the working section. The wastes would be allowed to close completely and indeed, be encouraged to do so. This was retreating with a vengeance, with the 'rapid' rate of 'advance' through the coal serving to keep the 'immediate' working face, trouble free!

The coalface commenced production at a face-line width of 180 metres, at datum level 9515 feet and would 'retreat' a full 425 metres uphill, to the 9750 feet datum height *{which was only 394 feet (120 metres) below ground-level at its termination}.* The 'retreat' commenced during the last week in August 1969, and progressed unhindered over clean ground, with nary a obstacle in sight. *(There was one small problem that the miners encountered and that was on the bottom right end of the run, where a narrow, double fault showed up around New Year's day 1970. It ran almost parallel to the main-gate for a period of approximately five weeks. It protruded a distance of 40 metres onto the bottom end of the face, but disappeared just as quickly as it had appeared!)* The rate of advance *(retreat)* on this last section was uniform, quick and largely unimpeded. The section operated through approximately 125 working days, where the face-line was advanced through 415 metres. The face-length slowly opened out to approximately 210 metres width after two months and continued at this length until termination. The height of the taken coal was approximately 40 inches in thickness *(one metre)*, that included two inches of stone.

The rate of advance on this final coal face in Lochhead pit, was a colossal 11.2 feet per 24 hours. This was by far, the fastest moving, coal face in the history of Lochhead colliery and using modern mechanisation. Whether it was because the miners *(Power Loaders)* had no need to worry about packing the wastes, or the certain knowledge that this was the very last coalface in Lochhead pit, - I simply don't know! I can surely hazard a certain guess, that the miners and everyone else concerned, knew that this was one, last, final fling and that there was nothing to be lost in 'going over the top!' *(There would be no more 'contracts' thrashed-out in this colliery ever again!)* Either way, the miners went 'all out' in one final dash, that would no doubt enhance their final wages for several months - and so what!? Willie Forbes the Colliery manager, was without doubt, getting more that his daily production quota's and that would keep him reasonably happy. Another reason why the coal-cutting on this run was never delayed or threatened, was that there was built at the south end of the **275 metres level** *(to the pit bottom)*, an enormous coal bunker, capable of holding several hundred tons of coal, which fed directly onto an endless conveyor belt and thence to the pit bottom.

The amount of coal produced from this No. 4 section over this relatively short period of time, would have exceeded 80,000 cubic metres. Surely, the most continuous volume of coal ever to surface at Lochhead Colliery!

\*

This No. 4 section within the Lower Dysart seam in the area of the Old Dook, was the last production face ever to yield coal from within Lochhead colliery. The 210 metres long, face-line, ergo, the No. 4 section, was officially terminated on 27th March 1970, which was the date that both Willie Forbes the Colliery manager, and the Chief Underground Surveyor, signed the abandoned plans for both the **Lower Dysart coal seam** and **Lochhead Colliery**. The deed was done and Lochhead colliery became yet another pit in the **Kingdom of Fife** to die an ignominious death. She had survived for approximately one hundred and seventy years (under two guises!) and partaken of all the coals from the Barncraig to the Lower Dysart seams, and in ever-increasing amounts. The mainstay of the pit had of course, been the prolific and ubiquitous Dysart Main, which had sustained the mine and its miners for a full sixty years and more, with very little of its coals ever being touched by modern machinery. The Coals had always flowed steadily from Lochhead pit and its Surface Dipping, but never more so, than since the introduction of the Anderson Boyes coal-cutting machines and latterly, the 'Panzer' Power-Loading equipment.

Greater quantities of coal had probably flowed from Lochhead pit during the last 20 years of her life, than during her entire lifetime - all due to increased mechanisation in her later life. 'Tis of great regret, that this coal-hungry machinery had not been available during the taking of the Dysart Main from 1890 to 1950 -

and then to 1960! It may well have been, that possibly up to fifty per cent or more of this great ubiquitous seam could have been developed and safely extracted, instead of the pitiful small percentage that ever saw the light of day.

The latter 'King' that was the Lower Dysart seam of coals was now also dead!

This day also saw the death of Lochhead colliery!

\*

# Chapter LVII

## Death of a Coalmine!
### *(The second Lochhead Pit!)*

The quietly unassuming, but proud 'Lady Lillian' had now reached the end of her tether and her resistance was all but gone. Her dying throes were not prolonged and her death came swiftly! It was also relatively painless! *(But not for her miners!)* She had in her lifetime, been repeatedly and comprehensively ravaged, not just once, but tentatively and then fiercely, many times in her younger years, and twice comprehensively in her middle and later years. She had been assaulted with a vengeance, she had been punched, kicked, drilled, bored and screwed, and she had even been driven asunder on numerous occasions. Her vast fortune in mineral jewels and precious wealth had been stolen, extricated, plundered, and clinically removed. She had suffered the gross indignity of having her internals eviscerated and extracted. She had experienced the intrusions of many and much, her vital organs were now in need of urgent repairs, which now, would never be countenanced. Her insides were falling apart, even though the surgery had been specific, detailed and intrusive. Her very vitals felt empty and used and she was tired. She was but a shell of the proud lady she had once been - and her time had come!

On the Twenty-Seventh day of March, in the year of Nineteen Hundred and Seventy, she took her last dying breaths and succumbed. She had been finally and ignominiously abandoned and her latter day carers were now vying to get shot of her. On that same day, her friend and carer William Forbes, signed the death certificate that was witnessed by the chief surveyor, and her official remains were eventually forwarded into the care of The Coal Authority, at the Mining Records Office at Burton on Trent in Stafford-shire.

Her steel, brick and tin superstructure was taken apart and removed, and nearly all trace of her once proud existence was obliterated, levelled and expurgated from the eyes of man. Her immediate innards were partially stopped and the in-roads to her very being topped-off. It was inevitable, that her remaining great emptiness, was to be allowed to slowly fill with the cold fresh and salt water solution, that would gather at her lowest extremities and ever so slowly, inch its way up through her multifarious, arterial routes and depleted veins, now bereft of their once throbbing life blood. This cold, salt and mineral impregnated water, would eventually fill every single, still open, capillary within her being, until it had equalised with Mother Nature's own water table. Very little of her outwards existence remains visible to the human eye, even though the discerning visitor can still see the slight, small, circular indentation on the small mound of earth and shale, that hides the concrete cap that sits seven metres below the surface of the ground and conceals what once was, the very productive and vibrant Lady Lilian.

The Lady Lilian, did of course, have her name changed to the much more impersonal name of Lochhead Colliery. Why? Well, I do know! But that is not for discussion in this narrative. I do know however, that she did have her deserved revenge, in that having been constantly ravaged and repeatedly assaulted over many years, by more and more sophisticated and savage machinery, her simple revenge was complete, inclusive, permanent and expensive in the extreme. She quite simply retained and buried forever, most of the diabolical instruments of her long ravishment and to a degree, depth and permanency, as to ever negate their recovery. She had not won, but merely held her own with the irretrievable burial of the instruments of her long torture, that were finally and deliberately abandoned and left to rust forever in her innards. She was irrevocably and permanently buried, with most of the instruments of her defilement, clutched to her enfeebled breast.

No flowers mark her grave, save the few yellow dandelions, daises and buttercups provided by her Earth Mother Nature. Now, she is all but forgotten, except by the few, old, Lochhead miners, who occasionally gaze rheumily in the direction of the resting place of a very proud old lady, who once, long-served the small, mining communities of East Wemyss, West Wemyss, and the Coaltown of Wemyss - and other small communities in the surrounding countryside.

**De mortuis nil nisi bonum!**

\*

# Chapter LVIII

## The Passing of Old Harry
### *A Man of his Time!*

Old Harry died as he had lived. . . . Quietly and with great dignity! - And mostly surrounded by his large extended family. He had attained the grand old age of 94 years and his time had finally come. His great frame was shrunken, his muscles had all but withered, and his memory had all but failed. His demeanour was still warm and pleasant, his smile was quick, his voice was soft and he hadn't had a cross word with anyone in many, long years. Old Harry was a coal miner and had been so for nigh on fifty years. He had worked underground as boy, youth, man and experienced oldster. He started in Lochhead pit at the age of fourteen, the day after he left school. His mentor was his brother-in-law by marriage, old Tot Smart, husband to his older sister Old Kate, who taken him at the age of fourteen when his own mother died, leaving a husband and fourteen, grown offspring, of whom young Hend (Harry) was the youngest. Old Tot took young Hend down Lochhead pit as his 'boy' in a 'pick-place', where young Hend's introduction to this underground world, was spluttering tallow lamps, jammie pieces, cold tea or water and mind-numbing, hard work.

Young Hend came up through this hard school as had many youngsters of his generation, whose only prospect was the coal-mines. Young Hend as he was still called, had by this time developed a very bad stammer, which often pitifully inhibited his slow speech, an impediment which followed him all through his long life, but did not stop Harry from achieving the life-time respect and admiration that was naturally and tacitly elicited, from both his fellow miners and from the colliery management. *(No one would ever dare to interrupt Harry, in his sometimes painstaking efforts to be understood.)*

Old Harry was a master craftsman of a miner, who applied his craft for seven days of every week. He was a contractor in brushing, back-brushing and developing, even though he always favoured the brushing contracts in his later years. He worked seven days a week every week and was forever being forced by the manager of the colliery *(Wullie Hampson)* to take time off, especially at the time of the annual holiday. Harry never travelled very far during his enforced holidays. Coal mines being the precarious occupation that they were, needed constant attention and maintenance and Harry knew that if there was trouble at the mine, he would be one of the few miners that Wullie Hampson would send for!

Harry would always turn out for anything, even in the middle of the night. I distinctly remember whilst I was still a pupil at Denbeath public school *(I was aged around 10 at the time),* there was a firm and insistent knock on the front door of our house, at 34 Barncraig Street in Buckhaven. The caller was none other than Wullie Hampson, the manager. Wullie Hampson was brief, if not articulate. "Mrs Moodie, - I need Harry at the Pit!" As my mother came back to warn Old Harry *(Wullie Hampson would not come inside, he returned to his small motor car),* she saw that Harry had risen without the least word of complaint or question. It was enough that Wullie Hampson had come to his front door in the middle of the night, requesting help! Old Harry dressed quickly, grabbed the few hurriedly made-up sandwiches and tin flask of cold tea, that would suffice for his day-shift piece. He then quickly left without fuss or explanation. *(I'll be back when I can!)* This had happened before and would no doubt happen again! This was the unspoken measure of Harry's skill, knowledge and expertise, which would be tested again, and again, over many years to come! Such was the relationship between the few miners of Old Harry's calibre and the colliery manager, who knew full well that the efficient running of a small pit such as Lochhead, depended on the hard-won knowledge and experience of men such as these.

Harry's long career at Lochhead pit was interrupted only once, when he worked at Wellsgreen pit for a few years, just after he got married and lived in Denbeath, then Buckhaven. Harry's knees then played-up, in that several 'beat-knees' in succession, determined that he get off his knees where possible and work on his two feet. This brought him into the world of brushing, stone mining and developing, where the work was even heavier, but the possible remuneration slightly better. Old Harry thrived in this environment and soon began to make a name for himself as a capable, dependable and competent miner. He soon gathered a small team of like-minded young miners and began vying for brushing contracts with the mine manager. *(First George Welsh, then Wullie Hampson.)* Small contracts soon became big contracts and after a few years, Harry joined the small and exclusive band of heavyweight contractors that helped develop Lochhead pit and Surface Dipping. This was when Harry decided that there would be no more day-shift for him and from now on, he would prefer to work back-shifts and night-shifts, and have some free time during the day to follow his main single hobby, which was building coal bunkers, garden sheds and garages and literally anything using wood.

He became very frustrated in later life after his family had grown up and left home, by the complaints of new young neighbours who objected to the noise made by Harry, with his constant sawing and hammering of nails into wood. *(Old Harry could drive a mean nail!)* This was quietly devastating to old Harry, a man who had never complained about anything in his life and simply could not understand the attitude of people, who probably had never done a hard day's work in their miserable lives!

Old Harry was also well known for his smoking of his short Stonehaven pipe. He took up the pipe sometime in his late thirties, just after I had stopped pinching his cigarettes. *(I think he did this to further deter me from cigarette smoking.)* He first of all started on 'thick black', then 'bogie roll', then 'condor twist', then 'condor plug' and finally, 'condor flake!' His short, stubby pipe was his trade-mark and Harry was never to be seen without it! When the coal-mines act determined that Lochhead pit would become a Part II mine, undeterred, Harry started back on 'condor plug' and began to chew it underground. A few of his work-mates did complain behind his back, which is not surprising, considering that a pick or shovel covered in tobacco gob - isn't the most pleasant of experiences!

During his time in the Surface Dipping and in the Pit, Old Harry had made many friends and acquaintances, and many miners considered it a honour and privilege to work in his team. This did not however, deter or inhibit his many co-workers from labelling him with the odd sobriquet or two *(but never to his face!)* during his long career. 'Harry Gimme' or 'Harry, gees the pick', or even 'Harry, haund me this', would be quite commonplace. Harry would merely hold out his hand without turning around, with the necessary request, and no matter as to what any other member of his team was doing, Harry's needs came first! And . . . Harry got away with it! He was the craftsman!

If there was a 'dirty', dangerous, unsafe, 'This is your life' sort of task to be done in any section in Lochhead pit, Old Harry was one of the few miners that would be entrusted with the job. This usually meant recovering expensive equipment that was in danger of being buried forever, when the roof finally decided to fall. An entirely unpredictable equation! Old Harry and miners such as Old Hugh Gilbert from the Coaltown, would be entrusted to 'go in' and recover such expensive equipment, knowing full well, that one false move or miscalculation would bury not only the equipment, but them as well. Even the mine engineers and electricians who were concerned for equipment under threat in such an environment, would express satisfaction, when miners of the calibre of Old Harry and Old Hugh came on the scene. This kind of undertaking took an especial kind of courage found only in the most confident of miners, whose faith in their own abilities outweighed thoughts of personal danger. Old Harry was like a few other close-mouthed, old miners. He was a precipitator. He was a 'practical mover and a continuous shaker' and always led by wordless example. These were the type of miners who always 'punched' at full volume and were worth their weight in gold to the management of any working colliery.

Old Harry was finally forced into retirement at the age of 63 years, at a time when the coal mines in East Fife were closing down. Lochead pit had only a year or two to go, when it too, would wind coal for the last time and its structure be raised to the ground. I cannot even begin to guess just how many miles of roadways, tunnels and mines that Old Harry was responsible for, or indeed, how much coal he actually dug.

Old Harry was a 'quiet' man, well liked, much respected, and regarded as somewhat of an institution at Lochhead Colliery, both down the Pit and in the Surface Dipping. He knew his inestimable worth as did all supremely confident men and felt little need to explain his actions, or the need to exploit other men in the doing thereof! He was a legend in his own life-time and his name remains well known even to this day, amongst the ever-dwindling band of old Lochhead colliers. Old Harry was but one of many competent miners who gave their 'all' to the management and mines of Lochhead Pit and the Surface Dipping - and who for the most part, are ill-remembered, often rejected, sometimes vilified and thoughtlessly disparaged as - only COALMINERS!

\*

HARRY MOODIE . . . . . Coalminer.   1904 - 1998.
Rest in Peace - old man, you deserve it!

THE TERMINATION

of

## A Bright and Shining Life!

and

## THE END
of
## A DARK and DISMAL STRIFE

## A Story about Coalminers!

\*

# The GÖTEBURG Experience

## Not in Here you Don't!
## – *A Sobering Emancipation!*

In the southern seaport city of Göteburg or Gothenburg in Sweden, around the middle of the nineteenth century, a limited experiment was initiated, whereby the granting of alcohol licenses was limited to publicly owned trusts or companies, whose remit was to exercise control of the price of spirits, curtail the sales of ales, wines and spirits *(where possible)* and generally operate the licence for the public good! One of the basic precepts of this experiment, was that profits over and above a fixed percentage, accrued or generated by the licence holders, would be returned to the local authorities, to be used to provide, build and maintain, community facilities such as public libraries, museums, craft centres and pleasure parks and indeed, anything, that in the opinion of the authorities could and would probably be a healthier alternative to drinking alcohol!
*(However, as any discerning reader may realise - and that which the city authority in their wisdom had already anticipated, that if the good folks of Gothenburg were to continue drinking, then, why should not the city benefit from the potentially large profits, that were to be made from the sale and consumption of alcohol.)*

These drinking establishments were also placed under the management of fixed salaried staff/managers, who were under no obligation whatsoever to promote the sale of alcohol. It was therefore quite immaterial to them, as to how much beer or spirits were sold to the general public. This was known as the concept of - 'Disinterested Management'! *(A moot point worth consideration regarding this 'experiment' was, that if the thirsty burgers of Gothenburg knew that all of the profits from their alcohol consumption were to be used to facilitate worthy local causes, might they then have drunk even more?)*

The initial result of this experiment *(the basis premise of which was to ostensibly curb the consumption of alcohol),* seemed to have the opposite effect, in that the Gothenburg city fathers were just a little startled *(delighted, would be an apt description!)* at the amount of monies that were beginning to accumulate in the City coffers.

Word of this experiment soon spread to other towns and cities on the large Scandinavian peninsula - and within a relatively short time, similar 'systems' became widespread both in Sweden and Norway. No doubt with the same measure of success!

Word, or knowledge of the 'Gothenburg System' as it became known, soon travelled to Great Britain *(there was an established trade route between the ports and harbours of Fife and the Scandinavian and Baltic States)* where it was brought to the attention of both the Temperance movement and that group of people known as Abolitionists, who of course, wished to ban the 'demon alcohol' in all of its forms. The public house system in being in Britain *(Scotland)* at that time, did not allow for this form of licensing, in that the licence was invariably invested in the person of one named worthy individual, whose name by law, appeared above the door of the main entrance of the establishment. However, the basic idea behind the system soon caught the attention of the trade union and socialist movement, whose own ideas about the regularisation of social drinking were very similar. The main stumbling block to the possible implementation of this system, was that the British Licensing laws of the period did not provide for the conferring of licenses on public trusts or bodies, or supposedly disinterested groups.

This was about to change with the passing of the Industrial and Providential Societies Act of 1893, which allowed the application of public house licenses to groups, bodies, trusts and established private companies. The Act was of special interest to some of the private coal-owners, who in the past, had viewed the hard-drinking habits of a great number of their workforce as an unnecessary evil and the root cause of many avoidable pit accidents - and unwanted absenteeism. *(A few colliers in the East Fife coal mines, had acquired what could have been described as a local reputation for their consummate and irresponsible, week-end binge-drinking, and actually 'appeared' at the mine/pit-head on a Monday morning, in a state of hung-over inebriation. Quite often, their state of residual intoxication was concealed from the banksman, in an attempt to get them underground to their place of work, thereby hopefully enabling them to sweat it out!)*

Many coal-owners had always opposed the licensing of public house within the limits of their privately owned and sometimes isolated, mining villages. They were all to aware of the inherent dangers of uncontrolled drinking, even though they realised all too well that much of it was obtained through covert shebeens and illegal drinking howfs.

Previous to this Act in 1893, private coal owners and companies could not, as incorporated bodies, be granted a public house licence. This had now changed, in that they could now appoint a body, or a trust, in which the licence could now be vested and operated. This Gothenburg system had the double attraction or incentive, that the coal-owners through their nominees (trusts), could strive to control the drinking habits and hopefully curb the excess drinking of their 'captive' coal miners, which of course, could result in less absenteeism and a safer working environment through reduced numbers of partially, inebriated miners. Moreover, the monies raised by this venture, could be used to provide extra basic facilities for the mining communities, by way of limited, additional education *(small libraries),* and recreational pleasure *(village halls, bowling greens, community centres. etc!),* which by right, should have been the committed prerogative of the coal-owners themselves! In a much darker vein, the coal-owners realised that they could probably exercise a much greater social control over their workers, in that within the confines of the 'Goths', they could 'keep an ear to the ground' and unobtrusively have watched and 'suspicioned', any possible subversive elements within the mining community. *(Certain elements in the East Fife coal fields were known for their undisguised left-leaning proclivities, both underground and socially!)* Seemingly, one of the reasons why in times past, that the coal-owners had been dead set against outsiders being granted licenses, was that the company writ or influence could not and would not prevail in what was after all, their 'company owned' or privately owned villages.

Now, with the advent of the proposed Gothenburg system, the trust or body that held the new licence could be quietly intimidated or effectively controlled by unseen manipulation, from an unscrupulous coal-owner cum coal-master. Either way, the granting of licences under the Gothenburg system, gave the coal-owners the means by where their writ could run large and continue to proliferate and hold sway, if they so desired!

In those small isolated mining communities and villages where the Company held a monopoly of housing, or indeed owned the complete village, the introduction of even one public house operating on the Gothenburg principle was bound to succeed, was bound to generate income and would inevitably, if allowed, generate good, unencumbered profits.

The Experiment was bound to start somewhere in Fife, and in 1885, two years after the promulgation of the Ind. & Prov. Societies Act, the Fife Coal Coy., in their own right, applied for, and during the following year, gained their first license to operate a public house on the Gothenburg principle. This first such venture in the Kingdom of Fife was originated in the small mining village of the Hill o' Beath, utterly owned by the Fife Coal Coy. This village was situated immediately to the south-west of Cowdenbeath and lay to the north-west of 'Andrew Carnegie's' town of Dunfermline. This actually became the first Gothenburg or 'Goth(s)' as they were popularly known in Scotland - and such was the initial success of this experiment and the monies raised, that very soon, the local mining populace were 'seen' to be illuminated, by the provision of electric lighting for the village. *(Surely, one of the earliest communities to be modernised by such a splendid innovation!)*

The experiment soon expanded, in that by the start of W.W.I, no fewer than twenty or more such licences had been granted in the Kingdom of Fife alone, with other coal-owners and private companies in different mining areas seeking to adopt the experiment. Thus, the Gothenburg experiment gained proponents and quickly proliferated! *(In Fife at least!)*

However, all was not sweetness and light within the system *(except for Hill o'
Beath!)*. The new structures were large to the point of being monolithic, very
imposing and severe, mostly built like substantial country houses and incorporating
much grey stone. Their size was impressive and certainly dwarfed many of the
miners' smaller house and cottage rows, but in spite of their forbidding looks, the
buildings soon became popular gathering places for the local population. The
rooms inside were often large, roomy, but with just the minimum of comfort by
way of sitting accommodation. The insides could usually be described as austere,
just barely equipped and maybe just a little overwhelming, especially after the
cramped conditions and indigent state of some of the miners cottages. The
attractions that drew most of the miners *(and some of their wives into the 'jug-bar')*
into the Goths, could only have been the reasonable quality of the beers and spirits
and the thought of some general, or even intimate conversations. It could not have
been the non-convivial atmosphere of recreation and games! There was so many
'don'ts' and 'muss'nts' within the confines of the buildings, that it was no small
wonder that any miner set foot inside at the first instance!

No card games! No dominoes! No games of chance! No darts! No
snooker or billiards. No music or dancing - and miners must not, under any
circumstances 'Ask for credit!' This was strictly TABOO! In other words, all of
the things that a normal, healthy and fit young miner would take for granted
outside of the confines of the colliery. But flourish they did, and for many years
the Gothenburg's were the sole centre of attraction in the life of many villagers.
Some large villages and towns were known to boast of more than one Gothenburg
- and in the town of Kelty, there was a total of four such establishments, which
gave some of the villagers so much of headache in trying to describe them, that
they eventually settled the problem by giving them numbers from 1 to 4! *(Just
like parallel levels down the coal-mines!)*

The Goths that were situated within the isolated mining communities and
smaller villages, did remarkably well in terms of monies generated for local
amenities, whilst those Goths that faced the competition of other non-involved
public-houses within the same town or village, did not fare so well! Given the
choice, the miners mostly preferred the much more convivial and unrestrained
atmosphere of other hostelries, where they felt that 'big-brother's' writ did not
prevail! The general feeling amongst some of the more independent-minded of
miners, was that they should throw-off the enveloping yoke of the coal-owners, as
soon as they had come up the pit!

The Gothenburg system of public houses were quite strategically placed
within the mining communities, especially where the village was completely, or
even partially owned by the Coal company. This was determined by the size of
the mining population within the potential catchment area, even to the instance of
more than one Gothenburg being built or opened in the same large village or

small town. Kinglassie had two, known as the Upper Goth and the Lower Goth. Cardenden had three. Kelty as we know, had four! In the East Fife area, there were three such Gothenburgs, each within a sizable mining community. One at the Tower-bar at Aberhill. One opposite the Wellesley colliery and now known as the Wellesley bar. One in the Coaltown of Wemyss that still stands in nearly its original state, but has reverted to the name of one of the most famous of the Wemyss lairds - The 'Great' second Earl, Lord David Wemyss of that Ilk!

*

Author's Note: - There is on record, a practice perpetrated on certain coal miners in the Stirling-shire area, where several of the underground contractors actually paid out the weekly wages of their encumbered miners, in one of the local Gothenburgs! *(No doubt, to the eye-watering strains of "Sixteen tons and deeper in debt!")*

*

# ADDENDUM

# I.  Tiddley-Winks Clubs

As to why they were named the tiddley-winks clubs, I know not know, nor have I come across any single person that can give me an answer.  I do know why they were formed, how they came about and how they were operated.  At that point of time after the 1926 General Strike, when the miners returned to the coal mines after being 'out' for nigh on ten months or more, the men actually returned to the same work for exactly one shilling a day less than the amount that they had 'struck*' for, and an increase of one hour on their working day!  Needless to say, after having lived on the bread-line and having had to submit to the indignity of being partly fed by 'soup-kitchens', the pitifully small amount of money coming in was sorely needed and mostly spoken for by each successive 'pay-day'.  There were a few traders and shopkeepers who continued to support and subsidise miners and their families in the long term, and those traders who were true to their customers were repaid by the loyalty of the miner's and their wives, in that they stayed with such traders as loyal customers - a trust not lightly given!  Most miners' somehow inherit long memories and small acts of kindness *(and unsolicited charity)* were well remembered by families, with signal memories uncluttered by frivolities.  Life was often much too grim and miners always remembered their friends.  Loyalties were most often very fierce!

Most wives will claim, that given a small, but reasonable sum of money, they would then be able to put it to better use on something more substantial in the way of the dire necessities, that always seemed to be out of their grasp.  To this end, one wife, or even a small group of wives got together and came up with an original inspiration.  They would, all of them, up to a fixed number of wives at any one time, pay a small sum of money into a common fund every week and this fund to be held by one person.  The weekly amount of money to be paid-in every week was to be one shilling - and if the number of wives in this co-operative were to number twenty, then the weekly pay-out would be twenty shillings.  This amount would of course, be guaranteed to a different wife every week, with no one woman getting out any more money than they had actually put in!  In order to ensure absolute fairness and produce an element of surprise, the 'draw' would be organised in one of two ways. One, the draw could be made on a weekly basis on a Friday, with the week's winner not known until the actual draw took place, or two, the number of women taking part in the draw at any one time, would have their 'number' placed in a hat and the numbers would be drawn out one at a time, so that each and every participant would know the exact week, on a scale of one to twenty, when she could expect her little 'bonanza! Either way, both methods worked extremely well, with all of the wives concerned conscientiously in-putting their single shilling on a weekly basis.

The continued success of the system was exemplified by the spirit of the women themselves, in maintaining their contributions to the fund, irrespective of when they themselves were paid out! The whole venture was based on absolute trust and many such schemes were abroad during these times. If, for any reason, a wife could not make her contribution on any given week, there was always a friend or neighbour who was willing pitch-in! They, each other knew, that it could quite easily be themselves in the same dire straights the following week!

These women were surely the very first Y.U.M.M.I.E.S on record!
(Young Upwardly Mobile Miner's Inherently Economical Spouses!)

*\* At the start of the 1926 General Strike, the miners came out with strong hearts, a firm belief that their cause was just and the rousing, sloganised, couplet ringing in their ears:-*

*Not a penny off the daily pay!*
*Not a minute on the working day!*

*Several months later, these same miners were forced to return to work with equally heavy hearts, a feeling of utter dejection and with a different couplet that seemed to mock their futile efforts:-*

*Another shilling off the pay!*
*Another hour on the day!*

\*

# II.  Miner's Wedding Cake.

This consumable comes in all shapes and sizes, it comes in various shades and textures, is to be found in different colours and consistencies and has permeated into all levels of society, from all walks of life!  From the platters of the most humble dwelling to tables of the rich and famous.

It is consumed by people both young and old and from all manner of background, from earthy peasants, to Kings and Queens.  It has a status all out of proportion to its lowly origins, for in itself, it is derived from the very 'salt' of the Earth!

Its simple ingredients are supplied by Mother Nature herself, quite unneedful of the 'help' of Man.  The One ingredient comes from a number of sources, all naturally occurring and re-occurring!, and obtained for the most part, for next to nothing in its original form.

This itself, a naturally forming product, is a stable food for many, is a 'must' for some and is used as a 'status' symbol by certain people in various lands.  Its production is taken to a very high degree in some quarters and raised to a status sometimes beyond recognition, in its name and price!  Yet, it is of the most simple and inexpensive produce in its basic presentation.

It is loved, hated, yearned for and rejected, it is beaten, whipped, ground, sliced, cut, broken, grated and crumbled.  It is boiled, fried, roasted, grilled, baked and even taken in its raw state.  It is soft, hard, fresh, stale, gooey and rock-like. It comes in colours like Red, Yellow, Orange, Brown and even Blue.  It is elevated to high esteem and given Royal patronage!  It has acquired a reputation all out of all proportion to its humble beginnings.

It has acquired the status of great cities, like Gloucester, Leicester, Bicester and Lanark.  It has even been given the accolade of County status, such as Devon, Cornwall and Lancashire!

It can be vile, smelly, dirty in colour and completely indigestible.  Many people can not abide its smell, it's taste or its texture.  It is the subject of many and vivid descriptions and has a following of hundreds of thousands, both for and against it's existence!  It is with nearly all of us every day of our waking lives.  It is welcomed and thoroughly enjoyed, it is rejected and utmostly reviled.  Its inestimable value was realised as a propaganda weapon during W.W.II., in that the nation's Coal-Miner's were given an extra ration of three-quarters of a 'pound' every week, just to maintain their moral.  It was regarded as a National Life-Saver!

It has taken on the mantle of great Countries such as Denmark, Holland and Switzerland.  In point of fact, French peasants have claimed it as their own and given it a new name:- Fromage!  It is produced in varieties and variations and concoctions that are almost unpalatable or inedible to other nationalities and creeds, and its fame transcends national borders!

It is calorific!  It is ubiquitous!  It is Universal!  It is a thundering great Paradox! It is CHEESE in all of its forms!!  The MINER'S Wedding Cake!

# III.  Recipe for 'Po Toast'!

Prologue: When two coal miners are sitting down together at 'piece-time', especially on the day-shift and one or both of them extract 'Fire place' buttered or jammed toast from their 'piece-box', they will usually look at one another without saying anything - and smile!  A very knowing and secret smile, each one knowing precisely the exact thoughts that are visualised in each other's minds' eye!

*Definition of Toast: -  Fire-place Toast is identified by the large double or treble 'tine' marks in the toasted bread, caused by the toasting fork (and thence the bread) being swung across the glowing coals.  Tine-holes can be single-ended or double-ended, depending on whether the toast is turned vertically or laterally. Vertically turned toast always has double tine holes.*

*Definition of a Po: -  Large, Bum-sized, porcelain, basin-like vessel having one or more carrying handles.  (Coal Miners wives' po's were usually made from Enamelled-Tin.)*

Tableau: *Miner's Cottage with inside cold running water, but having outdoor Water-closet!  Coal-miner on Day-shift, which means 'up' at 4.30am for the wife/young mother to prepare 'piece' for a husband.  The coal-fire 'range' was 'banked' the previous evening with one large lump of coal. (Miner's Fire-coal at One shilling per cwt.)*  A simple 'prod' with a strong poker will immediately 'break' the lump into many parts thereby giving an 'instant' hot blaze!  *(An essential requirement for 'rough' toast!)  Wife is wearing heavy long flannel nightdress.*

Operation: - *Wife gets out of bed, then cuts a half-dozen thick slices of bread, she then obtains two-foot long, toasting fork.  She then fetches the Po!, which is then strategically placed.  With the cut bread to one side and toasting fork in left hand, wifey then kneels before blazing fire, raises her nightgown and 'squats' for her morning 'Tinkle!'''.*  (This action is made all the more mysterious by the fact that 'all' is hidden under the voluminous /ubiquitous folds of her gown.)  *This operation lasts thro' the sequential production of six pieces of bread/toast until completion. The one single unmitigated 'factor' that needs to be realised is that all the way thro' this entire operation, wifey has only 'one free hand!'*  To add further 'spice' to this already tantalising 'offering', it must be remembered that *wifey* has yet to butter, jam or even 'honey' the toast - and then wrap and pack the delicacy in her man's 'piece-box'. *Then, and not before, shall wifey then re-occupy the still warm bed newly vacated by her spouse and 'that's* IT! *- all dutifully performed prior to her own morning ablutions when 'she' finally 'gets up!'*

Conclusion: *Readers'!..... "I give you:- PO-TOAST!" (Smile Please!)*

# IV.   Goodness Gracious Me!

Within the Coaltown of Wemyss at the time described in my narrative, there was no resident Doctor.  The nearest doctor, Dr. Khambatta by name, lived in the mining village of East Wemyss at Kingslaw House, East Brae.  He was quite a well known doctor, in that he had quite a large and far-flung practice that at least stretched from Denbeath, Methil to the village of West Wemyss.  His patients included many miners and their families, which, in the spirit of the times could be quite large.  Like most country doctors, he was forced by the sheer size of his out-flung practice and the demands made on his small vehicle, to limit his visits to given areas to perhaps once or twice a week.  This of course, did not take into account any emergency calls that required his immediate attention, if he could be found.  Telephones were few and far between, and just to be able to pinpoint the Doctor's whereabouts at any one time, could be a matter of pure speculation.  That was one of the reasons that he attempted to be in a given area, at loosely defined times.

As to how this problem of arranging a Doctor's visit was overcome in the Coaltown of Wemyss, was by a very simple and unique method of communication.  There was a designated house in the village, named 'Fernlea Cottage' at 86 Main Street, where access to the back scullery was immediate, through an unlocked back door.  On the wall of this room there hung a clean slate and a slate pencil!  This slate was there for the exclusive use of those patients, or anyone else for that matter, that required the services of Dr. Khambatta.  All they had to do, was to write their name and address on this slate, or get someone to do it for them.

This 'Call-House' was where Dr. Khambatta would make his first port of call.  He would note the names and addresses of all patients listed, then wipe the slate clean to show that he had visited!  He would then visit each patient in turn, without their having to visit his surgery in East Wemyss.  This house in the Coaltown of Wemyss is still known today as the Ca' Hoose!' *(Call house)* and is presently occupied by Brian Smart and his wife.

When Dr. Khambatta retired from practice and was succeeded by Dr. Meiklejohn, the same slate system was embraced by the said new doctor and was used very successfully for many years to come, in both the Coaltown of Wemyss and West Wemyss!

\*

# IN MEMORANDUM

The following is a list of some of those coal miners whose lives were lost within the Wemyss pits described in this narrative and over the past one hundred and fifty years. This includes both of the Lochhead collieries, the Victoria pit and several unnamed sinks around the village of West Wemyss. It does not include the names of those brave miners whose lives were lost at any of the other Wemyss Coal Company pits, or indeed, in any other Fife coal mine.

This list is by no means complete. There are many omissions for which I can only apologise. One of the main reasons for this, is the fact, that on or around Vesting Day at January 1st 1947, many of the Private Coal Company records were simply discarded or burnt, or otherwise disposed of! This includes many individual, handwritten records belonging to the Wemyss Coal Company at Lochhead and the Victoria pits.

By the time someone within the N.C.B structure realised what was happening, it was already too late! Those records were irretrievably lost forever!

| Name | Age | Colliery | Date of Death | Cause of Death or Fatal Injury *(if known)* |
|------|-----|----------|---------------|---------------------------------------------|
| Anderson, James | 18 | Lochhead | 2 Feb. 1903 | Fall of Side Coals |
| Arbuckle, James | ? | Lochhead | 27 Sept. 1948 | Overcome by Carbon Monoxide *(Whitedamp)* |
| Band, Stuart | 18 | Lochhead | 25 Mar. 1907 | Fall of Roof Coals |
| Baxter, John | 58 | Wemyss | 7 Dec. 1887 | Fell into Bell Crank |
| Birrel, Thomas | 19 | Wemyss | 15 Nov. 1873 | Fall of Roof |
| Brown, James | 61 | Lochhead | 10 Oct. 1907 | Fall of Roof |
| Buist, Thomas | ? | Lochhead | March 1920 | Unknown at this time! |
| Chalmers, Johnston | 82? | Lochhead | 24 Oct. 1906 | Fall of Coal |
| Christie, Alexander | 64 | Wemyss | 16 Feb. 1901 | Fall of Roof |
| Christie, James | 20 | Lochhead | 15 Aug. 1907 | Run over by Coal Hutch |
| Christie, Thomas | 59 | Wemyss | 20 May 1900 | Fall of Roof Coal |
| Dryburgh, George | 66 | Wemyss | 28 Nov. 1805 | Hit by Crane |
| Dryburgh, John | 22 | Wemyss | 11 Mar. 1890 | Fall of Head Coal |

| | | | | |
|---|---|---|---|---|
| Duff, Robert | 62 | Lochhead | 21 Dec. 1907 | Fall of Side Coals |
| Fairfull, Alex. | ? | Victoria | 20 Dec. 1922 | Unknown at this time! |
| Kilpatrick, John | 44 | Lochhead | 29 Dec. 1907 | Underground Fire |
| Kilpatrick, Thomas | ? | Victoria | 29 Dec. 1907 | Unknown at this time! |
| Kirk, William | 54 | Lochhead | 9 Sep. 1921 | Unknown at this time! |
| Malcolm, James | 19 | Lochhead | 29 May 1896 | Fall of Fireclay. (Soft Blaes?) |
| McCrae, Robert | 55 | Victoria | 4 May 1911 | Fall of Stone |
| McNeish, Samuel | 28 | Wemyss | 11 Aug. 1897 | Fall of Stone. |
| McRae, Robert | 55 | Lochhead | 4 May 1911 | Fall of Stone. |
| Melville, John | 14 | Wemyss | 11 July 1873 | Fall of Coal. |
| Laing, George | 56 | Wemyss | 18 Dec. 1897 | Fall of Coal |
| Mathieson, P. | 55 | Wemyss | 9 Aug. 1863 | Crushed by Pit Cage |
| McLaren, William | 15 | Victoria | 30 Jan. 1934 | Run over by Rake of moving Hutches |
| Morris, Andrew | 49 | Lochhead | 29 Dec. 1907 | Underground Fire |
| Mulholland, Stewart | 45 | Wemyss | 26 Oct. 1898 | Crushed by Tubs |
| Munro, John | 38 | Wemyss | 17 April 1902 | Fall of Side Coal |
| Murray, Alexander | 17 | Wemyss | 19 Jan. 1865 | Fall of Roof |
| Ness, Walter | ? | Wemyss | 27 Jan. 1859 | Falling Wood |
| Nicol, John | ? | Wemyss | 17 June 1855 | Fall of Coal |
| Nicol, Thomas | 52 | Lochhead | 23 April 1912 | Fell down Pit-Shaft |
| Parker, William | ? | Victoria | 15 Jan. 1935 | Unknown at this time! |
| Paterson, David | ? | Lochhead | 27 Sept. 1948 | Overcome by Carbon Monoxide (*Whitedamp*) |
| Pattison, David | ? | Lochhead | ?  ?  1955 | Eviscerated by Coal-Cutter in East Side Mine |
| Penman, John | 42 | Wemyss | 17 Nov. 1903 | Crushed |
| Penman, James | ? | Lochhead | 27 Sept. 1948 | Overcome by Carbon Monoxide (*Whitedamp*) |
| Pryde, Robert | 46 | Lochhead | 30 Nov. 1926 | Unknown at this time |
| Reid, David | 25 | Wemyss | 8 Sept. 1887 | Fall of Roof |

| | | | | |
|---|---|---|---|---|
| Rodger, Andrew | ? | Victoria | 19 Nov. 1851 | Unknown at this time |
| Rodger, David | ? | Lochhead | ?  ?  1918 | Run-away hutch on New Dook |
| Rodger, David | 50 | Lochhead | ?  ?  1941 | Fall of Side Coal |
| Rodger, John | 36 | Wemyss | 20 April 1888 | Mine Explosion |
| Russell, William | 24 | Lochhead | 15 May 1913 | Fall of Side Coal |
| Scott, R. | 44 | Wemyss | 29 Aug. 1864 | Fall of Roof Coal |
| Scott, William | ? | Victoria | 29 Dec. 1907 | Unknown at this time |
| Shepherd, William | 19 | Lochhead | 11 May 1912 | Fell down Pit-Shaft |
| Stenhouse, Albert | 17 | Wemyss | October 1925 | Fall of Coal |
| Stevenson, David | 45 | Lochhead | 30 June 1904 | Crushed by Wagons |
| Storrar, James | 60 | Lochhead | 11 Nov. 1904 | Machinery Accident |
| Watson, David | 19 | Lochhead | 6 Jan. 1913 | Fall of Side Wall, *(Probably coal)* |
| Welsh, Archie | 70 | Lochhead | 18 May 1910 | Fell from Coal Wagon |
| Welsh, Walter | 22 | Victoria | 29 July 1879 | Fall of Coal, *(Side or Roof Coal)* |
| Williamson, Andrew | 16 | Lochhead | 30 May 1911 | Fall of Coal, *(Side or Roof Coal)* |

\*

Author's Note: - **Currently available for viewing at the Public Library in Dunfermline (Historical Section) in the Kingdom of Fife, there is presently being compiled, a singular, leather-bound volume, containing the names of the known miners who lost their lives in one of the many Fife coal mines. This record is by no means complete and is being constantly reviewed, but requires the written input of members of the general public, who have sundry information with regard to those miners who were killed in the Fife coal mines and who's names do not appear in that memorial list!**

**This List with several additions has been extracted from that Memorial Roll!**

\*

Author's Comment: - **A close inspection of the above described Causes of Death and Fatal Injuries, shall surely reveal that the greatest percentage of** underground fatalities, both in Lochhead and the Victoria pits were caused by falling coal or stone, or by the silent dropping of standing coals from side walls. These falling side coals were usually a silent killer, with little or no warning being given or experienced. This was especially true within the Stoop and room workings of the Dysart Main seam, or indeed, with any tall stand of coal, where the slow, but inexorable, downwards crush will render a once solid, side wall of coal, into a crushed and broken entity within weeks. This process was sometimes exacerbated in part by a few desperate colliers, ever hungry for a 'quick' filling, defying mine convention by removing this 'easy' coal!

*

# BIBLIOGRAPHY

## Books and publications consulted, used and quoted,
## in the preparation of this book

A Glossary of Mining Terms used in Fife ........................ Edited by: R.D.Kerr, Fife Colleges 1980

Guide to the (Fife) Coalfields - 1961 ................ Edited by: G.R.Strong, M.I.Min.E.,& E.G.Corbin

Mine Gases & Explosions - 1908 ............................................................ by: J.T.Beard, C.E.,E.M.

The Coal Fields of Scotland - 1921 ........................................ R.W. Dron, F.R.S.Edin.,M.I.Min.E.

Economic Geology of the Fife Coalfields - Area I ........................................ H.M. St. Office 1931

Economic Geology of the Fife Coalfields - Area II .................................... H.M. St. Office 1934

Practical Coal Mining, Vols. I, II & III - 1908 ............ Edited by: Prof. W.S.Boulton B.Sc.,F.G.S.

The Lost Village of Buckhaven - Pub. 1993 ................................................................. Eric Eunson

Methil No More! - Pub. 1994 ................................................................................... Paul Murray

A Short History of the Scottish Coal-Mining Industry - 1958 .......... Pub. by: N.C.B. Scottish Div.

The Miners - 1976 ............................................................................................ Anthony Burton

Randolph G. E. Wemyss, Purves & Cunningham. Leven - 1909 .............. Andrew S. Cunningham

First Statistical Account of Scotland - 1791 Vol. X (Fife) ............. Compiled by: Sir John Sinclair

New (second) Statistical Account of Scotland - 1845 (Fife) ............. Pub. by: William Blackwood

Third Statistical Account of Scotland - 1952 (County of Fife) ............. Alexander R. Smith. M.A.

Mining in the Parish of Wemyss - (Short Precis held at Methil Public Library).......... Frank Rankin

Bygone Leven - 1991 ............................................................................................... Eric Eunson

Old Wemyss - 1995 ........................................................................................ Margaret Thomson

Chambers Scots Dictionary - (W. & R. Chambers, Ltd.) ......... Compiled by: Alex Warrack, M.A.

Memorials of the Family of Wemyss of Wemyss - Vol I, II & III – 1888 ........................................
........................................................................................by: Sir William Fraser, K.C.B.,LL.D.
*(Kindly loaned to this author by*: Mr Michael Wemyss of Wemyss of that Ilk)

The Gothenburg Experiment in Scotland - 1988 - A Short Precis ................ by: Charles McMaster

Memoirs of the Geological Survey of Scotland ............. Sir Archibald Geikie, F.R.S.,D.C.L.,D.Sc.

Coal Mining in Wemyss ...................... Compiled by: The Wemyss Environment Education Centre

The Fife Coal Company Ltd. - 1953 - A Short History ...................................... by: Augustus Muir

The Wemyss Private Railway - 1998 ............................................................... by: A.W.Brotchie

Mining in the Kingdom of Fife - 1913 ........................................................ Andy. S. Cunningham